T0212009

Lecture Notes in Computer Science 12951

More information about this subseries at http://www.springer.com/series/7407

Osvaldo Gervasi · Beniamino Murgante ·
Sanjay Misra · Chiara Garau ·
Ivan Blečić · David Taniar ·
Bernady O. Apduhan · Ana Maria A. C. Rocha ·
Eufemia Tarantino · Carmelo Maria Torre (Eds.)

Computational Science and Its Applications – ICCSA 2021

21st International Conference
Cagliari, Italy, September 13–16, 2021
Proceedings, Part III

 Springer

Editors
Osvaldo Gervasi ⓘ
University of Perugia
Perugia, Italy

Sanjay Misra ⓘ
Covenant University
Ota, Nigeria

Ivan Blečić ⓘ
University of Cagliari
Cagliari, Italy

Bernady O. Apduhan
Kyushu Sangyo University
Fukuoka, Japan

Eufemia Tarantino ⓘ
Polytechnic University of Bari
Bari, Italy

Beniamino Murgante ⓘ
University of Basilicata
Potenza, Potenza, Italy

Chiara Garau ⓘ
University of Cagliari
Cagliari, Italy

David Taniar ⓘ
Monash University
Clayton, VIC, Australia

Ana Maria A. C. Rocha ⓘ
University of Minho
Braga, Portugal

Carmelo Maria Torre ⓘ
Polytechnic University of Bari
Bari, Italy

ISSN 0302-9743 ISSN 1611-3349 (electronic)
Lecture Notes in Computer Science
ISBN 978-3-030-86969-4 ISBN 978-3-030-86970-0 (eBook)
https://doi.org/10.1007/978-3-030-86970-0

LNCS Sublibrary: SL1 – Theoretical Computer Science and General Issues

This Springer imprint is published by the registered company Springer Nature Switzerland AG
The registered company address is: Gewerbestrasse 11, 6330 Cham, Switzerland

Preface

These 10 volumes (LNCS volumes 12949–12958) consist of the peer-reviewed papers from the 21st International Conference on Computational Science and Its Applications (ICCSA 2021) which took place during September 13–16, 2021. By virtue of the vaccination campaign conducted in various countries around the world, we decided to try a hybrid conference, with some of the delegates attending in person at the University of Cagliari and others attending in virtual mode, reproducing the infrastructure established last year.

This year's edition was a successful continuation of the ICCSA conference series, which was also held as a virtual event in 2020, and previously held in Saint Petersburg, Russia (2019), Melbourne, Australia (2018), Trieste, Italy (2017), Beijing. China (2016), Banff, Canada (2015), Guimaraes, Portugal (2014), Ho Chi Minh City, Vietnam (2013), Salvador, Brazil (2012), Santander, Spain (2011), Fukuoka, Japan (2010), Suwon, South Korea (2009), Perugia, Italy (2008), Kuala Lumpur, Malaysia (2007), Glasgow, UK (2006), Singapore (2005), Assisi, Italy (2004), Montreal, Canada (2003), and (as ICCS) Amsterdam, The Netherlands (2002) and San Francisco, USA (2001).

Computational science is the main pillar of most of the present research on understanding and solving complex problems. It plays a unique role in exploiting innovative ICT technologies and in the development of industrial and commercial applications. The ICCSA conference series provides a venue for researchers and industry practitioners to discuss new ideas, to share complex problems and their solutions, and to shape new trends in computational science.

Apart from the six main conference tracks, ICCSA 2021 also included 52 workshops in various areas of computational sciences, ranging from computational science technologies to specific areas of computational sciences, such as software engineering, security, machine learning and artificial intelligence, blockchain technologies, and applications in many fields. In total, we accepted 494 papers, giving an acceptance rate of 30%, of which 18 papers were short papers and 6 were published open access. We would like to express our appreciation for the workshop chairs and co-chairs for their hard work and dedication.

The success of the ICCSA conference series in general, and of ICCSA 2021 in particular, vitally depends on the support of many people: authors, presenters, participants, keynote speakers, workshop chairs, session chairs, organizing committee members, student volunteers, Program Committee members, advisory committee members, international liaison chairs, reviewers, and others in various roles. We take this opportunity to wholehartedly thank them all.

We also wish to thank Springer for publishing the proceedings, for sponsoring some of the best paper awards, and for their kind assistance and cooperation during the editing process.

We cordially invite you to visit the ICCSA website https://iccsa.org where you can find all the relevant information about this interesting and exciting event.

September 2021

Osvaldo Gervasi
Beniamino Murgante
Sanjay Misra

Welcome Message from the Organizers

COVID-19 has continued to alter our plans for organizing the ICCSA 2021 conference, so although vaccination plans are progressing worldwide, the spread of virus variants still forces us into a period of profound uncertainty. Only a very limited number of participants were able to enjoy the beauty of Sardinia and Cagliari in particular, rediscovering the immense pleasure of meeting again, albeit safely spaced out. The social events, in which we rediscovered the ancient values that abound on this wonderful island and in this city, gave us even more strength and hope for the future. For the management of the virtual part of the conference, we consolidated the methods, organization, and infrastructure of ICCSA 2020.

The technological infrastructure was based on open source software, with the addition of the streaming channels on YouTube. In particular, we used Jitsi (jitsi.org) for videoconferencing, Riot (riot.im) together with Matrix (matrix.org) for chat and ansynchronous communication, and Jibri (github.com/jitsi/jibri) for streaming live sessions to YouTube.

Seven Jitsi servers were set up, one for each parallel session. The participants of the sessions were helped and assisted by eight student volunteers (from the universities of Cagliari, Florence, Perugia, and Bari), who provided technical support and ensured smooth running of the conference proceedings.

The implementation of the software infrastructure and the technical coordination of the volunteers were carried out by Damiano Perri and Marco Simonetti.

Our warmest thanks go to all the student volunteers, to the technical coordinators, and to the development communities of Jitsi, Jibri, Riot, and Matrix, who made their terrific platforms available as open source software.

A big thank you goes to all of the 450 speakers, many of whom showed an enormous collaborative spirit, sometimes participating and presenting at almost prohibitive times of the day, given that the participants of this year's conference came from 58 countries scattered over many time zones of the globe.

Finally, we would like to thank Google for letting us stream all the live events via YouTube. In addition to lightening the load of our Jitsi servers, this allowed us to record the event and to be able to review the most exciting moments of the conference.

Ivan Blečić
Chiara Garau

Welcome Message from the Organizers

Organization

ICCSA 2021 was organized by the University of Cagliari (Italy), the University of Perugia (Italy), the University of Basilicata (Italy), Monash University (Australia), Kyushu Sangyo University (Japan), and the University of Minho (Portugal).

Honorary General Chairs

Norio Shiratori	Chuo University, Japan
Kenneth C. J. Tan	Sardina Systems, UK
Corrado Zoppi	University of Cagliari, Italy

General Chairs

Osvaldo Gervasi	University of Perugia, Italy
Ivan Blečić	University of Cagliari, Italy
David Taniar	Monash University, Australia

Program Committee Chairs

Beniamino Murgante	University of Basilicata, Italy
Bernady O. Apduhan	Kyushu Sangyo University, Japan
Chiara Garau	University of Cagliari, Italy
Ana Maria A. C. Rocha	University of Minho, Portugal

International Advisory Committee

Jemal Abawajy	Deakin University, Australia
Dharma P. Agarwal	University of Cincinnati, USA
Rajkumar Buyya	University of Melbourne, Australia
Claudia Bauzer Medeiros	University of Campinas, Brazil
Manfred M. Fisher	Vienna University of Economics and Business, Austria
Marina L. Gavrilova	University of Calgary, Canada
Yee Leung	Chinese University of Hong Kong, China

International Liaison Chairs

Giuseppe Borruso	University of Trieste, Italy
Elise De Donker	Western Michigan University, USA
Maria Irene Falcão	University of Minho, Portugal
Robert C. H. Hsu	Chung Hua University, Taiwan
Tai-Hoon Kim	Beijing Jaotong University, China

Vladimir Korkhov   St. Petersburg University, Russia
Sanjay Misra   Covenant University, Nigeria
Takashi Naka   Kyushu Sangyo University, Japan
Rafael D. C. Santos   National Institute for Space Research, Brazil
Maribel Yasmina Santos   University of Minho, Portugal
Elena Stankova   St. Petersburg University, Russia

Workshop and Session Chairs

Beniamino Murgante   University of Basilicata, Italy
Sanjay Misra   Covenant University, Nigeria
Jorge Gustavo Rocha   University of Minho, Portugal

Awards Chair

Wenny Rahayu   La Trobe University, Australia

Publicity Committee Chairs

Elmer Dadios   De La Salle University, Philippines
Nataliia Kulabukhova   St. Petersburg University, Russia
Daisuke Takahashi   Tsukuba University, Japan
Shangwang Wang   Beijing University of Posts and Telecommunications,
                                 China

Technology Chairs

Damiano Perri   University of Florence, Italy
Marco Simonetti   University of Florence, Italy

Local Arrangement Chairs

Ivan Blečić   University of Cagliari, Italy
Chiara Garau   University of Cagliari, Italy
Alfonso Annunziata   University of Cagliari, Italy
Ginevra Balletto   University of Cagliari, Italy
Giuseppe Borruso   University of Trieste, Italy
Alessandro Buccini   University of Cagliari, Italy
Michele Campagna   University of Cagliari, Italy
Mauro Coni   University of Cagliari, Italy
Anna Maria Colavitti   University of Cagliari, Italy
Giulia Desogus   University of Cagliari, Italy
Caterina Fenu   University of Cagliari, Italy
Sabrina Lai   University of Cagliari, Italy
Francesca Maltinti   University of Cagliari, Italy
Pasquale Mistretta   University of Cagliari, Italy

Augusto Montisci University of Cagliari, Italy
Francesco Pinna University of Cagliari, Italy
Davide Spano University of Cagliari, Italy
Giuseppe A. Trunfio University of Sassari, Italy
Corrado Zoppi University of Cagliari, Italy

Program Committee

Vera Afreixo University of Aveiro, Portugal
Filipe Alvelos University of Minho, Portugal
Hartmut Asche University of Potsdam, Germany
Ginevra Balletto University of Cagliari, Italy
Michela Bertolotto University College Dublin, Ireland
Sandro Bimonte INRAE-TSCF, France
Rod Blais University of Calgary, Canada
Ivan Blečić University of Sassari, Italy
Giuseppe Borruso University of Trieste, Italy
Ana Cristina Braga University of Minho, Portugal
Massimo Cafaro University of Salento, Italy
Yves Caniou University of Lyon, France
José A. Cardoso e Cunha Universidade Nova de Lisboa, Portugal
Rui Cardoso University of Beira Interior, Portugal
Leocadio G. Casado University of Almeria, Spain
Carlo Cattani University of Salerno, Italy
Mete Celik Erciyes University, Turkey
Maria Cerreta University of Naples "Federico II", Italy
Hyunseung Choo Sungkyunkwan University, South Korea
Chien-Sing Lee Sunway University, Malaysia
Min Young Chung Sungkyunkwan University, South Korea
Florbela Maria da Cruz Polytechnic Institute of Viana do Castelo, Portugal
 Domingues Correia
Gilberto Corso Pereira Federal University of Bahia, Brazil
Fernanda Costa University of Minho, Portugal
Alessandro Costantini INFN, Italy
Carla Dal Sasso Freitas Universidade Federal do Rio Grande do Sul, Brazil
Pradesh Debba The Council for Scientific and Industrial Research
  (CSIR), South Africa
Hendrik Decker Instituto Tecnolčgico de Informática, Spain
Robertas Damaševičius Kausan University of Technology, Lithuania
Frank Devai London South Bank University, UK
Rodolphe Devillers Memorial University of Newfoundland, Canada
Joana Matos Dias University of Coimbra, Portugal
Paolino Di Felice University of L'Aquila, Italy
Prabu Dorairaj NetApp, India/USA
Noelia Faginas Lago University of Perugia, Italy
M. Irene Falcao University of Minho, Portugal

Cherry Liu Fang	Ames Laboratory, USA
Florbela P. Fernandes	Polytechnic Institute of Bragança, Portugal
Jose-Jesus Fernandez	National Centre for Biotechnology, Spain
Paula Odete Fernandes	Polytechnic Institute of Bragança, Portugal
Adelaide de Fátima Baptista Valente Freitas	University of Aveiro, Portugal
Manuel Carlos Figueiredo	University of Minho, Portugal
Maria Celia Furtado Rocha	Universidade Federal da Bahia, Brazil
Chiara Garau	University of Cagliari, Italy
Paulino Jose Garcia Nieto	University of Oviedo, Spain
Jerome Gensel	LSR-IMAG, France
Maria Giaoutzi	National Technical University of Athens, Greece
Arminda Manuela Andrade Pereira Gonçalves	University of Minho, Portugal
Andrzej M. Goscinski	Deakin University, Australia
Eduardo Guerra	Free University of Bozen-Bolzano, Italy
Sevin Gümgüm	Izmir University of Economics, Turkey
Alex Hagen-Zanker	University of Cambridge, UK
Shanmugasundaram Hariharan	B.S. Abdur Rahman University, India
Eligius M. T. Hendrix	University of Malaga, Spain/Wageningen University, The Netherlands
Hisamoto Hiyoshi	Gunma University, Japan
Mustafa Inceoglu	EGE University, Turkey
Peter Jimack	University of Leeds, UK
Qun Jin	Waseda University, Japan
Yeliz Karaca	University of Massachusetts Medical School, USA
Farid Karimipour	Vienna University of Technology, Austria
Baris Kazar	Oracle Corp., USA
Maulana Adhinugraha Kiki	Telkom University, Indonesia
DongSeong Kim	University of Canterbury, New Zealand
Taihoon Kim	Hannam University, South Korea
Ivana Kolingerova	University of West Bohemia, Czech Republic
Nataliia Kulabukhova	St. Petersburg University, Russia
Vladimir Korkhov	St. Petersburg University, Russia
Rosa Lasaponara	National Research Council, Italy
Maurizio Lazzari	National Research Council, Italy
Cheng Siong Lee	Monash University, Australia
Sangyoun Lee	Yonsei University, South Korea
Jongchan Lee	Kunsan National University, South Korea
Chendong Li	University of Connecticut, USA
Gang Li	Deakin University, Australia
Fang Liu	Ames Laboratory, USA
Xin Liu	University of Calgary, Canada
Andrea Lombardi	University of Perugia, Italy
Savino Longo	University of Bari, Italy

Tinghuai Ma	Nanjing University of Information Science and Technology, China
Ernesto Marcheggiani	Katholieke Universiteit Leuven, Belgium
Antonino Marvuglia	Research Centre Henri Tudor, Luxembourg
Nicola Masini	National Research Council, Italy
Ilaria Matteucci	National Research Council, Italy
Eric Medvet	University of Trieste, Italy
Nirvana Meratnia	University of Twente, The Netherlands
Giuseppe Modica	University of Reggio Calabria, Italy
Josè Luis Montaña	University of Cantabria, Spain
Maria Filipa Mourão	Instituto Politécnico de Viana do Castelo, Portugal
Louiza de Macedo Mourelle	State University of Rio de Janeiro, Brazil
Nadia Nedjah	State University of Rio de Janeiro, Brazil
Laszlo Neumann	University of Girona, Spain
Kok-Leong Ong	Deakin University, Australia
Belen Palop	Universidad de Valladolid, Spain
Marcin Paprzycki	Polish Academy of Sciences, Poland
Eric Pardede	La Trobe University, Australia
Kwangjin Park	Wonkwang University, South Korea
Ana Isabel Pereira	Polytechnic Institute of Bragança, Portugal
Massimiliano Petri	University of Pisa, Italy
Telmo Pinto	University of Coimbra, Portugal
Maurizio Pollino	Italian National Agency for New Technologies, Energy and Sustainable Economic Development, Italy
Alenka Poplin	University of Hamburg, Germany
Vidyasagar Potdar	Curtin University of Technology, Australia
David C. Prosperi	Florida Atlantic University, USA
Wenny Rahayu	La Trobe University, Australia
Jerzy Respondek	Silesian University of Technology Poland
Humberto Rocha	INESC-Coimbra, Portugal
Jon Rokne	University of Calgary, Canada
Octavio Roncero	CSIC, Spain
Maytham Safar	Kuwait University, Kuwait
Francesco Santini	University of Perugia, Italy
Chiara Saracino	A.O. Ospedale Niguarda Ca' Granda, Italy
Haiduke Sarafian	Pennsylvania State University, USA
Marco Paulo Seabra dos Reis	University of Coimbra, Portugal
Jie Shen	University of Michigan, USA
Qi Shi	Liverpool John Moores University, UK
Dale Shires	U.S. Army Research Laboratory, USA
Inês Soares	University of Coimbra, Portugal
Elena Stankova	St. Petersburg University, Russia
Takuo Suganuma	Tohoku University, Japan
Eufemia Tarantino	Polytechnic University of Bari, Italy
Sergio Tasso	University of Perugia, Italy

Ana Paula Teixeira	University of Trás-os-Montes and Alto Douro, Portugal
Senhorinha Teixeira	University of Minho, Portugal
M. Filomena Teodoro	Portuguese Naval Academy/University of Lisbon, Portugal
Parimala Thulasiraman	University of Manitoba, Canada
Carmelo Torre	Polytechnic University of Bari, Italy
Javier Martinez Torres	Centro Universitario de la Defensa Zaragoza, Spain
Giuseppe A. Trunfio	University of Sassari, Italy
Pablo Vanegas	University of Cuenca, Equador
Marco Vizzari	University of Perugia, Italy
Varun Vohra	Merck Inc., USA
Koichi Wada	University of Tsukuba, Japan
Krzysztof Walkowiak	Wroclaw University of Technology, Poland
Zequn Wang	Intelligent Automation Inc, USA
Robert Weibel	University of Zurich, Switzerland
Frank Westad	Norwegian University of Science and Technology, Norway
Roland Wismüller	Universität Siegen, Germany
Mudasser Wyne	National University, USA
Chung-Huang Yang	National Kaohsiung Normal University, Taiwan
Xin-She Yang	National Physical Laboratory, UK
Salim Zabir	National Institute of Technology, Tsuruoka, Japan
Haifeng Zhao	University of California, Davis, USA
Fabiana Zollo	University of Venice "Cà Foscari", Italy
Albert Y. Zomaya	University of Sydney, Australia

Workshop Organizers

Advanced Transport Tools and Methods (A2TM 2021)

| Massimiliano Petri | University of Pisa, Italy |
| Antonio Pratelli | University of Pisa, Italy |

Advances in Artificial Intelligence Learning Technologies: Blended Learning, STEM, Computational Thinking and Coding (AAILT 2021)

Alfredo Milani	University of Perugia, Italy
Giulio Biondi	University of Florence, Italy
Sergio Tasso	University of Perugia, Italy

Workshop on Advancements in Applied Machine Learning and Data Analytics (AAMDA 2021)

Alessandro Costantini	INFN, Italy
Davide Salomoni	INFN, Italy
Doina Cristina Duma	INFN, Italy
Daniele Cesini	INFN, Italy

Automatic Landform Classification: Spatial Methods and Applications (ALCSMA 2021)

Maria Danese	ISPC, National Research Council, Italy
Dario Gioia	ISPC, National Research Council, Italy

Application of Numerical Analysis to Imaging Science (ANAIS 2021)

Caterina Fenu	University of Cagliari, Italy
Alessandro Buccini	University of Cagliari, Italy

Advances in Information Systems and Technologies for Emergency Management, Risk Assessment and Mitigation Based on the Resilience Concepts (ASTER 2021)

Maurizio Pollino	ENEA, Italy
Marco Vona	University of Basilicata, Italy
Amedeo Flora	University of Basilicata, Italy
Chiara Iacovino	University of Basilicata, Italy
Beniamino Murgante	University of Basilicata, Italy

Advances in Web Based Learning (AWBL 2021)

Birol Ciloglugil	Ege University, Turkey
Mustafa Murat Inceoglu	Ege University, Turkey

Blockchain and Distributed Ledgers: Technologies and Applications (BDLTA 2021)

Vladimir Korkhov	St. Petersburg University, Russia
Elena Stankova	St. Petersburg University, Russia
Nataliia Kulabukhova	St. Petersburg University, Russia

Bio and Neuro Inspired Computing and Applications (BIONCA 2021)

Nadia Nedjah	State University of Rio de Janeiro, Brazil
Luiza De Macedo Mourelle	State University of Rio de Janeiro, Brazil

Computational and Applied Mathematics (CAM 2021)

Maria Irene Falcão	University of Minho, Portugal
Fernando Miranda	University of Minho, Portugal

Computational and Applied Statistics (CAS 2021)

Ana Cristina Braga	University of Minho, Portugal

Computerized Evaluation of Economic Activities: Urban Spaces (CEEA 2021)

Diego Altafini	Università di Pisa, Italy
Valerio Cutini	Università di Pisa, Italy

Computational Geometry and Applications (CGA 2021)

Marina Gavrilova   University of Calgary, Canada

Collaborative Intelligence in Multimodal Applications (CIMA 2021)

Robertas Damasevicius   Kaunas University of Technology, Lithuania
Rytis Maskeliunas   Kaunas University of Technology, Lithuania

Computational Optimization and Applications (COA 2021)

Ana Rocha   University of Minho, Portugal
Humberto Rocha   University of Coimbra, Portugal

Computational Astrochemistry (CompAstro 2021)

Marzio Rosi   University of Perugia, Italy
Cecilia Ceccarelli   University of Grenoble, France
Stefano Falcinelli   University of Perugia, Italy
Dimitrios Skouteris   Master-Up, Italy

Computational Science and HPC (CSHPC 2021)

Elise de Doncker   Western Michigan University, USA
Fukuko Yuasa   High Energy Accelerator Research Organization
                                  (KEK), Japan
Hideo Matsufuru   High Energy Accelerator Research Organization
                                  (KEK), Japan

Cities, Technologies and Planning (CTP 2021)

Malgorzata Hanzl   University of Łódż, Poland
Beniamino Murgante   University of Basilicata, Italy
Ljiljana Zivkovic   Ministry of Construction, Transport and
                                  Infrastructure/Institute of Architecture and Urban
                                  and Spatial Planning of Serbia, Serbia
Anastasia Stratigea   National Technical University of Athens, Greece
Giuseppe Borruso   University of Trieste, Italy
Ginevra Balletto   University of Cagliari, Italy

Advanced Modeling E-Mobility in Urban Spaces (DEMOS 2021)

Tiziana Campisi   Kore University of Enna, Italy
Socrates Basbas   Aristotle University of Thessaloniki, Greece
Ioannis Politis   Aristotle University of Thessaloniki, Greece
Florin Nemtanu   Polytechnic University of Bucharest, Romania
Giovanna Acampa   Kore University of Enna, Italy
Wolfgang Schulz   Zeppelin University, Germany

Digital Transformation and Smart City (DIGISMART 2021)

Mauro Mazzei   National Research Council, Italy

Econometric and Multidimensional Evaluation in Urban Environment (EMEUE 2021)

Carmelo Maria Torre   Polytechnic University of Bari, Italy
Maria Cerreta   University "Federico II" of Naples, Italy
Pierluigi Morano   Polytechnic University of Bari, Italy
Simona Panaro   University of Portsmouth, UK
Francesco Tajani   Sapienza University of Rome, Italy
Marco Locurcio   Polytechnic University of Bari, Italy

The 11th International Workshop on Future Computing System Technologies and Applications (FiSTA 2021)

Bernady Apduhan   Kyushu Sangyo University, Japan
Rafael Santos   Brazilian National Institute for Space Research, Brazil

Transformational Urban Mobility: Challenges and Opportunities During and Post COVID Era (FURTHER 2021)

Tiziana Campisi   Kore University of Enna, Italy
Socrates Basbas   Aristotle University of Thessaloniki, Greece
Dilum Dissanayake   Newcastle University, UK
Kh Md Nahiduzzaman   University of British Columbia, Canada
Nurten Akgün Tanbay   Bursa Technical University, Turkey
Khaled J. Assi   King Fahd University of Petroleum and Minerals,
                                        Saudi Arabia
Giovanni Tesoriere   Kore University of Enna, Italy
Motasem Darwish   Middle East University, Jordan

Geodesign in Decision Making: Meta Planning and Collaborative Design for Sustainable and Inclusive Development (GDM 2021)

Francesco Scorza   University of Basilicata, Italy
Michele Campagna   University of Cagliari, Italy
Ana Clara Mourao Moura   Federal University of Minas Gerais, Brazil

Geomatics in Forestry and Agriculture: New Advances and Perspectives (GeoForAgr 2021)

Maurizio Pollino   ENEA, Italy
Giuseppe Modica   University of Reggio Calabria, Italy
Marco Vizzari   University of Perugia, Italy

Geographical Analysis, Urban Modeling, Spatial Statistics (GEOG-AND-MOD 2021)

Beniamino Murgante	University of Basilicata, Italy
Giuseppe Borruso	University of Trieste, Italy
Hartmut Asche	University of Potsdam, Germany

Geomatics for Resource Monitoring and Management (GRMM 2021)

Eufemia Tarantino	Polytechnic University of Bari, Italy
Enrico Borgogno Mondino	University of Turin, Italy
Alessandra Capolupo	Polytechnic University of Bari, Italy
Mirko Saponaro	Polytechnic University of Bari, Italy

12th International Symposium on Software Quality (ISSQ 2021)

Sanjay Misra	Covenant University, Nigeria

10th International Workshop on Collective, Massive and Evolutionary Systems (IWCES 2021)

Alfredo Milani	University of Perugia, Italy
Rajdeep Niyogi	Indian Institute of Technology, Roorkee, India

Land Use Monitoring for Sustainability (LUMS 2021)

Carmelo Maria Torre	Polytechnic University of Bari, Italy
Maria Cerreta	University "Federico II" of Naples, Italy
Massimiliano Bencardino	University of Salerno, Italy
Alessandro Bonifazi	Polytechnic University of Bari, Italy
Pasquale Balena	Polytechnic University of Bari, Italy
Giuliano Poli	University "Federico II" of Naples, Italy

Machine Learning for Space and Earth Observation Data (MALSEOD 2021)

Rafael Santos	Instituto Nacional de Pesquisas Espaciais, Brazil
Karine Ferreira	Instituto Nacional de Pesquisas Espaciais, Brazil

Building Multi-dimensional Models for Assessing Complex Environmental Systems (MES 2021)

Marta Dell'Ovo	Polytechnic University of Milan, Italy
Vanessa Assumma	Polytechnic University of Turin, Italy
Caterina Caprioli	Polytechnic University of Turin, Italy
Giulia Datola	Polytechnic University of Turin, Italy
Federico dell'Anna	Polytechnic University of Turin, Italy

Ecosystem Services: Nature's Contribution to People in Practice. Assessment Frameworks, Models, Mapping, and Implications (NC2P 2021)

Francesco Scorza	University of Basilicata, Italy
Sabrina Lai	University of Cagliari, Italy
Ana Clara Mourao Moura	Federal University of Minas Gerais, Brazil
Corrado Zoppi	University of Cagliari, Italy
Dani Broitman	Technion, Israel Institute of Technology, Israel

Privacy in the Cloud/Edge/IoT World (PCEIoT 2021)

Michele Mastroianni	University of Campania Luigi Vanvitelli, Italy
Lelio Campanile	University of Campania Luigi Vanvitelli, Italy
Mauro Iacono	University of Campania Luigi Vanvitelli, Italy

Processes, Methods and Tools Towards RESilient Cities and Cultural Heritage Prone to SOD and ROD Disasters (RES 2021)

Elena Cantatore	Polytechnic University of Bari, Italy
Alberico Sonnessa	Polytechnic University of Bari, Italy
Dario Esposito	Polytechnic University of Bari, Italy

Risk, Resilience and Sustainability in the Efficient Management of Water Resources: Approaches, Tools, Methodologies and Multidisciplinary Integrated Applications (RRS 2021)

Maria Macchiaroli	University of Salerno, Italy
Chiara D'Alpaos	Università degli Studi di Padova, Italy
Mirka Mobilia	Università degli Studi di Salerno, Italy
Antonia Longobardi	Università degli Studi di Salerno, Italy
Grazia Fattoruso	ENEA Research Center, Italy
Vincenzo Pellecchia	Ente Idrico Campano, Italy

Scientific Computing Infrastructure (SCI 2021)

Elena Stankova	St. Petersburg University, Russia
Vladimir Korkhov	St. Petersburg University, Russia
Natalia Kulabukhova	St. Petersburg University, Russia

Smart Cities and User Data Management (SCIDAM 2021)

Chiara Garau	University of Cagliari, Italy
Luigi Mundula	University of Cagliari, Italy
Gianni Fenu	University of Cagliari, Italy
Paolo Nesi	University of Florence, Italy
Paola Zamperlin	University of Pisa, Italy

13th International Symposium on Software Engineering Processes and Applications (SEPA 2021)

Sanjay Misra	Covenant University, Nigeria

Ports of the Future - Smartness and Sustainability (SmartPorts 2021)

Patrizia Serra	University of Cagliari, Italy
Gianfranco Fancello	University of Cagliari, Italy
Ginevra Balletto	University of Cagliari, Italy
Luigi Mundula	University of Cagliari, Italy
Marco Mazzarino	University of Venice, Italy
Giuseppe Borruso	University of Trieste, Italy
Maria del Mar Munoz Leonisio	Universidad de Cádiz, Spain

Smart Tourism (SmartTourism 2021)

Giuseppe Borruso	University of Trieste, Italy
Silvia Battino	University of Sassari, Italy
Ginevra Balletto	University of Cagliari, Italy
Maria del Mar Munoz Leonisio	Universidad de Cádiz, Spain
Ainhoa Amaro Garcia	Universidad de Alcalà/Universidad de Las Palmas, Spain
Francesca Krasna	University of Trieste, Italy

Sustainability Performance Assessment: Models, Approaches and Applications toward Interdisciplinary and Integrated Solutions (SPA 2021)

Francesco Scorza	University of Basilicata, Italy
Sabrina Lai	University of Cagliari, Italy
Jolanta Dvarioniene	Kaunas University of Technology, Lithuania
Valentin Grecu	Lucian Blaga University, Romania
Corrado Zoppi	University of Cagliari, Italy
Iole Cerminara	University of Basilicata, Italy

Smart and Sustainable Island Communities (SSIC 2021)

Chiara Garau	University of Cagliari, Italy
Anastasia Stratigea	National Technical University of Athens, Greece
Paola Zamperlin	University of Pisa, Italy
Francesco Scorza	University of Basilicata, Italy

Science, Technologies and Policies to Innovate Spatial Planning (STP4P 2021)

Chiara Garau	University of Cagliari, Italy
Daniele La Rosa	University of Catania, Italy
Francesco Scorza	University of Basilicata, Italy

Anna Maria Colavitti   University of Cagliari, Italy
Beniamino Murgante   University of Basilicata, Italy
Paolo La Greca   University of Catania, Italy

Sustainable Urban Energy Systems (SURENSYS 2021)

Luigi Mundula   University of Cagliari, Italy
Emilio Ghiani   University of Cagliari, Italy

Space Syntax for Cities in Theory and Practice (Syntax_City 2021)

Claudia Yamu   University of Groningen, The Netherlands
Akkelies van Nes   Western Norway University of Applied Sciences,
                             Norway
Chiara Garau   University of Cagliari, Italy

Theoretical and Computational Chemistry and Its Applications (TCCMA 2021)

Noelia Faginas-Lago   University of Perugia, Italy

13th International Workshop on Tools and Techniques in Software Development Process (TTSDP 2021)

Sanjay Misra   Covenant University, Nigeria

Urban Form Studies (UForm 2021)

Malgorzata Hanzl   Łódź University of Technology, Poland
Beniamino Murgante   University of Basilicata, Italy
Eufemia Tarantino   Polytechnic University of Bari, Italy
Irena Itova   University of Westminster, UK

Urban Space Accessibility and Safety (USAS 2021)

Chiara Garau   University of Cagliari, Italy
Francesco Pinna   University of Cagliari, Italy
Claudia Yamu   University of Groningen, The Netherlands
Vincenza Torrisi   University of Catania, Italy
Matteo Ignaccolo   University of Catania, Italy
Michela Tiboni   University of Brescia, Italy
Silvia Rossetti   University of Parma, Italy

Virtual and Augmented Reality and Applications (VRA 2021)

Osvaldo Gervasi   University of Perugia, Italy
Damiano Perri   University of Perugia, Italy
Marco Simonetti   University of Perugia, Italy
Sergio Tasso   University of Perugia, Italy

Workshop on Advanced and Computational Methods for Earth Science Applications (WACM4ES 2021)

Luca Piroddi	University of Cagliari, Italy
Laura Foddis	University of Cagliari, Italy
Augusto Montisci	University of Cagliari, Italy
Sergio Vincenzo Calcina	University of Cagliari, Italy
Sebastiano D'Amico	University of Malta, Malta
Giovanni Martinelli	Istituto Nazionale di Geofisica e Vulcanologia, Italy/Chinese Academy of Sciences, China

Sponsoring Organizations

ICCSA 2021 would not have been possible without the tremendous support of many organizations and institutions, for which all organizers and participants of ICCSA 2021 express their sincere gratitude:

Springer International Publishing AG, Germany
(https://www.springer.com)

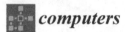

Computers Open Access Journal
(https://www.mdpi.com/journal/computers)

IEEE Italy Section, Italy
(https://italy.ieeer8.org/)

Centre-North Italy Chapter IEEE GRSS, Italy
(https://cispio.diet.uniroma1.it/marzano/ieee-grs/index.html)

Italy Section of the Computer Society, Italy
(https://site.ieee.org/italy-cs/)

University of Perugia, Italy
(https://www.unipg.it)

University of Cagliari, Italy
(https://unica.it/)

University of Basilicata, Italy
(http://www.unibas.it)

Monash University, Australia
(https://www.monash.edu/)

Kyushu Sangyo University, Japan
(https://www.kyusan-u.ac.jp/)

University of Minho, Portugal
(https://www.uminho.pt/)

Scientific Association Transport Infrastructures,
Italy
(https://www.stradeeautostrade.it/associazioni-e-
organizzazioni/asit-associazione-scientifica-
infrastrutture-trasporto/)

Regione Sardegna, Italy
(https://regione.sardegna.it/)

Comune di Cagliari, Italy
(https://www.comune.cagliari.it/)

Città Metropolitana di Cagliari

Cagliari Accessibility Lab (CAL)
(https://www.unica.it/unica/it/cagliari_
accessibility_lab.page/)

Referees

Nicodemo Abate	IMAA, National Research Council, Italy
Andre Ricardo Abed Grégio	Federal University of Paraná State, Brazil
Nasser Abu Zeid	Università di Ferrara, Italy
Lidia Aceto	Università del Piemonte Orientale, Italy
Nurten Akgün Tanbay	Bursa Technical University, Turkey
Filipe Alvelos	Universidade do Minho, Portugal
Paula Amaral	Universidade Nova de Lisboa, Portugal
Federico Amato	University of Lausanne, Switzerland
Marina Alexandra Pedro Andrade	ISCTE-IUL, Portugal
Debora Anelli	Sapienza University of Rome, Italy
Alfonso Annunziata	University of Cagliari, Italy
Fahim Anzum	University of Calgary, Canada
Tatsumi Aoyama	High Energy Accelerator Research Organization, Japan
Bernady Apduhan	Kyushu Sangyo University, Japan
Jonathan Apeh	Covenant University, Nigeria
Vasilike Argyropoulos	University of West Attica, Greece
Giuseppe Aronica	Università di Messina, Italy
Daniela Ascenzi	Università degli Studi di Trento, Italy
Vanessa Assumma	Politecnico di Torino, Italy
Muhammad Attique Khan	HITEC University Taxila, Pakistan
Vecdi Aytaç	Ege University, Turkey
Alina Elena Baia	University of Perugia, Italy
Ginevra Balletto	University of Cagliari, Italy
Marialaura Bancheri	ISAFOM, National Research Council, Italy
Benedetto Barabino	University of Brescia, Italy
Simona Barbaro	Università degli Studi di Palermo, Italy
Enrico Barbierato	Università Cattolica del Sacro Cuore di Milano, Italy
Jeniffer Barreto	Istituto Superior Técnico, Lisboa, Portugal
Michele Bartalini	TAGES, Italy
Socrates Basbas	Aristotle University of Thessaloniki, Greece
Silvia Battino	University of Sassari, Italy
Marcelo Becerra Rozas	Pontificia Universidad Católica de Valparaíso, Chile
Ranjan Kumar Behera	National Institute of Technology, Rourkela, India
Emanuele Bellini	University of Campania Luigi Vanvitelli, Italy
Massimo Bilancia	University of Bari Aldo Moro, Italy
Giulio Biondi	University of Firenze, Italy
Adriano Bisello	Eurac Research, Italy
Ignacio Blanquer	Universitat Politècnica de València, Spain
Semen Bochkov	Ulyanovsk State Technical University, Russia
Alexander Bogdanov	St. Petersburg University, Russia
Silvia Bonettini	University of Modena and Reggio Emilia, Italy
Enrico Borgogno Mondino	Università di Torino, Italy
Giuseppe Borruso	University of Trieste, Italy

Michele Bottazzi	University of Trento, Italy
Rahma Bouaziz	Taibah University, Saudi Arabia
Ouafik Boulariah	University of Salerno, Italy
Tulin Boyar	Yildiz Technical University, Turkey
Ana Cristina Braga	University of Minho, Portugal
Paolo Bragolusi	University of Padova, Italy
Luca Braidotti	University of Trieste, Italy
Alessandro Buccini	University of Cagliari, Italy
Jorge Buele	Universidad Tecnológica Indoamérica, Ecuador
Andrea Buffoni	TAGES, Italy
Sergio Vincenzo Calcina	University of Cagliari, Italy
Michele Campagna	University of Cagliari, Italy
Lelio Campanile	Università degli Studi della Campania Luigi Vanvitelli, Italy
Tiziana Campisi	Kore University of Enna, Italy
Antonino Canale	Kore University of Enna, Italy
Elena Cantatore	DICATECh, Polytechnic University of Bari, Italy
Pasquale Cantiello	Istituto Nazionale di Geofisica e Vulcanologia, Italy
Alessandra Capolupo	Polytechnic University of Bari, Italy
David Michele Cappelletti	University of Perugia, Italy
Caterina Caprioli	Politecnico di Torino, Italy
Sara Carcangiu	University of Cagliari, Italy
Pedro Carrasqueira	INESC Coimbra, Portugal
Arcangelo Castiglione	University of Salerno, Italy
Giulio Cavana	Politecnico di Torino, Italy
Davide Cerati	Politecnico di Milano, Italy
Maria Cerreta	University of Naples Federico II, Italy
Daniele Cesini	INFN-CNAF, Italy
Jabed Chowdhury	La Trobe University, Australia
Gennaro Ciccarelli	Iuav University of Venice, Italy
Birol Ciloglugil	Ege University, Turkey
Elena Cocuzza	Univesity of Catania, Italy
Anna Maria Colavitt	University of Cagliari, Italy
Cecilia Coletti	Università "G. d'Annunzio" di Chieti-Pescara, Italy
Alberto Collu	Independent Researcher, Italy
Anna Concas	University of Basilicata, Italy
Mauro Coni	University of Cagliari, Italy
Melchiorre Contino	Università di Palermo, Italy
Antonella Cornelio	Università degli Studi di Brescia, Italy
Aldina Correia	Politécnico do Porto, Portugal
Elisete Correia	Universidade de Trás-os-Montes e Alto Douro, Portugal
Florbela Correia	Polytechnic Institute of Viana do Castelo, Portugal
Stefano Corsi	Università degli Studi di Milano, Italy
Alberto Cortez	Polytechnic of University Coimbra, Portugal
Lino Costa	Universidade do Minho, Portugal

Alessandro Costantini	INFN, Italy
Marilena Cozzolino	Università del Molise, Italy
Giulia Crespi	Politecnico di Torino, Italy
Maurizio Crispino	Politecnico di Milano, Italy
Chiara D'Alpaos	University of Padova, Italy
Roberta D'Ambrosio	Università di Salerno, Italy
Sebastiano D'Amico	University of Malta, Malta
Hiroshi Daisaka	Hitotsubashi University, Japan
Gaia Daldanise	Italian National Research Council, Italy
Robertas Damasevicius	Silesian University of Technology, Poland
Maria Danese	ISPC, National Research Council, Italy
Bartoli Daniele	University of Perugia, Italy
Motasem Darwish	Middle East University, Jordan
Giulia Datola	Politecnico di Torino, Italy
Regina de Almeida	UTAD, Portugal
Elise de Doncker	Western Michigan University, USA
Mariella De Fino	Politecnico di Bari, Italy
Giandomenico De Luca	Mediterranean University of Reggio Calabria, Italy
Luiza de Macedo Mourelle	State University of Rio de Janeiro, Brazil
Gianluigi De Mare	University of Salerno, Italy
Itamir de Morais Barroca Filho	Federal University of Rio Grande do Norte, Brazil
Samuele De Petris	Università di Torino, Italy
Marcilio de Souto	LIFO, University of Orléans, France
Alexander Degtyarev	St. Petersburg University, Russia
Federico Dell'Anna	Politecnico di Torino, Italy
Marta Dell'Ovo	Politecnico di Milano, Italy
Fernanda Della Mura	University of Naples "Federico II", Italy
Ahu Dereli Dursun	Istanbul Commerce University, Turkey
Bashir Derradji	University of Sfax, Tunisia
Giulia Desogus	Università degli Studi di Cagliari, Italy
Marco Dettori	Università degli Studi di Sassari, Italy
Frank Devai	London South Bank University, UK
Felicia Di Liddo	Polytechnic University of Bari, Italy
Valerio Di Pinto	University of Naples "Federico II", Italy
Joana Dias	University of Coimbra, Portugal
Luis Dias	University of Minho, Portugal
Patricia Diaz de Alba	Gran Sasso Science Institute, Italy
Isabel Dimas	University of Coimbra, Portugal
Aleksandra Djordjevic	University of Belgrade, Serbia
Luigi Dolores	Università degli Studi di Salerno, Italy
Marco Donatelli	University of Insubria, Italy
Doina Cristina Duma	INFN-CNAF, Italy
Fabio Durastante	University of Pisa, Italy
Aziz Dursun	Virginia Tech University, USA
Juan Enrique-Romero	Université Grenoble Alpes, France

Annunziata Esposito Amideo	University College Dublin, Ireland
Dario Esposito	Polytechnic University of Bari, Italy
Claudio Estatico	University of Genova, Italy
Noelia Faginas-Lago	Università di Perugia, Italy
Maria Irene Falcão	University of Minho, Portugal
Stefano Falcinelli	University of Perugia, Italy
Alessandro Farina	University of Pisa, Italy
Grazia Fattoruso	ENEA, Italy
Caterina Fenu	University of Cagliari, Italy
Luisa Fermo	University of Cagliari, Italy
Florbela Fernandes	Instituto Politecnico de Braganca, Portugal
Rosário Fernandes	University of Minho, Portugal
Luis Fernandez-Sanz	University of Alcala, Spain
Alessia Ferrari	Università di Parma, Italy
Luís Ferrás	University of Minho, Portugal
Ângela Ferreira	Instituto Politécnico de Bragança, Portugal
Flora Ferreira	University of Minho, Portugal
Manuel Carlos Figueiredo	University of Minho, Portugal
Ugo Fiore	University of Naples "Parthenope", Italy
Amedeo Flora	University of Basilicata, Italy
Hector Florez	Universidad Distrital Francisco Jose de Caldas, Colombia
Maria Laura Foddis	University of Cagliari, Italy
Valentina Franzoni	Perugia University, Italy
Adelaide Freitas	University of Aveiro, Portugal
Samuel Frimpong	Durban University of Technology, South Africa
Ioannis Fyrogenis	Aristotle University of Thessaloniki, Greece
Marika Gaballo	Politecnico di Torino, Italy
Laura Gabrielli	Iuav University of Venice, Italy
Ivan Gankevich	St. Petersburg University, Russia
Chiara Garau	University of Cagliari, Italy
Ernesto Garcia Para	Universidad del País Vasco, Spain,
Fernando Garrido	Universidad Técnica del Norte, Ecuador
Marina Gavrilova	University of Calgary, Canada
Silvia Gazzola	University of Bath, UK
Georgios Georgiadis	Aristotle University of Thessaloniki, Greece
Osvaldo Gervasi	University of Perugia, Italy
Andrea Gioia	Polytechnic University of Bari, Italy
Dario Gioia	ISPC-CNT, Italy
Raffaele Giordano	IRSS, National Research Council, Italy
Giacomo Giorgi	University of Perugia, Italy
Eleonora Giovene di Girasole	IRISS, National Research Council, Italy
Salvatore Giuffrida	Università di Catania, Italy
Marco Gola	Politecnico di Milano, Italy

A. Manuela Gonçalves	University of Minho, Portugal
Yuriy Gorbachev	Coddan Technologies LLC, Russia
Angela Gorgoglione	Universidad de la República, Uruguay
Yusuke Gotoh	Okayama University, Japan
Anestis Gourgiotis	University of Thessaly, Greece
Valery Grishkin	St. Petersburg University, Russia
Alessandro Grottesi	CINECA, Italy
Eduardo Guerra	Free University of Bozen-Bolzano, Italy
Ayse Giz Gulnerman	Ankara HBV University, Turkey
Sevin Gümgüm	Izmir University of Economics, Turkey
Himanshu Gupta	BITS Pilani, Hyderabad, India
Sandra Haddad	Arab Academy for Science, Egypt
Malgorzata Hanzl	Lodz University of Technology, Poland
Shoji Hashimoto	KEK, Japan
Peter Hegedus	University of Szeged, Hungary
Eligius M. T. Hendrix	Universidad de Málaga, Spain
Edmond Ho	Northumbria University, UK
Guan Yue Hong	Western Michigan University, USA
Vito Iacobellis	Polytechnic University of Bari, Italy
Mauro Iacono	Università degli Studi della Campania, Italy
Chiara Iacovino	University of Basilicata, Italy
Antonino Iannuzzo	ETH Zurich, Switzerland
Ali Idri	University Mohammed V, Morocco
Oana-Ramona Ilovan	Babeş-Bolyai University, Romania
Mustafa Inceoglu	Ege University, Turkey
Tadashi Ishikawa	KEK, Japan
Federica Isola	University of Cagliari, Italy
Irena Itova	University of Westminster, UK
Edgar David de Izeppi	VTTI, USA
Marija Jankovic	CERTH, Greece
Adrian Jaramillo	Universidad Tecnológica Metropolitana, Chile
Monalisa Jena	Fakir Mohan University, India
Dorota Kamrowska-Załuska	Gdansk University of Technology, Poland
Issaku Kanamori	RIKEN Center for Computational Science, Japan
Korhan Karabulut	Yasar University, Turkey
Yeliz Karaca	University of Massachusetts Medical School, USA
Vicky Katsoni	University of West Attica, Greece
Dimitris Kavroudakis	University of the Aegean, Greece
Shuhei Kimura	Okayama University, Japan
Joanna Kolozej	Cracow University of Technology, Poland
Vladimir Korkhov	St. Petersburg University, Russia
Thales Körting	INPE, Brazil
Tomonori Kouya	Shizuoka Institute of Science and Technology, Japan
Sylwia Krzysztofik	Lodz University of Technology, Poland
Nataliia Kulabukhova	St. Petersburg University, Russia
Shrinivas B. Kulkarni	SDM College of Engineering and Technology, India

Pavan Kumar	University of Calgary, Canada
Anisha Kumari	National Institute of Technology, Rourkela, India
Ludovica La Rocca	University of Naples "Federico II", Italy
Daniele La Rosa	University of Catania, Italy
Sabrina Lai	University of Cagliari, Italy
Giuseppe Francesco Cesare Lama	University of Naples "Federico II", Italy
Mariusz Lamprecht	University of Lodz, Poland
Vincenzo Laporta	National Research Council, Italy
Chien-Sing Lee	Sunway University, Malaysia
José Isaac Lemus Romani	Pontifical Catholic University of Valparaíso, Chile
Federica Leone	University of Cagliari, Italy
Alexander H. Levis	George Mason University, USA
Carola Lingua	Polytechnic University of Turin, Italy
Marco Locurcio	Polytechnic University of Bari, Italy
Andrea Lombardi	University of Perugia, Italy
Savino Longo	University of Bari, Italy
Fernando Lopez Gayarre	University of Oviedo, Spain
Yan Lu	Western Michigan University, USA
Maria Macchiaroli	University of Salerno, Italy
Helmuth Malonek	University of Aveiro, Portugal
Francesca Maltinti	University of Cagliari, Italy
Luca Mancini	University of Perugia, Italy
Marcos Mandado	University of Vigo, Spain
Ernesto Marcheggiani	Università Politecnica delle Marche, Italy
Krassimir Markov	University of Telecommunications and Post, Bulgaria
Giovanni Martinelli	INGV, Italy
Alessandro Marucci	University of L'Aquila, Italy
Fiammetta Marulli	University of Campania Luigi Vanvitelli, Italy
Gabriella Maselli	University of Salerno, Italy
Rytis Maskeliunas	Kaunas University of Technology, Lithuania
Michele Mastroianni	University of Campania Luigi Vanvitelli, Italy
Cristian Mateos	Universidad Nacional del Centro de la Provincia de Buenos Aires, Argentina
Hideo Matsufuru	High Energy Accelerator Research Organization (KEK), Japan
D'Apuzzo Mauro	University of Cassino and Southern Lazio, Italy
Chiara Mazzarella	University Federico II, Italy
Marco Mazzarino	University of Venice, Italy
Giovanni Mei	University of Cagliari, Italy
Mário Melo	Federal Institute of Rio Grande do Norte, Brazil
Francesco Mercaldo	University of Molise, Italy
Alfredo Milani	University of Perugia, Italy
Alessandra Milesi	University of Cagliari, Italy
Antonio Minervino	ISPC, National Research Council, Italy
Fernando Miranda	Universidade do Minho, Portugal

B. Mishra	University of Szeged, Hungary
Sanjay Misra	Covenant University, Nigeria
Mirka Mobilia	University of Salerno, Italy
Giuseppe Modica	Università degli Studi di Reggio Calabria, Italy
Mohammadsadegh Mohagheghi	Vali-e-Asr University of Rafsanjan, Iran
Mohamad Molaei Qelichi	University of Tehran, Iran
Mario Molinara	University of Cassino and Southern Lazio, Italy
Augusto Montisci	Università degli Studi di Cagliari, Italy
Pierluigi Morano	Polytechnic University of Bari, Italy
Ricardo Moura	Universidade Nova de Lisboa, Portugal
Ana Clara Mourao Moura	Federal University of Minas Gerais, Brazil
Maria Mourao	Polytechnic Institute of Viana do Castelo, Portugal
Daichi Mukunoki	RIKEN Center for Computational Science, Japan
Beniamino Murgante	University of Basilicata, Italy
Naohito Nakasato	University of Aizu, Japan
Grazia Napoli	Università degli Studi di Palermo, Italy
Isabel Cristina Natário	Universidade Nova de Lisboa, Portugal
Nadia Nedjah	State University of Rio de Janeiro, Brazil
Antonio Nesticò	University of Salerno, Italy
Andreas Nikiforiadis	Aristotle University of Thessaloniki, Greece
Keigo Nitadori	RIKEN Center for Computational Science, Japan
Silvio Nocera	Iuav University of Venice, Italy
Giuseppina Oliva	University of Salerno, Italy
Arogundade Oluwasefunmi	Academy of Mathematics and System Science, China
Ken-ichi Oohara	University of Tokyo, Japan
Tommaso Orusa	University of Turin, Italy
M. Fernanda P. Costa	University of Minho, Portugal
Roberta Padulano	Centro Euro-Mediterraneo sui Cambiamenti Climatici, Italy
Maria Panagiotopoulou	National Technical University of Athens, Greece
Jay Pancham	Durban University of Technology, South Africa
Gianni Pantaleo	University of Florence, Italy
Dimos Pantazis	University of West Attica, Greece
Michela Paolucci	University of Florence, Italy
Eric Pardede	La Trobe University, Australia
Olivier Parisot	Luxembourg Institute of Science and Technology, Luxembourg
Vincenzo Pellecchia	Ente Idrico Campano, Italy
Anna Pelosi	University of Salerno, Italy
Edit Pengő	University of Szeged, Hungary
Marco Pepe	University of Salerno, Italy
Paola Perchinunno	University of Cagliari, Italy
Ana Pereira	Polytechnic Institute of Bragança, Portugal
Mariano Pernetti	University of Campania, Italy
Damiano Perri	University of Perugia, Italy

Federica Pes	University of Cagliari, Italy
Marco Petrelli	Roma Tre University, Italy
Massimiliano Petri	University of Pisa, Italy
Khiem Phan	Duy Tan University, Vietnam
Alberto Ferruccio Piccinni	Polytechnic of Bari, Italy
Angela Pilogallo	University of Basilicata, Italy
Francesco Pinna	University of Cagliari, Italy
Telmo Pinto	University of Coimbra, Portugal
Luca Piroddi	University of Cagliari, Italy
Darius Plonis	Vilnius Gediminas Technical University, Lithuania
Giuliano Poli	University of Naples "Federico II", Italy
Maria João Polidoro	Polytecnic Institute of Porto, Portugal
Ioannis Politis	Aristotle University of Thessaloniki, Greece
Maurizio Pollino	ENEA, Italy
Antonio Pratelli	University of Pisa, Italy
Salvatore Praticò	Mediterranean University of Reggio Calabria, Italy
Marco Prato	University of Modena and Reggio Emilia, Italy
Carlotta Quagliolo	Polytechnic University of Turin, Italy
Emanuela Quaquero	Univesity of Cagliari, Italy
Garrisi Raffaele	Polizia postale e delle Comunicazioni, Italy
Nicoletta Rassu	University of Cagliari, Italy
Hafiz Tayyab Rauf	University of Bradford, UK
Michela Ravanelli	Sapienza University of Rome, Italy
Roberta Ravanelli	Sapienza University of Rome, Italy
Alfredo Reder	Centro Euro-Mediterraneo sui Cambiamenti Climatici, Italy
Stefania Regalbuto	University of Naples "Federico II", Italy
Rommel Regis	Saint Joseph's University, USA
Lothar Reichel	Kent State University, USA
Marco Reis	University of Coimbra, Portugal
Maria Reitano	University of Naples "Federico II", Italy
Jerzy Respondek	Silesian University of Technology, Poland
Elisa Riccietti	École Normale Supérieure de Lyon, France
Albert Rimola	Universitat Autònoma de Barcelona, Spain
Angela Rizzo	University of Bari, Italy
Ana Maria A. C. Rocha	University of Minho, Portugal
Fabio Rocha	Institute of Technology and Research, Brazil
Humberto Rocha	University of Coimbra, Portugal
Maria Clara Rocha	Polytechnic Institute of Coimbra, Portugal
Miguel Rocha	University of Minho, Portugal
Giuseppe Rodriguez	University of Cagliari, Italy
Guillermo Rodriguez	UNICEN, Argentina
Elisabetta Ronchieri	INFN, Italy
Marzio Rosi	University of Perugia, Italy
Silvia Rossetti	University of Parma, Italy
Marco Rossitti	Polytechnic University of Milan, Italy

Francesco Rotondo	Marche Polytechnic University, Italy
Irene Rubino	Polytechnic University of Turin, Italy
Agustín Salas	Pontifical Catholic University of Valparaíso, Chile
Juan Pablo Sandoval Alcocer	Universidad Católica Boliviana "San Pablo", Bolivia
Luigi Santopietro	University of Basilicata, Italy
Rafael Santos	National Institute for Space Research, Brazil
Valentino Santucci	Università per Stranieri di Perugia, Italy
Mirko Saponaro	Polytechnic University of Bari, Italy
Filippo Sarvia	University of Turin, Italy
Marco Scaioni	Polytechnic University of Milan, Italy
Rafal Scherer	Częstochowa University of Technology, Poland
Francesco Scorza	University of Basilicata, Italy
Ester Scotto di Perta	University of Napoli "Federico II", Italy
Monica Sebillo	University of Salerno, Italy
Patrizia Serra	University of Cagliari, Italy
Ricardo Severino	University of Minho, Portugal
Jie Shen	University of Michigan, USA
Huahao Shou	Zhejiang University of Technology, China
Miltiadis Siavvas	Centre for Research and Technology Hellas, Greece
Brandon Sieu	University of Calgary, Canada
Ângela Silva	Instituto Politécnico de Viana do Castelo, Portugal
Carina Silva	Polytechic Institute of Lisbon, Portugal
Joao Carlos Silva	Polytechnic Institute of Cavado and Ave, Portugal
Fabio Silveira	Federal University of Sao Paulo, Brazil
Marco Simonetti	University of Florence, Italy
Ana Jacinta Soares	University of Minho, Portugal
Maria Joana Soares	University of Minho, Portugal
Michel Soares	Federal University of Sergipe, Brazil
George Somarakis	Foundation for Research and Technology Hellas, Greece
Maria Somma	University of Naples "Federico II", Italy
Alberico Sonnessa	Polytechnic University of Bari, Italy
Elena Stankova	St. Petersburg University, Russia
Flavio Stochino	University of Cagliari, Italy
Anastasia Stratigea	National Technical University of Athens, Greece
Yasuaki Sumida	Kyushu Sangyo University, Japan
Yue Sun	European X-Ray Free-Electron Laser Facility, Germany
Kirill Sviatov	Ulyanovsk State Technical University, Russia
Daisuke Takahashi	University of Tsukuba, Japan
Aladics Tamás	University of Szeged, Hungary
David Taniar	Monash University, Australia
Rodrigo Tapia McClung	Centro de Investigación en Ciencias de Información Geoespacial, Mexico
Eufemia Tarantino	Polytechnic University of Bari, Italy

Sergio Tasso	University of Perugia, Italy
Ana Paula Teixeira	Universidade de Trás-os-Montes e Alto Douro, Portugal
Senhorinha Teixeira	University of Minho, Portugal
Tengku Adil Tengku Izhar	Universiti Teknologi MARA, Malaysia
Maria Filomena Teodoro	University of Lisbon/Portuguese Naval Academy, Portugal
Giovanni Tesoriere	Kore University of Enna, Italy
Yiota Theodora	National Technical Univeristy of Athens, Greece
Graça Tomaz	Polytechnic Institute of Guarda, Portugal
Carmelo Maria Torre	Polytechnic University of Bari, Italy
Francesca Torrieri	University of Naples "Federico II", Italy
Vincenza Torrisi	University of Catania, Italy
Vincenzo Totaro	Polytechnic University of Bari, Italy
Pham Trung	Ho Chi Minh City University of Technology, Vietnam
Dimitrios Tsoukalas	Centre of Research and Technology Hellas (CERTH), Greece
Sanjida Tumpa	University of Calgary, Canada
Iñaki Tuñon	Universidad de Valencia, Spain
Takahiro Ueda	Seikei University, Japan
Piero Ugliengo	University of Turin, Italy
Abdi Usman	Haramaya University, Ethiopia
Ettore Valente	University of Naples "Federico II", Italy
Jordi Vallverdu	Universitat Autònoma de Barcelona, Spain
Cornelis Van Der Mee	University of Cagliari, Italy
José Varela-Aldás	Universidad Tecnológica Indoamérica, Ecuador
Fanny Vazart	University of Grenoble Alpes, France
Franco Vecchiocattivi	University of Perugia, Italy
Laura Verde	University of Campania Luigi Vanvitelli, Italy
Giulia Vergerio	Polytechnic University of Turin, Italy
Jos Vermaseren	Nikhef, The Netherlands
Giacomo Viccione	University of Salerno, Italy
Marco Vizzari	University of Perugia, Italy
Corrado Vizzarri	Polytechnic University of Bari, Italy
Alexander Vodyaho	St. Petersburg State Electrotechnical University "LETI", Russia
Nikolay N. Voit	Ulyanovsk State Technical University, Russia
Marco Vona	University of Basilicata, Italy
Agustinus Borgy Waluyo	Monash University, Australia
Fernando Wanderley	Catholic University of Pernambuco, Brazil
Chao Wang	University of Science and Technology of China, China
Marcin Wozniak	Silesian University of Technology, Poland
Tiang Xian	Nathong University, China
Rekha Yadav	KL University, India
Claudia Yamu	University of Groningen, The Netherlands
Fenghui Yao	Tennessee State University, USA

Contents – Part III

**International Workshop on Application of Numerical Analysis
to Imaging Science (ANAIS 2021)**

**International Workshop on Advances in information Systems
and Technologies for Emergency management, risk assessment
and mitigation based on the Resilience concepts (ASTER 2021)**

**International Workshop on Advances in Web Based Learning
(AWBL 2021)**

General Track 5: Information Systems and Technologies

General Book 3: Information Systems
and Technologies

Predicting Physiological Variables of Players that Make a Winning Football Team: A Machine Learning Approach

Alberto Cortez[1]([✉]) [iD], António Trigo[1] [iD], and Nuno Loureiro[2,3] [iD]

[1] Coimbra Business School Research Centre | ISCAC,
Polytechnic of Coimbra, Coimbra, Portugal
alberto.v.cortez@protonmail.com, aribeiro@iscac.pt
[2] Sport Sciences School of Rio Maior, 2040-413 Rio Maior, Portugal
nunoloureiro@esdrm.ipsantarem.pt
[3] Life Quality Research Centre (CIEQV), Polytechnic Institute of Santarem, Santarem, Portugal

Abstract. As football gained popularity and importance in the modern world, it been under greater scrutiny, so the industry must have better control of the training sessions, and most importantly, the football games and their outcomes. With the purpose of identify the physiological variables of the players that most contribute to winning a football match, a study based on machine learning algorithms was conducted on a dataset of the players GPS positions during the football matches of a team from the 2nd division of the Portuguese championship. The findings reveal that the most important players' physiological variables for predicting a win are Player Load /min, Distance m/min, Distance 0.3 m/s, Acceleration 0.2 m/s with an accuracy of 79%, using the XGBoost algorithm.

Keywords: Football · Winning · Physiological variables · Predicting · Machine learning · Classification

1 Introduction

Football is a team sport seen by several authors, as one of the most popular forms of sport in the world, as referred by Oliver et al. [1]: "football is the most popular sport in the world". As its popularity increases it comes under greater scrutiny, so the industry must have better control of training sessions, and most importantly, football games.

With the purpose of identifying which physiological variables of the players most contribute to winning a football match, a study was performed using Machine Learning (ML) algorithms on a dataset with the GPS positions of the players during football matches, sampled at 10 Hz (Playertek, Catapult Innovations, Melbourne, Australia [2]). The study aimed to both to determine the athletes' physiological variables, in a team, best contribute to the team's win, and the variables that most influence the wins by player position on the field.

Regarding this, was used this information, to apply ML algorithms. Various classification algorithms were used for understanding how these variables influence the victory

© Springer Nature Switzerland AG 2021
O. Gervasi et al. (Eds.): ICCSA 2021, LNCS 12951, pp. 3–15, 2021.
https://doi.org/10.1007/978-3-030-86970-0_1

in the game. For the ML process the Cross Industry Standard Process for Data Mining (CRISP-DM) [3] process was selected, since it is considered the most relevant and comprehensive approach for this kind of analysis [3].

2 Related Work

It is widely known that technology fosters learning with generalized impact on society, and in sports it is not different: "the future of the sports industry lies in the hand of technology" [4]. Because of that, "sports activities have many opportunities for intelligent systems" [5].

As it can be easy understandable, data-driven methods can effectively overcome the subjective limitations (manual analysis) of the match and offer better results for football clubs [6]. As Baboota and Kaur state, "whatever form of Artificial Intelligence (AI) is used, it is evident that football is a sport that benefits from technical integration" [7].

ML has been employed by several authors in different areas of knowledge, for example, Žemgulys et al. [8] used ML techniques to recognize the hand signals made by a basketball referee in recorded games and achieved a 95,6% of accuracy, regarding three of the hand signals.Yang [4] defends, ML offers a contemporary statistical approach where algorithms have been specifically designed to deal with imbalanced data sets and enable the modelling of interactions between a large number of variables. As Knauf et al. [9] affirm in their study, this field of research, known as ML, is a form of AI that uses algorithms to detect meaningful patterns based on positional data. ML is a relatively new concept in football, and little is known about its usefulness in identifying performance metrics that determine match outcome.

Nonetheless, there are some studies regarding ML in football, such as a study that explores injury risk [1], studies about the categorization of football players and football training sessions [4, 11], and even one study that tries to evaluate the football players regarding their market value [12]. Regarding injury risk, the study done by Oliver et al. [1] aims to understand whether the use of ML improved the ability of a neuromuscular screen to identify injury risk factors in elite male youth football players. In categorization, there are some studies. García-Aliaga et al. [11] aimed to find out, by using ML methods, whether the technical-tactical behaviours of players according to position could be identified by their statistics, without including spatial-temporal descriptors. In addition, the authors of this work wanted to check the capacity of ML to identify the most influential variables in each of the positions and to find groups of anomalous players. Another model regarding categorization is the one proposed by Yang [4] to evaluate football training using AI, through the analysis of the evaluation indexes of the running ability of athletes at different positions, showing that the evaluation efficiency is 24.12% higher than that of traditional artificial team, which proves the feasibility of this model. In [12], authors propose a novel method for estimating the value of players in the transfer market, based on the FIFA 20 dataset. The dataset was clustered using the APSO-clustering algorithm which resulted in detecting four clusters: goalkeepers, strikers, defenders, and midfielders. Then, for each cluster, an automatic regression method, able to detect the relevant features, was trained, and they were able to estimate the value of players with 74% accuracy. As illustrated by those examples, ML algorithms hold the potential to provide coaches and analysts with additional information to evaluate the game [10].

As GPS devices have become commonplace in professional football to track player performance. As some authors defend, the identification of internal and external load measures can provide answers about which variables to include in an integrated approach [13]. The amount of information about athletes collected by these GPS systems is increasing, making it difficult for users to analyse this information, so it is necessary to create analytical algorithms that effectively analyse it. Optimization methods, especially metaheuristic optimization algorithms, are powerful methods that can help the researchers to overcome the difficulties and challenges in different fields [12]. Data science has emerged as a strategic area that, supported by the great possibility of data production for analysis, allows knowledge discovery in sport science with the aim of filling some gaps that traditional statistical methods could not achieve [9]. The enormous advances in technology make it possible to process an exorbitant amount of data to draw extremely useful conclusions [5].One such area where predictive systems have gained a lot of popularity is in the prediction of football match results [5].

3 Methodology

In this work was used the CRISP-DM process, being especially useful in modelling. Laureano et al. [14] use this methodology to better model for predicting the waiting time for internments. This allowed them to identify the clinic attributes related to time of interment. Another study done in the medicine with this framework was done by Morais et al. [15] and they aim to better understand what characteristics of a new-born, brings the need of assistance for breathing at birth. This was done by analysing the characteristics of the mother and the pregnancy to prevent neonatal mortality.

The CRISP-DM framework has six phases [3]: business understanding, data understanding, data preparation, modelling, evaluation, and deployment. This study followed these phases which are presented in this section.

For this study, a dataset was compiled with the physiological variables of football players recorded by a GPS system during the playing of football games of a Portuguese team from the 2nd Regional Division of AF Santarém in the 2018/2019 season. Besides the physiological variables, the dataset also contains the results of the games that will be the label/target variables of the study.

During the 2018/2019 the team with 28 players with an average age of 22 years old played 14 games in the first phase of the championship, 10 games in the second phase and 2 extra games in Ribatejo Cup. A total of 33748 different episodes were registered, regarding the different players, and games they played. ML algorithms were applied to these records, which results are presented in Sect. 4.

3.1 Business Understanding

The business goal of this work is the prediction of the physiologic variables that have more influence in the game winning. For this was consider, all the variables recorded by the GPS tracking system, some environment characteristics, the rate of perceived exertion (RPE) and the final score of the game. This prediction can have influence in training analysis, because the coaching staff can understand where they should apply more focus in training sessions to have a better performance on game day.

3.2 Data Understanding

Table 1 presents the variables used in this study that are present in the dataset.

Table 1. Dataset variables

Variables	Description
Game	Is the number of the game in the Championship games sequence
Final Score	Represents the points that the team won in the game, which relates to a victory, draw or loss. 0 it's for a lost game, 1 for a draw and 3 for a win
Minutes	It is the number of minutes that the players are actively in the game
RPE	It is the rate of perceived exertion, made by a numeric estimate of someone's exercise intensity. It is a way to measure how hard a person is exercising, which ranges from 1 (no exertion) to 10 (extremely hard)
Heart Rate (HR)	The maximum heart rate calculated as HRmax = 220 − age. It is calculated in absolute and %
Player Load	Calculated based on the acceleration data that are registered by the triaxial accelerometers. This variable, considered as a magnitude vector, represents the sum of the accelerations recorded in the anteroposterior, medio-lateral and vertical planes. Represented in Total and in Arbitrary Units (U.A.) per minutes
Distance (Total and m/s)	The Total distance provides a good global representation of volume of exercise (walking, running) and is also a simple way to assess individual's contribution relative to a team effort. It is divided in five different speed zones: "walking/jogging distance, 0.0 to 3.0 m/s; running speed distance, 3.0 to 4.0 m/s; high-speed running distance, 4.0 to 5.5 m/s; very high-speed running distance, 5.5 to 7.0 m/s; and sprint distance, a speed greater than 7.0 m/s [13]
Work Ratio (WRRatio)	It is used to describe footballer's activity profiles, which is divided in two categories: 1) pause if the distance is travelled at a speed < 3.0 m/s; and as 2) work if the distance is travelled at a speed > 3.0 m/s

(*continued*)

Table 1. (*continued*)

Variables	Description
Acceleration (Accel.)	Categorized based upon the acceleration of the movement, which is thought to represent the "intensity" of the action. It is divided in "low intensity", 0.0 to 2.0 m/s^2; "moderate intensity", 2.0 to 4.0 m/s^2; and "high intensity", greater than 4.0 m/s^2 [13]
Deacceleration (Deacc.)	Categorized based upon the deacceleration of the movement, which is thought to represent the "intensity" of the action. It is divided in "low intensity", 0.0 to −2.0 m/s^2; "moderate intensity", −2.0 to −4.0 m/s^2; and "high intensity", greater than −4.0 m/s^2 [13]

3.3 Data Preparation

In this phase of the CRISP-DM process, the data was selected and prepared to be used by the ML models. First the dataset was scanned for null or inconsistent information.

After cleaning the data, a final transformation was done to support the two goals of the study, which created two different datasets. One dataset where the original dataset was grouped by game (computing the means for the variables), which held 676 lines and a second dataset where the original dataset was grouped by player position so that it would be possible to identify the best variables by player position on the field and was use a dataset of 10725 episodes.

3.4 Modelling

In the modelling phase, two models were created to study the importance of physiological variables of the athletes on the outcome of the football game. The first model evaluates the team as a whole, while the second model evaluates the players' positions on the field of play.

Given that the objective of our game focuses on winning the match, it does not matter to distinguish between draw and defeat, so the binary variable win/lose seems to us to be the most interesting, with draws being coded as defeat. Thus, a new label/target variable which stores the information regarding the Win of the game (1 – victory; 0 - other) was added to the original dataset. This variable was computed from the variable with the outcome of the game, "final score" (see Table 1).

3.4.1 Variables that Most Contribute to Game Victory

Several models with different features were hypothesized, both to determine which variables as a whole best contribute to the team's win and to determine which variables most influence the wins by player position on the field.

To better evaluate the variables and understanding which were the most relevant in predicting a Win, four models with different sets of variables where created (see Table 2). In the first model was chose to apply all the variables to have a better understanding how those variables all together will affect the game winning. For the second model, were removed all the variables that had big correlation between them, and those that show high intensity for the game. Those variables that were related to high intensity were used in the third model, to understand how those variables, influence the game result. For the fourth model, were chose the variables that showed more correlation between them and victory, and the variables defended by some authors [16], like Player Load/min.

Table 2. ML models

Model	Features
Model 1	`['RPE', 'HR', '%HR', ' < 60%HR', '60-74,9%HR', '75-89,9%HR', ' > 90%HR', 'Player Load', 'Player Load.UA/min', 'Distance Total', 'Distance.m/min', 'Distance.0-3', 'Distance.3.4', 'Distance.4-5.5', 'Distance.5.5-7', 'Distance. > 7', 'WRRatio', 'Accel.0-2', 'Accel.2-4', 'Accel. > 4', 'Deacc.0-2', 'Deacc.2-4', 'Deacc. > 4']`
Model 2	`['60-74,9%HR', '75-89,9%HR', 'Player Load', 'Distance.m/min', 'Distance.0-3', 'Distance.3.4', 'Distance.4-5.5', 'Distance.5.5-7', 'Accel.0-2', 'Accel.2-4', 'Deacc.0-2', 'Deacc.2-4']`
Model 3	`[' > 90% HR', 'Distance. > 7', 'WRRatio', 'Acceler. > 4', 'Deacc. > 4']`
Model 4	`['Player Load.UA/min', 'Distance.m/min', 'Distance.0.3', 'Acceler.0.2']`

3.4.2 Physiological Variables that Most Contribute to Game Victory by Player Position

To achieve this second goal, the dataset was divided by position on the field, and was used a Recursive Feature Elimination (RFE) [17] for each position, to rank the feature variables related to winning.

The variables selected by RFE allowed to understand the different effects of a football games in the different positions, and how these variables affect winning football games. The variables are different for each position, and they are related to the specific demands. The select variables for each position are presented in Table 3 for each player position on the field.

In the scope of this work the Goalkeeper position was excluded because of the specific physiological demands regarding the positions. In this work, the following positions were used, Central Defender (CD), Full Back (FB), Midfielder (MC), Offensive Midfielder (OM), Winger (W) and Forward (F).

Table 3. ML position models

Position	Features
CD	['Distance.m', 'Distance.0-3', 'Distance.3-4', 'Distance. > 7']
FB	[' < 60.0%HR', '60-74,9%HR', 'Player Load.UA/min', 'WRRatio']
MC	['Player Load.UA/min', 'Distance.m/min', 'WRRatio', 'Aceler. > 4']
OM	[<'60.0%HR', '60-74,9%HR', 'Player Load.UA/min', 'WRRatio']
W	[' < 60.0%HR', '60-74,9%HR', '75-89,9%HR', 'Player Load.UA/min']
F	['Distance.m', 'Distance.4-5.5', 'Distance. > 7']

3.5 Evaluation

This section presents the results obtained for the two models.

3.5.1 Variables that Most Contribute to Game Victory

Since the goal is to determine which variables contribute to the win of the game classifications algorithms were chose. Thus, six ML algorithms were selected: Naïve Bayes (NB), K-Nearest Neighbours (KNN), Random Forest (RF), Decision Tree (DT), Support Vector Machine (SVM) and Extreme Gradient Boosting (XgB).

For each ML algorithm the dataset was divided into training and testing. Training dataset was composed by 66% of all the data, the remaining was used for testing.

The performance of each ML algorithm was assessed through a confusion matrix, which shows the present results of True Positives, False Positives, True Negatives and False Negatives. With these, it was possible to calculate the sensitivity, specificity and accuracy of each algorithm, in order to evaluate their performance regarding the 4 models.

Table 4, 5 and 6, and Fig. 1 present the results of the ML algorithms for the 4 models with and without Cross-Validation (CV). For the results with CV was also calculated the Standard Deviation (StD). The best model was Model 4, which was presented with a chart to better understand the results (see Fig. 1). All the data was used for testing using 7-fold CV.

Model 1 (Table 4) has a mean accuracy of 0.476 without CV, and 0.560 with CV. The ML algorithms with the best accuracy without a CV were DT and NB with 57% and with CV were RF and SVM with 62% both.

Table 4. Results of ML algorithms for Model 1

Metrics/Algorithms	KNN	DT	RF	NB	SVM	XgB
Accuracy	**0.571**	**0.571**	0.429	**0.571**	0.286	0.429
Accuracy with CV	0.600	0.520	**0.620**	0.400	**0.620**	0.600
StD Accuracy with CV	0.330	0.140	0.080	0.330	0.280	0.220
Sensitivity	0.500	0.500	0.750	0.500	0.500	0.250
Specificity	0.333	0.667	0.667	0.667	0.667	0.667

Table 5 presents the results for Model 2 with a mean accuracy of 0.429 without CV, and 0.552 with CV. The best ML algorithm for this model regarding accuracy was KNN with a result of 57% without CV and XgB with a result of 76% with CV.

Table 5. Results of ML algorithms for Model 2

Metrics/Algorithms	KNN	DTC	RFC	NBC	SVMC	XgB
Accuracy	**0.571**	0.286	0.429	0.429	0.429	0.429
Accuracy with CV	0.600	0.620	0.400	0.450	0.480	**0.760**
StD Accuracy with CV	0.220	0.190	0.280	0.340	0.290	0.230
Sensitivity	0.500	0.000	0.750	0.250	0.250	0.500
Specificity	0.667	0.667	0.000	0.667	0.667	0.333

Model 3 has the worst mean results for accuracy (see Table 6) from all the models with 0.429 without CV and 0.340 with CV. The algorithms with the best accuracy without CV were RF and SVM 57% and with CV was NB with a result of 45%.

Table 6. Results of ML algorithms for Model 3

Metrics/Algorithms	KNN	DTC	RFC	NBC	SVMC	XgB
Accuracy	0.429	0.286	**0.571**	0.429	**0.571**	0.286
Accuracy with CV	0.290	0.210	0.360	**0.450**	0.400	0.330
StD Accuracy with CV	0.190	0.190	0.360	0.290	0.330	0.240
Sensitivity	0.200	0.000	0.800	0.600	0.400	0.000
Specificity	1.000	1.000	0.000	0.000	1.000	1.000

The last model used in this analysis was Model 4 (Fig. 1) that held the best overall results, and the best individual ML algorithm results. This model has a mean accuracy of 0.542 without CV and 0.675 with CV. The two algorithms with the best results were

DTC and XgB with a 63% of accuracy without CV and an accuracy of 79% with CV. This model reveals the most important variables that could help predict a football game win.

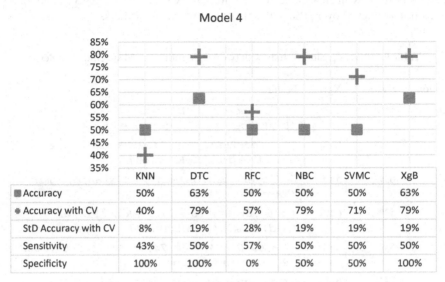

	KNN	DTC	RFC	NBC	SVMC	XgB
■ Accuracy	50%	63%	50%	50%	50%	63%
✦ Accuracy with CV	40%	79%	57%	79%	71%	79%
StD Accuracy with CV	8%	19%	28%	19%	19%	19%
Sensitivity	43%	50%	57%	50%	50%	50%
Specificity	100%	100%	0%	50%	50%	100%

Fig. 1. Plot of the ML algorithm results for Model 4

From the results obtained for all the four models using the ML algorithms, the algorithms that provided best mean results were SVMC and XgB (see Table 8).

Table 8. Average results of ML models with CV

Models	KNN	DTC	RFC	NBC	SVMC	XgB
Model 1	0.600	0.520	0.620	0.400	0.620	0.600
Model 2	0.600	0.620	0.400	0.450	0.480	0.760
Model 3	0.290	0.210	0.360	0.450	0.400	0.330
Model 4	0,400	0,790	0,570	0,790	0,710	0,790
Average with CV	47%	54%	49%	52%	**55%**	**62%**

Therefore, these two algorithms were chosen for the second model of the study, which objective is to detect which physiological variables most influence the game win by the player position on the field (see Sect. 3.4.2).

3.5.2 Physiological Variables that Most Contribute to Game Victory by Player Position

The best results obtained with the ML algorithms were relative to the models for the Forwards position with an accuracy of 74% for SVMC and 62% for XgB (Table 9). All the positions achieved over 50% with at least one of the algorithms.

Table 9. Players positions study with two ML algorithms and specific variables

Position	SVMC	XgB	Features
CD	66%	62%	['Distance.m', 'Distance.0-3', 'Distance.3-4', 'Distance. > 7']
FB	63%	61%	[' < 60.0%HR', '60-74,9%HR', 'Player Load.UA/min', 'WRRatio']
MC	63%	59%	['Player Load.UA/min', 'Distance.m/min', 'WRRatio', 'Aceler. > 4']
OM	52%	62%	[<'60.0%HR', '60-74,9%HR', 'Player Load.UA/min', 'WRRatio']
W	55%	56%	[' < 60.0%HR', '60-74,9%HR', '75-89,9%HR', 'Player Load.UA/min']
F	**74%**	**62%**	['Distance.m', 'Distance.4-5.5', 'Distance. > 7']

The results obtained were not very good because only one model obtained a good result that was the Forwards (F) model with a result around 70%. Nonetheless, all the results were higher than 50%, which shows that these models for the players' positions can be a contribution for the football coach.

4 Discussion

This study aimed to identify which physiological variables have the most influence on winning the game by analysing them, using ML algorithms. With that objective, several models with different features/variables were created.

These models were constructed based upon the correlation between all the variables and the understanding of the variables regarding the intensity applied in each game. For this, the first model was composed by all the variables measured, while for the construction of the second model the same set of variables was used excluding the ones with greater correlation between them and the ones used in the third model. The third model was constructed to understand if high intensity had a better association with the victory, and that was the model with the worst results. Finally, the fourth model was created with the variables defend by some authors [18] in previous studies and those that had some correlation with wining. This model was the one who got better results with an accuracy of 79% with three of the algorithms using a CV. An important finding of this study was that the high intensity is not a key factor in game winning, which is not in agreement with the results presented in the study by Altavilla, where it is claimed that that "high intensity covered distance has traditionally been identified as a key indicator of the physical performance during the matches" [18].

This study allowed to verify that teams that kept their players more active, in the end will have better results. It is also possible to verify that the greater the distance covered in this championship, the better the proportion of victories. These results are made by the average values of the team, so with this they can only be applied to a team, not to an individual player.

In the second part of study the physiological variables of the players by player position on the field were analysed, this has revealed that the physiological variables that contribute to winning games vary according to the players' positions on the field and differ from the team as a whole. Some of these variables were identified and confirmed in different studies, as the key variables, for the positions on the field. As was seen in this study, distance covered at max speed (*distance.* > 7) by Forwards (F) was one of the variables that has a bigger impact/influence in winning. Almulla et al. [19] and Baptista et al. [20] defended that the distance covered by forwarders at very high speed from the winning team was higher than the losing one. The variables chosen for the Forward position presented the best results in predicting a win in this context, with a 70% in SVMC and 62% in XgB.

One of the variables chosen to apply in the Central Midfielders (CM) *was distance covered per minutes,* which is backed by Borghi et al. and Altavilla et al. [18, 21] which refer that Central Midfielders (CM) have the highest value in total distance covered during games. In the same study Borghi et al. [21] affirm that Central Back (CB) covered the shorter distances during match, and this is important because, *Distance 0.3,* was one of the variables chosen to be used in this study for this position.

Another variable that was chosen by the RFE and was the *Player Load U.A./min* for the Full Back (FB) and Winger (W) but was not selected for the Central Back (CB). This was corroborated in a study done by Baptista et al., where they defended that Central Back (CB) had less turns per match than the Full Back (FB) and Winger (W) [20]. The same author also defended that FB covered more high intensity and sprinting distance than CB during the matches [20]. This was also revealed in our study because *Work-Ratio* was one of the variables selected for the Full Back (FB). The variables selected for the Full Back (FB). The variables selected for the FB position presented the second-best results in predicting a win in this context, with a 63% in SVMC and 61% in XgB.

5   Conclusions

Modern football competition has the characteristics of fierce confrontation, long duration, intensity of the game, large amount of exercise, and high technical and tactical requirements [5]. Because of these characteristics, it can be easily understandable that football is a highly complex sport. Selecting and evaluating properly the team players is an important and difficult task in multiplayer sports such as football that has many different aspects to be considered [5].

Therefore, studies in this area may play an important role in terms of evaluating the potential performance of teams and players. It is important to stress out that teams should be analysed as a whole in order to identify important variables that are correlated with a match winning outcome. The individual data, for instance, can be very important for predicting players performance and injuries with time, not in the scope of this work.

The results obtained in this study can be used by the coaching staff of a football team as they help to prepare the team, as a whole, in relation to the most important variables, also helping to identify by player's position on the pitch which are the most important variables for the players' performance. Bottom line, the use of ML applied to training and team selection can represent a major step forward in the evolution of the sport, providing possibly critical information to the training staff that will allow in the end to help to better understand the demands of the positions on the field.

This study was conducted in only one team, being also a reduced sample. This represents an important limitation to the study, but at the same time can serve as steppingstone to other studies to help a better understanding of the game.

In the future, it is intended to further analyse different teams and therefore improve our sample set. This will enable to get a more complete picture of what the most important variables are and correlate them with the tactical decisions of the teams. In addition, it is intended to perform this analysis in different competition contexts and verify if these variables still maintain the same importance, if some of the contextual variables changed.

Without a doubt that in the future AI will contribute to improve the results of football teams. Still, it is important to understand that when a human element is involved in sport, there will always be unpredictability and uncertainty that makes it fascinating and surprising for its viewers [22].

References

1. Oliver, J.L., Ayala, F., de Ste Croix, M.B.A., Lloyd, R.S., Myer, G.D., Read, P.J.: Using machine learning to improve our understanding of injury risk and prediction in elite male youth football players. J. Sci. Med. Sport **23**(11), 1044–1048 (2020). https://doi.org/10.1016/j.jsams.2020.04.021
2. Catapult Innovations, "Playertek" (2021). https://www.playertek.com/gb/. Accessed 12 Apr 2021
3. Jaggia, S., Kelly, A., Lertwachara, K., Chen, L.: Applying the CRISP-DM framework for teaching business analytics. Decis. Sci. J. Innov. Educ. **18**(4), 612–634 (2020). https://doi.org/10.1111/dsji.12222
4. Yang, Y.: Evaluation model of soccer training technology based on artificial intelligence. J. Phys. Conf. Ser. **1648**, 042085 (2020)
5. Maanijou, R., Mirroshandel, S.A.: Introducing an expert system for prediction of soccer player ranking using ensemble learning. Neural Comput. Appl. **31**(12), 9157–9174 (2019). https://doi.org/10.1007/s00521-019-04036-9
6. Kusmakar, S., Shelyag, S., Zhu, Y., Dwyer, D., Gastin, P., Angelova, M.: Machine learning enabled team performance analysis in the dynamical environment of soccer. IEEE Access **8**, 90266–90279 (2020). https://doi.org/10.1109/ACCESS.2020.2992025
7. Baboota, R., Kaur, H.: Predictive analysis and modelling football results using machine learning approach for English Premier League. Int. J. Forecast. **35**(2), 741–755 (2019). https://doi.org/10.1016/j.ijforecast.2018.01.003
8. Žemgulys, J., Raudonis, V., Maskeliūnas, R., Damaševičius, R.: Recognition of basketball referee signals from real-time videos. J. Ambient. Intell. Humaniz. Comput. **11**(3), 979–991 (2019). https://doi.org/10.1007/s12652-019-01209-1

9. Knauf, K., Memmert, D., Brefeld, U.: Spatio-temporal convolution kernels. Mach. Learn. **102**(2), 247–273 (2015). https://doi.org/10.1007/s10994-015-5520-1
10. Herold, M., Goes, F., Nopp, S., Bauer, P., Thompson, C., Meyer, T.: Machine learning in men's professional football: current applications and future directions for improving attacking play. Int. J. Sports Sci. Coach. **14**(6), 798–817 (2019). https://doi.org/10.1177/1747954119879350
11. García-Aliaga, A., Marquina, M., Coterón, J., Rodríguez-González, A., Luengo-Sánchez, S.: In-game behaviour analysis of football players using machine learning techniques based on player statistics. Int. J. Sports Sci. Coach. (2020). https://doi.org/10.1177/1747954120959762
12. Behravan, I., Razavi, S.M.: A novel machine learning method for estimating football players' value in the transfer market. Soft. Comput. **25**(3), 2499–2511 (2020). https://doi.org/10.1007/s00500-020-05319-3
13. Miguel, M., Oliveira, R., Loureiro, N., García-Rubio, J., Ibáñez, S.J.: Load measures in training/match monitoring in soccer: a systematic review. Int. J. Environ. Res. Public Health **18**(5), 2721 (2021)
14. Laureano, R.M.S., Caetano, N., Cortez, P.: Previsão de tempos de internamento num hospital português: Aplicação da metodologia CRISP-DM. RISTI - Revista Iberica de Sistemas e Tecnologias de Informacao (13), 83–98 (2014). https://doi.org/10.4304/risti.13.83-98
15. Morais, A., Peixoto, H., Coimbra, C., Abelha, A., Machado, J.: Predicting the need of neonatal resuscitation using data mining. Procedia Comput. Sci. **113**, 571–576 (2017). https://doi.org/10.1016/j.procs.2017.08.287
16. Soto, P.R., Nieto, D.C., Suarez, A.D., Ortega, J.P.: Player load and metabolic power dynamics as load quantifiers in soccer. J. Hum. Kinetics **69**, 259 (2019). https://doi.org/10.2478/hukin-2018-0072
17. Arndt, C., Brefeld, U.: Predicting the future performance of soccer players. Stat. Anal. Data Min. **9**(5), 373–382 (2016). https://doi.org/10.1002/sam.11321
18. Altavilla, G., Riela, L., di Tore, A.P., Raiola, G.: The physical effort required from professional football players in different playing positions. J. Phys. Educ. Sport **17**(3), 2007–2012 (2017). https://doi.org/10.7752/jpes.2017.03200
19. Almulla, J., Alam, T.: Machine learning models reveal key performance metrics of football players to win matches in qatar stars league. IEEE Access **8**(December), 213695–213705 (2020). https://doi.org/10.1109/ACCESS.2020.3038601
20. Baptista, I., Johansen, D., Seabra, A., Pettersen, S.A.: Position specific player load during matchplay in a professional football club. PLoS ONE **13**(5), 1–11 (2018). https://doi.org/10.1371/journal.pone.0198115
21. Borghi, S., Colombo, D., la Torre, A., Banfi, G., Bonato, M., Vitale, J.A.: Differences in GPS variables according to playing formations and playing positions in U19 male soccer players. Res. Sports Med. **29**(3), 225–239 (2020). https://doi.org/10.1080/15438627.2020.1815201
22. Rathi, K., Somani, P., Koul, A.V., Manu, K.S.: Applications of artificial intelligence in the game of football: the global perspective. Res. World **11**(2), 18–29 (2020). https://doi.org/10.18843/rwjasc/v11i2/03

A Framework for Supporting Ransomware Detection and Prevention Based on Hybrid Analysis

Alfredo Cuzzocrea[1]([✉]), Francesco Mercaldo[2], and Fabio Martinelli[2]

[1] University of Calabria, Rende, Italy
alfredo.cuzzocrea@unical.it
[2] IIT-CNR, Pisa, Italy
{francesco.mercaldo,fabio.martinelli}@iit.cnr.it

Abstract. Ransomware is a very effective form of malware, which recently raised a lot of attention since an impressive number of workstations was affected. This malware is able to encrypt the files located in the infected machine and block the access to them. The attackers will restore the machine and files only after the payment of a certain amount of money, usually given in bitcoins. In this paper we discuss an hybrid framework, combining static and dynamic analysis, exploiting APIs to prevent and mitigate ransomware threats. The evaluation, considering 1000 legitimate and ransomware applications, demonstrates that the hybrid API calls-based detection can be proved to be a promising direction in ransomware prevention and mitigation.

Keywords: Ransomware · Malware · Static analysis · Dynamic analysis · Hybrid analysis · Security

1 Introduction

Ransomware is a malware able to encrypts documents, essentially locking users out of their computers. Attackers in order to give back the original data try to extort money to restore access.

The main difference between ransomware and the other widespread malware families is represented by the exhibited behaviour. As a matter of fact, as demonstrated in [30], generic malware focuses on gathering and sending to the attackers user sensitive and private information. Moreover, malware is usually coded to be silent in order to continuously perpetuate its harmful actions. From the other side, ransomware is a threat in the malware landscape that aims at making users aware of its presence by locking the screen and, in most dangerous cases, ciphering data on the infected machine.

During the past few years, ransomware has become a relevant and emerging security problem, due to the rapid spread and the devastating consequences which can bring. Law enforcement agencies from all around Europe teamed up with antimalware vendors and other IT security companies to form the *No More*

© Springer Nature Switzerland AG 2021
O. Gervasi et al. (Eds.): ICCSA 2021, LNCS 12951, pp. 16–27, 2021.
https://doi.org/10.1007/978-3-030-86970-0_2

Ransom! project[1]. According to McAfee security analysts [28], 2016 might be remembered as "the year of ransomware", confirming the predictions of exponential growth of these kinds of attacks [25]. As an example, the WannaCry ransomware attack in May 2017 rendered unusable over 200,000 computers in more than 150 countries with ransom demands[2]. At the end of 2017, the first ransomware as a service frameworks (the so-called RaaS) appeared and this trend, as pointed out by Sophos experts, is growing up during 2018. RaaS allows attackers to rent ransomware infrastructure rather than develop it themselves[3]. The RaaS paradigm works in the following way: the attacker generally pays an upfront fee, and the author of the RaaS keeps a small percentage of each ransom paid. Generally, the rentee is allowed to set the ransom price and build the attack campaign.

From the defensive side, a weak point of the current antimalware solutions is their main operating mechanism: they mostly apply a signature-based detection; this approach requires the vendor's awareness of the malware, this is the reason why a threat must be widespread in order to be recognized.

For this reason current anti-malware software fails in identifying several ransomware threats, and could play no role in the last set of attacks (WannaCry, Petya, etc.), since new ransomware samples, whose binary signature is unknown, are continuously produced and spread in the wild as zero-day attacks. As a matter of fact, zero-day attacks are able to elude commercial anti-malware, requiring thus more advanced techniques to be detected. For instance, the Satan service claims to generate working ransomware samples and allows to set your own price and payment conditions, collects the ransom on the customer behalf and finally provides a decryption tool to victims who pay the ransom: its creators keep the remaining 30% of income, so if the victim pays a ransom worth 1 bitcoin, the customer receives 0.7 in bitcoin.

Considering that ransomware damages cost $5 billion in 2017 and $11.5 billion in 2019 (those figures are up from just $325 million in 2015[4]) ransomware detection represents a challenge also for scientific community.

Regarding the current state-of-the-art related to ransomware analysis, authors in [40] presented EldeRan, a model which dynamically monitors the initial activities of a sample, checking for specific characteristics and actions in order to classify this sample as ransomware or as benign. Unveil [27] is also a system, which dynamically tries to identify ransomware: the main functionality of this model is to monitor system's desktop and look for any changes which indicate the presence of a ransomware. In [6] authors present a SDN based analysis for detecting malware which utilizes asymmetric cryptography, i.e. CryptoWall or Locky. In their study they observe the network communication and the HTTP messages of the under examination samples and are able to detect malicious behavior. CryptoDrop is presented in [39] and it is a system which

---

[1] https://www.nomoreransom.org.
[2] http://wapo.st/2pKyXum?tid=ss_tw&utm_term=.6887a06778fa.
[3] goo.gl/N7PAjh.
[4] goo.gl/Gyrt1N.

monitors the activity of an application during its execution, aiming at detecting changes in a big amount of files within a system.

Behavioral analysis is the key in detecting and preventing the action of ransomware. In fact, even if they show different code, the ransomware behavior is very similar for all the samples falling in this category. Unfortunately, these common characteristics can be detected only at run-time. In this work we propose an hybrid static/dynamic framework for malware detection and effect mitigation, which exploit static analysis to identify features that can be monitored at run-time and exploited to prevent the malicious action. In the following we will introduce the framework design, the methodology and promising experimental results. Starting from the analysis of current literature, it emerges that there is no work aiming to prevent and mitigate ransomware behaviours (as a matter of fact, all the cited work are focused on ransomware detection), this is the reason why in this paper we design a framework with the aim to prevent and mitigate ransomware threats. A high-level design and a preliminary experiment about the effectiveness of the proposed framework are provided.

As regards future conceptual extensions, the topics we investigate in this paper has relevant correlations with the emerging big data trend (e.g., [17–21]). This synergy turns to be a critical aspect to be considered by future research efforts.

The paper proceeds as follows: in the next section the proposed methodology is described, in Sect. 3 the results of the experimental analysis is presented, Sect. 4 discusses the current state-of-the-literature in ransomware detection and, finally, conclusion and future work are drawn in the last section.

2   Methodology

This section presents the architecture of the proposed framework for ransomware mitigation and prevention. Two scenarios are being considered, one related to the in vitro analysis (depicted in Fig. 1), and the other one related to the in vivo analysis (depicted in Fig. 1).

Scenario I in Fig. 1 represents the workflow of the proposed framework aimed to ransomware identification.

As anticipated, a hybrid static/dynamic analysis is exploited in order to obtain information concerning the APIs invoked by the analyzed ransomware samples. More specifically:

– Through static analysis, we obtain from the executable under analysis a list of the APIs and libraries being invoked, using a reverse engineering process. Whether the gathered information match the ones stored into the SA knowledge repository (containing information related to APIs and libraries of widespread ransomware samples), the analyzed sample will be marked as "ransomware", otherwise it will be marked as "suspicious" and will be sent to the dynamic analysis module.

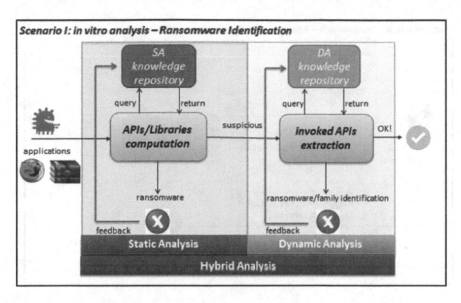

Fig. 1. Scenario I: identification of ransomware samples with static and dynamic analysis with knowledge-based feedbacks.

- From dynamic analysis, we obtain the frequency of the APIs invoked at runtime. As a matter of fact, techniques like dynamic loading (i.e., a mechanism by which a computer program is able at run time to load, execute and unload a library into memory) or reflection (i.e., the ability of a computer program to examine, introspect, and modify its own structure and behavior at runtime) can easily elude the static analysis. In addition, several obfuscation techniques are able to cipher the name of the invoked API functions, aiming by this way to evade the current signature-based antimalware mechanism. Whether the invoked APIs gathered from this module match the ones stored in the DA knowledge repository (containing information related to the APIs invoked by ransomware samples) the executable under analysis will be marked as "ransomware", otherwise, the evaluated application will be marked as "legitimate". Furthermore, in this step we plan to assign the sample in a specific malware family, if it is characterized as ransomware.

Moreover, the proposed framework provides a feedback system: once a ransomware is identified, the gathered features (from both static and dynamic analysis) are stored in the correspondent knowledge repository, in order to increase the performance of the identification and to reduce the response time of the framework (the feedback system is highlighted with red arrows in Fig. 1).

Figure 2 represents the ransomware mitigation and prevention process.

The Scenario II (the in-vivo analysis) starts when the user performs operations related to email attachment/file opening or when he/she browses an URL. A daemon installed on the user computer is in charge to collect APIs in real-time and to send the trace to the in vitro analysis for deep analysis. Whether the

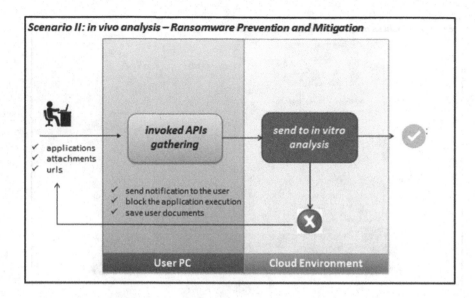

Fig. 2. Scenario II: ransomware prevention and mitigation

trace is labeled as ransomware, the deamon on the user computer performs tree different operations: (i) send notification about the maliciousness of the application to the user, (ii) block the application execution (prevention) and (iii) save user documents that could be encrypted in case the daemon is not able to kill the ransomware main process (mitigation).

3 Experimental Analysis

In this section a set of results is presented, aiming to prove the discrimination between ransomware and benign applications based on the APIs calls (provided by the static analysis) and the frequency of the invocations (provided by the dynamic analysis), as depicted in Fig. 1.

To perform the experimental evaluation we exploited the Cuckoo Sandbox [36], a widely used tool for malware analysis, which has the ability to analyze different malicious files and report back log files regarding a huge amount of information such as API calls, network traffic, behavioral and memory analysis. In this study Cuckoo is considered because of its widely use with regard to malware inspection. A number of studies and research works have also exploited this tool as for instance, in [34] the authors extend its functionality in order to execute malware samples multiple times, under different circumstances and study their overall behavior. In [35] the authors use Cuckoo in order to extract network indicators of compromise from malware samples. Another study is [11], where the authors used Cuckoo to analyze malware and dynamically reconfigure a SDN network, based on the behavior of the sample.

As depicted in Fig. 1 we compute libraries and APIs called by the applications under analysis (with regard to the static analysis) and the frequency of the API calls (with regard to the dynamic analysis). We considered APIs aimed to ciphering operations and network communication, since ransomware need usually to communicate with a Command & Control server, either to save the encryption key or use it for the ransom payment.

For this study, the libraries and the APIs taken into consideration are related mostly to ciphering operations and also to network communication, since usually a ransomware need to communicate with a Command & Control server, either to save the encryption key or use it for the ransom payment. We evaluated 500 real-world samples belonging to WannaCry family and 500 well-known legitimate applications (250 of them with ciphering abilities and the remaining 250 with no ciphering abilities). To ensure their trustworthiness, component that creates and manipulates the standard elements of the Windows user interface, such as the desktop, windows, and menus. It thus enables programs to implement a graphical user interface (GUI) Results are shown in Tables 1, 2, 3. More specifically, Table 1 presents the libraries being invoked from the WannaCry samples and the percentage of their invocation both from the ransomware samples and the benign applications. The outcome of this analysis is that in most of the ransomware samples there are calls to specific APIs, which are not used neither from application with no ciphering functionality, nor from applications with encryption capabilities i.e., *ws2_32.dll* (the Windows Sockets API used by most Internet and network applications to handle network connections), *iphlpapi.dll* (a dynamic link library) and *wininet.dll* (related to Internet functions used by Windows applications) libraries: this is symptomatic of the massive use of the network for the C&C communication usually performed by ransomware samples. Table 2, shows a list of the most common APIs used for encryption and network connection and the percentage of their presence in the under analysis samples. The interesting fact which can be observed in this Table is that the majority of the Wannacry samples uses the *LdrGetProcedureAddress* API which includes a number of encryption functions (i.e. CryptGenKey, CryptAcquireContextW, CryptAcquireContextW etc.) and the *LdrLoadDll* API, which gives the possibility to an application to drop its own .dll files to the system. Moreover, it can also been observed that the ransomware samples make use of APIs related to the connection with the internet, a behavior that is not presented by the benign applications. Finally, Table 3, presents the average frequency of the API calls, for WannaCry samples (*W* column in Table 3), Trusted Ciphering Applications (*TC* column in Table 3) and two Trusted Non Ciphering Applications (*TNC* column in Table 3). A first conclusion which can be extracted is that ransomware samples make in total less API calls, which is reasonable since their main functionality is the encryption of the files. In addition, regarding the two aforementioned APIs, we can also observe that they are being called more times in the WannaCry samples, rather than in the benign applications: this is symptomatic that ransomware performs its harmful behaviour continuously (as a matter of fact the encryption functions is called for each file to cipher).

Table 1. Invoked libraries

	Wannacry samples	Trusted ciphering apps	Trusted non ciphering apps
kernel.dll	100%	100%	90%
user32.dll	20%	80%	90%
advapi32.dll	90%	80%	70%
msvcrt.dll	90%	20%	10%
ws2_32.dll	80%	0%	0%
iphlpapi.dll	80%	0%	0%
wininet.dll	80%	0%	0%

Table 2. Invoked APIs

	Wannacry samples	Trusted ciphering apps	Trusted non ciphering apps
LdrGetProcedureAddress	90%	80%	40%
EncryptMessage	80%	20%	30%
DecryptMessage	10%	20%	30%
CryptExportKey	10%	0%	0%
CryptEncrypt	10%	0%	0%
CryptDecrypt	10%	0%	0%
LdrLoadDll	90%	80%	0%
InternetOpenA	80%	0%	0%
WSAStartup	80%	0%	10%
InternetOpenUrlA	80%	0%	0%
InternetCloseHandle	80%	0%	10%

Table 3. Frequency of API calls

	W	TC	TNC
Total API calls	893	37.812	125.284
LdrGetProcedureAddress	19.8%	4.57%	2.99%
LdrLoadDll	3.79%	0.32%	0.19%

The obtained results of this investigation demonstrate that the usage of API calls and invocations to discriminate between ransomware and legitimate applications seem to be a promising viable way for ransomware identification.

4   Related Work

In this section we present a review of the current literature regarding the area of ransomware detection, mainly based on the detection of malicious samples (i.e., malware detection) but also related to family identification and phylogenetic analysis.

Several studies try to propose solutions for the identification of a malware's family. More specifically, the authors of [33] present a solution which is based on the BIRCH clustering algorithm. After extracting static and dynamic analysis features from the malware samples, they construct the necessary vectors for the clustering algorithm and group the samples into families based on these information. Another study which uses the call graph clustering, is proposed in [26]. The authors base their analysis on graph similarities through the minimization of the graph edit distance. With respect to both of these studies, the proposed model of this paper uses the phylogenetic analysis and is targeting on ransomware family identification.

Another approach is that of [29], where the authors extract n-gram features from the binary code of the samples. These features were selected using the Sequential Floating Forward Selection (SFFS) method with three classifiers and the accuracy of the method reached the 96.64%. However, their methodology is different from the one presented in this paper and they do not provide results specifically referred to ransomware.

In [2] the authors make use of the PrefixSpan algorithm in order to create groups of malware families, based on sequence patterns. Although, these patterns are related only to network traffic packets and not to the overall behavior of the malware. In [43] ARIGUMA is proposed, a malware analysis system for family classification. The authors consider the number of functions being called in a function and the number of functions which call a specific function. Also, the local variables, the arguments and the instructions are taken into account. Even though, this method can also detect obfuscated APIs, the classification accuracy is only 61.6%. In [24] the authors present a malware family classification system which is based on the instruction sequence. The analysis is made through three steps: the feature extraction, the classification and the signature generator. The accuracy of their method reaches up to 79%, which is less than the one achieved in this paper.

Phylogenetic analysis has been used in a number of studies regarding malware. More specifically, in [8] the authors use the phylogenetic analysis in order to assign android malware samples in families. They consider the sequence of the system calls and based on these features they compute the similarity functions. The main difference between this work and the proposed one, is that the second focuses on the ransomware and is also able to distinguish trusted applications. Another study is that of [9], where the authors exploit phylogenesis so as to compute the relationship between 6 groups of malware. In [41] a method of extracting API calls out of malware samples and calculating the similarity amongst them, is presented. Although, this study is focused on analyzing the malware behavior and not on classifying the samples into families. A study which also uses phylogenetic trees is [42], where the authors analyze internet worms and network traffic and present families of worms can be related by their normalized compression distance. Martinelli and colleagues [30] adopt supervised classification algorithms for ransomware family identification. Moreover they consider the binary trees generated by these algorithms to infer the phylogenetic relationships. They experiment the proposed method on real-world ransomware applications belonging to three widespread families (i.e., petya, badrabbit and wannacry),

A phylogenetic analysis related to Android malware is proposed authors in [12,14] where authors consider model checking techniques for malware evolution tracking and for discovering ancestor and descendant relationship.

5   Conclusion and Future Work

Considering the massive damage recently caused by ransomware, the development of techniques and methodologies focused to the identification and mitigation are needed. In this paper we proposed a hybrid framework that exploiting API calls (by static analysis) and invocations (by dynamic analysis) is designed to prevent the ransomware threat. We evaluated, using the Cuckoo framework, the effectiveness of API calls and invocations to discriminate between ransomware and legitimate applications, obtaining encouraging results. As future work we plan to evaluate the proposed framework and evaluate their effectiveness on a larger set of applications, both benign and malicious. In addition we plan to consider the adoption of formal methods [1,37,38] for reaching better accuracy for the ransomware detection and mitigation tasks, considering that formal methods already showed their efficacy in other fields as, for instance, as biology [10,23], medicine [3–5] but also software security [7,13,15,16,22,31,32].

Acknowledgments. This work has been partially supported by MIUR - SecureOpen-Nets, EU SPARTA, CyberSANE and E-CORRIDOR projects.

References

1. Barbuti, R., De Francesco, N., Santone, A., Vaglini, G.: Reduced models for efficient ccs verification. Formal Methods Syst. Des. **26**(3), 319–350 (2005)
2. Boukhtouta, A., Lakhdari, N.E., Debbabi, M.: Inferring malware family through application protocol sequences signature. In: 2014 6th International Conference on New Technologies, Mobility and Security (NTMS), pp. 1–5, March 2014
3. Brunese, L., Mercaldo, F., Reginelli, A., Santone, A.: Formal methods for prostate cancer gleason score and treatment prediction using radiomic biomarkers. Magn. Reson. Imaging **66**, 165–175 (2020)
4. Brunese, L., Mercaldo, F., Reginelli, A., Santone, A.: Neural networks for lung cancer detection through radiomic features. In: 2019 International Joint Conference on Neural Networks (IJCNN), pp. 1–10. IEEE (2019)

5. Brunese, L., Mercaldo, F., Reginelli, A., Santone, A.: An ensemble learning approach for brain cancer detection exploiting radiomic features. Comput. Methods Programs Biomed. **185**, 105134 (2020)
6. Cabaj, K., Gregorczyk, M., Mazurczyk, W.: Software-defined networking-based crypto ransomware detection using HTTP traffic characteristics. Comput. Electr. Eng. **66**, 353–368 (2018)
7. Canfora, G., Mercaldo, F., Moriano, G., Visaggio, C.A.: Composition-malware: building android malware at run time. In: 2015 10th International Conference on Availability, Reliability and Security (ARES), pp. 318–326. IEEE (2015)
8. Canfora, G., Mercaldo, F., Pirozzi, A., Visaggio, C.A.: How i met your mother? In: Proceedings of the 13th International Joint Conference on e-Business and Telecommunications, pp. 310–317. SCITEPRESS-Science and Technology Publications, Lda (2016)
9. Carrera, E., Erdélyi, G.: Digital genome mapping - advanced binary malware analysis (2004)
10. Ceccarelli, M., Cerulo, L., Santone, A.: De novo reconstruction of gene regulatory networks from time series data, an approach based on formal methods. Methods **69**(3), 298–305 (2014)
11. Ceron, J.M., Margi, C.B., Granville, L.Z.: MARS: an SDN-based malware analysis solution. In: 2016 IEEE Symposium on Computers and Communication (ISCC), pp. 525–530, June 2016
12. Cimino, M.G.C.A., De Francesco, N., Mercaldo, F., Santone, A., Vaglini, G.: Model checking for malicious family detection and phylogenetic analysis in mobile environment. Comput. Secur. **90**, 101691 (2020)
13. Cimitile, A., Martinelli, F., Mercaldo, F., Nardone, V., Santone, A.: Formal methods meet mobile code obfuscation identification of code reordering technique. In: 2017 IEEE 26th International Conference on Enabling Technologies: Infrastructure for Collaborative Enterprises (WETICE), pp. 263–268. IEEE (2017)
14. Cimitile, A., Martinelli, F., Mercaldo, F., Nardone, V., Santone, A., Vaglini, G.: Model checking for mobile android malware evolution. In: 2017 IEEE/ACM 5th International FME Workshop on Formal Methods in Software Engineering (FormaliSE), pp. 24–30. IEEE (2017)
15. Cimitile, A., Mercaldo, F., Nardone, V., Santone, A., Visaggio, C.A.: Talos: no more ransomware victims with formal methods. Int. J. Inf. Secur. **17**(6), 719–738 (2018)
16. Ciobanu, M.G., Fasano, F., Martinelli, F., Mercaldo, F., Santone, A.: Model checking for data anomaly detection. Procedia Comput. Sci. **159**, 1277–1286 (2019)
17. Cuzzocrea, A.: Improving range-sum query evaluation on data cubes via polynomial approximation. Data Knowl. Eng. **56**(2), 85–121 (2006)
18. Cuzzocrea, A., Matrangolo, U.: Analytical synopses for approximate query answering in OLAP environments. In: Galindo, F., Takizawa, M., Traunmüller, R. (eds.) DEXA 2004. LNCS, vol. 3180, pp. 359–370. Springer, Heidelberg (2004). https://doi.org/10.1007/978-3-540-30075-5_35
19. Cuzzocrea, A., Moussa, R., Xu, G.: OLAP*: effectively and efficiently supporting parallel OLAP over big data. In: Cuzzocrea, A., Maabout, S. (eds.) MEDI 2013. LNCS, vol. 8216, pp. 38–49. Springer, Heidelberg (2013). https://doi.org/10.1007/978-3-642-41366-7_4
20. Cuzzocrea, A., Saccà, D., Serafino, P.: A hierarchy-driven compression technique for advanced OLAP visualization of multidimensional data cubes. In: Tjoa, A.M., Trujillo, J. (eds.) DaWaK 2006. LNCS, vol. 4081, pp. 106–119. Springer, Heidelberg (2006). https://doi.org/10.1007/11823728_11

21. Cuzzocrea, A., Serafino, P.: LCS-hist: taming massive high-dimensional data cube compression. In: Proceedings of the 12th International Conference on Extending Database Technology: Advances in Database Technology, pp. 768–779 (2009)
22. Martinelli, F., Mercaldo, F., Orlando, A., Nardone, V., Santone, A., Sangaiah, A.K.: Human behaviour characterization for driving style recognition in vehicle system (2018)
23. Francesco, N.D., Lettieri, G., Santone, A., Vaglini, G.: Grease: a tool for efficient "nonequivalence" checking. ACM Trans. Softw. Eng. Methodol. (TOSEM) **23**(3), 24 (2014)
24. Huang, K., Ye, Y., Jiang, Q.: ISMCS: an intelligent instruction sequence based malware categorization system. In: 2009 3rd International Conference on Anti-counterfeiting, Security, and Identification in Communication, pp. 509–512, August 2009
25. Infosec Institute: Evolution in the World of Cyber Crime. Technical report, Infosec Institute, June 2016
26. Kinable, J., Kostakis, O.: Malware classification based on call graph clustering. J. Comput. Virol. **7**(4), 233–245 (2011)
27. Kirda, E.: Unveil: a large-scale, automated approach to detecting ransomware (keynote). In: 2017 IEEE 24th International Conference on Software Analysis, Evolution and Reengineering (SANER), p. 1, February 2017
28. McAfee Labs: McAfee Labs Threats Report - December 2016. Technical report, McAfee Labs, August 2016
29. Liangboonprakong, C., Sornil, O.: Classification of malware families based on n-grams sequential pattern features. In: 2013 IEEE 8th Conference on Industrial Electronics and Applications (ICIEA), pp. 777–782, June 2013
30. Martinelli, F., Mercaldo, F., Michailidou, C., Saracino, A.: Phylogenetic analysis for ransomware detection and classification into families. In: ICETE, no. 2, pp. 732–737 (2018)
31. Martinelli, F., Mercaldo, F., Nardone, V., Santone, A.: Car hacking identification through fuzzy logic algorithms. In: 2017 IEEE International Conference on Fuzzy Systems (FUZZ-IEEE), pp. 1–7. IEEE (2017)
32. Mercaldo, F., Nardone, V., Santone, A., Visaggio, C.A.: Hey malware, i can find you! In: 2016 IEEE 25th International Conference on Enabling Technologies: Infrastructure for Collaborative Enterprises (WETICE), pp. 261–262. IEEE (2016)
33. Pitolli, G., Aniello, L., Laurenza, G., Querzoni, L., Baldoni, R.: Malware family identification with birch clustering. In: 2017 International Carnahan Conference on Security Technology (ICCST), pp. 1–6, October 2017
34. Provataki, A., Katos, V.: Differential malware forensics. Digit. Investig. **10**(4), 311–322 (2013)
35. Rudman, L., Irwin, B.: Dridex: analysis of the traffic and automatic generation of IOCs. In: 2016 Information Security for South Africa (ISSA), pp. 77–84, August 2016
36. Cuckoo Sandbox. Cuckoo Sandbox - Automated Malware Analysis (2018). https://cuckoosandbox.org/. Accessed 06 Mar 2018
37. Santone, A.: Automatic verification of concurrent systems using a formula-based compositional approach. Acta Inf. **38**(8), 531–564 (2002)
38. Santone, A.: Clone detection through process algebras and java bytecode. In: IWSC, pp. 73–74. Citeseer (2011)
39. Scaife, N., Carter, H., Traynor, P., Butler, K.R.: Cryptolock (and drop it): stopping ransomware attacks on user data. In: 2016 IEEE 36th International Conference on Distributed Computing Systems (ICDCS), pp. 303–312, June 2016

40. Sgandurra, D., Muñoz-González, L., Mohsen, R., Lupu, E.C.: Automated dynamic analysis of ransomware: Benefits, limitations and use for detection. arXiv preprint arXiv:1609.03020 (2016)
41. Wagener, G., State, R., Dulaunoy, A.: Malware behaviour analysis. J. Comput. Virol. 4(4), 279–287 (2008)
42. Wehner, S.: Analyzing worms and network traffic using compression. J. Comput. Secur. 15(3), 303–320 (2007)
43. Zhong, Y., Yamaki, H., Yamaguchi, Y., Takakura, H.: Ariguma code analyzer: efficient variant detection by identifying common instruction sequences in malware families. In: 2013 IEEE 37th Annual Computer Software and Applications Conference, pp. 11–20, July 2013

A Neural-Network-Based Framework for Supporting Driver Classification and Analysis

Alfredo Cuzzocrea[1(✉)], Francesco Mercaldo[2], and Fabio Martinelli[2]

[1] University of Calabria, Rende, Italy
alfredo.cuzzocrea@unical.it
[2] IIT-CNR, Pisa, Italy
{francesco.mercaldo,fabio.martinelli}@iit.cnr.it

Abstract. The proliferation of info-entertainment systems in nowadays vehicles has provided a really cheap and easy-to-deploy platform with the ability to gather information about the vehicle under analysis. The infrastructure in which these information can be used to improve safety and security are provided by Ultra-response connectivity networks with a latency of below 10 ms. In this paper, we propose a service-oriented architecture based on a fully connected neural network architecture considering position-based features aimed to detect in real-time: (i) the driver and (ii) the driving style, with the goal of providing an architecture for increasing security and safety in automotive context. The experimental analysis performed on real-world data shows that the proposed method obtains encouraging results.

Keywords: Automotive · Artificial Intelligence · Neural network · Machine learning

1 Introduction

The Internet of Things is represented by the network of physical devices, home appliances, vehicles and a plethora of devices embedded with actuators, sensors, software and, in general, with connectivity [33]. More attention has been recently obtained by the Internet of Vehicle (i.e., IoT related to automotive) [28]. In fact, due to this technology a vehicle can turn from an autonomous and independent system into a component of a more efficient cooperative one [15].

As a matter of fact, vehicles can send data between them or to an external server in order to adapt the driving style to the road condition; traffic lights can send information about the number of cars in a lane in order to suggest to other cars an alternative path to avoid traffic congestion [22]. In general, a very low end-to-end latency of below 10 ms is required for vehicle safety applications. Nevertheless, a latency less than 1 ms will be required considering the bi-directional exchange of data is required with these emerging applications.

© Springer Nature Switzerland AG 2021
O. Gervasi et al. (Eds.): ICCSA 2021, LNCS 12951, pp. 28–39, 2021.
https://doi.org/10.1007/978-3-030-86970-0_3

Nowadays, several useful features from vehicles while are traveling can be gathered by the proliferation of both mobile devices and info-entertainment systems in modern cars. This has boosted the research community to focus about safety and security in the automotive context. In this scenario, data mining algorithms can be useful to analyze features collected from vehicles with the aim to extract knowledge related to the driver behavioral analysis. For this reason, driver detection and driving style recognition are becoming the main research trends. In particular, the main aims regarding driver are to develop systems proposing ad-hoc policies for single driver (for insurance companies) and anti-theft system (for end users).

New anti-theft paradigms are necessary, considering that cars are equipped with many computers on board, exposing them to a new type of attacks [1,20]. As any software, the exposition to bug and vulnerabilities of the operating systems running on cars has to be considered [30]. Auto insurance markets is interested in driver detection. In fact, through these plethora of new data available and by measuring and rating a particular person's driving ability insurance may estimate the individual risk [4]. These paradigms can also be extended via different-but-related initiatives, such as the traditional the context of adaptive and flexible paradigms in web intelligence (e.g., [3,8]).

This idea is defined into the Usage Based Insurance (UBI) concept, introduced into the personal motor insurance market over a decade ago. Basically, two typical models: "pay as you drive" and "pay how you drive" are adopted. Moreover, time of usage, distance driven, driving behavior and places driven to are the main features impacting the premiums. Relating to driving style recognition, this trend is more related to safety than driver detection but currently there are few studies proposing methods to address this issue.

The features gathered from Controller Area Network (CAN) bus, a vehicle bus standard designed to allow micro-controllers and devices to communicate with each other in applications without a host computer are exploited from several methods aimed to the driver detection [24]. Brake usage and fuel consumption rate are just two examples of information passing though this bus.

Different methods proposed in the current literature consider several sets of features extracted from the CAN bus. Thus different vehicle manufacturers provides different features (apart from a small subset common to all manufacturers). Thus, it is possible that a method evaluated on a vehicle produced by a certain manufacturer may not work on a vehicle by another manufacturer or on a new version of the same vehicle.

Furthermore, for exploiting data from CAN bus the proposed methodologies must have access to these data. This results in possible communication delay, critical infrastructure errors, and in possible attacks, as discussed in [23]. Starting from these considerations, in this paper we propose a method aimed to (i) discriminate between the vehicle owner and impostors and (ii) detect the driving style. A fully connected neural network architecture exploiting a feature vector based on positioning systems is proposed. The goal is to detect the driver and the driving style without accessing to critical information CAN bus-related.

The paper is organized as follows: Sect. 2 present the state of the ark works in the field. Section 3 motivate the work and underline the main contribution.

Finally, Sect. 4 show the results of the experiments and Sect. 5 concludes the work and proposes new lines of research.

2 Related Work

Current literature related to the driver behavioral analysis in discussed in following section.

In [34], researchers by analyzing data from the steering wheel and the accelerator and by training a Hidden Markov model (HMM) classifies the driver characteristics. In particular, they build two models for each driver, one trained from accelerator data and one learned from steering wheel angle data. They achieve an accuracy of 85% in identifying the different drivers. Instead, in [32] the authors observe the signals of the driver following another vehicle in order to identify her behavior. They analyzed signals, as accelerator pedal, brake pedal, vehicle velocity, and distance from the vehicle in front, were measured using a driving simulator. The results achieved an identification rates of 73% for thirty real drivers and 81% for twelve drivers using a driving simulator.

For modeling driver behavior a HMM based similarity measure has been proposed in [14]. They employ a simulated driving environment to test the effectiveness of the proposed solution. A method based on driving pattern of the car has been proposed in [18]. In particular, they apply four classifiers: the k Nearest Neighbor, RandomForest, Multilayer Perceptron and Decision Tree to mechanical feature extracted from the CAN and they achieved performance of 84%, 96%, 74% and 93%, respectively. Researchers in [13] applies Temporal and Spatial Identifiers in order to control the road environment and driver behavior profiles to provide a common measure of driving behavior based on the risk of a casualty crash for assessing the effectiveness of a PAYD (i.e., pay-as-you drive) scheme on reducing driving risks. They prove that in order to induce significant changes, in several cases personalized feedback alone are sufficient. Nevertheless, if a financial incentives are applied, the risk reductions is much greater.

In [13] researcher proposed a biometric driver recognition system utilizing driving behaviors. Authors, using Gaussian mixture models, extract a set of features from the accelerator and brake pedal pressure were then used as inputs to a fuzzy neural network system to ascertain the identity of the driver. The effectiveness of relating the use of the FNN for real-time driver identification and verification has been proven in the experiments.

In [17], researchers consider data coming from the real-time in-car video in order to detect near vehicles for safety, autodriving, and target tracing. They propose an approach with the aim to localize target vehicles in video under various environmental conditions. The motion in the field of view is modeled according to the scene characteristic. Furthermore, a HMM is trained by the vehicle motion model with the goal of probabilistically separate target vehicles from the background and track them. By considering different light setting and street types they conclude that their method is robust with respect to side conditions. Working with these data is difficult because of privacy regulation; nevertheless, innovative techniques to deal with this topic have been developed [10].

Trasarti et al. [31] extract mobility profiles of individuals from raw GPS traces studying how to match individuals based on profiles. The profile matching problem is instantiated to a specific application context, namely proactive car pooling services. Then a matching criterion that satisfies various basic constraints obtained from the background knowledge of the application domain is developed.

Another way to collect data useful for observing driver behavior is by means of questionnaire as done in [19]. Despite being subjected to cognitive bias they provide a means for studying driving behaviors, which could be difficult or even impossible to study by using other methods like observations, interviews and analysis of national accident statistics. Their findings demonstrate that bias caused by socially desirable responding is relatively small in driver behavior questionnaire responses.

The proposal of an architecture able to identify the driver and the driving style exploiting features related to the accelerometer sensor available from the info-entertainment system or in a mobile device fixed in the vehicle is the main difference with regard to the elicited works. This represents the first attempt to detect the driver and the driving style with the same feature set, overcoming the state of the art literature in terms of performances. This innovation is the first attempt to infer both the identity of the driver and her driving style with the same feature set.

3   Modeling and Analyzing the Driver Behavior

In the following we describe the high-level architecture and the method we proposed aimed to: (i) identify the driver and (ii) detect the driving style. Table 1 shows the set of features available from the accelerometer.

Table 1. Features involved in the study.

Feature	Description	Measured in
F1	Acceleration in X	(Gs)
F2	Acceleration in Y	(Gs)
F3	Acceleration in Z	(Gs)
F4	Acceleration in X filtered by KF	(Gs)
F5	Acceleration in Y filtered by KF	(Gs)
F6	Acceleration in Z filtered by KF	(Gs)
F7	Roll	(degrees)
F8	Pitch	(degrees)
F9	Yaw	(degrees)

Those features are sampled by the inertial sensors (available on mobile phones and on info-entertainment systems) at a frequency 10 Hz. In particular, since mobile phone sampled with a frequency 100 Hz the time series is reduced by taking the mean of every 10 samples. With the aim to gather data, the mobile device is fixed on the windshield at the start of the travel, in this way the axes are the same during the whole trip. In the calibration process lateral axis of the vehicle is set to the Y axis(reflects turnings) while the longitudinal axis is set to the X axis (positive value reflects an acceleration, negative a braking).

A Kalman filter (KF) is applied to the observations [26]. KF is an algorithm exploiting a series of measurements observed over time, containing statistical noise and other inaccuracies, and produces estimates of unknown variables that tend to be more accurate than those based on a single measurement alone, by estimating a joint probability distribution over the variables for each timeframe. Guidance, navigation, and control of vehicles, particularly aircraft and spacecraft are common applications [12]. In order to test our method with all the possible noises in the real application we consider real-world data and not simulated one. This enables us to consider all the possible variables that are not considerable from simulated environments.

3.1   The Classification Process

We consider supervised techniques. To build models we consider the following three main steps: (*i*) data instances *pre-processing* (data cleaning and labeling), (*ii*) *training* the designed neural network and (*iii*) classification performance assessment (*Testing*). Cleaning of the raw instances of the feature by removing incomplete and wrong instances is the main process of *pre-processing*.

Figure 1 show the architecture of the *Training* phase: the inferred function is generated starting from the features gathered from the acceleration data. The inferred function, provided by the neural network, should be able to discriminate between accelerometer features belonging to several classes i.e., the function should define the boundaries between the numeric vectors belonging to several classes.

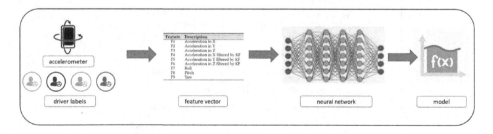

Fig. 1. The *Training* step.

Drivers and the driver behaviors are the two classes of labels used to discriminate. Those labels must be computed free from any bias for the training set in order to let the supervised machine learning techniques to infer the right function. The third step, once generated the model is the evaluation of its effectiveness: the *Testing* one, is depicted in Fig. 2.

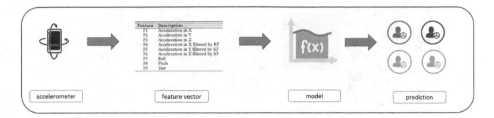

Fig. 2. The *Testing* step.

Well-know information retrieval metrics are used to measure the effectiveness of the models. Furthermore, instances not included into the training data must be used in order to evaluate the (real) performance of the built models. To assure this, we consider the cross validation: we split the full dataset in two parts and we used the first part as training dataset and the second one as testing dataset. In order to evaluate all the instances belonging to the full dataset, this process is repeated considering different instances for both training and testing dataset.

Two different classifiers are defined: the first one is used for the *driver* detection, while the second one is used for her *driving style*. It is worth noting that both classifiers considers the same accelerometer feature set. To train the classifiers, with regard to the *driver* detection, we defined T as a set of labeled traces *(M, l)*, where each M is associated to a label $l \in \{U_1, \ldots, U_n\}$ (where U_n represents the n-th driver under analysis i.e., *Driver #1, Driver #2, Driver #3, Driver #4, Driver #5 and Driver #6*). For the *driving style* classifier, the set of labeled traces *(M, l)* is defined as T. Where M is associated to a label $l \in \{U_1, \ldots, U_n\}$ with U_n representing different driving style i.e., *Normal, Drowsy* and *Aggressive*. A feature vector V_f is used as input for the classifier during training for each M.

In order to perform assessment during the training step, a k-fold cross-validation is used [29]: the dataset is subdivided into k subsets using random sampling. We use a subset as a validation dataset for assessing the performance of the trained model, while we use the other $k-1$ subsets for the training phase. We repeat this procedure $k = 5$ times. In each of the ten iterations, just one subset of the k defined has been used as validation set. To obtain a single reliable estimate, the final results are evaluated by computing the average of the results obtained during the ten iterations. As first step the dataset D is partitioned in k slices.

Then, the following steps are executed for each iteration i: (1) the training set $T_i \subset D$ is generated by selecting an unique set of k–1 slices from the dataset D; (2) we generate the test set $T_i' = D - T_i$ by considering the kth slice (it is evaluated as the complement of T_i to D); (3) T_i is used as training set for the classifier; (4) the trained classifier is applied to T_i' to evaluate accuracy.

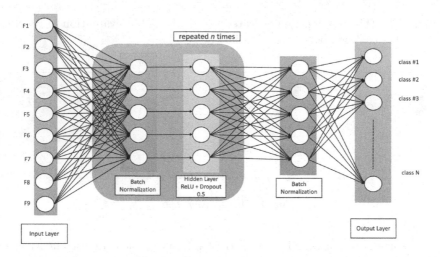

Fig. 3. The designed neural network.

Since $k = 5$, for each iteration i 80% of the dataset is used as training set (T_i), while the remaining 20% is used as a test set (T_i'). The final step has the goal of assessing the performances of the method by running the trained classifiers on real user data. The aim is to identify the driver under analysis (with the first model) and the driving style (with the second model) in real-time, the adopted features model and feeding with it the trained classifier.

3.2   The Neural Network Architecture

In order to build the two models related to the *driver detection model* and *driving style detection model* we propose the neural network architecture depicted in Fig. 3. Figure 3 shows the neural network architecture which number of hidden layers empirically set as equal to 25.

With the aim to generate the classifier model, the supervised classification approach considers a neural network architecture composed by the following sets: (*i*) *Input layer*: this layer is used as an entry point into the network, there is a node for each considered feature; (*ii*) *Hidden layers*: artificial neurons such as perceptrons composed it. The output of each artificial neuron is the application of an activation function to the weighted sum of its input. Usually the functions used are sigmoid, a ReLu or a soft plus function; (*iii*) *Dropout layer*: this layer implements a regularization technique, which aims to reduce the complexity of the model with the goal to prevent over-fitting. Several neurons are randomly deactivated following a Bernoulli distribution of parameter p. Then, the obtained network is trained and the deactivated node are set in their original position with their original weights. In the training stages, the "drop" probability for each hidden node is usually in the range [0,0.5] (for this study 0.5 was used because dropping a neuron with 0.5 probability gets the highest variance for the distribution); (*iv*) *Batch Normalization*: in order to improve the training of feed-forward neural networks, batch normalization is used [16]. It is a method that allows to improve the speed of learning and enables the usage of higher learning rates, more flexible parameter initialization, and saturating non-linearities. We adopted this approach since batch normalized models achieve higher accuracy on both validation and test, due to a stable gradient propagation within the network; (*v*) *Output layer*: the output of the networks are produced in this last layer. The layer used is a dense one using a linear operator in which each input is connected to each output.

Figure 3 shows the neural network architecture used in this work. Cross-entropy has been selected as loss function and the stochastic gradient descent method has been used for learning optimization [21].

3.3 Study Design

We design an experiment aimed to investigate whether the feature set we consider is useful to discriminate between different drivers and different driving style using the proposed neural network architecture. In particular, the experiments aims to verify if the considered features can provide a prediction of unseen driver instances, and unseen driving style. The neural network described in the previous section is used for classification by using the 9 features gathered from accelerometer as input neurons.

The evaluation consists of a classification analysis aimed at assessing whether the considered features are able to correctly discriminate between different drivers and driving styles instances. In order to identify a unique feature of an instance under analysis we use each layer in the neural network [27]. In particular, each layer pick out a particular feature in order to classify the instance. Due to this reason, we choose to not apply a procedure of feature selection. Nevertheless, descriptive statistics and hypothesis testing is used for providing evidence of statistical impact about the effectiveness of the proposed feature vector.

4   Experimental Assessment and Evaluation

In this section we describe the real-world dataset considered in the evaluation of the proposed method and the results of the experiment. The libraries *Tensorflow*[1] (an open source software library for high-performance numerical computation) and the library *Keras*[2] (a high-level neural networks API, able to run on top of TensorFlow) are used to implement the designed neural network architecture. In order to plot the results, *Matplot*[3] is used. We developed the network using the Python programming language. We use an Intel Core i7 8th gen, equipped with 2GPU and 16Gb of RAM to run the experiments.

---

[1] https://www.tensorflow.org/.
[2] https://keras.io/.
[3] https://matplotlib.org/.

4.1 The Dataset

The dataset considered in the evaluation was gathered from a smartphone fixed in the car using an adequate support [26]. The experiments considers the data collected from six drivers that were involved in the experiment. More than 8 h of observations have been monitored and stored in the dataset. Each observation has been assigned to a particular driver (i.e., *Driver #1, Driver #2, Driver #3, Driver #4, Driver #5* and *Driver #6*) and the driving style (i.e., *Normal, Drowsy* and *Aggressive*).

The three driving styles (*Normal, Drowsy* and *Aggressive*) were performed in two different routes, one is 25 km (round trip) in a motorway type of road with normally 3 lanes on each direction and 120 km/h of maximum allowed speed, and the other is around 16 km in a secondary road of normally one lane on each direction and around 90km/h of maximum allowed speed. At the link http://www.robesafe.com/personal/eduardo.romera/uah-driveset/ the interested reader can find the anonymized data used for the tests. The labels driver and the driving style are manually added to each instance of the dataset by researchers in [26].

4.2 Classification Analysis

In the follow we present the classification results obtained from both the neural network and machine learning classification approaches (considered as baseline) on the real-world dataset we evaluate. For the models considering driving style detection, the neural network has 25 hidden layers. Since the neural networks reached the stability with less than 100 epochs, the number of epochs is set to 100.

Figure 4 shows the performances (in terms of accuracy and loss) related to *Driver #1* and *Driver #2* detection. Due to length constraints the plots for the remaining drivers are not reported. Nevertheless, conclusions can be drawn.

The performances of the proposed neural network model for the driver detection problem are the following: (*i*) *Driver #1*: the obtained accuracy is equal to 0.99644 while the relative loss is equal to 0.40479; (*ii*) *Driver #2*: an accuracy of 0.99833 and a loss of 0.40272 is achieved by this driver; (*iii*) *Driver #3*: an accuracy of 0.99833 and a loss of 0.40307 is achieved by the design network for this driver; (*iv*) *Driver #4*: this driver exhibits an accuracy equal to 0.99833 and a loss of 0.40277; (*v*) *Driver #5*: the proposed methodology achieves accuracy equal to 0.99833 and a loss equal to 0.40274; (*vi*) *Driver #6*: an accuracy of 0.99833 and a loss of 0.40273 is achieved for this driver.

In detail we obtain following results: (1) *Normal Style*: the method achieves an accuracy equal to 0.98 and a loss of 0.18596; (2) *Drowsy Style*: for this driving style the accuracy achieved is equal to 0.92003 and the loss is equal to 0.28644; (3) *Aggressive Style*: for the aggressive style the accuracy reached by the designed network is equal to 0.98004 and the loss is equal to 0.17792.

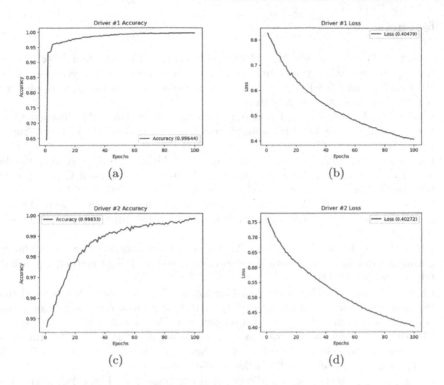

Fig. 4. Driver detection performances: (a) accuracy for *Driver #1*; (b) loss for *Driver #1*; (c) accuracy for *Driver #2*; (d) loss for *Driver #2*.

5 Conclusions and Future Work

In this paper we presented a neural network architecture aimed to provide a framework to analyze driver behaviors aimed to detect the driver and the driving style. The proposed method is evaluated on a real-world dataset using 6 different drivers and by obtaining the analysed feature vector from the accelerometer integrated in a fixed mobile device. The results on the instances considered show results of accuracy equal to 99% for the identification of the driver. Furthermore, the accuracy with respect the driver style detection has been proven to be between 92% and 98%.

As future work, we plan to investigate whether the application of formal methods can be useful for the driver/driving style identification, which have been already successfully used in other domains like, for instance, malware detection [2, 7, 25] and in system biology [5]. Finally, another interesting line of research consists in extending our proposed framework with features proper of the emerging big data trend (e.g., [6, 9, 11]).

References

1. Alheeti, K.M.A., Gruebler, A., McDonald-Maier, K.D.: An intrusion detection system against malicious attacks on the communication network of driverless cars. In: 2015 12th Annual IEEE Consumer Communications and Networking Conference (CCNC), pp. 916–921. IEEE (2015)
2. Canfora, G., Martinelli, F., Mercaldo, F., Nardone, V., Santone, A., Visaggio, C.A.: Leila: formal tool for identifying mobile malicious behaviour. IEEE Trans. Softw. Eng. **45**(12), 1230–1252 (2018)
3. Cannataro, M., Cuzzocrea, A., Pugliese, A.: XAHM: an adaptive hypermedia model based on XML. In: Proceedings of the 14th International Conference on Software Engineering and Knowledge Engineering, pp. 627–634 (2002)
4. Carfora, M.F., et al.: A "pay-how-you-drive" car insurance approach through cluster analysis. Soft. Comput. **23**(9), 2863–2875 (2018). https://doi.org/10.1007/s00500-018-3274-y
5. Ceccarelli, M., Cerulo, L., Santone, A.: De novo reconstruction of gene regulatory networks from time series data, an approach based on formal methods. Methods **69**(3), 298–305 (2014)
6. Chatzimilioudis, G., Cuzzocrea, A., Gunopulos, D., Mamoulis, N.: A novel distributed framework for optimizing query routing trees in wireless sensor networks via optimal operator placement. J. Comput. Syst. Sci. **79**(3), 349–368 (2013)
7. Cimitile, A., Mercaldo, F., Nardone, V., Santone, A., Visaggio, C.A.: Talos: no more ransomware victims with formal methods. Int. J. Inf. Secur. **17**(6), 719–738 (2017). https://doi.org/10.1007/s10207-017-0398-5
8. Cuzzocrea, A.: Combining multidimensional user models and knowledge representation and management techniques for making web services knowledge-aware. Web Intell. Agent Syst. Int. J. **4**(3), 289–312 (2006)
9. Cuzzocrea, A., De Maio, C., Fenza, G., Loia, V., Parente, M.: OLAP analysis of multidimensional tweet streams for supporting advanced analytics. In: Proceedings of the 31st Annual ACM Symposium on Applied Computing, pp. 992–999 (2016)
10. Cuzzocrea, A., De Maio, V., Fadda, E.: Experimenting and assessing a distributed privacy-preserving olap over big data framework: principles, practice, and experiences. In: 2020 IEEE 44th Annual Computers, Software, and Applications Conference (COMPSAC), pp. 1344–1350 (2020)
11. Cuzzocrea, A., Mansmann, S.: Olap visualization: models, issues, and techniques. In: Encyclopedia of Data Warehousing and Mining, Second Edition, pp. 1439–1446. IGI Global (2009)
12. Einicke, G.A.: Smoothing, filtering and prediction: estimating the past, present and future. Technical report, Intech, Rijeka (2012). ISBN 978-953-307-752-9
13. Ellison, A.B., Bliemer, M.C., Greaves, S.P.: Evaluating changes in driver behaviour: a risk profiling approach. Accid. Anal. Prev. **75**, 298–309 (2015)
14. Enev, M., Takakuwa, A., Koscher, K., Kohno, T.: Automobile driver fingerprinting. Proc. Priv. Enhancing Technol. **2016**(1), 34–50 (2016)
15. Fettweis, G.P.: The tactile internet: applications and challenges. IEEE Veh. Technol. Mag. **9**(1), 64–70 (2014)
16. Ioffe, S., Szegedy, C.: Batch normalization: accelerating deep network training by reducing internal covariate shift. In: Proceedings of the 32nd International Conference on International Conference on Machine Learning, ICML 2015, vol. 37, pp. 448–456. JMLR.org (2015)

17. Jazayeri, A., Cai, H., Zheng, J.Y., Tuceryan, M.: Vehicle detection and tracking in car video based on motion model. IEEE Trans. Intell. Transp. Syst. **12**(2), 583–595 (2011)
18. Kwak, B.I., Woo, J., Kim, H.K.: Know your master: driver profiling-based anti-theft method. In: PST 2016, pp. 211–218 (2016)
19. Lajunen, T., Summala, H.: Can we trust self-reports of driving? effects of impression management on driver behaviour questionnaire responses. Transport. Res. F: Traffic Psychol. Behav. **6**(2), 97–107 (2003)
20. Lyamin, N., Vinel, A., Jonsson, M., Loo, J.: Real-time detection of denial-of-service attacks in IEEE 802.11 p vehicular networks. IEEE Commun. Lett. **18**(1), 110–113 (2014)
21. Mannor, S., Peleg, D., Rubinstein, R.: The cross entropy method for classification. In: Proceedings of the 22nd International Conference on Machine Learning, ICML 2005, pp. 561–568. ACM, New York (2005)
22. Martinelli, F., Mercaldo, F., Nardone, V., Orlando, A., Santone, A.: Context-awareness mobile devices for traffic incident prevention. In: 2018 IEEE International Conference on Pervasive Computing and Communications Workshops (PerCom Workshops), pp. 143–148. IEEE (2018)
23. Martinelli, F., Mercaldo, F., Nardone, V., Santone, A.: Car hacking identification through fuzzy logic algorithms. In: 2017 IEEE International Conference on Fuzzy Systems (FUZZ-IEEE), pp. 1–7. IEEE (2017)
24. Martinelli, F., Mercaldo, F., Orlando, A., Nardone, V., Santone, A., Sangaiah, A.K.: Human behavior characterization for driving style recognition in vehicle system. Comput. Electr. Eng. **83**, 102504 (2020)
25. Mercaldo, F., Nardone, V., Santone, A., Visaggio, C.A.: Download malware? No, thanks. How formal methods can block update attacks. In: 2016 IEEE/ACM 4th FME Workshop on Formal Methods in Software Engineering (FormaliSE), pp. 22–28. IEEE (2016)
26. Romera, E., Bergasa, L.M., Arroyo, R.: Need data for driver behaviour analysis? Presenting the public UAH-DriveSet. In: 2016 IEEE 19th International Conference on Intelligent Transportation Systems (ITSC), pp. 387–392. IEEE (2016)
27. Schmidhuber, J.: Deep learning in neural networks: an overview. Neural Netw. **61**, 85–117 (2015)
28. Skibinski, J., Trainor, J., Reed, C.: Internet-based vehicle communication network. SAE Trans. 820–825 (2000)
29. Stone, M.: Cross-validatory choice and assessment of statistical predictions. J. Roy. Stat. Soc. Ser. B **36**, 111–147 (1974)
30. Taylor, A., Leblanc, S., Japkowicz, N.: Anomaly detection in automobile control network data with long short-term memory networks. In: 2016 IEEE International Conference on Data Science and Advanced Analytics (DSAA), pp. 130–139. IEEE (2016)
31. Trasarti, R., Pinelli, F., Nanni, M., Giannotti, F.: Mining mobility user profiles for car pooling. In: Proceedings of the 17th ACM SIGKDD International Conference on Knowledge Discovery and Data Mining, pp. 1190–1198. ACM (2011)
32. Wakita, T., et al.: Driver identification using driving behavior signals. IEICE Trans. Inf. Syst. **89**(3), 1188–1194 (2006)
33. Xia, F., Yang, L.T., Wang, L., Vinel, A.: Internet of things. Int. J. Commun. Syst. **25**(9), 1101–1102 (2012)
34. Zhang, X., Zhao, X., Rong, J.: A study of individual characteristics of driving behavior based on hidden Markov model. Sens. Transducers **167**(3), 194 (2014)

Web Document Categorization Using Knowledge Graph and Semantic Textual Topic Detection

Antonio M. Rinaldi[(✉)], Cristiano Russo, and Cristian Tommasino

Department of Electrical Engineering and Information Technologies,
University of Napoli Federico II, 80125 Via Claudio, 21, Napoli, Italy
{antoniomaria.rinaldi,cristiano.russo,cristian.tommasino}@unina.it

Abstract. In several contexts, the amount of available digital documents increases every day. One of these challenging contexts is the Web. The management of this large amount of information needs more efficient and effective methods and techniques for analyzing data and generate information. Specific application as information retrieval systems have more and more high performances in the document seeking process, but often they lack of semantic understanding about documents topics. In this context, another issue arising from a massive amount of data is the problem of information overload, which affects the quality and performances of information retrieval systems. This work aims to show an approach for document classification based on semantic, which allows a topic detection of analyzed documents using an ontology-based model implemented as a semantic knowledge base using a No SQL graph DB. Finally, we present and discuss experimental results in order to show the effectiveness of our approach.

1 Introduction

The widespread diffusion of new communication technologies, such as the Internet together with the development of intelligent artificial systems capable of producing and sharing different kinds of data, have led to a dramatic increasing of the number of available information. One of the main goal in this context is to transform heterogeneous and unstructured data into useful and meaningful information through the use of Big Data, deep neural networks and the myriad of applications that derive from their implementations. For this purpose, documents categorization and classification is an essential task in the information retrieval domain, strongly affecting user perception [16]. The goal of classification is to associate one or more classes to a document, easing the management of a document collection. The techniques used to classify a document have been widely applied to different contexts paying attention to the semantic relationships especially between terms and the represented concepts [27,28]. The use of semantics in the document categorization task has allowed a more accurate detection of topics concerning classical approaches based on raw text and meaningless label [24]. Techniques relying on semantic analysis are often based on the idea of

O. Gervasi et al. (Eds.): ICCSA 2021, LNCS 12951, pp. 40–51, 2021.
https://doi.org/10.1007/978-3-030-86970-0_4

semantic network (SN) [32]. Woods [37] highlighted the lack of a rigorous definition for semantic networks and their conceptual role. In the frame of this work, we will refer to a semantic network as a graph entity which contains information about semantic and/or linguistic relationships holding between several concepts. Lately, semantic networks have been often associated to ontologies which are now a keystone in the field of knowledge representation, integration and acquisition [26, 29–31]. Moreover, ontologies are designed to be machine-readable and machine-processable. Over the years, the scientific community provided many definitions of ontologies. One of the most accepted is in [11]. It is possible to represent ontologies into graphs and vice versa, with this duality making them interchangeable. The use of graphs and analysis metrics permits us to have a fast retrieval of information and for finding new patterns of knowledge hard to recognize. Topic detection and categorization are crucial task which allows quick access to contents in a document collections when used in an automatic way. A disadvantage of many classification methods is that they treat the categorization structure without considering the relationships between categories. A much better approach is to consider that structures, either hierarchical or taxonomic, constitute the most natural way in which concepts, subjects or categories are organized in practice [1].

The novelty of the proposed work has to be found in the way we combine statistical information and natural language processing. In particular, the approache uses an algorithm for word sense disambiguation based on semantic analysis, ontologies and semantic similarity metrics. The core is a knowledge graph which represents our semantic networks (i.e. ontology). It is used as a primary source for extracting further information. It is implemented by means of a NoSQL technology to perform a "semantic topic detection".

The paper is organized as follows: in Sect. 2 we provide a review of the literature related to Topic Modeling and Topic Detection techniques and technologies; Sect. 3 introduces the approach along with the general architecture of the system and the proposed textual classification methodology; in Sect. 4 we present and discuss the experimental strategy and results; lastly, Sect. 5 is devoted to conclusions and future research.

2   Related Works

This section analyzes relevant recent works related to textual topic detection, as well as the differences between our approach and the described ones. Over the years, the scientific community proposed several methodologies, hare grouped according to the main technique used. The goal of approaches based on statistics is to identify the relevance of a term based on some statistical properties, such as TF-IDF [33], N-Grams [8], etc. *Topic modeling* [20] instead is an innovative and widespread analytical method for the extraction of co-occurring lexical clusters in a documentary collection. In particular, it makes use of an ensemble of unsupervised text mining techniques, where the approaches are based on probabilities. The authors in [21] described a probabilistic approach for Web page classification, they propose a dynamic and hierarchical classification system that is capable of adding new categories, organizing the Web pages into a tree structure and classifying them by searching through only one path of the tree structure. Other approaches use *features* based on linguistic, syntactic, semantic, lexical properties. Hence, they are named linguistic approaches. Similarity functions are employed to extract representative keywords. Different machine learning techniques, such as Support Vector Machine [35], Naive Bayes [39] and others are used. The keyword extraction is the result of a trained model able to predict significant keywords. Other approaches attempt to combine the above-cited ones in several ways. Other parameters such as *word position, layout feature*, HTML tags, etc. are also used. In [13], the authors use an approach based on machine learning techniques in combination with semantic information, while in [18] co-occurrence is employed for the derivation of keywords from a single document. In [12], the authors use linguistic features to represent term relevance considering the position of a term in the document and other researches [25] build models of semantic graphs for representing documents. In [36], the authors presented an iterative approach for keywords extraction considering relations at different document levels (words, sentences, topics). With such an approach a graph containing relationships between different nodes is created, then the score of each keyword is computed through an iterative algorithm. In [2], the authors analyzed probabilistic models for topic extraction. Xu et al. [38] centered their research on topic detection and tracking but focusing on online news texts. The authors propose a method for the evolution of news topics over time in order to track topics in the news text set. First, topics are extracted with LDA (latent Dirichlet allocation) model from news texts and the Gibbs Sampling method is used to define parameters. In [34] an extended LDA topic model based on the occurrence of topic dependencies is used for spam detection in short text segments of web forums online discussions. Khalid et al. [14] use parallel dirichlet allocation model and elbow method for topic detection from conversational dialogue corpus. Bodrunova et al. [5] propose an approach based on sentence embeddings and agglomerative clustering by Ward's method. The Markov stopping moment is used for optimal clustering. Prabowo et al. [22] describe a strategy to enhance a system called ACE (Automatic Classification Engine) using ontologies. The authors focus on the use of ontologies for classifying Web pages concerning the Dewey Decimal Classification (DDC) and Library of Congress Classification (LCC) schemes using weighted terms in the Web pages and the structure of domain ontologies. The association between significant conceptual instances into their associated class representative(s) is performed using an ontology classification scheme mapping and a feed-forward network model. The use of ontologies is also explored in [17]. The authors propose a method for topic detection and tracking based on an event ontology that provides event classes hierarchy based on domain common sense.

In this paper, we propose a semantic approach for document classification. The main differences between our approach and the other presented so far are in the proposing of a novel algorithm for topic detection based on semantic information extracted from a general knowledge base for representing the user domains of interest and the fully automatization of our process without a learning step.

3 The Proposed Approach

In this section, we provide a detailed description of our approach for topic detection. The main feature of our methodology is its ability to combine both statistical information, natural language processing and several technologies to categorize documents using a comprehensive semantic analysis, which involves ontologies and metrics based on semantic similarity. To implement our approach, we follow a modular framework for document analysis and categorization. The framework makes use of a general knowledge base, where textual representation of semantic concepts are stored.

3.1 The Knowledge Base

We realized a general knowledge base using an ontology model proposed and implemented in [6,7]. The database is realized by means of a NoSQL graph technology. From an abstract, conceptual point of view, the model representation is based on *signs*, defined in [9] as *"something that stands for something, for someone in some way"*. These signs are used to represent concepts. The model structure is composed of a triple $<S, P, C>$ where S is the set of signs; P is the set of properties used to link signs with concepts; C is the set of constraints defined on the set P. We propose an approach focused on the use of textual representations and based on the semantic dictionary WordNet [19]. According to the terminology used in the ontology model, the textual representations are our signs. The ontology is defined using the DL version of the *Web Ontology Language*(OWL), a markup language that offers a high level of expressiveness preserving completeness and computational decidability. The model can be seen as a top-level ontology, since it contains a very abstract definition for its classes. The model and the related knowledge graph have been implemented in Neo4J *graph-db* using the *property-graph-model* [3].

Figure 1 shows a part of our knowledge graph to put in evidence the complexity of the implemented graph for a sake of clarity. It is composed of near 15,000 nodes and 30,000 relations extracted from our knowledge base.

3.2 The Topic Detection Strategy

Our novel strategy for textual topic detection is based on an algorithm called SEMREL. Its representation model is the classical *bag-of-words*. Once a document is cleaned, i.e. unnecessary parts are removed, the *tokenization* step allows to obtain a list of terms in the document. Such a list of terms is the input for a *Word Sense Disambiguation* step that pre-processes the list assigning the right meaning to each term. Then, Semantic Networks dynamically extracted

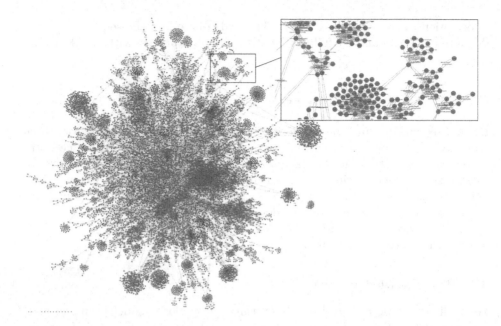

Fig. 1. Knowledge graph excerpt with 30000 edges and about 15000 nodes

from our knowledge base are generated for all the terms, in the *SN Extractor* step. Common nodes between an SN resulting from each concept and SNs from other concepts are used to compute their intersections. The common nodes correspond to the degree of representation of the concept considered with respect to the entire document. This measure is indicated as *Sense Coverage*. The latter factor would favor the more generic concepts and for this reason a scaling factor depending on the *depth* of the considered concept is used. It is computed as the number of hops to the root of our knowledge base considering only the hypernymy relationships. The *TopicConcept* is the one with the best trade-off between the *SenseCoverage* and the *Depth*. The formula used for calculating the topic concept of a given document is shown in Eq. 1.

$$TopicConcept = max(depth(C_i) * Coverage(C_i)) \tag{1}$$

where C_i is the *i-th* concept resulting from the WSD step. Only concepts in the *noun* lexical category are considered from the WSD list, because in the authors' opinion they are more representative to express the topic of a document.

In the Algorithm 1, we show the logic used to find the topic concept.

The WSD attempts to palliate the issue related to term *polysemy*. Indeed, it tries to "sense" the correct meaning of a term by comparing each sense of a term with all the senses of the others. The similarity between terms is calculated through a linguistic based approach and a metric computes their *semantic relatedness* [23].

Algorithm 1. Topic Concept Algorithm

```
1: procedure TopicConcept(ConceptList)
2:      BestConceptScore = 0
3:      for each concept Cᵢ in ConceptList (after WSD) do
4:          ScoreCᵢ = 0
5:          SN_Cᵢ = BuildSN(Cᵢ)
6:          CoverCᵢ = 0
7:          for each concept Cⱼ ≠ Cᵢ in ConceptList do
8:              SN_Cⱼ = BuildSN(Cⱼ)
9:              NumberOfCommonConcept = Match(SN_Cᵢ, SN_Cⱼ)
10:             CoverCᵢ = CoverCᵢ + NumberOfCommonConcept
11:         end for
12:         ScoreCᵢ = depth(Cᵢ) * CoverCᵢ
13:         if BestConceptScore < ScoreCᵢ then
14:             BestConceptScore = ScoreCᵢ
15:             TopicConcept = Cᵢ
16:         end if
17:     end for
18:     return TopicConcept
19: end procedure
```

This metric is based on a combination of the best path between pairs of terms and the depth of their Lowest Common Subsumer, expressed as the number of hops to the root of our knowledge base using hypernymy relationships.

The best path is calculated as follows:

$$l(w_1, w_2) = min_j \sum_{i=1}^{h_j(w_1,w_2)} \frac{1}{\sigma_i} \tag{2}$$

where l is the best path length between the terms w_i and w_j, $h_j(w_i, w_j)$ corresponds to the number of hops of the $j\text{-}th$ path and σ_i corresponds to the weight of the $i\text{-}th$ edge of the $j\text{-}th$ path. The weights σ_i are assigned to the properties of the ontological model described in Sect. 3.1 to discriminate the expressive power of relationships and they are set by experiments.

The depth factor is used to give more importance to specific concepts (low level and therefore with high depth) than generic ones (low depth). A non-linear function is used to scale the contribution of the sub-ordinates concepts in the upper level and increase those of a lower ones. The metric is normalized in the range $[0,1]$ (1 when the length of the path is 0 and 0 when the length go to infinite).

The Semantic Relatedness Grade of a document is then calculated as:

$$SRG(v) = \sum_{(w_i,w_j)} e^{-\alpha \cdot l(w_i,w_j)} \frac{e^{\beta \cdot d(w_i,w_j)} - e^{-\beta \cdot d(w_i,w_j)}}{e^{\beta \cdot d(w_i,w_j)} + e^{-\beta \cdot d(w_i,w_j)}} \tag{3}$$

where (w_i, w_j) are pairs of terms in v, $d(w_i, w_j)$ is the number of hops from the w_i, w_j subsumer to the root of the WordNet hierarchy considering the IS-A relation, α and β are parameters whose values are set by experiments.

The WSD process calculates the score for each sense of the considered term using the proposed metric. The best sense associated with a term is the one which maximizes the SRG obtained by the semantic relatedness between all terms in the document.

The best sense recognition is shown in the Algorithm 2.

Algorithm 2. Best Sense Algorithm

1: **procedure** BEST_SENSE(W_t)
2:     **for each sense** $S_{t,i}$ **of target word** W_t **do**
3:         set $Score\_St_{t,i} = 0$
4:         **for each word** $W_j \neq W_t$ **in windows of context do**
5:             **init array** $temp\_score$
6:             **for** $each sense S_{j,k} of W_j$ **do**
7:                 $temp\_score[j] = SRG(S_{t,i}, S_{j,k})$
8:             **end for**
9:             $Score\_S_{t,i} = Score\_S_{t,i} + MAX(temp\_score)$
10:         **end for**
11:         **if** $best\_sense\_score < Score\_S_{t,i}$ **then**
12:             $best\_sense\_score = Score\_S_{t,i}$
13:             $best\_sense = S_{t,i}$
14:         **end if**
15:     **end for**
16:     **return** $best\_sense$
17: **end procedure**

The best sense of a term is the one with the maximum score obtained by estimating the semantic relatedness with all the other terms of a given window of context.

3.3   The Implemented System

The system architecture is shown in Fig. 2. It is composed by multiple modules which are responsible of managing several tasks.

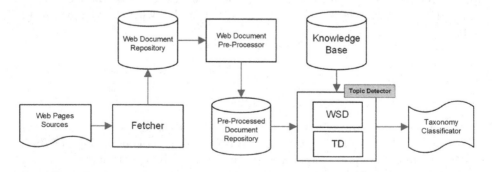

Fig. 2. The system architecture

The Web Documents can be fetched from different data sources by means of the Fetcher module and stored in the Web document repository. The textual information is first pre-processed. Cleaning operations are carried out by the Document Pre-Processor module. Such operations are: (i) tags removing, (ii) stop words deleting, (iii) elimination of special characters, (iv) stemming. The Topic Detection module uses an algorithm based on text analysis to address the correct topic of a document and our graph knowledge base. It is based on WSD and TD tasks based on the algorithm previously discussed. It is able to classify a document by the recognition of its main topic. Topic Detection result is the input of the Taxonomy Classificator used to create, with the help of our knowledge base, a hierarchy beginning from a concept. The proposed metric and approach have been compared with baselines and the results are shown in the next section.

4   Test Strategy and Experimental Results

In order to measure the performances of our framework we have carried out several experiments, which are discussed in the following. First we compare it with two reference algorithms widely used in the topic detection research field in order to have a more robust and significant evaluation: *LSA* [15] and *LDA* [4]. *Latent Semantic Analysis* (LSA), also known as *Latent Semantic Indexing* (LSI), is based on a vectorial representation of a document though the *bag-of-words* model. *Latent Dirichlet Allocation* (LDA) is a text-mining technique based on statistical models.

One of the remarkable feature of this system is that it is highly generalizable thanks to the development of autonomous modules. In this paper, we have used the textual content of *DMOZ* [10], one of the most popular and rich multilingual web directories with open content. The archive is made up of links to web content organized according to a hierarchy. The reason why we choose DMOZ lays into the fact that we want to compare our results with baselines. This way we can test against a real experimental scenario by using a public and well know repository. The category at the top level is the root of the DMOZ hierarchy. Since this is not informative at all, it has been discarded. Then we built a *ground truth* has been built considering a subset of documents from categories placed at the second level. These are shown in Table 1 together with statistics for the used test set.

The list of URLs is submitted to our fetcher to download the textual content. The restriction to a subset of DMOZ was necessary, due to the presence of numerous dead links and textual information. On a total of 12120 documents, we selected 10910 of them to create the topic modeling models used by LSA and LDA, while 1210 documents are used as test-set. The testing procedure employed in this paper uses our knowledge graph for the topic classification task. In order to have a fair and reliable comparison with all implemented algorithms, the same technique must be used, hence we need to perform a manual mapping of the used DMOZ categories to their respective WordNet synonyms. In this way, we create a ground truth using a pre-classified document directory (i.e., DMOZ) through a mapping with a formal and well-known knowledge

Table 1. DMOZ - URLs/category

DMOZ category	URLs/Category	URLs/Ground truth category
Arts	164 873	1 312
Business	171 734	1 208
Computers	78 994	1 189
Games	28 260	1 136
Health	41 905	1 011
News	6 391	1 264
Science	79 733	1 173
Shopping	60 891	1 430
Society	169 054	1 272
Sports	71 769	1 125
	Tot. URLs 3 573 026	12 120

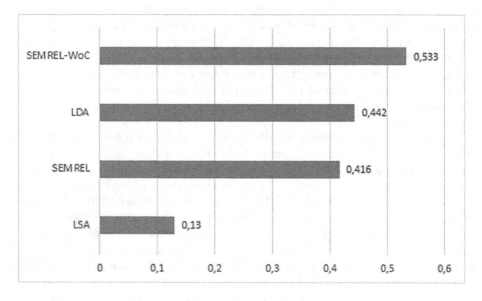

Fig. 3. Accuracy textual topic detection

source (i.e., WordNet). The annotation process also facilitates the classification of documents by other algorithms, e.g. LSA and LDA, because they give several topics that represent the main topics of the analyzed collection without dealings with the DMOZ categories. The central facet of our framework has been carefully evaluated to show the distinguishable performances of the proposed methodology. For the textual topic detection, the LSA and LDA models have been implemented and generated, as well as the proposed SEMREL algorithm in two variants. The first one consists in computing the (SRG) of a sense related to a term semantically compared with all the terms of the whole document. The second one is performed dividing a document in grammatical periods, defined by the punctuation marks dot, question mark and exclamation mark (i.e. wondows of context). The *semantic relatedness* of a concept is calculated considering each sense of a term belonging to its window of context. Figure 3 shows the obtained results accuracy.

We argue that these results depend to the impossibility of mapping some topics generated by LSA or LDA with the corresponding WordNet sysnset. This issue dosn't allow an accurate topic detection due to the dependency of these models to the data set. On the other hand, SEMREL have a better concept recognition taking out noise from specific datasets.

5   Conclusion and Future Works

In this paper, we have proposed a semantic approach based on a knowledge graph for web document textual topic detection. For this purpose, a word sense disambiguation algorithm has been implemented, and semantic similarity metrics have been used. The system has been fully tested with a standard web document collection (i.e. *DMOZ*). The design of the system allows the use of different collections of documents. The evaluation of our approach shows promising results, also in comparison with state-of-art algorithms for textual topic detection. Our method has some limitations due to the lack of knowledge in several conceptual domains in our knowledge base (i.e. WordNet). In future works, we are interested in the definition of automatic techniques to extend our knowledge base with additional multimedia information and domain specific ontologies. Moreover, we want investigate on the novel methodologies to improve the performance of the topic detection process exploiting multimedia data considering new metrics to compute semantic similarity. Other aspects to point out are the computational efficiency of our approach and additional testing with different document collections.

References

1. Albanese, M., Picariello, A., Rinaldi, A.: A semantic search engine for web information retrieval: an approach based on dynamic semantic networks. In: Proceedings of Sheffield SIGIR - Twenty-Seventh Annual International ACM SIGIR Conference on Research and Development in Information Retrieval (2004)
2. Alghamdi, A.: A survey of topic modeling in text mining. Int. J. Adv. Comput. Sci. Appl. IJACSA (2015)
3. Angles, R.: The property graph database model. In: AMW (2018)
4. Blei, D.M., Ng, A.Y., Jordan, M.I.: Latent dirichlet allocation. J. Mach. Learn. Res. **3**(Jan), 993–1022 (2003)
5. Bodrunova, S.S., Orekhov, A.V., Blekanov, I.S., Lyudkevich, N.S., Tarasov, N.A.: Topic detection based on sentence embeddings and agglomerative clustering with Markov moment. Future Internet **12**(9), 144 (2020)

6. Caldarola, E.G., Picariello, A., Rinaldi, A.M.: Experiences in wordnet visualization with labeled graph databases. In: Fred, A., Dietz, J.L.G., Aveiro, D., Liu, K., Filipe, J. (eds.) IC3K 2015. CCIS, vol. 631, pp. 80–99. Springer, Cham (2016). https://doi.org/10.1007/978-3-319-52758-1_6

7. Caldarola, E.G., Picariello, A., Rinaldi, A.M.: Big graph-based data visualization experiences: the wordnet case study. In: 2015 7th International Joint Conference on Knowledge Discovery, Knowledge Engineering and Knowledge Management (IC3K), vol. 1, pp. 104–115. IEEE (2015)

8. Cavnar, W.B., Trenkle, J.M., et al.: N-gram-based text categorization. In: Proceedings of SDAIR 1994, 3rd Annual Symposium on Document Analysis and Information Retrieval, vol. 161175. Citeseer (1994)

9. Danesi, M., Perron, P.: Analyzing Cultures: An Introduction and Handbook. Indiana University Press, Bloomington (1999)

10. DMOZ: Dmoz website. http://dmoz-odp.org/

11. Gruber, T.R.: Toward principles for the design of ontologies used for knowledge sharing? Int. J. Hum. Comput. Stud. **43**(5–6), 907–928 (1995)

12. Hu, X., Wu, B.: Automatic keyword extraction using linguistic features. In: Sixth IEEE International Conference on Data Mining Workshops, ICDM Workshops 2006, pp. 19–23. IEEE (2006)

13. Hulth, A.: Improved automatic keyword extraction given more linguistic knowledge. In: Proceedings of the 2003 Conference on Empirical Methods in Natural Language Processing, pp. 216–223. Association for Computational Linguistics (2003)

14. Khalid, H., Wade, V.: Topic detection from conversational dialogue corpus with parallel dirichlet allocation model and elbow method. arXiv preprint arXiv:2006.03353 (2020)

15. Landauer, T.K., Foltz, P.W., Laham, D.: An introduction to latent semantic analysis. Discourse Process. **25**(2–3), 259–284 (1998)

16. Liaw, S.S., Huang, H.M.: An investigation of user attitudes toward search engines as an information retrieval tool. Comput. Hum. Behav. **19**(6), 751–765 (2003)

17. Liu, W., Jiang, L., Wu, Y., Tang, T., Li, W.: Topic detection and tracking based on event ontology. IEEE Access **8**, 98044–98056 (2020)

18. Matsuo, Y., Ishizuka, M.: Keyword extraction from a single document using word co-occurrence statistical information. Int. J. Artif. Intell. Tools **13**(01), 157–169 (2004)

19. Miller, G.A.: Wordnet: a lexical database for English. Commun. ACM **38**(11), 39–41 (1995)

20. Papadimitriou, C.H., Raghavan, P., Tamaki, H., Vempala, S.: Latent semantic indexing: a probabilistic analysis. J. Comput. Syst. Sci. **61**(2), 217–235 (2000)

21. Peng, X., Choi, B.: Automatic web page classification in a dynamic and hierarchical way. In: Proceedings of 2002 IEEE International Conference on Data Mining, pp. 386–393. IEEE (2002)

22. Prabowo, R., Jackson, M., Burden, P., Knoell, H.D.: Ontology-based automatic classification for web pages: design, implementation and evaluation. In: Proceedings of the Third International Conference on Web Information Systems Engineering, WISE 2002, pp. 182–191. IEEE (2002)

23. Rinaldi, A.M.: An ontology-driven approach for semantic information retrieval on the web. ACM Trans. Internet Technol. (TOIT) **9**(3), 10 (2009)

24. Rinaldi, A.M.: Using multimedia ontologies for automatic image annotation and classification. In: 2014 IEEE International Congress on Big Data, pp. 242–249. IEEE (2014)

25. Rinaldi, A.M., Russo, C.: A novel framework to represent documents using a semantically-grounded graph model. In: KDIR, pp. 201–209 (2018)
26. Rinaldi, A.M., Russo, C.: A semantic-based model to represent multimedia big data. In: Proceedings of the 10th International Conference on Management of Digital EcoSystems, pp. 31–38. ACM (2018)
27. Rinaldi, A.M., Russo, C.: User-centered information retrieval using semantic multimedia big data. In: 2018 IEEE International Conference on Big Data (Big Data), pp. 2304–2313. IEEE (2018)
28. Rinaldi, A.M., Russo, C.: Using a multimedia semantic graph for web document visualization and summarization. Multimedia Tools Appl. **80**(3), 3885–3925 (2021)
29. Rinaldi, A.M., Russo, C., Madani, K.: A semantic matching strategy for very large knowledge bases integration. Int. J. Inf. Technol. Web Eng. (IJITWE) **15**(2), 1–29 (2020)
30. Russo, C., Madani, K., Rinaldi, A.M.: Knowledge acquisition and design using semantics and perception: a case study for autonomous robots. Neural Process. Lett. 1–16 (2020)
31. Russo, C., Madani, K., Rinaldi, A.M.: An unsupervised approach for knowledge construction applied to personal robots. IEEE Trans. Cogn. Dev. Syst. **13**(1), 6–15 (2020)
32. Sowa, J.F.: Principles of Semantic Networks: Explorations in the Representation of Knowledge. Morgan Kaufmann, Burlington (2014)
33. Sparck Jones, K.: A statistical interpretation of term specificity and its application in retrieval. J. Doc. **28**(1), 11–21 (1972)
34. Sun, Y.: Topic modeling and spam detection for short text segments in web forums. Ph.D. thesis, Case Western Reserve University (2020)
35. Suykens, J.A., Vandewalle, J.: Least squares support vector machine classifiers. Neural Process. Lett. **9**(3), 293–300 (1999)
36. Wei, Y.: An iterative approach to keywords extraction. In: Tan, Y., Shi, Y., Ji, Z. (eds.) ICSI 2012. LNCS, vol. 7332, pp. 93–99. Springer, Heidelberg (2012). https://doi.org/10.1007/978-3-642-31020-1_12
37. Woods, W.A.: What's in a link: Foundations for semantic networks. Read. Cogn. Sci. 102–125 (1988)
38. Xu, G., Meng, Y., Chen, Z., Qiu, X., Wang, C., Yao, H.: Research on topic detection and tracking for online news texts. IEEE Access **7**, 58407–58418 (2019)
39. Zhang, H.: The optimality of Naive Bayes. AA **1**(2), 3 (2004)

Understanding Drug Abuse Social Network Using Weighted Graph Neural Networks Explainer

Zuanjie Ma$^{(\boxtimes)}$, Hongming Gu, and Zhenhua Liu

Emerging Technology Research Division,
China Telecom Research Institute, Beijing, China
{guhm,liuzhh11}@chinatelecom.cn

Abstract. Relational data are universal in today's society, such as social media network, citation network, and molecules. They can be transformed into graph data consisting of nodes and edges. To classify the class of nodes in graph, GNNs is a state-of-the-art model in this domain. For example, anti-drug police rely on social data to identify hidden drug abusers. However, in the deployment of GNNs, it still require interpretability techniques to support. Because GNNs embed the node information and pass them through the adjacency network, resulting in the lack of transparency in its model. However, the common explanation technique used in other models are not directly applicable on GNNs. And those explanation methods in CNNs are not applicable either considering the topological variety of graph. In order to explain our GNN model on drug abuse social network, we proposed a framework to preprocess graph data, train GNN model, and apply our Maximal Mutual Information (MMI) method to help explain the prediction made by the GNN model. We evaluate the performance of the explanation model with evaluation metrics *consistency* and *contrastivity*. The explanation model performs impressively in both real world data and two synthetic data compared to other baselines. The results also show that our model can mimic GNN explanation made by human in node classification.

Keywords: Interpretability · Graph Neural Networks · Max mutual information

1 Introduction

Many real-world applications, including social networks, rely heavily on structured data as graphs. Graph is a powerful data structure to represent relational data. It maps the features of entities as node features and the relational information between them as edges. In order to take advantage of such data structure, Graph Neural Networks (GNNs) have emerged as state-of-the-art learning approach on these data to perform node classification task, such as drug abuser identification. The principle of GNNs is to recursively incorporate information

O. Gervasi et al. (Eds.): ICCSA 2021, LNCS 12951, pp. 52–61, 2021.
https://doi.org/10.1007/978-3-030-86970-0_5

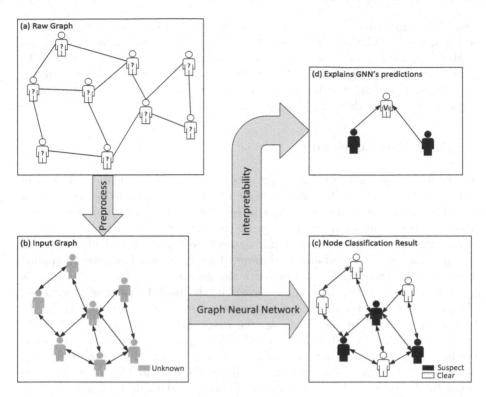

Fig. 1. The pipeline of the explanation framework. (a) represents the raw drug abuse graph to be preprocessed. (b) and (c) refer to the training and inference process of node classification task by GNN. (d) explains the prediction result of the trained model on drug abuse social data.

from neighboring nodes in graph, thus capturing both graph structure information and node or edge features. Despite their powerful representation ability, GNNs lack transparency for a human-friendly explanation for their predictions. Besides, the ability to explain the prediction of GNNs is important and useful because it can increase trust in GNNs model and transparency in decision-critical applications so that practitioners can improve its performance before deploying them in the real world. However, due to the complex data representation and non-linear transformations performed on the data, explaining decisions made by GNNs is a challenging problem. Therefore we formulate our explanation to identify the critical pattern in the input data used by the model to perform prediction. Although there are many explanation methods developed in deep learning models such as convolution neural network (CNN) [2, 7, 9, 12], those methods are not directly applicable to explain GNNs. Although some work have been done in explaining GNN [5, 8, 11, 13], but they fails in keeping the relational information in non-linear transformation or taking the edge weight information into account. Here we design a framework as denoted in Fig. 1 to preprocess, train, and explain drug abuse graph data based on GNNs model. We first extract feature from the raw graph to build the input graph for GNNs. Given the node classification results from GNNs, we apply a post-hoc graph feature explanation method to formulate an explanation. Our experiments on drug abuse data and synthetic data demonstrate that our the applied methods and pipeline can generate similar explanations and evidence as human interpretation.

2   Data Preprocessing

We are given a spatial-temporal heterogeneous drug abuse social data to perform node classification task. This data consist of social connections between nodes of suspect and clear collected in 17 consecutive months. There are four kinds of edges responding to different ways of contact between nodes. In order to extract information from this dynamic dataset, we apply sliding window based data preprocessing method on this dataset. The size of window here we applied is one month. Specifically, y_u represents the ground truth label of each node v_u, it is defined by the drug abuse record in the past 17 months, 0 for clear who never exist in abuse record, 1 for suspect who have been appeared in the drug abuse record. For each node v_u, the associated features, $X_u \in \mathbb{R}^d$ is the boolean indicator of the drug abuse record in the past 10 months. And the edge weight e_{uv} is the concatenation of original features consisting of the frequency of statistic values such as number of calls and call duration. There are 17 graphs in the final processed data, representing 17 months data. Each graph contains around 2000 nodes.

3   Graph Neural Networks

3.1   Data Representation

In this section, we introduce the notation and definitions of weighted and directed graphs. We denote the graph by $G = (V, \mathcal{E})$ where V is the set of nodes, $\mathcal{E}$ the set of edges linking the nodes and X the set of nodes' features. For connected nodes, $u, v \in V$, we denote by $e_{vu} \in \mathbb{R}$ the weight of the directed edge $\langle v, u \rangle \in \mathcal{E}$ linking them. We denote $E[v, u] = e_{vu}$, where $E \in \mathbb{R}^{|\mathcal{E}|}$. For each node, v_u, we associate a d-dimensional vector of features, $X_u \in \mathbb{R}^d$ and denote the set of all features as $X = \{X_u : u \in V\} \in \left(\mathbb{R}^d\right)^{|V|}$.

Edge features contains important features about graphs. For instances, the graph G may represent a drug abuse social network, where the nodes V represent suspects or clears and the edges E are the different kinds of social connections between them.

We consider a node classification task, where each node u is assigned a label $y_u \in I_C = \{0, \ldots, C - 1\}$. The two perspectives of the explanation correspond to the informative subset on E and X of the weighted graph.

3.2 Node Classification

GNN features its advantage by keeping the flow of information across neighbor-hoods and graphs. Besides, it is invariant to permutations on ordering. Different from the traditional GNN architectures, we focus on GNNs exploiting edge infor-mation on weighted and directed graphs. The per-layer update of our GNN model are as follows. At layer l, the update of GNN model Φ involves three key compu-tation steps. (1) First, the model computes neural messages between every pair of nodes. The message for node pair $\langle v_u, v_v \rangle$ is a function MSG of v_u's and v_v's representations h_u^{l-1} and h_v^{l-1} in the previous layer and of the weight of edge e_{uv} pointing from u to v: $m_{uv}^l = \text{MSG}\left(\mathbf{h}_u^{l-1}, \mathbf{h}_v^{l-1}, e_{uv}\right)$. (2) Second, for each node v_u, GNN aggregates messages from v_u's neighborhood $\mathcal{N}_{v_u}$ and aggregated mes-sage M_u via an aggregation method AGG: $M_u^l = \text{AGG}\left(\left\{\mathbf{m}_{uv}^l, e_{uv}\right\} \mid v_v \in \mathcal{N}_{v_u}\right)$, where $\mathcal{N}_{v_u}$ is neighborhood of node v_u whose definition depends on a particular GNN variant. (3) Finally, GNN takes the aggregated message M_u^l along with v_i's representation $\mathbf{h}_u^{l-l}$ from the previous layer, and it non-linearly transforms them to obtain v_u's representation $\mathbf{h}^l$ at layer l: $\mathbf{h}_u^l = \text{UPDATE}\left(M_u^l, \mathbf{h}_u^{l-1}\right)$. The final embedding for node v_u after L layers of computation is $\mathbf{z}_i = \mathbf{h}_i^L$. In summary, our explainer provides explanation for any GNN that follows the computations in terms of MSG, AGG, and UPDATE.

4 Graph Feature Explanation

Next we describe our explanation approach in forms of a two-stage pipeline.

First, given the input graph $G = \{V, \mathcal{E}\}$, node feature X and true nodes labels Y, we will get a trained GNN model Φ: $G \mapsto (u \mapsto y_u)$, where prediction $y_u \in I_C$.

Second, the explanation part will provide a subgraph and a subset of features retrieved from the k-hop neighborhoods of each node u based on the classification model Φ and the node u's true label where $k \in \mathbb{N}$ and $u \in V$. The subgraph and the subset of node features are theoretically the minimal set of information the trained model Φ needed to perform the node classification task.

Here, we define the subgraph of G as $G_S = (V_S, \mathcal{E}_S)$, where $G_S \in G$, $V_S \in V$, and $\mathcal{E} \in \mathcal{E}$.

4.1 Informative Components Detection

Following the Maximal Mutual Information (MMI) Mask method in [6]. For node u, mutual information (MI) quantifies the change in the probability of prediction when u's computational graph is limited to subgraph G_S and its node features are limited to X_S. The explainable subgraph is chosen to maximize the mutual information:

$$\max_{G_S} I\left(Y, (G_S, X_S)\right) = H(Y \mid G, X) - H\left(Y \mid G_S, X_S\right) \tag{1}$$

As the trained classifier Φ is fixed, the $H(Y)$ term of Eq. 1 is constant. Therefore, it is equivalent to minimize the conditional entropy $H(Y \mid G_S, X_S)$:

$$H(Y \mid G_S, X_S) = -\mathbb{E}_{Y|G_S, X_S} [\log P_\Phi(Y \mid G_S, X_S)] \tag{2}$$

In this way, the graph component explanation w.r.t node u's prediction $\hat{y}_u$ is the subgraph G_S that minimize. Some edges in node u's computational graph may not be useful in node prediction, while others are most informative edges to aggregate information from neighborhoods. We thus aim to find the most important subgraph component in computational graph to explain the useful edges in node prediction task. In order to optimize Eq. 2, we also extends the edge mask $\mathcal{M}_{vw}$ by considering edge weight e_{vw} and adding extra regularization to make sure a connected component to be formed in the subgraph. We add constraints to the value of mask before every iteration of optimization:

$$\begin{cases} \sum_w \mathcal{M}_{vw} e_{vw} = 1 \\ \mathcal{M}_{vw} \geq 0, \qquad \text{for } \langle v, w \rangle \in \mathcal{E}_c(u) \end{cases} \tag{3}$$

In this way, we optimize a relaxed adjacency matrix in our explanation rather than optimizing a relaxed adjacency matrix in graph. Thus we optimize a mask $\mathcal{M} \in [0,1]^Q$ on weighted edges, where there are Q edges in $G_c(u)$. Then the masked edge $E_c^\mathcal{M} = E_c \odot \mathcal{M}$, where $\odot$ is element-wise multiplication, is subject to the constraint that $E_c^\mathcal{M}[v,w] \leq E_c[v,w], \forall (v,w) \in \mathcal{E}_c(u)$. So the objective function can be written as:

$$\min_M -\sum_{c=1}^C \mathbb{I}[y = c] \log P_\Phi(Y \mid G_c = (V_c, E_c \odot \mathcal{M}), X_c) \tag{4}$$

4.2 Node Feature Importance

Node features also contribute to information aggregation in propagation of message layer. Thus we also take into account the node feature in our explanation. In addition to learn a mask on weighted graph, we follow [6] to learn a mask on node attribute to filter node features in subgraph G_S. The filtered node feature $X_S^T = X_S \odot M_T$, where M_T is the feature mask to be learned, is optimized by

$$\min_{M_T} -\sum_{c=1}^C \mathbb{I}[y = c] \log P_\Phi(Y \mid G_S, X_S \odot M_T) \tag{5}$$

The reparameterization on node attribute X is also applied in the optimization of M_T as in [10].

5   Evaluation Metrics

We can use metrics to evaluate accuracy of classification with our drug abuse data. But for these real world data, we can hardly know the ground truth of the explanation. However, we propose the evaluation metrics to quantify the explanation results.

In this paper, we define *consistency* and *contrastivity* to quantify the explanation result. We first introduce graph edit distance (GED) [1], which is a measure of similarity between two graphs. It is defined as the minimum number of graph edit operations to transform one graph to another. Thus, we follow the sense of classification that informative components detected for the node in the same class should be consist, we design *consistency* as the GED of informative components between nodes in the same class as whether the informative components detected for the node in the same class are consist. Besides, we design *contrastivity* as the GED of informative components of nodes across classes, as whether the informative components detected for the node in the different class are contrastive, to measure the contrastivity of informative components of nodes in different class.

Specifically, *consistency* is defined as below:

$$\frac{\sum_{a=1}^{N_i} \sum_{b=1}^{N_i} GED\left(G_a, G_b\right)}{N_i \times (N_i - 1)} \qquad a \neq b \tag{6}$$

where G_a and G_b are the informative components from node a and b in same class i, N_i is the number of nodes in class i. And *contrastivity* is defined as:

$$\frac{\sum_{a=1}^{N_i} \sum_{b=1}^{N_j} GED\left(G_a, G_b\right)}{N_i \times N_j} \tag{7}$$

where G_a is the informative component from node in class i, G_b the informative component from node in class j, N_i and N_j the number of nodes in class i and j.

6   Experiments

6.1   Alternative Baseline Approaches

We here consider the following alternative approaches that can provide explanation of prediction made by GNNs: (1) GRAD [3] is a gradient-based method. We compute gradient of the GNN's loss function w.r.t the model input. The gradient-based score can be used to indicate the relative importance of the input feature since it represents the change in input space which corresponds to the maximizing positive rate of change in the model output. (2) GNNEXPLAINER is also a MMI-based method to explain the prediction made by GNNs. However, it does not take the weight of edges into consideration. The experiments in this section of both real world and synthetic dataset reveals the outstanding performance of our proposed method by keeping relational information in non-linear transformation and taking edge weight information of graph into account.

6.2 Drug Abuse Social Network

In order to evaluate the performance of our explanation method, we perform node classification and different explanation methods. The node classification task on this dataset is to classify each node in graph as suspect or clear in drug abuse. Thus, we choose every pair of adjacent months to be train graph and test graph. In this way, we have 17 pair of train graph and test graph. Both node and edge in graph are associated with categorical features, and the edge features are normalized to [0, 1] as 1-D feature. We perform inductive learning on the train graph and generalize the trained GNN on test graph. The learning rate of GNN is 0.001 and we trained the model for 100 epochs until convergence. Based on the trained model, we performed the explanation methods to compare their performance in Table 1. We measured consistency and contrastivity of informative components within and between classes by selecting top 6 edges. The performance of explanation methods are averaged on all train and test graph pairs and repeated by 100 times. The lower consistency value shows that nodes in the same class have similar topology structure. The higher contrastivity value means GNN relies more on topological difference to classify nodes into different class. From the results in Table 1, GRAD performs worst since it fails to keep the structural information in propagation. While GNNExplainer performs worse than our MMI method, because it ignores the edge features information.

Table 1. Model performance

	Consistency	Contrastivity
MMI	0.42	1.24
GNNExplainer	0.47	1.18
GRAD	0.63	0.87

6.3 Synthetic Data

In order to evaluate the applicability of our model, following [10], We construct a synthetic dataset called $BA - SHAPES$ with a base Barabási–Albert (BA) [4] graph with 100 nodes and a set of 60 five-node house structured network motifs randomly attached to the nodes of the base graph. Besides, we add the extra $0.1N$ random edges to perturb the graph, where N is the number of nodes poorly connected. We assigned different weights for edges in graph as denoted in Fig. 2. We haved tried different combinations of edges weights, and as the weight of different roles of edges are sparse enough, the explanation result remains similar. The four classes of nodes are assigned according to their structure role: the top, middle, and bottom of the house and the nodes that don't belong to a house. Several natural and human-made systems, including the Internet, the world wide web, citation networks, and some social networks are thought to be approximately scale-free and certainly contain few nodes (called hubs) with unusually high degree as compared to the other nodes of the network. All the node feature X_i was designed the same constant to constrain the node label to be determined by motif only. We randomly split 60% of the nodes for training and the rest for testing. We performed informative component detection and compared the result with ground truth in Table 2. The GRAD performed worst in this case because it fails to keep the relational structure in graph data. The GNNExplainer performed worse in this case because it ignore the attribute of edges in graph so that the nodes of base graph may usually be included into the informative subgraphs.

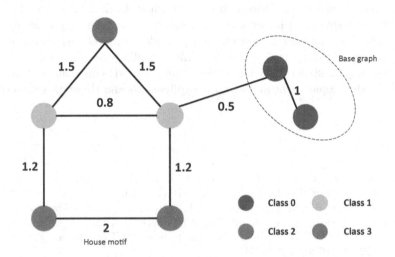

Fig. 2. The edge weights of corresponding edges in graph.

Table 2. Average AUC score of nodes in house shape

Method	MMI mask	GRAD	GNNExplainer
AUC	0.945	0.893	0.831

We also constructed a graph called $BA-FEAT$ with BA graph and designed the 2-D node attributes for each node in graph to constrain the node labels to be determined by node features only. We assigned the normal distribution $S_u \sim N(0,0,1)$ as the first entry of node u's feature, and $s_u + (y_u + 1)*0.2$ as the second entry, where y_u is the ground truth label of node u. We thus constructed a graph with 100 nodes with the node features above. And we followed the same training model and methods in BA-SHAPES. To measure the performance of the explanation for node importance, we calculated the accuracy of classifying the second entry of node features as the important features in Table 3.

Table 3. Node feature importance performance (averaged on 10 times)

Method	MMI mask	GNNExplainer	GRAD
ACC	1	1	1
MSE	0.23	0.27	0.33

7 Conclusion

In this paper, we introduce the framework to preprocess and train on graph data, formulate the MMI based explanation method to explain its predictions. In order to evaluate its explanation performance, we applied it on both real world social data and synthetic data with evaluation metrics and baselines. The experiment results show that our explanation method provides comprehensive and impressive explanation of data used by GNN, while revealing the important information in topological structure. This explanation methods can benefit debugging in GNNs model, deployment in real word applications and those decision-critical area.

References

1. Abu-Aisheh, Z., Raveaux, R., Ramel, J.Y., Martineau, P.: An exact graph edit distance algorithm for solving pattern recognition problems. In: ICPRAM 2015–4th International Conference on Pattern Recognition Applications and Methods, Proceedings, vol. 1, pp. 271–278. SciTePress (2015). https://doi.org/10.5220/0005209202710278. http://www.scitepress.org/DigitalLibrary/Link.aspx?doi=10.5220/0005209202710278
2. Baehrens, D., Schroeter, T., Harmeling, S., Kawanabe, M., Hansen, K., Mueller, K.R.: How to explain individual classification decisions. J. Mach. Learn. Res. **11**, 1803–1831 (2009). http://arxiv.org/abs/0912.1128
3. Baldassarre, F., Azizpour, H.: Explainability techniques for graph convolutional networks. Technical report (2019)
4. Barabási, A.L., Albert, R.: Emergence of scaling in random networks. Science **286**(5439), 509–512 (1999). https://doi.org/10.1126/science.286.5439.509. https://science.sciencemag.org/content/286/5439/509science.sciencemag.org/content/286/5439/509.abstract
5. Huang, Q., Yamada, M., Tian, Y., Singh, D., Yin, D., Chang, Y.: GraphLIME: local interpretable model explanations for graph neural networks (2020). http://arxiv.org/abs/2001.06216
6. Li, X., Saude, J., Reddy, P., Veloso, M., Morgan, J.P., Research, A.I.: Classifying and Understanding Financial Data Using Graph Neural Network. Technical report. www.aaai.org
7. Liu, X., Wang, X., Matwin, S.: Interpretable deep convolutional neural networks via meta-learning. In: Proceedings of the International Joint Conference on Neural Networks 2018-July, February 2018. http://arxiv.org/abs/1802.00560
8. Neil, D., Briody, J., Lacoste, A., Sim, A., Creed, P., Saffari, A.: Interpretable graph convolutional neural networks for inference on noisy knowledge graphs, December 2018. http://arxiv.org/abs/1812.00279

9. Simonyan, K., Vedaldi, A., Zisserman, A.: Deep inside convolutional networks: visualising image classification models and saliency maps. In: 2nd International Conference on Learning Representations, ICLR 2014 - Workshop Track Proceedings, December 2013. http://arxiv.org/abs/1312.6034

10. Ying, R., Bourgeois, D., You, J., Zitnik, M., Leskovec, J.: GNNExplainer: generating explanations for graph neural networks, March 2019. http://arxiv.org/abs/1903.03894

11. Yuan, H., Tang, J., Hu, X., Ji, S.: XGNN: towards model-level explanations of graph neural networks. virtual event, pp. 430–438, June 2020. https://doi.org/10.1145/3394486.3403085. http://arxiv.org/abs/2006.02587

12. Zhang, J., Lin, Z., Brandt, J., Shen, X., Sclaroff, S.: Top-down neural attention by excitation backprop. Int. J. Comput. Vis. **126**(10), 1084–1102 (2016). http://arxiv.org/abs/1608.00507

13. Zhang, Y., Defazio, D., Ramesh, A.: RelEx: a model-agnostic relational model explainer, May 2020. http://arxiv.org/abs/2006.00305

SemApp: A Semantic Approach to Enhance Information Retrieval

Sameh Neji[1]([✉]), Tarek Chenaina[2], Abdullah M. Shoeb[3], and Leila Ben Ayed[2]

[1] Faculty of Economics and Management, Sfax University, Sfax, Tunisia
neji.sameh@yahoo.fr
[2] National School of Computer Science, University of Manouba, Manouba, Tunisia
Tarek.chenaina@ensi.edu.tn, leila.benayed@ensi-uma.tn
[3] Faculty of Computers and Information, Fayoum University, Fayoum, Egypt
ams02@fayoum.edu.eg

Abstract. The present work proposed a semantic retrieval approach to treat the issues of semantic ambiguity of indexed terms, the uncertainty, and imprecision that is inherent in the information retrieval process. The proposed approach constitutes of three different phases. The query meaning was discovered in the first phase by formulating a set of candidate queries from possible contexts. A score for each alternative was calculated based on its semantic tree and inherent dispersion between its concepts. This score assesses the overall meaning of the alternative query. This phase was finished by selecting the candidate query that attains the highest score to be the best representative to the original query. A semantic index was built in the second phase exploiting the classic and semantic characteristics of the document concepts to finally assign a weight for each concept to estimate its relative importance. The third phase proposed a ranking model that utilizes the semantic similarities and relations between concepts to calculate the query-document relevance. This ranking model is based on a query likelihood language model and a conceptual weighting model. The validity of the proposed approach was evaluated through performance comparisons with the related benchmarks measured in terms of the standard IR performance metrics. The proposed approach outperformed the compared baselines and improved the measured metrics. A statistical significance test was conducted to guarantee that the obtained improvements are true enhancements and are not a cause of random variation of the compared systems. The statistical test supported the hypothesis that the obtained improvements were significant.

Keywords: Information retrieval · Semantic information retrieval · Query reformulation · Semantic indexing · Conceptual weighting · Semantic relations · Ranking model · Query language model

1 Introduction

The aim of IR is to respond effectively to users' queries by retrieving information that better meets their expectations. Information needs are dependent on the user defined

O. Gervasi et al. (Eds.): ICCSA 2021, LNCS 12951, pp. 62–78, 2021.
https://doi.org/10.1007/978-3-030-86970-0_6

context. Primarily, the level of domain expertise of the user directly influences the query richness. However, the information retrieval system (IRS) must be intelligent enough to identify information needs and respond effectively to meet the expected needs, regardless the level of user expertise level. This process is difficult and remains a major and open challenge in the domain of IR.

The present work fits into the general problem of searching for textual information. The huge increase in the amount of information accessible on the internet has resulted in an excessive demand for tools and techniques capable of semantically processing web data. The current practice of information retrieval is based mainly on keywords which are modeled with a bag of words. However, such a model lacks the actual semantic information in the text. To deal with this problem, ontologies are proposed for the representation of knowledge, which is the backbone of semantic web applications. Information retrieval processes can benefit from this metadata, which gives semantics to the text.

There is a semantic information search gap between the data, the application, and the used algorithms. Semantic Web technology has an intelligent role in bridging the semantic gap and retrieving relevant information through data mining algorithms for systematic knowledge incorporation. The model [1] was acquired by analyzing user behavior to record user information based on his/her interests.

Ontologies are the main tools for the representation of knowledge. They facilitate the classification as well as the cartography of concepts and their relations in a hierarchical structure. The literature shows that the use of conceptual knowledge (ontologies) in the information retrieval process has helped to overcome the main limitations of information retrieval. An ontology-based IRS is presented in [2]. In this model, the extraction of the ontology vocabulary is performed on the query terms. The terms of the documents are extracted by projection onto the ontology and they are presented by a vector space. The relevance between the concepts of the document and query is calculated using a correlation matrix between these concepts.

The main limitations concern the relations between the different concepts are the difficulty of inference of semantic information from concepts, the need to unify the semantic representations, and the mapping of knowledge from heterogeneous ontologies.

In [3], an OWL-based (Web Ontology Language) search engine was proposed. The proposed indexing algorithm offers better ontology maintenance and the retrieval algorithm facilitates the processing of user queries. The limitation here is that the authors manually tested the indexing algorithm considering a sample of ten OWL ontologies. In [4], the authors presented an IRS based on domain ontology. The meanings of user query concepts are inferred from the domain ontology and they are used for query expansion. The expansion of queries has positive effects on the search for relevant information. The goal of query expansion can be either to expand the set of returned documents by using terms similar to the query words, or to increase search accuracy by adding completely new terms to the query. Semantic similarity between inferred concepts and domain ontology concepts is obtained using structure-based measures [5].

After obtained the semantic knowledge and represented it via ontologies, the next step is to inquiring the semantic data to extract the relevant information. There are query languages designed for semantic enquiries, but these formal query languages are not intended to be used by end users because it requires knowledge of the ontology domain as well as the syntax of the language. Combining the ease of using key-word-based interfaces with the power of semantic technologies is considered among the most difficult areas of semantic search.

In our view, all efforts to increase retrieval performance while maintaining usability will ultimately improve semantic search with keyword-based interfaces. This is a difficult task because it requires answering complex queries with only a few natural language keywords. The present work introduces a novel approach that utilizes three successive phases to improve IR: query meaning extraction, conceptual weighting documents representation, and a hybrid ranking model based on a query likelihood language model.

2 Related Works

In natural language processing, the similarity between words is often distinct from the relations between them. Similar words are almost synonyms while related words are connected to each other through semantic relations. When interpreting the query, the system can integrate other information to better meet the information user's needs. The content of the user's query is not always optimal for effective search. To improve the search performance which mainly depends on this query, various methods seek to improve its content by coordinating between the query and an effective index. While other methods, are interested in studying the relevance between the query and the document. Other systems study the behaviors and interests of the user to extract relevant information. Several methods use expansion techniques to extend the query by enriching the content of this query with other information [6, 7]. The extension of the query can be done by referring to ontologies and thesauri to discover syntactic, lexical or semantic relationships between words [8]. Other methods use additional information to identify meanings of query concepts in a disambiguation task [9]. Although statistical methods are used, the principle techniques are based on term co-occurrence to generate correlations between pairs of terms [10]. In [11], the author evaluates the co-occurrence by considering the frequency of terms found only in the first retrieved document. Other relevance feedback methods use user feedback to assess the relevance of documents that will be used to extend the query. This technology is based on the semantics of words in order to recognize their meaning in a specific context.

Discovering the correct meaning of the query is insufficient on its own to cope with IR challenges. Optimizing the representations of documents and queries is very important to extract the semantic information they contain. Term weighting technique is used in indexing phase rather than in the search phase. Several indexing approaches are based on external semantic resources. The basic principle of these approaches is to extract the representative elements from the document and the query. To assign degrees of importance to indexed terms, weighting algorithms are used to reduce the feature space to more specific terms in the search domain using knowledge bases such as ontologies [12]. In particular, the authors in [13] have proposed a term weighting scheme which combines the semantic similarity between each term for a category of the text and its frequency in the document to obtain the characteristic vector of each document. This approach outperforms TF-IDF in case of training data shortage and in case of the content of documents is focused on well-defined categories.

In the weighting model described in [14], the importance of a concept is assessed by a measure of its semantic similarity with the other concepts of the document which is combined with the frequency of their occurrences in this document.

Another semantic indexing approach [15] was extracted concepts by a construction of a semantic network by projection on WordNet. A semantic weight, which is adapted from the TF-IDF model, is calculated for each concept. In [16], the authors propose an indexing approach which is based on the frequency of the concept in the document with estimation of the semantic distances between each concept and the other most frequent concepts in a document. The results show that the use of TF-IDF weighting in semantic indexing increases precision by more than 50%. However, using the proposed semantic weighting, the precision is still low.

Recent approaches have been proposed for semantic indexing. This work [17], tackled the problem of indexing documents in the Arabic language. The author uses the Universal WordNet ontology to extract the index's concepts. This method improves the capacity of the semantic vector space model VSM.

To customize the search according to the user's interests, the work [18], presents a model to disambiguate queries. This model highlights the behavior of users via the history of conducted searches which is limited and insufficient to detect new information needs. To solve this problem, the user profile detection method uses semantic relations of ontologies to deduce its interests according to a hierarchy of concepts [19, 20].

Indexing is therefore an effective solution which consists in extracting descriptive elements from a document. In this case, the document is converted to a smaller structure than the original full text. This new representation contains the same amount of information. To successfully perform IR, index descriptors must be well exploited in the search phase. Ranking retrieved documents is an important task that measures query-document relevance. Some search models have been proposed to obtain semantic information between formal concepts [21]. However, assigning equal importance to all terms can significantly reduce the quality of the IRS.

Several approaches using fuzzy concept networks based on fuzzy formal concept analysis (FFCA) [22–24] have been proposed to address this challenge. They adopt the partial order relation of the concepts to calculate the relation between them and return a list of documents associated with the given query. However, these methods neglect explicit semantic information between concepts.

In [25], the authors conceived the idea of an intelligent IRS based on semantic web. In this model, the semantic relations between web pages are used to estimate the similarity between queries and documents. In [26], the authors proposed a model of knowledge search process using semantic metadata and artificial intelligence tech-niques. The works [27, 28] were introduced an automatic derivation of concept hyponymies and meronymies relations from web search results. Its method was based on generating derivative features from web search data and applying the machine learning techniques.

Language models have been applied in various applications such as ad hoc research, expert research and social research. These models estimate the relevance of a document to a query using the probability that this query can be generated by this document [29]. This model is based on the maximum likelihood estimate. Using of a bag of words, which assumes the independence of words, is common in the approaches based on language models. However, natural language terms are often dependent.

Returning relevant results is the primary goal of IR. To achieve this goal, the proposed approach considers two main factors before displaying results. The first factor is the extraction of the information from the semantic processing which discovers the meaning of the query. The second factor, that directly influences the precision of retrieval system, is the estimation of query-document relevance exploiting semantic indexing of concepts the related to terms of documents and queries.

3   The Proposed Approach

The present work is based on general domain ontology for extraction and retrieval of semantic information. Previous works [30–33] were interested in developing individual steps to improve some parts of the IRS separately. The first phase of the proposed framework is to disambiguate the user's query to resolve the ambiguity caused by the multiple senses for each term in the query. The second phase is to disambiguate the queries and documents terms. They are replaced and indexed with a new format. A semantic weight is calculated for each concept that reflects the degree of importance of each unit in the index. This step improves the representation of documents and queries. The third phase measures the query-document relevance by calculating a semantic score for each document. This score exploits the conceptual relations between the documents, query terms, and the meanings of these terms in the context of the document. The search results will be ranked according to this relevance score.

Most IRSs use a keyword-based search mechanism. However, the proposed framework defines information in terms of concepts. The concept can be considered as a sequence of terms which identify a clear meaning of the context concerned. The present work introduces a complete semantic processing of information that encompasses the three major phases of information retrieval in a single approach. The originality of this work compared to previous works is that it presents a general integrated semantic approach for processing information and retrieving relevant documents with respect to a query introduced in natural language in the form of generally ambiguous keywords. The limitations that mentioned in the previous section show the need to a unified system that conducts all the necessary steps for an effective IR. The present work contribution is proposing an IR system which processes information in three consecutive phases, Fig. 1. This system uses ontologies and semantic relations to represent documents and define the meanings of terms. It is intelligent enough to retrieve documents that do not contain the terms of the query.

3.1   Discovering Query Meaning

The query disambiguation process leads the search of documents that are truly related to the user's information needs. The ambiguity can be lexical (polysemy, disambiguation), or structural (type of information expected). The ambiguity of the query can lead to heterogeneity of documents belonging to different contexts.

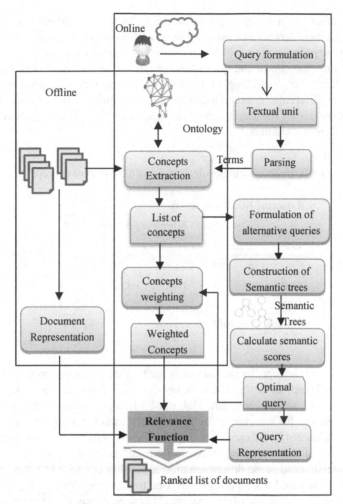

Fig. 1. The proposed approach outlines

Usually, the query introduced by the user is insufficient to return the most relevant documents. The query must be very clear to express the user's needs. The system is based on this query to retrieve the relevant documents. The search results will therefore be meaningless if the query is poorly expressed. Accordingly, the first step in the proposed technique for improving the performance of IRS is to discover the meaning expected by the user.

3.1.1 Extraction of Query Concepts

This task consists of identifying the concepts of the query. This is very important step to improve the performance of the system. Identifying concepts requires a word sense disambiguation step because of word ambiguity in natural language. A concept in the text can be represented by more than a term. The purpose of term disambiguation is therefore to select the most appropriate meaning. In specific domain, it is easier to define the meaning of terms since using more specialized terms in a specific domain can help to define the precise context than using more ambiguous terms in a general domain.

Consider the example of the query 'let the cat out of the bag'. The single word concepts are: 'let', 'cat', 'out', 'bag'. Table 1 shows the vast variety of possible meanings of this query in different contexts. Discovering the meaning of the query or otherwise the meaning expected by the user is a great challenge especially in cases where the query has several different meanings.

Table 1. Number of possible senses with the POS of concept

Concept	Number of senses				
	Noun	Verb	Adjective	Adverb	Total
Let	2	9	–	–	11
Cat	8	2	–	–	10
Out	1	3	10	3	17
Bag	9	5	–	–	14
Number of possible senses					26180

The multi-word concept 'let the cat out of the bag' has only one meaning as a verb. Its definition in WordNet is: "spill the beans, let the cat out of the bag, talk, tattle, blab, peach, babble, sing, babble out, blab out - (divulge confidential information or secrets; "Be careful – his secretary talks")". If the different meanings of each term were considered, it yields 26180 different meanings for the query in the case of separate words. So, the discovery of query meaning presents a difficult problem.

A very precise word sense disambiguation algorithm is needed to increase performance of IRS. Extended Lesk algorithm was used to disambiguate words by calculating a semantic score for each sense of the concept. This score is based on the extended measure of gloss overlap of different relations between synsets in WordNet. This algorithm returns better results for nouns and adjectives [34].

The original query was transformed to alternatives by a combination of all the possible concepts it contains. The same grammatical structure of the original query was kept so as not to deviate from the global meaning of the query.

3.1.2 Extraction of the Most Relevant Candidate Query

The retrieved documents may not meet the needs of the users because of wrong expression of the query. In this case, the level of user expertise and its direct effect on the accuracy of the results present a challenge to be tackled. Ambiguous words lead to a dispersion of the query context. To solve this problem, our approach extracts the closest context that maximizes the meaning of the query to help the user find relevant information. A semantic score is calculated for each candidate query. This score (*Can_Score*) reflects the semantic rapprochement with the user's expectations. The most relevant (optimal) query is selected to pass to the next phase.

$$Can\_Score = \cfrac{1}{D^{\alpha} \sum_{i=1}^{n} \frac{1}{(depth(c_i))^{\beta}} \sum_{\substack{c = c_i \\ c \in path(c_i, LCS)}} log(1 + (c, hyper(c)))} \quad (1)$$

Where n is the number of query concepts, c_i is the concept of rank i, LCS is least common subsumer of c_i, $hyperc$ is the hypernym of the concept c, δ $(c, hyper(c))$ is the similarity between the current concept c and its subsume, D is the coefficient of dispersion that calculate the number of branches containing at least one concept of the query, $depth(c_i)$ is the depth of the concept c_i with respect to LCS and α, β are two constants. Many experiments are conducted to set the values of α and β to finally assign to 2 and 1 respectively.

3.2 Indexing Phase

The use of semantic notion allows solving the problems of ambiguity of the indexed terms. Thus, exploitation of different semantic relations makes it possible to limit the disparity between the representation of documents and user's query. A semantic indexing technique based on semantic relations (is-a) and the synsets associated with the terms of the documents and the query was utilized. This technique makes it possible to represent the terms of documents and queries by their associated senses in the context in which they appear. An index based on the identified synsets was built. Representation of documents and queries was done by these synsets rather than the terms themselves.

3.2.1 Semantic Representation of the Document

The main problem with traditional IRSs is that they generally return too many results or irrelevant results. As keywords have multiple lexical characteristics or have multiple senses, in this case, relevant information may be lost because of not specifying the exact search's context. The resolution of the ambiguity problem by integration of the conceptual representation becomes necessary to represent the documents by a set of descriptors in the form of concepts belonging to a predefined semantic resource (ontology). Two issues that can be reported; the used external resource could be limited to a certain domain or it may miss certain terms. This representation brings new challenges because of the presence of synonymies and polysemy's in natural language, concerning the association of a term with the best concept.

A combined index based on terms and their senses was designed as the representation of data is a very important task in the IR process. The output of this phase is a new representation of the document and the query. Each term is replaced by the following format:

Term /POS/ Sens/ Spec/Tauto/ Weight

Where *Term* is the common term, *POS* is the part of speech of this term in the document (Part Of Speech), *Sens* is the concept sense associated with this term in WordNet,

Spec is estimated by the "depth" of the current concept in the ontology, *Tauto* is the cardinality of the concept's synsets in the document, *Weight* is the concept weighting. The system also indexes the terms that are not defined in the ontology.

3.2.2 Concepts Weighting

Term weighting is also an essential task in defining the importance of query and document terms. This weight is generally measured by a statistical method which is based on the presence of the term and its frequency in a text. Traditionally, the weighting score does not consider the senses of the terms. There is a problem with this method which neglects the notion of term's senses. This measure only considers the factor of existence of the concept in a document and is reflected by its frequency. This metric does not consider the semantic importance of the concept in the document. To remedy these problems, we utilize concept weighting model that calculates the importance of concepts according to different criteria that allow better presentation of relevant information.

Estimating the importance of each index item considers many factors that have improved the performance of the system. The index is built by weighted synsets. To give importance to each concept, we use the weighting model, Eq. (2).

$$w(s_t, d) = \frac{F + |Syn(s_t)|_d + \#Rel(hyper)(s_t, root) * \frac{1}{|C|} * \sum_{\forall t, POS(t)=NOUN} freq(t, C)}{\sum_{\forall s_i \in synset(s_t)} \sum freq(s_i, C) + \overline{POS(N, C)}}$$

$$(2)$$

$$F = \alpha freq(s_t, d) + (1 - \alpha) * (freq(t, d) - freq(s_t, d))$$

Where α is a parameter whose value is set empirically, $freq(s_t, d)$ is the frequency of the sense identified by a term t within a document d, $freq(t, d)$ is the frequency of the term t within a document d, and $freq(s_t, C)$ is the frequency of the sense identified by a term t throughout the collection C.

3.3 Ranking Phase

This process involves calculating a relevance measure for each document against a query. This measure reflects the semantic similarity between the query and the document. Unlike classical systems based on keywords, the semantic relevance formula uses the relations between the concepts. These relations are expressed in the used semantic resource.

An effective representation of documents and queries has been used to ensure effective retrieval of information expected by the user. The proposed approach benefits from the advantages of classical information retrieval. It combines between weighting formula and the language model which estimate the semantic similarities between the concepts. The document ranking score is defined by:

$$SemRank(D) = \left(\prod_{i=1}^{|Q|} P(q_i|D) \right)^{\gamma} * (Score_W(D))^{\delta}$$

Where γ and δ are two parameters to distinguish importance of each ingredients of the SemRank. The default values for these two constants are 1 and 3, respectively.

$$P(q_i|D) = \frac{\sum_{j=1}^{m} Sim(q_i, d_j)}{|D| * \max_{\substack{\forall d_k \in D' \\ D' \in C}} Sim(q_i, d_k)}$$

$Score_W(D)$

$$= \prod_{i=1}^{n} \left(\frac{\left(w(q_i, D) - \min_{\forall D' \in C} w(q_i, D') \right) * \left(\underset{\forall D' \in C}{Max} P(q_i|D') - \min_{\forall D' \in C} P(q_i|D') \right)}{\left(\max_{\forall D' \in C} w(q_i, D') - \min_{\forall D' \in C} w(q_i, D') \right)} + \min_{\forall D' \in C} P(q_i|D') \right)$$

$$w(q_i, D) = \frac{\beta * F + Tau(q_i, D) + Spec(q_i) * \overline{POS(N, C)}}{freq(q_i, C) + \overline{POS(N, C)}}$$

$$F = \alpha freq(q_i, D) + (1 - \alpha) * (freq(t_i, D) - freq(q_i, D))$$

This ranking model combines between classical and semantic search. The semantic similarity measure "Resnik" is used to find $Sim(q_i, d_j)$.

4 Evaluation

The aim of the experimental work is to evaluate the quality of the proposed approach as a function of the quality of its retrieved ranked list by comparing this result with other benchmarks. The plan is started by selecting a widely used dataset collection. Then, we choose the relative benchmarks to compare with. The effectiveness of each system is assessed through the standard IR measures.

TREC collection, one of the standard test collections, and its relevance judgements were chosen to conduct the evaluations. The proposed approach exploits the semantic and classic characteristics of the documents; therefore, we built a semantic and a classic index to the dataset. SemApp is compared to TFIDF, BM25, and QL (query likelihood language model). Each benchmark is implemented on both classic and semantic index and it will be indicated by a prefix C or S respectively before its name to indicate the underlying index. Hence, there are six models for comparison: C_TFIDF, S_TFIDF, C_BM25, S_BM25, C_QL, and S_QL. A set of forty TREC queries, and their qrels, was selected to conduct the experiments. Each model was run on every topic and the effectiveness measures are assessed and compared to each other. SemApp ranking part sets the following parameters: $\alpha = 0.9$, $\beta = 2$, $\gamma = 1$, and $\delta = 3$.

4.1 Evaluation of MAP, GMAP, Rprec, and MRR Metrics

In this experiment, four IR measures were assessed; mean average precision (MAP), geometric mean average precision (GMAP), precision after R (the number of relevant documents for a query) documents retrieved (Rprec), and mean reciprocal rank (MRR).

Figure 2 demonstrates a comparison among the seven models in the four measures. Also, Table 2 shows the percentage of improvements that achieved by SemApp.

The MAP, among of other measures, plays an important role in evaluating the IR systems because MAP, by its definition, assesses the gross precision of the examined system. SemApp attains the highest MAP among the other competitors. The improvement percentages are ranged from 15.5%, over C_QL, to 130.9%, over S_TFIDF. The descending order of systems performance is as follows: SemApp, C_QL, C_BM25, S_QL, S_BM25, C_TFIDF, and then S_TFIDF respectively. The benchmark model that built on classic index shows significant improvement over the corresponding one that built on semantic index. This result may happen because of change in frequency of words that appears in the classical index into several frequencies corresponding to the different meanings of these words in the semantic index. Also, word sense disambiguation algorithm may drift the meaning away from the correct sense.

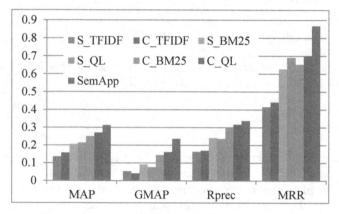

Fig. 2. Comparison of MAP, GMAP, Rprec, and MRR

Table 2. SemApp improvements percentage of MAP, GMAP, Rprec, and MRR

Improvement %	C_QL	S_QL	C_TFIDF	S_TFIDF	C_BM25	S_BM25
MAP	15.5%	46.6%	99.1%	130.9%	24.8%	52.2%
GMAP	46.0%	210.2%	483.3%	344.3%	63.5%	160.0%
R_{prec}	6.6%	42.9%	99.1%	106.5%	12.9%	40.2%
MRR	23.9%	25.4%	96.5%	109.0%	32.8%	38.3%

MAP is defined as the arithmetic mean of average precisions, so it assigns equal weight to each topic score change regardless of the relative size of the change. Alternatively, GMAP concentrates on the low performed topics; the topics that have low average precision. GMAP measures the impact of improvement in these special topics on the overall system performance. The experiment shows that the greatest GMAP appears in SemApp by a considerable gap with the other benchmarks. While C_QL and C_BM25 come in second and third places respectively, the worst behavior appears in C_TFIDF. This result shows that SemApp is capable of improving the low performed topics by 46% over C_QL and 63.5% over C_BM25.

Let R be the number of relevant documents, as judged in the qrels, of a query. The third assessment measure is Rprec which measures the precision after retrieve R documents whatever the relevancy of the retrieved documents. So, if an IR system returns x relevant documents within the first R retrieved documents; Rprec remains the same and independent of the location of the x documents within the top R-sized returned list. While SemApp comes first in Rprec results, the differences between SemApp and the nearest two competitors, C_QL and C_BM25 respectively, are not considerable. The improvement percentages over all benchmarks are ranged from 6.6% to 106.5%.

Reciprocal rank is defined as the multiplicative inverse of rank of the first retrieved relevant document. MRR is the arithmetic mean of reciprocal rank over all topics. By this definition, MRR is arranged the studied systems according to their first correct answer. MRR of SemApp exceeds those of C_QL, S_QL, C_BM25, S_BM25, C_TFIDF, and S_TFIDF by 23.9%, 25.4%, 32.8%, 38.3%, 96.5%, and 109% respectively. So, SemApp is the fastest system to find its first correct answer with considerable improvements with the other benchmarks.

4.2 Evaluation of P@x Metric

This experiment calculates the accuracy of each studied system at standard predefined steps of retrieved documents. Systems accuracy is measured after retrieve 5, 10, 15, 20, 30, and 100 documents. Figure 3 shows a comparison between the seven systems at these steps. Table 3 demonstrates the improvements that SemApp achieve over its competitors. The results show that SemApp rejects more irrelevant documents in each step compared to the other benchmarks. SemApp succeeds to enhance P@5 by 36.7% to 132.1%, P@10 by 24.4% to 137.5%, P@15 by 20.6% to 134.2%, and P@20 by 19.1% to 139% over the other competitors. In fact, the first two steps are very important because the user is usually interested to browse the top ten retrieved documents. So, the systems try to enhance their accuracy especially at this step. Commonly, the precision is inversely

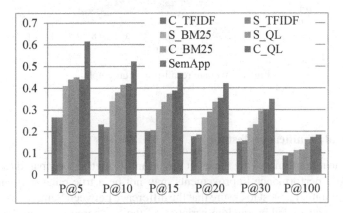

Fig. 3. Precision @ 5, 10, 15, 20, 30, and 100

proportional to the step because the system accepts more irrelevant documents in the higher steps. But, SemApp is still outperformed the other benchmarks in the rest of steps.

4.3 Evaluation of Recall-Precision Graphs

Recall-precision curves are means to compare the performance of IR systems. In this experiment, the recall-precision curve is plotted for each system. These curves are plotted by joining the average accuracy points at eleven recall steps. Figure 4 compares the seven graphs of the seven models. SemApp curve lies above the curves of the competitors which proves that SemApp is more accurate than the other benchmarks.

Table 3. SemApp improvements percentage of P@x

Improve%	P@5	P@10	P@15	P@20	P@30	P@100
C_QL	39.8%	24.4%	20.6%	19.1%	15.7%	5.6%
S_QL	39.8%	37.5%	39.8%	45.3%	50.7%	58.7%
C_TFIDF	132.1%	124.7%	134.2%	139.0%	129.0%	110.0%
S_TFIDF	132.1%	137.5%	128.4%	129.3%	122.9%	85.1%
C_BM25	36.7%	25.9%	25.4%	25.3%	18.4%	12.4%
S_BM25	50.0%	53.7%	57.0%	59.7%	61.8%	61.8%

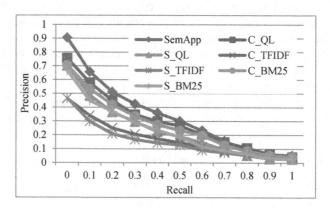

Fig. 4. Average recall-precision graph

4.4 Statistical Significance Testing

SemApp is capable of improving the most important IR metrics compared to baselines. However, these performance improvements may not be a true enhancement unless it goes through a statistical test to ensure that the difference between SemApp and the other benchmarks is not caused by random variation of the compared systems. The assumed characteristics to conduct the statistical test, like distributions normality and absence of outliers, were validated before performing the test. The null hypothesis assumes that there is no difference between SemApp and the benchmark to be tested. If the probability of the null hypothesis is less than the specified significance level, the alternative hypothesis is accepted which supports the superiority of the proposed approach.

Table 4 shows the metrics that have been subjected to paired two sample t-test: average precision (AP), Rprec, reciprocal rank (RR), P@5, P@10, P@15, P@20, and P@30 under 39 degrees of freedom. The listed 48 computed student t-statistics are compared to the t-critical one/two tail cases (2.429/2.712 in case of significance level 0.01 and 1.685/2.023 in case of significance level 0.05).

Table 4. Significance testing of SemApp metrics improvements

	C_QL	S_QL	C_TFIDF	S_TFIDF	C_BM25	S_BM25
t Stat of AP	3.038	3.646	5.528	5.354	3.873	3.968
t Stat of Rprec	1.533	3.945	5.500	5.277	2.537	3.407
t Stat of RR	2.916	2.333	5.650	6.684	3.793	3.181
t Stat of P@5	4.376	3.484	5.982	5.730	4.444	3.776
t Stat of P@10	3.295	3.439	5.378	5.381	5.047	4.149
t Stat of P@15	3.057	3.628	5.733	5.218	4.509	4.505
t Stat of P@20	3.004	3.787	5.683	5.333	4.260	4.646
t Stat of P@30	3.109	3.466	5.291	5.145	3.367	4.115

The first row of the table shows the computed t-statistic based on values of SemApp AP and values of each benchmark individually. All these values exceeds the threshold 2.712 which means that the AP improvements that achieved by SemApp over the other benchmarks are highly significant at significance level 0.01. The entries in the second row show some variations. First, the difference between SemApp and C_QL in Rprec metric is not statistically significant which means that C_QL and SemApp almost have the same performance in this metric. Secondly, the difference between SemApp and C_BM25 is statistically significant at significance level 0.05. The same result was happened between SemApp and S_QL in RR metric. To sum up: there are 45 out of 48 computed t-statistics show high statistically significant differences between SemApp and its competitors and 2 out of the 48 test show significant differences and only one test shows not statistical difference. These results lead to the superiority of the proposed approach over the other baselines.

5 Conclusions

The work presented in this paper is based on three integrated phases to enhance IR. First phase is devoted to query meaning extraction. Discovering the global meaning of the query is necessary to explore the true user's needs. A set of semantic trees corresponding to the possible query alternatives was extracted. After scoring each alternative (semantic tree), the optimal alternative was selected as the best query representative. An effective document and query representations were conducted in the second phase. This indexing phase utilize a conceptual weighting scheme to distribute importance effectively on the document concepts. In phase three, a hybrid ranking model was used to rank the documents. This model is based on a query likelihood language model and a conceptual weighting model.

The validity of the proposed approach has been tested through empirical experiments. The conducted evaluations assessed the standard IRS metrics to the proposed integrated approach and also compared them to the metrics of related benchmarks. The results show that the proposed integrated system is outperformed the other baselines. The proposed approach improves MAP by 15.5%, GMAP by 46%, Rprec by 6.6%, MRR by 23.9%, P@5 by 36.7%, P@10 by 24.4, P@15 by 20.6%, and P@20 by 19.1% over the nearest competitor in each metric. Statistical tests were performed to ensure that the obtained metric improvements are statistically significant and they are not caused by either random variations or coincidence.

The main limitation of this model is that the terms in the ontologies are limited and that the conceptual relations that connect the different concepts are not complete. To improve the performance of the proposed system, it may supported by more external semantic resources. Also, the formulae of weighting and ranking may be enhanced.

References

1. Nidelkou, E., Papastathis, V., Papadogiorgaki, M.: User Profile Modeling and Learning. in Encyclopedia of Information Science and Technology, 2nd edn, pp. 3934–3939. IGI Global, Hershey (2009)
2. Yibing, S., Qinglong, M.: Research of literature information retrieval method based on ontology. In: 2014 International Conference on Multisensor Fusion and Information Integration for Intelligent Systems (MFI), pp. 1–6 (2014)
3. Kumar, S., Singh, M., De, A.: OWL-based ontology indexing and retrieving algorithms for semantic search engine. In: 2012 7th International Conference on Computing and Convergence Technology (ICCCT), pp. 1135–1140 (2012)
4. Chauhan, R., Goudar, R., Sharma, R., Chauhan, A.: Domain ontology based semantic search for efficient information retrieval through automatic query expansion. In: 2013 International Conference on Intelligent Systems and Signal Processing (ISSP), pp. 397–402 (2013)
5. Ali, A., Bari, P., Ahmad, I.: Concept-based information retrieval approaches on the web: a brief survey. IJAIR 3, 14–18 (2011)
6. Zhou, D., Lawless, S., Liu, J., Zhang, S., Xu, Y.: Query expansion for personalized cross-language information retrieval. In: 2015 10th International Workshop on Semantic and Social Media Adaptation and Personalization (SMAP), pp. 1–5 (2015)
7. Carpineto, C., Romano, G.: A survey of automatic query expansion in information retrieval. ACM Comput. Surv. 44, 1–50 (2012)
8. Segura, A., Vidal-Castro, C., Ferreira-Satler, M., Salvador-Sánchez: Domain ontology-based query expansion: relationships types-centered analysis using gene ontology. In: Castillo, L., Cristancho, M., Isaza, G., Pinzón, A., Rodríguez, J. (eds.) Advances in Computational Biology. AISC, vol. 232, pp. 183–188. Springer, Cham (2014). https://doi.org/10.1007/978-3-319-01568-2_27

9. Laatar, R., Aloulou, C., Belguith, L.H.: Disambiguating Arabic words according to their historical appearance in the document based on recurrent neural networks. ACM Trans. Asian Low-Resource Lang. Inf. Process, **19**, 1–16 (2020)

10. Boughareb, D., Farah, N.: A query expansion approach using the context of the search. In: van Berlo, A., Hallenborg, K., Rodríguez, J., Tapia, D., Novais, P. (eds.) Ambient Intelligence - Software and Applications. AISC, vol. 219, pp. 57–63. Springer, Heidelberg (2013). https://doi.org/10.1007/978-3-319-00566-9_8

11. Kumar, N., Carterette, B.: Time based feedback and query expansion for twitter search. In: Serdyukov, P., et al. (eds.) ECIR 2013. LNCS, vol. 7814, pp. 734–737. Springer, Heidelberg (2013). https://doi.org/10.1007/978-3-642-36973-5_72

12. Nagaraj, R., Thiagarasu, V., Vijayakumar, P.: A novel semantic level text classification by combining NLP and thesaurus concepts. IOSR J. Comput. Eng. **16**, 14–26 (2014)

13. Luo, Q., Chen, E., Xiong, H.: A semantic term weighting scheme for text categorization. Expert Syst. Appl. **38**, 12708–12716 (2011)

14. Boubekeur, F., Azzoug, W., Chiout, S., Boughanem, M.: Indexation sémantique de documents textuels. In: 14e Colloque International sur le Document Electronique (CIDE14), Rabat, Maroc (2011)

15. Boughanem, M., Mallak, I., Prade, H.: A new factor for computing the relevance of a document to a query. In: International Conference on Fuzzy Systems, pp. 1–6 (2010)

16. Boubekeur, F., Boughanem, M., Tamine, L., Daoud, M.: Using WordNet for concept-based document indexing in information retrieval. In: Fourth International Conference on Semantic Processing (SEMAPRO), Florence, Italy (2010)

17. Al-Zoghby, A.: A new semantic distance measure for the VSM-based information retrieval systems. In: Shaalan, K., Hassanien, A.E., Tolba, F. (eds.) Intelligent Natural Language Processing: Trends and Applications. SCI, vol. 740, pp. 229–250. Springer, Cham (2018). https://doi.org/10.1007/978-3-319-67056-0_12

18. Luo, C., Liu, Y., Zhang, M., Ma, S.: Query ambiguity identification based on user behavior information. In: Jaafar, A., et al. (eds.) AIRS 2014. LNCS, vol. 8870, pp. 36–47. Springer, Cham (2014). https://doi.org/10.1007/978-3-319-12844-3_4

19. Hawalah, A., Fasli, M.: Dynamic user profiles for web personalisation. Expert Syst. Appl. **42**, 2547–2569 (2015)

20. Safi, H., Jaoua, M., Belguith, L.H.: Intégration du profil utilisateur basé sur les ontologies dans la reformulation des requêtes Arabes. In: ACTES DU COLLOQUE, p. 40 (2015)

21. Codocedo, V., Lykourentzou, I., Napoli, A.: A semantic approach to concept lattice-based information retrieval. Ann. Math. Artif. Intell. **72**(1–2), 169–195 (2014). https://doi.org/10.1007/s10472-014-9403-0

22. Formica, A.: Semantic web search based on rough sets and fuzzy formal concept analysis. Knowl.-Based Syst. **26**, 40–47 (2012)

23. Poelmans, J., Ignatov, D.I., Kuznetsov, S.O., Dedene, G.: Fuzzy and rough formal concept analysis: a survey. Int. J. Gen. Syst. **43**, 105–134 (2014)

24. Kumar, C.A., Mouliswaran, S.C., Amriteya, P., Arun, S.R.: Fuzzy formal concept analysis approach for information retrieval. In: Ravi, V., Panigrahi, B., Das, S., Suganthan, P. (eds.) FANCCO - 2015. AISC, vol. 415, pp. 255–271. Springer, Cham (2015). https://doi.org/10.1007/978-3-319-27212-2_20

25. Raj, T.F.M., Ravichandran, K.S.: A novel approach for intelligent information retrieval in semantic web using ontology. World Appl. Sci. J. **29**, 149–154 (2014)

26. Martín, A., León, C., López, A.: Enhancing semantic interoperability in digital library by applying intelligent techniques. In: 2015 SAI Intelligent Systems Conference (IntelliSys), pp. 904–911 (2015)

27. Damaševičius, R.: Automatic generation of concept taxonomies from web search data using support vector machine. In: Proceedings of 5th International Conference on Web Information Systems and Technologies, pp. 673–680 (2009)
28. Damaševičius, R.: Automatic generation of part-whole hierarchies for domain ontologies using web search data. In: 32nd International Convention Proceedings: Computers in Technical Systems and Intelligent Systems, vol. 3, pp. 215–220 (2009)
29. Tu, X., He, T., Chen, L., Luo, J., Zhang, M.: Wikipedia-based semantic smoothing for the language modeling approach to information retrieval. In: Gurrin, C., et al. (eds.) ECIR 2010. LNCS, vol. 5993, pp. 370–381. Springer, Heidelberg (2010). https://doi.org/10.1007/978-3-642-12275-0_33
30. Chenaina, T., Neji, S., Shoeb, A.: Query sense discovery approach to realize the user search intent. Int. J. Inf. Retr. Res. **12** (2022)
31. Neji, S., Jemni Ben Ayed, L., Chenaina, T., Shoeb, A.: A novel conceptual weighting model for semantic information retrieval. 07-Inf. Sci. Lett. **10**, (2021)
32. Neji, S., Chenaina, T., Shoeb, A., Ben Ayed, L.: HyRa: an effective hybrid ranking model. In: 25th International Conference on Knowledge Based and Intelligent information and Engineering Systems, vol. 10 (2021)
33. Neji, S., Chenaina, T., Shoeb, A., Ben Ayed, L.: HIR: a hybrid IR ranking model. In: International Computer Software and Applications Conference (2021)
34. Billami, M.B., Gala, N.: Approches d'analyse distributionnelle pour améliorer la désambiguïsation sémantique. Journées internationales d'Analyse statistique des Données Textuelles (JADT) (2016)

Family Matters: On the Investigation of [Malicious] Mobile Apps Clustering

Thalita Scharr Rodrigues Pimenta[1]([✉])[iD], Rafael Duarte Coelho dos Santos[2][iD],
and André Grégio[3][iD]

[1] Federal Institute of Paraná, Irati, Brazil
thalita.pimenta@ifpr.edu.br
[2] INPE, São José dos Campos, SP, Brazil
rafael.santos@inpe.br
[3] Federal University of Paraná, Curitiba, Brazil
gregio@inf.ufpr.br

Abstract. As in the classification of biological entities, malicious software may be grouped into families according to their features and similarity levels. Lineage identification techniques can speed up the mitigation of malware attacks and the development of antimalware solutions by aiding in the discovery of previously unknown samples. The goal of this work is to investigate how the use of hierarchical clustering on malware statically extracted features can help on explaining the distribution of applications into specific groups. To do so, we collected 76 samples of several versions from popular, legitimate mobile applications and 111 malicious applications from 11 well-known scareware families, produced their dendograms, and discussed the outcomes. Our results show that the proposed apporach is promising for the verification of relationships found between samples and their attributes.

Keywords: Clustering · Mobile malware · Lineage

1 Introduction

The development of signatures/heuristics to detect malicious mobile applications is a complex and time-consuming task, due to the large number of daily releases in app stores. According to McAfee Mobile Threat Report, more than 1.5 million variants of malicious code in mobile applications were detected in the first four months of 2019 [29]. Applying different types of analysis allows the classification of malware in groups or families and, consequently, the identification of suspicious behaviors and/or attributes extracted from mobile applications. Antiviruses (AVs) for the Android platform are based on package databases, file location in the operating system, and information from mobile applications [9]. Non-malicious and malicious applications use an extensive quantity of attributes like permissions, intent filters, external classes, and components. AV scanners check the packages and their locations before allowing the application installation. However, files added after installation may be unable to be monitored

© Springer Nature Switzerland AG 2021
O. Gervasi et al. (Eds.): ICCSA 2021, LNCS 12951, pp. 79–94, 2021.
https://doi.org/10.1007/978-3-030-86970-0_7

by AVs [9]. Therefore, external classes and the dynamic download of additional components may cause installed malware to escalate privileges.

Some types of malware, like Ransomware and Scareware call external components and classes to manipulate users. Scareware, which can be classified as a type of ransomware, takes advantage of violations related to the user experience [7]. Scareware tricks users with false applications that leverage security warnings on their devices. These fake treats intend to sell useless services, allowing the theft of bank information, such as the victim's credit card details. Installing these services makes it easier to prepare the device for Ransomware-type attacks, which encrypt files and require a ransom to be paid, usually with cryptocurrencies. For instance, in 2020, the 'COVID-2019 Scareware' used a fake data encryption threat and demanded ransom to terrify its victims [23]. Although not as harmful as traditional crypto-ransomware, this type of malware support scams and information theft, since this malicious code presents dialog boxes demanding ransom, without having genuine encryption features. During the COVID-19 pandemic, the terms related to the disease have been used for disseminating malware and fraudulent pages, as happened in Canada [5]. According to Norton UK [32], criminals have been using search engine optimization poisoning, a type of attack that associates malicious web pages with popular search terms. The malicious page is listed in the search results, looking legitimate to the end-user.

Due to the importance of mitigating this type of threat to the current cybercrime scenario, it is beneficial to develop alternative security solutions, considering device hardware limitations, operating system peculiarities, and end-user expertise. Cloud database-based solutions with attribute lists assigned to known families can be queried and updated by users, particularly enabled by pop-up windows and other components used by these plagues.

Concerning the window for identifying new threats that can last for months [3], we investigate the application of unique attributes and hierarchical clustering to recognize unknown variants of known malware families. In the literature, it is common to encounter hierarchical clustering analysis for malware classification and goodware. However, it is rare to find attempts to explain the divisions between families in the resulting clusters. Another little-explored sub-area is the investigation of the diversity in characteristics and behaviors found in samples from the same family. These variances can result in other branches or even other labels, depending on the antivirus consulted.

To be able to investigate the attributes usually used in mobile malware, it is interesting to compare with sets of elements of non-malicious applications. Thus, in conjunction with the malicious samples, versions of ten benign applications were analyzed. The goodware samples are from known apps present in the most downloaded general apps ranking from the Android system app store. In our experiments, a dataset of 111 samples of 11 Scareware families was used. The contributions of this paper are twofold: (i) we show and discuss the reasons behind the distribution of malicious and non-malicious app among their corresponding clusters, concluding that what differentiates them is the use of external

and unverified sources, peculiar strings, and uncommon URLs, also often called
to the same vulnerable attributes, principally permissions; (ii) we analyze which
of the extracted features unique to Android malware families aid in the identification of Scareware type, aiming to foster a base of external code lists and
critical strings, when found concomitantly with distinct attributes, for dynamic
monitoring of application operations.

The rest of the paper is organized as follows. In Sect. 2, we discuss the closest
available work and position ours, and a short description for Android features
and clustering is given. In Sect. 3, the proposed classification method is presented. The results discussions are explained in Sect. 4. Conclusions and future
directions for research are shown in Sect. 5.

2 Background and Related Work

In the context of malware analysis, variants are examples of a group or family,
which share similar properties, behaviors, and structures. Features are software
elements used to classify code as malicious or not.

According to the literature, there is a time window for new variants to be
noticed by security analysts and/or end-users. For Android malware analysis,
researches indicating that one can spend about three months to detect new
malware [3].

The malware classification problem has cumulative complexity due to the
increasing number of characteristics necessary to represent the application's
behavior exhaustively [19]. Certain permissions and services can be used by
normal or malicious applications. However, some attributes are more frequent in
illegal activities. For example, numerous event logs related to boot, SMS messages, phone call logs, and package management features are related to financial
malware [1].

The properties present in the application can be extracted from several methods, mainly through static (without code execution), dynamic (with code execution), or hybrid analysis techniques, combining static and dynamic methods
[12]. In addition to the extraction methods, several forms of representation of
the collected features are found in the literature. Some popular techniques are
listed below: *N-grams* [14], *opcodes* [27], sequences of function calls [30], flow
control graphs [2], data flow graphs [33], dangerous API calls [31] and network
communication data [24].

It is noteworthy that depending on the sample's complexity, it is necessary to
investigate the extraction methods for more appropriate characteristics, directly
influencing the quality of the results obtained after analysis.

2.1 Android Architecture and Analyzed Components

The AndroidManifest.xml file displays statements about the operating system
resources and hardware assets used in the application. As enabling usage permissions and functionality can be dangerous, this file is employed to detect anoma-

lous and unsafe behavior [10]. About the Android architecture, the values of the following components are extracted:

- **Intents:** These components provide data about operations and behaviors that can be activated by the application. Examples of resources for sharing data and services between applications include: *Activities, Intents, Broadcast Receivers* and *Services* [18].
- **Permissions:** These are authorizations to access the hardware resources and system functionalities. For example, some applications store user location data via GPS. Others may ask for information about the contact list. Thus, the control of relevant parts for security and privacy is carried out by the permission system, when the application is installed [11].
- **External Classes:** Applications for the Android platform, malicious or not, run external code. Still, these mechanisms can result in security issues, such as evasion of malware detectors and Code Injection attacks [26]. Consequently, the investigation of methods of external classes called by the application allows the verification of unsafe behaviors.

2.2   Clustering and Similarity

Due to the large number of variants that have been created automatically, which bypasses detection mechanisms of patterns and signatures, manual analysis has become a significant bottleneck in the control of virtual pests [4].

Several techniques (e.g., phylogeny, clustering) have been applied over the years to identify whether a mobile application is malicious or not. Clustering is an unsupervised learning technique for finding patterns in data. Given the difficulty of analyzing all the possibilities of groups in a large amount of data, several techniques were created to assist the formation of the clusters.

The use of these techniques intends to identify unknown malware mainly by calculating the similarity of a sample with previously grouped, already known families of malicious apps. While the similarity between two strongly related objects has a value of 1, the distance has a value of zero and vice versa [22]. Examples of popular indices include the Jaccard Index and Euclidean distance. Complementary equations and information can be found in the study of [6].

Grouping methods known as Hierarchical Clustering Algorithms create the hierarchy of relationships between objects. The central aspect of these algorithms is the pair selection based on a connection measure between the elements. The linkage measure uses similarity or distance indices to cluster similar objects. The most common forms of bonding are Complete Linkage, Simple Linkage, Average Linkage, Centroid Methods, Median Method, and Ward Method. Regarding the quality analysis of the generated clusters, internal and external indexes can be applied. External indexes assess inter-group variations and distances, while internal indexes assess intra-group variations. Examples of popular indices include Purity, Entropy, Precision, Recall and *F-Score* [28].

The quality of the clustering methods results depends on the chosen algorithm, parameters, and similarity measures [20].

2.3   Related Work

In the literature, several papers applied hierarchical clustering for Android malware detection and/or classification. However, we selected some related works focused on hierarchical clustering and relationships between families. Zhou et al. carried out characterization and evolution studies of more than 1200 samples from 49 different families, collected between 2010 and 2011. The results showed that 86% samples collected are legitimate applications packaged with malicious payloads [34].

Pfeffer et al. presented a system for analyzing malware and its attributes using genetic information, called MAAGI. From a bag-of-words model, three types of malware characteristics are used: two strings and one type of N-gram with a call procedure graph. The first strings are extracted from the binaries and the used libraries. In the clustering part, the Louvain method, based on the graphs grouping, was used together with K-means. Also, they present a method for extracting the lineage of malware using a probabilistic model. Thus, as a result, the authors believe that the system is the first to integrate clustering, lineage, and prediction of malware and families [25].

Hsiao et al. highlight that several threads are used to perform intrinsically complex tasks in Androids applications. Thus, it becomes difficult to analyze malicious applications without prior knowledge of their structures. Based on behavior analysis, the authors propose an investigation scheme, which involves phylogenetic trees, UPGMA, significant main components, and a matrix of points. For the experiments, more than 980 samples from different families were studied, making it possible to detect which code was shared between "sibling" applications and which was inherited from common ancestors [13].

In [17], the authors presented an Android malware clustering technique by mining malicious payloads. The proposed solution checks whether the applications share the same version of malicious payload. The authors created a method to remove code from legitimate Android application libraries while still keeping malicious payloads, even if they are injected under the name of popular libraries. In experiments carried out, the accuracy rate was 0.90 and the recall was 0.75 for the Android Gnome malware set.

Time is essential when dealing with thousands of samples to classify. Performing clustering methods allows screening apps to check for signs of hostility in the sample and groups with similar traits. As mentioned earlier, static analysis methods have strengths and weaknesses. Nevertheless, it is unnecessary to use a sandbox to run the sample for a while. Besides, it is not trivial to know the minimum necessary execution time to observe unlawful behavior.

However, our focus is on investigating the features that create a distinction in the grouping of malware and goodware, as well as the impacts of the classification of variants in families. Discovering the value of these attributes, the model can be used for datasets of different sizes. Unlike most articles, we aim to explain the divisions resulting from clustering, to interpret why some versions of the very application can be allocated in different groups. When analyzing malicious samples from the same family, this verification is even more important, due to the large number of modifications made from common code.

3   Experiments and Results

For the experiments, 111 Scareware applications available on the CICAnd-Mal2017 database of the UNB - University of New Brunswick [15] were analyzed. In addition to the malware samples, several samples of 10 benign applications (*goodware*) were analyzed for feature extraction and pattern determination. The graph of the Fig. 1 displays the names of applications and the distribution of malicious and non-malicious samples.

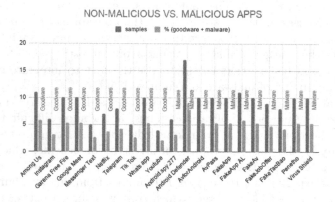

Fig. 1. Information about the non-malicious and malicious dataset. The blue column shows the number of samples of each application/family. The red column shows the percentage to the total dataset (malicious and non-malicious apps). (Color figure online)

We used Androguard [8] to preprocess/unpack the applications. The extracted *xml* files and features values were stored into a database for posterior analysis of similarity and clustering.

We applied the Levenshtein distance [16] to calculate the similarity of malware samples using strings extracted from attributes, set of permissions, activities, services, external classes, and receivers. Levenshtein distance employs the minimum number of insertions, deletions, and substitutions required to transform one string into another.

The Average Linkage algorithm was applied for clustering, a method in which the criterion is the distance of all individuals from one group to each group. For the implementation, we used the `sklearn` library and Python language, in combination with the Maria DB database.

This section presents charts and dendrograms of the groups resulting from sample clustering. Regarding the scale used on the vertical axis (distance), two identical samples have a value of 0. Consequently, the more divergent the two samples are, the greater this value is.

3.1   Analysis of Non-malicious Applications

Non-malicious individuals were grouped using the K-means algorithm (Fig. 3), by analyzing classes, permissions, and activities. It was observed that the applications Tik Tok and Among Us stand out for the greater and lesser amounts of these attributes, respectively. As not all applications had the use of services and receivers, these attributes were not employed in this first experiment.

The tree resulting from the goodware samples clustering is shown in Fig. 2. As can be observed in the dendrogram, the applications were properly grouped. However, copies of Google Meet have been divided into two separate groups (clusters G and H). The Group H samples have fewer attributes of all types when compared to more recent samples, placed in the other group.

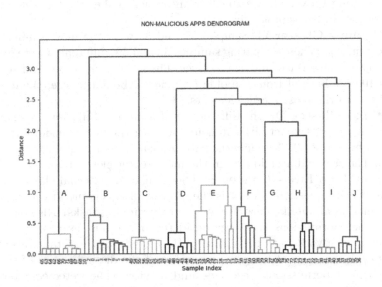

Fig. 2. Dendrogram of all goodware samples.

According to the analysis of distinct and common features, it was noticed that applications in the same cluster had dozens of identical features, and in separate groups, the applications had hundreds of different features. For example, Instagram and Whats app samples had 24 permissions, three services, nine classes, and two activities in common. Comparing the sample feature sets from Instagram and Tik Tok, we found differences in 355 activities, 1238 classes, 66 services, 32 receivers, and 53 permissions.

1. **Whats App - Cluster A**. As differentiators from the other communication applications analyzed, Whats App samples have permissions to receive maps, backup service, NFC, install packages, read phone numbers, and read and write settings. Comparing the 62 and 66 elements, the founded difference was 24 activities, 275 classes and three services, and one permission. The added permission in the last five samples was the android.permission.READ_PHONE_NUMBERS. Just by studying the total characteristics of each version, it would not be probable to regard this dissimilarity. There were several variations in size in the samples.

2. **Among Us - Cluster B**. Unlike Garena Free Fire, this game does noDue to space limitations, the list of analyzed samples, the number of features and additional trees weret have multiple attributes with questionable security. Among Us version v2020.6.9 is an outlier of the set. There were some alterations in the samples size.

3. **Telegram - Cluster C**. Samples 37 and 38 were the most similar in the Telegram set. When comparing samples 32 and 39, differences were found in 10 classes, one service, and one receiver. The set of permissions and activities was 100% inherited from the oldest version. About the size, there was an addition in the most recent elements.

4. **Netflix - Cluster D**. The difference in the totals of attributes seems succinct, however, the activities in common with the older version of the set, varied by about 50.5%, while the external classes reached 59% of the variation. The size had an addition in the most recent elements.

5. **Garena Free Fire - Cluster E**. The number of external classes is large. One security question is what is the interest of this game in getting device information about the currently or recently running tasks? Also, this game has other permissions that are considered dangerous, as recording audio, obtaining network status, and Bluetooth. Examining the total of attributes, it is evident a split of the groups because of the dissimilarities of the total numbers of permissions, receivers, and services. The detected alterations between the samples 51 and 60 were three permissions, eight activities, three receivers, seven services, and 62 classes.

6. **Tik Tok - Cluster F** Some differentiated permissions are external storage media reading and access Wifi state. Other several packages are used to read and modify the device's settings. The differences in the Tik Tok samples are perceptible in the feature quantities: an increment in classes, receivers, and activities. These alterations divided the five versions into two groups. When comparing samples 32 and 39, we detected differences of 28 activities, one receiver, 115 classes, and five services. The size had an addition in the most recent elements.

7. **Google Meet - Cluster G and Cluster H**. The divergences between the classes in common with the oldest version varied between 25 and 117 distinct classes. Comparing the elements 22 and 31, for example, we found differences in 12 services, 117 classes, 19 receivers, 38 activities, and 17 permissions. Also, there were some fluctuations in the sample sizes.

8. **Youtube - Cluster H.** The YouTube examples differ due to the growth in classes and decrease activities in the most recent samples. The elements of numbers 6 and 7 had differences of 217 classes, being the least similar pair in the group. The set of permission, service, and receivers was 100% inherited from the version YouTube.v16.01.34. For activity and class sets, despite changes in the total, the inheritance was higher than 83%. Regarding the size, there was a reduction in the most recent elements.

9. **Messenger - Cluster I.** This is the communication application with the highest number of attributes. Messenger elements have several Facebook permissions as write and receive SMS, and call phones. The five versions of Messenger were separated by the number of classes. The set of permissions and receivers was 100% inherited from the oldest version. However, when comparing samples 71 and 75, the discrepancies found were three activities, 607 classes, and one service. Regarding the size, there was an increase in the most recent elements.

10. **Instagram - Cluster J.** The separation of the six Instagram versions can be done based on the addition in the number of services and receivers and the decrease in the number of classes in samples 4 and 5. The added permission in Instagram.v174.0 was the `android.permission.READ_PHONE_NUMBERS` while the activity was the `com.instagram.rtc.activity.RtcGridSandbox Activity`. There were several variations in the number of external classes called in the samples. The three services added in the last two versions were related to background services and the alarm system. The size had an addition in the most recent elements, except in the last one.

Clustering Validation. For the 10 generated clusters, the Precision value was equal to 1. The Recall and F1-Score values were equal to 1 in almost all clusters, except in cluster H where Recall was 0.6 and F1-Score was 0.75.

3.2   Analysis of Malicious Applications

The result of clustering families using K-means is shown in Fig. 3. Due to space limitations, the list of analyzed samples, the number of features and additional trees are in the online Appendix[1]. It was noticed that there was confusion in the samples, both in K-means results and in the groups resulting from hierarchical clustering. In addition, information about unique attributes of each family is in Table 1.

(A) **Android spy 277.** The most appealing permissions are camera, audio recording, SMS messages, and authorization to disable the keylock and any associated password security. In the malware families clustering, the samples from this family were separated into purple and light blue clusters. Analyzing the total of features, samples 3 and 4 would be more similar to each other if compared to the pair 4–5. Meanwhile, samples 3 and 4 have only one activity and ten permissions in common, while the pair 4–5 shares

---

[1] Online Appendix is available at https://github.com/tsrpimenta/onlineappendix.

GOODWARE AND MALWARE K-MEANS CLUSTERING USING PERMISSIONS, CLASSES, AND ACTIVITIES

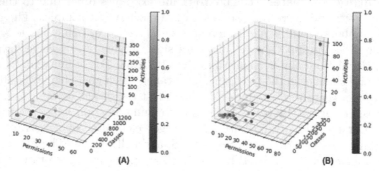

Fig. 3. Results of Goodware(A) and Malware (B) K-Means Clustering using Activities, Classes, and Permissions.

Table 1. Exclusive features of each family.

Family	Permissions	Activities	Receivers	Classes	Services
Android.spy.277	1	14	4	47	2
Android Defender	14	22	9	165	13
AvForAndroid	0	43	9	51	4
AvPass	0	13	2	74	2
FakeApp	0	1	0	1	0
FakeApp AL	2	1	1	26	1
FakeAV	4	5	4	6	7
FakeJobOffer	1	24	4	8	3
FaketaoBao	0	1	4	0	2
Penetho	0	3	0	2	0
Virus shield	3	7	12	59	15

one activity, seven classes, and ten permissions. We emphasize that there were several changes in the size of the samples.

(B) **Android Defender.** This family contains the most extensive number of samples and exclusive features (Table 1). After clustering just the samples from this family, we realized that three groups were formed. The most similar element pairs were 15–17, 8–20, and 9–10. In the clustering of all malware samples, 4 samples from this family were separated into a cluster, demonstrating that they are distinct from the others in this set. We found that this type of malware uses several services and classes with different names, especially for reading and modifying system configurations. This family contains samples with heterogeneous amounts of attributes and sizes.

(C) **AvForAndroid.** This family highlighted out with the highest number of exclusive activities (Table 1). In the clustering, 8 elements were in one cluster and two in the last group. It is interesting to compare samples from the same family to verify dissimilar behaviors. For example, sample 23 has permissions to change network state, record audio and another 19 different permissions. When, the sample 25 can install shortcuts, get user accounts and get the bookmarks list.

(D) **AVPass.** This family calls several exclusive external classes (Table 1). This family has numerous activities related to books and some related to sniffer actions. Examples are espionage on shopping activities, such as shopping cart, zoom in on products and click on promotions. All samples were in the same group. For the most similar samples of indexes 38, 36, and 39, we detected that what separated element 38 from pair 36–39 were changes in classes and activities.

(E) **FakeApp.** This group has the most simple set of unique attributes. Some of the permissions that stand out are `android.permission.KILL_BACK GROUND PROCESSES`, `com.android.browser.permission.WRITE_HISTORY_BOOKMARKS`, and `android.permission.BLUETOOTH_ADMIN`. Analyzing the elements of this family, it can be seen that sample 47 was an outlier, especially due to the increase in the number of permissions. There were several changes in the size of the samples.

(F) **FakeApp AL.** This family has the amplest set of unique classes (Table 1). The samples were divided into two clusters. In sample clustering, the most similar pairs were 54–59 and 60–61. Unusual founded permissions are battery status, access Bluetooth sharing, turn on the device's flashlight, camera, and execute calls.

(G) **FakeAV.** Element 70 was an outlier in this set because of the number of classes. Nevertheless, all samples remained in the same group. Concerning permissions found in this malware were
`android.permission.CLEAR_APP_CACHE`, `android.permission.INJECT_EVENTS`, `android.permission.DELETE_PACKAGES`, `android.permission.INSTALL_PACKAGES`.

(H) **FakeJobOffer.** These samples were grouped with elements of the Android Defender set. Simply by the number of attributes, we could have considered all the samples are equivalent. Nevertheless, the most related pairs were 86–88 and 90–91. The exclusive permission found in this family was the `com.saavn.android.permission.C2D_MESSAGE`.

(I) **Faketaobao.** This family presents few exclusives receivers, services, and activities. Some that stand out are related to boot broadcast, receiving SMS, and activities performing in the background. Although the samples are all in the same cluster, the most divergent elements are copies 83 and 87 and the most similar are pairs 84–89 and 88–90.

(J) **Penetho.** Despite being in the same cluster, the element 92 was separated by the difference in the number of permissions, activities and classes to the other samples. It is interesting to note that these samples do not use receivers and services.

(K) **Virus Shield.** This family has the most significant set of unique receivers (Table 1). Element 107 was an outlier in this set because of the number of classes, therefore being grouped in the same cluster. The pair 103–108 was the least dissimilar in the set. The number of features and the size of each sample were heterogeneous.

Clustering Validation. As shown in Table 2, the values of the validation of the clustering were varied, with the lowest precision values being in the clusters 6, 9, and 11.

Table 2. Calculated metrics (%) for malware clustering results (each cluster is denoted by C followed by its number).

Metric/Cluster	C1	C2	C3	C4	C5	C6	C7	C8	C9	C10	C11
Precision	100	90	89	64	70	45	80	0	37	100	0
Recall	100	100	100	100	100	100	57	0	75	25	0
F1-Score	100	94	94	78	82	62	66	0	50	40	0

3.3   Goodware and Malware Samples Clustering

The K-means clustering of all samples(malware + goodware dataset) is shown in Fig. 4. Due to the size of the dendrogram, the figure can be found in the Online Appendix.

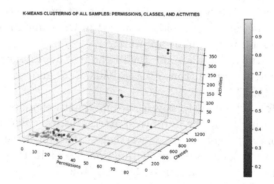

Fig. 4. K-means clustering of all samples.

Most benign apps were well separated, except for a few samples from Google Meet. About malicious samples, the classification of Penetho samples stands out with all samples in the same cluster. We observed that when comparing the familiar attributes between the Garena Free Fire and FakeApp_AL samples, we found thirteen unsafe permissions in common. These permissions have already been mentioned previously in the analyzes of these groups. Unfortunately, the chosen sets of attributes and the similarity measure were not satisfactory for separating families, yet they were capable of distinguishing between malicious and non-malicious samples.

3.4   Clustering Validation

Due to the limitation in the number of pages, the table for validating the clusters of all samples can be found in the Online Appendix. We observed that besides the clusters with values 0, the lowest values of precision, accuracy, and F1-score were those of groups 20, 22. The indices adopted to examine the results of the clusters are shown in the Table 3. According to the outcomes, it was clear that the clustering of non-malicious samples had the best indicators, reaching 0.95 of accuracy. Regarding the grouping of malicious samples families and the total dataset (malware and goodware), some measures revealed that the latter performed better. According to the results of the validation of clustering of malicious samples, we verified that the recall was not satisfying, indicating that although there are exclusive characteristics, the set of equal attributes is also large. This causes the samples to barely distinguishable, creating mixing between groups. Although the non-malicious samples are more distinguishable, the classification of the malicious ones ended up reducing the accuracy of the clustering of all dataset samples.

Table 3. Calculated indices (%) for the goodware, malware and goodware-malware clustering results

Clustering	Accuracy	Precision	Recall	F1-score	Homogeneity	V-Measure
Goodware	95	1	94	96	100	98
Malware	64	83	64	63	72	66
Goodware-Malware	67	80	67	71	89	85

4   Discussion

Concerning Goodware, some questions that remained about similar applications are in different groups. For example, *why did the communication apps stay in separate groups?* The samples of these applications have heterogeneous number of attributes, and some permissions are found in a few applications only (e.g., the Google Meet app has a number of conference-related receivers, such as meeting entry and exit, and data sharing). In the Whats app attribute analysis, we found several backup features, contact blocklists management, two-step authentication functionality, and payment actions. Another finding was that not all applications are allowed to use SMS messages, as noted in Google Meet and Telegram.

Malware vs Goodware. Examining the characteristics of the versions of the non-malicious applications tested, we verified that can be less confused to determine the lineage of the samples, due to a large number of features common among the samples.

Regarding ground truth and labeled datasets, building bases for from benign applications is less complex than for malicious code. The use of antivirus to check whether a file is malicious or not is different from the sample labeling process in families. In addition, in the analysis of malware, there is still the verification of whether a variable is new or modified code and repackaged.

Static Features. The classification of malware only through static features has limitations, as already mentioned in [21]. Still, permissions are useful tools for detecting dangerous and unwanted behavior (e.g., all analyzed malicious mobile applications had the KILL_BACKGROUND_PROCESSES, REAL_GET_TASKS, BLUETOOTH_ADMIN, and READ_LOGS permissions).

During the analysis of the attributes, it was evident that malicious and non-malicious applications use consents potentially harmful to the user's security. However, how do you know what the application intends when it requests permission for recording audio? Would this consent enables the sharing of audio messages among unauthorized people or spying the victim's conversations and use the audios for extortion? Other permissions considered critical appeared in diverse goodware:

- WRITE_EXTERNAL_STORAGE: Youtube, Whats app, Telegram, Messenger, Instagram, Garena Free Fire, Netflix, and Tik Tok.
- ACCESS_FINE_LOCATION: Youtube, Whats app, Telegram, Messenger, and Instagram.
- ACCESS_COARSE_LOCATION: Youtube, Whats app, Telegram, and Messenger.
- GET_TASKS: Whats app, Tik Tok, and Garena Free Fire.
- CHANGE_WIFI_STATE: Whats app, Messenger, and Garena Free Fire.
- READ_PHONE_STATE: Youtube, Whats app, Telegram, Messenger, Instagram, Garena Free Fire, and Netflix.

Some permissions appeared only in malicious code, such as DISABLE_KEYGUARD (FaketaoBao, Android.spy.277, FakeApp_AL, Android Defender, and AvForAndroid), and READ_HISTORY_BOOKMARKS (FakeApp and AndroidDefender).

Concerning other attributes, we found that families Android Defender and FakeApp_AL have battery monitoring services. The fakcAV and faketaoBao samples presented status and Wifi changes receivers. FakeJobOffer samples have attributes related to audio while AvForAndroid monitors readers of books and pdfs activities. Also, android.spy.277 and FakeApp_AL families are interested in the user's actions with images and videos. Virus Shield and AvPass elements presented payment activities and services.

As a limitation, the similarity measure and the features chosen were insufficient to distinguish malware families and some cases of goodware-malware too.

5   Conclusion

From the experiments presented, it was possible to conclude that malware variants from known families were grouped into separate groups. These results allow us to assume that the feature extraction model is still flawed and demands improvement. Based on these assumptions, the following future works are considered: use of hybrid analysis (static and dynamic features) to describe the most representative behaviors of malware; automation of the proposed model for large sets of samples acquired incrementally from real environments; and the use of unique attributes in clustering with weights.

References

1. A. F. A. Kadir, N.S., Ghorbani, A.A.: Understanding android financial malware attacks: taxonomy, characterization, and challenges. J. Cyber Secur. Mob. **7**, 1–52 (2018)
2. Alam, S., Traore, I., Sogukpinar, I.: Annotated control flow graph for metamorphic malware detection. Comput. J. **58**, 2608–2621 (2015)
3. Apvrille, A., Strazzere, T.: Reducing the window of opportunity for android malware Gotta catch'em all. J. Comput. Virol. **8**(1–2), 61–71 (2012)
4. Awad, R.A., Sayre, K.D.: Automatic clustering of malware variants. In: IEEE Conference on Intelligence and Security Informatics (ISI) (2016)
5. Burke, D.: Fake covid notification apps and websites aim to steal money and personal data. https://www.cbc.ca/news/canada/nova-scotia/covid-apps-phones-scammers-fraudulent-personal-data-1.5877496
6. Cha, S.H.: Comprehensive survey on distance/similarity measures between probability density functions. City **1**(2), 1 (2007)
7. Connolly, L.Y., Wall, D.S.: The rise of crypto-ransomware in a changing cybercrime landscape: taxonomising countermeasures. Comput. Secur. **87**, 101568 (2019)
8. Desnos, A., et al.: Androguard-reverse engineering, malware and goodware analysis of android applications. google.com/p/androguard 153 (2013)
9. Fedler, R., Schutte, J., Kulicke, M.: On the effectiveness of malware protection on android. Fraunhofer AISEC **45**, 53 (2013)
10. Feizollah, A., Anuar, N.B.R., Salleh, G.S.T., Furnell, S.: AndroDialysis: analysis of android intent effectiveness in malware detection. Comput. Secur. **65**, 121–134 (2017)
11. Felt, A.P., Chin, E., Hanna, S., Song, D., Wagner, D.: Android permissions demystified. In: Proceedings of the 5th ACM Symposium on Information, Computer and Communications Security, pp. 627–638. ACM (2011)
12. Firdaus, A., Anuar, N., Karim, A., Razak, M.F.A.: Discovering optimal features using static analysis and a genetic search based method for android malware detection. Front. Inf. Technol. Electron. Eng. **19**, 712–736 (2018)
13. Hsiao, S.W., Sun, Y.S., Chen, M.C.: Behavior grouping of android malware family. In: 2016 IEEE International Conference on Communications (ICC), pp. 1–6. IEEE (2016)
14. Karim, M.E., Walenstein, A., Lakhotia, A., Parida, L.: Malware phylogeny generation using permutations of code. J. Comput. Virol. **1**, 13–23 (2005)
15. Lashkari, A.H., Kadir, A.F.A., Taheri, L., Ghorbani, A.: Toward developing a systematic approach to generate benchmark android malware datasets and classification. In: 2018 International Carnahan Conference on Security Technology (ICCST), pp. 1–7 (2018)
16. Levenshtein, V.I.: Binary codes capable of correcting deletions, insertions, and reversals. In: Soviet Physics Doklady, pp. 707–710. Soviet Union (1966)
17. Li, Y., Jang, J., Hu, X., Ou, X.: Android malware clustering through malicious payload mining. In: Dacier, M., Bailey, M., Polychronakis, M., Antonakakis, M. (eds.) RAID 2017. LNCS, vol. 10453, pp. 192–214. Springer, Cham (2017). https://doi.org/10.1007/978-3-319-66332-6_9

18. Nauman, M., Khan, S., Zhang, X.: Apex: extending android permission model and enforcement with user-defined runtime constraints. In: Proceedings of the 5th ACM Symposium on Information, Computer and Communications Security, pp. 328–332. ACM (2010)

19. Martín, A., Fuentes-Hurtado, F., Naranjo, V., Camacho, D.: Evolving deep neural networks architectures for android malware classification. In: IEEE Congress on Evolutionary Computation (CEC) (2017)

20. Metz, J.: Análise e extração de características estruturais e comportamentais para perfis de malware. Master's thesis, Mestra em Ciências de Computação e Matemática Computacional - USP., São Carlos - SP (Junho 2006)

21. Moser, A., Kruegel, C., Kirda, E.: Limits of static analysis for malware detection. In: Twenty-Third Annual Computer Security Applications Conference (ACSAC 2007), pp. 421–430. IEEE (2007)

22. Nadeem, A.: Clustering malware's network behavior using simple sequential features. Master's thesis, University of Technology, Faculty of Electrical Engineering, Mathematics and Computer Science, September 2018

23. News, S.W.: The covid-19 hoax scareware. https://securitynews.sonicwall.com/xmlpost/the-covid-19-hoax-scareware/. Accessed 25 Mar 2021

24. Perdisci, R., W.Lee, Feamster, N.: Behavioral clustering of http-based malware and signature generation using malicious network traces. In: NSDI, vol. 10, p. 14 (2010)

25. Pfeffer, A., et al.: Malware analysis and attribution using genetic information. In: 7th International Conference on Malicious and Unwanted Software (2012)

26. Poeplau, S., Fratantonio, Y.A., Bianchi, C.K., Vigna, G.: Execute this! Analyzing unsafe and malicious dynamic code loading in android applications. In: NDSS Symposium 2014, pp. 23–26 (2014)

27. Rathore, H., Sahay, S.K., Chaturvedi, P., Sewak, M.: Android malicious application classification using clustering. In: Abraham, A., Cherukuri, A.K., Melin, P., Gandhi, N. (eds.) ISDA 2018 2018. AISC, vol. 941, pp. 659–667. Springer, Cham (2020). https://doi.org/10.1007/978-3-030-16660-1_64

28. Rendón, E., Abundez, I., Arizmendi, A., Quiroz, E.M.: Internal versus external cluster validation indexes. Int. J. Comput. Commun. 5(1), 27–34 (2011)

29. Samani, R.: Mcafee mobile threat report: Mobile malware is playing hide and steal. https://www.mcafee.com/enterprise/pt-br/assets/reports/rp-quarterly-threats-nov-2020.pdf. Accessed 25 Mar 2021

30. Schmidt, A.D., et al.: Static analysis of executables for collaborative malware detection on android. In: 2009 IEEE International Conference on Communications (2009)

31. Skovoroda, A., Gamayunov, D.: Review of the mobile malware detection approaches. In: 2015 23rd Euromicro International Conference on Parallel, Distributed, and Network-Based Processing (2015)

32. Team, N.: What is scareware and how can i avoid it? https://uk.norton.com/norton-blog/2015/09/what_is_scarewarean.html

33. Wüchner, T., Ochoa, M., Pretschner, A.: Malware detection with quantitative data flow graphs. In: Proceedings of the 9th ACM Symposium on Information, Computer and Communications Security (2014)

34. Zhou, Y., Jiang, X.: Dissecting android malware: characterization and evolution. In 2012 IEEE symposium on security and privacy (2012)

Experiments for Linking the Complexity of the Business UML Class Diagram to the Quality of the Associated Code

Gaetanino Paolone[1], Martina Marinelli[1], Romolo Paesani[1], and Paolino Di Felice[2]($\boxtimes$) (iD)

[1] Gruppo SI S.c.a.r.l, 64100 Teramo, Italy
{g.paolone,m.marinelli,r.paesani}@softwareindustriale.it
[2] Department of Industrial and Information Engineering and Economics, University of L'Aquila, 67100 L'Aquila, Italy
paolino.difelice@univaq.it

Abstract. A relevant goal of software engineering is to assure the quality of object oriented software from the conceptual modeling phase. UML Class diagrams constitute a key artifact at that stage. Among existing models for iterative and incremental development of software systems, Model Driven Architecture (MDA) has reached a leadership position. MDA enables model-driven software development which treats models as primary development artifacts. The present empirical study answers the following three research questions: (RQ1) are there available in the literature class complexity metrics that can be adopted at the business level? (RQ2) Is it possible to adopt those metrics (if any) to predict the quality of the code returned by xGenerator? The latter is a Java technology platform for the creation of MVC Web applications, which implements the model-driven approach. (RQ3) Is it possible to identify a threshold for the adopted metrics (if any) that might suggest when a business Class diagram should be refactored?

Keywords: UML Class diagram · Complexity metrics · Attribute complexity · Method complexity · Association complexity · Class complexity · Code quality

1 Introduction

At a coarse level of granularity, the lifecycle of the development of software systems includes the phases of: *business analysis, system design*, and *implementation*. Maciaszek [1] defines business analysis as "the activity of *determining* and *specifying* customer requirements". The output of the requirements determination phase is a requirement document, input of the requirements specification phase. The Unified Modeling Language (UML) is the de facto standard for implementing the Object-Oriented (OO) paradigm that today is largely

This research was funded by Software Industriale.

used in the development of complex software systems. UML describes notations for a number of different models that may be produced during analysis and design. UML *class diagrams* and *use-case diagrams* are the two most relevant specification techniques. These diagrams cover, respectively, the *structure* and the *behavior* of the system to be developed, as well as their *interactions*.

Among existing models for iterative and incremental development of software systems, Model Driven Architecture (MDA) has reached a leadership position. "MDA provides guidelines for structuring software specifications that are expressed as models. MDA separates business and application logic from the underlying platform technology."[1] MDA enables model-driven software development which treats models as primary development artifacts.

Figure 1 shows the three phases of the standard development of software systems as well as the alternative path in case of adoption of MDA. The same figure shows how MDA concepts link to the main phases of analysis, design, and implementation. In the right path, at the PSM level takes place a (manual, semi-manual, or automatic) transformation that generates the code.

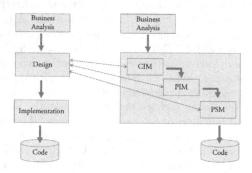

Fig. 1. Overview of two alternative software development lifecycles.

In [2] it is described an MDA-oriented software development process and a tool (`xGenerator`) which accomplishes the transformations between the levels. The latter is a proprietary Java technology platform for the creation of enterprise MVC Web applications. At a high level of abstraction, such a tool acts as a black box that receives as input an UML model and returns the Java code of the Web application.

[3] reports about the experiments carried out in order to investigate the quality of the code outputted by `xGenerator`. Through the `Springlint` tool,[2] it was possible to investigate the quality of the `Controller` and `Model` classes of the `ATMProject` by measuring a set of code metrics. The assessment of the quality of the code was based on the Software Architecture Tailored Thresholds proposed by Aniche and his colleagues in [4]. In such a paper, the authors take into account

---

[1] Sentences taken from: https://www.omg.org/mda/.

[2] http://www.github.com/mauricioaniche/springlint (accessed on June 20, 2020).

five different code metrics measuring coupling, cohesion, and complexity of OO code. Those metrics are able to distinguish the architectural role of the different layers of the MVC pattern. This feature is fundamental since `Controller` classes are different from `Model` classes. Accordingly, the threshold of the five metrics changes with regard to the class's architectural role. The numerical values of the five code metrics in [4] measured for the `ATMProject` were always below the threshold. Generalizing the findings of the experiments, Paolone et al. [3] stated that the classes returned by `xGenerator` are characterised by high cohesion, low coupling, and low complexity.

The present empirical study is motivated by answering the following three research questions:

(RQ1) are there available in the literature class complexity metrics that can be adopted at the business level? The business class diagram documents the internal structure of the business, at the same time it models the structure of the application to be developed; ultimately, it lays the foundation for the implementation work to be done. Therefore, monitoring the class diagram quality is a precondition to produce quality OO software.

(RQ2) Is it possible to adopt those metrics (if any) to predict the quality of the code returned by `xGenerator`? Anticipating the evaluation phase at the business level reduces the cost of the whole validation process. This is true in general and in MDE as well, where the models should be validated before being transformed, preventing the spread of defects introduced into the design.

(RQ3) Is it possible to identify a threshold for the adopted metrics (if any) that might suggest when a business Class diagram should be refactored? The knowledge of such a value is fundamental because it tells us when the refactoring process of the UML Class diagram should be started. UML model refactoring is a highly active area of research (see, for instance, [5]) and tools supporting the automation of the refactoring of UML Class diagrams are already available (e.g., [6,7]).

This study provides an answer to the three research questions listed above. To answer RQ1, we conducted a careful investigation of the state of the art. The decision was to adopt the complexity metrics proposed by Masmali and Badreddin [8,9] to measure the quality of the UML Class diagram of the `ATMProject`.

Concerning RQ2, the answer is affirmative but only for a portion of the whole generated code, namely the `Bean` classes which belong to the Model layer of the MVC pattern. The explanation is the following. First, as it is explained in [2], the `Bean` classes are in a 1-to-1 correspondence with the business classes in the UML Class diagram. Second, since `xGenerator` produces code in a deterministic way, once the quality of the Class diagram is proved, then it is correct to infer that the quality of the code that will be generated will be satisfactory as well. According to our long term experience in the development of enterprise Web applications for the bank sector, the `Bean` classes represent from 25 to 30% of the total number of lines of code of the whole application.

Concerning RQ3, the study allowed to determine empirically the upper bound of the value of the adopted complexity metrics (for the business classes), beyond which we cannot exclude that might be appropriate refactoring the UML Class diagram.

The present study is an integral part of our previous research on automatic generation of code [2,3].

The paper is structured as follows. Section 2 introduces the background underlying the present work. A real-life case study is the focus of Sect. 3, while Sect. 4 presents and discusses the results of the study. Section 5 mentions a survey paper about complexity metrics for UML Class diagrams. Section 6 concludes this work.

2   Background

This section presents a slight variation of the proposal by Masmali and Badreddin [8], followed by the recall of the code metrics in Aniche et al. [4]. Many alternative metrics applicable to the Class diagram are available today. Some of them will be part of our future studies devoted to evaluate the quality of the code returned by xGenerator. The reasons behind the adoption of the code metrics in Aniche et al. [4] are explained in [2].

2.1   UML Class Diagram Complexity Metrics

Masmali and Badreddin [8,9] proposed to assign to a UML class a score equal to the sum of the complexity rate of its components, namely the attributes, methods and associations. The class complexity metrics has been validated from a practical point of view by applying the metrics evaluation framework proposed by Kaner and Bond [10] and from a theoretical point of view by referring to the Weyuker's properties [11]. In the following we recall the fundamentals of the class complexity metrics by Masmali and Badreddin since it plays a relevant role in our study.

Equation 1 computes the class complexity ($Class_{comp}$) as the sum of the complexity of all the attributes ($\sum_{k=1}^{n} Attr_{comp}$) and methods ($\sum_{k=1}^{n} Meth_{comp}$) in the class and the associations ($\sum_{k=1}^{n} Asso_{comp}$) among the classes.

$$Class_{comp} = \sum_{i=1}^{n_1} Attr_{comp} + \sum_{j=1}^{n_2} Meth_{comp} + \sum_{k=1}^{n_3} Asso_{comp} \qquad (1)$$

Attribute Complexity. The complexity of each attribute ($Attr_{comp}$) is computed according to Eq. 2, where $Visibility_{score}$ and $Type_{score}$ are, in sequence, the complexity score of attribute visibility and attribute type. Table 1 and Table 2 show, respectively, the complexity score of the attribute visibility and attribute type. The values in the latter are not identical to those adopted in [8].

$$Attr_{comp} = Visibility_{score} + Type_{score} \qquad (2)$$

Table 1. Complexity score of the attribute visibility.

Visibility	Score
Private	1
Protected, Package	2
Public	3

Table 2. Complexity score of the attribute type. The names are as those of Java since xGenerator returns Java code.

Type	Score
byte, short, int, float, long, double, char, boolean	1
String	2
Array, Date, Map	3
Object, List	4

Method Complexity. The complexity of each method ($Meth_{comp}$) is computed as shown in Eq. 3, where $Visibility_{score}$ is the complexity score of the method visibility, $ReturnType_{score}$ is the complexity score of the method return type, and the third term denotes the complexity score of its parameters. The complexity score of the method visibility is identical to the complexity score of the attribute visibility (see Table 1), while the complexity score of the method's return type and parameters are identical to that of Table 2.

We count constructor methods as ordinary methods. McQuillan and Power [12] have pointed out that there is no uniformity on this point among the available proposals about metrics for OO software. This uncertainty becomes an obstacle when comparing different metrics.

$$Meth_{comp} = Visibility_{score} + ReturnType_{score} + \sum_{l=1}^{n_4}(Parameter_{score}) \quad (3)$$

Association Complexity. UML classes may be interrelated to each other through: Association, Directed Association, Reflexive Association, Aggregation, Composition, Inheritance, and Realization. In [8], authors took into account Association and Directed Association. In our study we focus on Association and Aggregation, the only two relationships that are used in the real-life case study of the next section. These two relationships are equivalent from the point of view of the complexity. We use the generic term of association complexity to refer to both. The association complexity ($Asso_{comp}$) of a class is computed as shown in Eq. 4, where $Link_{score}$ denotes the complexity score of the association link in the class diagram. The idea behind the equation is simple: the higher is the number of links of a class the higher is its complexity.

$$Asso_{comp} = \sum_{p=1}^{n_5} Link_{score} \tag{4}$$

Table 3 collects the complexity score of the associations. Our values are not identical to those in [8].

Table 3. Complexity score of the associations.

Multiplicity	Score
1..1	1
0..1 or 0..* or 1..*	2
..	3

2.2   Code Metrics

Code metric analysis is the common method for assessing the quality of a software system. In the metrics-based methods the false-positives cannot be eliminated until the *context* is taken into account, and this because one set of thresholds do not hold good necessarily in another context. Until few years ago, software tools and published proposals did not take the system architecture into account. Consequently, *all* classes of OO applications were treated as they were equal to each other and, hence, assessed in the same way, regardless of their specific *architectural role*. This approach is not satisfactory if applied to a MVC Web application, where Controller classes are quite different from Model classes simply because they play very different roles: the former are responsible for coordinating the flow between the View and the Model layers, while the latter implements business concepts.

Adding context to code metrics is a recent research topic. In Aniche et al. [4], the authors adapt the threshold of metrics to the "class's architectural role" with regard to the (Spring) MVC pattern; where the *architectural role* is defined as the particular role that a class plays in a given system architecture. Specifically, Aniche et al. focus on the server-side code, namely on the Controller and Model layers. [4] relies on the Chidamber and Kemerer metrics suite [13], as it covers different aspects of OO programming, such as *coupling*, *cohesion*, and *complexity*.

Table 4 shows the class level metrics taken from [13], while Table 5 shows the thresholds found by Aniche et al. [4] for the classes of the five architectural role belonging to the Controller and Model layers. Each metric varies from 0 or 1 to infinite. The triple $[v_1, v_2, v_3]$ denotes, in order, the *low*, *high* and *very high* thresholds corresponding to *moderate/high/very high risk*. Classes in which metric values range in the 70%–80% percentiles have "*moderate risk*", while from 80%–90% the risk is "*high*", and "*very high*" between 90%–100%.

Table 4. The reference class level metrics in [4].

Metric	Acronym	Meaning
Coupling between Object Classes	CBO	The number of classes a class depends upon. It counts classes used from both external libraries as well as classes from the project
Lack of COhesion in Methods	LCOM	The lack of cohesion in a class counts the number of intersections between methods and attributes. The higher the number, the less cohesive is the class
Number Of Methods	NOM	Number of methods in a class
Response For a Class	RFC	It is the count of all method invocations that happen in a class
Weighted Methods per Class	WMC	Sum of McCabe's cyclomatic complexity for each method in the class

Table 5. The thresholds for the reference class level metrics of Table 4.

Architectural role	CBO	LCOM	NOM	RFC	WMC
Controller	[26, 29, 34]	[33, 95, 435]	[16, 22, 37]	[62, 78, 110]	[57, 83, 130]
Repository	[15, 21, 31]	[36, 106, 351]	[19, 26, 41]	[50, 76, 123]	[60, 104, 212]
Service	[27, 34, 47]	[133, 271, 622]	[23, 32, 45]	[88, 123, 190]	[97, 146, 229]
Entity	[16, 20, 25]	[440, 727, 1844]	[33, 42, 64]	[8, 12, 25]	[49, 61, 88]
Component	[20, 25, 36]	[50, 123, 433]	[15, 22, 35]	[56, 81, 132]	[52, 79, 125]

The MVC pattern implemented by xGenerator is not identical to Spring MVC. The consequence is that the structure, and hence the content, of the three layers of the code are not identical in the two solutions. In [3] it has been shown that the Bean classes (part of the MVC Model layer implemented by xGenerator) match the Entity classes (part of the Spring MVC Model layer). Within this study the Bean classes are the only classes we refer to because they are in a 1-to-1 correspondence with the business classes in the UML Class diagram.

3   The Case Study

The case study is the ATMProject presented in [3] and recalled below, briefly. It concerns the development of a Web application. The example was carried out by following the steps of Software Development Process proposed in [2] and by making use of xGenerator.

The basic *business vocabulary* (i.e., the general concepts) of our example is composed of the following nouns and noun phrases: *Customer, ATM, Bank Account, PIN, Bancomat code, Check Balance, Amount, Deposit Money, Transfer Money, Transaction, Currency, Maintenance,* and *Repair*.

The *business rules* to be implemented were the following: (a) a customer can hold many bank accounts, while an account is owned by a single customer. (b) An account may be linked to many cards. (c) A customer must be able to make a cash withdrawal from any account linked to the card entered into the ATM;

moreover a customer must be able to make a deposit to any account linked to the card entered into the ATM. (d) Many transactions can be made on an account. (e) A transaction refers to one currency and takes place at a physical ATM. (f) Many maintenances and repairs may happen on ATMs.

Figure 2 shows the diagram of the Business Objects of the ATM Subsystem at the CIM level. To be able to apply the method proposed in [8], it was necessary to enrich such a diagram with the data type of attributes, the list of methods as well as the input/output parameters and return values of each method. Notice that in the MDA paradigm, the information about data types, input/output parameters and the return values are part of the PIM and PSM activities.

4   Results and Discussion

4.1   Measurements at the Code Level

Figure 3 shows the names of the classes that compose the Bean package of the Model layer of the ATMProject. As already said, the attention is restricted to the Bean classes because they correspond to the classes of the UML Class diagram. Table 6 collects the numerical values of the metrics of Table 4 returned by Springlint for the Bean classes. The empty boxes in these tables denote the zero value. The values of metrics NOM and WMC are identical because the McCabe's cyclomatic complexity equals 1 for each of the methods belonging to the Bean classes. The output of Springlint is an HTML file. We named it springlint-result ATMProject.html and made available at URL www.softwareindustriale.it/ATMProject.html ATMProject. The results are excellent. In fact, the values of the five metrics for the Bean classes are all below the smallest threshold of Table 5 (row "Entity"). Those values confirm that the classes returned by xGenerator are characterised by high cohesion and low coupling.

Table 6. The values of the metrics for the Bean classes.

	CBO	LCOM	NOM	RFC	WMC
ATM	8	56	12	1	12
BankAccount	12	120	17		17
Currency	7	11	6	1	6
Customer	9	91	15		15
Maintenance	12	45	11		11
MaintenanceCategory	7	7	6		6
Repair	11	28	9		9
Transaction	12	120	17		17

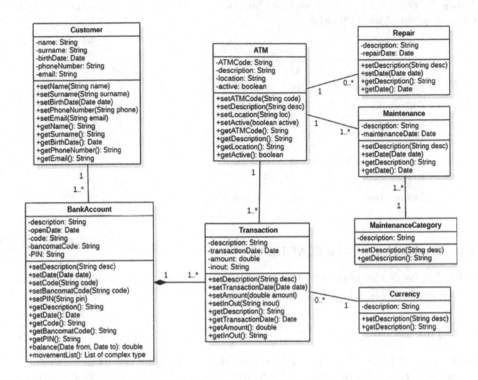

Fig. 2. The enhanced diagram of the Business Objects of the ATM Subsystem.

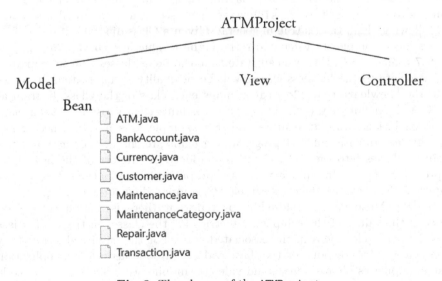

Fig. 3. The classes of the `ATMProject`

Table 7. The values of the complexity metrics for the business classes in the UML Class diagram of the `ATMProject`.

	$Attr_{comp}$	$Meth_{comp}$	$Link_{comp}$	$Class_{comp}$
ATM	11	38	6	55
BankAccount	16	69	3	88
Currency	3	10	2	15
Customer	16	52	2	70
Maintenance	7	22	2	31
MaintenanceCategory	3	10	2	15
Repair	7	22	1	30
Transaction	12	40	3	55

4.2 Evaluation at the CIM Level

Table 7 collects the numerical values of the complexity metrics of the business classes in the UML Class diagram of Fig. 2.

4.3 Discussion

As it is said in Sect. 4.1, the values of the five metrics for the `Bean` classes are below the smallest threshold of Table 5 (row "Entity"). This means that *all* these classes of the `ATMProject` are of good quality. This awareness implies that we can propagate backward the goodness of the code and state that the quality of the business classes in the UML Class diagram (from which the `Bean` classes come from) is good too. Exiting the `ATMProject` case study, now we are able to make the following claim that is true in general. Given a Class diagram, if none of its classes has a complexity value above the highest value in column $Class_{comp}$ of Table 7 (namely, 88), then we can state that the `Bean` classes that `xGenerator` can associate to the business classes will be of quality. This statement comes from the knowledge that `xGenerator` maps each class of the Class diagram at the CIM level uniquely (i.e., through a deterministic process) in a `Bean` class. This conclusion is not true in general, for example it is not true necessarily if programmers with different programming skills are charged for mapping the business classes into the code. Concrete evidence in support of the last claim is provided below. In summary, we are proposing the adoption of 88 as the empirical threshold for the class complexity computed according to Eq. 1.

In [15], Masmali and Badreddin proved that there is a high correlation between the values of the complexity metrics for the classes in the UML Class diagram and code smells in the associated code. For the empirical study, they investigated nine versions of the Java code of MobileMedia[3]: an application that manipulates photos, music, and video on mobile devices. Overall, the code

---

[3] https://github.com/julioserafim/MobileMedia.

(about 28,000 LOC) is composed of 252 classes and 997 methods. The third row of Table 8 (taken from [15]) shows the number of smelly classes detected by JDeodorant [16]. JDeodorant is an open source Eclipse plugin for Java, actively developed and well maintained[4]. The tool supports five code smells, namely: feature envy, type/state checking, long method, god class and duplicated code. JDeodorant has been widely investigated in the literature. For example, in [17] Paiva et al. conclude as follows a comparative study involving four smell tools: "JDeodorant identified most of the correct entities." An indirect evidence of the merits of such a tool is represented by the fact that it is used in several important studies (see, for instance, [18,19]). The examination of Table 8 shows that the percentage of smelly classes passes from 20% to 100% in correspondence with the values of the class complexity metrics that pass from the range 1–50 to 150–200. This data is a direct evidence that in absence of a deterministic transformation process of the business classes into Bean classes, the probability that the final code suffers of smells is very high.

Table 8. Correlation between the values of the complexity metrics for the classes in the UML Class diagram and code smells in the associated code of MobileMedia in absence of a deterministic mapping process between the two levels (source: [15]).

Class complexity	1–50	51–100	101–150	151–200
Number of classes	180	43	21	8
Number of smells	36 (20%)	15 (35%)	13 (62%)	8 (100%)

Table 9. The values of the NOM metrics applied to the business classes in the UML Class diagram of the ATMProject.

	NOM
ATM	8
BankAccount	12
Currency	2
Customer	10
Maintenance	4
MaintenanceCategory	2
Repair	4
Transaction	8

Overall, from the above considerations it follows that the three research questions in Sect. 1 have been answered.

---

[4] https://github.com/tsantalis/JDeodorant.

Fig. 4. The output of JDeodorant for the BankAccount class.

We conclude the discussion by adding three more comments, as indirect further evidences of the quality of the business classes of the ATMProject returned by xGenerator:

- of the five metrics for measuring the code quality (i.e., CBO, LCOM, NOM, RFC, and WMC), NOM is applicable also to the business classes of the Class diagram. Table 9 reports the NOM value for the eight business classes of the ATMProject. The values confirm the expectation, namely that, at the business class level, they are always lower than the values of the corresponding metrics applied to the code (see Table 6) and, moreover, their value is below the threshold of the code metrics (i.e., 33 of Table 5, row "Entity"). It follows that, if the value of NOM for the business classes is sufficiently below the 33 value, then this is a further evidence for predicting that the quality of the Bean classes returned by xGenerator will be good.
- We analysed the code of the Bean classes of the ATMProject by means of JDeodorant. No smells were detected by the tool (Fig. 4).
- In [3] the authors carried out experiments to measure the value of the six MVC smells part of the catalog in [14] for the Bean classes of the ATMProject. The values were all below the thresholds of the six metrics, which confirm that those classes are not smelly.

Threats to Validity

- *Internal validity* concerns factors that may impact the measurements' outcome. The only one factor that could negatively affect the internal validity of the results is about errors in the computation of the class complexity in absence of a supporting tool. In our case this eventuality did not occur given the small number of classes in the UML Class diagram and because we repeated the computation many times.

- *External validity* refers to the relevance of the results and their generalizability. We conducted the experiments on a Java software project returned by a proprietary tool (xGenerator); consequently, we cannot claim generalizability of our approach.
- *Conclusion validity* refers to threats that can impact the reliability of our conclusions. At the moment, we are not able to exclude that projects bigger than ATMProject might give rise to a bigger value of the upper bound of the complexity metrics of the business classes (namely, 88 in Table 7). To overcome this issue, more experiments are needed. The future experiments have to concern Class diagrams with almost all the relationships among classes. This constraint, in turn, implies that the complexity model we referred to in the present study (taken from [8]) has to be extended or, alternatively, that different methods, among those already appeared (see next section), are taken into consideration.

5   Metrics for UML Class Diagrams

Most studies aiming at predicting the quality of OO software systems (e.g., class fault proneness, code readability, maintainability, implementation time and cost of the software system under development, ...) use metrics extracted from the source code. A much smaller number of proposals use design metrics collected from design artifacts. In this paper we were interested in the adoption of "early" metrics for the prediction of the quality of the code outputted by the proprietary tool xGenerator by measuring the quality of the UML Class diagram. We adopted a slight variation of the complexity metrics described in [8].

Due to the reached page limits, this section first focuses on a published survey paper about metrics for UML Class diagrams; then it mentions two specific proposals which adopt two completely different approaches. We plan to adopt both in future studies.

[20] is the first survey about metrics for UML Class diagrams. Genero et al. compared twelve proposals along five dimensions: (a) the definition of the metrics, (b) the goals pursued by the metrics, (c) the theoretical validation of the metrics (if any), (d) the empirical validation of the metrics (if any), and (e) the availability of automatic tools supporting the metric calculation. The findings of the survey can be summarized as follows:

- the research on measures for UML Class diagrams at the business level is scarce and is not yet consolidated (both theoretical and empirical validation is needed);
- most metrics are related to classes while little emphasis has been put on the Class diagrams as a whole;
- limited efforts have been put on measures about the relationships among classes;

– last but not least, the identification of robust thresholds for the proposed metrics is still an open issue.

Marchesi [21] proposed one of the earliest sets of metrics for computing the complexity of UML Class diagrams. The suite is applicable to a single class, a package or an overall software system. The metrics for a single class concern: weighted number of responsibilities; weighted number of dependencies; depth of inheritance tree; number of immediate subclasses; and number of distinct classes depending on the class.

In [22] a Class diagram is modeled as a graph. The nodes of the graph denote classes while the relations are modeled as edges between the nodes. A weight is associate to each edge according to the type of relations it corresponds to in the class diagram. Ten different kinds of relations in UML Class diagrams are considered. The *entropy distance* is used to associate a complexity measure to the class diagram.

6   Conclusions

This study put in correspondence the complexity metrics of the business classes of the Class diagram of a real-life project (`ATMProject`) with the complexity metrics of the associated code. A tool (`xGenerator`) generated the code mapping it, in a deterministic way, from the business classes. The determinism of such a mapping implied that from the quality of the code we inferred the quality of the classes. In this way it was possible to set an upper bound for the value of the class complexity metrics. Our proposal is to keep this value as the threshold for the complexity metrics for the business classes of future projects. Staying below such a threshold, we can say that any code outputted by `xGenerator` will be of good quality.

It has been pointed out that this claim is true only for a portion of the whole generated code, namely the `Bean` classes which belong to the Model layer of the MVC pattern. According to our long term experience in the development of enterprise Web applications for the bank sector, the `Bean` classes represent from 25 to 30%percent of the total number of lines of code of the whole application.

The joint usage of the class complexity measurement method discussed in the paper and `xGenerator` makes it unnecessary to carry out the measurements of the quality of the generated code by means of any of the available suites (e.g., [13]).

It is worth emphasizing that the measurement of the complexity of the business classes should be mandatory if the software development process is manual. In fact, in such a scenario discovering that there are business classes with a too high complexity value imposes refactoring actions on the Class diagram before proceeding in the development, because very likely the code that will be produced will result smelly.

The next experiments we are going to perform will link the complexity of the code outputted by `xGenerator` to the complexity of the business classes computed by means of the suite proposed by Marchesi [21] and the graph-based approach in [22].

References

1. Maciaszek, L.A.: Requirements Analysis and System Design, 3rd edn. Addison Wesley, Harlow (2007)
2. Paolone, G., Marinelli, M., Paesani, R., Di Felice, P.: Automatic code generation of MVC web applications. Computers **9**(3), 56 (2020). https://doi.org/10.3390/computers9030056
3. Paolone, G., Paesani, R., Marinelli, M., Di Felice, P.: Empirical assessment of the quality of MVC web applications returned by xGenerator. Computers **10**(2), 20 (2021). https://doi.org/10.3390/computers10020020
4. Aniche, M., Treude, C., Zaidman, A., van Deursen, A., Gerosa, M.A.: SATT: tailoring code metric thresholds for different software architectures. In: IEEE 16th International Working Conference on Source Code Analysis and Manipulation (SCAM), Raleigh, NC, pp. 41–50 (2016). https://doi.org/10.1109/SCAM.2016.19
5. Misbhauddin, M., Alshayeb, M.: UML model refactoring: a systematic literature review. Empir. Softw. Eng. **20**, 206–251 (2015). https://doi.org/10.1007/s10664-013-9283-7
6. Nikulchev, E., Deryugina, O.: Model and criteria for the automated refactoring of the UML class diagrams. Int. J. Adv. Comput. Sci. Appl. (IJACSA) **7**(12), 76–79 (2016)
7. Deryugina, O., Nikulchev, E., Ryadchikov, I., Sechenev, S., Shmalko, E.: Analysis of the AnyWalker software architecture using the UML refactoring tool. Procedia Comput. Sci. **150**, 743–750 (2019). 13th International Symposium "Intelligent Systems" (INTELS'18)
8. Masmali, O., Badreddin, O.: Code complexity metrics derived from software design: a framework and theoretical evaluation. In: Proceedings of the Future Technologies Conference, 5–6 November, Vancouver, Canada (2020)
9. Masmali, O., Badreddin, O.: Comprehensive model-driven complexity metrics for software systems. In: The 20th IEEE International Conference on Software Quality, Reliability and Security Companion (QRS-C 2020) Macau, China, 11–14 Dic (2020). https://doi.org/10.1109/QRS-C51114.2020.00115
10. Kaner, C., Bond, W.P.: Software engineering metrics: what do they measure and how do we know?. In: Proceedings of the 10th International Software Metrics Symposium, 11–17 September, Chicago, IL, USA, pp. 1–12 (2004)
11. Weyuker, E.: Evaluating software complexity measures. IEEE Trans. Softw. Eng. **14**, 1357–1365 (1988)
12. McQuillan, J.A., Power, J.F.: On the application of software metrics to UML models. In: Kühne, T. (ed.) MODELS 2006. LNCS, vol. 4364, pp. 217–226. Springer, Heidelberg (2007). https://doi.org/10.1007/978-3-540-69489-2_27
13. Chidamber, S.R., Kemerer, C.F.: A metrics suite for object oriented design. IEEE Trans. Softw. Eng. **20**(6), 476–493 (1994). https://doi.org/10.1109/32.295895
14. Aniche, A., Bavota, G., Treude, C., Gerosa, M.A., van Deursen, A.: Code smells for model-view-controller architectures. Empir. Softw. Eng. **23**, 2121–2157 (2018). https://doi.org/10.1007/s10664-017-9540-2

15. Masmali, O., Badreddin, O., Khandoker, R.: Metrics to measure code complexity based on software design: practical evaluation. In: Arai, K. (ed.) FICC 2021. AISC, vol. 1364, pp. 142–157. Springer, Cham (2021). https://doi.org/10.1007/978-3-030-73103-8_9

16. Tsantalis, N., Chaikalis, T., Chatzigeorgiou, A.: JDeodorant: identification and removal of type-checking bad smells. In: 12th European Conference on Software Maintenance and Reengineering, Athens, Greece, pp. 329–331 (2008). https://doi.org/10.1109/CSMR.2008.4493342

17. Paiva, T., Damasceno, A., Figueiredo, E., Sant Anna, C.: On the evaluation of code smells and detection tools. J. Softw. Eng. Res. Dev. **5**, 7 (2017). https://doi.org/10.1186/s40411-017-0041-1

18. Nyamawe, A.S., Liu, H., Niu, Z., Wang, W., Niu, N.: Recommending refactoring solutions based on traceability and code metrics. IEEE Access **6**, 49460–49475 (2018). https://doi.org/10.1109/ACCESS.2018.2868990

19. Palomba, F., Panichella, A., Zaidman, A., Oliveto, R., De Lucia, A.: The scent of a smell: an extensive comparison between textual and structural smells. IEEE Trans. Softw. Eng. **44**(10), 977–1000 (2018). https://doi.org/10.1109/TSE.2017.2752171

20. Genero, M., Piattini, M., Calero, C.: A survey of metrics for UML class diagrams. J. Object Technol. **4**(9), 59–92 (2005). http://www.jot.fm/issues/issue_2005_11/article1

21. Marchesi, M.: OOA metrics for the unified modeling language. In: Proceedings of the Second Euromicro Conference on Software Maintenance and Reengineering, Florence, Italy (March 8–11, 1998), pp. 67–73 (1998). https://doi.org/10.1109/CSMR.1998.665739

22. Xu, B., Kang, D., Lu, J.: A structural complexity measure for UML class diagrams. In: Bubak, M., van Albada, G.D., Sloot, P.M.A., Dongarra, J. (eds.) ICCS 2004, Part I. LNCS, vol. 3036, pp. 421–424. Springer, Heidelberg (2004). https://doi.org/10.1007/978-3-540-24685-5_56

Probability-to-Goal and Expected Cost Trade-Off in Stochastic Shortest Path

Isabella Kuo[ID] and Valdinei Freire[✉][ID]

Escola de Artes, Ciências e Humanidades - Universidade de São Paulo,
São Paulo, Brazil
{15483114.ik,valdinei.freire}@usp.br

Abstract. Markov Decision Processes (MDPs) model problems where a decision-maker makes sequential decisions and the effect of decisions is probabilistic. A particular formulation of MDPs is the Shortest Stochastic Path (SSP), in which the agent seeks to accomplish a goal while reducing the cost of the path to it. Literature introduces some optimality criteria; most of them consider a priority of maximizing probability to accomplish the goal while minimizing some cost measure; such criteria allow a unique trade-off between probability-to-goal and path cost for a decision-maker. Here, we present algorithms to make a trade-off between probability-to-goal and expected cost; based on the Minimum Cost given Maximum Probability (MCMP) criterion, we propose to treat such a trade-off under three different methods: (*i*) additional constraints for probability-to-goal or expected cost; (*ii*) a Pareto's optimality by finding non-dominated policies; and (*iii*) an efficient preference elicitation process based on non-dominated policies. We report experiments on a toy problem, where probability-to-goal and expected cost trade-off can be observed.

Keywords: Shortest stochastic path · Markov Decision Process · Preference elicitation · Non-dominated policies · Goal and cost trade-off

1 Introduction

Markov Decision Process (MDP) models problems where an agent makes sequential decisions and the results of these actions are probabilistic [9]. The agent's objective is to choose a sequence of actions so that the system acts optimally according to previously defined criteria. Such a sequence of actions follows a policy, which maps each observed state into an action.

A special case of MDP is the Stochastic Shortest Path (SSP) [2], where a terminal goal state is considered. The agent is supposed to reach the goal state while minimizing the incurred cost. Two different theoretical scenarios appear whether there exists or there does not exist a proper policy. A proper policy guarantees that the goal state is reached with probability 1.

Usually, when there exists a proper policy, the objective of the agent is to minimize the expected accumulated cost. If a proper policy does not exist, the

© Springer Nature Switzerland AG 2021
O. Gervasi et al. (Eds.): ICCSA 2021, LNCS 12951, pp. 111–125, 2021.
https://doi.org/10.1007/978-3-030-86970-0_9

agent may get trapped in a dead-end; in this case, infinity cost may be incurred and the problem of minimizing the expected accumulated cost is not well-posed. There exist some approaches in the literature to overcome SSP problems with Unavoidable Dead-End.

The iSSPUDE (SSP with Unavoidable Dead-End) criterion [8] considers two opposite objectives: probability-to-goal and cost-to-goal; and gives priority for the first. Then, an optimal policy maximizes the chance of reaching the goal state, while minimizing the average cost incurred in traces that reaches the goal. Because only traces that reach the goal are considered, the objective cost-to-goal is finite and iSSPUDE is well-posed. Similarly, the MCMP (Minimum Cost given Maximum Probability) criterion [15] considers only policies that maximize probability-to-goal. An additional action of giving up is considered; if a dead-end state is reached, the agent gives up, guaranteeing a finite cost in every trace. Then, different from iSSPUDE, MCMP considers costs of traces that do not reach a goal state and pays the cost incurred until the agent reaches a dead-end.

As pointed out in [6], prioritizing probability-to-goal may incur in undesirable decisions such as trading off a huge cost increase for a tiny probability-to-goal increase. The GUBS criterion [6,7] proposes a semantic of goal prioritization based on the expected utility theory and shows that a more reasonable trade-off can be obtained under a rational approach. Other less refined criteria consist of penalizing traces with dead-ends, for example, by assigning a final penalty cost when giving up [8] or making use of discount factor [13,14].

Although criteria that allow a trade-off between cost-to-goal and probability-to-goal is necessary to avoid undesirable decision in real problems, they require some preference elicitation process to tune parameters in any formalization [1, 3]. Because the GUBS (Goal with UtilityBased Semantics) problem is based on Expected Utility theory, it may be more suitable to preference elicitation processes independent of the SSP dynamics and a trade-off specification can be obtained beforehand. However, setting appropriately the amount of penalty to giving up depends on the SSP dynamics, and requires an on-line preference elicitation based on the SSP problem at hand.

Based on the algorithms developed for MCMP problems and the final penalty cost to giving up, we contribute with an algorithm to obtained non-dominated policies regarding the trade-off of probability-to-goal and cost-to-goal. Moreover, we show how to use such an algorithm to support decisions by eliciting the preference of a decision-maker.

2   Background

2.1   Stochastic Shortest Path

In this article, we consider a special case of Markov Decision Process, the Stochastic Shortest Path (SSP) problem [2]. A SSP is defined by a tuple $\langle S, A, P, C, s_0, G \rangle$ where:

- $s \in S$ are possible states;
- $a \in A$ are possible actions;

- $\mathcal{P} : \mathcal{S} \times \mathcal{A} \times \mathcal{S} \to [0, 1]$ is the transition function;
- $\mathcal{C} : \mathcal{S} \times \mathcal{A} \to R^+$ is the cost function;
- $s_0 \in \mathcal{S}$ is the initial state; and
- $\mathcal{G}$ is the set of goal states.

A SSP defines a process of an agent interacting with an environment and at all time step t: (i) the agent observes a state s_t, (ii) the agent chooses an action a_t, (iii) the agent pays a cost c_t; and (iv) the process moves to a new state s_{t+1}. The process ends when a goal state $s \in \mathcal{G}$ are reached. Transitions and costs present Markov's property, i.e., they both depend only on the current state s_t and chosen action a_t.

The solution for a SSP consists of policies that describe which actions should be taken in each situation. Here we consider deterministic stationary policies $\pi : \mathcal{S} \to \mathcal{A}$ that maps each state for a unique action. A policy is proper if it guarantees that the goal is reached with probability 1 when the process starts in the state s_0.

Classically, only problems with proper policies are considered. Given a proper policy, the expected accumulated cost to any state $s \in \mathcal{S}$ is defined as:

$$V^\pi(s) = \mathrm{E} \left[\sum_{t=0}^{\infty} c_t \middle| s_0 = s, \pi \right] \tag{1}$$

and the optimal policy is defined by $\pi^*(s) = \arg\min_{\pi \in \Pi} V^\pi(s)$, where Π is the set of deterministic stationary policies.

When a proper policy does not exist, dead-ends occur, which are states from which a goal state cannot be reached. In the presence of dead-ends, the expected accumulated cost is infinite and $V^\pi(s)$ is uninformative. Therefore, criteria different from the classic one should be considered.

An alternative criterion consists in maximizing the probability of reaching a goal state in $\mathcal{G}$ from the initial state s_0 (probability-to-goal). While probability-to-goal can differentiate policies, such a criterion is not well-posed, since policies with completely different cost profile can present the same probability-to-goal. MCMP and iSSPUDE consider probability-to-goal as a first criterion and present some cost-dependent informative criterion as a tiebreaker [8,15].

2.2   Minimum Cost Given Maximum Probability (MCMP) Criterion

The MCMP criterion [15] considers policies that maximize probability-to-goal, while minimizing expected cost. To avoid infinite expected costs, the agent is allowed to giving up the process when he/she finds himself/herself in a dead-end.

Formalization. When following a policy, the process described by a SSP generates a random state-trace T, i.e., a sequence of states. Traces can be finite, when the agent reached a goal state, or infinite, otherwise. Consider the following concepts:

- T^i: the i-th state of the trace T,
- $\mathcal{T}_{\pi,s}$: the set of all traces obtained from s following a policy π, and
- $\mathcal{T}^{\mathcal{G}}_{\pi,s} \subseteq \mathcal{T}_{\pi,s}$: the subset of traces that reach the goal.

Note that, if a trace $T \in \mathcal{T}_{\pi,s}$ is finite, then T reaches a goal state. The occurrence probability for any finite trace $T \in \mathcal{T}^{\mathcal{G}}_{\pi,s}$ conditioned on $s_0 = s$ and policy π is given by:

$$P_{T,\pi} = \prod_{i=0}^{|T|-1} \mathcal{P}(T^{i+1}|T^i, \pi(T^i)). \tag{2}$$

The probability that a policy π will reach a goal state in $\mathcal{G}$ when starting in state $s \in \mathcal{S}$ is:

$$P^{\mathcal{G}}_{\pi,s} = \sum_{T \in \mathcal{T}^{\mathcal{G}}_{\pi,s}} P_{T,\pi},$$

and the accumulated cost of a finite trace is

$$C_{T,\pi} = \sum_{i=0}^{|T|-1} \mathcal{C}(T^i, \pi(T^i)).$$

To deal with infinite traces, the MCMP criterion allows a give up action, that is, when it is faced with a dead-end, it can give up and pay the cost spent up to that moment. To ensure that the probability of reaching the goal is maximized, the give up action is used only when a dead-end is found.

Consider the following auxiliary function that makes all traces of a policy finite:

$$\psi(sT) = \begin{cases} s & \text{, if } |T| = 0 \text{ or } P^{\mathcal{G}}_{\pi,s} = 0 \\ s\psi(T) & \text{, otherwise} \end{cases}. \tag{3}$$

If T is an infinite trace, then $\psi(T) = T'$ is a truncated version of T that ends up in the first dead-end in trace T when following π. If T is a finite trace, then $\psi(T) = T$ and T ends up in a goal state.

Consider $p_{max} = \max_{\pi \in \Pi} P^{\mathcal{G}}_{\pi,s_0}$, that is, the maximum probability of reaching the goal from the initial state s_0, an optimal policy under the MCMP criterion is defined by:

$$\pi^* = \arg \min_{\{\pi | P^{\mathcal{G}}_{\pi,s_0} = p_{max}\}} \sum_{T \in \mathcal{T}_{\pi,s_0}} C_{\psi(T),\pi} P_{T,\pi}. \tag{4}$$

Here, the expected cost is calculated over truncate versions of traces occurred when starting in the initial state s_0 and following policy π.

Linear Programming Solution. The MCMP criterion can be solved using two linear programming problems, LP1 and LP2 [15]. LP2 finds the maximum probability p_{max} to reach a goal state. LP1 finds the minimum expected cost policy that reaches a goal state with probability p_{max}.

Both problems consider variables $x_{s,a}$ for all $s \in \mathcal{S}, a \in \mathcal{A}$ that indicates an expected accumulated occurrence frequency for every state-action pair. The SSP dynamics restricts the solutions by specifying an $in(s)$ and $out(s)$ flow model for every state s. Every state but initial state and goal state must equalize $in(s)$ and $out(s)$. Initial state s_0 presents a unity in-out difference, while goal states in $\mathcal{G}$ has no output. For every state $s \in \mathcal{S}$, we have:

$$in(s) = \sum_{s' \in \mathcal{S}, a \in \mathcal{A}} x_{s',a} P(s|s', a) \quad \text{and} \quad out(s) = \sum_{a \in \mathcal{A}_{(s)}} x_{s,a}. \qquad (5)$$

The LP2 finds the p_{max} value by maximizing the accumulated input for all the goal states in $\mathcal{G}$ and by not considering at all accumulated cost. LP2 is described as follows:

$$\begin{aligned}
\text{LP2} \quad &\max_{x,a} \sum_{s_g \in \mathcal{G}} in(s_g) \\
&\text{s.t. } x_{s,a} \geq 0 && \forall s \in \mathcal{S}, a \in \mathcal{A}(s) \\
&\quad out(s) - in(s) = 0 && \forall s \in \mathcal{S} \setminus (\mathcal{G} \cup s_0) \\
&\quad out(s_0) - in(s_0) = 1
\end{aligned} \qquad (6)$$

The LP1 finds the minimum cost given maximum probability p_{max}. LP1 differs from LP2 regarding four aspects:

1. the flow that enters an state is allowed to be bigger than the flow that runs out, this characterize the give up action;
2. previous aspect is also applied to initial state s_0, i.e., give up action can be applied since the initial state;
3. the flow that enters goal state must be p_{max} and guarantees that the optimal policy maximize probability-to-goal; and
4. the objective minimizes the expected cost.

Implicitly, previous conditions guarantee Eq. 4. The linear programming LP1 is described as follows:

$$\begin{aligned}
\text{LP1} \quad &\min_{x_{s,a}} \sum_{s \in \mathcal{S}, a \in \mathcal{A}} x_{s,a} C(s, a) \\
&\text{s.t. } x_{s,a} \geq 0 && \forall s \in \mathcal{S}, a \in \mathcal{A}(s) \\
&\quad out(s) - in(s) \leq 0 && \forall s \in \mathcal{S} \setminus (\mathcal{G} \cup s_0). \\
&\quad out(s_0) - in(s_0) \leq 1 \\
&\quad \sum_{s_g \in \mathcal{G}} in(s_g) = p_{max}
\end{aligned} \qquad (7)$$

The optimal policy π^* to the MCMP criterion can be obtained from expected accumulated occurrence frequency $x_{s,a}$ in LP1 by:

$$\pi^*(s) = \arg \max_{a \in \mathcal{A}} x_{s,a} \quad \forall s \in \mathcal{S},$$

and π^* incurs in expected cost $c_{min} = \sum_{s \in \mathcal{S}, a \in \mathcal{A}} x_{s,a} C(s, a)$.

3   Probability-to-Goal and Average Cost Trade-Off

The MCMP and iSSPUDE criteria establish a preference system that prioritizes probability-to-goal, disregarding the associated cost. The great advantage of

MCMP and iSSPUDE is that both criteria have no parameter to be tuned. However, as highlighted by [6], a small reduction in the probability to reach the goal can bring a considerable reduction in the expected cost and decision-makers should be able to choose a better trade-off between probability-to-goal and expected cost. In the next subsections, we discuss and propose different strategy to obtain better trade-offs between Probability-to-Goal and Average Cost.

3.1 Trade-Offs Under Additional Constraint

Trade-offs can be easily founded by constraining one of the variable: probability-to-goal or expected cost. First, make use of LP2 and LP1 to find p_{max} and c_{min}. Then, ask for a decision-maker an additional equality constrain on probability-to-goal or expected cost with range on $[0, p_{max}]$ and $[0, c_{min}]$, respectively.

If a constraint in probability-to-goal $p \leq p_{max}$ is desired, minimize the expected cost under the constraint that $\sum_{s_g \in \mathcal{G}} in(s_g) = p$, i.e.,

$$
\begin{aligned}
\text{LP3} \quad &\min_{x_{s,a}} \ \sum_{s \in \mathcal{S}, a \in \mathcal{A}} x_{s,a} \mathcal{C}(s,a) \\
&\text{s.t. } x_{s,a} \geq 0 && \forall s \in \mathcal{S}, a \in \mathcal{A}(s) \\
&\quad out(s) - in(s) \leq 0 && \forall s \in \mathcal{S} \backslash (\mathcal{G} \cup s_0), \quad (8) \\
&\quad out(s_0) - in(s_0) \leq 1 \\
&\quad \sum_{s_g \in \mathcal{G}} in(s_g) = p
\end{aligned}
$$

and return expected cost $c = \sum_{s \in \mathcal{S}, a \in \mathcal{A}} x_{s,a} \mathcal{C}(s,a)$.

If a constraint in expected cost $c \leq c_{min}$ is desired, maximize the probability-to-goal under the constraint that $\sum_{s \in \mathcal{S}, a \in \mathcal{A}} x_{s,a} \mathcal{C}(s,a) = c$, i.e.,

$$
\begin{aligned}
\text{LP4} \quad &\max_{x,a} \ \sum_{s_g \in \mathcal{G}} in(s_g) \\
&\text{s.t. } x_{s,a} \geq 0 && \forall s \in \mathcal{S}, a \in \mathcal{A}(s) \\
&\quad out(s) - in(s) \leq 0 && \forall s \in \mathcal{S} \backslash (\mathcal{G} \cup s_0). \quad (9) \\
&\quad out(s_0) - in(s_0) \leq 1 \\
&\quad \sum_{s \in \mathcal{S}, a \in \mathcal{A}} x_{s,a} \mathcal{C}(s,a) = c
\end{aligned}
$$

and return probability-to-goal:

$$
p = \sum_{s_g \in \mathcal{G}} in(s_g) = \sum_{s_g \in \mathcal{G}} \sum_{s' \in \mathcal{S}, a \in \mathcal{A}} x_{s',a} \mathcal{P}(s_g | s', a).
$$

3.2 Non-dominated Policies

Although with additional constraints it is possible to obtain any trade-off for a SSP problem, it demands of decision-makers additional information which may not be at hand. Another alternative is to present Pareto optimal policies, i.e., non-dominated policies regarding probability-to-goal and expected cost. Then, decision-makers may choose which one is better regarding their preference. In this section we make use of a geometric approach to find non-dominated policies for SSPs [11].

Consider a policy π^α found by minimizing a linear combination between probability-to-goal and expected cost, i.e.,

LP5 $\min_{x,a} \sum_{s \in \mathcal{S}, a \in \mathcal{A}} x_{s,a} \mathcal{C}(s,a) - \alpha \sum_{s_g \in \mathcal{G}} in(s_g)$

$$
\begin{aligned}
\text{s.t. } & x_{s,a} \geq 0 && \forall s \in \mathcal{S}, a \in \mathcal{A}(s) \\
& out(s) - in(s) \leq 0 && \forall s \in \mathcal{S} \backslash (\mathcal{G} \cup s_0) \\
& out(s_0) - in(s_0) \leq 1
\end{aligned}
$$

(10)

Note that policy π^α is non-dominated. For $\alpha = 0$, LP5 returns probability-to-goal equal 0 and expected cost equal 0; and for $\alpha \to \infty$, LP5 returns probability-to-goal equal p_{max} and expected cost equal c_{min}.

Consider two non-dominated policies π' and π'' and their respective probabilities-to-goal and expected costs, (p', c') and (p'', c''). Then, it is possible to design a non-stationary probabilistic policy π^β that returns any interpolation between (p', c') and (p'', c''). At time step 0, policy π^β chooses policy π' with probability β and follows it thereafter, and chooses policy π'' with probability $1 - \beta$ and follows it thereafter.

In a finite SSP, i.e., when $|\mathcal{S}| < \infty$, we have a finite amount of stationary deterministic policies. In this case, the set of non-dominated policies are formed by: (i) a finite set of stationary deterministic policies ordered by their probability-to-goal returns; and (ii) interpolation between two consecutive stationary deterministic non-dominated policies. We introduce NDGC algorithm (Non-Dominated Probability-to-Goal and Expected Cost Trade-off), an anytime algorithm to find the set of stationary deterministic non-dominated policies Π_{ND}.

NDGC algorithm is based on the idea of non-dominated witness [10]. Given any two stationary deterministic non-dominated policies π' and π'', a non-dominated witness is a stationary deterministic policy π^w with respective return (p^w, c^w) that dominates some interpolation π^β of π' and π''.

Figure 1 shows the NDGC algorithm working. NDGC algorithm starts with two non-dominates policies: policy π^0, the policy that gives up at time step 0, and policy π^*, the optimal policy for MCMP criterion; and all interpolation between them. Call both policies generically π' and π'' with respective performance (p', c') and (p'', c''), NDGC algorithm finds a witness that is the farthest performance from any interpolation (see Fig. 1.a)); this is done by choosing $\alpha > 0$ such that $c' - \alpha p' = c'' - \alpha p''$. Then, the original interpolations are substituted by two new ones based on π' and π^w; and π^w and π''. NDGC algorithm continues by choosing one of the interpolations and finding a new witness (Fig. 1). NDGC algorithm stops when no witness is founded for any interpolation.

Algorithm 1 details the NDGC algorithm. Initial interpolation is found by solving MCMP problem (lines 1–3); then, the initial interpolation is added to a queue (lines 4–5). For each pair of policies in the queue, a witness policy is found if it does exist (lines 7–9). The witness policy is evaluated regarding the difference in cost between witness solution and interpolation (line 10). If the π^w improves on cost more than ϵ (line 11), π^w is added to $NonDominatedPolicies$ (line 12) and both of the new interpolations are added to $queue$ (lines 13–14). Note that

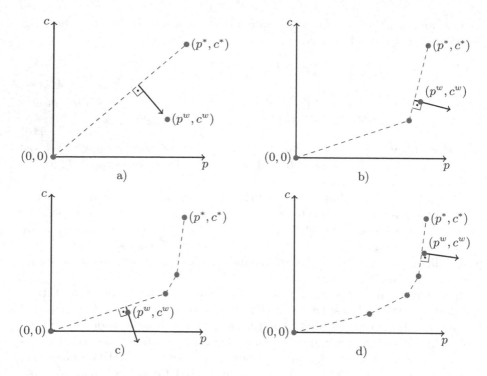

Fig. 1. Graphic demonstration of NDGC algorithm working.

after any iteration, we always have an approximation to the non-dominate set of policies.

3.3 Preference Elicitation

The NDGC algorithm finds out every non-dominated policy within an arbitrary error ϵ. Although it may inform a decision-maker a limited set of policies, it may generate too many ones. This can be a burden regarding computational complexity and also decision-maker cognition effort.

Based in [12], we adapt NDGC algorithm to a preference elicitation algorithm, PEGC algorithm, which elicits decision-maker preference. First, it decreases cognition effort on decision-makers by asking dichotomy questions, i.e., preference between two policies, both summarized in probability-to-goal and expected cost. Second, it decreases computational complexity by generating just a few policies to infer decision-maker preference, NDGC algorithm manages to do so by a search procedure that restricts the number of LP problems to be solved.

Algorithm 2 details the PEGC algorithm. PEGC algorithm considers that decision-maker evaluates policies linearly on probability-to-goal and expected cost, i.e., the value of a policy π^w is $V(\pi^w) = \alpha^* p^w - c^w$, where α^* is the unknown decision-maker trade-off.

Algorithm 1: NDGC algorithm (Non-Dominated Probability-to-Goal and Expected Cost Trade-off)

Input: SSP $\mathcal{M}$, threshold ϵ

1   $p_{max} \leftarrow \text{LP2}(\mathcal{M})$

2   $(\pi^*, p^*, c^*) \leftarrow \text{LP1}(\mathcal{M}, p_{max})$

3   $NonDominatedPolicies \leftarrow \{\pi^0, \pi^*\}$

4   $queue \leftarrow \emptyset$

5   $queue.Add((0,0), (p^*, c^*))$

6   **while** $priorityQueue$ *is not empty* **do**

7      $(p', c'), (p'', c'') \leftarrow priorityQueue.Remove()$

8      $\alpha \leftarrow \dfrac{c'' - c'}{p'' - p'}$

9      $(\pi^w, p^w, c^w) \leftarrow \text{LP5}(\mathcal{M}, \alpha)$

10     $d \leftarrow c' + \alpha(p^w - p')$

11   **if** $d > \epsilon$ **then**

12        $NonDominatedPolicies \leftarrow NonDominatedPolicies \cup \{\pi^w\}$

13        $queue.Add((p', c'), (p^w, c^w))$

14        $queue.Add((p^w, c^w), (p'', c''))$

15   **end**

16 **end**

Output: $NonDominatedPolicies$

Suppose that policy π' is preferred to policy π'', and remember that $\alpha^* > 0$, then:

$$V(\pi') > V(\pi'')$$
$$\alpha^* p' - c' > \alpha^* p'' - c''$$
$$\alpha^* < \frac{c'' - c'}{p'' - p'}$$

and, by choosing $\alpha = \frac{c''-c'}{p''-p'}$ for LP5, PEGC algorithm discovers a new non-dominated policy π^w that is better than π''. The same analysis can be made when policy π'' is preferred to policy π', in this case, we have that:

$$\alpha^* > \frac{c'' - c'}{p'' - p'}.$$

In PEGC algorithm, a pair of extreme policies is found by solving MCMP problem (lines 1–3). Then, the following steps are repeated:

1. a witness policy is founded (lines 5–6);
2. an observation of decision-maker preference is made (line 5–6); and
3. conditioned on decision-maker preference, the witness policy substitute π' or π'' (line 9–15).

The PEGC algorithm stops if the witness policy is too similar to the interpolation of π' and π''.

Algorithm 2: PEGC algorithm (Preference Elicitation Probability-to-Goal and Expected Cost Trade-off)

Input: SSP $\mathcal{M}$, threshold ϵ

1   $p_{max} \leftarrow \text{LP2}(\mathcal{M})$
2   $(\pi', p', c') \leftarrow (\pi^0, 0, 0)$
3   $(\pi'', p'', c'') \leftarrow \text{LP1}(\mathcal{M}, p_{max})$
4   **while** *true* **do**
5      $\alpha \leftarrow \dfrac{c'' - c'}{p'' - p'}$
6      $(\pi^w, p^w, c^w) \leftarrow \text{LP5}(\mathcal{M}, \alpha)$
7      $d \leftarrow c' + \alpha(p^w - p')$
8   ask decision maker between (p', c') and (p'', c'')
9   **if** (p', c') *is preferred to* (p'', c'') **then**
10        $(\pi'', p'', c'') \leftarrow (\pi^w, p^w, c^w)$
11        $\pi^* \leftarrow \pi'$
12   **else**
13        $(\pi', p', c') \leftarrow (\pi^w, p^w, c^w)$
14        $\pi^* \leftarrow \pi''$
15   **end**
16   **if** $d < \epsilon$ **then**
17   **break**
18   **end**
19 **end**

Output: π^*

---

4 Experiments

In this section we report experiments of NDGC and PEGC algorithms. We designed experiments on the River Domain [7] where trade-offs between probability-to-goal and expected cost can be easily observed.

4.1 The River Problem

The River problem, as described in [7], considers a grid world where an agent starts in one side of the river and must reach a goal state in the other side (Fig. 2).

The safest path consists in walking along the riverside until a bridge, walk by the bridge, and go back by the opposite riverside. On the other hand, the shortest path consists in crossing the river; however, actions in the river are influenced by the flow of the river and the agent may be dragged to the waterfall, which is a dead-end.

When walking on the riverside, there is a small probability of falling at the river; therefore, the river problem does not present proper policies for the initial state. When inside the river, instead of behaving deterministically, the river flow compels with a influence chance the agent to the waterfall. Depending on

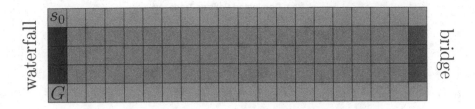

Fig. 2. 5 × 20 River problem.

whether the agent avoids the river, or where the agent chooses to enter the river incurs in different probability-to-goal and expected cost trade-offs.

We experiment on two sizes of River problems: 5 × 40 and 10 × 60. The chance of falling into the river was set to 0.1 and the influence chance of the river flow was set to 0.5 (see [7] for a complete description of the river problem).

4.2   Non-Dominated Policies

Figures 3 and 4 show the trade-off between probability-to-goal and expected cost for 5 × 40 and 10 × 60 river problems, respectively. We run NDGC algorithm with $\epsilon = 0.1$.

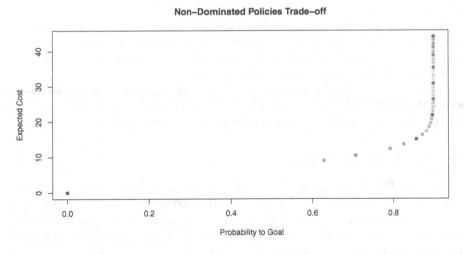

Fig. 3. Non-dominated policies for the 5 × 40 river problem. (Color figure online)

We plot trade-off points in order of occurrence in NDGC algorithm, from red (first) to green (last). In both scenarios, we can see that there exist many dondominated policies with very close probability-to-goal. Although the number of non-dominated policies can be large (more than 30 policies for small problems), we can see it is possible to have a good approximation with just a few policies.

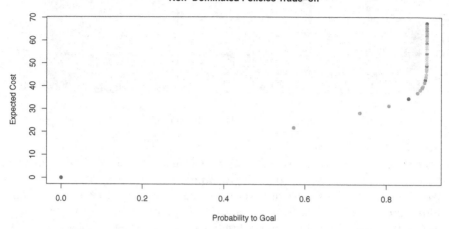

Fig. 4. Non-dominated policies for the 10×60 river problem. (Color figure online)

In the case of size 5×40 we have that $p_{max} = 0.9009$ and $c_{min} = 44.08$. If we allow a probability-to-goal $p = 0.8982$, we have expected cost $c = 21.74$, half of c_{min}. In the case of size 10×60 we have that $p_{max} = 0.9009$ and $c_{min} = 67.64$. If we allow a probability-to-goal $p = 0.8983$, we have expected cost $c = 43.96$, two thirds of c_{min}.

As already noted in [6], a tiny decrease in maximum probability-to-goal, in general, brings a large decrease in expected cost. This experiments corroborate the hypotheses that giving priority to probability-to-goal, as in iSSPUDE and MCMP, may incur on unnecessary spent in cost.

Note that a probability-to-goal and expected cost trade-off can be easily observed after the non-dominated policies are plotted in a graph. Knowing such graph, a decision-maker can make a informed choice over policies.

4.3   Preference Elicitation

To evaluate PEGC algorithm, we simulate a decision-maker whose preference equals MCMP criterion and test it in the 10×60 river problem. In this case, the decision-maker prefers policy π^* to any other. Note that this decision-maker represents the worst scenario for PEGC algorithm, since its preference is located in the most dense region in Fig. 4.

Figure 5 shows the trade-off between probability-to-goal and expected cost for all policies found during PEGC algorithm execution; 10 policies were necessary.

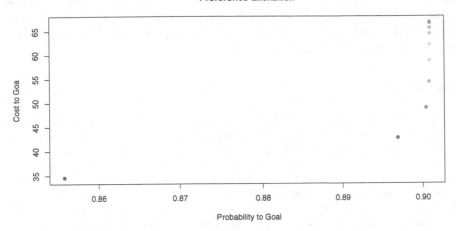

Fig. 5. Subset of non-dominated policies for the 10×60 river problem when executing PEGC algorithm.

5   Conclusion

In this work, were presented three methods to help a decision-maker to find a personal trade-off between probability-to-goal and the expected cost. In the first method, additional constraints, the decision-maker proposes an additional constraint to the SSP problem; to help the decision-maker, extremes trade-offs $(0,0)$ and (p^*, c^*) are informed beforehand. In the second method, non-dominated policies, the decision-maker is informed with a set of non-dominated policies, from which he/she can choose the one that best pleases him/her. In the third method, preference elicitation, dichotomy questions are formulated and asked decision-maker; in the end, the policies that best pleases him/her is returned.

Although we experiment on small toy problems, the conclusion is powerful. First, by using some criteria from the literature, such as MCMP or iSSPUDE, may generate a disastrous decision in real problems; a huge expected cost may be paid for a small increase in probability-to-goal. Second, the decision-maker must be in the loop if we want to avoid such a disastrous decision, we have offered three types of interaction between decision-maker and stochastic planning algorithms. Third, there is an inverse proportional relationship between the cognitive effort required from the decision-maker and computational effort.

We only considered the trade-off between probability-to-goal and the expected cost. Here, we considered the expected cost as defined in the MCMP criterion, which also considers the cost spent on traces that do not reach goal states. On the other side, iSSPUDE criterion considers only cost on traces that reach the goal. Presenting MCMP and iSSPUDE expected cost to decision-makers may reach a better-informed trade-off. Instead of considering expectation, we can also consider a distribution over cost, for example, by considering percentile, such as the one considered in CVaR MDP [4,5]. How to efficiently summarize policies into different aspects of costs and design an efficient preference elicitation process are the next steps of our research.

References

1. Benabbou, N., Leroy, C., Lust, T.: Regret-based elicitation for solving multi-objective knapsack problems with rank-dependent aggregators. In: The 24th European Conference on Artificial Intelligence (ECAI 2020). Saint Jacques de Compostelle, Spain (June 2020). https://hal.sorbonne-universite.fr/hal-02493998
2. Bertsekas, D.P., Tsitsiklis, J.N.: An analysis of stochastic shortest path problems. Math. Oper. Res. **16**(3), 580–595 (1991)
3. Branke, J., Corrente, S., Greco, S., Gutjahr, W.: Efficient pairwise preference elicitation allowing for indifference. Comput. Oper. Res. **88**, 175–186 (2017)
4. Carpin, S., Chow, Y.L., Pavone, M.: Risk aversion in finite Markov decision processes using total cost criteria and average value at risk. In: 2016 IEEE International Conference on Robotics and Automation (ICRA), pp. 335–342. IEEE (2016)
5. Chow, Y., Tamar, A., Mannor, S., Pavone, M.: Risk-sensitive and robust decision-making: a CVaR optimization approach. In: Advances in Neural Information Systems, pp. 1522–1530 (2015)
6. Freire, V., Delgado, K.V.: GUBS: a utility-based semantic for goal-directed Markov decision processes. In: Proceedings of the 16th Conference on Autonomous Agents and MultiAgent Systems, pp. 741–749 (2017)
7. Freire, V., Delgado, K.V., Reis, W.A.S.: An exact algorithm to make a trade-off between cost and probability in SSPs. In: Proceedings of the Twenty-Ninth International Conference on Automated Planning and Scheduling, pp. 146–154 (2019)
8. Kolobov, A., Mausam, Weld, D.S.: A theory of goal-oriented MDPs with dead ends. In: de Freitas, N., Murphy, K.P. (eds.) Proceedings of the Twenty-Eighth Conference on Uncertainty in Artificial Intelligence, pp. 438–447. AUAI Press (2012)
9. Mausam, A.K.: Planning with Markov decision processes: an AI perspective. Synth. Lect. Artif. Intell. Mach. Learn. **6**(1), 1–210 (2012)
10. Regan, K., Boutilier, C.: Robust policy computation in reward-uncertain MDPs using nondominated policies. In: Proceedings of the Twenty-Fourth AAAI Conference on Artificial Intelligence, AAAI 2010, pp. 1127–1133. AAAI Press (2010)
11. Freire da Silva, V., Reali Costa, A.H.: A geometric approach to find nondominated policies to imprecise reward MDPs. In: Gunopulos, D., Hofmann, T., Malerba, D., Vazirgiannis, M. (eds.) ECML PKDD 2011, Part I. LNCS (LNAI), vol. 6911, pp. 439–454. Springer, Heidelberg (2011). https://doi.org/10.1007/978-3-642-23780-5_38
12. Silva, V.F.D., Costa, A.H.R., Lima, P.: Inverse reinforcement learning with evaluation. In: IEEE International Conference on Robotics and Automation (ICRA 2006), pp. 4246–4251. IEEE, Orlando (May 2006)
13. Teichteil-Königsbuch, F.: Stochastic safest and shortest path problems. In: Proceedings of the Twenty-Sixth AAAI Conference on Artificial Intelligence (AAAI 2012), pp. 1825–1831 (2012)

14. Teichteil-Königsbuch, F., Vidal, V., Infantes, G.: Extending classical planning heuristics to probabilistic planning with dead-ends. In: Proceedings of the Twenty-Fifth AAAI Conference on Artificial Intelligence, AAAI 2011, San Francisco, California, USA, August 7–11, 2011 (2011)
15. Trevizan, F., Teichteil-Königsbuch, F., Thiébaux, S.: Efficient solutions for stochastic shortest path problems with dead ends. In: Proceedings of the Thirty-Third Conference on Uncertainty in Artificial Intelligence (UAI) (2017)

StoryTracker: A Semantic-Oriented Tool for Automatic Tracking Events by Web Documents

Welton Santos[1], Elverton Fazzion[1], Elisa Tuler[1], Diego Dias[1(⊠)],
Marcelo Guimarães[2], and Leonardo Rocha[1]

[1] Universidade Federal de São João del-Rei, São João del-Rei, Brazil
{welton,elverton,etuler,diegodias,lcrocha}@ufsj.edu.br
[2] Universidade Federal de São Paulo/UNIFACCAMP, São Paulo, Brazil
marcelo.paiva@unifesp.br

Abstract. Media vehicles play an essential role in investigating events and keeping the public informed. Indirectly, logs of daily events made by newspapers and magazines have been built rich collections of data that can be used by lots of professionals such as economists, historians, and political scientists. However, exploring these logs with traditional search engines has become impractical for more demanding users. In this paper, we propose *StoryTracker*, a temporal exploration tool that helps users query news collections. We focus our efforts (i) to allow users to make queries by adding information from documents represented by *word embbedings* and (ii) to develop a strategy for retrieving temporal information to generate *timelines* and present them using a suitable interface for temporal exploration. We evaluated our solution using a real database of articles from a huge Brazilian newspaper and showed that our tool can trace different *timelines*, covering different subtopics of the same theme.

Keywords: Timelines · Search engines · Word embeddings

1 Introduction

Search engines' evolution (e.g., Google Searching, Microsoft Bing, DuckDuckGo) has increased access to information. Individuals and institutions feed these mechanisms daily, which indirectly document history through records of daily events. These tools are based on keyword queries that aim to retrieve information by associating the database's information with a few keywords. This type of query returns broad information about a topic and does not focus on details. For example, an economist who needs to find news and information about the stock market will provide some keywords (e.g., variations on the stock market price) to the search engine. The result will be the latest news about events that caused the stock prices to oscillate (e.g., the fusion between two companies) [20].

Supported by CAPES, CNPq, Finep, Fapesp and Fapemig.

© Springer Nature Switzerland AG 2021
O. Gervasi et al. (Eds.): ICCSA 2021, LNCS 12951, pp. 126–140, 2021.
https://doi.org/10.1007/978-3-030-86970-0_10

Despite being efficient in retrieving recent and more comprehensive information, traditional search engines still have limitations for users with more demanding goals looking for more specific and temporally related themes. For example, a historian interested in the stock's behavior market over time from a particular event (e.g., the Mariana dam collapse in Brazil in 2015 or Brazilian President Dilma Rousseff's impeachment in 2016) will require costly and not always workable work. Therefore, it will force him to perform multiple queries, sort and filter large volumes of data. Thus, from the perspective of more specific searches, performed by more demanding users such as historians, we observe that there are still two particular challenges during the search process: (i) generating good queries to retrieve relevant and more specific documents in databases and; (ii) temporally organizing large numbers of documents (e.g., news, articles) in such a way that navigation through these data is efficient [1].

The present work aims to provide a tool for more demanding users to easily search, organize, and explore documents by presenting the results through timelines. Our proposal consists of dividing a traditional query into two parts, as described below. First, we perform a traditional search, in which the user defines the keywords of the topic s/he wants to search. It uses a traditional search engine, presenting several documents related to the topic sorted according to some criterion predetermined by the search algorithm (i.e., often related to commercial issues). In the second part, which corresponds to this work's first contribution, among the various documents returned by the search machine, the user chooses the one s/he considers most relevant (reference). This reference document makes it possible to construct a timeline through the algorithm we proposed, called Story Tracker (ST). From an initial date (t_0) determined by the user, the temporal coverage for the timeline (e.g., five months) and the information related to the query and the selected document, the ST algorithm divides the temporal coverage into subspaces, e.g., weeks (e.g., $t_1, t_2, ...t_n$) preserving the chronological order of the subspaces. For the time subspace t_0, the ST algorithm uses the representative vectors of the documents generated via *word embbedings* doc2vec model and identifies the documents most similar to the query and the reference document selected by the user, which will be presented for time t_1. Generally, to maintain consistency in the relationship between documents in an t_i interval with documents in the t_{i+1} interval, ST updates the reference document for each interval. This updated reference document in the t_{i+1} interval refers to the document most similar to query and the reference document in the t_i interval. This update process allows the user to force the tool to induce the retrieval of documents associated with more specific subjects addressed by the document and the query.

Our second contribution comprises a Web tool that uses the data retrieved by the process described above and presents them using an intuitive visual metaphor, which allows the user to explore the data in an organized way with convenience. Our tool organizes the time intervals in a sequence of sections (slides). Each section contains the documents from one interval to allow the user to explore the data after performing the query for building the timeline. The tool also provides document filters by categories (e.g., sports, market) to assist in the exploration process, which can be used in the originally retrieved data and in the timeline's construction, allowing specialized searches by categories. The tool also has a function that allows the user to define the number of documents he wants to view per interval. With this functionality, the user can generate more summarized and specific timelines, relating few documents per interval or more comprehensive timelines with a larger number of documents per interval.

We can integrate our tool with any search engine. However, to perform a controlled and consistent evaluation, we simulated a search engine using a database of the Brazilian newspaper Folha de São Paulo,[1] with documents from January 2015 to September 2017, composed of 167 thousand documents of different categories. To instantiate our tool, we built a vector representation for each document in the database considering the *word embbedings* doc2vec model. We also built the vector representation of the user query with the *word embbedings* model. We then select the most similar documents to the user query using cosine distance between these vectorial representations. The N documents most similar to the query are presented to the user, and within the N documents returned by the simulated search machine, it must choose a reference document.

To evaluate the functioning of the proposed tool, we simulated a search for the terms "the impeachment of Dilma Rousseff and its impacts," choosing different reference documents and then detailing the timelines built from them. Our results show that by varying the reference document resulting from the original query, the proposed tool can trace different timelines, covering different sub-topics of the impeachment process (e.g., lava jato investigation, public manifestations). It would not be possible through a single query in a traditional search engine. Also, we analyzed the impact of the query and the reference document for the construction of the timelines. We observe that, as we increased the importance of the reference document to the query, we retrieved documents with topics more correlated to the reference document. This result is a strong sign that the reference document works well as an information source. To the best of our knowledge, we have found no work in the literature that aims to perform more specific and temporally related topic searches. Moreover, our proposal is generic enough that any current search engine can apply and use it.

2 Related Works

2.1 Query Expansion

Query expansion is a strategy widely known in the literature and employed in search engines to optimize search results. Query expansion can change the user's original query by recalibrating the weight of terms or adding new terms. One of the main reasons for using query expansion is that users rarely construct good queries that are short on based on ambiguous words. Within the existing lines for query expansion, many works focus on identifying new terms from the analysis of an initial document ranking [3,6,17]. Basically, after an initial query of documents, the search engine returns a ranking of K documents, and the query expansion algorithm will use these documents to perform query enrichment. This approach is known as feature distribution of top-ranked documents (FD-TRD)

---

[1] https://www.kaggle.com/marlesson/news-of-the-site-folhauol.

and assumes that the K documents in the initial ranking are good sources of correlated query information because of their proximity to the words.

We can classify FD-TRD application into two fronts, Relevance Feedback (RF) and Pseudo-Relevance Feedback (PRF) [4]. The difference between the two strands comprises the user interaction with the initial document ranking, returned from the original query. PRF-based techniques assume that the initial document ranking is consistent with the query and use it as a source of information for its expansion. RF techniques rely on user interaction with the ranking documents and use this interaction to learn about the user's interests [17]. The system can learn about the user's interests in two ways, implicit or explicit. In the implicit form, the system analyzes a user's interaction with documents (e.g., the opened ones). In the explicit form, the user voluntarily shows the system which documents are relevant to him, and therefore should be used as a source of extra information. PRF-based techniques are popular because of the convenience of not depending on the user to collect information. However, that is hardly available, being more susceptible to querying drift, adding irrelevant information to queries, which is not suitable for real-time systems.

Since this work aims at users interested in conducting further research, and the user's explicitness about which documents are important is available, our work closely resembles works that apply PR to query expansion. Among the works that present approaches with PR, new proposals explore the potential of *word embbedings* models. [18] presents a KNN-based approach, which computes the distance between words using representation vectors based on the Word2Vec model [14,15]. In [23], the authors present that approaches with *word embbedings* trained on local data show better results than global approaches. Compared to other literature approaches, wembb-based techniques are equal or superior to traditional techniques (e.g., RM3) using RF [9]. Unlike all the work presented above, our approach is the first to add to the original user query the vector representations of documents.

2.2 Temporal Information Retrieval

Organizing and retrieving information in databases over time is a challenge addressed in several works [1,2,5,7,12,13,16,20]. In [2], the authors conduct a systematic review of time-focused information retrieval work and present the main issues inherent in existing challenges in distinct lines (e.g., temporal clustering, timelines, and user interfaces). In [7], the authors review temporal information retrieval techniques and classify the existing challenges into temporal indexing and query processing, temporal query analysis, and time-aware ranking. In literature, we noticed efforts aimed at two research fronts: (i) to develop models for using time as a variable to enhance classification, clustering, and information retrieval techniques; and (ii) to create suitable visual metaphors for users to easily explore the data returned by information retrieval models.

Following the first front, works in the literature introduce the time parameter into various text processing tasks. In [1], the authors develop an add-on for re-ranking search results from web content search engines (e.g., Google Searching). The authors process the data returned by the search engines and use the creation and update dates of the documents with temporal data present in the page's content to generate timelines in the users' browsers. Focusing on clustering and summarizing social network data, in [12], the authors propose a technique to extract and generate timelines from Twitter users. They develop a non-parametric topic modeling algorithm able to efficiently identifying tweets related to important events in users' lives (e.g., getting a university place) amid tweets with low importance (e.g., comment about a movie). With the model, the authors can trace ordinary users' timelines and more influential users (e.g., celebrities). In [5], the authors present a model for detecting patterns of behavior, called episodes, by analyzing social network user profiles. The authors propose a Bayesian model to generate summarized timelines with important episodes.

Some works direct their efforts toward presenting information to users. [13] presents the Time Explorer tool that explores graphical resources to generate intuitive timelines for users, aiming to generate timelines. The tool allows users to relate information extracted from the content of documents and make predictions about possible future events involving entities (e.g., celebrities, companies). More similar to our work, we found [20], which presents a topic coverage maximization tool focusing on generating timelines for historians. The authors argue that the way a historian seeks information differs from ordinary users. This type of user gets broad and general views of an event and not one-off information.

The works presented in this section focus on building timelines and exploratory research. Differently, we propose a complementary tool to traditional search engines. Our proposal is the only one focused on searching for more specific and temporally related topics with RF query expansion.

2.3   Document Representation

Many works largely use document vector representation in word processing because of its importance in document classification and clustering. Among the state-of-the-art works in the vector representation of documents, *word embeddings* models stand out. Traditional models based on Bag-Of-Words (BOW) have two inherent weaknesses in their representations: (i) loss of semantic information in the representations; and (ii) the vectors generated to represent documents have large dimensions. Usually, the amount of dimensions is equal to the size of the vocabulary of the document set.

Facing these problems, *word embbedings* models represent words in reduced vector spaces, preserving the semantics of words and documents. Among the existing works in *word embbedings*, [10,14,15] stand out. [14] proposes word2vec, a widely used model. The work proposes to use words in a multi-dimensional vector space so that a vector represents each word (usually 50 to 2000 thousand dimensions) in which semantically close words have similar vectors. Following the line of word representation, [10] extends word representation to sentence and document representation with the widely used doc2vec model.

Regarding document classification, [19] compares several representation methods for binary document classification based on data taken from clinical records to identify patients who are users of alternative medicines. In [11], the authors compare the doc2vec model with traditional representation models and with the word2vec model. They investigate the performance of the representation techniques for the task of classifying product categories from product descriptions. They show that traditional models perform similarly to word2vec, being better than doc2vec and suggesting that tasks, where semantic context preservation is not required *word embbeginds* algorithms do not promote improvement. In the paper [21], the authors use the doc2vec model and the HDBSCAN algorithm to identify duplicate use case tests and reduce software testing costs. Focusing on topic modeling, [8] uses vector representations with *words embbedings* to train classifiers and suggest to researchers topics for paper submission. Closer to our work, in [22] they train a doc2vec model with a Twitter database and propose a new technique to measure document distance in information retrieval tasks. Thus, although we have not found works that use *word embbedings* models to improve query results in search machines and organize them temporally, the excellent results in other areas motivated our proposal, detailed in the next section.

3   Story Tracker

In this section, we explain *Story Tracker* (ST), our proposal for performing query result refinements in search engines that aims to make the search easier, organize and explore documents by presenting the results using *timelines*. Figure 1 shows an overview of the tool, and we describe each of the steps next.

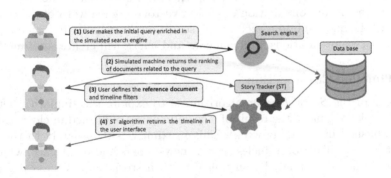

Fig. 1. Overview of user interaction in the process of *timelines* generation

3.1   Enriching the Query

The ST's first component uses query expansion to build *timelines* about specific subjects with the documents' information. The technique developed here comprises a variation of the Relevance Feedback technique by user interaction. Our goal is to allow the user to create *timelines* based on specific subjects from simple *queries*, exploiting information from few documents selected by the user. To do so, we show to the user a *ranking* of documents related to a *query* before building the *timeline* from the results of a search. Among the documents in the *ranking*, the user chooses a document that is closer to his interest. This document will be used as a source of information to retrieve documents next to the *query* in *timelines* creation (Sect. 3.2). We can see the process of choosing the reference document in Fig. 1 in steps one and two.

Given a search engine operating on a document base D relative to all indexed documents and a user-created *query* q with the keywords on the topic of interest, we construct a tuple (q, C_q, p_q, m_q), where C_q, p_q, m_q are also user-defined. C_q is a set of categories used for filtering where only documents $(d_i \in D)$ with categories in C_q are retrieved $(Cat(d_i) \in C_q)$. p_q consists of a base date for document exploration by our solution. Finally, m_q is a time period in months that is used as a search radius for base date p_q. From p_q and m_q, we define the interval $T_q = [(p_q - m_q), (p_q + m_q)[$. For example, for $p_q = 2016/04/01$ and $m_q = 2$, documents in the period $T_q = 2016/02/01$ to $2016/06/01$ will be analyzed. Thus, the scope of documents to be analyzed by the ST is given by $E_q \in D$ (Eq. 1).

$$E_q = \{d_i \in D | d_i \in T_q \wedge Cat(d_i) \in C_q\} \tag{1}$$

In this way, the new *query* E_q is submitted to a search engine, which will return a set of documents D_q sorted by some criteria pre-defined by the search algorithm (i.e., often related to commercial issues). Then, the user chooses a reference document in D_q to be used as an information source in the *timeline* building process described in Sect. 3.2. For each of the D_q documents, the ST will create a vector representation using the *word embbedings* doc2vec model. Similarly, the linear combination of all the vectors of each word of query q will transform the query q into a vector representation. The similarity calculation of these vector representations of documents and *query* will create the *timeline*.

3.2   Timeline Construction

The focus of the *Story Tracker* algorithm is to explore the similarity between documents belonging to the result of a specific query published in closer periods (e.g., published in the same week). This proximity is due mainly to the reactive nature of publication vehicles such as newspapers and magazines, which, to keep their audience updated, constantly publish stories about sub-events of a broader event. As sub-events are the causes of new sub-events, the relationship between them tends to be stronger, as does the similarity between the content of the stories that record them. Based on this assumption, the core of the algorithm proposed here consists of correlating documents temporally. Given an initial query and a reference document, documents published in neighboring time intervals can be retrieved and become sources of information to retrieve other documents. The application of the ST algorithm on the database is related to Steps 3 and 4 of the example in Fig. 1.

As input, the ST algorithm receives a tuple of parameters (q, d_r, D_q, T_q, s). d_r comprises the reference document chosen by the user, s refers to the granularity in days between the intervals that compose the *timeline* and q, D_q and

T_q are defined as before. From these parameters, the ST algorithm divides the construction of the *timeline* into two parts: past and future. Starting from the publication date of the reference document d_r $(Pub(d_r))$, the algorithm creates two sets of intervals (time subspaces) T_{pas} and T_{fut} referring to the past and future of $Pub(d_r)$, respectively. These interval sets comprise dividing the period T_q into T intervals of size s (in days). To generate the *timeline*, the expansion process described below applies to the interval sets T_{pas} and T_{fut} similarly. After an expansion of each interval set, the past and future outputs are concatenated and presented to the user as a *timeline*.

Expansion

The expansion process is presented in Algorithm 1. For each time interval $t \in T$, the ST algorithm creates subsets of documents S_t (lines 5–8) with the documents $d_t \in D_q$, whose publication date is within the t interval. For each document, its importance (semantic proximity) regarding the *query* and the reference document d_r (line 7) is calculated by the function $util(q, d_r, d_i)$ (Eq. 2). This function balances the importance between *query* q and reference document d_r, relative to any given document d_t through the parameter α, which varies in the range [0,1]. Higher α, the higher the importance of *query* and vice versa.

Algorithm 1: Story Tracker (ST) Expansion

---

1 **Input:** (q, d_r, D_q, T_q, s)
2 **Output:** Set S of documents for all intervals t
3 $T \leftarrow \frac{T_q}{s}$;
4 **foreach** *interval* $t \in T$ **do**
5    $S_t \leftarrow \{\}$;
6   **foreach** *document* d_t *where* $((d_t \in D_q)\&(Pub(d_t) \in t))$ **do**
7       $d_t.util \leftarrow util(q, d_r, d_t)$;
8       $S_t \leftarrow S_t \cup d_t$
9    $sort\_util(S_t, descending)$;
10    $d_r \leftarrow MAX(S_t)$;
11    $S \leftarrow S \cup S_t$;
12 **return** S

We control the amount of noise added in a reference document (d_r) using the α parameter since it can address several correlated subjects. Thus, using all the document information, we may correlate more documents than expected by the user, configuring the effect of *query drifting*, present in *query* expansion systems, where irrelevant information is added to the initial *query*. This way, with higher values of α, the user can reduce the impact of possible noisy information and leave the results closer to the original *query*.

$$util(q, d_r, d_i) = \alpha.cos(q, d_t) + (1 - \alpha).cos(d_r, d_t) \qquad (2)$$

Once the set S_t is built, we sorted it in descending order according to the value of the *util* function calculated for each of its documents (line 9). This set is concatenated to S, containing all the documents D_q sorted chronologically and by their relevance to the created timeline.

Updating the Reference Document

ST relies on a reference document to add information to the user's *query* and assist in retrieving subsets of documents that make up the *timeline*. Based on the assumption that documents in one interval tend to be more similar to documents in their neighboring intervals, ST updates the reference document for each subset of documents in each interval to search for more similar documents. Starting from the time interval t_0, ST retrieves the documents most closely related to *query q* and the reference document choose by the user, named d_{r0}. After constructing the subset of documents from the t_0 range, ST advances to the t_1 range. In constructing the subset of documents in this interval, the reference document d_{r0} is replaced by another document named d_{r1} (line 10), by means of the MAX function, which returns the most relevant document (*util* function) within the time t_0. We repeat this process for all remaining intervals so that the reference document used in the interval t_n comes from the interval t_{n-1}.

3.3   Graphical Interface

According to Fig. 1, the *timeline* construction uses two separated steps: *query* enrichment and *timeline* construction. The enrichment of the *query* comprises the user's first contact with the tool, interacting with the interface presented in Fig. 2. In this step, the user enriches the *query* by configuring the base date and the coverage of the *timeline* in Box 2 and Box 3, respectively. Figure 2 shows an example of the initial *ranking* with the cards containing the document title and a checkbox to mark the document as the reference document (Box 5). In Box 4, the user can define the number of documents in the initial *ranking*. In addition to the parameters present in the initial interface, the user can enrich their *query* with advanced parameters, found in the "Advanced Search" button, as highlighted in green in Fig. 2. The set of categories for document filters (Box 6), the granularity of time intervals (Box 7), and the balance between the importance of the *query* and the reference document (Box 8) are configured in this advanced tab.

After selecting a reference document and considering all the documents returned by the query, ST builds the timeline and returns it to the user. The interface depicts in Fig. 3. On this screen, the documents most associated with the *query* and the reference document in the interval are presented to the user. The title and date of the reference document are highlighted in blue in Box 1 and Box 2. The buttons used by the user to navigate through the *timeline* intervals are highlighted in Box 4 and Box 5. The interface offers another way to move through the *timeline* via the bottom bar (Box 6). Each link in the bar contains the title of the interval's reference document and a redirect function.[2]

---

[2] The code is available at https://github.com/warSantos/StoryTracker.git.

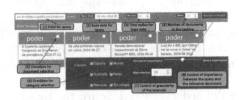

Fig. 2. Screen for building the initial document *ranking*

Fig. 3. Screen with documents from a *timeline* interval

4 Experimentation Evaluation

4.1 Experimental Setup

Models of *word embbedings* are famous for capturing semantic information between words and documents [10,14,15]. We explored the capacity of the doc2vec[3] model to find documents that are strongly correlated with each other. To train our models, we used the Folha de São Paulo Newspaper document base[4], which contains over 167,000 documents, from January 2015 to September 2017. Each document has the story title, textual content, publication date, category, and link to the page. We extracted the textual content of the document and removed the stopwords from each document. From this extraction, we trained a doc2vec representation model using the Paragraph-Vector Distributed Memory (PV-DM) algorithm, considering the default values for the algorithm parameters (100 dimensions, ten epochs, ...).

As described in the Sect. 3.1 and Sect. 3.2, our tool supports filtering documents by category. However, starting from the database without preprocessing, we found 43 different categories among the documents, with large overlapping topics and unbalance in the documents' volume among the categories. For this reason, we chose six main categories (sports, illustrated, market, world, power, and technology) before using a search engine. We redistributed the documents from the other categories into these six categories using a logistic regression model.[5] This process reduces the redundancy of categories and balances the volume of documents among them. For model training, we used the native documents from the main classes with the vector productions generated by the Paragraph-Vector Distributed Bag-of-Words (PV-DBOW) algorithm (i.e., using the algorithm's default parameters).

To evaluate our entire strategy in a more consistent and controlled way, we simulate a search engine over the database described in Sect. 4.1. Our search

---

[3] https://radimrehurek.com/gensim/models/doc2vec.html.

[4] https://www.kaggle.com/marlesson/news-of-the-site-folhauol.

[5] https://scikit-learn.org/stable/modules/generated/sklearn.linear_model.
LogisticRegression.html.

engine operates on a subset of documents from the database defined by the scope of an enriched user *query* (Sect. 3.1). It presents to the user the *ranking* of the K documents closest to the user *query*. To estimate the similarity between documents and the *query* q, we first create the vector representation v_q of q as a linear combination of the vectors of the words of q. The vector of each word is extracted from the PV-DM model. Words not existing in the model are ignored in the calculation. Given the vector v_q, we measure the cosine similarity of v_q with all documents in the scope of the user-defined *query* and return the set of K documents with the highest similarity score to the *query*.

4.2 Results

In this section, we discuss the impact of the proposed tool on *timelines* construction. We analyzed our method's ability to build different *timelines* correlating documents from a single input *query*, varying the reference document and the α threshold. For constructing *timelines*, we used as a case study the impeachment process of Brazil's former president, Dilma Rousseff. We chose this topic because the impeachment process lasted a long time, generating large volumes of news.

In our search engine input, we used the *query "Dilma Rousseff's impeach-ment and its impacts"*, with a base date 2016/08/01 (one month before the impeachment date), time radius of one month (total of two months, one in the future and one in the past) and the granularity of the intervals of 15 days. As a set of categories, we considered *power, market* and *world*. On these parameters, the search engine returned the *ranking* of documents contained in Fig. 4. The results are presented in Portuguese as the original documents of our dataset.

Fig. 4. *Ranking* generated by the search engine from the query "the impeachment of dilma rousseff and its impacts". In this Figure are the six documents most correlated to the user's query. The documents used as reference documents in the analyses in this section are highlighted in blue (Color figure online)

Given the *ranking* of news (Fig. 4), we took for analysis the document three (D3) (*"Should the Senate approve the impeachment of Dilma Rousseff? NO"* and document six (D6) (*"BC President says private interests hinder fiscal adjust-ment"*) and built a *timeline* related to each document. We chose these documents for dealing with topics that share the *"impeachment Dilma Rousseff"* theme, with D3 having more emphasis on the *impeachment* process with political content and D6 with emphasis on the impacts of the *impeachment* on the Brazilian economy. As the basis of the discussion of this paper, we generated four *timelines*

TM1 and TM2 with $\alpha = 0.5$ for documents D3 and D6, respectively; and TM3 and TM4 for document D6 with $\alpha = \{0.2, 0.5\}$. For analysis, we selected the top two documents from three intervals of the *timelines* and analyzed the variations of these as the reference document and α threshold change as shown in Fig. 5.

Figure 5 shows the documents from the TM1 *query* related to the D3. We observe that all the key documents of the three intervals deal with the main topic *impeachment*. Among the topics covered by the documents, we have themes such as "tax fraud", "corruption investigations", and "parliamentary coup". The themes share as their focus the process of *impeachment* using as background other processes involved. As an example, we can mention the second document in TM1 (*"Lula tells BBC that Dilma will expose herself to 'Judas' in the Senate"*). It deals with an interview of former president Lula to a newspaper, in which he defends himself against accusations regarding corruption involving his political party. Also, the third (*"Impeachment brings "didactic effect" to future leaders; see debate"*) and fourth (*"The judgment of history"*) documents deal with the event known as "pedaladas fiscais" (or pedaling fiscal). This event was used as an argument for the *impeachment* process. Comparing the documents in TM2 with those in TM1, it is noticeable that documents dealing with the *impeachment* process are still present. However, documents related to the economy show up as more relevant. As main examples of this change, we have the second (*"Odebrecht's pre-statement makes the dollar move forward; Stock Exchange goes up with Petrobras"*) and third (*"Fate of reforms proposed by Temer worries investors"*) documents in TM2. These documents deal with "variation in stock prices" and "worry of foreign investors" topics, respectively, which are topics that use the *impeachment* as background because of the economic instability caused in the country. Based on this analysis, we can see the impact caused by the alteration of the reference document, since with $\alpha = 0.5$ it was already possible to direct the construction of the *timeline* from the alteration of D3 to D6.

As described earlier, the second way to build different *timelines* is from the variation of the parameter α for the same tuple of *query* and reference document. Thus, the higher α, the greater the similarity of the documents to the *query* and vice versa. To analyze this parameter's impact, we use the D6 document to generate the T3 and T4 *timelines*. We chose this document to observe how much the variation of the α threshold distances and approximates the results of topics on economics. For testing, we set α to 0.2 and 0.8, so that the results are not radically dependent on the *query* or the reference document alone.

As shown in Fig. 5, varying α from 0.5 (TM2) to 0.8 (TM4), almost all the documents with the main topic related to the economy were replaced, leaving only one, the fourth document (*"Readjustment for STF ministers causes a divergence between PSDB and PMDB"*) in T4. Although this document is slightly related to economics because it contains several economic terms (e.g., salary, unemployment, debt), reading this document, we noticed that it focuses on the discussions between political parties caused by inappropriate salary adjustments during the impeachment process. On the other hand, varying α from 0.5 (TM2) to 0.2 (TM3), we can see that all the main documents of the intervals have an

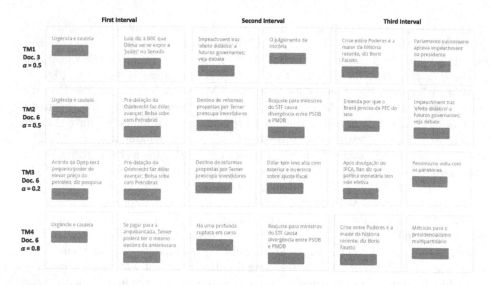

Fig. 5. Matrix with two most relevant documents by reference document and threshold intervals

economic focus, in general, on topics like "stock prices", "inflation", and "GDP growth". As a visible example of the favoring of economic topics tangential to the D6 document occasioned by the reduction of the α threshold, we can use the third (*"Odebrecht's pre-statement makes the dollar move forward; Stock Exchange goes up with Petrobras"*) document in TM2. This document became the most important document in its range in TM3. This article addresses the economic impacts occasioned by the suspected crimes of the Brazilian company Odebrecht[6] and its association with the Brazilian oil company Petrobras[7], Brazil's leading state-owned oil company. Investigations into the company at the time of the story's publication have caused distrust and oscillations in Petrobras' stock price, considerable economic impacts, which justifies the document's increased importance in its range.

Finally, comparing the main documents presented in TM3 with the documents in TM4, we can observe that both *timelines* do not share any main documents, even addressing constantly related topics. Observing the last document in TM3 (economic document), we observed it has no direct mention to the *impeachment* process. However, we noticed it brings related sentences, such as "The Congress may be in turmoil, running from the police" and "The public banks, stressed in the Dilma Rousseff years", despite the threshold $\alpha = 0.2$ drastically reducing the *query* importance. This behavior shows to be successful in capturing semantic information between documents by our models, allowing our method to correlate tangent documents from implicitly shared information. It also shows us the efficiency of the α threshold in directing the construction of *timelines*.

---

[6] https://en.wikipedia.org/wiki/Odebrecht.
[7] https://en.wikipedia.org/wiki/Petrobras.

5   Conclusion and Future Work

In this paper, we presented the StoryTracker tool, which can be integrated into any search engine so that users can easily organize and explore documents coming from a search by presenting the results through timelines. Our proposal is composed of three main parts: (1) Query Enrichment; (2) Timeline Construction; and (3) Summarization Interface. In the first part, we enrich the query by asking users to specify, besides the words, the document categories and a time interval. This query is then submitted to a search engine which returns several documents. The user chooses one s/he considers most relevant (reference document) to build a timeline. In the second part, our tool creates a vector representation using the *word embbedings* doc2vec model for all returned documents and the set of words that compose the query. The documents are then organized chronologically into different time subspaces. For each of these subspaces, we use the vectors representing the documents to identify the most similar documents to the query and the reference document, thus establishing a timeline. For the same set of documents returned by a search engine, the user can create different timelines, which present the same selected topic from different perspectives, varying the reference document. Finally, the third part comprises visual techniques to allow the user to explore the data.

To evaluate StoryTracker, we simulated a search engine using a database of articles published by the Brazilian newspaper Folha de São Paulo, from January 2015 to September 2017, composed of 167,000 documents. Simulating a search for the terms "the impeachment of Dilma Rousseff and its impacts", we build different timelines evaluating two different perspectives: political and economic. Our results show that our tool can trace different timelines using the original *query* and varying the reference document, covering different subtopics of the impeachment process (e.g., lava jato investigation, public manifestations). It would not be possible through a single *query* in a traditional search engine. As future work, our goal is to couple StoryTracker to different current search engines, performing an online evaluation of this combination under the usability perspective, considering different user profiles.

Acknowledgments. This project has partially supported by Huawei do Brasil Telecomunicações Ltda (Fundunesp Process # 3123/2020), FAPEMIG, and CAPES.

References

1. Alonso, O., Gertz, M., Baeza-Yates, R.: Clustering and exploring search results using timeline constructions. In: Proceedings of the 18th ACM Conference on Information and Knowledge Management, CIKM 2009, pp. 97–106. Association for Computing Machinery, New York (2009)
2. Alonso, O., Strötgen, J., Baeza-Yates, R., Gertz, M.: Temporal information retrieval: challenges and opportunities. In: TWAW Workshop, WWW, vol. 707, no. 01 (2011)
3. Attar, R., Fraenkel, A.S.: Local feedback in full-text retrieval systems. J. ACM **24**(3), 397–417 (1977)

4. Azad, H.K., Deepak, A.: Query expansion techniques for information retrieval: a survey. Inf. Process. Manage. **56**(5), 1698–1735 (2019)
5. Chang, Y., Tang, J., Yin, D., Yamada, M., Liu, Y.: Timeline summarization from social media with life cycle models. In: IJCAI (2016)
6. Jones, K.S., Walker, S., Robertson, S.E.: A probabilistic model of information retrieval: development and comparative experiments. In: Information Processing and Management, pp. 779–840 (2000)
7. Kanhabua, N., Anand, A.: Temporal information retrieval, pp. 1235–1238 (2016)
8. Karvelis, P., Gavrilis, D., Georgoulas, G., Stylios, C.: Topic recommendation using doc2vec. In: 2018 International Joint Conference on Neural Networks (IJCNN), pp. 1–6 (2018)
9. Kuzi, S., Shtok, A., Kurland, O.: Query expansion using word embeddings, pp. 1929–1932 (2016)
10. Le, Q.V., Mikolov, T.: Distributed representations of sentences and documents (2014)
11. Lee, H., Yoon, Y.: Engineering doc2vec for automatic classification of product descriptions on O2O applications. Electron. Commer. Res. **18**(3), 433–456 (2017). https://doi.org/10.1007/s10660-017-9268-5
12. Li, J., Cardie, C.: Timeline generation: tracking individuals on Twitter (2014)
13. Matthews, M., Tolchinsky, P., Blanco, R., Atserias, J., Mika, P., Zaragoza, H.: Searching through time in the New York times (2010)
14. Mikolov, T., Chen, K., Corrado, G., Dean, J.: Efficient estimation of word representations in vector space (2013)
15. Mikolov, T., Sutskever, I., Chen, K., Corrado, G., Dean, J.: Distributed representations of words and phrases and their compositionality (2013)
16. Qamra, A., Tseng, B., Chang, E.Y.: Mining blog stories using community-based and temporal clustering. In: Proceedings of the 15th ACM International Conference on Information and Knowledge Management, CIKM 2006, pp. 58–67. Association for Computing Machinery, New York (2006)
17. Rocchio, J.: Relevance feedback in information retrieval (1971)
18. Roy, D., Paul, D., Mitra, M., Garain, U.: Using word embeddings for automatic query expansion (2016)
19. Shao, Y., Taylor, S., Marshall, N., Morioka, C., Zeng-Treitler, Q.: Clinical text classification with word embedding features vs. bag-of-words features. In: 2018 IEEE International Conference on Big Data (Big Data), pp. 2874–2878 (2018)
20. Singh, J., Nejdl, W., Anand, A.: History by diversity. In: Proceedings of the 2016 ACM on Conference on Human Information Interaction and Retrieval - CHIIR 2016 (2016)
21. Tahvili, S., Hatvani, L., Felderer, M., Afzal, W., Bohlin, M.: Automated functional dependency detection between test cases using doc2vec and clustering (2019)
22. Trieu, L., Tran, H., Tran, M.-T.: News classification from social media using Twitter-based doc2vec model and automatic query expansion, pp. 460–467 (2017)
23. Wang, Y., Huang, H., Feng, C.: Query expansion with local conceptual word embeddings in microblog retrieval. IEEE Trans. Knowl. Data Eng. **33**(4), 1737–1749 (2019)

A Social Distancing-Based Facility Location Approach for Combating COVID-19

Suman Banerjee[1](✉), Bithika Pal[2], and Maheswar Singhamahapatra[3]

[1] Indian Institute of Technology, Jammu, India
suman.banerjee@iitjammu.ac.in
[2] Indian Institute of Technology, Kharagpur, India
bithikapal@iitkgp.ac.in
[3] School of Business, FLAME University, Pune, India

Abstract. In this paper, we introduce and study the problem of facility location along with the notion of *'social distancing'*. The input to the problem is the road network of a city where the nodes are the residential zones, edges are the road segments connecting the zones along with their respective distance. We also have the information about the population at each zone, different types of facilities to be opened and in which number, and their respective demands in each zone. The goal of the problem is to locate the facilities such that the people can be served and at the same time the total social distancing is maximized. We formally call this problem as the SOCIAL DISTANCING-BASED FACILITY LOCATION PROBLEM. We mathematically quantify social distancing for a given allocation of facilities and proposed an optimization model. As the problem is NP-Hard, we propose a simulation-based and heuristic approach for solving this problem. A detailed analysis of both methods has been done. We perform an extensive set of experiments with synthetic datasets. From the results, we observe that the proposed heuristic approach leads to a better allocation compared to the simulation-based approach.

Keywords: Facility location · Social distancing · Integer programming · COVID-19

1 Introduction

In recent times, the entire world is witnessing the pandemic Corona Virus Infectious Diseases (abbreviated as COVID-19) and based on the worldometers.com data, more than two hundred countries and territories are affected[1]. In this situation, every country tried with their best effort to combat this pandemic.

---

[1] https://www.worldometers.info/coronavirus/countries-where-coronavirus-has-spread/. Currently (around May 2021) the second wave has attacked countries like. As of now (dated 9th May, 2021), the number of deaths due to COVID-19 are 3304113, and the number of confirmed cases are 158863033.

© Springer Nature Switzerland AG 2021
O. Gervasi et al. (Eds.): ICCSA 2021, LNCS 12951, pp. 141–156, 2021.
https://doi.org/10.1007/978-3-030-86970-0_11

'Isolation', 'Quarantine', 'wearing of face mask' and 'Social Distancing' are the four keywords recommended by the *World Health Organization* (henceforth mentioned as WHO) to follow for combating this spreading disease. As per the topic of our study, here we only explain the concept of 'Social Distancing'. As per the WHO guidelines, the Corona virus spreads with the droplets that come out during the sneezing of an already infected person [27]. Hence, it is advisable to always maintain a distance of 1 meter (approximately 3 feet) from every other person.

Most of the countries (including India) were under temporary lock down and all the essential facilities including shopping malls, different modes of public transport, etc. were completely closed. However, for common people to survive some essential commodities (e.g., groceries, medicines, green vegetables, milk, etc.) are essential. It is better if the state administration can facilitate these essential services among the countrymen. However, as per the WHO guidelines, it is always advisable to maintain a distance of 1 meter from others whenever a person is out of his residence. Hence, it is an important question where to locate these facilities such that the demands of the customers can be served, and at the same time, social distancing among the people is maximized. Essentially, this is a facility location problem along with an additional measure 'social distancing'.

The study of the *Facility Location* problem goes back to the 17-th century and at that time this problem was referred to as the *Fermat-Weber Problem*[2]. This problem deals with selecting the location for placement of the facilities to best meet the demanded constraints of the customers [23]. Since its inception due to practical applications, this problem has been well studied from different points of view including algorithmic [9], computational geometry [6], the uncertainty of user demands [17], etc. However, to the best of our knowledge, the concept of social distancing is yet to be formalized mathematically, and also the facility location problem has not been studied yet with social distancing as a measure. It has been predicted that the third wave of COVID-19 is most likely to hit India sometime at the end of November or early December this year[3]. Hence, it is important to study the facility location problem considering social distancing as an important criterion.

In this paper, we study the problem of facility location under the social distancing criteria. The input to the problem is a road network (also known as graph) of a city where the nodes are small residential zone, edges are the streets connecting the zones. Additionally, we have how many different kinds of facilities are there and in which number they will be opened, population and demand of different kinds of facilities of every zone. The goal of this problem is locate the facilities in such a way that the basic amenities can be distributed to the people and the social distancing (mathematical abstraction is deferred until Sect. 3) is maximized. In particular we make the contributions in this paper.

---

[2] https://en.wikipedia.org/wiki/Geometric_median.

[3] https://www.indiatoday.in/coronavirus-outbreak/story/india-third-wave-of-covid-19-vaccine-prevention-1799504-2021-05-06.

- We mathematically formalize the notion of 'social distancing' and integrate it into the facility location problem. We formally call our problem as the SOCIAL DISTANCING-BASED FACILITY LOCATION Problem.
- We propose an optimization model for this problem which is a quadratic binary integer program.
- We propose two different approaches for solving the optimization problem. The first one is a simulation-based approach, and the other one is a heuristic solution.
- The proposed solution methodologies have been analyzed for time and space requirements.
- Finally, the proposed methodologies have been implemented with synthetic datasets and perform a comparative study between the proposed methodologies.

The remaining part of this paper is organized as follows: Sect. 2 presents some relevant studies from the literature. Section 3 describes our system's model, quantification of the 'social distancing', and introduce the SOCIAL DISTANCING-BASED FACILITY LOCATION PROBLEM. Section 4 describes the mathematical model of our problem. In Sect. 5, we propose one simulation-based approach and a heuristic solution approach for solving this problem. Section 6 contains the experimental evaluation, and finally, Sect. 7 concludes our study.

2   Related Work

The problem of facility location was initially originated from the operations research community [10] and has been investigated by researchers from other communities as well including computational geometry [6], graph theory [28], Management Sciences [25], Algorithm Design [9], Geographical Information Systems [18] and many more. Also, this problem has been studied in different settings such as Uncapasitated [14], Capasited [29], Multistage [7], Hierarchical [12], Stochastic Demand [17], Unreliable Links [21], Under Disruptions [1] and their different combinations [22]. The problem has got applications in different domains of society including health care [2], defense [15], etc. However, to the best of our knowledge, there does not exist any literature that studies the facility location problem considering the social distancing as a measure.

The variant of facility location introduced and studied in this paper comes under the facility location under disruption. There exists an extensive set of literature in this context. Recently, Afify et al. [1] studied the reliable facility location problem under disruption, which is the improvisation of the well-studied Reliable p-Median Problem [20] and Reliable Uncapacitated Facility Location Problem. They developed an evolutionary learning technique to near-optimally solve these problems. Yahyaei and Bozorgi-Amiri [32] studied the problem of relief logistics network design under interval uncertainty and the risk of facility disruption. They developed a mixed-integer linear programming model for this problem and a robust optimization methodology to solve this. Akbari et al. [3] presented a new tri-level facility location (also known as r-interdiction median model) model and proposed four hybrid meta-heuristics solution methodologies for solving the model. Rohaninejad et al. [24] studied a multi-echelon reliable capacitated facility location problem and proposed an accelerated *Benders Decomposition* Algorithm for solving this problem. Li and Zhang [16] studied the supply chain network design problem under facility disruption. To solve this problem they proposed a sample average approximation algorithm. Azizi [5] studied the problem of managing facility disruption in a hub-and-spoke network. They formulated mixed-integer quadratic program for this problem which can be linearized without significantly increasing the number of variables. For larger problem instances, they developed three efficient particle swarm optimization-based metaheuristics which incorporate efficient solution representation, short-term memory, and special crossover operator. Recently, there are several other studies in this direction [8,11]. As mentioned previously, there is an extensive set of literature on facility location problems on different disruption, and hence, it is not possible to review all of them. Look into [13,26] for a recent survey.

To the best of our knowledge, the problem of facility location has not been studied under the theme of social distancing, though this is the need of the hour. Hence, in this paper, we study the facility location problem with the social distancing criteria.

3   Preliminaries and the System's Model

In this section, we present the background and formalize the concept of social distancing. The input to our problem is the road network of a city represented by a simple, finite, and undirected graph $G(V, E, W)$. Here, the vertex set $V(G) = \{v_1, v_2, \ldots, v_n\}$ represents the small residential zones, the edge set $E(G) = \{e_1, e_2, \ldots, e_m\}$ denotes the links among the residential zones. W denotes the edge weight function that maps the distance of the corresponding street, i.e., $W : E(G) \longrightarrow \mathbb{R}^+$. We use standard graph-theoretic notations and terminologies from [31]. For any zone $v_j \in V(G)$, $N_{<d}(v_j)$ denotes the set of neighbors of v_j within the distance d. We denote the number of nodes and edges of G by n and m, respectively. Additionally, we have the following information. For each $v_j \in V(G)$, $P(v_j)$ denotes the people residing at zone v_j. In the rest of the paper, we use the term 'zone' and 'node', interchangeably. Let, $\mathcal{H} = \{h_1, h_2, \ldots, h_k\}$ denotes the k different kinds of facilities (say, $h_1 \equiv$ 'groceries' , $h_2 \equiv$ 'medicine' and so on) need to be opened. For every zone $v_j \in V(G)$ and every kind of facility $h_i \in \mathcal{H}$, we know the corresponding demand $\mathcal{D}_{v_j}(h_i)$. Hence, the total demand for the facility h_i is $\mathcal{D}(h_i) = \sum\limits_{v_j \in V(G)} \mathcal{D}_{v_j}(h_i)$.

Social Distancing. Now, we formally explain the meaning of the term 'social distancing' and express it as a mathematical expression. Suppose, a facility $h_i \in \mathcal{H}$ is opened for a particular duration $[t_1, t_2]$ in a day at the zone $v_j \in V(G)$. Let, $k_{i,j}^t$ denotes the number of people in the queue before this facility at time $t \in [t_1, t_2]$. From the geographical location, it is possible to determine the maximum number of people that can be present in the queue by maintaining WHO recommended distance to each other. For the zone $v_j \in V(G)$ and facility $h_i \in \mathcal{H}$, this number is denoted as $\gamma_{i,j}$. If the number of people in the queue is within the threshold then the social distancing should be maximum and when it crosses the threshold the social distancing factor decreases linearly or exponentially depending upon the number of people. We denote the social distancing function by $S()$. The following equation mentions this.

$$S(k_{i,j}^t) = \begin{cases} \text{positive constant,} & \text{if } k_{i,j}^t < \gamma_{i,j} \\ \text{decreases linearly or exponentially,} & \gamma_{i,j} \leq k_{i,j}^t < 2 \cdot \gamma_{i,j} \\ \text{decreases linearly or exponentially,} & k_{i,j}^t \geq 2 \cdot \gamma_{i,j} \end{cases}$$

Once $k_{ij}^t > \gamma_{i,j}$, the value of social distancing function decreases in two ways: linearly and exponentially, which is described below.

1. **Linear Social Distance Factor**: Here, $S(.)$ decreases linearly with the increasing value current number of people in the queue at a particular facility of certain zone. We model this using Eq. 1.

$$S(k_{i,j}^t) = A - b * max\{k_{i,j}^t - \gamma_{i,j}, 0\}, \quad \text{where } A \text{ and } b \text{ are positive constant.} \tag{1}$$

2. **Exponential Social Distance Factor**: Here, the $S(.)$ decreases exponentially with the linear increment of the current queue strength at a particular facility of certain zone. We model this using Eq. 2.

$$S(k_{i,j}^t) = A - b \cdot [\gamma_{i,j} + exp(k_{i,j}^t - 2 \cdot \gamma_{i,j})^{\frac{1}{4}}], \quad \text{where } A > 1 \tag{2}$$

The goal of the social distancing-based facility location problem is to assign facilities in the zones such that the total value of the social distancing function is maximized. Formally, we present this problem as follows.

SOCIAL DISTANCING-BASED FACILITY LOCATION PROBLEM

Input: The road network $G(V, E, W)$, Population at each zone $v_j \in V(G)$; i.e.; $P(v_i)$, Types of facilities $\mathcal{H} = \{h_1, h_2, \ldots, h_k\}$, Number of facilities of each type to be open; i.e.; $\{\ell_1, \ell_2, \ldots, \ell_k\}$, and their corresponding demand in each zone v_j; i.e.; $\{\mathcal{D}_{v_j}(h_1), \mathcal{D}_{v_j}(h_2), \ldots, \mathcal{D}_{v_j}(h_k)\}$ for all $v_j \in V(G)$.

Problem: Place maximum number of facilities to be opened of each type such that total social distancing is maximized.

Here, we note that each $\ell_i << n$. Symbols and notations used in this study are given in Table 1.

Table 1. Symbols with their interpretation

Symbol	Interpretation
$G(V, E, W)$	The road network under consideration
$V(G), E(G)$	Residential zones and road segments joining them
W	Edge weight function
$N_{<d}(v_j)$	Neighbors of the zone v_j within distance d
d_{max}	Maximum degree of a node in G
$P(v_j)$	Population at the zone v_j
P_{max}	Maximum population of a zone in $V(G)$
P	Total population of all zones of $V(G)$
$\mathcal{H}$	Set of different kinds of facilities
k	Number of different kinds of facilities; i.e.; $|\mathcal{H}| = k$
ℓ_i	Number of facilities opened of type h_i
ℓ_{max}	Maximum number of facilities among all types
$[t_1, t_2]$	Operation time window of the facilities
Δ	The difference between t_2 and t_1
$k_{i,j}^t$	Number of people in the queue at $v_j \in V(G)$, $h_i \in \mathcal{H}$, and $t \in [t_1, t_2]$
$\gamma_{i,j}$	Queue strength of the facility $h_i \in \mathcal{H}$ at $v_j \in V(G)$
$S(.)$	The social distancing function
$\mathcal{D}_{v_j}(h_i)$	Demand of facility type h_i at zone v_j

4   Mathematical Model Formulation

Now, we formulate a mathematical model of our problem. First, we define the decision variables involved in our formulation.

$$x_{i,j} = \begin{cases} 1, & \text{if the facility of type } h_i \text{ is opened at zone } v_j \\ 0, & \text{otherwise} \end{cases}$$

$$y_{i,j,p}^t = \begin{cases} 1, & \text{if the person } p \text{ visits the facility of type } h_i \text{ at zone } v_j \text{ at time } t \\ 0, & \text{otherwise} \end{cases}$$

Given the decision variables, total social distancing can be given by the following equation.

$$F = \sum_{h_i \in \mathcal{H}} \sum_{v_j \in V(G)} x_{i,j} \left(\sum_{t \in [t_1, t_2]} \sum_{p \in P(v_j)} S(k_{i,j}^t) \, y_{i,j,p}^t \right) \tag{3}$$

In Eq. 3, four summations are involved. The first one is to sum up for all the facilities. Similarly, the second one is to sum up for all the nodes. Next, one is used to sum up for all distinct time stamps within the operated time interval. Finally, the last one is for all the people of a node. The goal is to maximize the function mentioned in Eq. 3 subject to certain constraints. Now, we present the mathematical model, and next, we describe the meaning of each constraint.

Maximizing Social Distancing

$$max \sum_{h_i \in \mathcal{H}} \sum_{v_j \in V(G)} x_{i,j} \left(\sum_{t \in [t_1,t_2]} \sum_{p \in P(v_j)} S(k_{i,j}^t) \, y_{i,j,p}^t \right)$$

subject to,

$$\sum_{v_j \in V(G)} x_{i,j} \le \ell_i, \quad \forall h_i \in \mathcal{H} \tag{4}$$

$$\sum \boldsymbol{y}_{i,j,p}^{:T} \, \boldsymbol{Y}_{i,:,p}^{:} = 1, \quad \forall p \in \bigcup_{v_k \in V(G)} P(v_k), \quad \forall v_j \in V(G), \quad \forall h_i \in \mathcal{H} \tag{5}$$

$$\sum_{t \in [t_1,t_2]} \sum_{v_k \in \{v_j\} \cup N_{<d}(v_j)} y_{i,k,p}^t \ge 1, \forall h_i \subset \mathcal{H}, \forall v_j \subseteq V(G), \forall p \in P(v_j)$$
$$\tag{6}$$

$$y_{i,j,p}^t \left(\sum_{t' \in [t_1,t_2]} y_{i,k,p}^{t'} \right) \{ t - \sum_{t' \in [t_1,t_2]} y_{i,k,p}^{t'} t' \} \le 0, \forall h_i \in \mathcal{H}, \ \forall (j,k) \in E(G),$$

$$\forall p \in P(v_j), \forall t \in [t_1,t_2] \tag{7}$$

$$x_{i,j} \in \{0,1\}, \quad \forall h_i \in \mathcal{H}, \quad \forall v_j \in V(G) \tag{8}$$

$$y_{i,j,p}^t \in \{0,1\}, \quad \forall h_i \in \mathcal{H}, \quad \forall v_j \in V(G), \quad \forall t \in [t_1,t_2], \quad \forall p \in \bigcup_{v_k \in V(G)} P(v_k) \tag{9}$$

The first constraint in Inequation 4 enforces that the total number of open facilities of a particular type h_i should not be more than its allowance (which is ℓ_i). All the remaining constraints are on $y_{i,j,p}^t$. The assumption is a person can visit a facility location at once. Then, he or she can wait there for a certain period for being served or visit the nearest neighboring facility location to get the service. Now, the quantification of the continuous time domain is done in such a way that $y_{i,j,p}^t$ can be one, only once for the entire duration of $[t_1,t_2]$. Another physical condition implies that one person can not be in two different facility locations at the same time. Both the scenarios are capture in the second constraint described in Inequation 5. Here, $\boldsymbol{Y}_{i,:,p}^{:}$ is the matrix of shape total duration by the number of facility locations denoting the presence of a person p at any location in the entire time duration to avail h_i-type facility, and $\boldsymbol{y}_{i,j,p}^{:}$ is the one hot vector of the matrix $\boldsymbol{Y}_{i,:,p}^{:}$, which describes the presence at the particular location v_j. Now, the next constraint in Inequation 6 implies that the people should be served either at the location where they stay (e.g. v_j), or any nearby locations (e.g. $N_{<d}(v_j)$) within certain vicinity. The next constraint in Eq. 7 tells that if an arbitrary person p visits the locations at v_j and $v_k \in N_{<d}(v_j)$

to avail the facility type h_i then he first visits the place where he stays, then the neighboring ones. Finally, Inequation 8 and 9 tells that the decision variables x_{ij} and y^t_{ijp}, for all $h_i \in \mathcal{H}$, $v_j \in V(G)$, $t \in [t_1, t_2]$ take binary values.

Note: Observe that in this study we do not consider the cost of the facilities which is quite natural in case of traditional facility location problems [23]. This is due to the following reason. During the lock down period, a significant population are jobless, and hence government is providing the basic amenities free of cost. This is our assumption as mentioned in Sect. 1. We also assume that there is no dearth of supply.

5   Proposed Solution Approaches

As the facility location problem is NP-Hard, finding the optimal solution for our problem is also computationally intractable. Hence, we propose two different approaches to solve our problem. The first one is the simulation-based approach and the second one is a heuristic algorithm. It is important to note that none of these methods produces an optimal solution.

5.1   Simulation-Based Approach

A simulation-based approach has been adopted to obtain a solution for the developed model. Demand is generated for the people from all the nodes. As mentioned previously, a facility of type h_i can have at most ℓ_i many ($\ell_i \ll n$). People of a node get the service of a facility from the same node if it is available there, otherwise move to the neighbor node where a service facility is opened. People's arrival to the service facility follows *Poisson Distribution* and service time follows *Exponential Distribution* [4]. People get service based on the first come first serve basis. We have simulated these features of how people are arriving and getting the service and during this process how social distance is being maintained based on the number of people in the queue at a given time. People from several nodes may come and form a queue based on the arrival time and get the service based on first come first serve (FCFS) basis. Due to the space limitations we do not discuss anything related to the queuing theory and can be found at [4].

Now, the arrival time of people of all the nodes is generated based on the predefined arrival rate. It is implemented based on the inter-arrival time using poisson distribution which is reciprocal of arrival rate. Next, for each node, where the facility is not opened, the nearby open facility is found based on the least distance. Arrival time for all the people in a given and its nearby nodes are sorted. Service times are generated for the nodes where a given type of service facility is opened using the exponential distribution. Social distance is calculated at each open node using Eq. 3. This process is repeated for a defined number of simulation runs. Finally, the locations corresponding to the maximum value of the social distancing function are returned as a solution. This process is shown in the form of pseudo code in Algorithm 1.

Algorithm 1: Pseudo Code for the Simulation-Based Approach

Data: Underlying Road Network $G(V, E, W)$, Population and Respective Demand of Every Zone, Number of Different Types of Facilities, Number of facilities of each type

Result: Location of Different Facilities

1   $Simulation\_Run \longleftarrow 0$;

2   **while** $Simulation\_Run$ *not reached* **do**

3   **Step 1:** For every type of facility (say h_i) the allowed number of them (say ℓ_i)are opened randomly;

4   **Step 2:** Arrival time of peoples of all the nodes are generated based on the predefined arrival rate.;

5   **Step 3:** For each node, where facility is not opened, the nearest open facility is found based on the least distance.;

6   **Step 4:** For a node v_j, the people of of this zone and those who are assigned to this facility, their arrival times are sorted and queue is formed.

7   **Step 5:** Service times are generated for the nodes where a given type of service facility is opened using the exponential distribution of service time.;

8   **Step 6:** Based on the arrival time and service time, number of people in queue is derived. ;

9   **Step 7:** Social distance is calculated at each queue and summed up.;

10   **Step 8:** $Simulation\_Run$ is incremented by 1; i.e.; $Simulation\_Run = Simulation\_Run + 1$;

11 **end**

12 **Step 9**: Among all the simulation runs corresponding locations of the maximum is returned for locating the facilities.

Now, we analyze Algorithm 1 to understand its time and space requirement. Let, $\mathcal{R}$ denotes the number of simulation runs. Let, the number of different facilities are k and ℓ_{max} denotes the maximum number of facilities among all types. Now, it is easy to observe that the running time of Step 1 is of $\mathcal{O}(k \cdot \ell_{max})$. Let, $P(v_i)$ denotes the number of people of at node v_i and P denotes the number of people in all the nodes in G; i.e.; $P = \sum_{v_i \in V(G)} P(v_i)$. As, for a people generating the arrival time requires $\mathcal{O}(1)$ time, hence, time requirement of Step 2 requires $\mathcal{O}(P)$ time. The number of nodes where the facility is not opened will be greater than equals to $(n - \ell_{max})$ and in the worst case it will be of $\mathcal{O}(n)$. Let d_{max} denotes the maximum possible degree of a node in G; $d_{max} = argmax_{v \in V(G)} deg(v)$. For each node without facility to choose the nearest facility requires $\mathcal{O}(d_{max})$. So, for all the nodes and all types of facilities required computational time is of $\mathcal{O}(n \cdot k \cdot d_{max})$. This implies that the execution of Step 3 requires $\mathcal{O}(n \cdot k \cdot d_{max})$ time. Let, P_{max} denotes the maximum number of people residing at any zone, hence, $P_{max} = argmax_{v \in V(G)} P(v)$. As, in any facility the people of its neighbors may come, and hence, the total number of people in any facility is of $\mathcal{O}((d_{max}+1)P_{max})$. Sorting this people based on the arrival time will require

$\mathcal{O}((d_{max} \cdot P_{max} + P_{max}) \log(d_{max} \cdot P_{max} + P_{max}))$ time. This implies that the execution of Step 4 requires $\mathcal{O}((d_{max} \cdot P_{max} + P_{max}) \log(d_{max} \cdot P_{max} + P_{max}))$ time. Assuming that the generating service time requires $\mathcal{O}(1)$, executing Step 5 requires $\mathcal{O}(k \cdot \ell_{max})$ time. For a single queue, calculating the number of people in it requires $\mathcal{O}(1)$ time. Hence, executing Step 6 requires $\mathcal{O}(k \cdot (d_{max} + 1) \cdot P_{max})$.

Assume that $t_2 - t_1 = \Delta$. Hence, from the objective function it is easy to follow that computing the social distancing requires $\mathcal{O}(n \cdot k \cdot \Delta \cdot P_{max})$ time. If in each iteration of the while loop, we update the maximum value of the social distancing and the corresponding location of the facilities of different types then after the last iteration we obtain the solution of the proposed simulation approach. This requires $\mathcal{O}(k \cdot \ell_{max})$ time. So, the total time required by the simulation procedure is of $\mathcal{O}(\mathcal{R}(k \cdot \ell_{max} + P + n \cdot k \cdot d_{max} + (d_{max} \cdot P_{max} + P_{max}) \log(d_{max} \cdot P_{max} + P_{max}) + k \cdot \ell_{max} + k \cdot (d_{max} + 1) \cdot P_{max} + n \cdot k \cdot \Delta \cdot P_{max} + k \cdot \ell_{max}))$. This reduces to $\mathcal{O}(\mathcal{R}(k \cdot \ell_{max} + P + n \cdot k \cdot d_{max} + (d_{max} \cdot P_{max} + P_{max}) \log(d_{max} \cdot P_{max} + P_{max}) + k \cdot d_{max} \cdot P_{max} + n \cdot k \cdot \Delta \cdot P_{max}))$. Additional space required by the proposed simulation approach is to store the arrival time of the people which is of $\mathcal{O}(n \cdot P_{max})$, service time of the queues which is of $\mathcal{O}(k \cdot \ell_{max})$, location the facilities which is of $\mathcal{O}(k \cdot \ell_{max})$, social distancing function value at each queue which is of $\mathcal{O}(k \cdot \ell_{max})$. So, the total space requirement is of $\mathcal{O}(n \cdot P_{max} + k \cdot \ell_{max})$. Hence, Theorem 1 holds.

Theorem 1. *The running time and space requirement of the proposed simulation-based approach is of* $\mathcal{O}(\mathcal{R}(k \cdot \ell_{max} + P + n \cdot k \cdot d_{max} + (d_{max} \cdot P_{max} + P_{max}) \log(d_{max} \cdot P_{max} + P_{max}) + k \cdot d_{max} \cdot P_{max} + n \cdot k \cdot \Delta \cdot P_{max}))$ *and* $\mathcal{O}(n \cdot P_{max} + k \cdot \ell_{max})$, *respectively.*

5.2   Heuristic Solution

In this section, we propose a heuristic solution to solve our problem. First, we describe the working principle, and subsequently, we present it as a step-by-step procedure. As a first step, we sort all the nodes based on demand in descending order. Based on this heuristic for any type of facility $h_i \in \mathcal{H}$, ℓ_i many facilities will be placed at the top demand zones. Now, it allocates the people to the facility in the following way: For the high demand locations the people of that zones are allocated to the same zone, and subsequently, we allocate the people of other zone in the reverse order such that the entire population of $V(G)$ is distributed as uniformly as possible. Without loss of generality, assume that after sorting the nodes based on the demand, the order of the nodes are $\rho = < v_1, v_2, \ldots, v_n >$. For any zone $v_j \in \{v_1, v_2, \ldots, v_i\}$, the people of zone v_i are allocated to the facility at the same zone. Now, v_{i+1} is allocated to v_i, v_{i+2} is allocated to v_{i-1}, and v_{2i} to v_1. Similarly, v_{2i+1} is allocated to v_2, v_{2i+2} is allocated to v_3 and so on. Next from Step 2 to Step 7 of Algorithm 1 is executed. Pseudo code is given in Algorithm 2.

Now, we analyze Algorithm 2 to understand its time and space requirement. Sorting of the nodes based on the demand requires $\mathcal{O}(n \log n)$ time. It is easy to observe that the allocation of the facilities requires $\mathcal{O}(n)$ time. As described in

Algorithm 2: Pseudo code for the Heuristic Algorithm

Data: Underlying Road Network $\mathcal{G}(\mathcal{V}, \mathcal{E})$, Population and Respective Demand of Every Zone, Number of Different Types of Facilities, Number of facilities of each type

Result: Location of Different Facilities

1 **Step 1**: All the nodes of the network are sorted based on the demand of the nodes;

2 **Step 2**: People of nodes are allocated in the facilities as described.;

3 **Step 3**: Execute Step 2 to Step 7 of Algorithm 1.;

Sect. 5.1, the total execution time from Steps 2 to 7 is of $\mathcal{O}(P + n \cdot k \cdot d_{max} + (d_{max} \cdot P_{max} + P_{max}) \log(d_{max} \cdot P_{max} + P_{max}) + k \cdot \ell_{max} + k \cdot d_{max} \cdot P_{max} + n \cdot k \cdot \Delta \cdot P_{max})$ time. Hence, total time requirement of Algorithm 2 is of $\mathcal{O}(n \log n + P + n \cdot k \cdot d_{max} + (d_{max} \cdot P_{max} + P_{max}) \log(d_{max} \cdot P_{max} + P_{max}) + k \cdot \ell_{max} + k \cdot d_{max} \cdot P_{max} + n \cdot k \cdot \Delta \cdot P_{max})$. It is easy to observe that the space requirement of Algorithm 2 is same as Algorithm 1 which is of $\mathcal{O}(n \cdot P_{max} + k \cdot \ell_{max})$. Hence, Theorem 2 holds.

Theorem 2. *The running time and space requirement of the proposed heuristic approach is of $\mathcal{O}(n \log n + P + n \cdot k \cdot d_{max} + (d_{max} \cdot P_{max} + P_{max}) \log(d_{max} \cdot P_{max} + P_{max}) + k \cdot \ell_{max} + k \cdot d_{max} \cdot P_{max} + n \cdot k \cdot \Delta \cdot P_{max})$ and $\mathcal{O}(n \cdot P_{max} + k \cdot \ell_{max})$, respectively.*

6 Experimental Evaluation

In this section, we describe the experimental evaluation of the proposed methodology. Initially, we start by describing the datasets.

6.1 Dataset Description

As it is difficult to find datasets that contains all the required information in this study, hence we create some synthetic datasets. We perform our experiments with two different kinds of networks topology:

- **Complete Network**: In this kind of network, every vertex is connected to every other vertex of the network. It is easy to observe that an undirected complete network with n vertices will have $\frac{n(n-1)}{2}$ edges.
- **Rectangular Grid Network**: A rectangular grid network $G(V, E)$ of size $n \times n$ is defined with the vertex set $V(G) = [n] \times [n] = \{1, 2, \ldots, n\} \times \{1, 2, \ldots, n\}$ and two vertices (i, j) and (i', j') will be connected by an edge if both $|i - i'| \leq 1$ and $|j - j'| \leq 1$. Many real-world road networks are of grid structure [19, 30].

For the ease of understanding, we show a demo diagram of complete and grid network in Fig. 1.

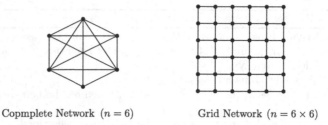

Copmplete Network $(n = 6)$ Grid Network $(n = 6 \times 6)$

Fig. 1. Figure of a complete and grid network of size $n = 6$ and $n = 6 \times 6$, respectively.

6.2   Experimental Setup

In this section, we describe the experimental setup. Several parameters are there, whose value needs to be set. Here, we describe them one by one.

- **Network Size**: As mentioned, we use networks of two different kinds, namely complete and grid. For complete network, we use $n = 60$ and $n = 100$. On the other hand for the grid network we consider $n = 60 \times 60$ and $n = 100 \times 100$.
- **Population at Each Zone**: This has been chosen uniformly at random from the interval $[1000, 2000]$ for each zone.
- **Demand**: For simplicity, we consider the uniform and demand for all the people.
- **Parameters Related to Queuing Theory**: We consider the inter arrival time and service time are 1 min and 0.7 min, respectively.
- **Parameters in the Social Distancing Function**: As mentioned previously, there are three parameters in the social distancing function. They are $\gamma_{i,j}, A,$ and b. In this study, we consider $\gamma_{i,j} = 4$ for all $h_i \in \mathcal{H}$ and $v_j \in V(G)$, $A = 10$, and $b = 0.5$.

We implement our both solution approaches in MATLAB (Version MATLAB 9.0 R2016a) and all the simulation codes and the synthetic datasets can be found at https://github.com/BITHIKA1992/Facility_location_Covid19.

6.3   Experimental Results

Now, we discuss the experimental results. Our goal here is to study the impact of number of facilities on social distancing and average queue length.

Impact on the Social Distancing. Here, we discuss a comparative study of two different solution methodologies on social distancing. Figure 2 shows the plots for the number of opened facilities vs. social distancing for both kinds of network topologies with two different network sizes. From the figure, it has been observed that for a fixed network size when the number of opened facilities increases, social distancing also increases. As an example For a complete network with 100 nodes, when the number of opened facilities is 20, the value of social distancing due to the allocation of facilities by the heuristic method is 2.4×10^6.

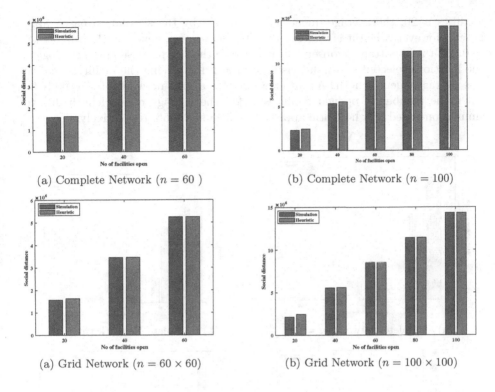

(a) Complete Network ($n = 60$)   (b) Complete Network ($n = 100$)

(a) Grid Network ($n = 60 \times 60$)   (b) Grid Network ($n = 100 \times 100$)

Fig. 2. Impact of the number of opened facilities on Social Distancing for Complete and Grid Network for two different sizes.

When the number of opened facilities are increased to 60, the value of social distancing also increased to 7.9×10^6.

The impact of algorithms on social distancing can also be observed from Fig. 2. We can conclude that when the number of opened facilities is more, it does not matter which algorithm is used to locate facilities the value of social distancing will not change much. However, if the number of opened facilities is much less, the location suggested by the heuristic approach leads to more value of the social distancing. As an example, in case of the 100×100 rectangular grid network when the number of opened facilities is 20, the value of social distancing function due to facility placement by simulation-based approach and heuristic approach is 2.08×10^6 and 2.29×10^6, respectively. Hence, the difference is approximately 210000.

Impact on Average Queue Length. Now, we discuss the impact of the number of opened facilities on the average queue length. Figure 3 shows the number of opened facilities vs. average queue length plots for both kinds of networks for two different sizes. From the figure, it has been observed that as the number of opened facilities increases naturally, the average queue length decreases. As an example, for a complete network with 60 nodes when the number of opened

facilities is 20, the average queue length is 2.58 by the simulation-based approach. However, when it is increased to 40 the average queue length drops down
to 0.68. It is important to observe that between the two proposed methodologies,
the locations selected by the heuristic approach for placing the facilities lead to
less average queue length. As an example, for a grid network of size 60×60
when the number of opened facilities is 20, the average queue length due to
simulation-based and heuristic approach is 2.98 and 1.56, respectively.

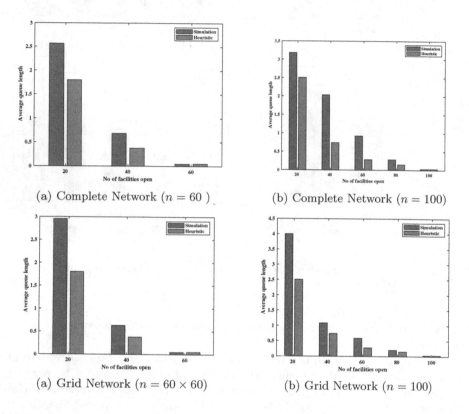

(a) Complete Network ($n = 60$)   (b) Complete Network ($n = 100$)

(a) Grid Network ($n = 60 \times 60$)   (b) Grid Network ($n = 100$)

Fig. 3. Impact of the number of opened facilities on average queue length for complete
and grid network for two different sizes.

From the experiments, we can observe that the allocation of facilities by the
heuristic approach leads to lesser queue length and consequently more value of
social distancing. This is due to the fact that the heuristic approach tries to
assign the people to the facilities in such a way that their loads are balanced.
Due to this reason, the social distancing achieved by the heuristic solution is
more.

7   Concluding Remarks

In this paper, we have studied the problem of the Social Distancing-Based Facility Location Problem and proposed two solution methodologies. The first one is a simulation-based approach and the second one is a heuristic solution. From the experiments with synthetic datasets, we observe that the heuristic solution leads to the allocation of facilities causing more social distancing. Our study can be extended in several directions. One can consider more realistic scenarios like unknown demand, supply constraint, and many more. We can also extend our study for competitive situation where there will be more than one agency for providing the facilities of each type and their goal is to maximize their profit at the same time the social distancing needs to be maintained.

References

1. Afify, B., Ray, S., Soeanu, A., Awasthi, A., Debbabi, M., Alloucho, M.: Evolutionary learning algorithm for reliable facility location under disruption. Expert Syst. Appl. **115**, 223–244 (2019)
2. Ahmadi-Javid, A., Seyedi, P., Syam, S.S.: A survey of healthcare facility location. Comput. Oper. Res. **79**, 223–263 (2017)
3. Akbari-Jafarabadi, M., Tavakkoli-Moghaddam, R., Mahmoodjanloo, M., Rahimi, Y.: A tri-level r-interdiction median model for a facility location problem under imminent attack. Comput. Ind. Eng. **114**, 151–165 (2017)
4. Asmussen, S.: Applied Probability and Queues, vol. 51. Springer, New York (2008)
5. Azizi, N.: Managing facility disruption in hub-and-spoke networks: formulations and efficient solution methods. Ann. Oper. Res. **272**(1–2), 159–185 (2019)
6. Bespamyatnikh, S., Kedem, K., Segal, M., Tamir, A.: Optimal facility location under various distance functions. Int. J. Comput. Geom. Appl. **10**(05), 523–534 (2000)
7. Biajoli, F.L., Chaves, A.A., Lorena, L.A.N.: A biased random-key genetic algorithm for the two-stage capacitated facility location problem. Expert Syst. Appl. **115**, 418–426 (2019)
8. Fotuhi, F., Huynh, N.: Reliable intermodal freight network expansion with demand uncertainties and network disruptions. Netw. Spatial Econ. **17**(2), 405–433 (2017)
9. Hassin, R., Ravi, R., Salman, F.S., Segev, D.: The approximability of multiple facility location on directed networks with random arc failures. Algorithmica **82**, 2474–2501 (2020)
10. Hassin, R., Tamir, A.: Improved complexity bounds for location problems on the real line. Oper. Res. Lett. **10**(7), 395–402 (1991)
11. Hatefi, S.M., Jolai, F.: Robust and reliable forward-reverse logistics network design under demand uncertainty and facility disruptions. Appl. Math. Model. **38**(9–10), 2630–2647 (2014)
12. Helber, S., Böhme, D., Oucherif, F., Lagershausen, S., Kasper, S.: A hierarchical facility layout planning approach for large and complex hospitals. Flexible Serv. Manufact. J. **28**(1–2), 5–29 (2016)
13. Ivanov, D., Dolgui, A., Sokolov, B., Ivanova, M.: Literature review on disruption recovery in the supply chain. Int. J. Prod. Res. **55**(20), 6158–6174 (2017)
14. Krishnaswamy, R., Sviridenko, M.: Inapproximability of the multilevel uncapacitated facility location problem. ACM Trans. Alg. (TALG) **13**(1), 1–25 (2016)

15. Lessin, A.M., Lunday, B.J., Hill, R.R.: A bilevel exposure-oriented sensor location problem for border security. Comput. Oper. Res. **98**, 56–68 (2018)
16. Li, X., Zhang, K.: A sample average approximation approach for supply chain network design with facility disruptions. Comput. Ind. Eng. **126**, 243–251 (2018)
17. Marinakis, Y., Marinaki, M., Migdalas, A.: A hybrid clonal selection algorithm for the location routing problem with stochastic demands. Ann. Math. Artif. Intell. **76**(1–2), 121–142 (2016)
18. Miller, H.J.: GIS and geometric representation in facility location problems. Int. J. Geograph. Inf. Syst. **10**(7), 791–816 (1996)
19. Miyagawa, M.: Optimal hierarchical system of a grid road network. Ann. Oper. Res. **172**(1), 349 (2009)
20. Mladenović, N., Brimberg, J., Hansen, P., Moreno-Pérez, J.A.: The p-median problem: a survey of metaheuristic approaches. Eur. J. Oper. Res. **179**(3), 927–939 (2007)
21. Narayanaswamy, N.S., Nasre, M., Vijayaragunathan, R.: Facility location on planar graphs with unreliable links. In: Fomin, F.V., Podolskii, V.V. (eds.) CSR 2018. LNCS, vol. 10846, pp. 269–281. Springer, Cham (2018). https://doi.org/10.1007/978-3-319-90530-3_23
22. Ortiz-Astorquiza, C., Contreras, I., Laporte, G.: Multi-level facility location problems. Eur. J. Oper. Res. **267**(3), 791–805 (2018)
23. Owen, S.H., Daskin, M.S.: Strategic facility location: a review. Eur. J. Oper. Res. **111**(3), 423–447 (1998)
24. Rohaninejad, M., Sahraeian, R., Tavakkoli-Moghaddam, R.: An accelerated benders decomposition algorithm for reliable facility location problems in multi-echelon networks. Comput. Indus. Eng. **124**, 523–534 (2018)
25. Ross, G.T., Soland, R.M.: Modeling facility location problems as generalized assignment problems. Manag. Sci. **24**(3), 345–357 (1977)
26. Snyder, L.V., Atan, Z., Peng, P., Rong, Y., Schmitt, A.J., Sinsoysal, B.: OR/MS models for supply chain disruptions: a review. IIE Trans. **48**(2), 89–109 (2016)
27. Tahir, M., Shah, S.I.A., Zaman, G., Khan, T.: Stability behaviour of mathematical model MERS corona virus spread in population. Filomat **33**(12), 3947–3960 (2019)
28. Tamir, A.: The K-centrum multi-facility location problem. Discrete Appl. Math. **109**(3), 293–307 (2001)
29. Tran, T.H., Scaparra, M.P., O'Hanley, J.R.: A hypergraph multi-exchange heuristic for the single-source capacitated facility location problem. Eur. J. Oper. Res. **263**(1), 173–187 (2017)
30. Watanabe, D.: A study on analyzing the grid road network patterns using relative neighborhood graph. In: The Ninth International Symposium on Operations Research and Its Applications, pp. 112–119. Citeseer (2010)
31. West, D.B., et al.: Introduction to Graph Theory, vol. 2. Prentice Hall, Upper Saddle River (2001)
32. Yahyaei, M., Bozorgi-Amiri, A.: Robust reliable humanitarian relief network design: an integration of shelter and supply facility location. Ann. Oper. Res. **283**(1), 897–916 (2019)

Using Mathematica and 4Nec2 to Design and Simulation of an Antenna in 2100 and 3500 MHz Bands

Ricardo Velezmoro-León⬭, Carlos Enrique Arellano-Ramírez⬭,
Daniel Alonso Flores-Córdova⬭, and Robert Ipanaqué-Chero(✉)⬭

Universidad Nacional de Piura, Urb. Miraflores s/n, Castilla, Piura, Peru
{rvelezmorol,carellanor,dflores,ripanaquec}@unp.edu.pe

Abstract. The design of an efficient antenna allows the delocalization of electromagnetic energy making possible the transmission of wireless energy. In this paper, we describe the design of the model of an antenna with an efficiency of 99.06% and an SWR of 1.01 after the impedance matching process at the 3500 MHz frequency. This antenna is generated from control points on a curve lying on a lemniscatic torus and it's projected on the XY plane. To carry out the design we used the Mathematica symbolic calculation system. In addition, we propose two frequencies for simulation purposes with the assistance of the free software 4Nec2, which will allow us to see its main characteristics, one of 2100 MHz (0.1428 m wavelength) and another of 3500 MHz (0.1428 m wavelength).

Keywords: Design · Simulation · Antenna · Lemniscatic torus

1 Introduction

An antenna is a transitional structure between free space and a guiding device [1]. The importance of the design of an antenna lies in its efficiency since the design of an efficient antenna allows the delocalization of electromagnetic energy making possible the transmission of wireless energy [4]. The use of computer software to perform modeling and simulation of various objects is gaining current popularity due to the development of computers and the constant updating, as well as the development, of specialized software [9].

In this paper, we describe the design of the model of an efficient antenna generated from control points on a curve lying on lemniscatic torus [3] and its projection on the XY plane. To carry out the design we use the Wolfram Language [10] of the Mathematica v.11.0 symbolic calculation system [11]. In addition, we propose two frequencies for simulation purposes with the assistance of the free software 4Nec2 [2], which will allow us to see its main characteristics, one of 2100 MHz (0.1428 m wavelength) and another of 3500 MHz (0.1428 m wavelength).

Supported by Fondo Especial de Desarrollo Universitario (FEDU), created by Law No. 25203 published on February 24, 1990.

We believe it convenient to mention the importance of having a guide for the correct writing of research articles in disciplines related to ICT. This paper has been prepared following the recommendations proposed in [6].

The structure of this paper is as follows: Sect. 2 introduces the mathematical definition of some surfaces and curves based on which the antenna proposed in this paper will be designed. Then, Sect. 3 introduces the antenna design proposed in this paper. Here, in the Wolfram Language, the lemniscatic torus, its projection, and the curve on which are the control points for the antenna design are defined. The simulation of the designed antenna with the 4Nec2 software is performed in Sect. 4. Finally, Sect. 5 closes with the main conclusions of this paper and some further remarks.

2   Mathematical Preliminaries

2.1   Surfaces of Revolution

Let C be a curve lying on a plane $P \subset \mathbb{R}^3$, and let A be a line that also lies on P but that does not meet C. When this profile curve C is revolved around axis A, it sweeps out a surface of revolution M in $\mathbb{R}^3$ [7].

2.2   Torus of Revolution

The torus of revolution is the surface of the revolution obtained when the profile curve C is a circle, according to Subsect. 2.1 [8].

2.3   Lemniscate of Bernoulli

The lemniscate of Bernoulli is defined as the locus of points of a plane such that the product of distances from two fixed points, of the same plane, $(-A, 0)$ and $(A, 0)$, $A > 0$, is a constant A^2. The Cartesian equation obtained from this definition is $(x^2 + y^2)^2 = 2A^2(x^2 - y^2)$. The parametric equations for the lemniscate with half-width A are

$$x(t) = \frac{A \cos t}{1 + \sin^2 t}; \; y(t) = \frac{A \sin t \cos t}{1 + \sin^2 t}; \; 0 < t < 2\pi \quad [5].$$

2.4   Lemniscatic Torus

The parametric equations for Bernoulli's lemniscate lying on the $x = B$, $B \in \mathbb{R}$, plane are

$$x(v) = B; \; y(v) = \frac{A \cos v}{1 + \sin^2 v}; \; z(v) = \frac{A \sin v \cos v}{1 + \sin^2 v}; \; -\frac{\pi}{2} < v < \frac{3\pi}{2}.$$

Given

$$R(u) = \begin{pmatrix} \cos u & -\sin u & 0 \\ \sin u & \cos u & 0 \\ 0 & 0 & 1 \end{pmatrix},$$

the lemniscatic torus (Fig. 1) is defined by

$$LT(u, v) = R(u) \cdot (x(v), y(v), z(v))^{\top}; \; 0 < u < 2\pi; \; -\frac{\pi}{2} < v < \frac{\pi}{2},$$

explicitly

$$LT(u, v) =$$
$$\left(B\cos(u) - \frac{A\sin(u)\cos(v)}{\sin^2(v) + 1}, \frac{A\cos(u)\cos(v)}{\sin^2(v) + 1} + B\sin(u), \frac{A\sin(v)\cos(v)}{\sin^2(v) + 1} \right);$$
$$0 < u < 2\pi; \; -\frac{\pi}{2} < v < \frac{\pi}{2}.$$

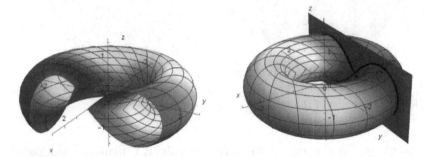

Fig. 1. Profile curve and the torus generated by a lemniscate (left), the lemniscate is also observed (right).

The families of basic curves inherent to the lemniscatic torus are the parallels, circumferences; and the meridians, lemniscate leaves.

2.5 Curves Lying on the Lemniscatic Torus

Curves lying on the lemniscatic torus are obtained by substituting the parameters u and v by differentiable functions $\alpha_1(t)$ and $\alpha_2(t)$, respectively. This is $LT(\alpha_1(t), \alpha_2(t))$.

3   Design of an Antenna Model with Mathematica

First, we define the lemniscatic torus in Mathematica:

```
LT[{u_,v_},{A_,B_}]:={B Cos[u]-(A Cos[v] Sin[u])/(1+Sin[v]^2),
  B Sin[u]+(A Cos[u] Cos[v])/(1+Sin[v]^2),
  (A Cos[v] Sin[v])/(1+Sin[v]^2)}
```

After compiling this definition it is possible to use the built-in `Parametric Plot3D` and `Show` functions to plot the lemniscatic torus and the curves lying on it (Fig. 2, left).

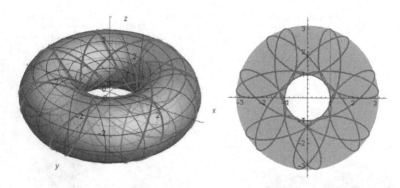

Fig. 2. A lemniscatic torus, LT[{u,v},{3,1}], and a curve, LT[{3t,7t},{3,1}], lying on it (left). Projections of the previous objects (right).

Next, we project the lemniscatic torus onto the XY plane. In Mathematica we have:

```
LTP[{u_,v_},{A_,B_}]:={B Cos[u]-(A Cos[v] Sin[u])/(1+Sin[v]^2),
  B Sin[u]+(A Cos[u] Cos[v])/(1+Sin[v]^2)}
```

After compiling this definition it is possible to use the built-in `ParametricPlot` and `Show` functions to plot the projections of the lemniscatic torus and the curves lying on it (Fig. 2, right).

Next, to facilitate later calculations, we define the curve, `cur2d`, as follows:

```
cur2d[t_]:=LTP[{3t,7t},{3,1}]
```

The domain of the curve is the interval $[0, 2\pi]$, we partition it into increments of $\frac{\pi}{35}$ and find the corresponding points on the curve:

```
m=cur2d/@Table[i,{i,0,2 Pi-Pi/35,Pi/35}];
```

We rearrange the point's list, m, and divide the coordinates of its elements by 100, taking into account that `m[[i]]` is the ith element of m.

```
data=m[[#]]&/@{46,45,63,39,60,67,18,64,12,11,66,65,13,59,
    10,17,38,14,32,31,16,15,33,9,30,37,58,34,52,51,36,35,53,
    29,50,57,8,54,2,1,56,55,3,49,70,7,28,4,22,21,6,5,23,69,20,
    27,48,24,42,41,26,25,43,19,40,47,68,44,62,61}/100;
```

See Table 1.

Table 1. Control points, stored in **data**, duly ordered and arranged in columns.

$(-0.0040, -0.031)$	$(-0.014, 0.015)$	$(0.031, 0.031)$
$(-0.0043, -0.020)$	$(-0.0077, 0.0081)$	$(0.021, 0.00026)$
$(-0.0014, -0.011)$	$(-0.0061, 0.0093)$	$(0.011, 0.0010)$
$(-0.0034, -0.011)$	$(-0.013, 0.016)$	$(0.011, -0.00095)$
$(-0.0066, -0.020)$	$(-0.017, 0.027)$	$(0.021, -0.0021)$
$(-0.011, -0.017)$	$(-0.010, 0.030)$	$(0.019, -0.0070)$
$(-0.0062, -0.0093)$	$(-0.0048, 0.020)$	$(0.010, -0.0040)$
$(-0.0077, -0.0080)$	$(-0.0035, 0.011)$	$(0.0095, -0.0057)$
$(-0.013, -0.016)$	$(-0.0015, 0.011)$	$(0.018, -0.0092)$
$(-0.022, -0.023)$	$(-0.0025, 0.020)$	$(0.027, -0.016)$
$(-0.027, -0.016)$	$(0.0025, 0.020)$	$(0.022, -0.023)$
$(-0.018, -0.0092)$	$(0.0015, 0.011)$	$(0.013, -0.016)$
$(-0.0095, -0.0057)$	$(0.0035, 0.011)$	$(0.0077, -0.0080)$
$(-0.010, -0.0040)$	$(0.0048, 0.020)$	$(0.0062, -0.0093)$
$(-0.019, -0.0070)$	$(0.010, 0.030)$	$(0.011, -0.017)$
$(-0.021, -0.0021)$	$(0.017, 0.027)$	$(0.0066, -0.020)$
$(-0.011, -0.00095)$	$(0.013, 0.016)$	$(0.0034, -0.011)$
$(-0.011, 0.0010)$	$(0.0061, 0.0093)$	$(0.0014, -0.011)$
$(-0.021, 0.00026)$	$(0.0077, 0.0081)$	$(0.0043, -0.020)$
$(-0.031, 0.0031)$	$(0.014, 0.015)$	$(0.0040, -0.031)$
$(-0.030, 0.011)$	$(0.018, 0.011)$	
$(-0.019, 0.0087)$	$(0.0096, 0.0057)$	
$(-0.010, 0.0039)$	$(0.010, 0.0039)$	
$(-0.0096, 0.0057)$	$(0.019, 0.0087)$	
$(-0.018, 0.011)$	$(0.030, 0.011)$	

The points in the *data* list are the control points for designing the antenna. Finally, we join the points of list **data** using straight lines and obtain the antenna design (Fig. 3).

```
lab=Table[Subscript[P,i],{i,Length[data]}];
g1=Graphics[{Blue,Thick,Line[data]}];
g2=Graphics[{Cyan,AbsolutePointSize[7],Point[data]}];
g3=Graphics[{MapThread[Text[#1,#2]&,{lab,data}]}];
Show[g1,g2,g3]
```

See Fig. 3 (left).

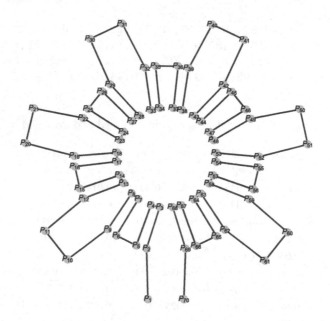

Fig. 3. Control points and antenna design.

4   Simulation of an Antenna Model with 4Nec2

The control points (Table 1) have been entered into the 4Nec2 software. With these points and in free space, the modeled antenna will have the appearance shown in Fig. 4.

4.1   Standing Wave Ratio (SWR) at 2100 MHz

Having the variables initialized, according to the points in Table 1, at a frequency of 2100 MHz the antenna presents an SWR = 13.2 (standing wave ratio) and an impedance of 52.3 Ω.

Figure 5 shows the simulation result for the 2000 MHz–2200 MHz range, with the same previous parameters. It is observed that, in a large part of the simulated range, the antenna could be adapted using some impedance matching technique; since the SWR (Standing Wave Ratio) increases with increasing frequency.

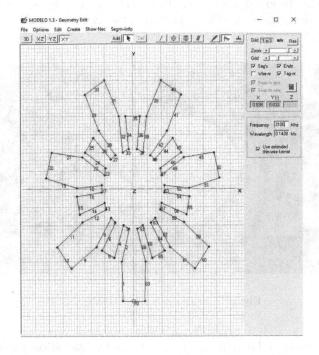

Fig. 4. Antenna modeled on 4Nec2.

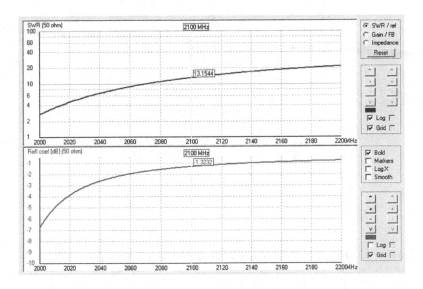

Fig. 5. Standing Wave Ratio at 2100 MHz in free space.

Antenna Gain. In Fig. 6 it is observed that the antenna presents a gain of 3.09 dBi and a relatively omnidirectional radiation pattern.

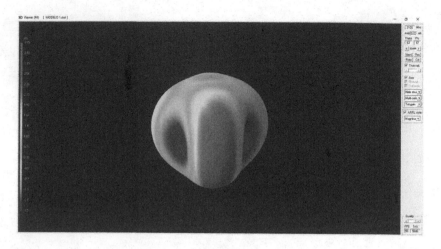

Fig. 6. Antenna gain at 2100 MHz.

Antenna Radiation Diagrams. Fig. 7 shows the radiation patterns of the antenna at 2100 MHz and Figs. 8 and 9 show the radiation patterns of the same antenna under free-space conditions.

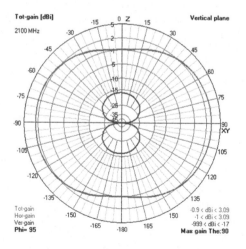

Fig. 7. Antenna radiation patterns at 2100 MHz.

Fig. 8. 3D radiation pattern, of the antenna at 2100 MHz, in the H plane.

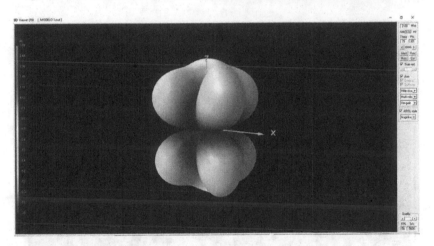

Fig. 9. 3D radiation pattern, of the antenna at 2100 MHz, in the V plane.

4.2   Standing Wave Ratio (SWR) at 3500 MHz

Having the variables initialized, according to the points in Table 1, at a frequency
of 3500 MHz the antenna presents an SWR = 10.81 with an impedance of 136 Ω.

Figure 10 shows the simulation result for the 3400 MHz–3600 MHz range,
with the same previous parameters. It is observed that, in a large part of the
simulated range, the antenna could be adapted using some impedance matching
technique, since it has a fairly high SWR, ideally being the SWR closer to unity
to avoid losses due to reflected power.

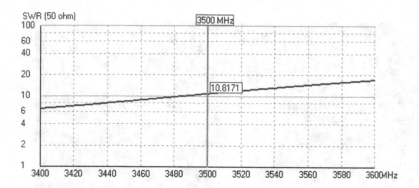

Fig. 10. Standing Wave Ratio at 3500 MHz in free space.

Antenna Gain. In Fig. 11 it is observed that the antenna presents a gain of 5.85 dBi.

Fig. 11. Antenna gain at 3500 MHz.

Antenna Radiation Diagrams. Figure 12 shows the radiation patterns of the antenna at 3500 MHz and Figs. 13 and 14 show the radiation patterns of the same antenna under free-space conditions.

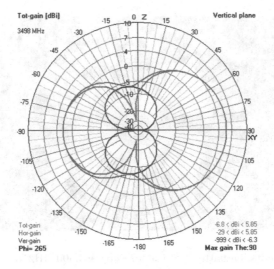

Fig. 12. Antenna radiation patterns at 3500 MHz.

Fig. 13. 3D radiation pattern, of the antenna at 3500 MHz, in the H plane.

Antenna Impedance Matching. Because the impedance found in the previous simulation is 136 Ω, it is necessary to perform the impedance matching simulation between the transmission line, which we assume 50 Ω, and the experimental antenna tested.

In Fig. 15 we can see the adaptation using 4Nnec2.

Fig. 14. 3D radiation pattern, of the antenna at 3500 MHz, in the V plane.

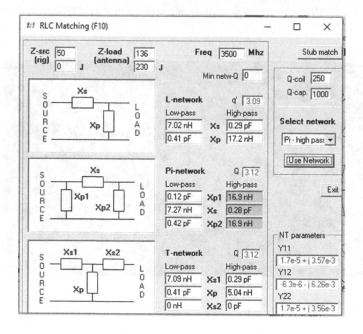

Fig. 15. Impedance matching with RLC Matching.

Figures 16 and 17 show the results after adapting the antenna to a frequency of 3500 MHz. It can be seen that the impedance of the antenna almost equals the impedance of the line, achieving an SWR of 1.01. This indicates that this antenna has high efficiency of 99.06%.

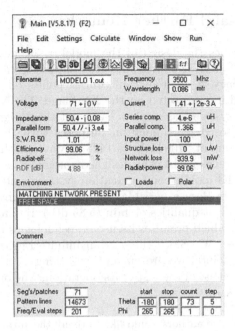

Fig. 16. Antenna parameters after matching with RLC Matching.

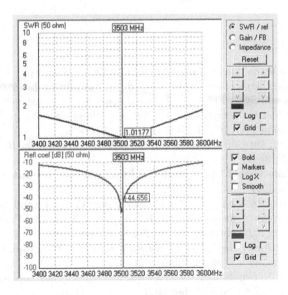

Fig. 17. SWR after adaptation with RLC Matching.

5   Conclusions and Further Remarks

This document describes the design of the model of an antenna, with an efficiency of 99.06% and a SWR of 1.01 after the impedance matching process at the 3500 MHz frequency, generated from control points on a curve lying on a lemniscatic torus and it's projected on the XY plane. The Mathematica v.11.0 symbolic calculation system was used to carry out the design. In addition, two frequencies were proposed for simulation purposes with the assistance of the free software 4Nec2, one of 2100 MHz (0.1428 m wavelength) and the other of 3500 MHz (0.1428 m wavelength). The designed antenna has qualities to be used in fourth-generation 4G mobile communication systems, and especially in fifth-generation 5G, being in Peru a possible frequency for its implementation. The antenna presents excellent qualities of gain (5.85 dBi), radiation pattern, adaptation, and efficiency; which promises its use in technologies in full development such as 4G and future 5G. Since there is an infinite range of possibilities to obtain more control points, we are currently working on the design of other antenna models; as well as, in the test of its operation. The obtained results will be reported elsewhere.

Acknowledgements. The authors would like to thank the authorities of the Universidad Nacional de Piura for the acquisition of the Mathematica v.11.0 license, to the developers of 4nec2 free software, and the reviewers for their valuable comments and suggestions.

References

1. Balanis, C.: Antenna Theory. 4th edn. John Wiley & Sons, Inc., Hoboken (2016)
2. Nec2 Homepage. https://www.qsl.net/4nec2/. Accessed 12 Jan 2021
3. Iglesias, A., et al.: Symbolic computational approach to construct a 3D torus via curvature. In: Fourth International Conference Ubiquitous Computing and Multimedia Applications, pp. 19–24. Xian (China) (2013)
4. López, J., et al.: Introducción al análisis y diseño de antenas. Rev. Escuela Fís. **2**(2), 82–98 (2019)
5. MathWorld Homepage. https://mathworld.wolfram.com/Lemniscate.html. Accessed 5 Jan 2021
6. Misra, S.: A step by step guide for choosing project topics and writing research papers in ICT related disciplines. In: ICTA 2020. CCIS, vol. 1350, pp. 727–744. Springer, Cham (2021). https://doi.org/10.1007/978-3-030-69143-1_55
7. ScienceDirect Homepage. https://www.sciencedirect.com/topics/mathematics/surface-of-revolution. Accessed 5 Jan 2021
8. ScienceDirect Homepage. https://www.sciencedirect.com/topics/mathematics/torus-of-revolution. Accessed 5 Jan 2021
9. Shobayo, O., Abayomi-Alli, O., Odusami, M., Misra, S., Safiriyu, M.: Modeling and simulation of impedance-based algorithm on overhead power distribution network using MATLAB. In: Sengodan, T., Murugappan, M., Misra, S. (eds.) Advances in Electrical and Computer Technologies. LNEE, vol. 672, pp. 335–345. Springer, Singapore (2020). https://doi.org/10.1007/978-981-15-5558-9_31

10. Wolfram, S.: An Elementary Introduction to the Wolfram Language. 2nd edn. Wolfram Media, Inc., Champaign (2017)
11. Wolfram, S.: The Mathematica Book. 5th edn. Wolfram Media, Inc., Champaign (2003)

An Article-Oriented Framework
for Automatic Semantic Analysis
of COVID-19 Researches

Antonio Pedro[1(✉)], Antônio Pereira[1], Pablo Cecilio[1], Nayara Pena[1],
Felipe Viegas[2], Elisa Tuler[1], Diego Dias[1], and Leonardo Rocha[1]

[1] Universidade Federal de São João del-Rei, São João del Rei, Brazil
{antoniopedro,antoniopereira,cecilio,nayara.p.pena}@aluno.ufsj.edu.br,
{etuler,diegodias,lcrocha}@ufsj.edu.br
[2] Universidade Federal de Minas Gerais, Belo Horizonte, Brazil
frviegas@dcc.ufmg.br

Abstract. In this work, we propose an article-oriented framework for
automatically extract semantic topics from scientific articles related to
COVID-19 researches. Our framework has four key building blocks (i) preprocessing, (ii) topic modeling, (iii) correlating topics, authors, and institutions, and (iv) a summarization interface. The first one corresponds to
apply traditional textual pre-processing strategies in the texts extracted
from articles and constructing their data representations. The topic modeling block aims at finding semantic topics from articles based on the
most relevant words of each discovered topic. The third block correlates
these discovered topics with the articles, authors (researchers), and institutions. The summarization interface provides an intuitive visualization
for these results. Our evaluation shows that our framework can automatically extract relevant features from the articles, identifying the key topics
covered by them, as well as the contribution of researchers, institutions,
and countries to the topics. Our framework can help research institutions
and companies to form multidisciplinary teams and funding agencies to
identify more promising research approaches regarding COVID-19.

Keywords: COVID-19 · Machine leaning · Topic modeling

1 Introduction

Technological advances in recent decades have considerably increased the
processing power and storage capacity of computers. Associated with these
advances, the Internet has presented a significant expansion and populariza-
tion in recent years. Every day new applications are created, generating and
using a higher amount of data (the expectation of reaching 44 ZiB within five
years [22]) of the most diverse types. As stated in [2], "too much information is
as bad as none", and one of the major challenges in this scenario is to provide
tools that can perform intelligent and automatic analyzes on these data to find
the information resources to meet the needs of users practically.

Supported by CAPES, CNPq, Finep, Fapesp and Fapemig.

O. Gervasi et al. (Eds.): ICCSA 2021, LNCS 12951, pp. 172–187, 2021.
https://doi.org/10.1007/978-3-030-86970-0_13

We can mention different scenarios where new techniques are being proposed to extract relevant information effectively. One example that illustrates these scenarios has to do with the so-called App Stores, such as Google Store and Apple Store, which provide over 2 million applications (each) so that users can search, buy, share and develop applications for mobile devices. As one could expect, both platforms produce useful data as textual comments and star ratings. Recent studies have shown that we can explore such information to leverage application development, with a clearer and more user-centered definition of requirements and improvements [19]. Currently, an important scenario that affects all segments of society is scientific production, which has been presenting an exponential expansion in terms of the number of articles, the number of researchers, and ongoing researches. The last detailed study on the *Google Scholar*[1] platform, in 2014, estimated an amount of 160 million documents in its database [37]. More recently, this scenario has been receiving even more prominence in consequence of the intense work of researchers to stop the advance of COVID-19, caused by a virus of the coronavirus family. Organizing semantically the information provided by the scientific articles published for this family of viruses can be extremely important to point out more promising research directions, to identify approaches that are still little explored, and to suggest possible collaborations of researchers with prominence in different areas, which corresponds to the application scenario of the present work.

In this work, we propose a general framework that allows filtering, summarizing, and analyzing articles written by researchers and published on different digital platforms. Our framework extracts relevant features from the articles automatically, identifying the key topics covered by them. The main goal of our proposal is to allow researchers and funding agencies to identify well-studied topics and little-explored approaches. Based on this information, take a more promising research direction to not only manage financial resources more efficiently but also to achieve even more innovative results. Based on the topics covered by articles, it is possible to establish the key topics studied by the researchers, authors of articles, and their institutions. This information can encourage researchers that are dealing with the same problem by different approaches to join efforts.

Our framework has four main building blocks (i) preprocessing, (ii) a topic modeling strategy, (iii) correlating topics, authors, and institutions, and (iv) a summarization interface. Our goal is to deal with the texts extracted from scientific articles, a proper data preprocessing allows the framework to achieve better effective results. From the raw textual content of articles, it is possible to apply different preprocessing strategies, such as lower-casing conversion, punctuation, and stopwords removal. The topic modeling strategy aims at finding semantic topics from articles. It is the ability to represent the set of articles as a bag-of-words matrix and decompose the represented dataset matrix properly (design matrix) into matrices that capture the latent relationships between terms and articles. It is also responsible for extracting the most relevant words of each discovered topic. Several strategies exist for this matter, such as non-negative matrix factorization

---

[1] https://scholar.google.com.

(NMF), latent dirichlet allocation (LDA), and its probabilistic counterpart, to name a few. Once the key topics of the articles have been identified, the third block consists of associating these topics with the articles and their authors and, its respective institutions. The summarization interface provides an intuitive visualization of the research topics and their associations with articles, authors, and institutions. Thus, the desiderata for our proposed framework is to offer, for each of its fundamental building blocks, methods that can perform their tasks losing no useful information, allowing fully automatic operation.

To evaluate our proposal, we instantiated the framework and provided an extensive analysis considering a dataset composed of articles published regarding COVID-19, SARS-CoV-2 and other coronaviruses, available at Kaggle [1]. Although this dataset contains complete works written in several languages and dated at least 20 years ago, we considered only the title and abstract of the 26 thousand articles published in 2020. We instantiated the first block of our framework applying some traditional preprocessing strategies: lower-casing conversion, punctuation, stopwords removal and named entity recognition (NER) approaches [13,25]. In the second block, we adopted the topic modeling strategy NMF [17] to infer different topics. For the third block, we proposed a strategy that consists to manipulate the matrices provide by NMF that correlates topics and articles, introducing in those the information of the articles' authors and their institutions and countries. Finally, we presented a visual interface proposal that summarizes all the gathered information, highlighting the main topics related to each article and author of our dataset. This interface is available at `labpi.ufsj.edu.br/covpapers/`[2]. Analyzing the results achieved by the instantiation of our framework, we were able to answer several questions such as: (i) *Which authors contribute the most to a certain topic?*; (ii) *Which lines of research are most explored by researchers at an institution?*; (iii) *Which countries lead research in a given area?*; and, (iv) *Which topics are most explored by a given author?*. Besides, with the definition of topics related to a set of articles, we discuss how it is possible to assist: (i) authors in the definition of descriptive tags about their research; (ii) companies to form multidisciplinary teams based on the researchers who most contribute to the desired area; (iii) government agencies, focused on Education and Research, to provide support in determining areas which are more, and less, studied, allowing a more equal redirection of financial resources.

2   Related Work

The proposal of strategies that aim to provide mechanisms for filtering, summarizing, and analyzing a specific area's scientific production is not new [7,8,29,36]. However, all these works concentrate on considering just metadata to provide this type of analysis, such as title, keywords, and tags, arguing that they are enough to provide an efficient data summarization. In this article, we raised some questions about these works: *To what extent are only metadata good descriptors*

---

[2] username: user, password: covid-19.

broadly? Are the keywords, alone, good descriptors about a large repository of articles? Can a researcher who wants to keep up to date on the scientific production in their fields be able to rely only on keywords to get this information? We believe that the answer to all these questions is "**No!**", motivating us to present the framework described and evaluated in this article, providing all these analyses considering the text that compose the articles.

Although potentially able to provide rich details information, using textual information is also more challenging and essentially requires preprocessing strategies application to minimize textual noise, selecting only data able to contribute to further steps of a machine learning processing, such as Topic Modeling (TM) [35]. In [5] the authors provided a complete analysis regarding different combinations of preprocessing strategies for text classification tasks. Regarding TM, in [30] the authors show that despite there is no correct, generic and unique sequence of preprocessing techniques, lower-casing conversion, punctuation and stopwords removal can achieve better effective results.

Another step that is considered in the textual preprocessing process is the data representation. It is necessary to build a $m \times n$ article-term matrix representation (or some latent term encoding) of the dataset. The representation widely adopted of this matrix in text mining exploits the term frequency-inverse document frequency (TF-IDF) paradigm (and its variants) [28]. The major problems with the TF-IDF representation have to do with its high dimensionality, sparseness and its lack of useful information such as context. Other alternative representations that aim at producing a more compact space in terms of latent dimensions (e.g., distributional word embeddings) exist, such as [24,27]. These strategies are based on co-occurrence statistics of textual datasets, representing words as vectors such that their similarities correlate with semantic relatedness. They usually use contextual information, such as terms adjacent to a target word. There is a consensus on literature regarding the contribution of these representations to TM tasks [34], especially the one presented in [32,33], called by the authors CluWords. CluWords correspond to clusters of semantically related word embeddings [24] built employing distance functions and filtering mechanisms. More than simple groups of (filtered) related words, CluWords are coupled with specific weighting schemes to capture their importance to a specific task.

A set of strategies commonly used to summarize, organize and analyze large volumes of textual data are those related to TM [12,14], which addresses the problem of discovering relationships between documents (D) and topics (Z), and relationships between the words (W) that compose the documents and topics. Each topic $z_i \in D$ comprises a probabilistic distribution between words that co-occur frequently, and the documents are represented by probabilistic distributions among the topics. They can be divided into two approaches: supervised and unsupervised. In supervised approaches, prior knowledge about the documents previously organized into semantically related groups is used so that the assignment of new documents to these groups can be performed [10]. With the ever-increasing and extremely dynamic data volume, the adoption of such strategies in real-world scenarios has become practically infeasible. Unsupervised approaches have been increasingly applied in the literature [19]. The unsupervised techniques are classified into probabilistic and non-probabilistic models. One of the most cited probabilistic techniques in the literature is LDA, which is based on a three-level hierarchical Bayesian model [3]. To use it, it is necessary to adjust and refine some parameters, among them the number of topics.

From the applicability point of view, we found in the literature several works that make use of probabilistic TM strategies to summarize and understand a large volume of textual data. In [16], the authors proposed the LDA as a topic modeling technique to analyze large journalistic repositories and check for trends and patterns in news across time. Whereas, in [4], the authors analyzed the topics raised in the posts of super users of Internet Support Groups (ISGs), concluding that a large part of the posts is related to mental health and recovery. In [31], meanwhile, the authors aimed to understand through LDA the various meanings that the term "Big Data" takes on in the biomedical literature. Tourism is also an area of great possibilities for using topic modeling techniques, especially with LDA, as presented in the works of [15] and [20]. LDA summarizes potential attractions content for tourist recommendation systems, assisting the used operators in recommending and deciding tourists. In the second, LDA has applied to reviews made at Disney parks, returning a series of topics discussed and commented on by park visitors. This allows the parks' management to have a better understanding of visitor perception, driving marketing campaigns.

Despite the diverse use of LDA, the work of [26] showed that, although topic modeling by LDA is an excellent technique for qualitative text analysis, the content obtained in these topics may not be understandable to humans due to multiple parameter settings. In this sense, non-probabilistic TM algorithms stand out, especially those based on matrix decomposition, which has good interpretability and few parameter settings. Despite the popularity of (probabilistic) LDA, the TM study began with an article defining a non-probabilistic algorithm for the TM task. In [6], the authors proposed the latent semantic analysis (LSA), an algorithm that decomposes the A word document matrix into three other matrices employing the singular value decomposition (SVD) technique. Despite being pioneering, the resulting SVD matrices are still uninterpretable, mainly by returning negative values. Thus, another matrix decomposition technique has gained prominence. In [18], the authors proposed the Non-Negative Matrix factorization strategy (NMF), which is a non-probabilistic method to retrieve topics from documents. This technique, which needs only the number of topics to work, decomposes $m \times n$ article-term matrix into two other matrices, W (of documents by topics) and H (of topics by terms). From the point of view of the work proposed here, the decomposed matrices are very valuable for further analysis. In [19], for example, the authors used these decomposed matrices to perform a sentiment analysis on topics obtained from reviews performed on mobile apps in the Google PlayStore. In [11], on the other hand, the authors exploited the NMF to evaluate the political agenda and topics of speeches of European parliamentarians through time (1999–2014).

3   Proposed Work

It is a fact that the huge amount of data present on the Internet makes extracting information a challenging task. The difficulty increases when the data to be considered is textual. However, with the help of data mining techniques, especially Topic Modeling, it is possible to combine automation and formalization with the quality of consistent results in the understanding about researchers. Currently, it is common to use metadata for papers published in scientific databases, such as e *tags*, to summarize the lines of research in which a researcher works, or even, the affinity with certain topics. However, these metadata may not be as efficient in the task of summarizing and categorizing researchers, either because of their limitations or because they are filled in by the researcher himself. To ensure more assertive analyzes of these researchers, in this work, we propose the use of the textual information present in these articles (the text itself). To this end, we present a general framework that allows filtering, summarize and analyze articles written by researchers and published on different digital platforms. More specifically, our framework can automatically extract relevant features from the articles, identifying the main topics covered by them, as well as, the main topics related to the authors and research institutions. In Fig. 1 we present an overview of our framework and in the following sections, we detail how we instantiate each one of the blocks that composed it.

Fig. 1. Proposed framework overview

3.1   Preprocessing

Adopting preprocessing techniques is indispensable to give data consistency. We propose the instantiation of the preprocessing block through the use of four strategies: (i) uppercase to lowercase conversion, (ii) stopwords removal, (iii) tokenization, and (iv) removal of some entities identified in the named entity recognition (NER) process.

As shown by [30], converting all characters to lowercase ensures better results for text classification algorithms, corresponding to our first task. Then, we remove the stopwords from the text, which are words that contribute little to relevant textual analysis [9]. Since such a list of words varies for each context, our proposal consists of using the list of stopwords from the spaCy library[3], adding some other noisy words specific to the dataset context to be analyzed.

Based on a more concise text, it is possible to move on to the third strategy: tokenization. This is the process of transforming words in the text into single tokens

---

[3] https://github.com/explosion/spaCy/blob/master/spacy/lang/en/stop_words.py.

- the conversion from uppercase to lowercase makes it easier to group words that refer to the same token, such as "Experiment" and "experiment". These tokens can be generated in several ways, e.g., using whitespaces that appear between words. Although efficient, for large volumes of texts, where contractions and compound words are present in excess, more sophisticated tokenization strategies are important. In this sense, we made use of a package from the NLTK library for Python, which performs tokenization based on Twitter data[4], leaving words closer to their original meaning, i.e., it doesn't decompose compound words like "infectious disease" into two distinct tokens, but into a single token.

The next step in this preprocessing process is to employ a strategy widely used in the literature to categorize people, places, organizations, and other entities of interest in the text [13]. These are recognized in the process of NER. Our work aims to deal with academic databases, composed by published scientific articles, so it is common to use numerical expressions, periods, descriptions of people or physical spaces, such as "in the **second week** the results were better" or "the experiment was performed in the **Turing lab** all week". Therefore, we apply NER to identify textual entities which add little semantic information to algorithms for automatic text manipulation and remove them.

Finally, the last step of this block consists into represent the set of articles as a bag-of-words (BOW) matrix $m \times n$ (article-term matrix). Thus, we adopt the CluWords presented in [32]. Basically, the CluWords transform the traditional BOW representation of documents to include semantic information related to the terms present in the documents. The semantic context is obtained by means of a pre-trained word representation, such as Word2Vec [23], FastText [24]. Each document has a new representation where original words are replaced by the CluWords representation [33]. This transformation is composed of two phases: (1) for each term t of the dataset is computed its corresponding CluWord. A CluWord for a term t is a set of terms in the vocabulary which word vectors are most similar to term t; (2) a modified version of the TF-IDF weighting scheme is calculated for the CluWords on which it is possible to exploit these new terms as a richer representation of documents of the collection. The best results reported in the literature point out the use of CluWords as the state-of-the-art for text representations for TM tasks [32,33].

3.2   Topic Modeling

The TM block is based on NMF, a strategy that simultaneously performs dimension reduction and clustering, with successful application in topic modeling. NMF [17] produces a "part-based" decomposition of latent relationships of a non-negative design matrix $A \in \mathbb{R}^{n \times m}$, where n is the number of documents and m the number of terms. The goal is to find a k-dimensional approximation of A ($k \ll m$) in terms of non-negative factors $H \in \mathbb{R}^{n \times k}$ and $W \in \mathbb{R}^{k \times m}$. H encodes the relationship between documents and topics while W encodes the relationship between terms and topics. The key idea behind NMF is to approximate the

---

[4] https://www.nltk.org/api/nltk.tokenize.html#module-nltk.tokenize.casual.

column vectors of A by non-negative linear combinations of non-negative basis vectors (columns of H) and the coefficients given by the columns of W. The Fig. 2 illlustrate this process.

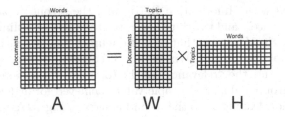

Fig. 2. Decomposition process performs by NMF

3.3 Correlating Topics, Authors and Institutions

Finding out which actors (authors and institutions) contribute more to each topic is a key point to enrich our analysis. Let's consider that the two previous steps of our framework have been applied to a database containing scientific articles related to the Computer Science area. Thus, at the end of the first two steps, we have the matrices H and W, that relate topics to articles and topics by words, respectively. Following the example, consider that the $i-th$ article in the H matrix deals mostly with the topic "Machine Learning", while the $j-th$ article deals with the topic "Sensor Networks". A pertinent question that can be asked is: *Who are the authors of the article that deal mostly with the topic "Machine Learning"?* It is evident that, for this example, each article has one or more authors, so it is possible, through the relations between articles and found topics, to highlight which authors most contribute to the topics obtained by TM. Similarly, as the authors of an article belong to research institutions, it is also possible to highlight the institutions by topics, also considering the relationship between articles and topics. Therefore, for the third block, we propose a strategy that consists to manipulate the matrices provided by NMF that correlates topics and articles, introducing the information of the articles' authors and their institutions. We will demonstrate below how this manipulation happens through an example.

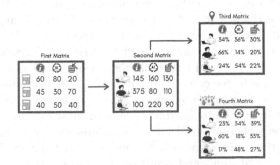

Fig. 3. Calculating author contributions to topics

Let us follow the same example above, considering that we generated the matrices H and W for three topics. First, you can identify what each topic is about by analyzing the H matrix and finding which words are most strongly associated with each topic. Let's assume the example where the first topic is mostly associated with "Information Retrieval", the second topic is related to "Artificial Intelligence", and the third one is related to "Data Mining". The next step is to analyze the W matrix that relates documents and topics. For this, let's take as an example the first matrix in Fig. 3, containing three articles, where each position presents the "relevance" of the topic for the document. From this matrix, we can group and sum the topic values achieved for articles that belong to the same author, leading us to the second matrix in Fig. 3. Assuming that the three articles in the first matrix belong to the first author in the second matrix, we can infer the "relevance" of this author for each topic: 145 (60+45+40) for the "Information Retrieval" topic, 160 (80+30+50) for the "Artificial Intelligence" topic, and 130 (20+70+40) for the "Data Mining" topic. Considering the complete W matrix, this same process can be applied to all collection authors.

From these data, we can perform two different analyses: (i) calculate for each author the distribution among the topics by which their works are related; or (ii) evaluate among all authors those who have their works most correlated to each topic. The first analysis is local information, so it is based only on what has been written by the same author. Thus, by performing a normalization on the rows representing each author in the second matrix of Fig. 3, it is possible to know how much the research of each author is related to each topic. We illustrate the result in the third matrix of Fig. 3. Analyzing the first row that represents the first author, we can observe that his articles are 34% related to the topic "Information Retrieval", 36% to the topic "Artificial Intelligence", and 30% to the topic "Data Mining". We can say that this author's articles deal more with the "Artificial Intelligence" topic. We based the second analysis on global information and consider everything that has been written for the same topic to distinguish those authors who have their articles more aligned with the topics got. In the fourth matrix of Fig. 3 we normalized each of the columns, which represents the topics got from the second matrix. Thus, following our example, we can observe that the first author, despite having his articles dealing mostly with "Artificial Intelligence" is not the one who has the greatest alignment with this topic, the third author has the greatest alignment. Even though the first author has their articles less related to the topic "Data Mining", among the authors evaluated, they are presenting articles more aligned with this topic. Considering that each author registers in their articles the research institution to which they are affiliated and that each institution is in a country, we can do an analogous analysis to the one presented above for institutions and countries.

Following the strategy presented by us, it is possible to perform several analyses, such as identifying the topics that have been most explored by researchers or by a specific researcher, evaluate how institutions and/or countries have been exploring each of the topics identified, and many other analyses. These analyses can help research institutions and companies to form multidisciplinary teams, and funding agencies to identify areas that need more resources.

3.4 Summarization Interface

We have developed a visualization interface that intuitively summarizes the topics found and their associations with researchers and institutions. This way, all the analyses proposed in Sect. 3.2 can be performed quickly and efficiently. We present the data related to topics and their correlations with articles, researchers, and institutions through a large set of visual metaphors in a web application. Charts, heatmaps, tables, and interactive search methods are available, allowing the same result to be viewed from different perspectives. This Web App is available at our website[5]. In Fig. 4 we present our application with one of the visual metaphors provided, the one detailing the topics related to a researcher. Other visual metaphors will be presented in the Sect. 4.1 when we will also discuss their potential to perform the possible analyses discussed in the previous section.

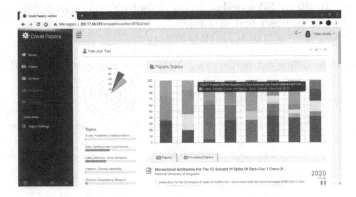

Fig. 4. Visualization of author in COVID's base

4 Experiments and Results

4.1 Dataset

To validate the proposed framework, we considered a dataset of published articles related to COVID-19, SARS-CoV-2 and, other viruses of the coronaviruses group. It is available at Kaggle [1], by the Allen Institute for AI in partnership with Chan Zuckerberg Initiative, Georgetown University Center for Security and Emerging Technologies, Microsoft Research, IBM, US National Library of Medicine, and in coordination with the White House Office of Science and Technology Policy. The dataset contains articles published in several languages, such as, Portuguese, English, French, German and, others. However, in our experimental evaluation, we only considered articles written in English. The dataset has several articles distributed in dates that mark various moments in history

---

[5] `labpi.ufsj.edu.br/covpapers/`
   username: user
   password: covid-19.

(avian flu in 2001, swine flu in 2009, and COVID-19 in 2020). Thus, we separated only articles published in the year 2020, resulting in more than 16,000 documents, for which we extracted the title and abstracts for a specific analysis on COVID-19.

4.2   Preprocessing and Topic Modeling

We apply the preprocessing steps highlighted in Sect. 3.1 (including the text representation using CluWords, the state-of-the-art in semantic representation [32,33]). The pre-processed dataset is then forwarded to the step our tool which performs the modeling of topics (as detailed in Sect. 3.2, we use NMF [17]). In the summarization interface, it is possible to define how many topics it should generate (**10, 20 or 30 topics**) and how many words should characterize each of the topics (**10, 20 or 30 words**). In our experimental analysis, we exploit 10 topics with 10 words, described in Table 1.

Table 1. Topic Modeling extraction using NMF, considering 10 topics and 10 words

Topic	Word 1	Word 2	Word 3	Word 4	Word 5
1	Infection	Viruses	Pathogen	Diseases	Parasite
2	Antigen	Immunogen	Crossreactivity	Hybridomas	Monoclonal
3	Economic	Social	Political	Governmental	Education
4	Cells	Neuroprogenitor	Fibroblast	Keratinocytes	Epithelium
5	Patients	Inpatient	Outpatient	Hospital	Clinic
6	Antimalarial	Drugs	Dosage	Antidepressants	Statins
7	Coatings	Alloy	Metals	Oxides	Conductive
8	Proteins	Cysteine	Enzyme	Nucleotides	Ubiquitins
9	Approximation	Parameters	Simulations	Algorithm	Equations
10	Venous	Catheterization	Carotid	Artery	Endovascular
Topic	Word 6	Word 7	Word 8	Word 9	Word 10
1	Contagions	Bacterial	Rhinovirus	Transmissible	Parvovirus
2	Epitope	Toxoid	Autoantibodies	Polyclonal	Sera
3	Cultural	Initiatives	Financial	Institutions	Personal
4	Macrophages	Tissues	Apoptosis	Intercellular	Osteoblasts
5	Doctors	Diagnoses	Nurses	Physicians	Oncologists
6	Immunosuppressant	Aspirin	Methadone	Quinine	Psychoactive
7	Superalloy	Nanocomposite	Thermoplastics	Electrodeposition	Polymer
8	Ribosome	Genes	Residues	Aminoacids	Peptide
9	Probabilistic	Submodel	Methodology	Computation	Model
10	Embolectomy	Thoracotomy	Aneurysm	Sternotomy	Endarterectomy

4.3   Analysing the Correlation Between Topics, Authors and Institutions

A researcher could identify how the proposed framework defines their research areas or which topics their research comes closest to. To illustrate this analysis, we considered the Dutch researcher Marion Koopmans, a virologist who specialized in molecular epidemiology at the Erasmus University Medical Center in Rotterdam, Netherlands. In Fig. 5, we present the topic distribution for her four articles available in the dataset (right graph) and an overall analysis of the researcher (left graph). We can see that her research aimed at 34% on Topic 1 (dealing with virology); 22% on Topic 2 (immune system); and 14% on Topic 3 (socioeconomic effects of the pandemic). This association, automatically identified by our framework, can be corroborated by the author's description in her institutional page[6]. The same association can also be observed in a recent report in the scientific journal Nature [21], which highlights a list of researchers who have been leading the research on the origin of COVID-19.

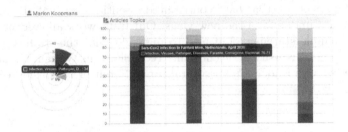

Fig. 5. Relationship between topics and articles by Marion Koopmans

The second analysis evaluates, among all the authors, those whose work is the most correlated to a topic. Returning to our previous example, despite Marion Koopmans' research having Topic 1 as her primary focus, the framework shows another researcher standing out in this area. He is the Japanese researcher Hiroshi Nishiura. In Fig. 6, we illustrate the topic distribution of his article entitled "*Incubation Period And Other Epidemiological Characteristics Of 2019 Novel Coronavirus Infections With Right Truncation: A Statistical Analysis Of Publicly Available Case Data*", which has 52.89% of its content linked to Topic 1. Just one article does not define the association of a researcher for a specific research topic, but several published articles mostly related to one research topic lead the framework to infer the researcher's contribution to this topic - Hiroshi Nishiura. He emerges as the researcher who has the highest correlation with Topic 1 (Fig. 7).

---

[6] https://www.erasmusmc.nl/en/research/researchers/koopmans-marion.

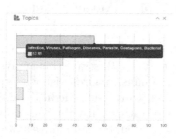

Fig. 6. Topics Related to a article

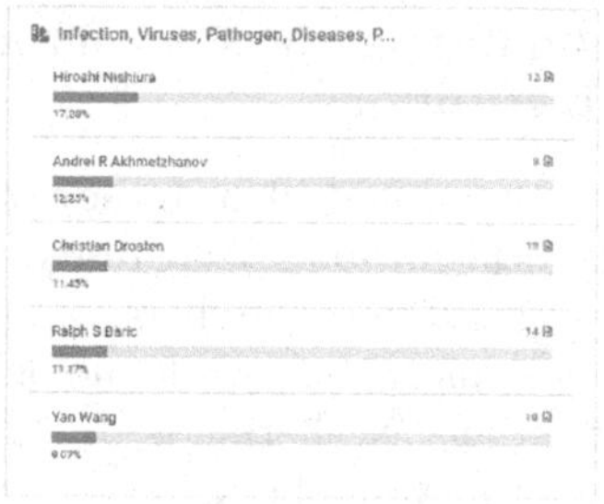

Fig. 7. Authors who contribute the most to a topic

Besides the analyses performed on researchers and articles, research institutions committee might be interested in finding out what their institutions study and how their studies affect the global scenario facing COVID-19. For example, the city of Wuhan, the epicenter of the new coronavirus pandemic, has some educational institutions that have contributed essentially so that the rest of the world could also understand the new virus. Our framework allowed not only to confirm this simple statement but also to highlight that the studies performed at the Huazhong University of Science and Technology, in Wuhan, China, placed them in the research leadership for several topics (5 out of 10 considered topics), as shown in Fig. 8.

Fig. 8. Correlation per institution

Fig. 9. Correlation per country

Finally, governments and research foundations interested in visualizing how their countries are facing the various topics related to the new coronavirus can use our framework to find out which areas of study have been prioritized and if the studies have had a significant impact when compared to other countries. The countries that have invested the most in fighting the pandemic or have suffered from it are the ones that lead the listed topics. Following the basic principle of science, they report, study, and propose solutions, and highlight potential problems that can surge elsewhere. Thus, the participation of China, the United States, Italy, India, the Netherlands, France, and Spain in the topics got by our method stands out as shown in Fig. 9.

5   Conclusion and Future Works

In this work, we proposed a novel framework that analyzing scientific articles published on different digital platforms and offers to researchers and funding agencies a more detailed and informative view of the well-studied topics and little-explored approaches considered on COVID-19 researches, pointing out more promising research direction. This framework has four key building blocks that automatically extract semantic topics from the papers, identifying the primary subjects covered by them; correlate these topics with the papers, authors (researchers), and institutions, and summarize all information in an intuitive interface. Analyses on a dataset composed of papers published regarding COVID-19 and other corona viruses [1] demonstrated that we can: (1) identify correctly the topics referred to an article, (2) identify the contribution of researchers to different topics, and (3) evaluate how institutions and/or countries have been contributing to each topic. In sum, the main implication is that this new framework can help research institutions and companies to form multidisciplinary teams and funding agencies to identify less and more explored areas to direct resources to more promising research approaches regarding COVID-19. As future work, we plan to instantiate different configurations regarding topic modeling and data representation. We also intend to propose and evaluate dynamic and interactive methods in the visualization interface and apply them to other databases.

Acknowledgements. This study was financed in part by the Coordenação de Aperfeiçoamento de Pessoal de Nível Superior - Brasil (CAPES) - Finance Code 001 and National Council for Scientific and Technological Development – CNPq.

References

1. Ai, A.I.F.: Covid-19 open research dataset challenge (cord-19) (2020)
2. Auden, W.H.: The Complete Works of WH Auden: Prose, vol. 2. Princeton University Press, Princeton, April 2002
3. Blei, D.M., Ng, A.Y., Jordan, M.I.: Latent dirichlet allocation. J. Mach. Learn. Res. **3**(null), 993–1022 (2003)
4. Carron-Arthur, B., Reynolds, J., Bennett, K., Bennett, A., Griffiths, K.M.: What's all the talk about? topic modelling in a mental health internet support group. BMC Psychiatry **16**(1), 367 (2016)
5. Cunha, W., et al.: Extended pre-processing pipeline for text classification: on the role of meta-feature representations, sparsification and selective sampling. Inform. Process. Manag. **57**(4), 102263 (2020)
6. Deerwester, S., Dumais, S., Landauer, T., Furnas, G., Harshman, R.: Indexing by latent semantic analysis. J. Am. Soc. Inf. Sci. **41**, 391–407 (1990)
7. Dutta, P.S., Tahbilder, H., et al.: Prediction of rainfall using data mining technique over Assam. Indian J. Comput. Sci. Eng. (IJCSE) **5**, 85–90 (2014)
8. Fathalla, S., Vahdati, S., Auer, S., Lange, C.: Metadata analysis of scholarly events of computer science, physics, engineering, and mathematics. In: TPDL (2018)

9. Gerlach, M., Shi, H., Amaral, L.A.N.: A universal information theoretic approach to the identification of stopwords. Nat. Mach. Intell. **1**(12), 606–612 (2019)
10. Ghosh, S., et al.: Temporal topic modeling to assess associations between news trends and infectious disease outbreaks. Sci. Rep. **7**, 40841 (2017)
11. Greene, D., Cross, J.P.: Exploring the political agenda of the european parliament using a dynamic topic modeling approach (2016)
12. Griffiths, T.L., Steyvers, M.: A probabilistic approach to semantic representation. In: Proceedings of the Annual Meeting of the Cognitive Science Society, vol. 24 (2002)
13. Gudivada, V.N., Arbabifard, K.: Chapter 3 - open-source libraries, application frameworks, and workflow systems for nlp. In: Gudivada, V.N., Rao, C. (eds) Computational Analysis and Understanding of Natural Languages: Principles, Methods and Applications, volume 38 of Handbook of Statistics, pp. 31–50. Elsevier (2018)
14. Hofmann, T.: Probabilistic latent semantic analysis. In: Proceedings of the Fifteenth Conference on Uncertainty in Artificial Intelligence, pp. 289–296. Morgan Kaufmann Publishers Inc. (1999)
15. Huang, C., Wang, Q., Yang, D., Xu, F.: Topic mining of tourist attractions based on a seasonal context aware lda model. Intell. Data Anal. **22**, 383–405 (2018)
16. Jacobi, C., van Atteveldt, W., Welbers, K.: Quantitative analysis of large amounts of journalistic texts using topic modelling. Digit. Journalism **4**(1), 89–106 (2016)
17. Lee, D.D., Seung, H.S.: Learning the parts of objects by non-negative matrix factorization. Nature **401**(6755), 788–791 (1999)
18. Lee, D.D., Seung, H.S.: Algorithms for non-negative matrix factorization. In: Proceedings of the 13th International Conference on Neural Information Processing Systems, NIPS'00, pp. 535–541. MIT Press, Cambridge, MA, USA (2000)
19. Luiz, W., et al.: A feature-oriented sentiment rating for mobile app reviews. In: Proceedings of the 2018 World Wide Web Conference, WWW 2018, pp. 1909–1918. International World Wide Web Conferences Steering Committee, Republic and Canton of Geneva, CHE (2018)
20. Luo, J.M., Vu, H.Q., Li, G., Law, R.: Topic modelling for theme park online reviews: analysis of disneyland. J. Travel Tourism Market. **37**(2), 272–285 (2020)
21. Mallapaty, S.: Meet the scientists investigating the origins of the covid pandemic (2020)
22. Marr, B.: 20 fatos sobre a internet que você (provavelmente) não sabe (2015)
23. Mikolov, T., Grave, E., Bojanowski, P., Puhrsch, C., Joulin, A.: Advances in pre-training distributed word representations. CoRR, abs/1712.09405 (2017)
24. Mikolov, T., Grave, E., Bojanowski, P., Puhrsch, C., Joulin, A.: Advances in pre-training distributed word representations. In: LREC 2018 (2018)
25. Nadeau, D., Sekine, S.: A survey of named entity recognition and classification. Lingvisticæ Investigationes **30**(1), 3–26 (2007)
26. Nikolenko, S.I., Koltcov, S., Koltsova, O.: Topic modelling for qualitative studies. J. Inf. Sci. **43**(1), 88–102 (2017)
27. Pennington, J., Socher, R., Manning, V.: Glove: global vectors for word representation. In: EMNLP (2014)
28. Qaiser, S., Ali, R.: Text mining: use of tf-idf to examine the relevance of words to documents. Int. J. Comput. Appl. **181**(1), 25–29 (2018)
29. Sundar, N.A., Latha, P.P., Chandra, M.R.: Performance analysis of classification data mining techniques over heart disease database, vol. 2, pp. 470–478. Citeseer (2012)
30. Uysal, A.K., Gunal, S.: The impact of preprocessing on text classification. Inform. Process. Manag. **50**(1), 04–112 (2014)

31. van Altena, A.J., Moerland, P., Zwinderman, A., Olabarriaga, S.: Understanding big data themes from scientific biomedical literature through topic modeling. J. Big Data **3**, 1–21 (2016)

32. Viegas, F., et al.: Cluwords: exploiting semantic word clustering representation for enhanced topic modeling, pp. 753–761 (2019)

33. Viegas, F., Cunha, W., Gomes, C., Pereira, A., Rocha, L., Goncalves, M.: CluHTM - semantic hierarchical topic modeling based on CluWords. In: Proceedings of the 58th Annual Meeting of the Association for Computational Linguistics, pp. 8138–8150. Association for Computational Linguistics, Online, July 2020

34. Viegas, F., et al.: Semantically-enhanced topic modeling. In: Proceedings of the 27th ACM International Conference on Information and Knowledge Management, CIKM 2018, pp. 893–902. Association for Computing Machinery, New York, NY, USA (2018)

35. Vijayarani, S., Ilamathi, M.J., Nithya, M.: Preprocessing techniques for text mining-an overview. Int. J. Comput. Sci. Commun. Netw. **5**, 7–16 (2015)

36. Yethiraj, N.G.: Applying data mining techniques in the field of agriculture and allied sciences. Int. J. Bus. **001**, 40–42 (2012)

37. You, J.: Just how big is google scholar? ummm..., (2014)

Using Quadratic Discriminant Analysis by Intrusion Detection Systems for Port Scan and Slowloris Attack Classification

Vinícius M. Deolindo[1], Bruno L. Dalmazo[2]([⊠]) [iD], Marcus V. B. da Silva[3],
Luiz R. B. de Oliveira[1], Allan de B. Silva[1], Lisandro Zambenedetti Granville[3],
Luciano P. Gaspary[3] [iD], and Jéferson Campos Nobre[3] [iD]

[1] University of Vale do Rio dos Sinos, São Leopoldo, Brazil
{vinicius.deolindo,luizbertoldi,fallanbs}@unisinos.br
[2] Federal University of Rio Grande, Rio Grande, Brazil
dalmazo@furg.br
[3] Federal University of Rio Grande do Sul, Porto Alegre, Brazil
{mvbsilva,granville,paschoal,jcnobre}@inf.ufrgs.br

Abstract. Identify and classify attacks through Intrusion Detection Systems is one constant challenge for security professionals. Computer networks are one of the significant IT components that support classification operations. Machine Learning (ML) techniques can aid in this process by providing methods capable of making decisions based on previously known information. In light of this, literature shows that Quadratic Discriminant Analysis (QDA) is barely explored as a classification method for IDS. To fill this gap, this study aims to create a new classifier able to distinguish legitimate network traffic from an attack by adopting ML techniques and QDA algorithms for identifying Port Scan and DoS Slowloris attacks.

Keywords: Intrusion Detection System · Machine Learning · Denial of Service · Quadratic Discriminant Analysis

1 Introduction

Currently, computer networks provide a means of communication between the most diverse devices, providing services to users and businesses. Consequently, networks also become targets of attacks that seek to compromise their security. The rapid detection of a potential attack and the type of attack in question is essential so that the necessary measures can be chosen for each case.

Services such as Intrusion Detection System (IDS) provide to the computer networks a tool capable of identifying events that may endanger the security of the environment, compromising the integrity, availability, and confidentiality of resources and services [10]. An attack identification, coupled with defense strategies and incident response, can provide a basis for network perimeter protection.

© Springer Nature Switzerland AG 2021
O. Gervasi et al. (Eds.): ICCSA 2021, LNCS 12951, pp. 188–200, 2021.
https://doi.org/10.1007/978-3-030-86970-0_14

However, the IDS must count on complementary detection mechanisms to assist with rapid detection.

Machine Learning is an area of knowledge that aims to develop techniques that allow computer programs to acquire information and knowledge in an automated way. This learning process is built on specific techniques that enable a system to make decisions based on previously evaluated experiences. Making systems able to precisely predicting occurrences is one of the challenges that Data Scientists face in Machine Learning [1,7]. For instance, defining a model capable of assertively predicting information depends on a number of factors and detailed preliminary analysis.

The adoption of Machine Learning techniques for recognizing attack patterns – which can be characterized as an anomaly detection approach – can help to determine the type of attack suffered, aiding decision-making for reaction measures [3]. Automated techniques are ideal for IDS as they allow you to monitor and correlate a large number of patterns, signatures, and anomalies information [8]. As a novelty, this work applies a classification algorithm, namely, Quadratic Discriminant Analysis (QDA) jointly with an IDS in order to perform anomaly detection. QDA is one of the standard approaches to data classification algorithms and performs well when applied to large amounts of data for factoring [15].

Traditional techniques applied by IDS based on signatures can present problems in identifying certain types of attacks. The main reason is an attack may suffer slight variations in its pattern of action. Thus changing its default features and hiding this new behaviour to the system. The combination between IDS, machine learning, along with the classification of network traffic through classifiers using the QDA algorithm, can support in the correct and quick decision making looking for anomalies in the network traffic [5]. To deal with these lacks, this study aims to create a new classifier able to distinguish legitimate network traffic from an attack by adopting ML techniques and QDA algorithms. In this context, the scope of this work is bounded to scenarios for *Port Scan* and *DoS Slowloris* attacks.

The remainder of the paper is organized as follows. Section 2 presents the theoretical basis of the methods and systems adopted in this study. Section 3 describes the steps of a machine learning project and challenges to overcome, whilst Sect. 4 presents the evaluation and discusses the results. Section 5 covers some of the most prominent related work. Section 6 concludes with some final remarks and prospective directions for future research.

2   Background

This section presents the theoretical framework for the technologies and methods used in this study, namely, Machine Learning (ML), Cross Industry Standard Process for Data Mining (CRISP-DM), Intrusion detection System (IDS), the explored attacks and Quadratic Discriminant Analysis (QDA).

2.1 Machine Learning

Machine Learning (ML) belongs to the artificial intelligence field and consists of a set of tools, techniques, and methods for extracting knowledge from a dataset. ML aims to build systems that can automatically learn by analyzing previous making decisions based on acquired information and knowledge. In other words, ML seeks to extract knowledge from one dataset so that this learning can be applied to other similar datasets [7].

2.2 Cross Industry Standard Process for Data Mining

Also referenced by the acronym CRISP-DM, it alludes to a methodology intended for data mining related projects. Developed through a consortium consisting of *DaimlerChryrler, SPSS and NCR*, the framework presents six steps that make up the lifecycle of a Machine Learning project [12].

The expected activities and deliverables in each of the steps of the CRISP-DM framework are as follows.

- *Business Understanding:* The first phase of the project focuses on understanding the goals and requirements as well as business expectations for the delivery. The challenge is to convert doubts and expectations into a clear definition of the problem so that only then can a plan for the project progress can be drawn up.
- *Data Understanding:* Understanding the collected data, assessing the initial data collection, preselecting information sources and identifying the quality of the records are examples of activities that should be performed at this stage of the project. By following this order, each of the information brings more familiarity and better knowledge regarding the whole process, facilitating to perform the first preliminary analyzes of the collected data.
- *Data Preparation:* The third stage of the project is designed for the formulation of the previously evaluated data, in this phase should be built the datasets that will be used in the future, it is also essential to establish the portion that will be used in the training phase and the portion intended to the classification model testing.
- *Modeling:* In this step, the models and techniques available for applying the data prepared previously are evaluated. This step is characterized by using different techniques to identify and assess the benefits and drawbacks of each algorithm.
- *Evaluation:* The fifth step of the project has two goals: to evaluate the models created in the previous stage and define the final model for implementation. Also, if necessary, it is possible to return to the Business Understanding step for data redefinition. This process may occur if the results of model evaluations are not satisfactory.
- *Deployment:* The last step of the cycle, the final model is created and implemented. Creating the data model is not necessarily the end of the project. Here, it is common presenting the project to the stakeholders and sponsors.

All knowledge built throughout the project should be documented and delivered along with the final results so that they can be available as learning in future projects.

2.3 Intrusion Detection System

Intrusion Detection Systems (IDSs) are systems that seek to identify potential threats and attacks that may be occurring on a device or a network. In general, an IDS may detect intrusions using two approaches: through anomalies or an attack signature identification. Anomaly detection seeks to identify standard deviations in the network traffic or device behaviour based on normal patterns to identify outliers. On the other hand, signature-based IDS uses known patterns (part of code or protocols behaviour, for instance) of various types of attacks to try to identify threats.

- Network-based Intrusion Detection System (NIDS): It operates at capturing network packets to protect a system from network-based threats. A NIDS reads all inbound packets and searches for any suspicious patterns.
- Host-based Intrusion Detection System (HIDS): It runs on individual hosts or devices on the network. A HIDS monitors the inbound and outbound packets from the device only and will alert the user or administrator if suspicious activity is detected.
- Hybrid Intrusion Detection Systems: It consists of using network-based and host-based systems to control and monitor the computational security of an environment.

2.4 Attacks

For this study, we consider two well-known attacks frequently triggered in networks. Each of the attacks has different characteristics and objectives, as described below:

- *Port Scan:* This attack performs a port scan, that is a method for discovering communication channels (ports) that can be employed in future attacks. The technique consists of investigating the communication ports of a given target and then evaluating which exploiting methods are the most appropriate. Although some authors do not consider Port Scan as an attack (since there is no intrusion), there is a consensus that it is a first stage in identifying vulnerabilities that could lead to future attacks [11].
- *DoS Slowloris:* Refers to a type of attack performed using the Slowloris tool, which targets HTTP servers. The method consists of requesting multiple connections to a given target server, in order to attempt exhausting the server's ability to respond to requests. The consecutive attacker connection requests seek for maintaining established connections over a long period. Usually, the IDS based on anomaly detection presents some difficulty to detect this attack due to the low volume traffic generated. Also, different from a Denial of Service, the connection is held as much possible to simulating a legitimate user behaviour [4]. Figure 1 illustrates the Slowloris attack based on the dataset [6].

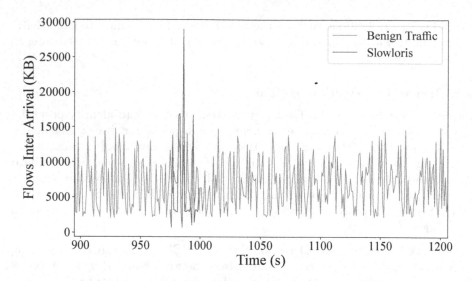

Fig. 1. Attack performed by DoD Slowloris

2.5 Quadratic Discriminant Analysis

Quadratic Discriminant Analysis (QDA) is one of the most used models for machine learning using the supervised learning technique. This method seeks to model the probability of each class based on a Gaussian Distribution, i.e., it is assumed that the measurements from each class are normally distributed. Then, after modeling the classes with the supervised method, the QDA uses a normal distribution to make predictions [15]. Although the Gaussian Model is simple, the QDA presents a limitation: it does not perform well from a small data sample [15].

3   Port Scan and DoS Slowloris Classification

This section presents an alternative for classifying Port Scan and DoS Slowloris attacks by collecting metrics generated by an IDS and machine learning. Intrusion detection plays a vital role in computer network defense processes by enabling the generation of various types of alerts for different malicious behaviours that may occur in the environment [13].

Sharafaldin *et al.* [13] created a complete dataset containing various types of attacks. Then enabling to extract the patterns of attacks adopted for the accomplishment of this work. Mapping of the most relevant information for each type of attack allows to establish patterns that made possible the use of machine learning algorithms for classification and consequent identification of Port San and DoS Slowloris attacks.

CRoss Industry Standard Process for Data Mining (CRISP-DM) was adopted in the context of this work. This framework provides standards for carrying out the activities of a machine learning project [9]. In this section, the data preparation and the proposed model for the classification of the attacks are presented. The environment for modelling this work was based on the RStudio, which is an open-source project that covers several components of R statistical.

3.1 Business Understanding

The business understanding phase can be summarized in the problem that the project seeks to solve. What challenges can be met and how to develop a solution that addresses the issue at hand. In this case, the main problem that this study seeks to solve is the identification of attack patterns drawn for Port Scan and DoS Slowloris jointly with the adoption of the QDA classifier. Therefore, it is necessary to evaluate which data sources are available and which of them may provide attack patterns to perform this study.

At this stage, several IDS datasets and their features were evaluated, so that it was possible to assess which sets would be useful for the study. In this sense, the Sharafaldin's paper [13] brings a complete dataset with valuable information about attacks collected by an IDS. Also, the attacks were performed in a controlled environment using recent techniques. Although using features as the source, destination, and duration of attacks, this dataset present a high quality of the classification labels of each attack, essential information that facilitates the future creation of a QDA model. In the next step was necessary to request the datasets for the Canadian Institute for Cybersecurity. This organization conducted the study around the datasets.

3.2 Data Understanding

Datasets were divided into days, as shown in Table 1, each day contains information from different attacks. As the purpose of this study is to evaluate Port Scan and DoS Slowloris attacks, we used the Friday files, the day in which these attacks were performed.

Several metrics in the dataset archives are available, but due to the diversity of parameters in each set, this study considers just the most relevant metrics, as detailed in [13]. Also, we evaluated the variety of classification labels of each attack contained in the files. Having a consistent diversity of information is a crucial point to create the model using QDA.

Table 1. File organization [13]

Days	Labels
Monday	Benign
Tuesday	BForce, SFTP and SSH
Wednesday	DoS and Hearbleed Attacks Slowloris, Slowhttptest, Hulk and GoldenEye
Thursday	Web and Infiltration Attacks Web BForce, XSS and Sql Injection Infiltration Dropbox Download and Cool disk
Friday	DDoS LOIT, Slowloris ARES, Port Scans (sS, sT, sF, sX, sN, sP, sV, sU, sO, sA, sW, sR, sL and B)

Table 2. Organization of the labels in the dataset

Port Scan		DoS Slowloris	
Var1	*Freq*	*Var1*	*Freq*
BENIGN	127537	*BENIGN*	440031
PortScan	158930	DoS GoldenEye	10293
		DoS Hulk	231073
		DoS Slowhttptest	5499
		DoS slowloris	5796
		Heartbleed	11

The data understanding allows to realize some metrics presented in the datasets did not present the type set on. This gap may impact the model construction, depending on the amount of data used. The files presented 85 columns with different features. According to Table 2, we observe that there are 127,537 data classified as *BENIGN* and 158,930 *Port Scan*, in other words, for *Port Scan* more than half of the data is classified as attack. This amount indicates that there are data enough for the model creation. Due to the similarity among the file structure, the parsing procedures for the data from both attack sets could be the same. It is worth noticing that the amounts of *labels* for each attack are the result of an initial assessment, because during the *Data Preparation* stage, the data changes according to the preparation criteria.

3.3   Data Preparation

Data preparation is a key point for creating the model; if it is not done correctly, the results and performance suffer an impact in a negative way. Usually, in case of having a lot of information in the initial dataset, you can build an "auxiliar" dataset. This technique of creating or transforming resources into a new set can assist in more accurate classifiers in later steps [2].

The quality of the assessed data is critical to the success of the project. In the data preparation, all information that will not be used in the future are removed, as most learning algorithms use knowledge extracted from the data without the use of external sources. In order to set up the datasets of each type of attack, we used as guide the study performed by [13]. Table 3 presents the most relevant metrics for each type of attack.

Table 3. Relevant metrics for each type of attack [13]

Label	Feature	Weigh
Port Scan	Init Win F.Bytes	0.0083
	B.Packets/S	0.0032
	PSH Flag Count	0.0009
DoS Slowloris	Flow Duration	0.0431
	F.IAT Min	0.0378
	B.IAT Mean	0.0300
	F.IAT Mean	0.0265

At this stage, we also performed the evaluation of the information whose origin was the internal network itself. In other words, during the creation of the datasets, we introduce all network traffic analyzed by the IDS without removing any information from the internal hosts. This approach could deform the information for analysis, as an attack usually originates from the external perimeter of the network. So, we remove all records that came from the internal network (LAN). However, removing the data affected the amount of benign data of the sets. The updated data are shown in Sect. 4.

After completing the data preparation step, only the columns used to build the models remain. Also, the overall data was reduced at the end of this step, disregarding all traffic from the internal network. In this sense, it was necessary to establish the data types for each column, which helps in the performance of model creation. The data generated during this process is subject to an evaluation to attest to its quality. Depending on the quality of the evaluated data and the results obtained, it could be necessary to return to the previous steps to ensure the data present the required level of quality. This step is essential to avoid future problems in the evaluation of the results.

3.4 Modeling

As previously described, this study seeks to create models that allow the classification and identification of *Port Scan* and *DoS Slowloris* attacks using QDA. In the Modeling phase occurs the tests for the creation of the templates. Firstly, in order to be able to create the model in QDA, it was necessary to revisit some steps of data preparation. In this context, 60% of the data was reserved for the training of each model, the remaining 40% used for evaluating the results.

To create the datasets, we select the records randomly according to the established percentages, not collecting sequence data, which could affect the model performance. For both models (training and testing), the same percentages were adopted, regardless of the label distribution of each dataset. The definition of the percentages for training and evaluation was defined based on similar studies in the literature. The size distribution of each dataset is shown in Table 4.

Table 4. Size distribution of each dataset

	Port Scan	DoS Slowloris
Training	106531	44397
Testing	71022	29599

The classification models for each type of attack start to work right after creating the training and testing datasets. The process consists of passing the parameters that the QDA uses to construct each model and which variable should be learned or predicted. In this case, the goal is that the models can predict the label according to each type of attack, distinguishing the regular traffic from the attack. Next, we present the performance evaluation of each case.

4   Evaluation

To evaluate our proposal, we used a dataset produced by the Canadian Institute for Cybersecurity. More specifically, the dataset provides a range of information regarding the main types of attacks detected by IDS. The dataset contains more than 80 metrics available for network traffic evaluation, as well as a description of the configurations applied in the environments and components used in this work, thus allowing the complete identification of the topology used for the instantiation of the scenarios.

During the evaluation of the proposal, several types of attacks presented in the datasets were evaluated. However, not all attacks contained enough information available for the investigation and creation of the classification model. The following metrics were individually evaluated for each of the two models created: Receiver Operating Characteristics (ROC), Accuracy, and Recall. All these metrics are commonly used for evaluating classification algorithms.

- Accuracy: It is one of the most common metrics for evaluating machine learning accuracy. It refers to a statistical measure of how well a binary classification test correctly identifies or excludes a condition. Accuracy is calculated using the total hits made by the classifier (true positive and true negative) on the total number of objects that were classified, as illustrated in Eq. 1.

$$Accuracy = \frac{TruePositive + TrueNegative}{TotalSamples} \tag{1}$$

When the variety of data for the model creation is not large, the Accuracy may not be the best metric. It is crucial to evaluate the available class balance so that the classification standards can be established. Accuracy refers to the number of correct ratings, but it is imperative to evaluate the percentage of correctly rated true positive and true negative. Table 5 presents the results for Accuracy, Sensitivity, and Specificity.

Table 5. Statistics of the classification models

	Accuracy	Sensitivity	Specificity
Port Scan	0.9996	0.9996	0.9996
DoS Slowloris	0.9832	0.9934	0.8616

The values shown in Table 5 range from 0 to 1. Where 1 means the best result possible. *Accuracy* is the overall accuracy, *i.e.*, the percentage that the model is correct. *Sensitivity* is the classifier's ability to identify benign traffic correctly. *Specificity* allows to measure the performance of the classification the attack; it means how often the algorithm hits. Observing results from the evaluation, it is worth noting that all metrics present values higher than 85%, highlighting the scenario for the *Port Scan*, with a hit rate of more than 99%.

- *Receiver Operating Characteristics* (ROC): It refers to a graphical method that aims to evaluate metrics and select them for systems – originally designed for radar signal detection. Currently, it is used for evaluating a wide range of activities, from psychology, medicine, economics, weather, and machine learning. Figure 2 presents the ROC graphs of the models created for *Port Scan* and *DoS Slowloris* respectively.

In Fig. 2, we observe there is no curve in the ROC for *Port Scan*, there is a straight line from (0.0; 0.0) to (1.0; 1.0), unlike the ROC for *DoS Slowloris*. Analyzing the different results, we can notice that for both models created, the positive class is the "Benign", but the amount of "Benign" data in both datasets for each attack is what differentiates the results. Table 6 presents the amount of data for each attack type according to the distribution of the datasets.

Table 6. Characterization of the data according to the datasets

	Benign	Attack
Port Scan	18.623	158.930
DoS Slowloris	68.200	5.796

From the analysis of different distributions for each dataset presented in Table 6, we observe that for *Port Scan* attack the number of records classified as Attack is far superior in comparison to Benign, which means that the model labeled with the Negative class has more data than the Positive

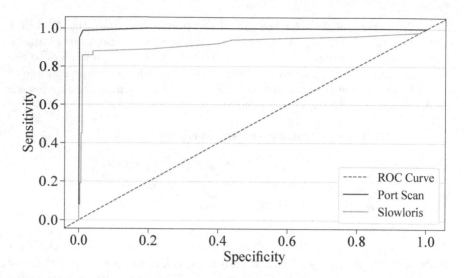

Fig. 2. ROC graph from the models

class. From the point of view of the *Dos Slowloris* dataset, this scenario is reversed. In which case, there is much more data classified as "Benign", which explains the difference in the curves of each of the graphs in Fig. 2. About 89% of data for *Port Scan* is classified as an attack against about 7% for *DoS Slowloris*.

- *Recall:* This metric is also referenced by *True Positive Rate*, which briefly answers the following question: When should the algorithm predict the Positive Class: how often does it hit? Other metrics such as Sensitivity and Specificity can evidence the performance of a classifier [16]. This relationship is presented in Fig. 2.

The *Recall* results are shown in Table 5 by the column Sensitivity. The lower percentage for Port Scan attack compared to DoS Slowloris can be explained by the low number of "Benign" records. This characteristic confuses the model once there are more attacks than normal network traffic. However, there was a high hit rate for *Recall* in identifying attacks. Recall and accuracy are related metrics, and in some cases, changing parameters to increase one may imbalance the other. The best case is looking for a trade-off where both achieve a suitable result.

5   Related Work

There are several studies related to the use of Machine Learning techniques for identifying attack patterns. On the other hand, usually, generalist methods for identifying and classifying attacks are proposed. Thus, ignoring specific behaviours of each type of attack, which can often make the difference in the result. Sinclair *et al.* [14] present a method based on the Decision Tree algorithm to identify anomalies in network layers. However, the result does not rank the attack or malicious traffic according to the technique used, only suspicious traffic is identified.

Osareh and Shadga [10] conducted a study using a public dataset, KDD, to evaluate the performance of different classification methods. The dataset used for the study is from the year 1998, which makes it challenging to evaluate its efficiency faced with current attacks. Similar to Sinclair *et al.* [14] work, the same methods, and variables were used for traffic classification and identification of attack types. The study explored an algorithm that, on average, could meet the general classification rate. However, this approach does not achieve a classification rate higher than 50% when applied individually for each type of attack. This perspective reinforces the importance and relevance of each specific measurement metric for each type of attack.

Sharafaldin *et al.* [13] developed a complete dataset containing several types of attacks identified by an IDS. This study presents the construction systems, topology overview, and methodology used to construct the available datasets, as well as containing several metrics distributed in more than 80 columns. The variety, amount of data available, as well as all the information used for the construction and execution of the study, allowed us to conduct our research. The diversity of data and the previous identification of the most relevant metrics for a given attack enabled the creation of efficient classification models.

6   Conclusion

Face the relevance of the attack scenarios considered in this study, classification models for assisting in the identification of risks of the network perimeters were proposed. It is worth observing the algorithm proposed is a complement to identification and classification mechanisms already existing in the IDS, and it may support in doubt scenarios or even "double-check" for the definition of defense strategies.

In light of this, the present work proposed a new approach for using QDA for identifying and classifying *Port Scan*, and *DoS Slowloris* attacks, allowing distinguish "benign" traffic from "malicious" traffic. From a test environment and based on detailed datasets, a QDA classification model obtained results with a high percentage of assertiveness in the classification of attacks. By observing the results, it is clear the influence of the correct metrics on the efficiency of each of the models created.

Despite the encouraging results depicted, as future work, we intend to evaluate other types of attacks not considered in this study. Also, to propose a methodology to compare the performance with different classifiers for several types of attacks. It is worth noticing the features used to construct the models were made available in the datasets available in [6]. Finally, an approach for on-the-fly collecting information must be designed in order to evaluate and perform a fair comparison among other classifiers.

References

1. Aksu, D., Aydin, M.A.: Detecting port scan attempts with comparative analysis of deep learning and support vector machine algorithms. In: 2018 International Congress on Big Data, Deep Learning and Fighting Cyber Terrorism (IBIGDELFT), pp. 77–80. IEEE (2018)
2. Aziz, A.S.A., Sanaa, E., Hassanien, A.E.: Comparison of classification techniques applied for network intrusion detection and classification. J. Appl. Log. **24**, 109–118 (2017)
3. Boutaba, R., et al.: A comprehensive survey on machine learning for networking: evolution, applications and research opportunities. J. Internet Serv. Appl. **9**(1), 16 (2018)
4. Choi, J., Park, J., Heo, S., Park, N., Kim, H.: Slowloris DoS Countermeasure over WebSocket. In: Choi, D., Guilley, S. (eds.) WISA 2016. LNCS, vol. 10144, pp. 42–53. Springer, Cham (2017). https://doi.org/10.1007/978-3-319-56549-1_4
5. Dalmazo, B.L., Vilela, J.P., Simoes, P., Curado, M.: Expedite feature extraction for enhanced cloud anomaly detection. In: NOMS 2016–2016 IEEE/IFIP Network Operations and Management Symposium. pp. 1215–1220, April 2016. https://doi.org/10.1109/NOMS.2016.7502990
6. Dalmazo, B.L., Deolindo, V.M., Nobre, J.C.: Public dataset for evaluating Port Scan and Slowloris attacks (2019)
7. Harrington, P.: Machine Learning in Action. Manning Publications Co. (2012)
8. Liao, H.J., Lin, C.H.R., Lin, Y.C., Tung, K.Y.: Intrusion detection system: a comprehensive review. J. Netw. Comput. Appl. **36**(1), 16–24 (2013)
9. Nadali, A., Kakhky, E.N., Nosratabadi, H.E.: Evaluating the success level of data mining projects based on CRISP-DM methodology by a fuzzy expert system. In: 2011 3rd International Conference on Electronics Computer Technology (ICECT), vol. 6, pp. 161–165. IEEE (2011)
10. Osareh, A., Shadgar, B.: Intrusion detection in computer networks based on machine learning algorithms. Int. J. Comput. Sci. Netw. Secur. **8**(11), 15–23 (2008)
11. Sangkatsanee, P., Wattanapongsakorn, N., Charnsripinyo, C.: Practical real-time intrusion detection using machine learning approaches. Comput. Commun. **34**(18), 2227–2235 (2011). https://doi.org/10.1016/j.comcom.2011.07.001
12. Shafique, U., Qaiser, H.: A comparative study of data mining process models (KDD, CRISP-DM AND SEMMA). Int. J. Innov. Sci. Res. **12**(1), 217–222 (2014)
13. Sharafaldin, I., Lashkari, A.H., Ghorbani, A.A.: Toward generating a new intrusion detection dataset and intrusion traffic characterization. In: Proceedings of Fourth International Conference on Information Systems Security and Privacy, ICISSP (2018)
14. Sinclair, C., Pierce, L., Matzner, S.: An application of machine learning to network intrusion detection. In: 15th Annual Computer Security Applications Conference, (ACSAC 1999) Proceedings, pp. 371–377. IEEE (1999)
15. Srivastava, S., Gupta, M.R., Frigyik, B.A.: Bayesian quadratic discriminant analysis. J. Mach. Learn. Res. **8**, 1277–1305 (2007)
16. Su, W., Yuan, Y., Zhu, M.: A relationship between the average precision and the area under the roc curve. In: Proceedings of the 2015 International Conference on The Theory of Information Retrieval, ICTIR 2015, pp. 349–352. Association for Computing Machinery, New York (2015). https://doi.org/10.1145/2808194.2809481

Systematic Literature Review on Service Oriented Architecture Modeling

Khouloud Zaafouri, Mariam Chaabane[(✉)], and Ismael Bouassida Rodriguez

ReDCAD Laboratory, University of Sfax, 3038 Sfax, Tunisia
mariam.chaabane@redcad.org

Abstract. Context: With the recent trend of shifting from traditional architectures towards Service Oriented Architectures (SOA), an enterprise can create, choreograph new business functions, deploy and integrate multiple services that communicate with each other using service interfaces to pass messages from one service to another. So, to build an SOA that accommodate business scalability, and flexibility and facilitate ongoing and changing needs of business, an SOA modeling language is required.

Objective: The purpose of this work is to determine the current state of the art in the field of SOA modeling shedding light on techniques have been used to model a SOA, its importance and domains where it is applied.

Method: In order to fulfill the objective of the research, the method which we choose was a Systematic Literature Review (SLR). This served in collecting and structuring the information that exists in the field of SOA modeling.

Results: Service oriented applications have been modeled with different modeling languages. Choosing a suitable modeling language is depending on the criteria for what is being modeled; if is it structural view point or behavioral aspects or even both.

Conclusion: Summing up the results, it can be concluded firstly that modeling mitigates the complexity of huge and complicated data of service oriented applications. Secondly, the research that has been performed has shown that the Unified Modeling Language (UML), Business Process Modeling Notation (BPMN), Service Component Architecture (SCA) and Event-B are the most modeling languages used on large scale.

Keywords: Systematic Literature Review · SLR · Modeling · Service Oriented Architecture · SOA

1 Introduction

This paper presents an SLR on SOA modeling. A systematic review is a means of identifying, evaluating and interpreting all available research relevant to a particular research question (RQ), topic area, or phenomenon of interest in an unbiased and repeatable manner [13]. There are a number of different reasons why we should be employ systematic reviews. Some common ones are: to provide

© Springer Nature Switzerland AG 2021
O. Gervasi et al. (Eds.): ICCSA 2021, LNCS 12951, pp. 201–210, 2021.
https://doi.org/10.1007/978-3-030-86970-0_15

a detailed summary of current literature relevant to a RQ; to identify where there are limitations in current research in order to help determine where further study might be needed and to analyse how far a given hypothesis is supported or contradicted by the available proof.

The goal of this review is to aggregate the published empirical knowledge on the topic of SOA modeling. With the recent trend of shifting from traditional architectures towards SOA, services can be founded and reused to assemble new applications faster than in the traditional architectures [24], thanks to service interfaces that utilize standard network protocols in such a way that they can be rapidly integrated into new softwares without having to perform deep integration all time. So that, the structure and behavior of the architecture can be changing constantly [12] whenever the business process was changed.

This change becomes a difficult issue, due to complexity and strict requirements of such systems [16]. So, the need to model becomes increasingly apparent. For this reason, this review sheds light on the topic of SOA modeling.

The remainder of this paper is structured as follows: the research method is presented in Sect. 2, the results are presented in Sect. 3, the forth section discusses some possible limitations of the research, a conclusion and potential directions for the future are given in the last section.

2   Research Method

This section describes the RQ and search strategy aiming to find all relevant primary studies related to model SOA.

2.1   Research Questions

The RQ states the specific issue or problem that the research project will focus on. It also outlines the task that will need to complete. The process of developing our RQ follows several steps: Firstly, We choose a broad topic which is SOA. Secondly, preliminary reading was done to find out about topical issues. Then, we narrow down a specific field that we want to focus on which is SOA modeling language. The last step was to formulate the RQ taking into consideration such types of questions: Descriptive and Comparative RQ.

The following RQ were established:

RQ1: Which techniques have been used to model a SOA?
RQ2: What are the interests of modeling languages for SOA?
RQ3: Is the choice of modeling language particular for a type of system?

2.2   Search Strategy

The search strategy combines the key concepts of our search question in order to retrieve accurate results. Based on our RQ we create a list of seed keywords that our audience might be searching on. All we need to do it is describe our offering as simple as possible and brainstorm how other people might search for it.

As a result, our search strategy is an organized structure of key words, related to the broad concepts of our topic, which are Service Oriented Architecture and Modeling languages, used to search a database. Then, we added synonyms, variations, and related terms for each keyword. A Boolean operator (AND and OR) allow us to try different combinations of search terms. The final search string is (Service-Oriented Architecture OR Service Oriented Architecture OR SOA) AND (modeling OR model OR models OR design OR graphical notation OR graphical representation) AND (language OR method OR methodology OR procedure OR technique).

3   Search Results

This section presents the manner in which papers were identified, examined and selected, and the method used to extract and analyze data from papers deemed as relevant.

3.1   Data Collection

The search process was a manual search on specific online databases. The selected sources are shown in Table 1. Those sources identified as the most important and likely to contain relevant material, were searched for relevant content. The covered timeframe caught between a lower bound that was set from the beginning of 2010 and an upper bound that was set to the end of 2020.

Table 1 shows the number of papers found in each source.

Table 1. Search results by resource

Resource	Number of papers
Springer https://link.springer.com/	83
IEE Xplore Digital Library https://www.ieee.org/	55
ACM Digital library https://www.acm.org/	128
Google Scholar https://scholar.google.com/	100
Science Direct https://www.sciencedirect.com/	32
Hyper Articles en Ligne (HAL) https://hal.archives-ouvertes.fr/	14
Total	412

3.2　Data Extraction and Analysis

Once the papers within the scope were identified, the next step was to extract the relevant information (data) from them by reading abstracts and after that by reading introductions.

Table 2 shows the number of relevant papers founded.

Table 2. Filtered search results

Irrelevant and duplicates	2
Incomplete and not related to RQ, Excluded by reading title and abstract	279
File not found	19
Total for introduction readingr	112
Not related to RQ, Excluded by reading introduction	49
Total for reading	63

Table 3 gives an overview of the relevant papers, indicating their source.

Table 3. Filtered search results by resource

Resource	Number of papers
Springer	8
IEE Xplore Digital Library	8
ACM Digital library	27
Google Scholar	13
Science Direct	4
Hyper Articles en Ligne (HAL)	3

3.3　Overview of Results

Out of 412 papers, 112 selected for introduction reading and were thus opened and inspected, yielding 63 relevant papers.

RQ1: Which Techniques Have Been Used to Model a SOA? This section deals with the first RQ and looks at empirical work what technique have been used to model SOA. We found 32 over 63 papers that dealt with this RQ.

Figure 1 reveals that a large majority of the papers contribute to UML (39%). On second place, BPMN is addressed by a quarter of the papers (26%). 16% of addressed papers are modeled using SCA and 10% of studies using Event-B method. Finally, each one of the modeling languages: Knowledge Acquisition in Automated Specification (KAOS), Open Archival Information System (OAIS) and Behavior-Interaction-Priority (BIP) are addressed by (3%) of the papers.

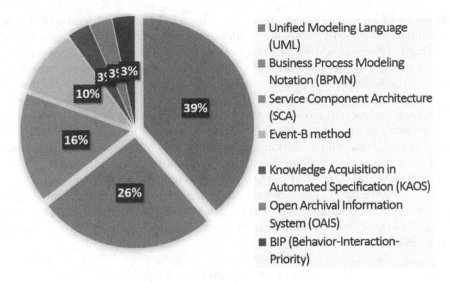

Fig. 1. Breakdown by empirical technique

RQ2: What Are the Interests of Modeling Languages for SOA? This section discusses papers that contribute to the understanding of the interests of modeling languages for SOA. In the set of 63 papers, 32 of them highlighted the motivation of modeling languages for SOA.

In terms of UML's effect on modeling SOA, three observations have been made on the topic: The peculiarity of UML is that a standard and understandable visual modeling language [25] that represent a natural, intuitive and expressive way to forge the behavioral and structural aspects of a system [2,4], well-designed software requirements definitions have a great impact on final product quality [6] and reduce costs and minimize development time [15]. Due to this, it can be concluded that UML is a visual modeling language applicable to design the structural and behavioral aspects of the system.

In terms of the interesting of BPMN to model SOA two perceptions have been made on the subject: To help the business and system stakeholders to align their business requirements and IT system implementations [8], and to more flexibly execute business transactions [5,21] thanks to the rich set of elements and attributes which mostly emphasize on illustrating the system behavior and its abstraction [17] using BPEL language [3]. So, BPMN as a visual modeling language can be readily used to design the behavioral view point of a system.

Regarding SCA's interest in SOA modeling, two visualizations have been made on the topic: Thanks to SCA design primitives [20], SCA provide graphical representation of components structure and of components assemblies, those Components can be reused to create composite components [9]. Therefore, it could be interesting to define a component that manages scenario execution [9]. On the other hand, the combination of SCA and ASM [19,20] tackles "the complexity of service oriented applications by offering a high degree of design and validation at early development phases". Unlike BPMN, SCA as a visual modeling language is addressed to model the structural aspects of a system.

As for Event-B method's interest to model SOA two observations have been made on the topic: It is a formal method. The key features of it are the use of set theory as a modelling notation, the use of refinement to represent systems at different abstraction levels and the use of mathematical proof to verify consistency between refinement levels. Event-B method is identical with UML on modeling structural and behavioral view points [11] of a system. Nevertheless, Event-B, which is a formal language is not a visual modeling language.

We discuss the interest of the remainder techniques in SOA modeling. For KAOS as a systematic approach to model SOA applications using goal-models [22], this approach can guide the service designer in identifying a set of services during SOA specification thanks to the use of a set of heuristics and model transformations.

Table 4. Breakdown by criterion

Modeling language	Visual modeling language	Behavioral aspects	Structural aspects
UML	X	X	X
BPMN	X	X	
SCA	X		X
Event-B method		X	X
KAOS	X	X	
OAIS			X
BIP		X	

As for OAIS as a reference model [14] was developed to be applicable in any organisational context in which digital content is managed for the long term. It is used in order to provide a set of policies and procedures in the implementation of content preservation programs [18]. Most organisations that manage digital collections have indicated an intention to design and implement their digital repositories in accordance with the OAIS. But, it is not intended to promote or discuss specific details or technical recommendations in the different software implementations.

Regarding BIP [23], thanks to a powerful executable modeling language with formally defined operational semantics and mathematically proven expressiveness properties [1], BIP supports innovative formal analyses by reasoning on the model structure thus avoiding as much as possible the scalability limitations of model checking techniques.

Graiet et al. [10] proposes a sound approach to formalize Web services composition mediation with the ACME ADL using Armani, which provides a powerful predicate language in order to ensure service execution reliability.

Table 4 summarizes the criterion of each Modeling Language, blank cells indicate that the topic was not addressed.

In the light of the criterion mentioned in Table 4, we deduced that the modeling languages UML, BPMN and SCA are a visual modeling language, in contrast of Event-B which is a formal method.

The modeling language BPMN models the behavioral aspects of the system, while SCA models the structural aspects by defining components and wires to represent a network.

Whereas, both UML and Event-B forge the behavioral and the aspects of the system. Even though, SCA out-performs UML by explicitly defining Services and References with various bindings [7].

Table 5. Breakdown by domain

Domain	UML	BPMN	SCA	Event-B method	KAOS	OAIS	BIP
Education	X	X					
Healthcare	X						
Transport	X	X			X		
Marketing	X						
Industry		X					
Unknown			X	X		X	X

RQ3: Is the Choice of Modeling Language Particular for a Type of System? Modelling language is applied on large scale, depending on the criteria for what is being modeled; if is it structural view point or behavioral aspects or even both. Regarding our research we extract five domains where modelling language was applied (Education, Healthcare, Transport, Marketing and Industry).

Table 5 shows that the most applied modelling language are UML and BPMN. UML is used in education, healthcare, transport and marketing field. While BPMN is applicable on the field of education, transport and industry.

From the outcome of our research it is possible to conclude that there are strong chances that BPMN will be adopted by the industry as a modeling language for modeling the integration of SOAs. While UML is intended to use mainly in areas like web applications, commercial applications and embedded systems.

4   Threats to Validity

The purpose of this section is to analyze the threats to validity regarding the conclusions of our SLR.

The main threats to the validity of our review we have identified are these:

Choice of Papers: relevant papers may not be identified because they do not match to the search criteria. To resolve this problem, we have taken into consideration the recommendations of the experts at each stage of our SLR to avoid this threat to validity.

Unbalanced Results: Impartiality in such papers can affect the interpretation of the results. In fact, the rather low number of existing research materials about Event-B method, KAOS, OAIS and BIP and the increased prevalence of modeling with UML, BPMN and SCA. Thus, the papers analyzed for those modeling language were not entirely numerically balanced. Nevertheless, the fact that our choice of papers is validated by experts, ensures that our interpretation reflects the reality of the literature.

UML Limitation: A potential limitation of the literature study is that: Regarding the findings we have obtained, we conclude that UML is the most modeling language that used in order to model softwares that follow SOA. Whereas De Troyer et al. [7] shows that SCA out-performs UML by explicitly defining Services and References with various bindings. Nevertheless, UML 2.x version could be very adequate to model Service Oriented application while replacing UML, SCA and BPMN to model both structural and behavioral aspects of SOA. In addition, it could model different instances of the modeled Web Services using the object diagram.

5   Conclusion and Further Research

The goal of this review is to aggregate the published empirical knowledge on the topic of SOA modeling. We started by developing our RQ aiming to find all relevant primary studies related to model SOA. When our RQ were established, we create a list of seed keywords that our audience might be searching on. Based on those words a search strategy was created. Using the search strategy that we established, we search manually on specific online databases. Out of 412 papers, 112 selected for introduction reading and were thus opened and inspected, yielding 63 relevant papers. Only 32 over 63 papers deals with our RQ.

From the research that has been carried out, we conclude that Service oriented applications have been modeled with various modeling languages: UML, BPMN, SCA, Event-B method, KAOS, OAIS and BIP. We analyzed those seven main languages for SOA modeling; having an evaluation between them, based on a defined set of criteria.

The main conclusions can be drawn is that UML, BPMN, SCA and Event-B are the most modeling languages used to model applications that follow a SOA. BPMN for designing the behavioral aspect of the system and SCA for modeling the structural view point of the system. Moreover, to support both modeling capabilities (behavioral and structural) a formal modeling language was proposed which is Event-B method, while UML is popular for software systems modeling and widely accepted as the standard for this, thanks to the visual modeling language.

The study can be continued by taking a more practical approach by modeling "Smart City" of SOA integration using those modeling language, and trying to identify the benefits and limitations resulting from the concrete practical modeling.

Acknowledgement. This work was partially supported by the LABEX-TA project MeFoGL: "Méthodes Formelles pour le Génie Logiciel".

References

1. Bliudze, S., Sifakis, J.: A notion of glue expressiveness for component-based systems. In: van Breugel, F., Chechik, M. (eds.) CONCUR 2008. LNCS, vol. 5201, pp. 508–522. Springer, Heidelberg (2008). https://doi.org/10.1007/978-3-540-85361-9_39
2. Booch, G., Rumbaugh, J., Jacobson, I.: The unified modeling language reference manual (1999)
3. Chaabane, M., Bouassida Rodriguez, I., Drira, K., Jmaiel, M.: Mining approach for software architectures' description discovery. In: 14th IEEE/ACS International Conference on Computer Systems and Applications, AICCSA 2017, Hammamet, Tunisia, 30 October–3 November, 2017, pp. 879–886 (2017). https://doi.org/10.1109/AICCSA.2017.169
4. Chaabane, M., Krichen, F., Bouassida Rodriguez, I., Jmaiel, M.: Monitoring of service-oriented applications for the reconstruction of interactions models. In: Computational Science and Its Applications - ICCSA 2015–15th International Conference, Banff, AB, Canada, 22–25 June, 2015, Proceedings, Part I, vol. 9155, pp. 172–186 (2015). https://doi.org/10.1007/978-3-319-21404-7_13
5. Cimino, M.G., Palumbo, F., Vaglini, G., Ferro, E., Celandroni, N., La Rosa, D.: Evaluating the impact of smart technologies on harbor's logistics via BPMN modeling and simulation. Inf. Technol. Manage. **18**(3), 223–239 (2017)
6. Davis, A., Dieste, O., Hickey, A., Juristo, N., Moreno, A.M.: Effectiveness of requirements elicitation techniques: empirical results derived from a systematic review. In: 14th IEEE International Requirements Engineering Conference (RE 2006), pp. 179–188. IEEE (2006)
7. De Troyer, O., Bauzer Medeiros, C., Billen, R., Hallot, P., Simitsis, A., Van Mingroot, H. (eds.): ER 2011. LNCS, vol. 6999. Springer, Heidelberg (2011). https://doi.org/10.1007/978-3-642-24574-9
8. Elvesæter, B., Panfilenko, D., Jacobi, S., Hahn, C.: Aligning business and it models in service-oriented architectures using BPMN and SOAML. In: Proceedings of the First International Workshop on Model-Driven Interoperability, MDI 2010, pp. 61–68. Association for Computing Machinery, New York (2010). https://doi.org/10.1145/1866272.1866281
9. Faure, M.: Management of Scenarized User-centric Service Compositions for Collaborative Pervasive Environments. Theses, Université Montpellier II - Sciences et Techniques du Languedoc (2012). https://tel.archives-ouvertes.fr/tel-00790156
10. Graiet, M., Maraoui, R., Kmimech, M., Bhiri, M.T., Gaaloul, W.: Towards an approach of formal verification of mediation protocol based on web services. In: Proceedings of the 12th International Conference on Information Integration and Web-Based Applications & Services, iiWAS 2010, pp. 75–82. Association for Computing Machinery, New York (2010). https://doi.org/10.1145/1967486.1967502
11. Hoang, T.S.: An introduction to the event-b modelling method. Industrial Deployment of System Engineering Methods, pp. 211–236 (2013). https://doi.org/10.1007/978-3-642-33170-1
12. Jia, X., Ying, S., Zhang, T., Cao, H., Xie, D.: A new architecture description language for service-oriented architec. In: Sixth International Conference on Grid and Cooperative Computing (GCC 2007), pp. 96–103. IEEE (2007)

13. Kitchenham, B.: Procedures for performing systematic reviews. Keele, UK, Keele University **33**(2004), 1–26 (2004)
14. Lee, C.A.: Open archival information system (OAIS) reference model. Encyclopedia of library and information. Sciences **3**, 1–11 (2010)
15. Loniewski, G., Insfran, E., Abrahão, S.: A systematic review of the use of requirements engineering techniques in model-driven development. In: Petriu, D.C., Rouquette, N., Haugen, Ø. (eds.) MODELS 2010. LNCS, vol. 6395, pp. 213–227. Springer, Heidelberg (2010). https://doi.org/10.1007/978-3-642-16129-2_16
16. Loukil, S., Kallel, S., Zalila, B., Jmaiel, M.: AO4AADL: aspect oriented extension for AADL. Central Eur. J. Comput. Sci. **3**(2), 43–68 (2013)
17. Soleimani Malekan, H., Afsarmanesh, H.: Overview of business process modeling languages supporting enterprise collaboration. In: Shishkov, B. (ed.) BMSD 2013. LNBIP, vol. 173, pp. 24–45. Springer, Cham (2014). https://doi.org/10.1007/978-3-319-06671-4_2
18. Neto, A.J.R., Borges, M.M., Roque, L.: Preliminary study about the applicability of a service-oriented architecture in the OAIS model implementation. In: Proceedings of the 5th International Conference on Technological Ecosystems for Enhancing Multiculturality, TEEM 2017. Association for Computing Machinery, New York (2017). https://doi.org/10.1145/3144826.3145381
19. Riccobene, E., Potena, P., Scandurra, P.: Reliability prediction for service component architectures with the SCA-ASM component model. In: 2012 38th Euromicro Conference on Software Engineering and Advanced Applications, pp. 125–132 (2012)
20. Riccobene, E., Scandurra, P., Albani, F.: A modeling and executable language for designing and prototyping service-oriented applications. In: 2011 37th EUROMICRO Conference on Software Engineering and Advanced Applications, pp. 4–11 (2011)
21. de Souza, A.P., Rabelo, R.J.: A dynamic services discovery model for better leveraging BPM and SOA integration. Int. J. Inf. Syst. Serv. Sect. (IJISSS) **7**(1), 1–21 (2015)
22. Souza, E., Moreira, A.: Deriving services from KAOS models. In: Proceedings of the 33rd Annual ACM Symposium on Applied Computing, SAC 2018, pp. 1308–1315. Association for Computing Machinery, New York (2018). https://doi.org/10.1145/3167132.3167273
23. Stachtiari, E., Mentis, A., Katsaros, P.: Rigorous analysis of service composability by embedding WS-BPEL into the BIP component framework. In: 2012 IEEE 19th International Conference on Web Services, pp. 319–326 (2012)
24. Todoran, I., Hussain, Z., Gromov, N.: SOA integration modeling: an evaluation of how SoaML completes UML modeling. In: 2011 IEEE 15th International Enterprise Distributed Object Computing Conference Workshops, pp. 57–66. IEEE (2011)
25. Zhang, H., Liu, J., Zheng, L., Wang, J.: Modeling of web service development process based on MDA and procedure blueprint. In: 2012 IEEE/ACIS 11th International Conference on Computer and Information Science, pp. 422–427 (2012)

A Deep Learning Solution for Integrated Traffic Control Through Automatic License Plate Recognition

Riccardo Balia[1], Silvio Barra[2], Salvatore Carta[1], Gianni Fenu[1], Alessandro Sebastian Podda[1(✉)], and Nicola Sansoni[1]

[1] University of Cagliari, DMI, 09124 Cagliari, Italy
{r.balia,salvatore,fenu,sebastianpodda,n.sansoni}@unica.it
[2] University of Naples "Federico II", DIETI, 80125 Napoli, Italy
silvio.barra@unina.it

Abstract. Nowadays, Smart Cities applications are becoming steadily popular, thanks to their main objective of improving people daily habits. The services provided by the aforementioned applications may be either addressed to the entire digital population or narrowed towards a specific kind of audience, like drivers and pedestrians. In this sense, the proposed paper describes a Deep Learning solution designed to manage traffic control tasks in Smart Cities. It involves a network of *smart lampposts*, in charge of directly monitoring the traffic by means of a bullet camera, and equipped with an advanced System-on-Module where the data are efficiently processed. In particular, our solution provides both: i) a risk estimation module, and ii) a license plate recognition module. The first module analyses the scene by means of a Faster R-CNN, trained over an ad-hoc set of synthetically videos, to estimate the risk of potential traffic anomalies. Concurrently, the license plate recognition module, by leveraging on YOLO and Tesseract, is active for retrieving the plate number of the vehicles involved. Preliminary experimental findings, from a prototype of the solution applied in a real-world scenario, are provided.

Keywords: Deep Learning · Smart Cities · Anomalies detection · License plate recognition

1 Introduction

Smart Cities services and applications are becoming ubiquitous in the modern life, positively influencing the daily habits their citizens. Currently, most of these applications fall in the areas of physical and logical security [12], targeted both at pedestrian [15] and drivers [3], including privacy aspects [12].

These brand new services are commonly provided by means of *Internet of Things* (IoT) sensors, spread in specific districts and roads or, in an increasing number of cases, all over the city. Such sensors are instrumental in sensing the environment, to infer new information and to provide people-oriented solutions

© Springer Nature Switzerland AG 2021
O. Gervasi et al. (Eds.): ICCSA 2021, LNCS 12951, pp. 211–226, 2021.
https://doi.org/10.1007/978-3-030-86970-0_16

[38]. Nevertheless, their applications nowadays pave the entire range of the road safety and facility topics, not comprehensively:

- Smart Parking [10,11];
- Intelligent Transportation Systems [33];
- Smart Traffic lights [4];
- Automated Vehicles and Pedestrian Safety [17].

In this context, several scientific and industrial approaches have been proposed, relying on the mere vehicle/object tracking [9,24], as well as aimed at addressing related issues like camera calibration [30], scene segmentation [16], just to mention a few. More recently, with the advent of *learning approaches* [37], these approaches have been boosted towards a more generalist scenario, involving heterogeneous data sources and fusions [23], scene understanding proposals [6,8], video captioning techniques [7,18].

In view of the foregoing, this paper aims to propose a novel solution, based on Deep Learning and Computer Vision methods, to automatically monitor traffic flows for recognizing possible abnormal traffic situations and identifying the license plates number of the vehicles involved. The main benefit provided is to drastically optimise the recording and storing of such events for subsequent audits, as well as lay the foundations for the integration with automatic sanctioning or traffic analysis systems. Indeed, current solutions in literature or in industry are either not focused on scalability, or they require expensive hardware and software. In this sense, the proposed system is cost-effective and leverages on open-source techniques only.

Specifically, the original contributions of this work are:

- the description of a novel distributed architecture for the automatic monitoring of traffic flows in Smart Cities;
- the definition of a taxonomy of traffic anomalies, so that they can be recognised by an intelligent automatic system, and the specification of a monitoring module based on Faster R-CNN, trained on synthetic and real videos, to estimate the probability that such anomalous events have occurred;
- the design of an original and efficient automatic number plate recognition approach, based on YOLO and Tesseract OCR;
- the development of a prototype implementing the above approaches, that supports the real-time execution on NVIDIA-based System-on-Modules to be embedded on Smart City lampposts;
- the validation of the proposed solution through preliminary experimental simulations, and by providing an in-depth analysis of the results obtained.

The remainder of this document is organised as follows. Section 2 briefly introduces relevant related work in the literature, whereas Sect. 3 describes in detail the proposed system and the approach on which it is based. Section 4 presents the evaluation results and discusses findings and limitations of our proposal. Finally, Sect. 5 ends the paper with conclusions and future directions where we are headed.

2   Related Works

The challenges tackled by this work have been topics of research from the 90s [29], falling within the broad scope of the anomaly detection, which took its first steps in the network security area [13,34] and then expanded to the wider field of surveillance [2]. Indeed, nowadays, the problems related to automatizing video surveillance and traffic monitoring tasks in Smart Cities [1], in particular those aimed at improving the safety of pedestrian and vehicles, are still open and remain vastly studied [21]. While earlier works mainly employed techniques based on pattern recognition and image manipulation [40] to extract useful data gathered from the sensors, today the developments in Internet of Things [19], Edge Computing [22], Machine/Deep Learning [31] and Computer Vision [26] are the prominent focus of the research.

In this context, Appathurai et al. [5] exploited an artificial neural network, with weights optimally selected through an oppositional gravitational search optimization, to implement a moving vehicle detection system. Through their approach, it is hence possible to gather data about vehicle tracking, vehicle normal speeds, movement examinations and vehicle classification. Makhmutova et al. [28] highlighted the usefulness of improving existing closed-circuit television systems with the application of intelligent information processing and data analysis. They employed computer vision methods to perform object detection and tracking and neural networks for real-time classification and detection of anomalies. They also tested the proposed system on real video streams provided by the city of Kazan.

Similarly, in [2], the authors focused on vehicle identification and counting, with the aim of identifying traffic congestions. They used blob analysis, background subtraction, a dynamic autoregressive moving average model, a boundary block detection algorithm and vehicle tracking to perform automatic vehicle counting. Chakraborty et al. [14] proposed an artificial vision-based smart city solution to the problems concerning traffic and transportation. They also implemented a human activity detection system to improve safety and security of Smart Cities.

This work aims to take a step forward in this direction, by exploiting different Deep Learning techniques in a coherent fashion, and by proposing a structure for the development and deployment of a comprehensive system that handles the problem of traffic safety in a Smart City environment.

3   The Proposed Solution

The proposed solution is designed for automatic traffic monitoring in Smart Cities, through the application of Deep Learning and Computer Vision techniques. The primary purpose of this solution is to carry out video surveillance of urban traffic, by implementing the distributed system shown in Fig. 1.

The solution consists of a network of *smart* lampposts, installed on the public street and equipped with a bullet camera and a general purpose *System-on-Module* (SoM), specifically a *NVIDIA Jetson AGX Xavier* board (performing a

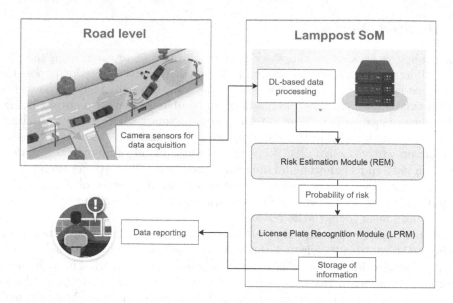

Fig. 1. The proposed solution.

Volta GPU with Tensor Core, an ARMv8 CPU and 32 GB of RAM), particularly suitable for running Neural Networks (NN) and Deep Learning (DL) algorithms on such embedded configuration.

Through this board, the lamppost is able to run the two modules based on Deep Learning and Computer Vision techniques, which perform automatic traffic monitoring functions: (i) a risk estimation module (REM), that automatically marks the segments of the entire observed video stream which are characterised by abnormal traffic conditions (i.e., anomalies are present); and, (ii) a license plate recognition module (LPRM), responsible for registering the license plate numbers of the vehicles present in each segment. In this way, the system is able to store only the information relevant for the traffic analysis, guaranteeing better storage space occupation and a longer time of data archiving. Moreover, such a specification is also effective where the public authority intends to permanently certify digests of these data in a distributed context, for instance by exploiting metadata storage on public blockchains [25, 27, 39]. In the remainder of this section, the approach adopted and the detailed functioning of these two modules are outlined.

3.1 Risk Estimation Module (REM)

The first of the two modules that make up the proposed solution, i.e. the one dedicated to the estimate the risk of abnormal traffic conditions, is described below.

Categorisation of the Anomalies. In order to build an effective risk detection module, part of the preliminary work on the development of this solution was devoted to categorise the different typologies of anomalies that can occur in the urban environment and that can be (potentially) captured through the exploitation of Deep Learning techniques. Specifically, the taxonomy of anomalies employed in this study, illustrated in what follows, specifies 8 different macro-categories, each of them consisting of a set of subcategories, with the aim of highlighting similarities and differences among them. This categorisation is presented in Table 1.

It is important to emphasise that such a categorisation is not intended to cover all possible road anomalies. In the specific case of this study, it is in fact the result of an early analysis within the scope of the project of which this work is part. However, the rationale adopted was to mainly include, as previously pointed out, anomalies that were potentially dangerous to the movement of vehicles and other road users, as well as violations that could cause traffic disruption (such as parking in prohibited areas).

In addition, the anomalies taken into account are those suitable for automatic recognition by artificial intelligence techniques based on computer vision, such as those based on Deep Learning and exploited in this study. In this context, the macro-categories defined in Table 1 need to be further discriminated between *static* and *dynamic* ones, with regard to how time factors into the detection process. In detail, static anomalies are those than can be identified by looking at a static image, as they are mainly characterized mainly by the position and appearance of the involved entities (i.e., vehicles, pedestrians, *etc.*) in the scene; in contrast, dynamic ones necessarily requires to observe the behaviour and movements of such entities in a protracted interval, thus implying the time component. In particular, the macro-categories can be subdivided as follows:

- **Static**: traffic, entity on the road, collision, prohibited stop;
- **Dynamic**: failure to give way, illegal lane change, invasion, reckless driving.

Also note that the risk estimation module of the proposed solution is currently focused only on identifying static anomalies, through frame-by-frame analysis (starting from the video acquired by the lamppost camera) by exploiting a Convolutional Neural Network (CNN). On the other hand, the part concerning the analysis of dynamic anomalies is still under development and is not the subject of this paper.

Faster R-CNN and Ad-hoc Dataset Generation. A supervised learning approach was adopted to estimate the risk associated with the currently observed traffic situation. To do so, an artificial intelligence algorithm, specifically a neural network, was trained by exploiting a dataset of static anomalies (identified through the categorisation described in the previous paragraph) and in order to enable the recognition of moderate/high risk traffic situations.

As previously stated, the goal of this approach is not that of estimating the traffic risk with high accuracy, but rather to improve the performance of the traffic control system by marking as *relevant* those clips – acquired from the continuous loop recording – whose associated risk is estimated as higher than

Table 1. Taxonomy of anomalies.

Macro-category	Subcategory	Description
Traffic	Congestion	A large number of vehicles in the monitored area is proceeding at a very low speed
Entity on the road	Animal	An animal invaded the roadway
	Person	A person invaded the roadway (in the absence of pedestrian crossings)
	Object	An object in the roadway constitutes an obstacle to vehicles
Failure to give way	Red light violation	A vehicle engaged the crossroads with a red traffic light
	Intersection	A vehicle engaged the intersection while another vehicle with the right of way is approaching
	Roundabout	A vehicle engaged a roundabout while another vehicle with the right of way is approaching
Illegal lane change	U-turn	A vehicle has performed a U-turn
	Wrong way	A vehicle was driving on the wrong side of the road
	Continuous line	A vehicle crossed the continuous line of the road
Collision	Head-on	Some vehicles have been involved in a frontal impact
	Sideswipe	Some vehicles collided sideways with each other
	Rear-end	A vehicle impacted the rear of another vehicle
	Obstacle	A vehicle impacted an obstacle, a wall or another obstacle in the environment
	Pedestrian investment	A vehicle impacted with a pedestrian
Invasion	Bike lane	A vehicle invaded a bike lane
	Sidewalk	A vehicle invaded a sidewalk or a pedestrian area
	Off-road	A vehicle was driving off-road but neither on a bike lane or on a pedestrian area
Reckless driving	Hard braking	A vehicle halted in a sudden way
	Zigzagging	A vehicle is zigzagging or changing lanes abruptly
	High speed	A vehicle is driving at very high speed, above the permitted limits
Prohibited stop	No parking zone	A vehicle is parked on-road, in a no parking zone
	Illegal parking	A vehicle is parked off-road, where not allowed

normal. These clips are therefore stored for a longer period of time and remain available for possible subsequent human inspection; moreover, the number plate recognition algorithm is executed on them (in a subsequent non-real time phase) in order to identify the vehicles involved.

Given the performance improvement purpose of the risk estimation module, the approach is agnostic with respect to the type of neural network used to identify static anomalies. However, in the implementation of the experimental prototype of this work, the selected neural network was a *Faster R-CNN*, built on top of *ResNet* [20]. In our preliminary tests, in fact, the Faster R-CNN showed a discrete ability to identify static anomalies, thanks to its peculiarity of being able to analyse individual regions of each frame acquired. However, instead of the Faster R-CNN, it is also possible to leverage on a lighter *single shot detector* (SSD), e.g. *YOLO* [36], properly trained to detect the targeted anomalies, by sacrificing the accuracy of the risk estimation in favour of a higher supported frame rate. The architecture of the Faster R-CNN adopted in our module is shown in Fig. 2.

In order to train the network to recognize the static anomalies defined by the taxonomy in Table 1, we decided to create an *ad-hoc* training dataset, composed by about 800 synthetic videos, manually realized in order to cover each of the target categories with an adequate number of samples (in widely varying scenarios, shots, lighting and traffic conditions).

These synthetic videos have been generated through a Rockstar Games' editor, that exploits the traffic simulator engine of the video game *Grand Theft Auto V*, produced by the same company. The main reason behind this choice falls in the absence of literature datasets that include all the kind of anomalies that do not belong to the category of *collisions*, as well as the impracticality of creating a balanced and comprehensive dataset by capturing real video, in a reasonable amount of time. However, the simulations mimic the real-world scenario at a good level (i.e., traffic dynamics and abnormal events are realistic, while the consequences, such as impact damage, are less so, although they are not considered in this analysis), and even rare events can thus be added to the training set very efficiently. Sample footage depicting normal behaviour, i.e. without anomalies, is also generated and used so as to minimize the amount of false positives. A part of the dataset has also been used as a test set (out-of-sample), together with a small set of real videos, for the validation of the obtained model.

Finally, *Python3* and the `detecto` library are used to train and the deploy the Faster R-CNN, while the tool *LabelImg* to manually annotate images.

3.2 License Plate Recognition Module (LPRM)

The license plate recognition module (LPRM) acts downstream of the risk estimation module (REM), recognising and storing, fully automatically, the number plates of vehicles involved in a situation of potential anomaly or risk to traffic.

This module operates through a pipeline consisting of three fundamental steps: i) the plate detection; ii) the character segmentation (which includes rotation, threshold and blobs rejection); and iii) the character recognition. Such a

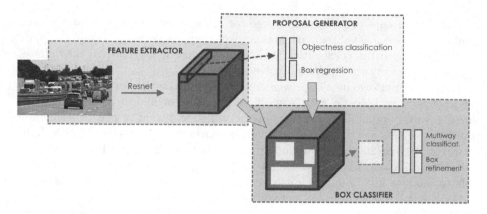

Fig. 2. Architecture of the Faster R-CNN used in the Risk Estimation Module.

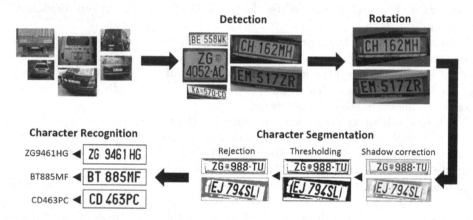

Fig. 3. The pipeline of the License Plate Recognition Module (LPRM).

pipeline is summarised in Fig. 3. The three steps are then described in the following of this section.

Plate Detection Phase. The plate detection phase is responsible for the identification and extraction of the plates within the video frames received by the REM. This step is performed by means of *ImageAI*[1], a Python library that acts as a wrapper for several deep learning algorithms and easily allows to train a custom model to detect objects from images or videos. In the proposed approach, the YOLOv3 algorithm has been used, trained on a dataset of license plates (described below) as to obtain a custom model.

Then, the detection algorithm takes as input an image and the generated model. Since YOLO can detect multiple objects of the same class within an

---

[1] https://github.com/OlafenwaMoses/ImageAI/releases/tag/essential-v4.

image, the plate with the highest probability is selected (with respect to the probability assigned by YOLO).

Figure 4 shows the output of the detection phase.

Fig. 4. Input (on the left) and output (on the right) of the detection phase.

Character Segmentation. Given the crop of the license plates, this step of LPRM pipeline aims to return an image containing only the characters of the license plate. This stage is performed through handcrafted pre-processing methods and it is divided, in turn, into four sub-phases: a) image rotation; b) image segmentation; c) anomalous blobs rejection; d) final refinements for the character recognition optimization.

The image rotation algorithm exploited by the LPRM represents an improved version of the one presented in [35]. First, an automatic thresholding is applied to generate a binary image, then edges are detected with the Canny algorithm. Finally, the module leverages on the Hough transform for the detection of the lines and their inclination, so as to straighten the image. Some improvements are applied to the Hough parameter space, i.e. a smoothing filter is used to remove outliers and the maxima pattern generated by the Hough transformation, making the angle detection more accurate. Furthermore, the rotation generates some empty space to entirely contain the final image and avoid cropping. An example of this rotation algorithm is depicted in Fig. 5.

After the image is rotated, the segmentation is performed. To do so, firstly the image is converted to grayscale and the height fixed to 100 pixels; then shadows are removed, by subtracting from the original image a copy of itself in which dilatation and smoothing filters are applied.

Subsequently, an adaptive thresholding based on the *Otsu's method* [32] is applied to extract the darker details. However, since several dark regions may have been lost due the removal of shadows, some isolated regions – whose shape may be similar to a character – can circumvent the blobs rejection. To overcome this issue, a map of darker areas is derived through a linear threshold and, secondly, dilation and closing filters, to make sure that the detected dark regions do not overlap with the characters. The final segmentation is thus obtained by subtracting the darker areas map from the processed license plate image.

From the previous step, the LPRM obtains a binary image containing the license plate characters fully separated from all the other regions. Then it

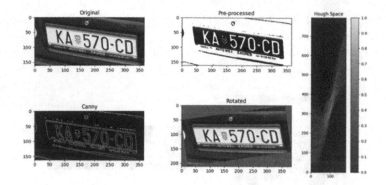

Fig. 5. An example of automatic image rotation.

needs to remove such anomalous regions using ad-hoc *blobs rejection* algorithms. Hence, three algorithms have been developed, based on *a priori* knowledge about the characteristic of the image; specifically, it is worth to note that the characters have a similar size and intensity to each other, and have a specific size within the image. Figure 6 graphically summarises the above steps.

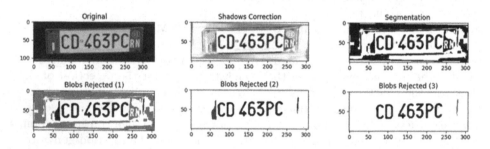

Fig. 6. The character segmentation sub-phases of the LPRM pipeline.

The last sub-phase of the character segmentation performs some refinements, aimed at optimizing the following recognition stage: all the white spaces are cut (substituted with a fixed padding), the image is resized, the height is fixed and the angle for a fine rotation adjustment is computed. Finally, a smoothing filter is applied.

Character Recognition. This is the final stage of the LPRM pipeline. It returns the extracted licence plate number by exploiting a majority voting approach. This stage leverages on an additional deep learning layer, based on the *Tesseract OCR*[2]. It is based on a Long-Short Term Memory (LSTM), lightweight and easy

---

[2] https://github.com/tesseract-ocr

to use, which (compared to CNNs) required shorter training times while still ensuring high levels of accuracy.

```
F  D  5  8  0  L  D  G  A  6  N         DU 879 BZ 304 E 726  IL 111 AT S  747 EA

7  1  3  G  A  E  P  4  5  Z  K         ST 377 I  SK 314 CK ZG 4497 P  ZG 348 CR

2  B  T  S  9  P  Y  C  3  X  H    4052 AC  DU 986 BM  RI 238  L HA D 677  ZG 9267 AD
```

Fig. 7. Example of the images created to train Tesseract.

Over 60 images (Fig. 7) were tailored and used for training the LSTM, obtaining a custom model. In addition, the most accurate models in the Tesseract OCR repository were considered. However, despite of the optimizations, the reading of the images varied between configurations. To tackle this last issue, a majority voting system was built in order to combine the decisions of the individual models and thus to maximise the final accuracy.

4 Results and Discussion

In this section, the preliminary experimental results for the proposed solution are shown and discussed, describing also the current limitations and possible future developments. The modules of the solution were developed and trained on a personal desktop configuration (Intel Core i7-8750H CPU @ 2.20 GHz with 32 GB of RAM and a GeForce GTX 1060 graphic card), while the smart lamppost operational execution has been simulated by providing, as input, a continuous stream of recorded videos to the Nvidia Jetson board.

In particular, a subset consisting of 80 test videos from the synthetic dataset (i.e., that created by means of the Rockstar editor), and 20 real footage videos, all of which have an average duration of between 20 and 120 s, were used to validate the effectiveness of the prototype and the proposed approach. We manually annotated the static anomalies in each video, with some videos having no anomalies. Moreover, we focus only on precision and recall as metrics, as the aim is to observe the effectiveness of identifying events in the anomalies class only (i.e., (the overall accuracy of the system in correctly identifying even *normal* behaviours is not of interest) (Figs. 8 and 9).

For what concerns the risk estimation module (REM), the proposed implementation showed, during the validation process, to have good recall values (>90%) but relative low precision (~35%). However, as the function of the risk estimation module is only to filter the data that the license place recognition module has to parse and the system has to store, it is required to avoid the presence of false negatives, even if at the cost of having a high number of false positives. Indeed, this behavior is desirable: not detecting an anomalous event would have much more impact on the reliability of the system than the parsing of extra data has. On the other hand, with this solution, we achieve high system performance and scalability, as the amount of interesting data parsed is maxed and the amount of useless data is significantly reduced when compared to the alternative (i.e., storing the full video stream).

(a) Traffic collision;   (b) traffic congestion.

Fig. 8. Examples of static anomalies detected in real footage videos.

With regard to the license plate recognition module (LPRM), the prototype has been validated in two modalities: an operational validation (from the video buffers of abnormal events, extrapolated by the REM module from the 80 videos described above), and a static one (using a set of 164 real-world license plate images, mainly downloaded from *PlatesMania*[3]). Globally, we obtained an over-all accuracy of 87.21%.

There are several causes that led to a failure in the recognition. For example, detection errors may be induced by crops that are too wide or narrow, or that do not fully include all the characters of the license plate. Segmentation errors are mainly due to characters that are not isolated correctly, while (very rare) reading errors may occur even if the image contains only well-segmented and isolated valid characters.

Notably, no errors has emerged during the object detection phase, in both datasets. Also the segmentation step showed to be susceptible to few errors, in part due to occlusions (usually not problematic, as the same number plate can be detected, and therefore recorded, in an earlier or later frame of the video stream) and in part to the connection with anomalous regions, even if for a few pixels (in this case, morphological operations with wider filters could solve the problem, but with the possible drawback of separating characters on damaged plates and consequent failure of the rejection sub-phase). Reading errors represent the main problem in the pipeline. Images with perspective distortions are particularly sensitive to errors, but also there exist some (infrequent) issues due to ambiguity between characters and numbers – e.g., for pairs like (S-5), (A-4), (Z-7), (G-C), (O-0) – that the voting system did not completely solve. However, as mentioned above, during the video stream it is still possible to detect the correct licence

---

[3] http://platesmania.com/.

(a) Object on the road;   (b) traffic collision;

(c) prohibited stop;   (d) traffic congestion.

Fig. 9. Examples of static anomalies detected in the synthetic test videos.

plate number in a different frame, so combinations that do not appear in a sufficient number of frames can easily be excluded (Fig. 10).

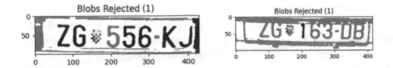

Fig. 10. Example of errors that may occur during the segmentation stage, which led to the rejection of valid characters. On the left, a character is linked to an invalid region; on the right, an occlusion disrupts the character separation.

5 Conclusions

In this paper, we have proposed a novel taxonomy of traffic anomalies, as a reference for artificial intelligence detection approaches, and we have presented an integrated solution for traffic control through automatic license plate recognition, based on Deep Learning and Computer Vision techniques, that leverages on the aforementioned taxonomy. The proposed solution consists of a network of smart lampposts, each equipped with a bullet camera and a System-on-Module (i.e., an NVIDIA Jetson board), capable of autonomously running two intelligent software components: a risk estimation module (REM) and a number plate detection module (LPRM). The solution aims at monitoring Smart City traffic automatically through a system of cameras placed on road lampposts, filtering out normal traffic flows and marking as relevant only those in which - by means of a Faster R-CNN - an above-average risk index is detected, i.e. a static anomaly has occurred. Then, the license plate recognition approach, exploiting YOLO and Tesseract, enables the automatic identification of vehicles in the scene. Such a solution is, firstly, efficient, as it allows only relevant traffic events to be stored and processed, optimising space usage and allowing longer archiving; secondly, it is easily scalable, thanks to its decentralised nature. At this stage, however, there are still some open problems, towards which our future research work is oriented. First, the risk estimation module currently takes into account static anomalies only; the detection of dynamic anomalies is indeed more complex and, to deal with this limitation, a set of heuristics is currently being studied. Second, the use of a single camera makes each observation point susceptible to problems of occlusion, glare and perspective. In this sense, the combined processing of two parallel video streams, from different observation points, would improve the overall detection accuracy.

References

1. Ahad, M.A., Paiva, S., Tripathi, G., Feroz, N.: Enabling technologies and sustainable smart cities. Sustain. Cities Soc. **61**, 102301 (2020)
2. Al-Heety, A.T., et al.: Moving vehicle detection from video sequences for traffic surveillance system. ITEGAM-JETIA **7**(27), 41–48 (2021)
3. Al-Turjman, F., Lemayian, J.P.: Intelligence, security, and vehicular sensor networks in internet of things (IoT)-enabled smart-cities: an overview. Comput. Electr. Eng. **87**, 106776 (2020)
4. Albatish, I.M., Abu-Naser, S.S.: Modeling and controlling smart traffic light system using a rule based system. In: 2019 International Conference on Promising Electronic Technologies (ICPET), pp. 55–60 (2019). https://doi.org/10.1109/ICPET.2019.00018
5. Appathurai, A., Sundarasekar, R., Raja, C., Alex, E.J., Palagan, C.A., Nithya, A.: An efficient optimal neural network-based moving vehicle detection in traffic video surveillance system. Circ. Syst. Signal Process. **39**(2), 734–756 (2020)
6. Atzori, A., Barra, S., Carta, S., Fenu, G., Podda, A.S.: Heimdall: an AI-based infrastructure for traffic monitoring and anomalies detection. In: 2021 IEEE International Conference on Pervasive Computing and Communications Workshops and other Affiliated Events (PerCom Workshops), pp. 154–159. IEEE (2021)
7. Barra, S., Bisogni, C., De Marsico, M., Ricciardi, S.: Visual question answering: which investigated applications? arXiv preprint arXiv:2103.02937 (2021)
8. Barra, S., Carta, S.M., Giuliani, A., Pisu, A., Podda, A.S., et al.: FootApp: an AI-powered system for football match annotation. arXiv preprint arXiv:2103.02938 (2021)
9. Barra, S., De Marsico, M., Cantoni, V., Riccio, D.: Using mutual information for multi-anchor tracking of human beings. In: Cantoni, V., Dimov, D., Tistarelli, M. (eds.) Biometric Authentication, pp. 28–39. Springer International Publishing, Cham (2014). https://doi.org/10.1007/978-3-319-13386-7_3
10. Bock, F., Di Martino, S., Origlia, A.: Smart parking: using a crowd of taxis to sense on-street parking space availability. IEEE Trans. Intell. Transp. Syst. **21**(2), 496–508 (2020). https://doi.org/10.1109/TITS.2019.2899149

11. Bock, F., Di Martino, S.: On-street parking availaibilty data in San Francisco, from stationary sensors and high-mileage probe vehicles. Data Brief **25**, 104039 (2019)

12. Braun, T., Fung, B.C., Iqbal, F., Shah, B.: Security and privacy challenges in smart cities. Sustain. Cities Soc. **39**, 499–507 (2018)

13. Carta, S., Podda, A.S., Recupero, D.R., Saia, R.: A local feature engineering strategy to improve network anomaly detection. Future Internet **12**(10), 177 (2020)

14. Chakraborty, M., Pramanick, A., Dhavale, S.V.: MobiSamadhaan—intelligent vision-based smart city solution. In: Gupta, D., Khanna, A., Bhattacharyya, S., Hassanien, A.E., Anand, S., Jaiswal, A. (eds.) International Conference on Innovative Computing and Communications. AISC, vol. 1165, pp. 329–345. Springer, Singapore (2021). https://doi.org/10.1007/978-981-15-5113-0_24

15. Cho, Y., Jeong, H., Choi, A., Sung, M.: Design of a connected security lighting system for pedestrian safety in smart cities. Sustainability **11**(5) (2019). https://doi.org/10.3390/su11051308, https://www.mdpi.com/2071-1050/11/5/1308

16. Choi, S., Kim, J.T., Choo, J.: Cars can't fly up in the sky: improving urban-scene segmentation via height-driven attention networks. In: Proceedings of the IEEE/CVF Conference on Computer Vision and Pattern Recognition (CVPR), June 2020

17. Combs, T.S., Sandt, L.S., Clamann, M.P., McDonald, N.C.: Automated vehicles and pedestrian safety: Exploring the promise and limits of pedestrian detection. Am. J. Prev. Med. **56**(1), 1–7 (2019)

18. Deng, J., Li, L., Zhang, B., Wang, S., Zha, Z., Huang, Q.: Syntax-guided hierarchical attention network for video captioning. IEEE Trans. Circ. Syst. Video Technol. (2021, in press)

19. Dhingra, S., Madda, R.B., Patan, R., Jiao, P., Barri, K., Alavi, A.H.: Internet of things-based fog and cloud computing technology for smart traffic monitoring. Internet Things, p. 100175 (2020). https://doi.org/10.1016/j.iot.2020.100175, https://www.sciencedirect.com/science/article/pii/S2542660519302100

20. He, K., Zhang, X., Ren, S., Sun, J.: Deep residual learning for image recognition. In: Proceedings of the IEEE Conference on Computer Vision and Pattern Recognition (CVPR), June 2016

21. Sri Jamiya, S., Esther Rani, P.: An efficient method for moving vehicle detection in real-time video surveillance. In: Suresh, P., Saravanakumar, U., Hussein Al Salameh, M.S. (eds.) Advances in Smart System Technologies. AISC, vol. 1163, pp. 577–585. Springer, Singapore (2021). https://doi.org/10.1007/978-981-15-5029-4_47

22. Khan, L.U., Yaqoob, I., Tran, N.H., Kazmi, S.M.A., Dang, T.N., Hong, C.S.: Edge-computing-enabled smart cities: a comprehensive survey. IEEE Internet Things J. **7**(10), 10200–10232 (2020). https://doi.org/10.1109/JIOT.2020.2987070

23. Khan, M.A., et al.: Human action recognition using fusion of multiview and deep features: an application to video surveillance. Multimedia Tools Appl. 1–27 (2020)

24. Malik, K.: Fast vehicle detection with probabilistic feature grouping and its application to vehicle tracking. In: Proceedings Ninth IEEE International Conference on Computer Vision, vol. 1, pp. 524–531 (2003). https://doi.org/10.1109/ICCV.2003.1238392

25. Li, W., Guo, H., Nejad, M., Shen, C.C.: Privacy-preserving traffic management: a blockchain and zero-knowledge proof inspired approach. IEEE Access **8**, 181733–181743 (2020)

26. Li, Y., et al.: Multi-granularity tracking with modularlized components for unsupervised vehicles anomaly detection. In: Proceedings of the IEEE/CVF Conference on Computer Vision and Pattern Recognition (CVPR) Workshops, June 2020

27. Longo, R., Podda, A.S., Saia, R.: Analysis of a consensus protocol for extending consistent subchains on the bitcoin blockchain. Computation **8**(3), 67 (2020)
28. Makhmutova, A., Anikin, I., Minnikhanov, R., Bolshakov, T., Dagaeva, M.: Detection of traffic anomalies for a safety system of smart city. In: Information Technology and Nanotechnology (ITNT-2020), pp. 638–645 (2020)
29. Michalopoulos, P.G.: Vehicle detection video through image processing: the autoscope system. IEEE Trans. Veh. Technol. **40**(1), 21–29 (1991). https://doi.org/10.1109/25.69968
30. Neves, J.C., Moreno, J.C., Barra, S., Proença, H.: Acquiring high-resolution face images in outdoor environments: a master-slave calibration algorithm. In: 2015 IEEE 7th International Conference on Biometrics Theory, Applications and Systems (BTAS), pp. 1–8 (2015). https://doi.org/10.1109/BTAS.2015.7358744
31. Nguyen, K.T., Dinh, D.T., Do, M.N., Tran, M.T.: Anomaly detection in traffic surveillance videos with GAN-based future frame prediction. In: Proceedings of the 2020 International Conference on Multimedia Retrieval, pp. 457–463 (2020)
32. Otsu, N.: A threshold selection method from Gray-level histograms. IEEE Trans. Syst. Man Cybern. **9**(1), 62–66 (1979)
33. Pagliara, F., Mauriello, F., Di Martino, S.: An analysis of the link between high speed transport and tourists' behaviour. Tourism Int. Interdisc. J. **67**(2), 116–125 (2019)
34. Pang, G., Shen, C., Cao, L., Hengel, A.V.D.: Deep learning for anomaly detection: a review. ACM Comput. Surveys (CSUR) **54**(2), 1–38 (2021)
35. Piccinelli, L.: Raddrizzare il contenuto di un'immagine, November 2016. https://luca-picci.medium.com/raddrizzare-il-contenuto-di-unimmagine-37f9bbc16207
36. Redmon, J., Divvala, S., Girshick, R., Farhadi, A.: You only look once: Unified, real-time object detection. In: Proceedings of the IEEE Conference on Computer Vision and Pattern Recognition (CVPR), June 2016
37. Sreenu, G., Durai, M.S.: Intelligent video surveillance: a review through deep learning techniques for crowd analysis. J. Big Data **6**(1), 1–27 (2019)
38. Suzuki, L.R.: Smart cities IoT: enablers and technology road map. In: Rassia, S.T., Pardalos, P.M. (eds.) Smart City Networks. SOIA, vol. 125, pp. 167–190. Springer, Cham (2017). https://doi.org/10.1007/978-3-319-61313-0_10
39. Yang, Y.T., Chou, L.D., Tseng, C.W., Tseng, F.H., Liu, C.C.: Blockchain-based traffic event validation and trust verification for VANETs. IEEE Access **7**, 30868–30877 (2019)
40. Fu, Z., Hu, W., Tan, T.: Similarity based vehicle trajectory clustering and anomaly detection. In: IEEE International Conference on Image Processing 2005, vol. 2, pp. II-602 (2005). https://doi.org/10.1109/ICIP.2005.1530127

Secure Machine Intelligence and Distributed Ledger

Dmitry Arseniev[1]([envelope]) [iD], Dmitry Baskakov[2] [iD], and Vyacheslav Shkodyrev[2] [iD]

[1] Saint-Petersburg State University, Saint Petersburg, Russia
[2] Peter the Great St. Petersburg Polytechnic University, St. Petersburg, Russia

Abstract. Modern machine and deep learning systems are becoming part of high-performance cloud services and technologies. It is extremely important to understand that in systems such as recommendation systems, data is stored on local machines, and the trained system (matrix) is located in the cloud vendor, for example, in AWS or Google Cloud. Data on local machines can be updated or deleted. Local machines are often networked, which requires the use of synchronization methods and specialized protocols for the exchange of such information. And the central server is used as a single computer center for machine learning tasks. At the same time, it is necessary to control both the integrity of local data and their relevance with respect to other local machines. An important aspect is that the data center in the cloud should not know about our data, that is, we must be able to transmit them in encrypted form. At the same time, the deep learning model should be able to work with such encrypted data and send us the answers in encrypted form too. All this should be calculated in polynomial time, that is, quickly enough. For encryption purposes, it is proposed to use homomorphic algorithms. This report attempts to combine two promising modern paradigms for solving similar problems: machine intelligence and distributed ledger. For the purposes of distributed deep learning in relation to recommender systems, this symbiosis shows very serious practical prospects.

Keywords: Deep learning · Differential privacy · Homomorphic encryption · Machine learning · Secure computation

1 Introduction

Machine learning techniques are widely used in practice to produce predictive models for use in medicine, banking, recommendation services, threat analysis, and authentication technologies. The popularity and relevance of cloud machine learning has grown significantly. Moreover, often in such projects, in our opinion, due attention is paid specifically to issues of confidentiality and data security. In this paper, we consider a problem in which some neural network is located in the provider's cloud. This network is trained on some of its data and provides a service, for example, classification on images or customer data. The purpose of this work is to show and identify possible problems of

© Springer Nature Switzerland AG 2021
O. Gervasi et al. (Eds.): ICCSA 2021, LNCS 12951, pp. 227–239, 2021.
https://doi.org/10.1007/978-3-030-86970-0_17

such interaction, as well as to demonstrate the possibility of building secure protocols that would provide the ability to obtain classification results by a client without revealing their data to a neural network.

In the past years, deep neural networks (DNNs) have achieved remarkable progress in various fields, such as computer vision, natural language processing, and medical images analysis [1]. Consider a distributed recommendation system. Several data sources, for example, medical, send data from each clinic to the cloud for automatic diagnosis. In the cloud is a decision-making system based on a deep learning system [2]. In the general case, the look of this service using the Amazon SageMaker implementation example is as follows (Fig. 1):

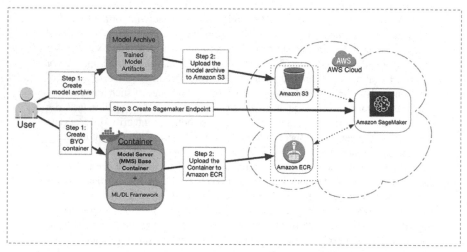

Fig. 1. Deep learning framework to Amazon SageMaker (https://aws.amazon.com/sagemaker/).

If this architecture is used, a problem arises when transferring data in *step 2* when loading the model or, much more often, just data in S3[1] or Amazon ECR[2]. Let's assume that it happens very often that we do not want to at least somehow provide the cloud service with access to our source data due to, for example, legislative restrictions. Suppose this is data both of a personal nature (personal) and commercial, which we cannot send in the usual way without encryption. In this case, we have the right to talk about *Private Distributed Recommender Systems* (businesses with proprietary consumer data would like to build recommender systems which can leverage data across all the businesses without compromising the privacy of any party's data) [3]. It should be understood that at this stage we continue to face all the features of machine learning that are traditionally inherent in this industry. Simple problem where standard deep earning either [4]:

---

[1] https://aws.amazon.com/s3/?nc=sn&loc=0.
[2] https://aws.amazon.com/ecr/

- Does not work well

  - Requires prior knowledge for better architectural/algorithmic choices
  - Requires other than gradient update rule
  - Requires to decompose the problem and add more supervision
  - Requires more data

- Does not work at all

  - No «local-search» algorithm can work
  - Even for «nice» distribution and well-specified models
  - Even with over-parameterization (a.k.a. improper learning)

In order to provide a secure exchange of data with cloud services, in addition to encryption, one or another protocol should be used that will make such data transactions more secure (Secure Deep Learning Inference, SDLI) [5]. It seems very promising to consider the possibility of using homomorphic encryption together with the use of blockchain technology to control both data integrity and to track the chain of possible data changes taking into account the distributed structure, which we will discuss later [6]. Deep learning as a service (DLaaS) has emerged as a promising to further enable the widespread use of DNNs in industry/daily-life. Google[3], Amazon[4], and IBM[5] have all launched DLaaS platforms in their cloud services. Using DLaaS, a client sends its private data to the cloud server. Then, the server is responsible for performing the DNN inference and sends the prediction results back to the client. Obviously, if the private client data, are not protected, using DLaaS will cause potential privacy issues. A curious server may collect sensitive information contained in the private data (i.e. client's privacy) [5].

To address this privacy issue, researchers have employed the homomorphic encryption to perform various DNN operators on encrypted client data [7]. As a result, the cloud server only serves as a computation platform but cannot access the raw data from clients. However, there exist two major obstacles in applying these approaches. First, some common non-linear activation functions, such as ReLU and Sigmoid, are not cryptographically computable [8]. Second, the inference processing efficiency is seriously degraded by thousands of times. To tackle these problems, a recent work proposes using an interactive paradigm. A DNN inference is partitioned into linear and non-linear computation parts. Then, only the linear computations are performed on the cloud server with encrypted data. The nonlinear computations are performed by the client with raw data. However, in such an interactive paradigm, the intermediate features extracted by the linear computations are directly exposed (sent back) to the client. Thus, a curious client can leverage these features to reconstruct the weights of the DNN model held by the cloud. This issue is called the leakage of server's privacy [5].

In fact, a practical solution for secure DNN inference should protect both client's privacy and server's privacy. In addition, it should support DNNs with all types of

---

[3] https://cloud.google.com/products/ai.

[4] https://aws.amazon.com/machine-learning/

[5] https://www.ibm.com/analytics/machine-learning.

non-linear activation functions. Unfortunately, there still lacks an effective approach in literature. Our key strategy is to combine deep learning, homomorphic encryption and distributed ledger.

The traditional approach of using distributed ledger involves the use of this technology in artificial intelligence systems, for example, in robotic systems. For example, cyberphysical systems (Robotics) exchange some information among themselves, the integrity of which should be controlled. Such systems learn, store this data, pass it to other nodes, and so on. There are even some implementations of such an approach and such solutions on the market [9]. There were other concepts for using distributed ledger in artificial intelligence. For example, there were prerequisites and ideas for building decentralized platforms that would allow the user to create, organize joint participation and monetize artificial intelligence systems using blockchain technologies [9]. And of course, the use of distributed ledger in conjunction with artificial intelligence systems was not at all limited to such solutions, of which there were, in fact, quite a lot. One of the curious examples is the use of machine learning for joint decision making or forecasting of certain processes [10]. The use of a distributed machine learning system along with blockchain technologies has been successful enough to predict the market prices of certain assets [10].

2 Materials and Methods

We introduce a few basic primitives that we will need for our solution, namely:

- Deep Neural Network (DNN), which is located at the cloud provider.
- Homomorphic encryption (HE). Additive homomorphic encryption (AHE).
- Secret Sharing (SS) and Garbled Circuit (GC).
- Oblivious ROM (OROM).
- Data Aggregation (DA).
- Differential Privacy (DP).

2.1 Deep Neural Network

DNN, for example, Convolutional neural network (CNN) in cloud provider. Suppose that a cloud provider or we, as the owners of this service, would not want to discover both the weights of this neural network, its hyperparameters, and the models that we used in the training process (Fig. 2):

In fact, in the provider's cloud, we store a matrix with the weights of a neural network of the form:

$$
\begin{bmatrix}
\omega_{1,1} & \cdots & \omega_{1n} \\
\vdots & \ddots & \vdots \\
\omega_{m,1} & \cdots & \omega_{m,n}
\end{bmatrix}
\tag{1}
$$

That is, we enter a certain vector with signs at the input, it is multiplied by matrices and at the output we get again the response vector and belonging to some class (Fig. 3):

It is important that the neural network does not know about the input data and their structure, and, preferably, does not remember the output data or stores it in its memory.

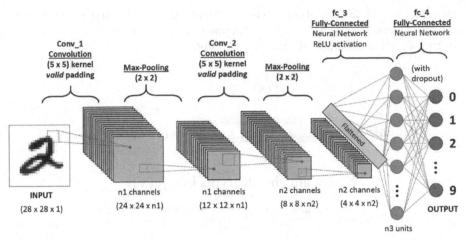

Fig. 2. Convolutional neural network (https://towardsdatascience.com/a-comprehensive-guide-to-convolutional-neural-networks-the-eli5-way-3bd2b1164a53).

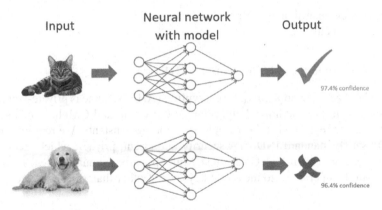

Fig. 3. Deep learning inference

2.2 Homomorphic Encryption (HE)

Homomorphic encryption is a form of encryption that allows computation on ciphertexts, generating an encrypted result which, when decrypted, matches the result of the operations as if they had been performed on the plaintext. Homomorphic encryption can be used for privacy-preserving outsourced storage and computation. This allows data to be encrypted and out-sourced to commercial cloud environments for processing, all while encrypted [11]. Fully Homomorphic Encryption (FHE), is an encryption method that allows anyone to compute an arbitrary function f on an encryption of x, without decrypting it and without knowledge of the private key [12]. Using just the encryption of x, one can obtain an encryption of $f(x)$. The major bottleneck for these techniques, notwithstanding these recent developments, is their computational complexity. But recent efforts, both theory and in practice have given us large results in the performance of homomorphic scheme [13, 14] (Fig. 4)

Enc (x)

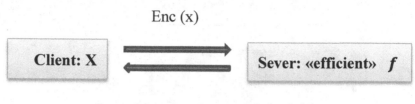

Something that decrypts into $f(x)$

Fig. 4. Homomorphic encryption

It is very important to use an effective homomorphic encryption model. What is efficient?

- Small low-degree arithmetic circuit.
- Small Boolean circuit.

2.3 Additive Homomorphic Encryption (AHE)

A (private-key) additive homomorphic encryption (AHE) scheme is private-key encryption scheme with three additional algorithms Add; CAdd and CMult, which supports adding two ciphertexts, and addition/multiplication by constants. We require our AHE scheme to satisfy standard IND-CPA security and circuit privacy, which means that a ciphertext generated from Add, CAdd and CMult operations should not leak more information about the operations to the secret key owner, other than the decrypted message [15].

2.4 Secret Sharing and Garbled Circuit

Gabled circuit (GC) is a cryptographic protocol that enables two-party secure computation in which two mistrusting parties can jointly evaluate a function over their private inputs without the presence of a trusted third party. In the garbled circuit protocol, the function has to be described as a Boolean circuit. The invention of garbled circuit was credited to Andrew Yao, as Yao introduced the idea in the oral presentation of his paper [16]. Improvements to GC have been proposed in literature, for example, free-XOR and half-gates [17]. Using Advanced Encryption Standard (AES) as the block cipher, we leverage Intel AES instructions for faster garbling procedure [15]. Yao's garbled circuits [16] and the secret-sharing based Goldreich-Micali-Wigderson (GMW) protocol [18] are two leading methods for the task of two-party secure computation (2PC). Now, after three decades theoretical and applied work about improving and optimizing these protocols, we have modern and efficient implementations [19].

One of the main problems of all these protocols and techniques until recently was their communication complexity[6]. Indeed, three recent works followed the garbled circuits paradigm and designed systems for secure neural network inference: the Secure ML system [20], the Mini ONN system [21], the Deep Secure system [14, 22]. Actually, it is not so obvious to use only Secret sharing or Garbled circuit in machine learning security problems. Our vision is that compromises should be sought, in which case it is better to use both technologies depending on the tasks.

We can use Hybrid protocols:

- mix homomorphic encryption and garbled circuits via secret sharing
- Homomorphic Encryption for linear operations and Garbled Circuits for non-linear operations
- Homomorphic Encryption for fully-connected layers and Garbled Circuits for ReLu-activation [15].

3 Oblivious RAM

Consider the simplest model of our calculations and secure compute $a[i]$ (Fig. 5):

Fig. 5. Securely compute a.

The key problem is that the network often stores the data of our access to it, information about transactions, other data. Ideally, I would like the network to somehow know how to forget all this and not even store encrypted data accessing it. Oblivious RAM (ORAM) algorithms, first proposed by Goldreich and Ostrovsky [23], allow a client to conceal its access pattern to the remote storage by continuously shuffling and re-encrypting data as they are accessed. An adversary can observe the physical storage locations accessed, but the ORAM algorithm ensures that the adversary has negligible probability of learning anything about the true (logical) access pattern. Since its proposal, the research community has strived to find an ORAM scheme that is not only theoretically interesting, but also practical [24, 25].

In this case ORAM, distributed machine learning is characterized by the following types of possible threats and attacks (Fig. 6):

At the very key factor of all these attacks is the ability Data Aggregation.

4 Data Aggregation

Here, we introduce the most prominent data privacy preserving mechanisms. Not all these methods are applied to deep learning, but we briefly discuss them for the sake of comprehensiveness. These methods can be broadly divided into two groups of context-free

---

[6] https://en.wikipedia.org/wiki/Communication_complexity.

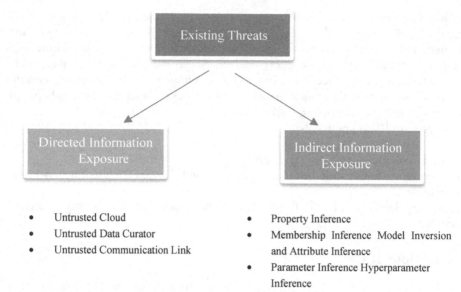

Fig. 6. Existing threats deep learning.

privacy and context-aware. Context-free privacy solutions, such as differential privacy, are unaware of the specific context or the purpose that the data will be used for. Whereas context-aware privacy solutions, such as information-theoretic privacy, are aware of the context where the data is going to be used, and can achieve an improved privacy-utility tradeoff.

4.1 Naïve Data Anonymization

What we mean by naive anonymization in this survey is the removal of identifiers from data, such as the names, addresses, and full postcodes of the participants, to protect privacy. This method was used for protecting patients while processing medical data and has been shown to fail on many occasions [26].

4.2 K-Anonymity

A dataset has k-anonymity property if each participant's information cannot be distinguished from at least $k - 1$ other participants whose information is in the dataset. K-anonymity means that for any given combination of attributes that are available to the adversary (these attributes are called quasi-identifiers), there are at least k rows with the exact same set of attributes. K-anonymity has the objective of impeding re-identification [26].

4.3 Semantic Security and Encryption

Semantic security (computationally secure) is a standard privacy requirement of encryption schemes which states that the advantage (a measure of how successfully an adversary can attack a cryptographic algorithm) of an adversary with background information should be cryptographically small.

5 Distributed Ledger

Actually, there are not so many works and studies where qualitative integration of two such well-known paradigms as machinery intelligence and blockchain was given or offered one way or another. Existing works clearly, of course, solve certain problems facing the industry and researchers, but this is clearly not enough if we are talking about the integration of such important modern technologies [9, 27, 28]. In our approach to distributed machine learning as a service, the use of Blockchain technology is due to a number of important circumstances, namely:

- Oblivious RAM implies a limit on the number of transactions N, where N must necessarily be bounded above by some integer variable:

$$N \leq K \tag{2}$$

- All participants in the distributed computer network of machine learning, in the case of access to the cloud provider (AWS, Google Cloud), save the most important parameters of the access to the database in the blockchain, for example, the number of requests or transactions n_i, where:

$$N \approx \sum_{i=1}^{m} n_i \tag{3}$$

- When the counter of transactions or hits N becomes greater than or equal to K $N \geq K$, then controlled deletion of data from the neural network should take place, or, as an option, reconfiguration of its parameters, hyperparameters which in theory should entail the removal of data from the network [29].
- A temporary restriction $t \leq T$ on the use of this neural network if, for some reason, users have stopped changing the counter settings or even stopped accessing the services of a cloud provider.

Thus, the general algorithm of work will consist in the fact that each time the service provider is accessed, the client stores data on the number of transactions in the distributed ledger, after exceeding which the saved user data on the neural network are reset, as well as the possible setting of new parameters in accordance with the concept Oblivious ROM [30].

6 Differential Privacy

Definition 5.1. $\epsilon - $ ***Differential Privacy***$(\epsilon - DP)$**.** For $\epsilon \geq 0$, an algorithm A. satisfies $\epsilon - DP$ [31] if and only if for any pair of datasets D and D' that differ in only one element:

$$P[A(D) = t] \leq e^{\epsilon} P\left[A(D') = t\right] \forall t \tag{4}$$

Where, $P[A(D) = t]$ denotes the probability that the algorithm A outputs t. In this setup, the quantity bel is named the *privacy loss:*

$$\ln \frac{P[A(D) = t]}{P[A(D') = t]} \tag{5}$$

DP tries to approximate the effect of an individual opting out of contributing to the dataset, by ensuring that any effect due to the inclusion of one's data is small. One of the widely used DP mechanisms when dealing with numerical data is the Laplace mechanism [32].

Definition 5.2. ***Laplace Mechanism.*** Given a target function f and fixed $\epsilon \geq 0$, the randomizing algorithm $A_f(D) + x$ where x is perturbation random variable drawn from Laplace distribution:

$$Lap\left(\mu, \frac{\Delta_f}{\epsilon}\right) \tag{6}$$

Is called the Laplace Mechanism and is $\epsilon - DP$. Here, Δ_f is called global sensitivity of function f, and is defined as:

$$\Delta_f = \sup\left|f(D) - f(D')\right| \tag{7}$$

For all the dataset pair (D, D') that differ in only one element. Finding this sensitivity is not always trivial, specifically if the function f is a deep neural network, or even a number of layers of it [33]. Differential privacy satisfies a composition property that states when two mechanisms with privacy budgets ε_1 and ε_2 are applied to the same datasets, together they use a privacy budget $\varepsilon_1 + \varepsilon_2$. As such, composing multiple differentially private mechanisms consumes a linearly increasing privacy budget. It has been shown that tighter privacy bound for composition can be reached, so that the privacy budget decreases sub-linearly (Fig. 7):

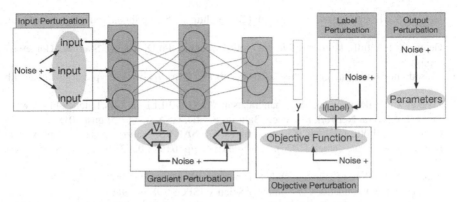

Fig. 7. How work differential privacy in deep learning [32]

7 Conclusion and Discussion

In this paper, we examined modern approaches and concepts that relate to distributed machine learning. We in no way claimed to be an exhaustive exposition of this area, which is impossible in principle. Nevertheless, we showed that it is extremely important to use a distributed ledger in order to control both the integrity of the data in the cloud, and forcibly delete data or reconfigure the parameters of the neural network. In the case of joint and remote operation of several nodes at once, in the case of joint and remote operation of several nodes at once, these requirements become extremely important.

References

1. TRADI: Tracking deep neural network weight distributions (2020). https://arxiv.org/pdf/1912.11316v3.pdf. Accessed 03 May 2020
2. Buniatyan, D.: Hyper: distributed cloud processing for large-scale deep learning tasks (2019). https://arxiv.org/pdf/1910.07172v1.pdf. Accessed 25 May 2020
3. Tsikhanovich, M., Magdon-Ismail, M., Ishaq, M., PD-ML-Lite: private distributed machine learning from lightweight cryptography (2019). https://arxiv.org/pdf/1901.07986v2.pdf. Accessed 27 Apr 2020
4. Failures of Deep Learning (2017). https://simons.berkeley.edu/talks/shai-shalev-shwartz-2017-3-28. Accessed 10 Apr 2020
5. BAYHENN: Combining Bayesian deep learning and homomorphic encryption for secure DNN inference (2019). https://arxiv.org/pdf/1906.00639v2.pdf. Accessed 01 May 2020
6. Boddeti, V.N.: Secure face matching using fully homomorphic encryption (2018). https://arxiv.org/pdf/1805.00577v2.pdf. Accessed 2020 Apr 15
7. Chialva, D., Dooms, A.: Conditionals in homomorphic encryption, and machine learning applications (2019). https://arxiv.org/pdf/1810.12380v2.pdf. Accessed 11 Mar 2020
8. Thaine, P., Gorbunov, S., Penn, G.: Efficient evaluation of activation functions over encrypted data (2020). http://www.cs.toronto.edu/~pthaine/EEAFED.pdf. Accessed 17 Mar 2020
9. Lopes, V., Alexandre, L.A.: An overview of blockchain integration with robotics and artificial intelligence (2018). https://arxiv.org/pdf/1810.00329.pdf. Accessed 24 Feb 2020
10. Craib, R., Bradway, G., Dunn, X., Krug, J.: White paper: Numeraire: a cryptographic token for coordinating machine intelligence and preventing overfitting (2017)

11. Halevi, S.: Homomorphic encryption (2017). https://shaih.github.io/pubs/he-chapter.pdf. Accessed 17 Apr 2020
12. Gentry, C.: A fully homomorphic encryption scheme. Ph.D. Thesis, Stanford University (2009)
13. Palisade homomorphic encryption software library. https://palisade-crypto.org/software-library/
14. Juvekar, C., Vaikuntanathan, V., Chandrakasan, A.: GAZELLE: A Low Latency Framework for Secure Neural Network Inference, Baltimore, MD, USA, 15–17 August 2018
15. Chen, H., Chillotti, I., Dong, Y., Poburinnaya, O.: SANNS: scaling up secure approximate K-nearest neighbors search (2020). https://arxiv.org/pdf/1904.02033.pdf. Accessed 04 Apr 2020
16. Yao, A.C.-C.: How to generate and exchange secrets (extended abstract). In: 27th Annual Symposium on Foundations of Computer Science (SFCS 1986) (1986)
17. Zahur, S., Rosulek, M., Evans, D.: Two Halves make a whole. In: Oswald, E., Fischlin, M. (eds.) EUROCRYPT 2015. LNCS, vol. 9057, pp. 220–250. Springer, Heidelberg (2015). https://doi.org/10.1007/978-3-662-46803-6_8
18. Goldreich, O., Micali, S., Wigderson, A.: How to plait any mental game (1987)
19. Demmler, D., Schneider, T., Zohner, M.: ABY – a framework for efficient mixed-protocol secure two-party computation. In: 22nd Annual Network and Distributed System Security Symposium, NDSS 2015, San Diego, California, USA (2015)
20. Mohassel, P., Zhang, Y.: SecureML: a system for scalable privacy-preserving machine learning. In: Conference: 2017 IEEE Symposium on Security and Privacy (SP) (2017)
21. Liu, J., Juuti, M., Lu, Y., Asokan, N.: Oblivious neural network predictions via minionn transformations. In: 2017 ACM SIGSAC Conference on Computer and Communications Security, CCS 2017, Dallas (2017)
22. Ouhani, B.D., Riazi, M.S., Koushanfar, F.: Deepsecure: scalable provably-secure deep learning (2017). https://arxiv.org/ftp/arxiv/papers/1705/1705.08963.pdf. Accessed 14 Mar 2020
23. Goldreich, O., Ostrovsky, R.: Software protection and simulation on oblivious rams. In: ACM (1996)
24. Stefanovy, E., van Dijkz, M., Shi, E.: Path ORAM: an extremely simple oblivious RAM protocol (2013). https://eprint.iacr.org/2013/280.pdf
25. Stefanovy, E., van Dijkz, M., Shi, E.: Path ORAM: an extremely simple oblivious RAM protocol (2013). https://eprint.iacr.org/2013/280.pdf
26. Sweeney, L.: "k-anonymity: A model for protecting. Int. J. Uncertain. Fuzz. Knowl. Based Syst. **10**, 557–570 (2002)
27. Chen, F., Wany, H., Caiz, H., Cheng, G.: Machine learning in/for blockchain: future and future and challenges (2020). https://arxiv.org/pdf/1909.06189v2.pdf. Accessed 07 May 2020
28. Zheng, Z., Dai, H.-N., Wu, J.: Blockchain intelligence: when blockchain meets artificial intelligence (2020). https://arxiv.org/pdf/1912.06485v3.pdf. Accessed 03 May 2020
29. Serizawaa, T., Fujita, H.: Optimization of convolutional neural network using the linearly decreasing weight particle swarm optimization (2020). https://arxiv.org/ftp/arxiv/papers/2001/2001.05670.pdf. Accessed 11 Mar 2020
30. Liu, J., Tai, X.-C., Luo, S.: Convex shape prior for deep neural convolution network based eye fundus images segmentation (2020). https://arxiv.org/pdf/2005.07476v1.pdf. Accessed 13 May 2020
31. Dwork, C., McSherry, F., Nissim, K., Smith, A.: Calibrating noise to sensitivity in private data analysis. In: Third Conference, Berlin (2006)
32. Mireshghallah, F., Taram, M., Vepakomma, P., Singh, A., Raskar, R., Esmaeilzadeh, H.: Privacy in deep learning: a survey. https://arxiv.org/pdf/2004.12254v3.pdf. Accessed 11 May 2020

33. Lecuyer, M., Atlidakis, V., Geambasu, R., Hsu, D., Jana, S.: Certified robustness to adversarial examples with differential privacy (2019). https://arxiv.org/abs/1802.03471. Accessed 07 Feb 2020
34. Lopes, V., Alexandre, L.A.: An overview of blockchain integration with robotics and artificial intelligence (2018). https://arxiv.org/pdf/1810.00329v1.pdf. Accessed 29 Apr 2020

On the Data Security of Information Systems: Comparison of Approaches and Challenges

Farah Abdmeziem[1]([✉])([iD]), Saida Boukhedouma[1],
and Mourad Chabane Oussalah[2]

[1] LSI Laboratory, University of Science and Technology Houari Boumediene,
Algiers, Algeria
{fabdmeziem,sboukhedouma}@usthb.dz
[2] LS2N Laboratory, University of Nantes, Nantes, France
Mourad.oussalah@univ-nantes.fr

Abstract. Information Systems (IS) are in constant evolution to cope with new technologies like cloud computing and Big data. Nevertheless, these new technologies bring several security issues, some caused by the numerous access points which make companies prone to an increasing number of attacks. The architecture of an information system can be structured in three main layers: data layer, application layer, and technology layer. Security must be reinforced in each of the previous layers to prevent internal and external attacks. In this paper, we focus on the security of the data layer, thus we survey the research work dealing with security related to this layer. For that, we propose a data meta-model with a high level of abstraction that shows the relationship between data, security issues, and security mechanisms covering the three aspects of security (confidentiality, integrity, availability) and considering different types of data and novel technologies (big data and cloud computing). We use the main concepts of this meta-model to classify security mechanisms according to security goals, to compare the different approaches proposed in the literature, and to identify open challenges in the data security field. Our data meta-model can be easily extended and serves as a reference to compare other approaches in the literature.

Keywords: Information system architecture · Security · Data layer · Security mechanisms · Security issues · Meta-model

1 Introduction

Information System (IS) is an essential asset of companies. Its main objectives consist in collecting, storing, and sharing data between the stakeholders of the companies. Nowadays, information systems are becoming more and more collaborative and dynamic to meet business needs. New technologies like Cloud

© Springer Nature Switzerland AG 2021
O. Gervasi et al. (Eds.): ICCSA 2021, LNCS 12951, pp. 240–255, 2021.
https://doi.org/10.1007/978-3-030-86970-0_18

Computing, Internet of Things (IoT), Big data, and Blockchain foster the collaborative work of companies where data and business processes are exchanged and shared between stakeholders of the same company or other companies.

Moreover, information System Architecture (ISA) attracts the interest of a lot of researchers and several works have been proposed. The most known frameworks for ISA design are: *the ZACHMAN framework* [44], *the TOGAF*[1], and *the DODAF*[2] frameworks. Nevertheless, and despite the researchers' efforts, ISA lacks a reference model. Overall, ISA can be divided into three levels [41]: the **data level** that supports business, the **application level** where the applications needed for data management and business support are defined and the **technological level** that represents the main technologies used for IS implementation. It is worth noting that these three layers are found, among others, in the ISA frameworks mentioned above. Furthermore, security is one of the most important non-functional requirements of information systems and in particular collaborative information systems. Specifically, it is about securing the data and the processes handling this data. Each layer of the ISA can be prone to security attacks, such as the data layer where we can count several known attacks including *SQL injection* [5] and *Man in the Middle* [21]. These attacks are not without consequences on the application layer too as these two layers are strongly linked and this can induce considerable losses for companies. It is becoming then crucial to secure the companies' information systems from the design phase onwards. In this paper we dig into the data layer of ISA by proposing a meta-model that gives our perception of IS' data security. The meta-model highlights the main concepts attached to data, security issues, and security mechanisms. The proposed meta-model is used to classify security mechanisms according to security goals; and also to compare the research work surveyed in this paper, according to a set of identified criteria. Notice that our meta-model can be easily used/extended to consider other approaches not mentioned in this survey. The remainder of this paper is structured as follows: Sect. 2 introduces our global ISA and security meta-model. Section 3 presents a generic data meta-model followed by the most known data security issues and security mechanisms. In the same Section, we present a classification of security mechanisms relative to security goals and finally the data security meta-model. In Sect. 4, we use our data security meta-model to compare research works discussed in Sect. 3 and identify open challenges in data security. Finally, Sect. 5 concludes the paper and gives our future work directions.

2   Information System and Security

Starting from the global Information System Architecture (ISA), we present in Fig. 1 a meta-model that shows the link between information system layers and security. The meta-model depicts the three main layers of an ISA that consist of

---

[1] The Open Group Architecture Framework. https://www.opengroup.org/togaf.

[2] Federal Enterprise Architecture Framework. https://dodcio.defense.gov/Library/DoD-Architecture-Framework/.

(i) the data layer containing data used/produced by business applications, (ii) the application layer where the applications for business support are defined, (iii) the technological layer that represents the main technologies used in application's implementation, data storage/access and the infrastructures that provide an environment for IS deployment (for instance, building services using web services technology or Cloud Computing). Information systems can carry some weaknesses and security vulnerabilities which make them prone to attacks and security issues like unauthorized accesses and data loss. These security issues compromise the security of the three layers of the IS. Security is defined through three main concepts called the CIA (Confidentiality, Integrity, Availability):

- Confidentiality: is described as the assurance that the IS assets are kept secret;
- Integrity: refers to the inability to alter or destroy IS assets by accident or malfeasance;
- Availability is the ability to access the IS assets whenever it is needed.

Security mechanisms supported by tools or frameworks are proposed to mitigate IS security issues and hence ensure security. However, these solutions have an impact on the IS non-functional properties; this will be detailed in Sect. 3.5.

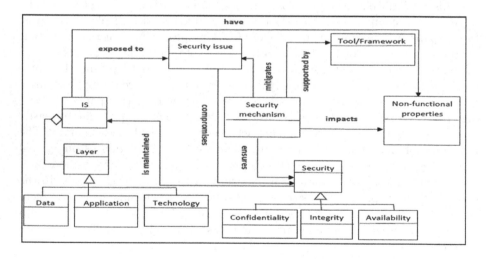

Fig. 1. IS and security

3 Data Security

Data is at the heart of companies and represents their key asset. Indeed, among other benefits, it helps companies to make better decisions and improve their processes. However, data is also the main target of cyber-attacks that cause considerable damages to companies. In this section, we begin by presenting our data meta-model followed by a brief description of the security issues that the data layer could be prone to. Then, we give an overview of the proposed security mechanisms in the literature to ensure data security and their impact on the IS non-functional properties. Finally, we present our data security meta-model.

3.1 Data Meta-model

Figure 2 depicts the main concepts associated with the data layer. Data can be of different types mainly structured, semi-structured or unstructured. It is stored, processed and, accessed via databases that can be SQL or NoSQL databases. These databases can be located on premise (on the organisation's site) or in a cloud environment. When data with its different types is with high volume and velocity, we talk about Big data. Data with its different types can be in two states: *at rest* when it is stored in the databases or *in motion* when it is travelling across a network.

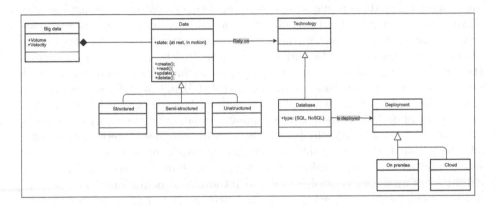

Fig. 2. Data meta-model

3.2 Data Security Issues

Data is exposed to security issues like unauthorised access, tempering of sensitive data, leakage, and data breach. Some of the security issues are carried with the used technologies like cloud computing. Those security issues compromise the confidentiality, integrity, and availability of data and come with an immense cost to companies. Thus it is becoming urgent to ensure data security by proposing a set of data security mechanisms.

3.3 Data Security Mechanisms: Overview of the Literature

Many researchers have worked on the issue of data security; in this Section, we provide an overview of the proposed solutions. We choose to review the well-known models on data security and the works that deals with the cloud computing, blockchain and big data technologies.

Access Control and Encryption Mechanisms: Access control and encryption are well-known and used techniques to ensure data confidentiality and integrity. Encryption [34], on one hand, involves transforming data to make it unreadable to those who don't have the key to decrypt it. Access control, on the other hand, is the decision to permit or deny a subject access to system objects [12]. Objects in our work scope refer to data.

Access Control (AC). Is achieved through models that provide a framework and a set of conditions on which objects, subjects, operations, and rules can be combined to generate and enforce an access control decision. AC can be static or dynamic i.e. whether the access control mechanism uses contextual information or is adjustable and adaptable to environmental changes. The most known access control models are Role-Based Access Control (RBAC) [8] and Attribute-Based Access Control (ABAC). Overall, RBAC employs pre-defined roles that carry a specific set of privileges associated with them and to which subjects are assigned. ABAC controls access to objects by evaluating rules against the attributes of entities (subject and object), operations, and the environment relevant to a request [13]. However, these mechanisms are not well suited for highly dynamic contexts like cloud computing or to deal with big data characteristics (the 3V). Hence, new access control mechanisms have been proposed. In [45] authors suggest Content-Based Access Control (CBAC) model, in which access control decisions are made using data content. More precisely CBAC makes access control decisions based on the content similarity between user credentials and data content dynamically. In [14] the authors propose a methodology with six phases for deploying reliable and efficient access control policies. The proposed solution aims to maintain coherence and compliance between the high-level definition of an access control policy and its low-level definition. To do so, the authors use a formal verification of the specified access control policy before proceeding to its implementation and a reverse engineering method to extract the enforced policy. On another note, [16] propose a generic and dynamic architecture for an access control meta-model that is composed of 6 levels. The proposed solution enables to derive different access control models and to combine them to create hybrid models. Last but not least, *the Bell-LaPadula (BLP) model*, a state machine-based multilevel security [3]. This model is based on subjects (users) and labelled objects from the most sensitive to the least sensitive. A user is restricted from reading up an object with higher label and writing down in object with lower label, it is the No Read Up, No Write Down principle.

Encryption Mechanism. Is the process of encoding information in a way that only authorized parties can access it. This mechanism is used for example in [4], where the authors propose to split the data into several parts and store each part on different cloud providers. Furthermore, the authors point out the fact that big data owners need to consider the cost of encryption in term of money and time. Hence, to deal with this issue they suggest to encrypt only storage path of big data. Finally, to improve availability, the proposed scheme stores more copies of the data parts and their index information on the cloud provider storage. To achieve a secure data transmission between users and cloud storage services, the authors in [18] propose an approach based on the combination of asymmetric (RSA) and symmetric (AES) encryption techniques. The symmetric technique was used to send files from user to the cloud service storage and the asymmetric technique was used to retrieve files from the cloud service storage. The idea behind this approach greatly strengthen the security of file transmission in cloud environment. However, it should be noticed that this technique poses a problem concerning the number of the generated keys for each file stored in the cloud, a number that will keep increasing in context of big data. In the same vein, Sandeep [36] propose an approach that combines three security parameters: data classification, index and encryption, MAC (Message Authentication Code) and authentication techniques to secure data in motion and at rest in a cloud environment. The solution consists of, first to divide data into three sections according to a sensitivity rating value, afterwards, data owner defines keywords to data and their corresponding indexes. Next, data, keywords and indexes are encrypted using the 128-bit SSL encryption algorithm and sent to the cloud. An authentication mechanism is also defined to access data stored in the cloud. The combination of security techniques as proposed in this approach can give successful results in regard to data security, however in case of massive amount of data, index management can become a cumbersome task and will cause performance problems. *Attribute-Based Encryption* (ABE) introduced by Cloud security Alliance (CSA) is another technique that consists on encrypting data as part of an access policy, so that only the user who has permission to access the data can decrypt it [31]. However, according to [43], policy updating poses heavy computation burden on data owner because this latter has to decrypt the data then re-encrypt it again according to the new access policy before sending it to the cloud. Hence the authors propose a dynamic policy updating.

Data Anonymization Mechanism: Data anonymization and data encryption both serve the purpose of securing data that is in motion or stored and can, sometimes, be considered as the same technique. However, as mentioned in [34], encryption can be a useful tool for anonymization, but it is neither necessary nor sufficient for anonymization, and encrypted data is not necessarily anonymized. Data anonymization is the process in which changes are made to data that will be used or in a way that will prevent the identification of key information [15]. The authors of [24] discussed and compared five anonymization techniques: suppression, generalization, swapping, masking, and distortion. The pros and cons of each technique have been discussed and the suppression technique that is "removing an entire part of data by changing the value to one value that doesn't have meaning [24]" was found as the most efficient one [24]. The challenge with anonymization is, first, to find the most suitable technique to apply to sensitive data as this latter depends on many other criteria [24] like the type of data and the appropriate anonymization level. Second, it is important to know when to stop anonymizing data to balance between privacy and data utility [6]. K-anonymity is a key concept of anonymization; it was introduced in [40]. A dataset containing sensitive information is transformed using K-anonymity in a way that guarantees protection against identity disclosure. However, to be efficient, this technique requires some conditions and is vulnerable to attacks like the background knowledge attack and the homogeneity attack. Furthermore, authors in [34] conduct a Proof Of Concept (POC) to show that anonymization is a viable technique to secure data stored in the public cloud. The POC principle is storing anonymized logs in public cloud-based Software as a Service (SaaS) application and then analysing logs to gather security and performance data. Virtual Machines (VMs) send the anonymized data and the de-anonymization takes place in a secure enclave. Successful results have been obtained on the fact that anonymization enhances public cloud security. [11] propose an efficient data anonymization scheme where anonymization and deanonymization are performed by a secure enclave rather than at the client's premises so that computational power is saved. The main drawback of the proposed solution is that it depends on the security of the enclave entity. In [26], Prakash et al. propose a personalized anonymization approach which consists of a top-down greedy algorithm that splits the data into partitions according to a chosen value. Then, a second algorithm is proposed to check if the partitioning violates the privacy requirement. Experimental results have shown that the proposed personalized approach gives better efficiency than other anonymization techniques like k-anonymity.

Mirroring, Checksumming, and Digital Signature: Mirroring, Checksumming, and Digital signature are well-known techniques to ensure data integrity. Mirroring [35] principle is to create two or more copies of data and the integrity checks can be made by comparing the copies. Checksumming is another technique based on checksums (also called hashes) computed using hash functions including MD5 and SHA1. Message Authentication Code (MAC) [20] is an application of the checksumming technique which involves generating cryptographic checksum on data to ensure that data has not been modified while in transit. Digital signature [17] is a mechanism that employs encryption, hashing, and digital signature algorithms to guarantee data integrity. In [25] the authors propose data integrity as a service (DIaaS) to address the integrity issue in cloud storage services. The proposed solution is based on the checksumming technique. Sandeep in [36] proposes a general framework to secure data in the cloud. One of the framework's phases is to generate a Message Authentication Code (MAC) to ensure data integrity. On another note, researchers investigate blockchain technology to ensure data integrity. Indeed, blockchain is a public ledger that works like a log by keeping a record of all transactions in chronological order, secured by an appropriate consensus mechanism and providing an immutable record. Its exceptional characteristics include immutability, irreversibility, decentralization, persistence and anonymity [28]. The integrity and persistence of the blocks in the blockchain are ensured using the hash principle and cryptographic signature as well as a consensual protocol. These characteristics make the blockchain technology an appropriate tool to guarantee data integrity. In [9] and [46] authors tackle the data integrity issue and use the blockchain characteristics to propose a blockchain-based database that ensures data integrity.

Data Backup and Recovery: Regarding the third concept of security: availability, backup plans and recovery mechanisms are proposed. Data backups and recovery mechanisms refer to all the activities and policies implemented to recover resources after an outage event [38] (in the scope of this work we are interested only in data recovery). This outage event can be due to a malign human intervention or a catastrophic event [1]. One of the most known techniques used to backup data are *Full backup, incremental backup and data de-duplication. Full backup* consists of making a copy of all data to a storage device. Such backup has to be made while a database is off-line and unavailable to its users, and would take more time and costs as data become larger [29]. To overcome the disadvantages of *Full backup*, the *Incremental backup* have been proposed and which consist of making copies of only files which have changed or are new since a prior backup [29]. With regard to the last technique: *Data de-duplication* [37,42] is used to find duplicate data and store a sole copy, then uses a pointer to replace the rest of duplicate data [39]. De-duplication concept reduces storage needs, costs and the amount of energy and processing power used to run the storage system [10]. Rahumed et al. in [30] propose *FadeVersion*, a secure cloud backup system based on version control technique and provide the property of assured deletion. The version control technique is based on the concept of de-duplication and the assured deletion property is guaranteed employing two-layer encryption. In [27] the authors present a simple recovery service from disaster based on multiple cloud service providers. Backing up data in different sites in the cloud gives maximum availability and greater reliability at the expense of cost and resource consumption. Sambrani et al. propose in [32] a seed block algorithm that aims to provide a secure backup and recovery process in a cloud environment. Experiments was conducted to show the ability of the proposed solution in maintaining the data without any lost during the recovering process.

3.4   What Security Mechanism for What Security Goal?

Based on the security mechanisms surveyed in Sect. 3.3, we propose Table 1 that classifies the security mechanisms according to security concepts (CIA) they tackle and data type (structured and big data; we consider that semi-structured and unstructured types are covered in big data). This work allowed us to come up with the following conclusions: (i) Confidentiality is the most addressed concept by the proposed security mechanisms despite the fact that data integrity and availability are equally important. (ii) We were not able to find a security mechanism that satisfies the three concepts of security all at once. (iii) Moreover, Big data security mechanisms lack standard models unlike the security mechanism for structured data like RBAC, CBAC... (4i) Finally, the same security mechanisms could be applied for structured data and big data (like

Table 1. Security mechanisms for security goals

Data type/security goal	Confidentiality	Integrity	Availability
Structured data	- RBAC, - CBAC, -BLP, - ABE, - RSA, - AES, -Data classi-fication, - Indexes, - K-anonymity, - Generalization, - Suppression, - Swapping, - Masking, - Distortion	- MAC, - DIaaS - Mirroring, - Digital signature, - Blockchain- based databases	- Full backup, - Incremental backup, de- duplication, - FadeVersion, - Backing up data in different sites in the cloud.
Big data	- CBAC, - Dynamic policy updating for ABE, - A top-down greedy algorithm	- MAC, - Mirroring, - Digital signature, - DIaaS	- Incremental backup, de- duplication, - Backing up data in different sites in the cloud

the mirroring technique to ensure data integrity) however, a security mechanism that suits structured data could impact the IS non-functional properties in case of big data (see Sect. 3.5).

3.5   Impact of Security Mechanisms on Non-functional IS Properties

One of the important criteria to consider when implementing security mechanisms is the impact of these solutions on IS non-functional properties. In our work, we consider four IS non-functional properties, namely: the management of the security solution, data transmission time, monetary cost of the solution, and the computational burden caused by the solution. Indeed, some of the proposed security solutions can be hard to manage/maintain and have negative impacts on the information system's performances causing heavy computation burden or extending the response time for data access requests, especially in the case of big data. Moreover, it is also important to evaluate the security mechanism with regard to the cost price.

3.6   Data Security Meta-model

Based on the security approaches surveyed in Sect. 3.3 and the work done in Sect. 3.4 and Sect. 3.5 we propose in Fig. 3 below, our data security meta-model which is made of three main components: Data that was detailed via our data meta-model in Sect. 2, security with its three main goals (CIA) and finally the proposed security mechanisms and their impact on the IS non-functional properties. The main objectives of our meta-model are as follow:

- Identify the security techniques used to ensure security goals/needs (CIA);
- Serves as a referential to compare data security mechanisms according to data type, security goal, and the impact of the solution on the IS non-functional properties.
- Easy to extend with other existing security mechanisms and non-functional properties;
- Use the meta-model to link between data type, security goal, and security mechanisms used to achieve this goal. This will help to find research directions not yet covered in the literature. In addition to these points, our meta-model is not bound to a single technology as is the case of [7,19,23] where the proposed meta-model is closely linked to a given technology.

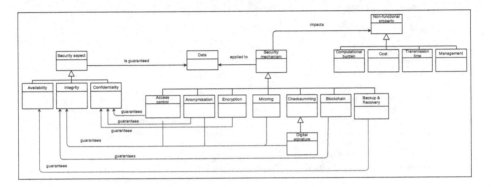

Fig. 3. Data security meta-model

4   Comparison of Approaches and Open Questions

In this section, we use our data security meta-model to compare the approaches discussed in Sect. 3.3. For this purpose, we distinguish the three security goals (CIA) and present three tables: Table 2, Table 3, and Table 4 that tackle respectively data confidentiality, data integrity, and data availability. We criticise the approaches in the drawback column in each table according to the non-functional properties presented in the meta-model. Lastly, regarding the tool/framework column we are interested in whether the paper in question gives implementation details about the proposed approach or proposes the solution only as a conceptual framework. The tables' cells with the "-" mark mean that we were unable to find any information in the paper to provide a conclusion or appreciation about the element.

Table 2. Security mechanisms for data confidentiality

Approach	Data type	Security mechanism	Technology	Tool/Framework	Drawback
[8]	Structured at rest	Static access control	On premise	Framework	- Cause "role explosion" and become hard to manage and maintain
[13]	Structured at rest	Dynamic access control	On premise	Framework	- Efforts to define attributes - Cause "attributes explosion" and become hard to manage
[45]	Big data at rest	Dynamic access control	On premise	Tool	- The experiments show that CBAC generates overhead
[14]	Structured, semi-structured and unstructured data at rest	Dynamic access control	On premise	Tool	–
[16]	Structured at rest	Dynamic access control	On premise	Framework	–
CSA	Structured in motion	Static access control and encryption.	On premise		- Policy updating poses heavy computational burden
[43]	Big data at rest and in motion	Dynamic access control and encryption.	Cloud	Tool	–
[34]	Structured at rest	Anonymization	Cloud	Tool	- Could pose computational burden in the case of big data
[11]	Structured at rest	Anonymization		Tool	–
[4]	Big data	Encryption	Cloud	Tool	–
[18]	Structured in motion	Encryption	Cloud	Tool	- Could become hard to manage the generating encryption keys; - Poses computational burden in case of big data
[36]	Structured at rest and in motion	Encryption	Cloud	Framework	- Data transmission time could increase; - Computational burden due to the index management in case of large amount of data
[40]	Structured in motion	Anonymization	On premise and cloud	Framework	–
[26]	Structured and big data at rest	Anonymization	On premise	Tool	–
[33]	Big data	Access control	Cloud	Framework	–
[3]	Structured and big data at rest	Access control	On premise and cloud	Framework	- A subject can write up but cannot read what he just wrote

Table 3. Security mechanisms for data "integrity"

Approach	Data type	Security mechanism	Tool / Framework	Technology	Drawback
[35]	Structured and big data at rest	Mirroring	Tool	On premise and cloud	- More copies equal higher costs
[20]	Structured and big data at rest	MAC	Tool	On premise and cloud	- Costly in time in case of big data; - Consumes more bandwidth in the case of big data [46]
[17]	Structured and big data at rest	Digital signature	Tool	On premise and cloud	- Costly in money
[25]	Structured and big data at rest	DIaaS	Tool	Cloud	–
[9, 46]	Structured at rest	Blockchain	Tool	-	- Blockchain can't deal with big data; - Performance issues due to the consensus protocol

Table 4. Security mechanisms for data "availability"

Approach	Data type	Security mechanism	Technology	Tool / Framework	Drawback
[37, 42]	Structured, Semi-structured, Unstructured and big data	De-duplication	On premise and cloud	Tool	- For data that must be encrypted, the de-duplication cannot play its optimization role [39]
[30]	Structured at rest	Backup	Cloud	Tool	- Adds performance overhead in terms of monetary cost and upload time
[27]	Structured, Semi-structured and unstructured at rest	Backup	Cloud	Tool	- As data amount increases the monetary cost increases accordingly; - The recovery time is not taken into account in the experimentation
[32]	Structured, Semi-structured and unstructured at rest	Backup	Cloud	Tool	–
[36]	Structured at rest and in motion	Combined mechanisms	Cloud	Framework	- Data transmission time could increase; - Computational burden could be noticed due to the index management in case of a large amount of data

From our comparison tables we have come to the following conclusions:

– **Security goals:** The proposed approaches don't consider all three security goals. Hybrid approaches that combine two or more security mechanisms could be investigated to ensure data confidentiality, integrity, and availability;
– **Encryption mechanism:** Approaches proposing encryption as a security mechanism to protect data don't give much attention to *key management* i.e. the process of protecting cryptographic keys, as this latter could be the target of security attacks. Besides, the proposed algorithms (symmetric

and asymmetric) to encrypt structured data can be used to encrypt unstructured and big data, however, in this case, these solutions induce higher computational burden and cost;

- **Data type:** Data type plays a crucial role in choosing the right security mechanism. As noticed in the comparison tables, a data security approach could give satisfying results when securing structured data but presents computational burden and performance problems when dealing with big data;
- **Big data security:** The existing security solutions for big data are still not mature compared to security solutions for structured data. Big data security lacks standard tools and models. Moreover, patterns could be proposed to help the automatic detection of sensitive information in unstructured data and big data as this becoming a challenge due to the diversity of content in these types of data;
- **Dynamic data:** One of the main characteristics of unstructured and big data is the constant changing of data and therefore, the security mechanisms and frameworks should be as dynamic as possible to cope with these changes.

5   Conclusions and Future Work Directions

Security is one of the most important properties of information systems. Specifically, it is about securing the data and the processes handling this data. In this paper, we began by proposing a general meta-model describing the three main layers of an information system linked with security aspects. Then, we surveyed research works dealing with the security of the data layer and we classified the proposed security approaches according to the data type and security goals. Finally, we put forward a data security meta-model made of three main aspects: data, security issues, security mechanisms, and the impacted non-functional properties (cost, management, computational burden, and data transmission time). We used our meta-model as a reference to compare different approaches in the literature and to identify the shortcomings of each of them. A number of criteria should be taken into consideration when proposing solutions to secure data, namely its type, its state, the technology used, and the non-functional properties impacted by the solution. It is about finding a balance between security mechanisms and the system's non-functional properties. Thus, the purpose of our work is to present a conceptual framework allowing to compare the existing approaches in the literature, and we plan as future work to extend our meta-model with dynamic security mechanisms as proposed in [2,22]. Moreover, we intend to present our work to professionals, in order to validate our meta-model on reel case scenarios, and evaluate the trade-off of each security solution. We are currently working on this meta-model and its application to the e-health domain in a Cloud context. Otherwise, we are about to propose another meta-model that covers the security of the application layer. Our goal is to build a generic meta-model to be used as a reference to secure the three layers of an information system.

References

1. Alshammari, M.M., Alwan, A.A., Nordin, A., Al-Shaikhli, I.F.: Disaster recovery in single-cloud and multi-cloud environments: Issues and challenges. In: 2017 4th IEEE International Conference on Engineering Technologies and Applied Sciences (ICETAS), pp. 1–7. IEEE (2017)
2. Brindha, K., Jeyanthi, N.: A novel approach to secure cloud data storage and dynamic data auditing in a cloud. In: Proceedings of the Second International Conference on Internet of things, Data and Cloud Computing, pp. 1–5 (2017)
3. Cankaya, E.C.: Bell-LaPadula Confidentiality Model, pp. 71–74. Springer, Boston (2011). https://doi.org/10.1007/978-1-4419-5906-5_773
4. Cheng, H., Rong, C., Hwang, K., Wang, W., Li, Y.: Secure big data storage and sharing scheme for cloud tenants. China Commun. **12**(6), 106–115 (2015)
5. Clarke-Salt, J.: SQL Injection Attacks and Defense. Elsevier, Amsterdam (2009)
6. El Emam, K., Rodgers, S., Malin, B.: Anonymising and sharing individual patient data. BMJ **350**, 1–6 (2015)
7. Erraissi, A., Belangour, A.: A big data security layer meta-model proposition. Adv. Sci. Technol. Eng. Syst. J. **4**(5), 409–418 (2019)
8. Ferraiolo, D., Kuhn, D.R., Chandramouli, R.: Role-Based Access Control. Artech House, Boston (2003)
9. Gaetani, E., Aniello, L., Baldoni, R., Lombardi, F., Margheri, A., Sassone, V.: Blockchain-based database to ensure data integrity in cloud computing environments (2017)
10. Geery, D.: Reducing the storage burden via data deduplication. Computer **41**(12), 15–17 (2008)
11. George, R.S., Sabitha, S.: Data anonymization and integrity checking in cloud computing. In: 2013 Fourth International Conference on Computing, Communications and Networking Technologies (ICCCNT), pp. 1–5. IEEE (2013)
12. Hu, V.C., et al.: Guide to attribute based access control (abac) definition and considerations (draft). NIST Spec. Publ. **800**(162), 1–54 (2013)
13. Hu, V.C., Kuhn, D.R., Ferraiolo, D.F., Voas, J.: Attribute-based access control. Computer **48**(2), 85–88 (2015)
14. Jaïdi, F., Labbene-Ayachi, F., Bouhoula, A.: Advanced techniques for deploying reliable and efficient access control: application to e-healthcare. J. Med. Syst. **40**(12), 262 (2016)
15. Karle, T., Vora, D.: Privacy preservation in big data using anonymization techniques. In: 2017 International Conference on Data Management, Analytics and Innovation (ICDMAI), pp. 340–343. IEEE (2017)
16. Kashmar, N., Adda, M., Atieh, M., Ibrahim, H.: Deriving access control models based on generic and dynamic metamodel architecture: industrial use case. Procedia Comput. Sci. **177**, 162–169 (2020)
17. Kaur, R., Kaur, A.: Digital signature. In: 2012 International Conference on Computing Sciences, pp. 295–301. IEEE (2012)
18. Khanezaei, N., Hanapi, Z.M.: A framework based on RSA and AES encryption algorithms for cloud computing services. In: 2014 IEEE Conference on Systems, Process and Control (ICSPC 2014), pp. 58–62. IEEE (2014)
19. Kritikos, K., Massonet, P.: An integrated meta-model for cloud application security modelling. Procedia Comput. Sci. **97**, 84–93 (2016)
20. Liu, D., et al.: Chapter 3 - an introduction to cryptography. In: Next Generation SSH2 Implementation, pp. 41–64. Syngress, Burlington (2009)

21. Mallik, A.: Man-in-the-middle-attack: Understanding in simple words. Cyberspace: Jurnal Pendidikan Teknologi Informasi **2**(2), 109–134 (2019)
22. Mathur, N., Bansode, R.: Aes based text encryption using 12 rounds with dynamic key selection. Procedia Comput. Sci. **79**, 1036–1043 (2016)
23. Menzel, M., Meinel, C.: A security meta-model for service-oriented architectures. In: 2009 IEEE International Conference on Services Computing, pp. 251–259. IEEE (2009)
24. Murthy, S., Bakar, A.A., Rahim, F.A., Ramli, R.: A comparative study of data anonymization techniques. In: 2019 IEEE 5th Internatioinal Conference on Big Data Security on Cloud (BigDataSecurity), IEEE Internatioinal Conference on High Performance and Smart Computing, (HPSC) and IEEE Internatioinal Conference on Intelligent Data and Security (IDS), pp. 306–309. IEEE (2019)
25. Nepal, S., Chen, S., Yao, J., Thilakanathan, D.: Diaas: Data integrity as a service in the cloud. In: 2011 IEEE 4th International Conference on Cloud Computing, pp. 308–315. IEEE (2011)
26. Prakash, M., Singaravel, G.: An approach for prevention of privacy breach and information leakage in sensitive data mining. Comput. Electr. Eng. **45**, 134–140 (2015)
27. Prathyakshini, M., Ankitha, K.: Data storage and retrieval using multiple cloud interfaces. Int. J. Adv. Res. Comput. Commun. Eng. **5**(4), 1–4 (2016)
28. Puthal, D., Malik, N., Mohanty, S.P., Kougianos, E., Das, G.: Everything you wanted to know about the blockchain: its promise, components, processes, and problems. IEEE Consum. Electron. Mag. **7**(4), 6–14 (2018)
29. Qian, C., Huang, Y., Zhao, X., Nakagawa, T.: Optimal backup interval for a database system with full and periodic incremental backup. JCP **5**(4), 557–564 (2010)
30. Rahumed, A., Chen, H.C., Tang, Y., Lee, P.P., Lui, J.C.: A secure cloud backup system with assured deletion and version control. In: 2011 40th International Conference on Parallel Processing Workshops, pp. 160–167. IEEE (2011)
31. Sahafizadeh, E., Nematbakhsh, M.A.: A survey on security issues in big data and nosql. Adv. Comput. Sci. Int. J. **4**(4), 68–72 (2015)
32. Sambrani, Y., Rajashekarappa: Efficient data backup mechanism for cloud computing. Int. J. Adv. Res. Comput. Commun. Eng. **5**(7), 1–4 (2016)
33. Sarkar, B.K.: Big data for secure healthcare system: a conceptual design. Complex Intell. Syst. **3**(2), 133–151 (2017)
34. Sedayao, J., Enterprise Architect, I.I.: Enhancing cloud security using data anonymization. White Paper, Intel Coporation (2012)
35. Sivathanu, G., Wright, C.P., Zadok, E.: Ensuring data integrity in storage: techniques and applications. In: Proceedings of the 2005 ACM Workshop on Storage Security and Survivability, pp. 26–36 (2005)
36. Sood, S.K.: A combined approach to ensure data security in cloud computing. J. Netw. Comput. Appl. **35**(6), 1831–1838 (2012)
37. Storer, M.W., Greenan, K., Long, D.D., Miller, E.L.: Secure data deduplication. In: Proceedings of the 4th ACM International Workshop on Storage Security and Survivability, pp. 1–10 (2008)
38. Suguna, S., Suhasini, A.: Overview of data backup and disaster recovery in cloud. In: International Conference on Information Communication and Embedded Systems (ICICES2014), pp. 1–7. IEEE (2014)
39. Sun, G.Z., Dong, Y., Chen, D.W., Wei, J.: Data backup and recovery based on data de-duplication. In: 2010 International Conference on Artificial Intelligence and Computational Intelligence, vol. 2, pp. 379–382. IEEE (2010)

40. Sweeney, L.: k-anonymity: a model for protecting privacy. Int. J. Uncertainty Fuzziness Knowl. Based Syst. **10**(05), 557–570 (2002)
41. Vasconcelos, A., da Silva, M.M., Fernandes, A., Tribolet, J.: An information system architectural framework for enterprise application integration. In: . Proceedings of the 37th Annual Hawaii International Conference on System Sciences 2004, p. 9 IEEE (2004)
42. Xia, W., et al.: A comprehensive study of the past, present, and future of data deduplication. Proc. IEEE **104**(9), 1681–1710 (2016)
43. Yang, K., Jia, X., Ren, K.: Secure and verifiable policy update outsourcing for big data access control in the cloud. IEEE Trans. Parallel Distrib. Syst. **26**(12), 3461–3470 (2014)
44. Zachman, J.A.: A framework for information systems architecture. IBM Syst. J. **26**(3), 276–292 (1987)
45. Zeng, W., Yang, Y., Luo, B.: Access control for big data using data content. In: 2013 IEEE International Conference on Big Data, pp. 45–47. IEEE (2013)
46. Zikratov, I., Kuzmin, A., Akimenko, V., Niculichev, V., Yalansky, L.: Ensuring data integrity using blockchain technology. In: 2017 20th Conference of Open Innovations Association (FRUCT), pp. 534–539. IEEE (2017)

Inherent Discriminability of BERT Towards Racial Minority Associated Data

Maryam Ramezanzadehmoghadam[1], Hongmei Chi[1(✉)], Edward L. Jones[1], and Ziheng Chi[2]

[1] Department of Computer and Information Sciences, Florida A&M University, Tallahassee, FL 32307, USA
{maryam1.ramezanzade,hongmei.chi,edward.Jones}@famu.edu
[2] Department of Electronic and Information Engineering, Hong Kong Polytechnic University, Hong Kong, China

Abstract. AI and BERT (Bidirectional Encoder Representations from Transformers) have been increasingly adopted in the human resources (HR) industry for recruitment. The increased efficiency (e.g., fairness) will help remove biases in machine learning, help organizations find a qualified candidate, and remove bias in the labor market. BERT has further improved the performance of language representation models by using an auto-encoding model which incorporates larger bidirectional contexts. However, BERT's underlying mechanisms that enhance its effectiveness, such as tokenization, masking, and leveraging the attention mechanism to compute vector score, are not well understood. This research analyzes how BERT's architecture and its tokenization protocol affect the low number of occurrences of the minority-related data using the cosine similarity of its embeddings. In this project, by using a dataset of racially and gender-associated personal names and analyzing the interactions of transformers, we present the unfair prejudice of BERTs' pre-trained network and autoencoding model. Furthermore, by analyzing the distance of an initial word's token and its MASK replacement token using the cosine similarity, we will demonstrate the inherent discriminability during pre-training. Finally, this research will deliver potential solutions to mitigate discrimination and bias in BERT by examining its geometric properties.

Keywords: BERT · Machine learning fairness · Natural language processing · Racial minority · Pre-trained model · Recruitment

1 Introduction

Our last names and given names are our first social labels that correspond with our gender identity, race and ethnicity, or country of origin. However, with AI being increasingly adopted in screening resumes in the human resources

Supported by organization x.

(HR) industry for purposes such as predicting employee attrition and background verification, women and individuals from minority ethnic groups (compared to majority groups) are facing unfavorable labor market outcomes in many economies [13]. In addition, AI-based recruitment platforms are widely adopted in some industries recruiting new employees. However, the potential discrimination in recruitment algorithms is hard to assess [28]. In this paper, we check on name discrimination only and review BERT's inherent bias.

Machine learning is being used in a wide range of applications. With machines being trained on different data sets to make decisions, suggestions, and recommendations, it is genuinely essential to take the "fairness" of these decisions into account. In the context of decision-making, fairness is the "absence of any prejudice or favoritism toward an individual or a group based on their inherent or acquired characteristics" [18]. A model's output, if not fair, will result in decisions that can be biased against certain groups with a set of characteristics.

With the widespread use of AI systems and their human and computer interaction applications processing natural language, it is crucial to consider system fairness while designing and engineering them. Unfortunately, there are many different sources of bias when dealing with Big Data, leading to unfair biases towards other groups. [18] in their work have introduced and discussed some of the most general and essential sources of bias introduced in [29] and [21]. Even though all of the introduced bias sources can be applied to the field of natural language processing, but Historical Biases, Representational Biases, Popularity Biases, and Social Biases are the main focus of this work [19].

Historical Biases are the already existing bias and socio-technical issues in the world which are unrelated to the policy under study and can permeate the data generation process even given a perfect sampling and feature selection which can have a significant effect on the policy's hoped for outcome [29]. Representational Biases are the cause of wrongfully defining and sampling from a population. Popularity Biases come from the following underlying issue: More popular Items tend to be exposed more. When studying natural language, popularity metrics are ubiquitous to be manipulated based on the existing historical biases towards minority groups available in linguistic corpora. Social bias happens when other people's actions or content coming from them affect systems judgment. Considering the existing historical biases towards minority-related groups and ethnicities (African America, Hispanic, Asian, Middle Eastern), Representational, Popularity, and Social biases are easy and probable to happen in linguistic corpora, leading to Indirect systematic discrimination.

Indirect discrimination happens when individuals appear to be treated based on seemingly neutral and non-protected attributes; however, protected groups or individuals still get to be treated unjustly as a result of implicit effects from their protected attributes [31]. Systematic discrimination happens as patterns of behavior, policies, or practices that are part of the culture or structures of an organization, and which create or perpetuate disadvantage for racialized persons [10].

In various applications, pre-trained natural language processing (NLP) models run the risk of exploiting and reinforcing the societal biases that are present in the underlying data [8]. Therefore, it is crucial to understand the unjust bias inherited by them and ensure that the decisions do not reflect discriminatory behavior toward certain groups or populations.

BERT has further improved the performance of deep learning and neural network architecture by using an auto-encoding model which incorporates larger bidirectional contexts [24]. However, the underlying mechanisms of BERT that are improving its effectiveness, such as tokenization, masking, and leveraging the attention mechanism to compute vector scores, are not well understood. This paper analyses how BERT's architecture and its tokenization protocol affect the low occurrences of the minority-related data via analyzing BERT's vocabulary, tokenization, and embeddings using WEAT and cosine similarity test.

The rest of this paper is organized as follows. Section 2 presents an overview of the Natural Language Processing, context-free and contextual pre-trained language models, as well as BERT. Related works will be discussed in the Sect. 3. A detailed description of the datasets that have been used is provided in Sect. 4. Details on the methodology and performed analysis is provided in Sect. 5. Preliminary results are presented in Sect. 6. Finally, Sect. 7 provides the conclusion and future works.

2 Background

Natural Language Processing is a theoretically motivated range of computational techniques for analyzing and representing naturally occurring texts at one or more levels of linguistic analysis to achieve human-like language processing for a range of tasks or applications [17]. Measurable property, in machine learning, is a piece of information that can be observed as a characteristic when solving a problem via selecting relative, quantitative and qualitative. Selected feature, is a critical step in model construction since it can simplify the model, shorten the training time, deliver a practical algorithm, and enhanced results.

2.1 Language Representations

Word embeddings are used to converting each word in text data into vectors which are a numerical representation of words, and vectors are used to extract the best features from the content of texts. These vector representations capture the underlying words in relation to the other words present in a sentence. Words with similar meanings will be grouped closer together in a hyperplane that distinct them from words that are position further in that hyperplane. This step could be done by either creating a corpus-specific word embedding based and trained on the selected features that will be introduced from the training data set, or it could be done by using a pre-trained word embedding. The problem with the embedding layer is that its features will be relevant for only a specific issue. There won't be enough words introduced to it to map the encoded categorical variables. Pre-trained representations can either be context-free or contextual:

- Context-free models such as Word2Vec [9] or GloVe [22] generate a single "word embedding" representation for each word in the vocabulary no matter the position of the word in the text.
- Contextual models such as ELMO [23], Universal Language Model Fine-tuning (ULMFit) [14] and BERT, generate a representation of each word based on the other words in the text.

Contextual representations can further be unidirectional or bidirectional. ELMO, ULMFit, and previously pre-trained models were unidirectional, which means that each word is only contextualized using the words to its left (or right). However, BERT has a bidirectional architecture.

2.2   BERT

BERT is designed to pre-train deep bidirectional representations from an unlabeled text by jointly conditioning on both left and right context in all layers. As a result, the pre-trained BERT model can be fine-tuned with just one additional output layer to create state-of-the-art models for a wide range of tasks, such as question answering and language inference, without substantial task-specific architecture modifications [12].

BERT with it's bidirectional architecture claims to obtain state-of-the-art results on eleven natural language processing tasks [12].

3   Related Works

Several studies on identification and mitigation of bias in statistical context-free word embeddings such as GloVE and Word2Vec by [5] and Skip-Gram embeddings trained on various data sets and FastText embeddings by [7] and [15]. However, there have been only a few studies [32] and [2] on the fairness of contextual language models (CLMs) which are being leveraged in a variety of domains. Studies on BERT's fairness are mainly focused on gender bias, and there have been only a few studies that analyzed fairness of BERT's word embedding towards racial minority groups.

Zhao et al. (2019) [32] have quantified, analyzed and mitigated gender bias exhibited in ELMo's contextualized word vectors, and demonstrated that training data for ELMo contains significantly more male than female entities, ELMo's embeddings systematically encode gender information, and its co-reference system inherits its bias and presents a significant bias on the WinoBias probing corpus.

Bhardwaj, Majumder and Poria (2020) [4] analyze the gender bias that CLMs induce in different downstream tasks related to emotion and sentiment intensity prediction. They have trained a simple regressor utilizing BERT's word embeddings for each task and evaluated the gender bias in regressors using an equity evaluation corpus. Using this method, they proved a significant dependence of BERT's predictions on gender-particular words and phrases.

Kurita et al. (2019) [16] argued that contextual word embeddings, such as BERT are optimized to capture the statistical properties of training data. They also tend to pick up and amplify social stereotypes present in the data. They proposed a template-based method to quantify bias in BERT, which is generalizable to unveiling other biases in multi-class settings, such as racial and religious biases. Our work leverages their approach to demonstrate the racial bias in BERT.

Bartl, Nissim and Gatt [1] measured BERT's gender bias by studying associations between gender-denoting target words and names of professions in English and German, comparing the findings with real-world workforce statistics. Their results demonstrated that male person terms seem to have a more stable position in BERT, which could cause their probabilities in the model to not vary much depending on the context and make them less susceptible to change through fine-tuning. Our work uses the list of professions provided by them to demonstrate the racial bias in BERT.

Sharma, Dey and Sinha (2020) [27] demonstrated in their work that three models (BERT, RoBERTa, and BART) trained on MNLI and SNLI data-sets are significantly prone to gender-induced prediction errors and using a debiasing technique, such as augmenting the training data set to ensure that it is a gender-balanced data can help reduce such bias in some instances.

Podkorytov, et al. (2020) [24] presented experimental results illustrating that BERT does not produce "effective" contextualized representations for words. Instead, its improved performance may mainly be due to fine-tuning or classifiers that model the dependencies explicitly by encoding syntactic patterns in the training data.

4   Dataset

Bias in pre-trained contextualized word representation models is commonly measured via sentence templates. For this purpose, we have used two different corpora (EEC and BEC-Pro) developed by [15] and [1] and a data set from the Field Experiment on Labor Market Discrimination (LMD) designed by [3].

The Equity Evaluation Corpus (EEC) data set analyzes bias in NLP systems using the context's connection with emotions. The BEC-Pro data set analyzes bias in NLP systems using the connection of the context with professions. The LMD data set provides a list of names associated with two different races of African American and European American and their sex. The details of the mentioned data set will be discussed in Sect. 4.1, 4.2 and 4.3.

4.1   Equity Evaluation Corpus

The Equity Evaluation Corpus (EEC) is a data set of thousands of sentences to determine whether automatic systems consistently give higher (or lower) sentiment intensity scores to sentences involving a particular race or gender [15]. This corpus includes 11 sentence templates divided into two groups of emotion and non-emotion words with two variables of <person> and <emotion word>. First, we have used this template to generate sentences by instantiating the <person> noun phrases with the African American and European American first names associated with both genders of female and male. Next, we have instantiated the <emotion word> with the emotional state words correspond to pleasantness/unpleasantness, rich/poor, and happy/sad following the work of [6] and [15].

4.2   BEC-Pro

The BEC-Pro data set includes sentence templates containing a gender-denoting noun phrase <person> and a <profession>. The professions on this template were formed by three groups of 20 professions from the U.S. Bureau of Labor Statistics (2020) [20] with the following categories:

- Group 1 with highest female participation (88.3%–98.7%)
- Group 2 with lowest female participation (0.7%–3.3%)
- Group 3 with a roughly 50-50 distribution of male and female employees (48.5%–53.3%)

We have used the five-sentence templates of [1] that include the <person> noun phrases and the carries explicit gender information and a profession term. In addition, we have instantiated the <person> noun phrases with the African American and European American first names associated with both genders.

4.3   Labor Market Discrimination

The Field Experiment on Labor Market Discrimination (LMD) data set was constructed by [3] when studying race in the labor market. They posted fictitious resumes to help-wanted ads in Boston and Chicago newspapers. Marianne Bertrand and Sendhil Mullainathan's [3] research used name frequency data calculated from birth certificates of all babies born in Massachusetts between 1974 and 1979. They tabulated these data by race to determine which names are distinctively European American and distinctively African-American. Distinctive names were those that had the highest ratio of frequency in one racial group to the frequency in the other racial group. A survey was also conducted to check the distinctiveness of their data. We have used their data set of racially distinct names and combined it with the two previously mentioned corpora to perform a racial bias analysis on the BERT algorithm.

5   Methodology

We have performed two different methods (WEAT and Cosine Similarity) to analyze bias in the BERT algorithm, which will be discussed in more detail in the following sections. Section 5.1 gives an overview of BERT's pre-trained language model. Section 5.2 discusses the vocabulary data set of BERT that is used for word embeddings. Section 5.3 explains the Word Embedding Association Test (WEAT), one of the methods used to analyze bias in BERT. Finally, Sect. 5.4 presents the cosine similarity method used to analyze bias towards word embeddings generated by BERT. We have used BERT pre-trained auto tokenizer and auto model for Tensorflow from the Hugginface transformer library to perform these tests. **BERT Base Cased** has been used for all tests since it was relevant with using names as a noun in sentence pairs.

5.1 BERT Algorithmic Discriminability

BERT's architecture for the training phase first forms a bidirectional representation of the sentence and then masks the target words with a special [MASK] token. This makes the prediction of the target word more difficult for the model. The masked language model (MLM) randomly masks some of the tokens from the input, and the objective is to predict the original vocabulary id of the masked word based only on its context. Even though this approach produces a bidirectional pre-trained model but introducing the [MASK] token during training creates a mismatch between the pre-training and fine-tuning input, which is mitigated by choosing only 15% of the token positions at random for prediction. [16] estimated the probability of a masked, gendered target word's association with an attribute word in a sentence using the MLM. However, the results in [24] show that the BERT models do not produce "effective" contextualized representations for words, and its improved performance can be the cause of fine-tuning or classifiers that model the dependencies explicitly by encoding syntactic patterns in the training data. Therefore if BERT is not introduced to certain words, such as Racial minority data (African American Names in our case) during its pre-training or training phase, it can not provide not only an effective but also a correct contextualized representation of the words.

5.2 BERT Vocabulary

BERT's pre-training corpus is developed by the BooksCorpus (800M words) [33] and English Wikipedia (2,500M words). BERT uses WordPiece [30] embeddings with a 30,000 token vocabulary. Our work used BERT-Base cased model, which includes a vocab file to map WordPiece to word id. After pursuing the vocab, we have looked for the names of African Americans along with the European Americans of both male and female genders. Section 6 will discuss the Notable results of this analysis. When BERT is introduced with a new word in a sentence that is not included in its pre-training vocabulary, it will use a word piece model that breaks the word embedding down into multiple sub-words, and looks up their embeddings in WordPiece, and then treats the word as separate embeddings of the subwords. However, if the introduced word does not have any sub-words, BERT will choose each alphabetical character of the word as a subword and finds its embedding from the WordPeice to create a representation of that word. We have analyzed the word embedding of all African American, and European

American names are using BERT's tokenization method. The notable results of this analysis are discussed in Sect. 6.

5.3   WEAT Test

Implicit Association Test (IAT) is a measure of the strength of associations between concepts, evaluations, and stereotypes [25]. The Word-Embedding Association Test (WEAT) is a statistical test analogous to the IAT, which is widely applied to the semantic representation of words in AI, and termed word embedding. Word embeddings represent each word as a vector in a vector space of about 300 dimensions, based on the textual context in which the word is found. In WEAT, the distance between a pair of vectors (their cosine similarity score) is used as analogous to reaction time in the IAT to compare a set of target concepts (e.g., Caucasian and African American names) denoted as $\mathbf{X}$ and $\mathbf{Y}$ (each of equal size N), with a set of attributes to measure bias over social attributes and roles (e.g., career/family words) denoted as $\mathbf{A}$ and $\mathbf{B}$. The degree of bias for each target concept $\mathbf{t}$ is calculated as shown below, where $\mathbf{sim}$ stands for cosine similarity:

$$s(t, A, B) = [mean_{a \epsilon A} sim(t, a) - mean_{b \epsilon B} sim(t, b)] \tag{1}$$

To measure bias using this method, target noun phrases are the names of Racially classified groups and the attributes are professions and sentimental phrases. In this method, Model aim's in masking tokens to create unbiased setting and then predict them. An example of a masking sentence is shown in Table 1.

Table 1. Masked sentence set example for WEAT

Type	Example
Original sentence	Keisha is a scientist
Target masked	[MASK] is a scientist
Attribute masked	Keisha is a [MASK]
Target & attribute masked	[MASK] is a [MASK]

Following [1] and [16] who use WEAT, we measure the influence of the attribute $\mathbf{A}$, which can be a profession or emotion, on the likelihood of the target (T), which denotes people of Caucasian and African American racial groups person: P(T—A). Overall assumptions to perform such analysis on BERT and expect valid results is as follow:

– likelihood of a token is influenced by all other tokens in the sentence
    – Thus likelihood of Target will be different based on presence of the Target:

$$P(T) \neq P(T—A)$$

– Therefore, likelihood of Target will be different depending on different racially classified groups:

$$P(T_{female}|A) \neq P(T_{male}|A)$$

Details of calculating the WEAT probability scores following [1] and [16] is as follow:

1. Take the Original sentence with Target and Attribute words (Demonstrated in **Original Sentence** row of Table 1)
2. Mask the Target word (Demonstrated in **Target Masked** row of Table 1)
3. Obtain the probability of target word in the sentence:

$$p_T = P(Keisha = [MASK]|sentence)$$

4. Mask both target and attribute word (Demonstrated in **Target and Attribute Masked** row of Table 1)
5. Obtain the probability of target word when the attribute is masked:

$$p_{prior} = P(Keisha = [MASK]|maskedsentence)$$

6. Calculate the association by dividing the target probability by the prior and take the natural logarithm

5.4   Cosine Similarity Test

Contextual embeddings are mostly transformations of embedding at layer 1 (having no contextual information) to layer 12 (pick up more contextual information with each layer once deeper into the network). We have used the output of the last layer as the word embedding in a sentence with Female and Male Names of the Caucasian and African American racial groups as the noun phrase and different attribute sets of Actor/Actress, Pleasant/Unpleasant, Rich/Poor and Wrong/Write. These sentences' cosine similarity has been measured between the gender and racial groups, and the results are demonstrated in Fig. 2 and 3.

6   Preliminary Results

The result of each of the tests ran on the data sets will be discussed in the following sections.

6.1   BERT's Vocabulary

As it has been discussed before, BERT decomposes each unknown word into subwords and character tokens that it can generate embeddings for so that it can average the sub word embedding vectors to generate an approximate vector and retain the contextual meaning of an original word. BERT's vocabulary contains:

- Whole words
- Sub-words occurring at the front of a word or in isolation ("em" as in "embeddings" is assigned the same vector as the standalone sequence of characters "em" as in "go get em")
- Sub-words not at the front of a word, which are preceded by '##' to denote this case
- Individual characters

The result of our analysis on the BERT's vocabulary and LMD's set of names indicates that Caucasian Names for both Female and Male genders are 100% included in BERT's vocabulary, and Only 11% of Female and 33% of Male African American Names are included in BERT's vocabulary shown in Fig. 1. In [16] study, the Words that are absent in the BERT vocabulary were removed from the targets, and Lack of racial minority-related names, i.e., "Targets" in BERT's vocabulary, can lead to not only social bias but also algorithmic bias in the architecture.

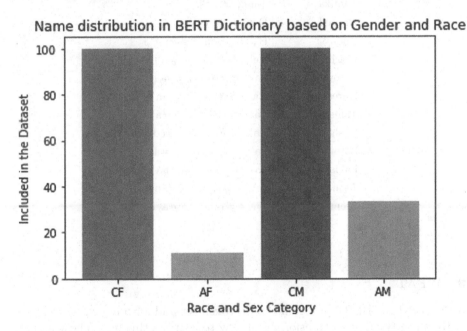

Fig. 1. Distribution of Caucasian and African American Names in BERT's vocabulary. C (Caucasian), A (African American), F (Female) and M (Male)

6.2 BERT's Algorithmic Discriminability

As shown in Table 1, you can observe that, since BERT does not include the African American names, it's treating them as unknown words and decomposing them into sub-words. The sub word embedding vectors to retain the contextual meaning of an original word which in the case of names it does not bring the proper context by having single character's and words like "ish" in "Keisha", therefore this will change the overall meaning and context of a sentence that includes a name that's not included in BERT's vocabulary like Social Minority groups name. Table 2 represents the sub-word decomposition of African American names. This result accords with [24] work, proving that the complex interaction of "context dependent representation" and "reconstruction of input sentence" via transformers lead to geometric properties of the embeddings and subsequently affect the inherent discriminability of the resulting representations.

Table 2. Sub word decomposition of African American names

Original name	BERT's tokenization
Aisha	['Ai', '##sha']
Hakim	['Ha', '##kim']
Tamika	['Tam', '##ika']
Darnell	['Dar', '##nell']
Ebony	['E', '##bon', '##y']
Keisha	['Ke', '##ish', '##a']
Latoya	['La', '##to', '##ya']
Kareem	['Ka', '##ree', '##m']
Tanisha	['Tan', '##ish', '##a']
Rasheed	['Ra', '##she', '##ed']
Latonya	['La', '##tony', '##a']
Lakisha	['La', '##kis', '##ha']
Jermaine	['Je', '##rma', '##ine']
Tremayne	['T', '##rem', '##ayne']

6.3   WEAT

We have used the BERT pre-trained auto tokenizer and auto model for Tensorflow from the Hugginface transformer library to perform this test. The sentence templates BEC-Pro provided by [1] which we have used and replaced the gender-related pronouns to African American and Caucasian related names, were used to measure the association of target and attributes in sentences. The associations were measured using the templates shown in Table 1. As shown in the Second row of Table 1, we masked the "Target", applied the softmax function to the unscaled output of earlier layers predicted for the "Target" position in the sentence to obtain the probability distribution of the "Target" over the BERT vocabulary which using its vocabulary index the assassination probability value was obtained.

As shown in Table 3 there is a negative association between a target and an attribute for African American associated names, which means the probability of the target decreased through the combination with the attribute. However, there is a positive association value between the target and attribute for Caucasian associated names, indicating that the probability of the target increased through the combination with the attribute, with respect to its association probability. For both racial groups, the female gender has less increment in the probability of target combined with the attribute. This indicates that not only BERT is biased towards Racial minority groups but also it is biased towards gender minority groups, as it was illustrated in [1] work.

Table 3. WEAT results for classified racial and gender groups

Race	Gender	Mean
African American	Male	−0.09
	Female	−0.12
Caucasian	Male	0.10
	Female	0.04

6.4   Cosine Similarity

It has been discussed by [11] that BERT was not optimized to get a well-shaped sentence embedding from it for unsupervised tasks, and it has the best performance on a supervised task where the classifier learns which dimension contribute and what the "scale" of those are. It has been further explained by [26] that BERT was not designed to produce helpful word/sentence embeddings that can be used with cosine similarities Because Cosine-similarity treats all dimensions equally, which puts high requirements for the created embed dings. We have analyzed the similarity of not just the sentences including both African American and Caucasian Targets and a sentimental Attribute such wrong/write, Actor/Actress, Pleasant and Unpleasant and the results demonstrated that for Caucasian names, two classes of female and male genders could be classified from each other using a boundary divider shown in Fig. 2 (A). Still, for the African American associated names, the two genders can not be classified into two different groups as if there is no similarity in their embeddings shown in Fig. 2 (B). Results also indicate that the word embedding for sentences including Caucasian male names is more biased towards positive sentiments such as pleasant and correct, shown in Fig. 3. This result is along with [4] observation that indicates having directional components from the embeddings that keep gender/racial features, causes into obtaining directions with a low magnitude of cosine similarity with actual gender/racial directions.

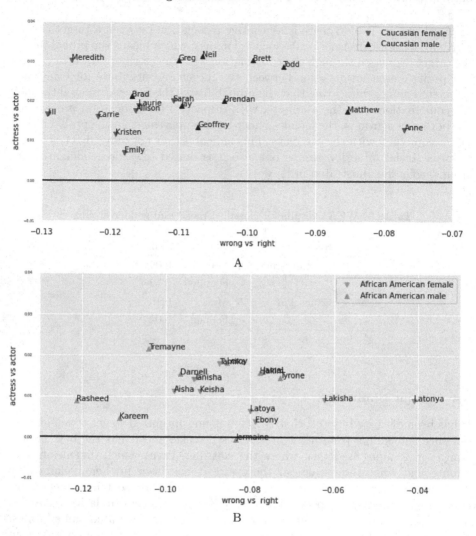

Fig. 2. Cosine similarity of Caucasian (A) and African American (B) associated names with sentimental attributes

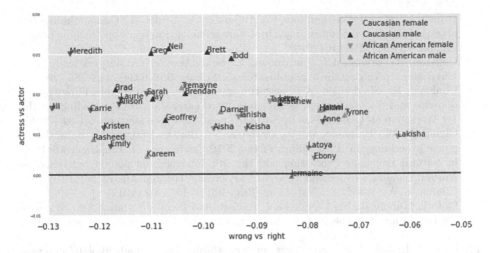

Fig. 3. Combined cosine similarity of both Caucasian and African American associated names with sentimental attributes

7 Conclusions

Among African Americans born in the last two decades, names provide a strong signal of socioeconomic status. AI recruitment platforms are widely adopted in industrial employee recruitment. The discrimination in recruitment algorithms affects these inequalities, which are difficult to assess. In this paper, we analyzed racial discrimination that can occur on such systems due to employee names. The results of our work illustrated that there exists an inherent discriminability that occurs during pre-training of BERT's algorithm. Future works and research on this study can analyze some mitigation strategies that can be applied to BERT and its architecture, such as having a more compiled vocabulary for the algorithm, training BERT with unique tokens for named entities and making them out from the raw text, using named entity recognition techniques to filter the names and replace them with an unknown token before feeding them into the pre-trained model.

References

1. Bartl, M., Nissim, M., Gatt, A.: Unmasking contextual stereotypes: measuring and mitigating Bert's gender bias. arXiv preprint arXiv:2010.14534 (2020)
2. Basta, C., Costa-Jussà, M.R., Casas, N.: Evaluating the underlying gender bias in contextualized word embeddings. arXiv preprint arXiv:1904.08783 (2019)
3. Bertrand, M., Mullainathan, S.: Replication data for: are Emily and Greg more employable than Lakisha and Jamal? A field experiment on labor market discrimination (2019)
4. Bhardwaj, R., Majumder, N., Poria, S.: Investigating gender bias in Bert. arXiv preprint arXiv:2009.05021 (2020)

5. Brunet, M.-E., Alkalay-Houlihan, C., Anderson, A., Zemel, R.: Understanding the origins of bias in word embeddings. In: International Conference on Machine Learning, pp. 803–811. PMLR (2019)
6. Caliskan, A., Bryson, J.J., Narayanan, A.: Semantics derived automatically from language corpora contain human-like biases. Science **356**(6334), 183–186 (2017)
7. Chaloner, K., Maldonado, A.: Measuring gender bias in word embeddings across domains and discovering new gender bias word categories. In: Proceedings of the 1st Workshop on Gender Bias in Natural Language Processing, pp. 25–32 (2019)
8. Chang, K.-W., Prabhakaran, V., Ordonez, V.: Bias and fairness in natural language processing. In: Proceedings of the 2019 Conference on Empirical Methods in Natural Language Processing and the 9th International Joint Conference on Natural Language Processing (EMNLP-IJCNLP): Tutorial Abstracts (2019)
9. Church, K.W.: Word2vec. Nat. Lang. Eng. **23**(1), 155–162 (2017)
10. Ontario Human Rights Commission: Racism and racial discrimination: Systemic discrimination (fact sheet). Ontario Human Rights Commission. Accessed 17 July 2018
11. Devlin, J.: Question: why Bert underperforms issue #80 ukplab/sentence-transformers (November 2018)
12. Devlin, J., Chang, M.-W., Lee, K., Toutanova, K.: Bert: pre-training of deep bidirectional transformers for language understanding. arXiv preprint arXiv:1810.04805 (2018)
13. Hangartner, D., Kopp, D., Siegenthaler, M.: Monitoring hiring discrimination through online recruitment platforms. Nature **589**(7843), 572–576 (2021)
14. Howard, J., Ruder, S.: Universal language model fine-tuning for text classification. arXiv preprint arXiv:1801.06146 (2018)
15. Kiritchenko, S., Mohammad, S.M.: Examining gender and race bias in two hundred sentiment analysis systems. arXiv preprint arXiv:1805.04508 (2018)
16. Kurita, K., Vyas, N., Pareek, A., Black, A.W., Tsvetkov, Y.: Measuring bias in contextualized word representations. arXiv preprint arXiv:1906.07337 (2019)
17. Liddy, E.D.: Natural Language Processing (2001)
18. Mehrabi, N., Morstatter, F., Saxena, N., Lerman, K., Galstyan, A.: A survey on bias and fairness in machine learning. arXiv preprint arXiv:1908.09635 (2019)
19. Mozafari, M., Farahbakhsh, R., Crespi, N.: Hate speech detection and racial bias mitigation in social media based on Bert model. PLoS ONE **15**(8), e0237861 (2020)
20. Bureau of Labor Statistics: Employed persons by detailed occupation, sex, race, and Hispanic or Latino ethnicity. U.S. Bureau of Labor Statistics (January 2021)
21. Olteanu, A., Castillo, C., Diaz, F., Kıcıman, E.: Social data: biases, methodological pitfalls, and ethical boundaries. Front. Big Data **2**, 13 (2019)
22. Pennington, J., Socher, R., Manning, C.D.: GloVe: global vectors for word representation. In: Proceedings of the 2014 Conference on Empirical Methods in Natural Language Processing (EMNLP), pp. 1532–1543 (2014)
23. Peters, M.E., et al.: Deep contextualized word representations. arXiv preprint arXiv:1802.05365 (2018)
24. Podkorytov, M., Biś, D., Cai, J., Amirizirtol, K., Liu, X.: Effects of architecture and training on embedding geometry and feature discriminability in Bert. In: 2020 International Joint Conference on Neural Networks (IJCNN), pp. 1–8. IEEE (2020)
25. Harward Project Implicit. https://www.projectimplicit.net
26. Reimers, N.: Why cosine similarity of BERT, ALBERT, Robert is so big, almost near 1.0? · issue #2298 · huggingface/transformers (December 2019)
27. Sharma, S., Dey, M., Sinha, K.: Evaluating gender bias in natural language inference (2021)

28. Steed, R., Caliskan, A.: Image representations learned with unsupervised pre-training contain human-like biases. arXiv preprint arXiv:2010.15052 (2020)
29. Suresh, H., Guttag, J.V.: A framework for understanding unintended consequences of machine learning. arXiv preprint arXiv:1901.10002 (2019)
30. Wu, Y., et al.: Google's neural machine translation system: bridging the gap between human and machine translation. arXiv preprint arXiv:1609.08144 (2016)
31. Zhang, L., Wu, Y., Wu, X.: A causal framework for discovering and removing direct and indirect discrimination. arXiv preprint arXiv:1611.07509 (2016)
32. Zhao, J., Wang, T., Yatskar, M., Cotterell, R., Ordonez, V., Chang, K.-W.: Gender bias in contextualized word embeddings. In: Proceedings of the 2019 Conference of the North American Chapter of the Association for Computational Linguistics: Human Language Technologies, Volume 1 (Long and Short Papers), Minneapolis, Minnesota, June 2019, pp. 629–634. Association for Computational Linguistics (2019)
33. Zhu, Y., et al.: Aligning books and movies: towards story-like visual explanations by watching movies and reading books. In: Proceedings of the IEEE International Conference on Computer Vision, pp. 19–27 (2015)

The Knowledge Base for Automating the Architecting of Software Systems

Gleb Guskov(iD), Anton Romanov(iD), and Aleksey Filippov(✉)(iD)

Ulyanovsk State Technical University, Ulyanovsk 432027, Russia
al.filippov@ulstu.ru
https://ulstu.ru

Abstract. This article describes an approach to knowledge base (KB) formation. KB may be used for automating the software systems (SS) architecting based on the experience of previous projects.

Software architecting is the presentation of software systems in the form of design artifacts set and their architecture. It is possible to improve the quality of a new SS by attracting the experience of previous projects in a new SS developing process. The experience of previous projects is successful architectural solutions that contain the design organization KB.

That KB should be formed in the analyzing process of design artifacts extracted from previous projects: source code, project diagrams, data models, structured text resources, etc.

This article describes the design organization KB model and the model of the 1C: Enterprise 8 platform solution (1C application) as an example of a design artifact. Also, the article presented the method for KB fragments forming based on analyzing the process of 1C application and a method for generating use-case diagrams based on the KB content.

The set of experiments was executed to evaluate the adequacy of the proposed models and methods. Precision and recall metrics were used to evaluate the quality of the generated use-case diagrams. Precision is the presence of ex-pert diagram elements in the generated diagram. Recall is the presence of elements in the generated diagram that are absent in the expert diagram. According to the results of the experiments, the average value of accuracy is 0.875, and the completeness is 0.6.

Keywords: Architecting · Software systems · Experience of previous projects · Design artifacts · Design experience units

1 Introduction

Currently, the complexity of developing and maintaining software systems has much increased. Therefore, it is necessary to apply the principles and procedures of the architecting [1]. Architecting is the software system (SS) architecture creation based on understanding the essence and basic properties of a SS. The

O. Gervasi et al. (Eds.): ICCSA 2021, LNCS 12951, pp. 272–287, 2021.
https://doi.org/10.1007/978-3-030-86970-0_20

properties of a SS determine the behavior of the system, its structure, and development approaches. These characteristics influence the stability, usefulness, and viability of the system [2–4]. An architecture description of a SS obtains during the architectural design process. An architecture description is used to coordinate the stakeholders involved in the development, implementation, support, and operation of the system.

Thus, requirements for the system are formed in the architectural designing of a SS for all stages of the life cycle: development, implementation, maintenance. The characteristics of the processes of development, implementation, operation, and maintenance of SS are highlighted.

A SS architecture can be considered as a concept of the primary properties of the system. A SS architecture can be represented as multiple representations: business representation, physical representation, technical representation [1]. An architecture description of a SS is defined in this study as a set of design artifacts obtained in a SS development. Design artifacts are the most primitive constructs that form an architecture description of a SS.

A SS architecture as a set of design artifacts allows systematizing architecture descriptions in different subject areas. The context of a SS defines many architectural requirements and constraints that determine the structure and characteristics of design artifacts [5]:

- components and elements of a SS;
- principles to the organization and interaction of an SS element;
- a SS structure;
- principles of a SS project;
- principles of a SS development within a life cycle.

Thus, an architecture description quality of a SS influences the development process and the SS quality. In the later stage of SS development are design errors discovered, that more resources are required to correct them [6].

It needs to use artificial intelligence methods to minimize errors in a SS at the early stages. Artificial intelligence methods and algorithms [7–9] involve extracting knowledge from design artifacts to build a knowledge base (KB) of a software company.

An architecture description of the current SS project can also be obtained by processing an architecture description of the previous projects [1,4]. Successful design decisions extracted from previous projects should be used in the process of working on new projects.

2 Knowledge Base of a Software Company

A modern large software company has a significant amount of completed projects. Each successfully implemented project is represented by a set of design artifacts obtained in the process of development. Design artifacts contain many experience units (EU) of highly qualified specialists who have been developing and designing SSs for many years. In this study, EU are understood as the implementation of domain entities and their business processes.

The EUs of previous projects is stored in various design artifacts:

- project documentation presented in the form of a weakly structured text in a natural language;
- set of design diagrams, usually presented in UML notation;
- source code;
- set of datamodels describes objects of a problem area, presented in the form of entities and relationships of relational databases;
- features of the project environment, usually presented in the form of wiki resources.

The complexity of the search for the necessary information depends on the amount of heterogeneous unstructured data. The experience of previous projects contained in design artifacts is not used on time. The use of design artifacts decreases the time of a SS development cycle and increases the quality of the developed SS.

The quality of a SS depends on the quality of execution of the development stages [6].

The central stage in the new SS development is the design stage. An architecture description and model of the developed SS are formed on the design stage. The quality of functioning of the whole complex of tasks assigned to a new SS depends on the quality of a new SS project. The following tasks should be solved at the design stage:

1. Search for key project participants.
2. Formation of the final architecture description of a new SS.
3. Object modeling and design of the basic elements of a new SS.

The formats and storage methods for various types of design artifacts are different, which complicates their analysis and the formation of KB. KB of a software company can automate the following tasks:

1. Best practices accumulation.
2. Extraction of frequently used EUs.
3. Evaluation of extracted EUs.
4. Definition of favorites in the choice of the EUs by designers, architects, and developers.
5. Determination of the frequency dynamics of various EUs usage.
6. Prototyping a project based on a set of EUs recorded in KB, etc.

2.1 Literature Review

Works [10–15] noted that using the experience of the previous projects at the initial stages of new project development is crucial. The solution to this problem can be based on intelligent methods and algorithms of design artifacts analysis to build the KB of a software company.

The specifics of the design experience located in the design artifacts of the previous projects require the formation of the KB of a software company with the structure based on an ontology system [16].

Works [17–21] note the importance of using ontological engineering in the design and development of SSs. In works [17,20] describes an approach to using ontologies instead of traditional modeling languages, such as UML, since ontologies allow controlling the logical integrity and consistency of the resulting model.

However, existing methods of forming a KB need experts in the problem area and knowledge engineering. This need leads to significant time costs for the formation of such a KB.

A large number of researchers are working in the field of extracting entities from program code [22–24]. Entities are understood as structural elements of the source code: files, classes, methods, variables, etc. The entities extraction is usually based on the abstract syntax tree analysis with a set of rules. In some cases, machine learning methods and neural networks are used for entity extraction.

In this study, entities are understood as objects of some problem area described in source code classes. Methods (business processes) in which entities are parameters or return value types are also extracted from the source code.

Therefore, the development of models, methods, and algorithms for building the KB of a software company in the automated analysis process of design artifacts is relevant.

The main difficulty of this approach is the necessity to unify design artifacts. Also, necessary to include in the KB information about the various subject areas within which previous projects were developed. The main objective of this research is to study the possibilities and limitations of data analysis and knowledge engineering methods for building KB for automating the process of the architectural design of SS. The methods for extraction of EUs in the KB content analyzing will be considered in subsequent papers.

Let's consider the ontology of the KB of a software company in more detail.

2.2 Domain Ontology

The domain ontology allows adapting the KB of a software company to the points of the subject areas. The domain ontology is the basis for creating a single set of concepts that unifies the entities of various subject areas.

The domain ontology is:

$$O^{dom} = \{O_1^{dom}, O_2^{dom}, \ldots, O_i^{dom}, \ldots, O_n^{dom}\}, \tag{1}$$

where n is the number of subject areas within which previous projects were developed;

O_i^{dom} is i-th ontology containing a description of the i-th subject area:

$$O_i^{dom} = \langle C^{dom}, T^{dom}, R^{dom}, F^{dom} \rangle, \tag{2}$$

where C^{dom} is a set of concepts;
T^{dom} is a set of terms that describe concepts;
F^{dom} is a set of interpretation functions that define a set of relationships R^{dom};
R^{dom} is a set of relationships:

$$R^{dom} = \{R_{CC}^{dom}, R_{CT}^{dom}\},$$

where R_{CC}^{dom} is a set of generalization relations that determine the hierarchy of concepts;
R_{CT}^{dom} is a set of association relationships between concepts and terms describing these concepts.

2.3   Ontology of Design Artifacts

The ontology of design artifacts contains formalized descriptions of design artifacts in the form of KB fragments. The central elements of this ontology are:

- entities are objects of a specific subject area, modeled using a design artifact, for example, a class, table, etc.;
- entity attribute is a property of an entity, such as a class field, table column, etc.;
- constraints of an entity attribute are properties of an entity attribute, for example, the type of a class field, the size of a class field, etc.;
- business processes are processes of a specific subject area, in which some entity is involved, for example, a class method, a use case, etc.

The ontology of design artifacts is:

$$O^{exp} = \langle E^{exp}, A^{exp}, C^{exp}, P^{exp}, R^{exp}, F^{exp} \rangle, \tag{3}$$

where E^{exp} is a set of entities;
A^{exp} is a set of entities attributes;
C^{exp} is a set of constraints on entities attributes;
P^{exp} is a set of business processes which entities are involved;
F^{exp} is a set of interpretation functions that define a set of relationships R^{exp};
R^{exp} is a set of relationships:

$$R^{exp} = \{R_{EE}^{exp}, R_{EA}^{exp}, R_{AC}^{exp}, R_{EP}^{exp}\},$$

where R_{EE}^{exp} is a set of generalization relationships defining a hierarchy of entities;
R_{EA}^{exp} is a set of association relationships defining the relationship between the entity E_i^{exp} and its attribute A_j^{exp};
R_{AC}^{exp} is a set of association relationships defining the relationship between the entity attribute A_i^{exp} and its constraint C_j^{exp};
R_{EP}^{exp} is a set of association relationships defining the relationship between the entity E_i^{exp} and business process P_j^{exp}.

2.4   Knowledge Base Model

The previously considered ontologies form the KB of a software company.
KB model is:

$$O = \langle O^{dom}, O^{exp}, R, F \rangle,$$

where O^{dom} is the ontology for describing various subject areas, which used on previous projects (Eq. 1);
O^{exp} is the ontology for representing design artifacts (Eq. 3). This ontology is formed in the process of analyzing design artifacts;
R is a set of association relations between concepts of domain ontology and design artifacts;
F is a set of interpretation functions that define a set of relationships R.

3   Using Knowledge Base to Architecting of Software Systems

3.1   Formation of KB Fragments in Design Artifacts Analyzing

The 1C: Enterprise 8 platform [25] application (1C application) is used as an example of a design artifact in this paper.

The 1C platform is widely used in CIS countries to automate various types of accounting. 1C applications are developed for each type of accounting.

Thus, the 1C applications are the design artifacts that contain the experience and knowledge of a large number of highly qualified specialists.

A 1C application can be represented as the following expression:

$$App = \{S_1^{Conf}, S_2^{Conf}, \ldots, S_n^{Conf}\},$$

where S_i^{Conf} is an i-th subsystem of a 1C application. The subsystem is used to group application objects within a certain context: "Orders", "Tools", "Employees", etc.

Subsystem S_i^{Conf} can be represented as the following expression:

$$S_i^{Conf} = \langle Name, E^{Conf}, P^{Conf} \rangle, \tag{4}$$

where $Name$ is unique subsystem name;
$E^{Conf} = \langle C^{Conf}, N^{Conf} \rangle$ is a set of application entities. A set of catalogs C^{Conf} and enumerations N^{Conf} are used to represent the entities of the problem area in a 1C application;
$P^{Conf} = \langle D^{Conf}, R^{Conf} \rangle$ is set of application business processes, in which the entities are involved. Set of business processes P^{Conf} contains set of documents D^{Conf} and set of registers R^{Conf}. Documents are used to change the state of the application entities. Registers are used to store the history of entity states. Register entries are formed based on documents data.

Each element of the sets C^{Conf} and D^{Conf} can be represented as the following expression:

$$C_i^{Conf}, D_j^{Conf} = \langle Name, V^{Conf}, T^{Conf} \rangle,$$

where $Name$ is unique element name;
$V^{Conf} = \{V_1^{Conf}, V_2^{Conf}, \ldots, V_m^{Conf}\}$ is a set of element attributes;
$V_k^{Conf} = \langle Name, Type, Constraint \rangle$ is an attribute of an element. $Name$ is attribute name, $Type$ is attribute data type and $Constraint$ is attribute constraints. The attribute data type $Type$ can contain a simple data type and a link to another application object (entity or business-process);
$T^{Conf} = \{T_1^{Conf}, T_2^{Conf}, \ldots, T_m^{Conf}\}$ is a set of application object tables. Each application object can have zero, one or more tables. Each table contains a unique name and set of attributes $T_l^{Conf} = \langle Name, V^{Conf} \rangle$.

Element of the set R^{Conf} represents as:

$$R_i^{Conf} = \langle Name, V^{Conf} \rangle.$$

Element of the set N^{Conf} represents as:

$$N_i^{Conf} = \langle Name \rangle.$$

Table 1 contains correspondence between objects of an i-th subsystem of a 1C application and entities of the ontology of design artifacts O^{exp}.

Table 1. Correspondence between objects of an i-th subsystem of a 1C application and entities of the ontology of design artifacts.

1C application	Ontology O^{exp}
$S_i^{Conf}.Name$	E_i^{exp}
$C_i^{Conf}.Name$	$E_i^{exp}, R_{EE_i}^{exp}$
$N_i^{Conf}.Name$	$E_i^{exp}, R_{EE_i}^{exp}$
$T_i^{Conf}.Name$	$E_i^{exp}, R_{EE_i}^{exp}$
$D_i^{Conf}.Name$	$P_i^{exp}, R_{EE_i}^{exp}$
$R_i^{Conf}.Name$	$P_i^{exp}, R_{EE_i}^{exp}$
$V_i^{Conf}.Name$	$A_i^{exp}, R_{EA_i}^{exp}$
$V_i^{Conf}.Type, V_i^{Conf} \in C_j^{Conf}$	$(C_i^{exp}, R_{EC_i}^{exp})$
$V_i^{Conf}.Type, V_i^{Conf} \in D_j^{Conf}$	$R_{EP_i}^{exp}$
$V_i^{Conf}.Constraint$	$C_i^{exp}, R_{AC_i}^{exp}$

Consider the algorithm for the formation of KB fragments using the example of the analysis of the 1C application of the following structure:

- subsystems: 'Trade'.
- catalogs: 'Products', 'Units', 'Warehouses', 'Suppliers'.
- 'Products' catalog attributes: 'Id' (string), 'Name' (string), 'Unit' (link to the element of 'Units' catalog).
- 'Units' catalog attributes: 'Id' (string), 'Name' (string).

- 'Warehouses' catalog attributes: 'Id' (string), 'Name' (string).
- 'Suppliers' catalog attributes: 'Id' (string), 'Name' (string).
- documents: 'Products supply'.
- 'Products supply' document attributes: 'Warehouse' (link to the element of 'Warehouses' catalog), 'Supplier' (link to the element of 'Suppliers' catalog), 'Products' (table).
- 'Products' table attributes: 'Products' (link to the element of 'Product' catalog), 'Price' (numeric), 'Quantity' (numeric), 'Amount' (numeric).

The algorithm for the formation of KB fragments based on the analysis of a 1C application can be presented as following steps:

1. Form the entity of the ontology of design artifacts for each subsystem of a 1C application:

$$S_i^{Conf} \rightarrow E_i^{exp},$$

for example:

$$\text{Trade} \in E^{exp}.$$

2. Create an entity of the ontology of design artifacts for each entity of the current subsystem, taking into account the relations between them:

$$C^{Conf} \rightarrow \hat{E}^{exp} \in E^{exp}, \hat{R}_{EE}^{exp} \in R_{EE}^{exp},$$

$$N^{Conf} \rightarrow \hat{E}^{exp} \in E^{exp}, \hat{R}_{EE}^{exp} \in R_{EE}^{exp},$$

for example:

$$\text{Products} \in E^{exp}, \text{Units} \in E^{exp}, \text{Warehouses} \in E^{exp}, \text{Suppliers} \in E^{exp},$$

$$\text{Products } R_{EE}^{exp} \text{ Trade}, \ldots, \text{Suppliers } R_{EE}^{exp} \text{ Trade}.$$

3. If the current entity of a 1C application is a catalog, then create a set of attributes, attribute constraints, and relationships for this entity of ontology:

$$V^{Conf} \in C_i^{Conf} \rightarrow \hat{A}^{exp} \in A^{exp}, \hat{R}_{EA}^{exp} \in R_{EA}^{exp},$$

$$V^{Conf} \in C_i^{Conf} \rightarrow \hat{C}^{exp} \in C^{exp}, \hat{R}_{EC}^{exp} \in R_{EC}^{exp},$$

for example:

$$\text{Id} \in A^{exp}, \text{Name} \in A^{exp}, \text{Unit} \in A^{exp},$$

$$\text{Products } R_{EA}^{exp} \text{ Id}, \text{Products } R_{EA}^{exp} \text{ Name}, \text{Products } R_{EA}^{exp} \text{ Unit}, \ldots,$$

$$\text{Suppliers } R_{EA}^{exp} \text{ Id}, \text{Suppliers } R_{EA}^{exp} \text{Name},$$

$$\text{string} \in C^{exp}, \text{numeric} \in C^{exp}, \text{link} \in C^{exp},$$

$$\text{Id } R_{AC}^{exp} \text{ string}, \text{Name } R_{AC}^{exp} \text{ string}, \text{Unit } R_{AC}^{exp} \text{ link}.$$

4. If the current entity of a 1C application has tables (is catalog), then, form a set of attributes, constraints of attributes, and relations for this entity of ontology for each table:

$$T^{Conf} \in C_i^{Conf} \rightarrow \hat{E}^{exp} \in E^{exp}, \hat{R}_{EE}^{exp} \in R_{EE}^{exp},$$

$$V^{Conf} \in T_j^{Conf}, T_j^{Conf} \in C_i^{Conf} \rightarrow \hat{A}^{exp} \in A^{exp}, \hat{R}_{EA}^{exp} \in R_{EA}^{exp},$$

$$V^{Conf} \in T_j^{Conf}, T_j^{Conf} \in C_i^{Conf} \rightarrow \hat{C}^{exp} \in C^{exp}, \hat{R}_{EC}^{exp} \in R_{EC}^{exp}.$$

5. Create a business process of ontology for each business process of the current subsystem, taking into account the connection between them:

$$D^{Conf} \rightarrow \hat{P}^{exp} \in P^{exp}, \hat{R}_{EP}^{exp} \in R_{EP}^{exp},$$

$$R^{Conf} \rightarrow \hat{P}^{exp} \in E^{exp}, \hat{R}_{EP}^{exp} \in R_{EP}^{exp},$$

for example:

$$\text{Products supply} \in P^{exp}, \text{Products supply } R_{EE}^{exp} \text{ Trade.}$$

6. Create a set of relationships for the current business process and the entities of the ontology based on the analysis of the attributes of the current business process of a 1C application:

$$V^{Conf} \in D_i^{Conf} \rightarrow \hat{R}_{EP}^{exp} \in R_{EP}^{exp},$$

$$V^{Conf} \in R_i^{Conf} \rightarrow \hat{R}_{EP}^{exp} \in R_{EP}^{exp},$$

for example:

$$\text{Suppliers } R_{EP}^{exp} \text{ Products supply}, \text{Warehouses } R_{EP}^{exp} \text{ Products supply.}$$

7. If the current business process of a 1C application has tables (is a document), then form a set of relationships to connect the current business process and entities of ontology for each table:

$$V^{Conf} \in T_j^{Conf}, T_j^{Conf} \in D_i^{Conf} \rightarrow \hat{R}_{EP}^{exp} \in R_{EP}^{exp},$$

for example:

$$\text{Products } R_{EP}^{exp} \text{ Products supply.}$$

Also, the following domain ontology is formed for the above 1C application:

$$\text{Catalog} \in C^{dom}, \text{Edit information about} \in T^{dom},$$

$$\text{Catalog } R_{CT}^{dom} \text{ Edit information about,}$$

$$\text{Document} \in C^{dom}, \text{Create document for} \in T^{dom},$$

$$\text{Document } R_{CT}^{dom} \text{ Create document for,}$$

3.2   Generating Use Case Diagrams Based on the Knowledge Base Content

A use case diagram in UML notation can be represented as the following expression:

$$UML = \langle A^{UML}, P^{UML}, R^{UML} \rangle, \tag{5}$$

where A^{UML} is a set of actors;
P^{UML} is a set of use cases;
R^{UML} is a set of relationships:

$$R^{UML} = \{R_C^{UML}, R_I^{UML}, R_E^{UML}\},$$

where R_C^{UML} is a set of generalization relationships;
R_I^{UML} is a set of inclusion relationships;
R_E^{UML} is a set of expansion relationships.

Thus, to generate of use case diagram in UML notation based on the content of the KB is need to transform the axioms of ontology O^{exp} (Eq. 3) into use case diagram UML elements (Eq. 5).

Table 2 contains correspondence between entities of ontology O^{exp} and the use case diagram structural components.

Table 2. Correspondence between entities of the ontology of design artifacts and the use case diagram structural components.

Ontology	Use-case diagram
Business process P_i^{exp}	Use case P_i^{UML}
Business processes $\hat{H}^P$	Inclusion relationship R_I^{UML}

As you can see from Table 2, the expansion and generalization relations are not formed. Additional work is required to improve the methods for generating and analyzing the content of the ontology O^{exp} in use UML case diagram creation.

The algorithm for generating the use cases diagram in UML notation based on the content of the KB consists of the following steps:

1. Selecting business processes of ontology as the use case diagram basis. For example, Products supply $\in P^{exp}$.
2. Analysis of the relationships R_{EP}^{exp} between selected business processes $\hat{P}^{exp}$ and other concepts $\hat{E}^{exp}$ of ontology to create a list of entities L^E for use case diagram generation. For example, Products $\in E^{exp}$, Units $\in E^{exp}$, Warehouses $\in E^{exp}$, Suppliers $\in E^{exp}$.
3. Formation of a hierarchy of entities H^E from the list of entities L^E based on analysis of relationships R_{EE}^{exp} of ontology. For example, Products R_{EE}^{exp} Trade, ..., Suppliers R_{EE}^{exp} Trade.

4. Replace the entities of the resulting hierarchy H^E with related business processes H^P if such a business process is not present in the diagram.
5. Form a use case P_i^{UML} for each node. Create inclusion relations R_I^{UML} between use cases.
6. Create an actor and link it to the root use case.

The set of commands is generated for the PlantUML system [20] after executing this algorithm. For example:

@startuml
* :user: − (Trade)*
* (Trade)..>(Products supply):include*
* (Products supply)..>(Edit information about Suppliers):include*
* (Products supply)..>(Edit information about Warehouses):include*
* (Products supply)..>(Edit information about Products):include*
* (Products supply)..>(Create document for Products supply):include*
@enduml

Also, the domain ontology considered above is used in the process of generating a use case diagram.

The resulting use-case diagram is shown in Fig. 1.

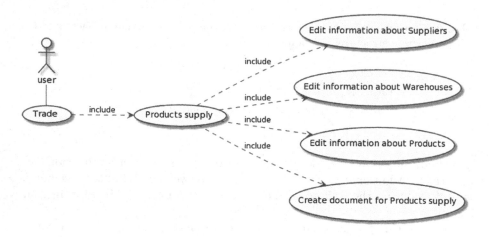

Fig. 1. The example of a generated use case diagram

4   Experiments Results

4.1   Plan of Experiments

A series of experiments were carried out to evaluate the quality of the proposed in this paper methods and models.

The experiments plan includes:

1. Formation of the KB in the form of OWL ontology in the process of analyzing a 1C application.
2. Validation of the obtained OWL ontology for correctness using the reasoner in Protege [27].
3. Generation of a set of PlantUML system command to generate use case diagrams in UML notation.
4. Evaluating the quality of generated design artifacts by comparing them with similar design artifacts created by an expert.

The following metrics were used to evaluate the quality of generated design artifacts:

1. Precision:
$$Precision = \frac{|Gen \cap Exp|}{|Exp|},$$

where $|Gen \cap Exp|$ is the number of matching elements in the generated design artifact and the design artifact created by an expert;
$|Exp|$ is the number of elements in the design artifact created by an expert.
2. Recall:
$$Recall = \frac{|Gen - Exp|}{|Gen|},$$

where $|Gen - Exp|$ is the number of elements in the generated design artifact that are missing in expert design artifact;
$|Gen|$ is the number of elements in the generated design artifact.

Precision and recall metrics used in evaluating the quality of design artifacts generated using the proposed approach are not similar to quality metrics of information retrieval. Precision and recall are not mutually exclusive metrics.

4.2 Evaluation of the Quality of the Generated Use Case Diagram in UML Notation

The "Trade Management 11.2" 1C application was used in this experiment as the design artifact to form the KB content.

The following objects from subsystem "Discrepancies" of application for the "Trade Management 11.2" 1C application were selected:

- subsystems "Purchase" (parent subsystem) and "Discrepancies";
- accumulation register "ShipmentSchedules";
- catalogs "Organizations", "Currencies", "Partners", "Counteragents", "Users", "Nomenclature", "Appointments", "Warehouses";
- Documents "DivergenceActAfterProductsMovement", "PostAcceptanceDiscrepancyAct", "PostDispatchAct".

Verification of the generated OWL ontology in the Protege editor was successful.

Also, the domain ontology O^{dom} (Eq. 1 and 2) was created. The ontology O^{dom} contains alternative names for objects of the "Trade Management 11.2" 1C application used as names of use cases. For example, for the concept "Purchase" set alternative name "Purchase Products", for the concept "Warehouses" set alternative name "Get info about warehouses", for the concept "DivergenceActAfterProductsMovement" – "Get divergence act after products movement".

The generated use case diagram (Fig. 2) contains 14 use cases. Nine use cases are not contained in the diagram created by the expert. The number of use cases in the diagram created by the expert (Fig. 3) is 5. Thus, values of quality metrics are: $Precision = 5/5 = 1$, and $Recall = 9/14 = 0.643$.

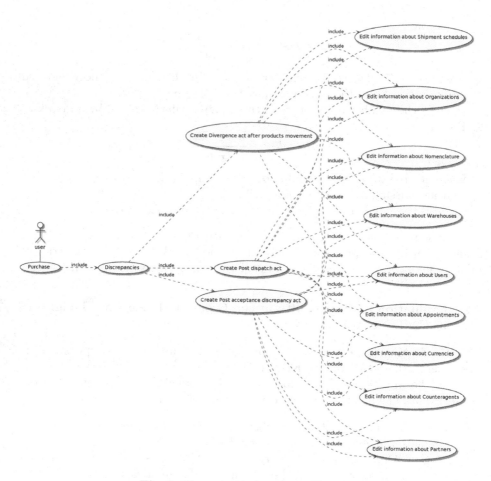

Fig. 2. The generated use case diagram

The time spent on the automatic generation of the use cases diagram was 3 min (without domain ontology O^{dom} formation time). An expert spent 17 min on a similar task.

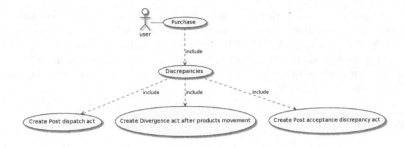

Fig. 3. The use case diagram built by the expert

Table 3 contains the results of experiments on the formation of use case diagrams in UML notation based on the KB content. KB was formed based on the analysis of various 1C applications.

Table 3. Experimental results.

1C application	Subsystem	The number of use cases			Precision	Recall
		Generated	Created by expert	Matching elements		
Trade management 11.2	Discrepancies	14	5	9	1	0.643
Salary and HR management 3.1	Personnel	8	4	4	1	0.5
Complex automation 2	Warehouse	12	6	4	0.667	0.667
Trade management 11.2	Bank and cash desk	13	6	5	0.833	0.615

The experimental results show that the generated design artifacts have a high precision rate and contain more structural elements, which increases the rate of recall. The high value of recall is related to a set of connections between the objects of the analyzed 1C application obtained during their life cycle.

5   Conclusion

This paper presents an approach to automating the architectural design process of a SS based on the KB of a software company. The KB contains ontologies that store knowledge about points of the subject area and successful design and architectural solutions extracted from previous projects. The new model of the KB of a software company is presented.

An application for the 1C platform is used in this paper as an example of a design artifact.

The paper also presents the experiments to evaluate the adequacy of the proposed models and methods by comparing the precision and recall values indicators for generated design artifacts and design artifacts created by the expert.

Acknowledgments. The authors acknowledge that the work was supported by the framework of the state task of the Ministry of Science and Higher Education of the Russian Federation No. 075-00233-20-05 "Research of intelligent predictive multimodal analysis of big data, and the extraction of knowledge from different sources".

The reported study was funded by RFBR and Ulyanovsk region, project numbers 19-47-730005 and 19-47-730006.

References

1. ISO/IEC/IEEE 42010:2011 Systems and Software Engineering - Architecture Description. http://docs.cntd.ru/document/1200139542. Accessed 22 Apr 2021
2. Van Heesch, U., Avgeriou, P., Hilliard, R.: A documentation framework for architecture decisions. J. Syst. Softw. **85**(4), 795–820 (2012). https://doi.org/10.1016/j.jss.2011.10.017
3. Hilliard, R.: Using aspects in architectural description. In: Moreira, A., Grundy, J. (eds.) EAW 2007. LNCS, vol. 4765, pp. 139–154. Springer, Heidelberg (2007). https://doi.org/10.1007/978-3-540-76811-1_8
4. Sosnin, P.: Substantially evolutionary theorizing in designing software-intensive systems. Information **9**(4), 91 (2018). https://doi.org/10.3390/info9040091
5. Sosnin, P., Pushkareva, A.: Ontological controlling the lexical items in conceptual solution of project tasks. In: Gervasi, O., et al. (eds.) ICCSA 2017. LNCS, vol. 10409, pp. 31–46. Springer, Cham (2017). https://doi.org/10.1007/978-3-319-62407-5_3
6. McConnell, S.: Code Complete. Pearson Education (2004)
7. Novák, V., Perfilieva, I., Jarushkina, N.G.: A general methodology for managerial decision making using intelligent techniques. In: Rakus-Andersson, E., Yager, R.R., Ichalkaranje, N., Jain, L.C. (eds.) Recent Advances in Decision Making. Studies in Computational Intelligence, vol. 222. Springer, Heidelberg (2009). https://doi.org/10.1007/978-3-642-02187-9_7
8. Yarushkina, N., Moshkin, V., Andreev, I., Klein, V., Beksaeva, E.: Hybridization of fuzzy inference and self-learning fuzzy ontology-based semantic data analysis. In: Abraham, A., Kovalev, S., Tarassov, V., Snášel, V. (eds.) Proceedings of the 1st International Scientific Conference on Intelligent Information Technologies for Industry (IITI'16). AISC, vol. 450, pp. 277–285. Springer, Cham (2016). https://doi.org/10.1007/978-3-319-33609-1_25
9. Yarushkina, N., Filippov, A., Moshkin, V.: Development of the unified technological platform for constructing the domain knowledge base through the context analysis. In: Kravets, A., Shcherbakov, M., Kultsova, M., Groumpos, P. (eds.) CIT&DS 2017. CCIS, vol. 754, pp. 62–72. Springer, Cham (2017). https://doi.org/10.1007/978-3-319-65551-2_5
10. Sosnin, P.: A way for creating and using a theory of a project in designing of a software intensive system. In: 17th International Conference on Computational Science and Its Applications (ICCSA). IEEE (2017). https://doi.org/10.1109/iccsa.2017.7999646
11. Sosnin, P.: Conceptual experiments in automated designing. In: Projective Processes and Neuroscience in Art and Design, pp. 155–181. IGI Global (2017). https://doi.org/10.4018/978-1-5225-0510-5.ch010
12. Henninger, S.: Tool support for experience-based software development methodologies. In: Advances in Computers, pp. 29–82. Elsevier (2003). https://doi.org/10.1016/s0065-2458(03)59002-7

13. Abioye, T.E., Arogundade, O.T., Misra, S., Akinwale, A.T., Adeniran, O.J.: Toward ontology-based risk management framework for software projects: an empirical study. J. Softw. Evol. Process **32**(12), e2269 (2020)
14. Júnior, A.A.C., Misra, S., Soares, M.S.: A systematic mapping study on software architectures description based on ISO/IEC/IEEE 42010:2011. In: Misra, S., et al. (eds.) ICCSA 2019. LNCS, vol. 11623, pp. 17–30. Springer, Cham (2019). https://doi.org/10.1007/978-3-030-24308-1_2
15. Oluwamayowa, A., Adedeji, A., Sanjay, M., Faith, A.: Empirical framework for tackling recurring project management challenges using knowledge management mechanisms. In: Gervasi, O., et al. (eds.) ICCSA 2020. LNCS, vol. 12254, pp. 954–967. Springer, Cham (2020). https://doi.org/10.1007/978-3-030-58817-5_67
16. Shaaban, A.M., Gruber, T., Schmittner, C.: Ontology-based security tool for critical cyber-physical systems. In: Proceedings of the 23rd International Systems and Software Product Line Conference 2019, pp. 207–210. ACM Press (2019). https://doi.org/10.1145/3307630.3342397
17. Bhatia, M.P.S., Kumar, A., Beniwal, R.: Ontologies for software engineering: past, present and future. Indian J. Sci. Technol. **9**(9), 1–16 (2016). https://doi.org/10.17485/ijst/2016/v9i9/71384
18. Falbo, R.A., et al.: An ontology pattern language for service modeling. In: Proceedings of the 31st Annual ACM Symposium on Applied Computing 2016, pp. 321–326. ACM Press (2016). https://doi.org/10.1145/2851613.2851840
19. Ilyas, Q.M.: Ontology augmented software engineering. In: Software Development Techniques for Constructive Information Systems Design, pp. 406–413. IGI Global (2013). https://doi.org/10.4018/978-1-4666-3679-8.ch023
20. Isotani, S., Ibert Bittencourt, I., Francine Barbosa, E., Dermeval, D., Oscar Araujo Paiva, R.: Ontology driven software engineering: a review of challenges and opportunities. IEEE Lat. Am. Trans. **13**(3), 863–869 (2015). https://doi.org/10.1109/tla.2015.7069116
21. Pan, J.Z., Staab, S., Aßmann, U., Ebert, J., Zhao, Y.: Ontology-Driven Software Development. Springer, Heidelberg (2013). https://doi.org/10.1007/978-3-642-31226-7
22. Godfrey, M.W., Zou, L.: Using origin analysis to detect merging and splitting of source code entities. IEEE Trans. Softw. Eng. **31**(2), 166–181 (2005)
23. Ali, N., Sharafi, Z., Guéhéneuc, Y.-G., Antoniol, G.: An empirical study on the importance of source code entities for requirements traceability. Empir. Softw. Eng. **20**(2), 442–478 (2014). https://doi.org/10.1007/s10664-014-9315-y
24. Savić, M., Rakić, G., Budimac, Z., Ivanović, M.: A language-independent approach to the extraction of dependencies between source code entities. Inf. Softw. Technol. **56**(10), 1268–1288 (2014)
25. What is 1C:Enterprise? Source, https://1c-dn.com/1c_enterprise/what_is_1c_enterprise/. Accessed 22 Apr 2021
26. PlantUML. UML Diagram Generator. https://plantuml.com. Accessed 22 Apr 2021
27. Protege. Free, open-source ontology editor and framework for building intelligent systems. https://protege.stanford.edu. Accessed 22 Apr 2021

Human Behavior Prediction for Risk Analysis

Adam Szekeres$^{(\boxtimes)}$ ⓘ and Einar Snekkenes ⓘ

Department of Information Security and Communication Technology,
Norwegian University of Science and Technology - NTNU,
Gjøvik, Norway
{adam.szekeres,einar.snekkenes}@ntnu.no

Abstract. The Conflicting Incentives Risk Analysis (CIRA) method makes predictions about human decisions to characterize risks within the domain of information security. Since traditional behavior prediction approaches utilizing personal features achieve low prediction accuracies in general, there is a need for improving predictive capabilities. Therefore, the primary objective of this study is to propose and test a psychological approach for behavior prediction, which utilizes features of situations to achieve improved predictive accuracy. An online questionnaire was used for collecting behavioral and trait data to enable a comparison of approaches. Results show that the proposed behavior prediction approach outperforms the traditional approach across a range of decisions. Additionally, interrater reliabilities are analyzed to estimate the extent of objectivity in situation evaluations, providing an indication about the potential performance of the approach when a risk analyst needs to rely on unobtrusive assessment of action-desirability.

Keywords: Risk analysis · Information security · Human motivation · Behavior prediction · Perception of situations

1 Introduction

Risk management aims to make predictions about potential future events, assess their consequences and mitigate undesirable outcomes by implementing appropriate controls. Within the Conflicting Incentives Risk Analysis (CIRA) method [25] risks result from conscious human decisions made by individuals. A decision-maker needs to consider how costs and benefits are allocated among the parties affected by the transaction: the decision-maker him/herself, other stakeholders, various groups and organizations. In the simplest case a decision has consequences only for the decision-maker. More often a personal choice impacts other stakeholders such as other individuals or an organization. In certain situations known as **threat risks** within CIRA personal benefits result in losses for other parties: Smart electricity meters have been reprogrammed for a fee by the employees of utility companies generating great financial losses for the organization [15]. Distribution system operators in the energy sector have to develop policies which balance the social costs of supply interruptions when creating policies for dealing with supply-demand imbalances [3]. Sexual harassment at workplaces causes trauma for the victims along with direct organizational costs (e.g. turnover and recruitment, investigating the complaint and legal penalties, damage to reputation paid by

© Springer Nature Switzerland AG 2021
O. Gervasi et al. (Eds.): ICCSA 2021, LNCS 12951, pp. 288–303, 2021.
https://doi.org/10.1007/978-3-030-86970-0_21

the organization) [20]. In another class of situations, a personal cost (i.e. time, money, resources, freedom, etc.) has to be borne by an individual to provide a benefit for others. The lack of perceived incentives to act in a desirable manner gives rise to **opportunity risks** within CIRA: whistleblowers motivated by moral concerns take a significant personal risk for the benefit of society to uncover questionable practices within their organization [4]. An individual developing a web-service to provide free access to pay-walled materials for everyone, risks personal freedom while publishers are faced with lost revenue [9]. Healthcare professionals have to deal with several costs (e.g. compliance with local and national regulations, delays in approval, complicated trial processes) to include patients in clinical trials for their benefit [13]. Spreading of malware can be prevented by a single individual sacrificing some resources (i.e. CPU, memory, storage) for the benefit of others. Risk management is becoming increasingly important as more and more domains of life become digitized and the number of people in interdependent relationships increases. Since organizational practices and legal systems identify individuals as responsible, accountable and punishable subjects for their actions [23,30], it is important to investigate how risks resulting from human decision-makers can be predicted at the individual level for the purpose of improving the CIRA method.

1.1 Problem Statement and Research Questions

To date, there is a lack of empirical tests investigating how a novel behavior prediction method performs within the CIRA method which takes into account differences in perception of situations among decision-makers. Thus, this study aims to empirically investigate the utility of a behavior prediction approach for improving the CIRA method, based on the following research questions: **RQ 1:** To what extent can predictive accuracy increase using an approach which takes into account differences in the perception of situational influences among decision-makers compared to a traditional approach utilizing personal features only? **RQ 2:** To what extent is the proposed method feasible when the situation perceptions need to be estimated unobtrusively by a risk analyst?

The paper is organized as follows: Sect. 2 presents existing results relevant to the paper's topics. Section 3 presents the characteristics of the sample, the instruments and procedures used for data collection as well as data processing steps. Results are presented in Sect. 4. A discussion of the results, their implications and the limitations of the study are discussed in Sect. 5. The key findings are summarized in Sect. 6 along with directions for further work.

2 Related Work

This section presents the risk analysis method under development; several types of risks in information security arising from conscious human behavior; and research results on the perception of situations providing the foundation for the proposed prediction approach.

2.1   Human Motivation at the Centre of Risk Analysis

The Conflicting Incentives Risk Analysis (CIRA) method developed within the domain of information security and privacy focuses on stakeholder motivation to characterize risks. The game-theoretic framework identifies two classes of stakeholders in an inter-dependent relationship [25]: **Strategy owners** are in position to select a certain strategy to implement an action based on the desirability of the available options, while **risk owners** are exposed to the actions or inactions of the strategy owners in a one shot game setting. Misaligned incentives can give rise to two types of risks: *threat risk* refers to undesirable outcomes for the risk owner, when the strategy owner selects a strategy based on its perceived desirability; *opportunity risk* refers to desirable outcomes for the risk owner, which fail to get realized, since the strategy owner has no incentives to act or would have to bear losses in order to provide a benefit for the risk owner. Motivation is represented by the expected change in overall utility after the execution of a strategy. The combination of several utility factors (i.e. personal attributes) con-tribute to each stakeholders' overall utility, and stakeholders are assumed to be utility maximisers. The method relies on the unobtrusive assessment of the strategy owner's relevant utility factors by a risk analyst, which is necessary since strategy owners are assumed to be adversarial (reluctant to interact with an analyst) and may be tempted to subvert the analysis by reporting false information. Previous work used the theory of basic human values (BHV) to operationalize a set of psychologically important util-ity factors [35], and investigated the extent to which observable features can be used for constructing the motivational profiles of strategy owners [37]. The theory of BHVs [29] distinguishes between 10 motivationally distinct desirable end-goals which form four higher dimensions as shown in Fig. 1. The trade-off among opposing values drives decision-making. Using a subject's profile information, the analyst makes a subjective assessment about action-desirability from the perspective of the strategy owner to pre-dict future behavior [33].

Fig. 1. Basic human value structure based on [29].

2.2 Risks Arising from Human Behavior Within Information Security

Insiders are trusted individuals who break information security policies deliberately. The purpose of a policy is to prescribe expected behaviors and to specify the consequences of undesirable behavior. Rule-breaking can be motivated by a variety of reasons (e.g. financial gain, curiosity, ideology, political, revenge) [19]. A recent systematic literature review revealed that most of the insider threat prediction applications focus on patterns of online activity as key features, whereas psychological approaches are less prevalent [6]. Some empirical work focuses on the psychological attributes of insiders, identifying and assessing personal or behavioral characteristics which are valid and reliable indicators of potential incidents. The insider threat prediction model prepared for the U.S. Department of Energy identifies 12 unique psychosocial behavioral indicators (e.g. disgruntlement, stress, absenteeism, etc.) that may be indicative of an insider threat [8]. The U.S. Department of Homeland Security [22] advises organizations to focus on a combination of personal attributes (e.g. introversion, greed, lack of empathy, narcissism, ethical flexibility) and behavioral indicators (e.g. using remote access, interest in matters outside of scope of duties, risk-taking behavior, etc.) to detect insider threats. Some of the most useful personal features for predicting insider threats are personality traits (e.g. Big Five, Dark Triad traits, sensation-seeking, etc.), emotions (e.g. hostility, anger) and mental disorders (e.g. paranoia, depression, etc.) [6].

Organizations are exposed to risks arising from negligent security practices by employees. Several empirical studies investigate conscious security-related behaviors using the Theory of Planned Behavior (TPB), which is one of the most widely used behavior prediction framework in psychology [1]. Using the theory (non-) compliant behavior can be predicted from the combination of the factors, and interventions can be targeted at specific factors to motivate desirable behavior. Most of the results included in [17] and [2] showed that all main constructs of TPB have significant associations with behavioral intentions, providing support for the model's utility. However, it is important to note that most studies measure behavioral intentions, as opposed to actual behavior. Prediction accuracy is generally measured by the R^2 metric averaging around $R^2 = 0.42$ across studies for intentions; but when actual behavior is measured accuracy may decrease to as low as $R^2 = 0.1$ [31], indicating that new models are needed to improve predictive capabilities.

A large number of risks are attributed to malicious external stakeholders. A taxonomy which categorizes hackers according to their properties (motivations, capabilities, triggers, methods) can be used for defense planning and forensic investigations. Key psychologically relevant motivations included in the taxonomy are: revenge, curiosity, financial gain, and fame [10].

2.3 Perception of Situations

The immediate situation in which decision-making takes place has important influence on behavior. It has been suggested, that behavior is a function of personal and situational attributes; known as the person-situation interactionist approach [18] and that the subject's perception of the situation determines their actions [14]. A simple process model developed in [26] explains how subjects perceive and process situational information and generate different behavioral outcomes. Cues of the situation (e.g. persons, objects) are perceived by the subjects and processed according to their specific stable and variable personal features (e.g. traits, roles, mental states, etc.), giving rise to the unique experience of the situation. The actions selected are dependent on the experience, and two persons' action will match to the extent that they attend to the same cues of the environment, and process the cues similarly due to their similarities in terms of personal features. The literature of situations still lacks consensus on how to conceptualize, define and measure situations. This is mainly attributed to the complex and multifaceted nature of situations [21]. Attempts have been made to address this gap by developing taxonomies of situational features as perceived by the individual in the situation, assuming that behavioral incentives are subjective rather than objective [38]. Since the Dark Triad traits (narcissism, Machiavellianism, psychopathy) are frequently associated with harmful workplace behavior [34], a taxonomy was developed which identifies situational triggers contributing to the behavioral tendencies associated with these traits [24], which can be used for the development of situational interventions to mitigate risks. While descriptive situation taxonomies exist, applications for the purpose of behavior prediction are lacking.

3   Materials and Methods

This study used an online questionnaire for collecting data from subjects. Two types of behavioral data was collected using dilemmas which operationalized threat and opportunity risks: **subjective ratings** capturing perception of situations; and **explicit choices** among the two options of each dilemma. Additionally, the following personal features were collected: basic demographic information and motivational profiles operationalized as BHVs. Sections were presented in the following order to maximize the number of questions between the two behavioral tasks:

1. Perception of situations (i.e. subjective ratings of dilemma-options).
2. Basic demographic information and motivational profiles (i.e. BHVs).
3. Choice between two options of each dilemma.

The questionnaire was completely anonymous, no personally identifiable information was collected, participants were required to express consent before starting the questionnaire. The questionnaire was implemented in Limesurvey and was hosted on servers provided by the Norwegian University of Science and Technology (NTNU).

3.1   Sample

Relying on the sample size recommendations for logistic regression analyses, data collection aimed at a minimum of 50 fully completed questionnaires [39]. First, a randomly selected sample of university students received an e-mail invitation to fill the online questionnaire, generating 22 fully completed surveys. Next, 40 additional respondents were recruited through the Amazon Mechanical Turk (MTurk) online workplace, where subjects receive payment for completing various human intelligence tasks (HITs). Each respondent who completed the questionnaire received $4 net compensation through the MTurk system, which equals to an hourly rate of $12–16. In addition to the higher-than average payment [12], additional options were selected to ensure data quality: the survey was available only for MTurk workers with a HIT Approval Rate greater than 90%, and only to Masters (using MTurk's quality assurance mechanism). Questionnaires below 9 min of completion time were entirely removed to increase the quality of the dataset. Thus, the final convenience sample contains data from 59 respondents (27 females and 32 males) with a mean age of 34 years (S.D. = 10.44). Citizenship of the respondents is as follows: 53% U.S., 25% Norway, 14% India, 8 % other. Most respondents have bachelor's degree (46%), followed by a completed upper secondary education (36%), master's degree (17%) and lower secondary education (2%).

3.2 Behavioral Data

Dilemmas. The stimuli for collecting two types of behavioral data (i.e. perception of situations and explicit choices) were dilemmas which operationalized threat risks as moral dilemmas and opportunity risks as altruistic dilemmas, based on a taxonomy of situations [36]. The dilemmas are based on real events and cover a broad range of behaviors resulting in threat or opportunity risks to the specified risk owners to increase the ecological validity of the stimuli. Dilemmas were presented as riskless choices (i.e. consequences are specified with certainty). Table 1 provides a summary about the theme of the nine dilemmas used in the questionnaire and the first dilemma is presented in full detail in the Appendix due to space limitations. Each dilemma comprised of a story (setting the context as shown in Item 1), and two mutually exclusive options as presented in Item 2 and Item 3 in the Appendix. Respondents answered the questions as strategy

Table 1. Short description of the main theme of the dilemmas included in the survey.

No	Theme of dilemmas
1	Kill an injured person to save rest of crew?*
2	Approach employees with a sexual offer looking for promotion?
3	Distribute electricity to residents instead of hospital during electricity crisis?
4	Reprogram customer's Smart Meters for a personal gain against the rules?
5	Inform customers of your employer about security issues identified in your products despite sure prosecution?
6	Include a patient in clinical trial despite significant personal costs?
7	Create paywall bypassing website to make research results freely available at the expense of personal freedom?
8	Sacrifice personal resources to run a virus to protect colleagues?
9	Accept Firm B's offer? Firm A: $100.000 salary + 14 days holiday vs. Firm B: $50.000 salary + 16 days holiday.*

Note. Dilemmas marked with * were taken from [7]

owners. The dilemmas were used at the beginning of the questionnaire to collect subjective ratings about the perception of situations, and at the end of the questionnaire to collect explicit choices from participants (explained below).

Perception of Situations. To capture the perceived losses/benefits from a particular choice, subjects were required to rate separately both options of all dilemmas (see Item 2 and Item 3) along five dimensions as shown in Fig. 4 in the Appendix. Ratings were collected by continuous sliding scales ranging from negative 100 through 0 to positive 100, with textual anchor labels at the two endpoints and at the mid-point of the scale (i.e. -100: Maximum possible decrease; 0: No impact; 100: Maximum possible increase). For each option an overall utility score was calculated as follows: $U_{total} = U_{socialstatus} + U_{careforothers} + U_{excitement} + U_{stability} + U_{pleasure}$, using an unweighted version of the multi-attribute utility theory (MAUT) [5] implemented in CIRA [25]. This step enables the comparison of the desirability of both options and provides a basis for checking the internal consistency of choices. A choice was considered internally consistent when the explicitly chosen option (in section 3 of questionnaire) received a higher utility score than the unchosen option of the same dilemma based on ratings (section 1 of the questionnaire). This metric gives an indication of data validity (i.e. whether respondents were following instructions) and subject rationality (i.e. making choices according to stated preferences).

Choice Between Dilemma-Options. Figure 5 in the Appendix shows an example of the final section of the questionnaire, requiring participants to make an explicit choice between two dilemma options as strategy owners. The proposed prediction method uses the subjective ratings (section 1 of questionnaire) corresponding to the chosen option (section 3 of questionnaire) for each subject for all dilemmas. For example, if a subject selected *option NO* in the choice task on *dilemma_1*, the ratings for *option NO* of *dilemma_1* were used as predictors, even if the total utility was higher for *option YES* of *dilemma_1*.

3.3 Personal Features: Demographics and Motivational Profile

Basic demographic data included age, sex, nationality, level of education. Individual motivational profiles were created using the Portray Value Questionnaire (PVQ-21), which is a 21-item questionnaire designed for self-assessment [28]. The instrument captures ten BHVs, which were computed according to the instructions in [28]. Five higher dimensions were created by computing the mean of the corresponding values as follows: *Self-enhancement:* power and achievement, *Self-transcendence:* universalism and benevolence, *Openness to change:* self-direction and stimulation, *Conservation:* security, tradition, conformity, while *Hedonism* was treated as a separate dimension, since "hedonism shares elements of both Openness and Self-Enhancement" [27].

4 Results

All analyses were conducted in IBM SPSS 25. Figure 2 presents descriptive statistics for all dilemmas. Red bars indicate the percentage of affirmative choices provided by subjects in the third section of the questionnaire. Blue bars represent the percentage of internally consistent choices. Dilemma 9 was included as a control question. It received a high number of internally consistent (signifying that evaluations provided by subjects were valid), correct responses (selecting the option with higher utility), indicating that most of the respondents were following instructions properly, thus results can be considered valid.

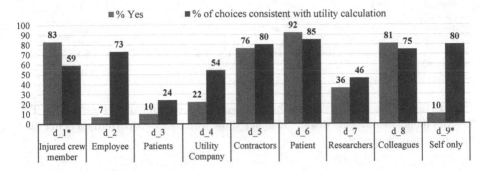

Fig. 2. Descriptive statistics for nine dilemmas. Percentage of affirmative choices and percentage of choices consistent with utility calculations derived from subjective situation ratings across dilemmas. Risk owners are identified under the dilemma numbers. The YES options refer to threat risks realized and opportunity risks avoided for the risk owners. Threat risks: d_1–d_4; Opportunity risks: d_5–d_8. A rational dilemma (d_9) was included as a control question. Dilemmas marked with * were taken from [7].

4.1   RQ 1: Comparison of Prediction Approaches

In order to assess the differences between the traditional (i.e. utilizing personal features only) and the proposed (i.e. utilizing perception of situations) approaches two separate sets of analyses were conducted. First, nine binary logistic regression models were built (one for each dilemma) using the motivational profiles as independent variables and explicit choices as dependent variables. The second set of analyses aimed at exploring the extent of potential improvements that can be expected when subjective perceptions of situations are used to predict the same choices. In the first case five of the nine predictive models were significantly better than the intercept-only models. In case of the proposed approach seven predictive models were significantly better than the intercept-only models. Table 2 presents a summary of all the logistic regression models' predictive performance. The proposed approach ("Subjective ratings of situation" columns) outperformed the traditional approach ("Personal features only" columns) in seven cases out of nine total cases, according to all performance metrics (percentage of correct classifications and the Nagelkerke R^2 - total variance explained).

Table 3 and Table 4 in Appendix presents the details for each model with the regression coefficients and corresponding tests of significance for each predictor. Predictive performance for each model is assessed by two variants of the R^2 metric: Cox & Snell

Table 2. Comparison of the two approaches for predicting identical outcomes.

	% of overall correct classification		% of variance explained (Nagelkerke's R^2)	
	Personal features only	Subjective ratings of situation	Personal features only	Subjective ratings of situation
Dilemma_1	86.4	**89.8***	30	**50***
Dilemma_2	93.2	91.5	44	41
Dilemma_3	86.4	**94.9***	36	**58***
Dilemma_4	74.6	**81.4***	24	**50***
Dilemma_5	76.3	**91.5***	11	**70***
Dilemma_6	89.8	93.2	28	31
Dilemma_7	71.2	**91.5***	29	**76***
Dilemma_8	81.4	**86.4***	30	**45***
Dilemma_9	91.5	**93.2***	35	**40***

Note. * improvement in predictive accuracy from traditional approach for models which were significantly better than the intercept-only model

R^2 and Nagelkerke R^2, measuring the total variances explained by the models. The "Overall model evaluation" row of the tables show which models performed better than the intercept-only models.

4.2   RQ 2: Practical Feasibility of the Proposed Predictive Approach

The second research question is concerned with exploring the potential accuracy which can be expected when ratings must be assessed unobtrusively by a risk analyst (i.e. subjective evaluations are not available). The level of agreement about the ratings between subjects indicates the level of objectivity in situation perceptions. In order to analyze the extent of objectivity, data was prepared as follows: as dilemma-options represent independent objects which were rated by subjects, for each dilemma-option (18 in total) a separate dataset was created, thus dilemma-options represent the unit of analysis. Each respondent was entered in the columns and situation ratings were entered as rows in each dataset following the guidelines in [32]. Intraclass correlations (ICC) are used as estimates of inter-rater reliability, a technique which is useful for understanding the proportion of reliable ("real") estimates provided by independent raters about a construct or a combination of constructs [16]. As respondents represent a sample from the population of potential respondents and all dilemma options were evaluated by all raters a two-way random analysis was selected ICC(2), which assumes random effects for raters as well as for the constructs being rated. Two types of reliability scores (consistency and absolute agreement) were computed to assess the accuracies of the ratings using ICC(2, 1). The difference between the consistency and absolute agreement measures is that "if two variables are perfectly consistent, they don't necessarily agree" [16]. Absolute agreement represents a more restrictive measure of inter-rater reliability. Figure 3 presents the intraclass correlation scores for all dilemmas. Consistency (red bars) refers to the extent of agreement about the directions of the ratings using a randomly selected analyst. Absolute agreement (blue bars) represents the expected accuracy when a single analyst estimates the exact scores. "An interrater reliability estimate of 0.80 would indicate that 80% of the observed variance is due to true score variance or similarity in ratings between coders, and 20% is due to error variance or differences in ratings between coders" [11].

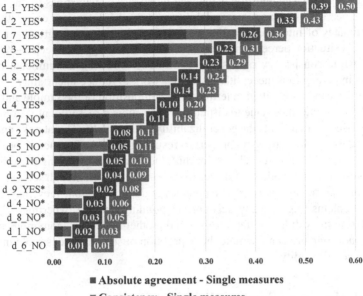

Fig. 3. Interrater reliability estimates across all dilemma-options. Red bars indicate the consistency of ratings, blue bars represent absolute agreement. Dilemma-options marked with * are statistically significant at $p \leq 0.05$. (Color figure online)

5 Discussion

Predictability is the essence of security. The primary purpose of the study was to test a proof of concept related to human behavior prediction using a novel approach. The first research question focused on assessing the extent of predictive improvement when subjective situation perceptions are used as predictors compared to a traditional prediction approach (personal features only). Two sets of analyses were conducted to enable an unbiased comparison between the two approaches using the same number of predictors. The overall percentage of correct classifications ranges between 71.2%–93.2% when only personal features are utilized, and between 81.4%–94.9% using the proposed approach. Predictive performance in terms of Nagelkerke's R^2 performance metric consolidates the findings for the proposed method's superiority. Nagelkerke's R^2 scores for the traditional models range between: 11%–44%, while for the proposed approach, predictive performance ranges between: 31%–76%. In sum, the proposed approach is able to predict more choices, as well as the models based on the proposed approach explain more variance in the outcomes.

The second research question aimed at exploring the extent of objectivity inherent in situation perceptions. The question is relevant when a risk analyst has to unobtrusively estimate the perceptions, due to unavailability of subjects. Intraclass correlations were used as estimates of interrater reliability to explore the extent of agreement between subjects about situation perceptions. Absolute agreement ranges between 2–39% across dilemmas, while consistency ranges between 3–50% showing a significant extent of subjectivity in perceptions, necessitating further work to reduce uncertainty.

Limitations can be identified in terms of the composition of the sample and the ecological validity of the study. Due to difficulties with subject recruitment, a convenience sample was used, which limits the generalizability of the results. Threat risks and opportunity risks are complex, emergent properties resulting from the interaction between the stakeholders. The dilemmas developed for this study aimed at capturing these risk types and were presented as hypothetical stories to respondents. Even though self-report questionnaires are the most-widely used methods for data collection, they may be prone to several problems (e.g. socially desirable responding, lack of experimental control, etc.) hampering the validity of the results. On the other hand, the anonymous nature of the online questionnaire can facilitate the expression of socially undesirable intentions, increasing overall validity.

6   Conclusions and Further Work

Risks attributed to human decisions are increasingly prevalent in all domains of life, necessitating better models and predictive capabilities for risk analysis. Therefore, this study aimed at contributing to the field by enhancing the predictive capabilities of the CIRA method, which focuses on decision-makers' motivation when characterizing risks. A behavior prediction approach was proposed and evaluated, which relies on how situations are perceived by decision-makers. The proposed approach consistently outperforms the traditional approach, but further work is needed to improve its utility in settings where unobtrusive assessment is the only option for a risk analyst. Future studies need to explore whether it is possible to increase the accuracy of situation perceptions by observers. This could be achieved by training analysts in situation-assessment. Furthermore, the development of automated situation-assessments would be necessary to increase reliability. The present study used dilemmas in which "the decision's impact on other people" can be considered the most salient feature of the environment, thus future studies could explore how this salience impacts decision-makers in real-world organizations.

Appendix

1. **Story of dilemma_1:** You are the captain of a military submarine travelling underneath a large iceberg. An onboard explosion has caused you to lose most of your oxygen supply and has injured one of your crew who is quickly losing blood. The injured crew member is going to die from his wounds no matter what happens. The remaining oxygen is not sufficient for the entire crew to make it to the surface. The only way to save the other crew members is to shoot dead the injured crew member so that there will be just enough oxygen for the rest of the crew to survive.

2. **NO option of dilemma_1:** By considering the consequences please rate how **NOT shooting the injured crew member** would influence each of the following factors from your perspective compared to your state before the decision!

3. **YES option of dilemma_1:** By considering the consequences please rate how **shooting the injured crew member** would influence each of the following factors from your perspective compared to your state before the decision!

Fig. 4. The rating scales used to collect the subjective perceptions about each situation (i.e. dilemma-option).

As the captain of the submarine in this situation would you shoot the injured crew member to save the rest of the crew?

Fig. 5. Task requiring an explicit choice between the two options of dilemma_1.

Table 3. Summary of nine binary logistic regression models for each dilemma. Each model uses personal features (BHVs) of subjects as independent variables to predict choices.

Predictor		d_1	d_2	d_3	d_4	d_5	d_6	d_7	d_8	d_9
Constant	β	9.04	−12.83	−5.53	−4.39	−1.67	7.74	−7.94	5.58	−10.72
	SE β	4.01	7.42	3.34	2.54	2.14	4.64	2.86	3.31	5.30
	p	**0.02***	0.08	0.10	0.09	0.44	0.10	**0.01***	0.09	**0.04***
Self-enhancement	β	−0.36	3.12	1.31	0.44	−0.02	−0.61	0.45	−1.48	1.01
	SE β	0.65	1.90	0.92	0.52	0.45	0.91	0.44	0.64	0.87
	p	0.58	0.10	0.15	0.40	0.97	0.50	0.30	**0.02***	0.24
Self-transcendence	β	−0.62	−0.47	−1.13	−0.75	0.67	1.18	0.17	0.11	−0.51
	SE β	0.74	1.34	0.79	0.60	0.52	1.11	0.58	0.79	1.11
	p	0.40	0.73	0.15	0.21	0.20	0.29	0.76	0.89	0.65
Openness to change	β	0.53	−2.18	−0.63	−0.01	0.39	−1.03	0.33	−0.26	0.83
	SE β	0.71	1.80	0.93	0.62	0.49	1.25	0.53	0.76	1.14
	p	0.46	0.23	0.50	0.99	0.43	0.41	0.54	0.73	0.47
Conversation	β	−1.49	1.43	0.30	1.00	−0.06	−1.06	0.71	−0.19	0.49
	SE β	0.67	1.45	0.71	0.52	0.40	0.82	0.42	0.47	0.66
	p	**0.03***	0.32	0.67	**0.05***	0.87	0.19	0.09	0.69	0.46
Hedonism	β	0.27	1.11	1.38	0.32	−0.39	0.01	0.25	0.43	0.46
	SE β	0.39	0.98	0.74	0.39	0.36	0.66	0.34	0.48	0.67
	p	0.49	0.26	0.06	0.41	0.27	0.99	0.47	0.37	0.49
Overall model evaluation	χ^2	11.80	11.14	11.15	10.08	4.51	7.86	14.17	11.89	10.90
	df	5								
	p	**0.04***	**0.05***	**0.05***	0.07	0.48	0.16	**0.02***	**0.04***	0.06
Goodness-of-fit-tests:										
Cox and Snell R^2		0.18	0.17	0.17	0.16	0.07	0.13	0.21	0.18	0.17
Nagelkerke R^2		0.30	0.44	0.36	0.24	0.11	0.28	0.29	0.30	0.35

Note. ***p ≤ 0.05.**

Table 4. Summary of nine binary logistic regression models for each dilemma. Each model uses subjective ratings of situations as independent variables to predict choices.

Predictor		d_1	d_2	d_3	d_4	d_5	d_6	d_7	d_8	d_9
Constant	β	0.90	−4.66	−2.87	−2.41	−0.63	0.14	−1.46	1.04	−3.31
	SE β	0.64	1.72	0.90	0.66	0.72	0.96	0.60	0.48	1.11
	p	0.16	**0.01***	**0.00***	**0.00***	0.38	0.89	**0.02***	**0.03***	**0.00***
S_1 Social status	β	−0.04	0.02	−0.01	0.01	−0.02	−0.02	0.02	−0.02	−0.06
	SE β	0.01	0.02	0.02	0.02	0.02	0.03	0.03	0.02	0.03
	p	**0.01***	0.34	0.45	0.75	0.40	0.55	0.48	0.23	**0.03***
S_2 Care for others	β	0.02	0.00	0.04	0.03	0.06	0.05	−0.03	0.02	0.03
	SE β	0.01	0.02	0.02	0.01	0.02	0.03	0.03	0.01	0.02
	p	0.08	0.99	**0.02***	**0.02***	**0.00***	0.10	0.26	0.12	0.23
S_3 Excitement	β	−0.02	0.06	0.00	−0.01	0.00	0.01	0.06	0.00	0.08
	SE β	0.01	0.03	0.02	0.02	0.03	0.02	0.03	0.01	0.03
	p	0.32	**0.05***	0.97	0.43	0.99	0.81	**0.01***	0.91	**0.01***
S_4 Stability	β	0.02	−0.03	0.01	−0.03	0.00	0.01	−0.03	0.03	0.00
	SE β	0.01	0.02	0.02	0.01	0.02	0.02	0.02	0.01	0.02
	p	0.12	0.19	0.48	**0.03***	0.82	0.68	**0.03***	**0.03***	0.96
S_5 Pleasure	β	0.01	−0.02	−0.01	0.04	0.00	0.01	0.02	−0.02	−0.02
	SE β	0.01	0.02	0.02	0.02	0.02	0.02	0.02	0.01	0.02
	p	0.36	0.37	0.69	**0.04***	0.92	0.65	0.29	0.22	0.28
Overall model evaluation	χ^2	21.00	10.40	19.45	22.97	37.13	8.50	47.65	19.15	12.64
	df	5								
	p	**0.00***	0.07	**0.00***	**0.00***	**0.00***	0.13	**0.00***	**0.00***	**0.03***
Goodness-of-fit-tests:										
Cox and Snell R^2		0.30	0.16	0.28	0.32	0.47	0.13	0.55	0.28	0.19
Nagelkerke R^2		0.50	0.41	0.58	0.50	0.70	0.31	0.76	0.45	0.40

Note. *$p \leq 0.05$. S_1-5: subjective ratings of situational features collected in section 1 of the questionnaire.

References

1. Ajzen, I.: The theory of planned behavior. Organ. Behav. Hum. Decis. Process. **50**(2), 179–211 (1991)
2. Cram, W.A., Proudfoot, J., D'Arcy, J.: Seeing the forest and the trees: a meta-analysis of information security policy compliance literature. In: Proceedings of the 50th Hawaii International Conference on System Sciences, pp. 4051–4060 (2017)
3. De Nooij, M., Lieshout, R., Koopmans, C.: Optimal blackouts: empirical results on reducing the social cost of electricity outages through efficient regional rationing. Energy Econ. **31**(3), 342–347 (2009)
4. Dungan, J.A., Young, L., Waytz, A.: The power of moral concerns in predicting whistleblowing decisions. J. Exp. Soc. Psychol. **85**, 103848 (2019)
5. Fischer, G.W.: Experimental applications of multi-attribute utility models. In: Wendt D., Vlek C. (eds.) Utility, Probability, and Human Decision Making. Theory and Decision Library. An International Series in the Philosophy and Methodology of the Social and Behavioral Sciences, vol. 11. Springer, Dordrecht (1975). https://doi.org/10.1007/978-94-010-1834-0_2
6. Gheyas, I.A., Abdallah, A.E.: Detection and prediction of insider threats to cyber security: a systematic literature review and meta-analysis. Big Data Anal. **1**(1), 6 (2016)

7. Greene, J.D., Sommerville, R.B., Nystrom, L.E., Darley, J.M., Cohen, J.D.: An fMRI investigation of emotional engagement in moral judgment. Science **293**(5537), 2105–2108 (2001)
8. Greitzer, F.L., Kangas, L.J., Noonan, C.F., Dalton, A.C.: Identifying at-risk employees: a behavioral model for predicting potential insider threats. Technical note PNNL-19665, Pacific Northwest National Lab (PNNL), Richland, WA (2010)
9. Greshake, B.: Looking into Pandora's Box: the content of sci-hub and its usage. F1000Research **6**, 541 (2017)
10. Hald, S.L., Pedersen, J.M.: An updated taxonomy for characterizing hackers according to their threat properties. In: 2012 14th International Conference on Advanced Communication Technology (ICACT), pp. 81–86. IEEE (2012)
11. Hallgren, K.A.: Computing inter-rater reliability for observational data: an overview and tutorial. Tut. Quant. Meth. Psychol. **8**(1), 23 (2012)
12. Hara, K., Adams, A., Milland, K., Savage, S., Callison-Burch, C., Bigham, J.P.: A data-driven analysis of workers' earnings on amazon mechanical turk. In: Proceedings of the 2018 CHI Conference on Human Factors in Computing Systems, pp. 1–14 (2018)
13. James, S., Rao, S.V., Granger, C.B.: Registry-based randomized clinical trials-a new clinical trial paradigm. Nat. Rev. Cardiol. **12**(5), 312–316 (2015)
14. Kihlstrom, J.F.: The person-situation interaction. In: The Oxford Handbook of Social Cognition, pp. 786–805. Oxford University Press (2013)
15. Krebs, B.: FBI: Smart Meter Hacks Likely to Spread—Krebs on Security (April 2012). https://krebsonsecurity.com/2012/04/fbi-smart-meter-hacks-likely-to-spread. Accessed 14 Feb 2020
16. Landers, R.N.: Computing intraclass correlations (ICC) as estimates of interrater reliability in SPSS. Winnower **2**, e143518 (2015)
17. Lebek, B., Uffen, J., Neumann, M., Hohler, B., Breitner, M.H.: Information security awareness and behavior: a theory-based literature review. Manage. Res. Rev. **37**, 1049–1092 (2014)
18. Lewin, K.: Principles of Topological Psychology. McGraw-Hill (1936)
19. Maasberg, M., Warren, J., Beebe, N.L.: The dark side of the insider: detecting the insider threat through examination of dark triad personality traits. In: 2015 48th Hawaii International Conference on System Sciences, pp. 3518–3526. IEEE (2015)
20. McDonald, P.: Workplace sexual harassment 30 years on: a review of the literature. Int. J. Manag. Rev. **14**(1), 1–17 (2012)
21. Meyer, R.D.: Taxonomy of Situations and Their Measurement (2015)
22. N.A.: Combating the insider threat. Technical report, Department of Homeland Security (May 2014). https://www.us-cert.gov/sites/default/files/publications/Combating%20the%20Insider%20Threat_0.pdf
23. Naffine, N.: Who are law's persons? From Cheshire cats to responsible subjects. Mod. Law Rev. **66**(3), 346–367 (2003)
24. Nübold, A., et al.: Developing a taxonomy of dark triad triggers at work-a grounded theory study protocol. Front. Psychol. **8**, 293 (2017)
25. Rajbhandari, L., Snekkenes, E.: Using the conflicting incentives risk analysis method. In: Janczewski, L.J., Wolfe, H.B., Shenoi, S. (eds.) SEC 2013. IAICT, vol. 405, pp. 315–329. Springer, Heidelberg (2013). https://doi.org/10.1007/978-3-642-39218-4_24
26. Rauthmann, J.F., Sherman, R.A., Funder, D.C.: Principles of situation research: towards a better understanding of psychological situations. Eur. J. Pers. **29**(3), 363–381 (2015)
27. Schwartz, S.: Value priorities and behavior: applying a theory of intergrated value systems. In: The Psychology of Values: The Ontario Symposium, vol. 8, pp. 119–144 (2013)
28. Schwartz, S.: Computing scores for the 10 human values (2016). https://www.europeansocialsurvey.org/docs/methodology/ESS_computing_human_values_scale.pdf. Accessed 02 Mar 2021

29. Schwartz, S.H.: An overview of the Schwartz theory of basic values. Online Read. Psychol. Cult. **2**(1), 1–20 (2012)
30. Selznick, P.: Foundations of the theory of organization. Am. Sociol. Rev. **13**(1), 25–35 (1948)
31. Shropshire, J., Warkentin, M., Sharma, S.: Personality, attitudes, and intentions: predicting initial adoption of information security behavior. Comput. Secur. **49**, 177–191 (2015)
32. Shrout, P.E., Fleiss, J.L.: Intraclass correlations: uses in assessing rater reliability. Psychol. Bull. **86**(2), 420 (1979)
33. Snekkenes, E.: Position paper: privacy risk analysis is about understanding conflicting incentives. In: Fischer-Hübner, S., de Leeuw, E., Mitchell, C. (eds.) IDMAN 2013. IAICT, vol. 396, pp. 100–103. Springer, Heidelberg (2013). https://doi.org/10.1007/978-3-642-37282-7_9
34. Spain, S.M., Harms, P., LeBreton, J.M.: The dark side of personality at work. J. Organ. Behav. **35**(S1), S41–S60 (2014)
35. Szekeres, A., Snekkenes, E.A.: Predicting CEO misbehavior from observables: comparative evaluation of two major personality models. In: Obaidat, M.S. (ed.) ICETE 2018. CCIS, vol. 1118, pp. 135–158. Springer, Cham (2019). https://doi.org/10.1007/978-3-030-34866-3_7
36. Szekeres, A., Snekkenes, E.A.: A taxonomy of situations within the context of risk analysis. In: Proceedings of the 25th Conference of Open Innovations Association FRUCT, pp. 306–316. FRUCT Oy. Helsinki, Finland (2019)
37. Szekeres, A., Snekkenes, E.A.: Construction of human motivational profiles by observation for risk analysis. IEEE Access **8**, 45096–45107 (2020)
38. Ten Berge, M.A., De Raad, B.: Taxonomies of situations from a trait psychological perspective: a review. Eur. J. Pers. **13**(5), 337–360 (1999)
39. VanVoorhis, C.W., Morgan, B.L.: Understanding power and rules of thumb for determining sample sizes. Tut. Quant. Meth. Psychol. **3**(2), 43–50 (2007)

A Deep Learning Approach to Forecast SARS-CoV-2 on the Peruvian Coast

I. Luis Aguilar[1]([✉])(iD), Miguel Ibáñez-Reluz[2](iD), Juan C. Z. Aguilar[3](iD), Elí W. Zavaleta-Aguilar[4](iD), and L. Antonio Aguilar[5](iD)

[1] Department of Mathematics, National University of Piura Castilla s/n, Piura, Peru
laguilari@unp.edu.pe
[2] Medicine Faculty, Cesar Vallejo University, Av. Victor Larco 1770, Trujillo, Peru
mibanezr@ucvvirtual.edu.pe
[3] Department of Mathematics and Statistics, Universidade Federal de São João del-Rei C.P. 110, 36301-160 São João del-Rei, MG, Brazil
jaguilar@ufsj.edu.br
[4] São Paulo State University (Unesp), Campus of Itapeva Rua Geraldo Alckmin 519, 18409-010 Itapeva, SP, Brazil
eli.zavaleta@unesp.br
[5] Artificial Intelligent Research, KapAITech Research Group, Condominio Sol de Chan-Chan, Trujillo, Peru
antonio@kapaitech.com

Abstract. The current spreading of the SARS-CoV-2 pandemic had put all the scientific community in alert. Even in the presence of different vaccines, the active virus still represents a global challenge. Due to its rapid spreading and uncertain nature, having the ability to forecast its dynamics becomes a necessary tool in the development of fast and efficient health policies. This study implements a temporal convolutional neural network (TCN), trained with the open covid-19 data set sourced by the Health Ministry of Peru (MINSA) on the Peruvian coast. In order to obtain a robust model, the data was divided into validation and training sets, without overlapping. Using the validation set the model architecture and hyper-parameters were found with Bayesian optimization. Using the optimal configuration the TCN was trained with a test and forecasting window of 15 days ahead. Predictions on available data were made from March 06, 2020 until April 13, 2021, whereas forecasting from April 14 to April 29, 2021. In order to account for uncertainty, the TCN estimated the 5%, 50% and 95% prediction quantiles. Evaluation was made using the MAE, MAD, MSLE, RMSLE and PICP metrics. Results suggested some variations in the data distribution. Test results shown an improvement of 24.241, 0.704 and 0.422 for the MAD, MSLE and RMSLE metrics respectively. Finally, the prediction interval analysis shown an average of 97.886% and 97.778% obtained by the model in the train and test partitions.

Keywords: Deep learning · SARS-CoV-2 · Temporal convolutional neural networks · Time series data

© Springer Nature Switzerland AG 2021
O. Gervasi et al. (Eds.): ICCSA 2021, LNCS 12951, pp. 304–319, 2021.
https://doi.org/10.1007/978-3-030-86970-0_22

1   Introduction

On December 2019 a novel virus was reported in Wuhan city, Hubei province, China. It was latter identified as SARS-CoV-2; a virus belonging to a family of coronavirus strains responsible for causing severe acute respiratory syndrome [7]. Initial clinical studies revealed the presence of thrombocytopenia, elevated D-dimer, prolonged time, and disseminated intravascular coagulation in SARS-CoV-2 patients [10]. Also, comorbidities were associated as high risk factors in SARS-CoV-2 patients [12]. SARS-CoV-2 shown a fast infection rate which resulted in a global pandemic. In order to better understand the virus spread, researchers over the world began to implement a wide variety of models. These initial models required relatively small data sets. As such, the SIR model offered a fast approach to estimate the SARS-CoV-2 infection tendencies. The SIR [2] model is composed by a system of differential equations involving three epidemiological parameters: susceptible (S), infectious (I) and removed (R). Other variants, like the SEIR [2] and SIRD [3] included additional parameters over the base SIR model such as exposition (E) and deceased (D) cases respectively.

In the following months, new data set repositories [9] allowed the application of deep learning models. Due to the evolving dynamics observed in SARS-CoV-2 cases, a time series approach was used. Therefore, recurrent neural networks (RNN) like long short term memory (LSTM) [15] and its variants were applied. In [8] a LSTM model was used to forecast SARS-CoV-2 cases in Canada. A 80-20% data split strategy was implemented for train and test respectively. The RMSE metric was used to evaluate the model performance. The trained LSTM model was able to estimate a peak in cases within a two week period.

Besides recurrent models, other non-recurrent architectures have shown successfully results in forecast SARS-CoV-2 cases. For example, in [25] a system of two neural networks were used in real time to estimate SARS-CoV-2 trends at country and regional levels. These models used a window of 12 and 8 days for each level respectively. The models were built using a global data set repository [9]. Margin error variations shown consistent accuracy results in several countries, such as Peru and Brazil.

In a comparative study, a Variational auto encoder (VAE) shown better forecasting performance than four recurrent architectures including a: RNN, GRU, LSTM and Bi-LSTM [26]. In a similar fashion, convolutional neural networks (CNN) also reported good performance in forecasting SARS-CoV-2 trends. In [19] a CNN model was applied to forecast the daily, cumulative, recoveries and deceased SARS-CoV-2 cases alongside with hospitalizations (with or without artificial ventilation). The study was executed on France at regional and national levels. Evaluation results for the CNN model reported by the MAE, RMSE, and R2 metrics shown a consistent forecasting performance. Furthermore, CNN variations such as temporal convolutional neural networks (TCN) have been able to forecast using prediction intervals with a multi output setting [19].

Lastly Bayesian optimization techniques have shown a positive impact in learning, when applied to LSTM and CNN models. In [1] a multi-head attention, LSTM and CNN models were developed to predict confirmed SARS-CoV-2 cases.

This study was applied to a short and long horizon using two data sets of several countries such as Peru and Brazil. CNN results in the short horizon were superior than the LSTM model according with the SMAPE, MAPE and RMSE metrics. In the long horizon, the CNN model was superior in Peru and Brazil.

As observed, deep learning models are capable of achieve good forecasting performance across different geographical areas. In this study, a TCN deep learning model is developed to forecast daily SARS-CoV-2 cases. The data used encompass the 3 geographical regions of Peru. Among them, the Peruvian coast was selected due to its high population density areas in contrast with the highlands and amazon rain-forest [16]. The model architecture and hyper-parameters were optimized using a Bayesian approach. To avoid any data leaking. a hold out method was used. Our model was able to predict from March 30 to April 16, 2021 and forecasting 15 days ahead of the available data until April 27, 2021. Finally, a set of evaluation metrics were applied to evaluate the model performance. The following sections will describe the study in more detail.

2   Temporal Convolutional Neural Networks (TCN)

A Temporal convolutional neural network (TCN) is a variation of a CNN model, which can deal with sequence data. A TCN model applies the convolution operation over an input sequence s. Since s contains events in a period of time, convolutions are applied over the time dimension. A TCN is able to estimate future events of s based on past information. To achieve this, a TCN model modifies the standard convolution operation present in a CNN with casual and dilated convolutions. To avoid vanish gradients, skip residual connections are usually applied between TCN layers.

In a casual convolution, the kernel operation maintains the natural order of the input sequence s. This process is illustrated in Fig. 1a, where instead of applying the convolution directly over s, the kernel is moved from left to right. This process forces the model to only relay on past data to make its predictions. This mechanism also prevents any data leaking from the future, thus the name casual. However, a casual convolution is unable to maintain a long horizon history, which is essential in order to capture past patterns. Therefore, another variation called dilated convolutions are applied. Dilated convolution relay on dilatations to increase the receptive field [4]. This effect can be observed in Fig. 1b, where an increasing dilation allows next layers to expand the input information. Formally for an input sequence $s = \{x_0, ..., x_T\}$, where T denotes the sequence length; a casual dilated convolution $\mathcal{F}(s)$, is defined as:

$$\mathcal{F}(s) = (x *_d f)(s) = \sum_{i=1}^{k} f(i) * x_{s-d*i} \tag{1}$$

where the dilation factor d express the amount of variation over the receptive field. Next, a set of kernels f of size k are applied over the elements in s, such that $x_s - d * i$ constrains the process to only considered past information.

The last component of a TCN is a skip residual connection. This mechanism was introduced by [13] in the Residual Network architecture. Skip residual connections create a shortcut in the information flow (gradient). The residual is skipped over a certain number of layers. Then, it is combined in the main information flow. This operation allows to create deeper models and deal with the gradient vanish problem. Figure 1c., shows a standard residual connection scheme.

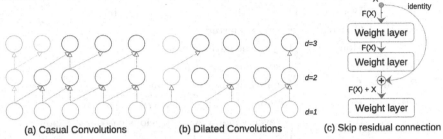

| (a) Casual Convolutions | (b) Dilated Convolutions | (c) Skip residual connection |

Fig. 1. Mechanisms in a TCN: (a) casual convolutions, (b) dilated convolutions and (c) skip residual connections.

3 Data Set

This study used the open covid-19 data set [18], created by the Ministry of Health in Peru (MINSA). The data set contained the daily reports of SARS-CoV-2 cases across the country. In its raw format, it was composed by 9 features and 155,649 observations, corresponding from March 03, 2020 until April 13, 2021 (**accessed on 16.04.2021**). Some features were not relevant for this study and were removed, therefore, a feature engineering process was applied to create new features which were used by the model. Although the data set had information for 25 regions, only 9 belonging to the coast were selected. These regions included: Lima, Ica, Callao, Piura, Lambayeque, La Libertad, Tumbes, Tacna and Moquegua, with a total of 63,735 samples. The coastal regions were selected due to its high population density [16]. Also, coastal regions shown a much larger number in infections than the highlands and amazon regions. As such determining the infection impact on the coast can help to mitigate the spreading to the remaining regions.

3.1 Feature Engineering

A total of five features were selected from the original covid-19 data set [18]. Missing observations were imputed using rolling windows. Next, a set of 9 features were engineered to capture: time dependencies, cases in lower administrative divisions (province and districts) and a seven day record of past SARS-CoV-2 cases per region. As a last pre-processing step, the data was grouped at regional level. This new data set was used by the TCN model for validation and training, containing a total of 10 features, which are described in Table 1.

Table 1. Feature engineering process description.

Features		Description
Selected	Engineering	
Date	Month	Current month, values from 1 to 12
	Day of year	Ongoing day in the year, values from 1 to 365
	Week	Week day, from 0 to Monday until 6 to Sunday
·	Weekday	Week of the year, values from 1 to 52
	Quarter	Current quarter
	Month day	Day of the current month
Region	–	Region name used as ID
Province	I_p	Number of infected provinces in a region
District	I_d	Number of infected districts in a region
	W_7	A seven window with previous cases per region
Cases	–	Daily cases variable used as target

3.2 Data Preparation

Since the obtained data set described in Table 1 was in tabular format, a preparation step was necessary in order to be used by the TCN model. First, W_7 was arranged into a fix sequence, where the eight day was used as target y (Fig. 2a). To obtain a fix sequence, all features were padding with zeros to match the length of the longest sequence, which corresponded to Lima. To keep the time correlation, all features were sorted by date.

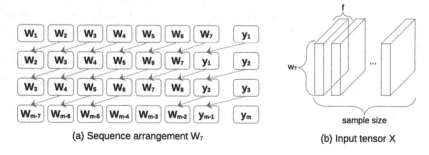

(a) Sequence arrangement W_7 (b) Input tensor X

Fig. 2. A sequence window W_7 arrangement (a) was used to represent a past seven-day input to the model, using the eight day as y. Then W_7 (a) was combined with the remaining 9 features, resulting in the X input tensor (b) used by the model.

Next, all features were combined into a 3D input tensor X, with dimensions: $samples \times sequence_7 \times features_{10}$ (Fig. 2b). To normalize the values in X, all features were scaled individually, such that the maximal absolute value of each one was set to 1.0.

3.3   Data Partition

After completing the previous process (subsection 3.2), the data was divided into validation and training sets without overlapping date periods. This avoided any data leaking between the training and validation sets, which could lead to overfitting. Then, each set was split into training and test partitions respectively. In both data sets, the training partition was used for learning, while the test partition for evaluation. Model architecture and hyper-parameters optimization were performed through Bayesian optimization on the validation data set. Table 2 shown the validation data set in more detail.

Table 2. Validation data set division used to build and optimize the TCN model.

Partition	Features	Samples	Data range	Days
Training	10	3366	Mar. 06, 2020–Mar. 14, 2021	374
Test	10	135	Mar. 15, 2021–Mar. 29, 2021	15

The training set was used to create the final TCN model, using the optimal hyper-parameter configuration found in the validation data set. Table 3 shown the training set details.

Table 3. Training data set division used to train the TCN model.

Partition	Features	Samples	Data range	Days
Training	10	3501	Mar. 06, 2020–Mar. 29, 2021	389
Test	10	135	Mar. 30, 2021–Apr. 13, 2021	15

A 15 day ahead window was selected for test and forecasting. We based this value according with the average 14 day incubation period of SARS-CoV-2 [7]. Next, in order to account for possible shifting variations in incubation periods, an extra day was added. This allowed the model to estimate SARS-CoV-2 variations more consistently across regions.

4   TCN Model

A TCN model was developed to estimate the daily SARS-CoV-2 cases in the Peruvian coast. Estimations were categorized into predictions and forecasting. Predictions were made using all the available data set [18], whereas forecasting used a multi-step approach without any available data. A set of 9 regions were selected from the data set [18] including: Lima, Ica, Callao, Piura, Lambayeque, La Libertad, Tumbes, Tacna and Moquegua. A predictive window of 15 days ahead were used for prediction and forecasting. The model was built and optimized using a Bayesian approach. The following sections will describe the model architecture as well as the optimization process in more detail.

4.1 Model Architecture

In order to determine the total number of casual convolution (DC-Conv) layers, an empirical approach was applied. In this setting, a single DC-Conv layer was added to the base model until no further improvements were reported on the validation set (Table 3). This process was repeated until the optimal number of DC-Conv layers were set to 5. Each DC-Conv layer applied the kernel operation described in Eq. 1.

DC-Conv layers were followed by a batch normalization layer [17], a ReLU [20] activation and a dropout layer [14] to avoid overfitting. These configuration was maintained with the exception of the output block. In the output block a 3 DC-Conv layers were implemented to learn the 5%, 50% and 95% prediction intervals. The output layers were followed by a ReLU activation to avoid convergence to negative values during training. As the last TCN component, two skip residual connections [13] were created from blocks 1–3 and 2–4. The model was implemented using Pytorch [21]. Figure 3 shown the detailed architecture.

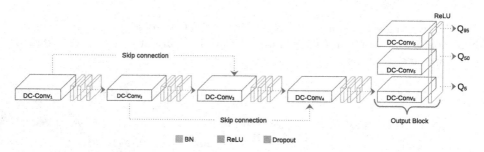

Fig. 3. Temporal Convolutional Neural Network (TCN) architecture. Blue blocks denotes casual dilated convolutions (DC-Conv). Batch normalization (green), relu activation (orange) and dropout (gray) are applied to blocks 1–4. A set of three Dc-Conv layers learns 5%, 50% and 95% prediction intervals. To avoid negative outputs, a relu layer is applied at the end of the output block.

4.2 Bayesian Optimization

In order to obtain the optimum architecture and hyper-parameters, a Bayesian search was performed on the validation data set using the BoTorch [6] library. Minimization of the quantile loss defined in Eq. 3 was used as objective. An initial list of hyper-parameters were build to start the search. Optimization was performed over the validation training data (Table 2) using the Sobol+GPEI strategy present in the Ax [5] library. The hyper-parameters were evaluated against the objective using the test partition from the validation data set (Table 2). Due to the sensitivity in the hyper-parameters, some configurations lead to non-optimal results. In such cases, the last well know configuration was restored. This cycle was repeated until no new improvements in the objective were found.

The search was executed using an AMD Ryzen 7-3700x CPU. Tables 4 and 5 shown the best configuration found for the architecture and hyper-parameters respectively.

Table 4. DC-convs configuration obtained after Bayesian optimization over the validation data set.

Block	Input ch.	Output ch.	Kernel size	Padding	Dilation	Dropout ratio
DC-Conv$_1$	10	32	3×3	8	4	3.00E-01
DC-Conv$_2$	32	32	3×3	16	8	6.58E-01
DC-Conv$_3$	32	32	3×3	16	8	4.66E-01
DC-Conv$_4$	32	32	3×3	20	10	3.62E-01
DC-Conv$_5$	32	1	2×2	8	14	–

Table 5. Optimal hyper-parameter configuration obtained with Bayesian optimization on the validation data set. This configuration was used to built the final model on the training data set.

Hyper-param	Lr	Max Lr	Momentum	Decay	Optim	Epochs	Batch
Value	5.18E-03	6.00E-03	5.27E-01	3.02E-06	RMSProp	7598	60

4.3   Training Configuration

Using the optimum values from Table 5, a new TCN model was trained on the training data set (Table 3). The model output was configured to learn the $Q_5\%$, $Q_{50}\%$ and $Q_{95}\%$ quantiles, which are defined as follows:

$$\mathcal{L}(\xi_i|q) = \begin{cases} q\xi_i, & \xi_i \geq 0 \\ (q-1)\xi_i, & \xi_i < 0 \end{cases} \tag{2}$$

where $q \in [0, 1]$ and ξ_i is defined as the difference between the real values y and predictions $\hat{y}$, such that: $\xi_i = y_i - \hat{y}_i$. Then, the total quantile error is averaged during training over the samples m using the following loss:

$$\mathcal{L}(y, \hat{y}|q) = \frac{1}{m} \sum_{i=1}^{m} \mathcal{L}(\xi_i|q) \tag{3}$$

The loss defined in Eq. 3 was optimized using the resilient backpropagation algorithm (RMSProp) [24]. To avoid any memorization during training, each batch was shuffled. In order to smooth the optimization area, the parameters were initialized using Xavier [11] with a normal distribution. To avoid any sub-optimal optimization, a cycle learning [23] strategy with a step size of 2000 was used during training.

Due to the different magnitude of the target variable y between regions, a normalization step was perform during training using: $log(y+1)$. This operation was reversed with $e(\hat{y}+1)$ to compute the estimates. Finally, in order to apply forecasting beyond the test data, a multi step strategy of 1 day ahead was used.

5   Evaluation Strategy

In order to measure the model performance, an evaluation strategy composed by two levels: global and locally was implemented. In the global level, the training partitions were used to evaluate the model learning capabilities as well as improvement variations between validation and training. On the other hand, the test partitions were used to measure the generalization to new unseen data, which offered a forecasting estimate error. Then at local level, the evaluation was applied per each region.

This strategy allowed us to draw a more complete evaluation of the model performance. Next, a set of metrics were selected to measure the inference $\hat{y}$ error against the real targets y for a sample size m. These metrics are: the mean absolute error (MAE), median absolute error (MAD), mean squared logarithmic error (MSLE) and root mean squared logarithmic (RMSLE). The following equations describe them in more detail:

$$MAE = \frac{1}{m}\sum_{i=1}^{m}|y_i - \hat{y}_i| \tag{4}$$

$$MAD = median(|y_i - \hat{y}_i|) \tag{5}$$

$$MSLE = \frac{1}{m}\sum_{i=1}^{m}[log(\hat{y}_i + 1) - log(y_i + 1)]^2 \tag{6}$$

$$RMSLE = \sqrt{\frac{1}{m}\sum_{i=1}^{m}[log(\hat{y}_i + 1) - log(y_i + 1)]^2} \tag{7}$$

Finally, to measure the prediction intervals robustness we used the percentage of real data points y which are between by the lower (L) and upper (U) intervals, such that: $L \leq y \leq U$. This metric is defined as the Prediction Interval Coverage Probability (PICP):

$$PICP = \frac{1}{m}\sum_{i=1}^{m}u_i, \qquad u_i(y) = \begin{cases} 0 & \text{if } L \geq y \geq U \\ 1 & \text{if } L \leq y \leq U \end{cases} \tag{8}$$

6   Results

This section analyzes the prediction and forecasting results using the evaluation strategy defined in Sect. 5. During the analysis the training and validation data sets were used. Predictions were computed on the training data set (Table 3). Forecasting error estimations were approximated from the test partition (Table 3). Finally, the model was evaluated using the metrics defined in Sect. 5. The following sub-sections will describe these analysis in more detail.

6.1   Prediction Analysis

Predictions were computed using the training and test partitions (Table 3). This process was performed in every coastal region. The training period started on March 06, 2020 and concluded on March 29, 2021, having more than one year of data. Meanwhile, the test period considered 15 days ahead of the training one, starting on March 30 and concluding on April 13, 2021. Analysis of the results from Fig. 4 shown that each coastal region have its particular trend, which are well captured by the model predictions $\hat{y}$. Furthermore, prediction intervals help us to drawn possible variations of daily SARS-CoV-2 cases in every region.

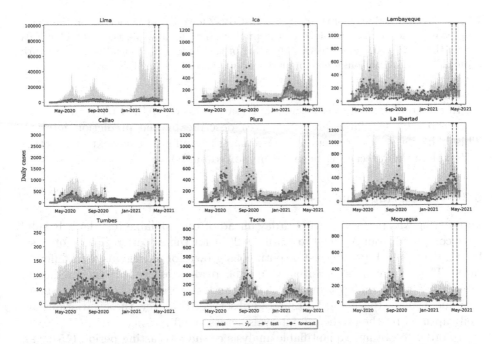

Fig. 4. Daily SARS-CoV-2 predictions from March 06, 2020 to March 29, 2021. The orange line represents the predictions $\hat{y}$, while the light blue are shown the 95% prediction interval. The green line denotes the beginning of the test period (March 30 to April 13, 2021). Meanwhile, the purple line denotes the start of forecasting from April 14 to April 29, 2021. Prediction and forecasting used a 15 day ahead window.

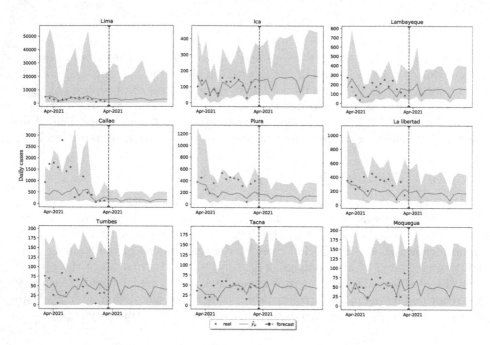

Fig. 5. An enlarged view of the daily SARS-CoV-2 forecasting tendencies in the Peruvian coast. The orange line denotes the prediction $\hat{y}$. Meanwhile the light blue area shown the 95% prediction interval. The start of the forecasting period is denoted with a purple line. Estimations were computed from April 14 to April 29, 2021.

Results on Lima suggest a large daily variations of SARS-CoV-2 cases. This is reflected by the large prediction area describe by the prediction interval. This behavior is less present in the remaining regions. The next sub-section will explore the forecasting behavior in more detail.

6.2   Forecasting Analysis

Forecasting was performed with 15 day ahead of the available test data. Results were computed from April 14 to April 29, 2021 for the 9 coast regions. Forecasting tendencies in Fig. 5 suggest a continuous grown for all regions with different rates. This behavior is accentuated by the peaks observed in the prediction intervals. Also an underestimation of $\hat{y}$ in the test period is observed for: Lambayeque, Callao, Piura and La Libertad observed. However, these variations are well captured by the prediction intervals across the 9 regions.

In order to obtain a quantifiable analysis of the forecasting period (15 days), the maximum and mean tendencies were averaged. Figure 6 shown a possible maximum of 22,937 cases for Lima. This represents the maximum expected variation. The remaining regions shown a more moderate estimate. It is important to note that this variations could change drastically with more recent data.

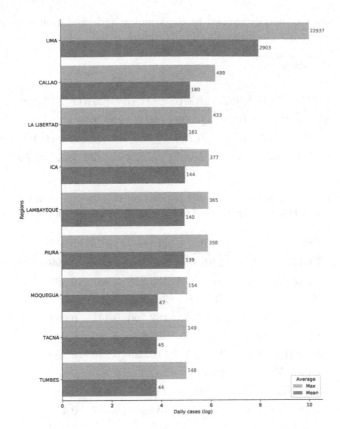

Fig. 6. An average estimated of the mean (orange) and maximum (light blue) $\hat{y}$ cases observed over the 15 day forecast for each coastal region. (Color figure online)

Nevertheless, factors like SARS-CoV-2 variants [22] combined with other components could explain the variations shown in Lima. For example, the Peruvian presidential elections took place on April 11, 2021, just 3 days before the star of the forecasting period on April 14, 2021. In this event, citizens mobilize in Lima, which contains the highest population density of the country. The previous factors could be related with the high forecasting variations observed.

6.3   Evaluation Metrics Analysis

Evaluation results from the validation and training data sets (global) shown in Table 6, indicates a slightly error increase in the training partition. These results suggest that, some data distributions were only present in the training set. On the other hand, the model shown an improvement on the test partition with a consistent decrease for the MAD, MSLE and RMSLE metrics. These results also serves as estimated errors to forecasting. Results from Table 7 shown the estimate error per each region (locally). Error variations are observed across regions,

Table 6. Evaluation metrics applied to the training and validation sets. In bold are shown the improvements from the final TCN model.

Data set	Partition							
	Train				Test			
	MAE	MAD	MSLE	RMLSE	MAE	MAD	MSLE	RMSLE
Validation	85.061	21.189	0.358	0.599	241.295	70.324	1.093	1.045
Training	87.423	25.536	0.514	0.717	245.211	46.082	0.389	0.624
Difference	↑2.362	↑4.347	↑0.156	↑0.119	↑3.260	**↓24.241**	**↓0.704**	**↓0.422**

Table 7. Metric scores per each coastal regions on the test partition (Table 3). The test period spanned from March 30 to April 13, 2021, with a window of 15 days.

Region	MAE	MAD	MSLE	RMSLE
LIMA	1102.096	973.231	0.249	0.499
CALLAO	674.262	480.318	0.931	0.965
ICA	30.554	26.634	0.154	0.393
PIURA	154.217	182.766	0.441	0.664
LAMBAYEQUE	76.616	89.297	0.400	0.633
LA LIBERTAD	112.487	116.535	0.240	0.490
TUMBES	29.418	25.229	0.786	0.887
TACNA	9.816	9.052	0.098	0.314
MOQUEGUA	17.435	13.856	0.202	0.449

Table 8. Prediction interval coverage probability (PICP) results for each coastal regions on the training and test partitions (Table 3). In bold are shown the average PICP estimates for both partitions.

Region	PICP-Train (%)	PICP-Test (%)
LIMA	100.000	100.000
CALLAO	96.915	86.667
ICA	98.715	100.000
PIURA	98.201	93.333
LAMBAYEQUE	96.658	100.000
LA LIBERTAD	97.943	100.000
TUMBES	95.373	100.000
TACNA	98.715	100.000
MOQUEGUA	98.458	100.000
AVERAGE	97.886	**97.778**

suggesting that some tendencies are more complex to learn than others. MAE and MAD metrics shown large variations, whereas MSLE and RMSLE shown 0.931 and 0.965 as their larger errors. PICP results in Table 8 shown an average of 97.886% and 97.778% for training and test respectively. Individually, almost all regions shown values above 90% for training and test with the exception of Callao (86.667% in test). These results shown the model robustness when addressing variations in the SARS-CoV-2 cases among regions.

7   Conclusions

In this research a TCN model was built to predict and forecasting the daily SARS-CoV-2 cases in the Peruvian coast. Bayesian optimization was used to determine the model architecture and best hyper-parameters. Due to the complexity of the problem, the model generated prediction intervals, allowing to capture variations in the daily SARS-CoV-2 cases per region. Maximum average forecast values suggested a very high estimate for Lima (22,937 cases) within 15 days. The presence of a new SARS-CoV-2 variant as well as the past presidential elections may contributed to the observed results in Lima. Evaluation metric analysis on the training partitions point out a possible difference in distribution between validation and training data sets. However, despite the difference, the model reported improvements over the test partition in three metrics: MAD, MSLE and RMSLE with a decrease of 24.241, 0.704 and 0.422 respectively. On the other hand, regional error analysis on the test partition suggested that some SARS-CoV-2 distributions like Lambayeque, Callao, Piura and La Libertad are harder to learn than others. Nonetheless, PICP results shown values greater than 90.0% (train) and 97.778 % (test) for almost all regions, with the exception of Callao (86.667%-test). On average the model prediction intervals captured about 97.886% (training) and 97.778% (test). The presented model can be used as a tool to estimate tendencies in daily SARS-CoV-2 cases, aiding in the decision making of health policies in each region. This study could also by applied to other countries. As such, it is recommended to add more variables to improve performance.

References

1. Abbasimehr, H., Paki, R.: Prediction of COVID-19 confirmed cases combining deep learning methods and Bayesian optimization. Chaos Solitons Fractals **142**, 110511 (2021). https://doi.org/10.1016/j.chaos.2020.110511, https://linkinghub.elsevier.com/retrieve/pii/S0960077920309036
2. Abou-Ismail, A.: Compartmental models of the COVID-19 pandemic for physicians and physician-scientists. SN Compre. Clin. Med. **2**(7), 852–858 (2020). https://doi.org/10.1007/s42399-020-00330-z, https://link.springer.com/10.1007/s42399-020-00330-z
3. Anastassopoulou, C., Russo, L., Tsakris, A., Siettos, C.: Data-based analysis, modelling and forecasting of the COVID-19 outbreak. PLOS ONE **15**(3), e0230405 (2020). https://doi.org/10.1371/journal.pone.0230405, https://dx.plos.org/10.1371/journal.pone.0230405

4. Bai, S., Kolter, J.Z., Koltun, V.: An empirical evaluation of generic convolutional and recurrent networks for sequence modeling (2018). http://arxiv.org/abs/1803.01271

5. Bakshy, E., et al.: AE: a domain-agnostic platform for adaptive experimentation (2018)

6. Balandat, M., et al.: BoTorch: a framework for efficient monte-carlo bayesian optimization (2019). http://arxiv.org/abs/1910.06403

7. Bchetnia, M., Girard, C., Duchaine, C., Laprise, C.: The outbreak of the novel severe acute respiratory syndrome coronavirus 2 (SARS-CoV-2): a review of the current global status. J. Infect. Public Health **13**(11), 1601–1610 (2020). https://doi.org/10.1016/j.jiph.2020.07.011, https://linkinghub.elsevier.com/retrieve/pii/S1876034120305918

8. Chimmula, V.K.R., Zhang, L.: Time series forecasting of COVID-19 transmission in Canada using LSTM networks. Chaos Solitons Fractals **135**, 109864 (2020). https://doi.org/10.1016/j.chaos.2020.109864, https://linkinghub.elsevier.com/retrieve/pii/S0960077920302642

9. Dong, E., Du, H., Gardner, L.: An interactive web-based dashboard to track COVID-19 in real time. Lancet Infect. Dis. **20**(5), 533–534 (2020). https://doi.org/10.1016/S1473-3099(20)30120-1, https://linkinghub.elsevier.com/retrieve/pii/S1473309920301201

10. Giannis, D., Ziogas, I.A., Gianni, P.: Coagulation disorders in coronavirus infected patients: COVID-19, SARS-CoV-1, MERS-CoV and lessons from the past. J. Clin. Virol. **127**, 104362 (2020). https://doi.org/10.1016/j.jcv.2020.104362, https://linkinghub.elsevier.com/retrieve/pii/S1386653220301049

11. Glorot, X., Bengio, Y.: Understanding the difficulty of training deep feedforward neural networks. In: Teh, Y.W., Titterington, M. (eds.) Proceedings of the Thirteenth International Conference on Artificial Intelligence and Statistics. Proceedings of Machine Learning Research, vol. 9, pp. 249–256. PMLR, Chia Laguna Resort (2010). http://proceedings.mlr.press/v9/glorot10a.html

12. Harrison, S.L., Fazio-Eynullayeva, E., Lane, D.A., Underhill, P., Lip, G.Y.H.: Comorbidities associated with mortality in 31,461 adults with COVID-19 in the United States: a federated electronic medical record analysis. PLOS Med. **17**(9), e1003321 (2020). https://doi.org/10.1371/journal.pmed.1003321, https://dx.plos.org/10.1371/journal.pmed.1003321

13. He, K., Zhang, X., Ren, S., Sun, J.: Deep residual learning for image recognition. In: 2016 IEEE Conference on Computer Vision and Pattern Recognition (CVPR), pp. 770–778. IEEE (2016). https://doi.org/10.1109/CVPR.2016.90, http://ieeexplore.ieee.org/document/7780459/

14. Hinton, G.E., Srivastava, N., Krizhevsky, A., Sutskever, I., Salakhutdinov, R.R.: Improving neural networks by preventing co-adaptation of feature detectors (2012). http://arxiv.org/abs/1207.0580

15. Hochreiter, S., Schmidhuber, J.: Long short-term memory. Neural Comput. **9**(8), 1735–1780 (1997). https://doi.org/10.1162/neco.1997.9.8.1735, https://direct.mit.edu/neco/article/9/8/1735-1780/6109

16. INEI: Perú - Censos Nacionales 2017: XII de Población, VII de Vivienda y III de Comunidades Indígenas (2017). http://webinei.inei.gob.pe/anda_inei/index.php/catalog/674

17. Ioffe, S., Szegedy, C.: Batch normalization: accelerating deep network training by reducing internal covariate shift. In: Bach, F., Blei, D. (eds.) Proceedings of the 32nd International Conference on Machine Learning. Proceedings of Machine Learning Research, vol. 37, pp. 448–456. PMLR, Lille (2015). http://proceedings.mlr.press/v37/ioffe15.html
18. MINSA: Casos positivos por COVID-19 - [Ministerio de Salud - MINSA] (2020). https://www.datosabiertos.gob.pe/dataset/casos-positivos-por-covid-19-ministerio-de-salud-minsa
19. Mohimont, L., Chemchem, A., Alin, F., Krajecki, M., Steffenel, L.A.: Convolutional neural networks and temporal CNNs for COVID-19 forecasting in France. Appl. Intell. (2021). https://doi.org/10.1007/s10489-021-02359-6, https://link.springer.com/10.1007/s10489-021-02359-6
20. Nair, V., Hinton, G.E.: Rectified linear units improve restricted boltzmann machines. In: Proceedings of the 27th International Conference on International Conference on Machine Learning, ICML'10, Omnipress, Madison, WI, USA, pp. 807–814 (2010)
21. Paszke, A., et al.: PyTorch: an imperative style, high-performance deep learning library. In: Wallach, H., Larochelle, H., Beygelzimer, A., d' Alché-Buc, F., Fox, E., Garnett, R. (eds.) Advances in Neural Information Processing Systems, vol. 32, pp. 8024–8035. Curran Associates, Inc. (2019). http://papers.neurips.cc/paper/9015-pytorch-an-imperative-style-high-performance-deep-learning-library.pdf
22. Reuters: Peru hits new COVID-19 case record as Brazilian variant spreads (2021). https://www.reuters.com/article/us-health-coronavirus-peru-idUSKBN2BG3CL
23. Smith, L.N.: Cyclical learning rates for training neural networks. In: 2017 IEEE Winter Conference on Applications of Computer Vision (WACV), pp. 464–472. IEEE (2017). https://doi.org/10.1109/WACV.2017.58, http://ieeexplore.ieee.org/document/7926641/
24. Tieleman, T., Hinton, G.: Lecture 6.5-rmsprop: divide the gradient by a running average of its recent magnitude. COURSERA Neural Netw. Mach. Learn. 4(2), 26–31 (2012)
25. Wieczorek, M., Siłka, J., Połap, D., Woźniak, M., Damaševičius, R.: Real-time neural network based predictor for cov19 virus spread. PLOS ONE 15(12), e0243189 (2020). https://doi.org/10.1371/journal.pone.0243189, https://dx.plos.org/10.1371/journal.pone.0243189
26. Zeroual, A., Harrou, F., Dairi, A., Sun, Y.: Deep learning methods for forecasting COVID-19 time-Series data: a comparative study. Chaos Solitons Fractals 140, 110121 (2020). https://doi.org/10.1016/j.chaos.2020.110121, https://linkinghub.elsevier.com/retrieve/pii/S096007792030518X

Analysis of Ontology Quality Dimensions, Criteria and Metrics

R. S. I. Wilson[1,2(✉)] (iD), J. S. Goonetillake[2] (iD), W. A. Indika[3] (iD), and Athula Ginige[4] (iD)

[1] Uva Wellassa University, Badulla, Sri Lanka
shyama@uwu.ac.lk
[2] University of Colombo School of Computing, Colombo 07, Sri Lanka
jsg@ucsc.cmb.ac.lk
[3] University of Ruhuna, Matara, Sri Lanka
waindika@dcs.ruh.ac.lk
[4] Western Sydney University, Penrith, NSW, Australia
a.Ginige@westernsydney.edu.au

Abstract. Ontology quality assessment needs to be performed across the ontology development life cycle to ensure that the ontology being modeled meets the intended purpose. To this end, a set of quality criteria and metrics provides a basis to assess the quality with respect to the quality requirements. However, the existing criteria and metrics defined in the literature so far are messy and vague. Thus, it is difficult to determine what set of criteria and measures would be applicable to assess the quality of an ontology for the intended purpose. Moreover, there are no well-accepted methodologies for ontology quality assessment as the way it is in the software engineering discipline. Therefore, a comprehensive review was performed to identify the existing contribution on ontology quality criteria and metrics. As a result, it was identified that the existing criteria can be classified under five dimensions namely syntactic, structural, semantic, pragmatic, and social. Moreover, a matrix with ontology levels, approaches, and criteria/metrics was presented to guide the researchers when they perform a quality assessment.

Keywords: Ontology quality assessments · Quality criteria · Quality metrics · Quality dimensions · Ontology evaluation

1 Introduction

Ontology is *"a formal, explicit specification of a shared conceptualization"* [1] that has numerous capabilities such as *analyzing domain knowledge, making available implicit knowledge explicit, sharing a common understanding of the structure of information among people or software agents* [2]. Thus, ontologies are incorporated in information systems as a component to manage heterogeneous information and high-volume data in domains like medicine, agriculture, defense, and finance. This supports information consumers or software agents to make the right decisions to tackle practical problems. However, the right decisions depend on the quality of the information provided, in the case of ontology-based information systems, the quality of the ontology.

© Springer Nature Switzerland AG 2021
O. Gervasi et al. (Eds.): ICCSA 2021, LNCS 12951, pp. 320–337, 2021.
https://doi.org/10.1007/978-3-030-86970-0_23

Ontology quality is a promising research area in the semantic web that has been discussed under ontology evaluation [3]. Ontology quality assessment is useful for ontology consumers to select a suitable ontology from a set of ontologies or assess the fitness of an ontology for an intended purpose [4, 5]. Moreover, quality assessment should not be limited to evaluate the product at the final stage. Thus, an ontology is also needed to be evaluated across the entire ontology life cycle. An ontology consists of levels namely syntactic, vocabulary (i.e., terminology), architecture, semantic, and context, also named as layers [6, 7]. The evaluation of ontology can be considered with respect to each ontology level/layer to reduce the complexity of the overall ontology quality assessments.

However, building a good quality ontology is not straightforward as it requires to consider several aspects such as logic, reasoning, structure, domain knowledge to be modeled concerning the specified tasks [8, 9]. To this end, ontology quality criteria are important to assess the components of an ontology that may have several measures (i.e., metrics) which provide an objective and a quantitative basis insight for ontology developers. Then, they can understand the areas to be revised to achieve a good quality of an ontology. Firstly, the authors of the article [6] have proposed a significant set of ontology quality criteria (i.e., characteristics) such as *correctness, soundness, consistency, completeness, conciseness, expandability,* and *sensitiveness*. Later, many more criteria have been added to this list by scholars from different points of view [3, 10–14]. For instance, the scholars of the article [3] have proposed a set of criteria by considering ontology as a software artifact. They have adopted the standard ISO/IEC 25000:2005 titled SQuaRE [15] and have suggested the criteria: *functional adequacy, reliability, performance efficiency, operability, maintainability, capability, transferability,* and *structural*. OntoQA is an approach that describes eleven (11) quality measures that can be used to evaluate the quality of an ontology at schema and data (i.e., knowledge bases) levels [14]. Schema level measures are *the richness of relationships, inheritance,* and *attributes*. The measures: *Class richness, average population, cohesion, fullness, connectivity* are classified as data level measures. Furthermore, OntoMetric is another web-based tool that assesses ontology quality under five criteria: *basic, schema, graph, knowledge,* and *class* which include 160 measures [13].

At present, many quality criteria and measures have been defined to assess the quality of an ontology from different perspectives such as ontology perspectives (i.e., inherent ontology quality), real-world perspectives, and users' perspectives [8, 13, 14, 16]. However, all these criteria and measures defined in the literature so far are messy and vague. For example, it is difficult to understand the quality criteria and thus the quality measures relevant to a given criterion in most cases as the terminology has been used inattentively. There is no distinction made between the two concepts quality criteria and quality measure from an ontology quality point of view. Moreover, the ontology has many levels as described in Sect. 4.1 and the quality of each level is significant for the overall quality of the ontology. Even if the quality of a certain level has been discussed in the literature in a very ad hoc way none of the existing definitions or approaches have defined quality criteria or measures in a methodical way for all of the levels. Consequently, no proper guidelines exist so far for ontology quality evaluation as the way it is with software engineering. For instance, when an ontology is evaluated through an application-based approach, it is necessary to understand what quality criteria to be adopted at the contextual level, semantic level, and structural level. Currently, ontology researchers and practitioners limit the quality evaluation only to a certain set of criteria namely *expressiveness* and *usefulness* due to the nonexistence of proper guidelines [16].

Nevertheless, our effort to analyze the quality criteria and measures that have been identified in the previous studies and synthesize them to provide an overview with ontology levels and approaches in order to produce a good quality ontology. To achieve this aim, data quality theories have been adopted. This would guide ontology developers and researchers to understand what quality criteria are to be assessed in each level (i.e., layer) and what the possible approaches would be to evaluate the ontologies.

2 Related Work

We analyzed the existing survey studies which have focused on ontology evaluation criteria, metrics, and approaches. Among them, a countable number of survey studies [16–20] were reviewed the related works comprehensively or systematically. However, none of them have provided a model or matrix, or overview among quality criteria, approaches, and ontology levels. This has caused a difficulty for researchers to gain insight on what quality criteria to be considered when performing ontology evaluation and what criteria would be more appropriate to assess each level of an ontology.

The author of the article [17] has highlighted the important quality criteria: *consistency, completeness, conciseness, expandability*, and *sensitiveness* through theoretical analysis and based on her experience. Then, a set of errors that can be made by ontology developers, have been classified under each quality criterion except *expandability* and *sensitiveness*. Finally, it has presented the ways of detecting *inconsistency, incompleteness*, and *redundancy*. Moreover, the work has also highlighted the requirement of developing language-dependent evaluation tools and the importance of documenting ontology quality with criteria. The research article [18] has considered the automatic, domain- and task-independent ontology evaluation as the scope of the study and has also focused on a set of ontology quality criteria that have been explained in five articles including [11, 17]. Furthermore, the evaluation of each ontology level with related measures has been described. For instance, the structure level can be evaluated by considering sub-graphs: *depth, breadth*, and *fan-outness*, and the context level can be assessed with competency questions, or through unit tests. Nevertheless, there is no clear comparison or discussion of ontology criteria and how those can be associated in each level when evaluating an ontology.

There are fifty-one structural quality measures that have been explored in [19]. Since the definition in natural language can be interpreted by different researchers from different perspectives and the paper has constructed formal definitions for each measure of quality criteria to provide a common understanding. Thus, the authors of the article [19] have addressed that issue by introducing formal definitions based on the Ontology Definition Model (ODM) and have presented the formal definitions for the quality measures: *Richness, Cohesion, Class Importance, Fullness, Coupling, Class Connectivity, Class Readability*. These formal definitions support researchers to compare the definitions and intents of measures when evaluating the structure of the ontology.

As the ontology quality assessment is required in each stage of the ontology life cycle, it is vital to aware of what criteria to be considered in each stage. Thus, the researcher of the article [16], has explored the quality criteria which are relevant for the evaluation of design and implementation stages. For that, a systematic review has been performed by retrieving articles from two reputed journals: the Journal of Web Semantics and the Semantic Web Journal. As the author has explored, *accuracy, adaptability, cognitive adequacy, completeness, conciseness, consistency, expressiveness*, and *grounding* as relevant criteria for evaluating the ontology in the design stage. To evaluate the quality of an ontology in the implementation stage, the criteria: *computational efficiency, congruency, practical usefulness, precision*, and *recall* have been recommended. Moreover, it has been revealed that few quality criteria such as *expressiveness* and *practical usefulness* have been used in practice though there are many quality criteria defined in theoretical approaches [16].

As a diverse set of ontology quality criteria exist, it is difficult for researchers to find a suitable set of quality criteria for assessing a particular ontology based on the intended purpose. To mitigate this issue, scholars have adopted well-defined theories and standards in the software engineering discipline [3, 20]. In the article [20], the authors have conducted a systematic review to identify the ontology quality criteria and grouped the measures of quality criteria into categories namely Inherent and Inherent-System, which have been defined in ISO/IEC 25012 Data Quality Standard. The adapted inherent quality criteria from this standard are *accuracy, completeness, consistency*, and *currentness*. The inherent-system criteria are *compliance, understandability, availability*. Under these criteria, the ontology measures identified through the survey have been mapped. For instance, the *accuracy* criterion includes the measures: *Incorrect Relationship, Hierarchy Over-specialization, Class Precision, Number of Deprecated Classes* and *Properties*, etc. and the *completeness* criterion includes the measures: *Number of Isolated Elements, Missing Domain or Range in Properties, Class Coverage*, and *Relation Coverage*. However, this classification can be applied to compare the quality of two or more ontologies in a similar domain, but it is not sufficient to assess a single ontology to get an idea on which components (i.e., levels) of an ontology have good quality and which are needed to be improved.

Moreover, the scholars [7, 21–25] have discussed several ontology evaluation approaches, criteria and ontology levels to be focused on when assessing ontology quality. Only the researchers of the articles [22, 23] have attempted to provide a comparison between the ontology quality approaches and criteria. However, the comparisons are abstract and difficult to interpret. The authors in [23] have stated that it is difficult to associate criteria with ontology approaches and ontology levels due to their diversity [23]. According to our study, a reason for having several criteria is due to the availability of different definitions to the same criteria, or vice versa (i.e., two or more closely related criteria may have the same definitions). This issue is further discussed in Sect. 4.2. Thus, we made an effort to carefully analyze these different definitions and to specify possible ontology quality criteria related to approaches, and levels (i.e., layers). To this end, a comprehensive theoretical analysis was conducted on ontology quality criteria and metrics (i.e., measures).

3 Methodology

To address the gap highlighted in Sects. 1 and 2, we performed a theoretical review by following the procedure proposed in [26]. To this end, the relevant background and gaps to be addressed have been explained in the previous sections. As the next step, the search terms to find the relevant papers from the databases: ACM Digital Library, IEEE Xplore Digital Library, Science Direct, Springer Link, and the search engine: Google Scholar, were defined. They are *ontology, ontology quality criteria, measures, metrics, quality assessment*, and *ontology evaluation*. Then, the general search strategy was developed to perform a search on the databases which are "*[ontology AND [Quality OR Evaluation OR Assessment] AND [Criteria OR Measures OR Metric]]*". At this stage, the articles were filtered purposefully by analyzing titles and abstracts as the study intention is not to explore the state-of-art in ontology quality assessment, but to analyze the ontology criteria which have been covered through the ontology levels, possibly with approaches. To reduce the searching results and to retrieve quality studies, the inclusion criteria such as;

- studies in English,
- studies published during (2010–2021),
- peer-reviewed,
- full-papers, and
- studies focused on quality assessment, criteria, and measures

have been applied. Finally, the relevant articles were downloaded through the reference management tool (i.e., Mendeley). Moreover, few potential articles were retrieved by looking up the references of the filtered articles. Thereafter, we selected the articles which are [3, 6, 10, 11, 27–35] for the analysis.

4 Data Analysis and Synthesis

4.1 Prerequisite

In previous studies, the following terms have been used interchangeably in their explanations of ontology quality. In this study, mostly we use the terms: criteria (i.e., characteristics), metrics (i.e., measures), dimensions, ontology levels, and ontology approaches to describe the theories in order to maintain consistency.

Criteria (i.e., Characteristics)
Ontology criteria (i.e., characteristics) describe a set of attributes. An attribute is a measurable physical or abstract property of an entity [36, 37]. In ontology quality, an entity can be a set of concepts, properties, or an ontology.

Metric (i.e., Measure)
Metric (i.e., measure) describes an attribute quantitatively or defines an attribute formally [38]. In other words, the ontology quality metric is used to measure the characteristic of an ontology that can be represented formally.

For instance, the conceptual complexity is a criterion (i.e., characteristic) that is used to evaluate an ontology (i.e., entity) and it can be quantitatively measured by using the metric: size, which may have measurements such as number of concepts and properties in the structure, number of leaf concepts and number of attributes per concepts.

Dimension (i.e., Aspects)

Dimensions (i.e., aspects) have been defined to classify several criteria/attributes based on different views. For example, if a dimension describes the content of an ontology, that may include a set of criteria related to the content assessments. The criteria like graph complexity, modularity, and graph consistency can be grouped into the structural dimension.

Therefore, similar to the software data quality, dimensions are qualitative and associate with several characteristics (i.e., attributes) that can be directly or indirectly measured through quantitative metrics.

Hereinafter, we use terms: *criteria* and *metric* instead of using the terms *ontology quality criteria* and *ontology quality metrics* respectively.

Ontology Levels (i.e., Layers)

In the ontology quality assessment, initially, three levels (also known as layers) to be focused on have been proposed in [6], namely: *content, syntactic & lexicon*, and *architecture*. Later, this was expanded by including *structural* and *context* [7, 18]. These levels/layers focus on different aspects of ontological information. The syntactic level considers the features related to the formal language that is used to represent the ontology. The lexicon level is also named vocabulary or data layer that takes into account the vocabulary that is used to describe concepts, properties, instances, and facts. The structural level/architectural layer focuses on the is-a relationship (i.e., hierarchical) which is more important in the ontology modeling against the other relations. Moreover, it considers the design principles and other structural features required to represent the ontology. Other non-hierarchical relationships and semantic elements are considered under the semantic level. The context level concerns the application level that the ontology is built for. It is important to assess whether the ontology confirms the real application requirements as a component of an information system or a part of a collection of ontologies [7, 39].

Ontology Evaluation Approaches (i.e., Methods, Techniques)

Mainly, ontology evaluation has been conducted under four approaches: application-based, data-driven-based, golden standard-based, and human-based [7]. In brief, the application-based approach: assesses the ontology when it is attached with the application and used in practice [39]. The data-driven approach: assesses the ontology against the data source (i.e., corpus) that is used for the ontology modeling. The golden-standard approach: compares the candidate ontology with the ontology that has the agreed quality or assesses the ontology with a benchmark/a vocabulary defined by experts. The human-based approach: assesses the ontology with the intervention of domain experts and ontology engineers based on the set of criteria, requirements, and standards [6, 7].

Table 1. The existing quality models for ontologies

Citation	Dimensions	Metric/attributes
Burton-Jones et al. 2005 [10]	Syntactic	Lawfulness, richness
	Semantic	Interpretability, consistency, clarity
	Pragmatic	Comprehensiveness, accuracy, relevance
	Social	Authority, history
Gangemi et al. 2006 [11]	Structural	Size, modularity, depth, breadth, tangledness, etc.
	Functional	precision, recall (coverage), Accuracy
	Usability-related	recognition, efficiency, and interfacing annotations
Zhu et al. 2017 [30]	Content	Correctness of entities, Semantic coverage, Vocabulary coverage
	Presentation	Size, relation, modularity: cohesion/coupling, non-redundancy
	Usage	Search efficiency, description complexity, definability, extendibility, tailorability, etc.

4.2 Ontology Quality Dimensions, Criteria, and Metrics

The ontology quality evaluation throughout the ontology life cycle ensures that good quality ontology is being developed. However, a major issue is the unavailability of an agreed methodology for it. As a result, several criteria and metrics have been defined without a strong theoretical foundation. According to our analysis, we were able to identify a set of criteria and metrics that were presented in Table 2. The related measurements have not been mentioned as all together hundreds of measurements are available. Thus, only the relevant citations have been provided for further references. Moreover, the metrics definitions similar to the ones in [11] and [3] can also be found in [12] and [40] respectively.

There are significant attempts in the literature to introduce generalized dimensions to classify criteria and metrics (see Table 1). The authors of the article [10] have introduced quality dimensions: *syntactic, semantic, pragmatic,* and *social* by adopting a semiotic framework named the semiotic metric suit. In the article [11], the researchers have classified metrics into three dimensions: *structural, functional,* and *usability-related.* In addition to that, the scholars in the research [30] have introduced a set of dimensions namely *presentation, content,* and *usage* considering the web service domain. When analyzing the metrics defined under each dimension, it has been recognized that the proposed dimensions are overlapping. For instance, the criteria: *modularity, size: concept/relations* defined under the structural dimension in [11] also appear in the presentation dimension in [30]. Moreover, the criteria in the dimensions: *semantic* [10], *functional* [11], and *content* [30] have been defined concerning the domain that the ontology being modeled. The *pragmatic* [10], *usability-related* [11], and *usage* [30] dimensions consider the quality when an ontology is at the application level and to this end, criteria such as *functional accuracy, relevancy, adaptability, efficiency,* and *comprehensibility* have been considered. Based on that, we have identified five main distinguish dimensions that the criteria can be grouped such as *syntactic, structural, semantic, pragmatic,* and *social.* This can be seen as an extended version of the semiotic metric suit [10]. Then, taking this as a basis, the identified criteria and metrics in the literature were mapped by analyzing the given definitions (see Table 2).

Classification of Criteria and Metrics
After a thorough analysis, we identified fourteen main ontology criteria namely *syntactic correctness, cognitive complexity, conciseness, modularity, consistency, accuracy, completeness, adaptability, applicability, efficiency, understandability, relevance, usability,* and *accessibility* which are classified as follows (the possible metrics are in the italic format).

– Syntactic: describes the conformance to the rules of the language that the ontology is written [10, 30]

   • Syntactic correctness: *lawfulness, richness*

– Structural: describes the topological and logical properties of an ontology [11]

   • Cognitive Complexity: *size, depth, breadth, fan-outness,* Modularity: *cohesion, coupling,* Internal Consistency: *tangledness, circularity, partition*

– Semantic: describes the characteristics related to the semantic (meanings) of an ontology [10, 34].

   • Conciseness: *precision,* Coverage: *recall,* External Consistency: *clarity, interpretability*

– Pragmatic: describes the appropriateness of an ontology for an intended purpose/s

   • Functional Completeness: *competency questions, precision,* Accuracy, Adaptability, Applicability, Efficiency, Understandability, Relevance, Usability: *ease of use,* Accessibility

– Social: describes the characteristics related to ontology quality in use (user-satisfaction/ social acceptance)

   • *recognition, authority, history*

Few of the criteria can be further decomposed into sub-criteria related to different perspectives: inherent to the ontology (i.e., ontology perspective), domain-depend (i.e., real-world perspective), and user-depend (i.e., user perspective). For instance, there are two types of consistency: *internal consistency* and *external consistency* [30].

Table 2. Associated ontology quality dimensions of criteria, and metrics

Dimensions	Criteria (i.e., characteristics)	Metrics	Citations	*Levels
Syntactic	Correctness: syntactic correctness	Lawfulness/well-formedness	[10, 30]	SY
Syntactic	Correctness: syntactic correctness	Richness	[10]	SY
Structural	Cognitive complexity: comprehensiveness	Size	[27, 29]	ST, L
Structural	Modularity	Coupling	[30, 35]	ST
Structural	Modularity	Cohesion	[3, 30, 35]	ST, S
Structural	Cognitive complexity	Depth	[12]	ST
Structural	Cognitive complexity	Breadth	[12]	ST
Structural	Cognitive complexity	Fan-outness	[12]	ST
Structural	Internal consistency	Tangledness	[12]	ST, S
Structural	Internal consistency	Circularity	[12, 17, 27]	ST, S
Structural	Internal consistency	Partition error	[6]	ST, S
Semantic	Conciseness	Precision	[3, 30, 33]	ST, S
Semantic	External consistency: semantic correctness	Clarity	[10, 17, 34]	S
Semantic	External consistency: semantic correctness	Interpretability	[17, 33]	S
Semantic	Coverage: semantic completeness	Inferred vocabulary coverage (recall)	[30]	S
Semantic	Coverage: syntactic completeness	Vocabulary coverage (recall)	[30]	L
Pragmatic	Functional completeness	Competency questions/precision	[34]	C
Pragmatic	Accuracy	–	[10, 34]	C
Pragmatic	Adaptability	–	[30, 34]	C
Pragmatic	Applicability	–	[30]	C
Pragmatic	Efficiency	–	[30]	C
Pragmatic	Understandability	–	[30]	C
Pragmatic	Relevance	–	[10, 34]	C
Pragmatic	Usability	Ease of use	[3, 34]	C
Pragmatic	Commercial accessibility: fitness	-	[18]	

(continued)

Table 2. (*continued*)

Dimensions	Criteria (i.e., characteristics)	Metrics	Citations	*Levels
Social	–	Recognition	[34]	C
Social	–	Authority	[10]	C
Social	–	History	[10]	C

*Levels are represented in symbols: syntactic - SY, structural - ST, lexical - L, semantic - S and context – C

Internal consistency is an inherent characteristic of ontologies that considers whether there is any self-contradiction within the ontology (i.e., ontology perspective) [18, 30]. In the article [17], the authors have classified three inconsistencies in this regard as *circularity errors, partition errors*, and *semantic errors*. Circularity and partition errors describe the logical inconsistencies related to the structure and the relations of an ontology, thus, both are inherent to the ontology. Tangledness that has been described in [11, 12] also considers as a measure of internal consistency of the structure as tangledness occurs when a class has multiple parent classes.

To determine the *semantic correctness* (i.e., semantic errors), it is necessary to consider the domain knowledge that the ontology used to specify the conceptualization. Thus, it comes under the *external inconsistency* that considers the consistency from the real-world perspective. Furthermore, *clarity* and *interpretability* can also be considered as metrics of *semantic correctness* [33].

Moreover, the definitions are given for the criteria: *completeness, coverage*, and *expressiveness* are closely related (see Table 3). Based on the *completeness* definition given in [41, 42] for Data Quality, we identified that ontology completeness also can be further decomposed as *coverage: syntactic completeness, coverage: semantic completeness*, and *functional completeness* considering different perspectives. For instance, *coverage* (or the semantic completeness) describes the completeness from the real-world perspective that determines the degree of covered entities (i.e., concepts, relation, attribute, instances) in the domain [17, 28, 33]. Moreover, the measures: *missing instances, missing properties, isolated relations*, and *incomplete formats* are also used to assess *coverage* from the ontological inherent point of view, which can be detected without domain knowledge. Thus, it is named *syntactic completeness*. In addition to that, we defined the criteria: *functional completeness* concerning the user perspective, in which the completeness is measured considering whether the ontology provides complete answers for the users' queries (i.e., competency questions), which is more subjective and difficult to measure. In the case of data-driven ontologies, functional completeness is measured against the corpus that the ontology to be covered.

Table 3. The definition is given for the criteria and metrics in the previous studies.

Criteria (C)/Metrics (M)	Definition
Coverage (C)	*"Measures how well the candidate ontology covers the terms extracted from the corpus"* [28]
Correctness (C)	*"Whether the ontology accurately represents the knowledge of a subject domain"* [30]
Syntactically correct (C)	*"Evaluates the quality of the ontology according to the way it is written"* [33]
Completeness (C)	*"Whether the ontology represents all of the knowledge of a subject domain"* [30] *"Refers to the extension, degree, amount or coverage to which the information in a user-independent ontology covers the information of the real world"* [17]. *"the level of granularity agreed"* [33]
Compatibility (C)	*"Whether the ontology contains junk, i.e., contents not in the subject domain"* [30]
Internal consistency (C)	*"Whether there is no self-contradiction within the ontology. (cannot measure)"* [30] *"the formal and informal definition have the same meaning"*. [6]
External consistency (C)	*"Whether the ontology is consistent with the subject domain knowledge"* [30]
Syntactic completeness (C)	*"How much the vocabulary of the ontology matches exactly that of the standard"* [30]
Semantic completeness (C)	*"How much the vocabulary of the standard can be derived from the ontology"* [30]
Expressiveness (C)	*"Refers to an ontology's degree of detail"* [9]
Well formedness (M)	*"Syntactic correctness with respect to the rules of the language in which it is written"* [30]
Conciseness (C)	*"The key attribute for this is the lack of redundancy within the ontology"* [30] *"All the knowledge items of the ontology must be informative, so that non-informative items should be removed"* [3] *"An ontology is concise if it does not store any unnecessary or useless definitions if explicit redundancies do not exist between definitions, and redundancies cannot be inferred using other definitions and axioms"* [17, 33]
Structural complexity (C)	The cardinality of ontology elements: classes, individuals, properties

(continued)

Table 3. (*continued*)

Criteria (C)/Metrics (M)	Definition
Modularity (C)	*"How well the ontology is decomposed into smaller parts, to make it easier to understand, use, and maintain"* [30]
Cohesion (M)	*"The degree of interaction within one module or ontology"*. [30]
Coupling (M)	*"The degree of cross-referencing"* [30]
Tangledness (M)	*"This measures the distribution of multiple parent categories"* [3]
Cycles (M)	*"The existence of cycles through a particular semantic relation"* [3], *"occurs when a class is defined as a specialization or generalization of itself"* [6]
Applicability (C)	*"This relates to whether the ontology is easy to apply for a specific task, which in this case is the description of web service semantics"* [30]
Adaptability (C)	*"How easily the ontology can be changed to meet the specific purposes of developing a particular web service"* [30]
Efficiency (C)	*"How easily semantic information can be processed for various purposes"* [30]
Comprehensibility (C)	*"Whether human readers can easily understand the semantic description"* [30]
Representation correctness (C)	*"Evaluate the quality of mappings of entities, relations, and features into the elements of the ontology were specified"* [33]
Structural accuracy (C)	*"It accounts for the correctness of the terms used in the ontology"* [3]
Lawfulness (M)	*"Correctness of syntax -the degree to which ontology language rules have complied"* [10]
Richness (M)	*"Refers to the proportion of features in the ontology language that have been used in an ontology"* [10]
Interpretability (M)	*"The knowledge provided by the ontology should map into meaningful concepts in the real world"* [33]
Consistency (C)	*"Whether terms have a consistent meaning in the ontology. E.g. if an ontology claims that X is a subclass_of Y, and that Y is a property of X, then X and Y have inconsistent meanings and are of no semantic value"* [10]
Clarity (M)	*"Whether the context of terms is clear. E.g.: in the context of academics, chair is a person, not furniture)"* [10]
Comprehensiveness (C)	*"Number of classes and properties - a measure of the size of the ontology"* [10]

(*continued*)

Table 3. (*continued*)

Criteria (C)/Metrics (M)	Definition
Accuracy (C)	*"Whether the claims an ontology makes are true"* [10]
Relevance (C)	*"Whether the ontology satisfies the agents specific requirements"* [10]
Authority (C)	*"Extent to which other ontologies rely on it (define terms using other ontology definitions)"* [10]
History (M)	*"Number of times ontology has been used"* [10]
Adaptability (C)	*"Whether the ontology provides a secure foundation that is easily extended and flexible enough to react predictably to small internal changes"* [34]
Ease of use (M)	*"Assess the level of documentation in the form of annotations included"* [34]
Recognition (M)	*"Assess the level of use it has received within its community"* [34]

5 Results and Discussion

When mapping the criteria with respect to the defined dimensions in our study, few deviations have been observed in the literature. For instance, in the article [34], the scholars have defined *structural complexity: the number of subclasses in the ontology* as a syntactic characteristic [34]. Although it describes the static property of the ontology structure, it is not a property that reflects the syntactic feature as defined in [34]. Importantly, the authors in [32] show that the structural metric: *cohesion* can be adopted to measure the semantics instead of the structural features of an ontology. This implies that the metrics do not strictly attach to one dimension and they can be measured in different ways to achieve the desired quality objectives. Thus, the measures would influence many dimensions through several criteria, which could be mapped with several ontology levels as shown in Table 2.

Moreover, in Table 4, we represented the metrics with levels and evaluation approaches that give an overview for researchers to identify possible quality metrics to be considered in each level with a suitable approach. Moreover, it has been mentioned that whether those metrics can be assessed manually or semi-/automatically. Based on our analysis, the metrics related to the structural, and syntactic levels can be automated as they are domain-independent. The metrics that come under the semantic and lexicon level also are automatable, however, need extra effort as those are domain depended. To assess the context level criteria, manual methods are mostly required as those are relative to the users and may not have specific quantitative metrics.

The metric: *formalism* describes the capabilities of the language that the ontology is written such as *machine understanding, reasoning,* and *defining required features* (i.e., entities, properties, relations). This has not been included under the proposed dimensions as it is considered before the ontology is modeled. However, the metric *formalism* is

Table 4. Matrix of ontology quality metrics, levels, and approaches.

Approaches	Layers				
	Context	Semantic	Lexicon	Structural	Syntactic
Application/Task/Users-based	Completeness: CQs (M) Accuracy (M) Ease of use (M) Relevancy (M) Efficiency (M) Applicability (M) Adaptability (M) Accessibility (M) History (M) Recognition (A)		*Clarity*		
Human-based (i.e., domain experts, ontology engineer)-	Accuracy (SA) Authority (M)	Conciseness: Precision (M) Comprehensibility (M)	*Cohesion (A)*	Size (A), Depth (A), Breadth (A), Fan-out (A), Tangledness (A), Circularity (A), Coupling (A), Cohesion (A), Partition (A), *Disjointsness*	Lawfulness (A) Richness (A) Formalization (M) *Flexibility*
Data-driven		Clarity (A) Interpretability (A)	Clarity (A)		
Golden-standard based	Accuracy (A)	Inferred Vocabulary coverage: (recall) (A), Clarity (A) Interpretability (A)	Vocabulary coverage(recall) (A)	Cohesion (A)	

*automate-A, semi-automate-SA, manual-M

noteworthy to consider when selecting a suitable ontology language for modeling. Thus, we included it in the matrix that can be assessed manually by ontology engineers.

To measure each metric, at least one approach is available. If many approaches are applicable, a suitable approach could be selected based on the purpose and availability of resources (i.e., time, experts, type of users, standards). For example, accuracy can be assessed through expert interventions, application-based or golden standard-based approaches [10]. However, in most cases, the golden standard (i.e., standard ontology, vocabularies, rules) is not available for comparison, and definitely, it is necessary to go with one of the other two approaches. If it is hard to evaluate an ontology in a real environment (i.e., application-based) with naïve users then the experts-based methods are acceptable.

When identifying the possible metrics with respect to the levels, we ignored the metrics defined in [3] since they have been defined by assuming ontology is a software artifact. As a result, the provided metrics are more subjective, and it is hard to match them with the ontology levels except the metrics defined under the structural dimension. Moreover, the authors in [22], have provided a criteria selection framework, without differentiating the metrics and the criteria. However, we adopted few metrics from it and have been included them in the matrix according to their classification (i.e., which are in the italic format in Table 4).

6 Conclusion and Future Work

A comprehensive theoretical analysis was performed on the ontology quality assessment criteria with the aim of providing an overview of criteria and possible metrics. The outcome of this analysis can be used to assess each ontology level with possible approaches as this domain has not been covered in the previous studies. To this end, we analyzed the definitions provided in the research works and clarified the vaguely defined definitions by studying theories in [36–38, 41–44]. Consequently, we were able to identify fourteen ontology quality criteria namely *syntactic correctness, cognitive complexity, conciseness, modularity, consistency, accuracy, completeness, adaptability, applicability, efficiency, understandability, relevance, usability,* and *accessibility*. These criteria have been classified under five dimensions namely: *syntactic, structural, semantic, pragmatic,* and *social*. Finally, a matrix was constructed that presents the association among the ontology levels, approaches, and criteria/metrics (see Table 4). This would become useful to gain an insight for researchers when dealing with the ontology quality assessment. Moreover, the absence of empirical evidence on the ontology quality assessment has limited the use of criteria in practice, and finding a methodological approach to derive ontology quality criteria with respect to the users' requirements (i.e., fit for the intended purpose) remains an open research problem.

References

1. Studer, R., Benjamins, V.R., Fensel, D.: Knowledge engineering: principles and methods. Data Knowl. Eng. **25**, 161–197 (1998)
2. Natalya, F.N., Deborah, L.M.: What is an ontology and why we need it. https://protege.stanford.edu/publications/ontology_development/ontology101-noy-mcguinness.html. Accessed 09 May 2021
3. Duque-Ramos, A., Fernández-Breis, J.T., Stevens, R., Aussenac-Gilles, N.: OQuaRE: a SQuaRE-based approach for evaluating the quality of ontologies. J. Res. Pract. Inf. Technol. **43**, 159–176 (2011)
4. Staab, S., Gomez-Perez, A., Daelemana, W., Reinberger, M.-L., Noy, N.F.: Why evaluate ontology technologies? Because it works! IEEE Intell. Syst. **19**, 74–81 (2004)
5. Duque-Ramos, A., et al.: Evaluation of the OQuaRE framework for ontology quality. Expert Syst. Appl. **40**, 2696–2703 (2013)
6. Gómez-Pérez, A.: Towards a framework to verify knowledge sharing technology. Expert Syst. Appl. **11**, 519–529 (1996)
7. Brank, J., Grobelnik, M., Mladenić, D.: A survey of ontology evaluation techniques. In: Proceedings of the Conference on Data Mining and Data Warehouses, pp. 166–170 (2005)
8. Ontology Summit 2013. http://ontolog.cim3.net/OntologySummit/2013/index.html. Accessed 09 May 2021
9. McDaniel, M., Storey, V.C.: Evaluating domain ontologies: clarification, classification, and challenges. ACM Comput. Surv. **52**, 1–44 (2019)
10. Burton-Jones, A., Storey, V.C., Sugumaran, V., Ahluwalia, P.: A semiotic metrics suite for assessing the quality of ontologies. Data Knowl. Eng. **55**, 84–102 (2005)
11. Gangemi, A., Catenacci, C., Ciaramita, M., Lehmann, J.: Modelling ontology evaluation and validation. In: Sure, Y., Domingue, J. (eds.) ESWC 2006. LNCS, vol. 4011, pp. 140–154. Springer, Heidelberg (2006). https://doi.org/10.1007/11762256_13
12. Gangemi, A., et al.: Ontology evaluation and validation: an integrated formal model for the quality diagnostic task, **53** (2005)
13. Lantow, B.: OntoMetrics: putting metrics into use for ontology evaluation. In: KEOD, pp. 186–191 (2016)
14. Tartir, S., Arpinar, I.B., Moore, M., Sheth, A.P., Aleman-Meza, B.: OntoQA: metric-based ontology quality analysis (2005)
15. ISO/IEC 25000:2005. https://www.iso.org/standard/35683.html. Accessed 09 May 2021
16. Degbelo, A.: A snapshot of ontology evaluation criteria and strategies. In: Proceedings of the 13th International Conference on Semantic Systems, pp. 1–8. Association for Computing Machinery, New York (2017)
17. Gómez-Pérez, A.: Ontology Evaluation. In: Staab, S., Studer, R. (eds.) Handbook on Ontologies. International Handbooks on Information Systems, pp. 251–273. Springer, Heidelberg (2004). https://doi.org/10.1007/978-3-540-24750-0_13
18. Vrandečić, D.: Ontology evaluation. In: Staab, S., Studer, R. (eds.) Handbook on ontologies. IHIS, pp. 293–313. Springer, Heidelberg (2009). https://doi.org/10.1007/978-3-540-92673-3_13
19. Reynoso, L., Amaolo, M., Vaucheret, C., Álvarez, M.: Survey of measures for the structural dimension of ontologies. In: 2013 IEEE 12th International Conference on Cognitive Informatics and Cognitive Computing, pp. 83–92 (2013)
20. Mc Gurk, S., Abela, C., Debattista, J.: Towards ontology quality assessment. In: MEPDaW/LDQ@ ESWC, pp. 94–106 (2017)
21. Pak, J., Zhou, L.: A framework for ontology evaluation. In: Sharman, R., Rao, H.R., Raghu, T. S. (eds.) WEB 2009. LNBIP, vol. 52, pp. 10–18. Springer, Heidelberg (2010). https://doi.org/10.1007/978-3-642-17449-0_2

22. Hooi, Y.K., Hassan, M.F., Shariff, A.M.: Ontology evaluation: a criteria selection framework. In: 2015 International Symposium on Mathematical Sciences and Computing Research (iSMSC), pp. 298–303 (2015)
23. Raad, J., Cruz, C.: A survey on ontology evaluation methods: In: Proceedings of the 7th International Joint Conference on Knowledge Discovery, Knowledge Engineering and Knowledge Management, pp. 179–186. SCITEPRESS - Science and and Technology Publications, Lisbon (2015)
24. Verma, A.: An abstract framework for ontology evaluation. In: 2016 International Conference on Data Science and Engineering (ICDSE), pp. 1–6 (2016)
25. Reyes-Pena, C., Tovar-Vidal, M.: Ontology: components and evaluation, a review. Res. Comput. Sci. **148**, 257–265 (2019)
26. Kitchenham, B.: Procedures for undertaking systematic reviews. Joint Technical report, Computer Science Department, Keele University (TR/SE-0401) and National ICT Australia Ltd. (0400011T.1) (2004)
27. Gavrilova, T.A., Gorovoy, V.A., Bolotnikova, E.S.: Evaluation of the cognitive ergonomics of ontologies on the basis of graph analysis. Sci. Tech. Inf. Process. **37**, 398–406 (2010)
28. Demaidi, M.N., Gaber, M.M.: TONE: a method for terminological ontology evaluation. In: Proceedings of the ArabWIC 6th Annual International Conference Research Track, pp. 1–10. Association for Computing Machinery, New York (2019)
29. Cross, V., Pal, A.: Metrics for ontologies. In: NAFIPS 2005 - 2005 Annual Meeting of the North American Fuzzy Information Processing Society, pp. 448–453 (2005)
30. Zhu, H., Liu, D., Bayley, I., Aldea, A., Yang, Y., Chen, Y.: Quality model and metrics of ontology for semantic descriptions of web services. Tsinghua Sci. Technol. **22**, 254–272 (2017)
31. Evermann, J., Fang, J.: Evaluating ontologies: towards a cognitive measure of quality. Inf. Syst. **35**, 391–403 (2010)
32. Ma, Y., Jin, B., Feng, Y.: Semantic oriented ontology cohesion metrics for ontology-based systems. J. Syst. Softw. **83**, 143–152 (2010)
33. Rico, M., Caliusco, M.L., Chiotti, O., Galli, M.R.: OntoQualitas: a framework for ontology quality assessment in information interchanges between heterogeneous systems. Comput. Ind. **65**, 1291–1300 (2014)
34. McDaniel, M., Storey, V.C., Sugumaran, V.: Assessing the quality of domain ontologies: metrics and an automated ranking system. Data Knowl. Eng. **115**, 32–47 (2018)
35. Oh, S., Yeom, H.Y., Ahn, J.: Cohesion and coupling metrics for ontology modules. Inf. Technol. Manag. **12**, 81 (2011)
36. IEEE Standard for a Software Quality Metrics Methodology: In IEEE Std 1061-1992, pp. 1–96 (1993)
37. Kitchenham, B., Pfleeger, S.L., Fenton, N.: Towards a framework for software measurement validation. IEEE Trans. Softw. Eng. **21**, 929–944 (1995)
38. Basili, V.R., Caldiera, G., Rombach, H.D.: The goal question metric approach. Encyclopedia Softw. Eng. **2**, 528–532 (1994)
39. Porzel, R., Malaka, R.: A task-based approach for ontology evaluation. In: Proceedings of the Workshop on Ontology Learning and Population at the 16th European Conference on Artificial Intelligence (ECAI), (2004)
40. OQuaRE: A SQUaRE based quality evaluation framework for Ontologies. http://miuras.inf.um.es/evaluation/oquare/. Accessed 09 May 2021
41. Wang, R.Y., Strong, D.M.: Beyond accuracy: what data quality means to data consumers. J. Manag. Inf. Syst. **12**, 5–33 (1996)
42. Mouzhi Ge., Markus, H.: A review of information quality research. In. Proceedings of the 2007 International Conference on Information Quality, pp. 76–91 (2007)

43. Uschold, M., Gruninger, M.: Ontologies: principles, methods and applications. Knowl. Eng. Rev. **11**, 93–136 (1996)
44. Nigel, B.: Human-computer interaction standards. In: Anzai, Y., Ogawa, K.(eds.) Proceedings of the 6th International Conference on Human Computer Interaction, vol. 20, pp. 885–890. Elsevier, Amsterdam (1995)

Semantic Improvement
of Question-Answering System
with Multi-Layer LSTM

Hanif Bhuiyan[1]([⊠])(iD), Md. Abdul Karim[2](iD), Faria Benta Karim[3](iD),
and Jinat Ara[4](iD)

[1] Data61, CSIRO, CARRS-Q, Queensland University of Technology, Queensland,
Australia
hanif.bhuiyan@data61.csiro.au
[2] Department of Computer Science and Engineering, University of Asia Pacific,
Dhaka, Bangladesh
[3] Department of Electrical and Electronic Engineering, University of Asia Pacific,
Dhaka, Bangladesh
[4] Department of Computer Science and Engineering, Jahangirnagar University,
Savar, Dhaka, Bangladesh

Abstract. Question answering (QA) is a popular method to extract
information from a large pool of data. However, effective feature extraction and semantic understanding, and interaction between question and
answer pairs are the main challenges for the QA model. Mitigating these
issues, here, we propose a Long-Short-Term-Memory (LSTM) model
using the recurrent neural network. This model aims to extract the effective interaction between questions-answers pairs using the Softmax and
Max-pooling function. The experiment performs on a publicly available
Wiki-QA dataset to identify the effectiveness of the proposed model. The
evaluation performs by comparing the result with other existing models.
The comparison shows that the proposed approach is significant and
competitive for QA and achieves 83% training and 81% testing accuracy
on the Wiki-QA dataset.

Keywords: Question answering · Long-short-term-memory ·
Recurrent neural network · Information extraction

1 Introduction

An intelligent question-answer (QA) is a popular human machine interaction
approach to satisfy the user by retrieving correct answers or information of a
specific question [1]. The rapid increase of web data allows a web user to store
data and make it available for the community for future use [2,3]. However, this
massive storage data makes the searching process is a difficult task [4]. This difficulty allows NLP researchers to develop new tools such as Question Answer to
find the correct information from structured and unstructured corpus data [5].

© Springer Nature Switzerland AG 2021
O. Gervasi et al. (Eds.): ICCSA 2021, LNCS 12951, pp. 338–352, 2021.
https://doi.org/10.1007/978-3-030-86970-0_24

In the past few years, artificial intelligence and knowledge base methods were popular in question-answer systems for many NLP researchers [6,7]. These methods were performed through several text analysis algorithms to analyze and identify the most similar keywords [8] for considering the appropriate answer to the question [9]. Surprisingly, these methods have a lack of semantic meaning. The accuracy of the QA system mostly depends on the QA sentence embedding and mapping. It helps to generate a sequential representation of the context. Generally, sequentially QA sentence mapping is difficult and hard to get good accuracy. Besides, it addresses the shortcoming of knowledge-based techniques [10]. However, a fully functional question-answer system is relatively challenging,and several researchers have tried to build a more systematic process to improve accuracy [11,12].

In recent days, to minimize the shortcoming of knowledge-based techniques, deep learning provides several methods and techniques [12]. The majority of NLP solutions have replaced by deep learning techniques to improve accuracy and produce accurate results (especially the abstract question-answering approaches) [13]. This method performs through an artificial neural network and showed significant improvement in the question-answer system. Therefore, deep learning uses widely and accepted by the NLP researcher to discover the distributed data representation process [14]. Among several deep learning methods, recurrent neural network (RNN) is a significant innovation and frequently used method to map the question and answer sentences [11]. It performs through continuously identifying the meaning of each word of a sentence to represent the entire meaning of the sentences and mapping with the other sentence word.

Recently researchers take advantage of recurrent neural network models for semantic analysis of contextual data [12,15–18]. Feng et al. [15] and Wang and Nyberg [16] used recurrent neural networks and Long Short-Term Memory Networks. These works are limited to identify short text semantics of small corpus data. They didn't consider predicting the full-sentence semantics of corpus data, resulting in a poor semantic understanding. Zhang and Ma [19] proposed a deep learning-based QA model with a single LSTM network concept. They didn't consider a multi-layer LSTM network that might reduce accuracy and produces inaccurate results.Although these methodologies are effective in QA systems,the implemented techniques still bring low accuracy.

In this paper, we implement a multilayer LSTM network through RNN architecture for the single sentence question-answer domain. This work aims to solve the need for semantic understanding and improve the overall accuracy of corpus data. Besides, an attention mechanism implements to improve the semantic understanding of the proposed model. It helps to increases the interaction between the question-answers pairs. An optimization technique uses to improve the performance of the proposed model. A comprehensive experiment conducted to determine the proposed model's effectiveness on the publicly available Wiki-QA data set.The experiment result shows that neural network-based multilayer LSTM architecture is more effective for question-answer tasks than single LSTM architecture. The comparison result of existing model considers evaluating the proposed model. The compared result shows that the proposed model is significant and achieved higher accuracy than the existing system (83% and 81% accuracy for training and testing, respectively). The contributions of this paper summarized as follows:

(a) Multilayer LSTM recurrent neural network architecture is implemented to identify the semantic meaning and effectively map the input question-answer sentence vector.

(b) The proposed model incorporates the ReLU activation function and softmax attention mechanism to obtain the most semantic representation of the QA pair.

(c) Max-pooling and Adam optimizer reconciled to calculate the accuracy for improving the accuracy rate.

The rest of the paper presents as follows. In Sect. 2, represent a brief overview of related work. In Sect. 3, presents the proposed LSTM-RNN architecture for the question answering framework. In Sect. 4, added the details about the implementation procedure of the proposed QA model. In Sect. 5, conclude the experimental result with performance analysis. Finally, a detailed conclusion or discussion presents in Sect. 6.

2 Related Work

To understand the insight of corpus data and retrieving the desired outcome from a massive number of semantic web data, QA is a popular approach. There are several techniques to perform QA tasks [6,7,11,15,16]. Larson et al. [20] proposed work enhances the importance of POS tagging and tf-idf technique in the question-answering system. Sharma et al. [21] proposed a QA model considering artificial intelligence algorithms. It shows the significant importance of the AI technique in QA approach. Both of these techniques are promising, however these fail to represent semantic understanding effectively and resulting low accuracy. Therefore, NLP researcher has started to implement deep learning methods for structure and unstructured data analysis. Motivated by deep learning techniques, Cai et al. [22] implement a neural network structure using a stacked Bidirectional Long Short-Term Memory (BiLSTM) model for question-answer tasks. Rohit et al. [23] implemented a neural network architecture using a single-layer LSTM architecture and memory network for closed domain question-answer analysis. The experimental result of these two proposed approaches [22,23] addressed the necessity of neural network architecture in the QA approach. Though these works showed impressive results, they only considered single layer LSTM architecture and didn't focus on the attention mechanism and optimization technique. Less consideration of the LSTM layer, attention mechanism, and optimization technique could lead to a lack of semantic understanding and reduce overall accuracy.

Therefore, Li et al. [26] emphasized the importance of multi-layer LSTM architecture. Furthermore, Yin et al. [18] added the importance and significance of the neural network technique or architecture to improve the Question Answering system. Here, we implement four LSTM architecture layers using the recurrent neural network (RNN) architecture with an attention mechanism and optimization algorithm to extract higher semantic accuracy.

3   Question Answer (QA) Framework

In this section, we describe the proposed deep learning-based question answering system (shows in Fig. 1). The proposed architecture design into three phases: input processing and embedding phase, LSTM neural network phase, and QA feature mapping phase. At first, the input processing and embedding phase performs the question and answer pair's vector representation. Then, a multilayer LSTM neural network implements to extracts the hidden feature of the input question-answer sentence. It acts as an encoder through the activation function (ReLU). Next, the attention mechanism (Softmax) considers for QA feature mapping. Max-Pooling uses to identify the higher semantic feature. Adam

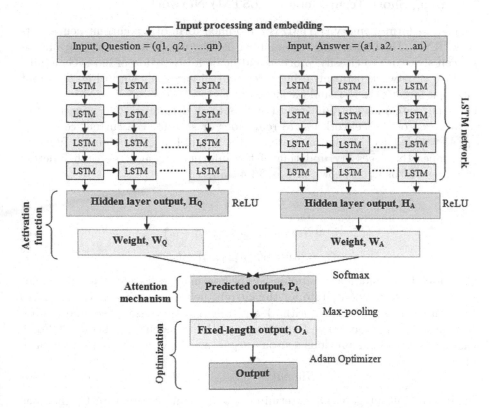

Fig. 1. Proposed LSTM question answer neural network architecture

optimizer applies for probability distribution to calculate the accuracy of the proposed model.

3.1 Input Processing and Embedding

To automatic question answering system, a set of questions and answers is taken as input first. Here, Question set, $Q = (q_1, q_2, q_3,.....q_m)$ and Answer set, $A = (a_1, a_2, a_3,....a_n)$, where m is the length of the question and n is the length of the answer. After feeding the sets of questions and answers into the embedding layer, run four layers of the LSTM network over the questions and answer embedding to obtain their hidden state vector. Mathematical formula of hidden state vector denoted by H_Q and H_A where H_Q is the question hidden state and H_A is the answer hidden state shown in Eqs. 1 and 2.

$$H_Q = [h_{q1}, h_{q2}, ..., h_{qn}] \subset R^{d*n} \tag{1}$$

$$H_A = [h_{a1}, h_{a2}, ..., h_{am}] \subset R^{d*m} \tag{2}$$

3.2 Long Short Term Memory (LSTM) Network

Long short term memory (LSTM) is a modified form of a recurrent neural network (RNN). Hochreiter and Schmidhuber [24] developed it to introduce a memory cell structure to classify, process, and make a prediction of data. Generally, the LSTM network performs using three control gates (shown in Fig. 2) : input gate (i_t), forget gate (f_t), and output gate (O_t).

The input gate allows incoming data or signal and determines which input information (x_t) of the network has to remember or save for the current cell state, C_t, and added to the neuron state. In the input layer, the sigmoid function determines the necessary updating of the information, and the tanh function generates vector C_t as shown in Eqs. 3 , 4 , 5.

$$C_t = f_t * C_{t-1} + i_t * C_t \tag{3}$$

$$i_t = Sigma(W_i.[h_{t-1}, x_t] + b_i) \tag{4}$$

$$C_t = tanh(W_c.[h_{t-1}, x_t] + b_c) \tag{5}$$

Then, forget gate modulate states and allow how many cell states have to remember or forget its previous state for the current neuron state C_{t-1}. It only saves the verified data (by the algorithm) and removes the unsatisfied information through the forget gate. The forgetting probability calculation shows in Eq. 6, where h_{t-1} represents previous neuron output, and X_t represents current neuron input.

$$f_t = Sigma(W_f.[h_{t-1}, X_t] + b_f) \tag{6}$$

Finally, the output gate, O_t determines the amount of cell state C_t that has to transmit in the current output h_t, which shows in Eq. 7 and 8, where σ is a

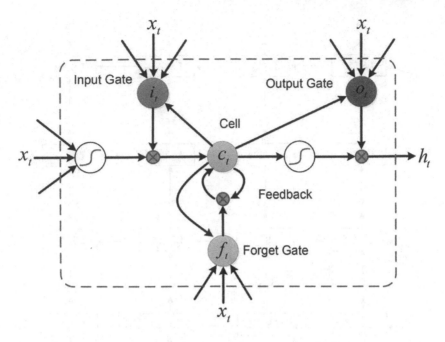

Fig. 2. LSTM basic cell architecture [18]

logistic sigmoid function, x_t is t-t^h word vector of the sentence, and h_t refers to hidden state.

$$O_t = Sigma(W_o.[h_{t-1}, x_t] + b_o) \tag{7}$$

$$h_t = O_t * tanh(C_t) \tag{8}$$

Moreover, here we define four layer forward LSTM network where input sequence of layer A, is $x_A = (x_{A1}, x_{A2}, \ldots, x_{An})$ which produce the output sequence, $y_A = (y_{A1}, y_{A2}, \ldots, y_{An})$. The output of layer A become the input of layer B and produce output sequence, $y_B = (y_{B1}, y_{B2}, \ldots, y_{Bn})$. Similarly output of layer B considered as the input of layer C and produce output sequence, $y_C = (y_{C1}, y_{C2}, \ldots, y_{Cn})$. Finally output of layer C considered as input of layer D and produce the forward hidden sequence $h_t = (h_1, h_2, h_3, \ldots h_n)$. The four layer forward LSTM structure is shown in Fig. 3.

4   QA Feature Mapping and Sequencing

To feature mapping, and QA pair sequencing, activation function, attention mechanism, and optimization technique considers and describes through the following subsections. The activation function considers extracting the appropriate feature of question-answer pairs. Then the attention mechanism uses for semantically mapping question-answer tokens. Finally, an optimization algorithm increases the overall accuracy. A detailed overview is described through the following subsections.

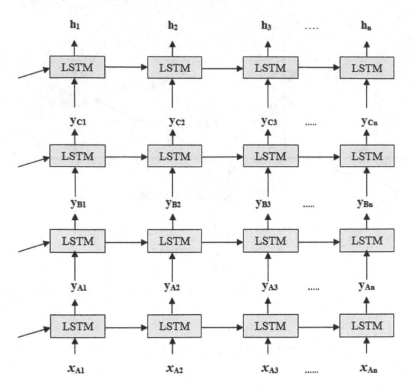

Fig. 3. Forward LSTM network architecture

4.1   Activation Function

In terms of the proposed model, we applied the activation function (showed in Fig. 1) to determine the necessary information which should add or delete and pass the selective information to the next hidden state. To determine whether information passes or not, the proposed model uses the Rectified Linear Unit[1] (ReLU) activation function on the output of hidden layer neurons, which is popular in deep learning techniques (shown in Eq. 9 , 10). The ReLU function provides a real number output between $[0, 1]$. It represents the weight of the information. It only active the neuron if the result of the linear transformation is greater than 0. It refers that for a negative input value, it produces the result 0 that indicates the deactivated status of the neuron. Generally, the ReLU function activates a certain number of neurons simultaneously. So it is more efficient than the sigmoid and tanh function and provides better performance by reducing the negative sampling or result.

$$F(x) = max(0, x) \tag{9}$$

$$R(z) = 1, z > 0; 0, z \leq 0 \tag{10}$$

---

[1] https://en.wikipedia.org/wiki/Rectifier_(neural_networks).

4.2   Attention Mechanism

In the proposed model, the attention mechanism reduces the amount of data loss of the LSTM layer. The attention mechanism is appropriate for inferring the mapping of different data (especially for question-answer tokens). Here, a popular attention mechanism known as a Softmax function[2] applies to mapping information from the questions and answers words to make answer predictions. Softmax is the most effective attention mechanism in dealing with probability distribution problems. The calculation process of the Softmax function shows in Eq. 11.

$$Softmax(x_i) = exp\frac{(x_i)}{\sum(x_j)} \tag{11}$$

4.3   Optimization

To increase the accuracy of the proposed model, here we implement max-pooling technique and the Adam algorithm. Maximum pooling uses to identify the highest weight and transform the input into a fixed-length output. Max-pooling is the most conventional pooling method to calculate the weight of arbitrary information [25]. Max-pooling is denoted by Eq. 12, where M is the size of the pooling region or context.

$$Output(i) = max(x_j); 1 \le j \le M * M \tag{12}$$

Besides, Adam optimizer[3] implements to improve the performance of the proposed model. It is a well-known optimization algorithm developed by Diederik Kingma and Jimmy Ba for predicting the outcome of a large scale of data. Generally, it uses natural language and computer vision problems to improve performance by continuously updating the parameters.

5   Implementation Details

This section describes the implementation process of the neural network model in detail. The implementation process performs into the data set, pre-processing, training, and LSTM network. Here we train the model and observe each phase of testing accuracy rate to identify the highest accuracy of the model.

5.1   Dataset

In this step, we introduce a public data set Wiki-QA Data set with its source, characteristics, number of pairs in detail. Here, the Wikipedia Question Answer

---

[2] https://en.wikipedia.org/wiki/Softmax_function.
[3] https://machinelearningmastery.com/adam-optimization-algorithm-for-deep-learning/.

dataset[4] is considered for the experiment. Carnegie Mellon University and the University of Pittsburgh launched this data set in 2010 for academic research. This Wiki-QA dataset includes 1165 pairs of questions and answers with both incorrect and missing/null answers. This Wiki-QA dataset focus on different types of question and answer, for instance, single supporting fact, yes/no, personal information, historical information, time, numbers, etc. We used this data set for both training and testing purposes (considering 80% for training and 20% for testing purposes). The experiment performs on this data set to evaluate the performance and effectiveness of the proposed model.

5.2   Preprocessing

The Wiki-QA dataset is a natural language question-answer dataset with numbers of notation, missing values, and unformed structure. To convert this unstructured data into a suitable data structure and make it human-readable, pre-processing performs as follows:

Removing Notation: At first, we remove the HTML and punctuation tags from the obtained text or string. Then all the single characters and spaces have been eliminated.

Splitting: Next, splitting the sentences into questions and answers to feed into the proposed model.

Tokenize: The third step is pre-processing for tokenizing the sentences into different words. For performing tokenizing, we used the Keras tokenizer. For example, a sample question: Was Alessandro Volta a professor of chemistry?. The tokenize form of this sample question is: ["Alessandro", "Volta", "a", "professor", "of", "chemistry", "?"].

Indexing: To make the equal length of questions and answer sentences, we perform post-padding and post-truncation operations as various length questions and answers introduced in the dataset. The ambiguous question-answer length ratio might lead to poor results. Therefore, the maximum length of the question and the answer converts into 14 and 30 (<30 might reduce accuracy and >30 might introduce over-fitting problem) respectively to increase accuracy and minimize the over-fitting problem. Those QA pair tokens that exceed this range will automatically eliminate.

5.3   Training Details

Here, 80% of the Wiki-QA dataset is used for the training phase. It helps to identify the optimal architecture of the proposed model. The proposed model trains until 100 epochs. For training, the batch size sets to 100. The memory

---

[4] www.cs.cmu.edu/~ark/QAdata/data/Question_Answer_Dataset_v1.2.tar.gz.

size is limited to the first 100 sentences. Padding operation performs to minimize the fixed-length problems to obtain fixed-size data. The embedding of the null symbol replaces with zero. For the final answer prediction, Softmax uses in the memory layer. Sometimes model performance was not significant. Therefore, we repeated the process with different initialization or epoch (30, 50, and 100) to identify the highest and lowest accuracy (shown in Fig. 5).

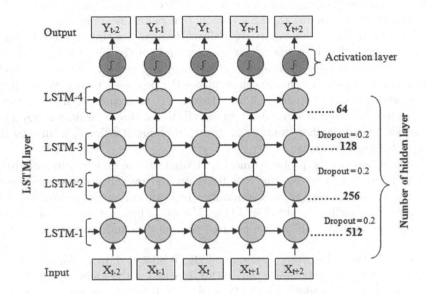

Fig. 4. Basic structure of multi layer forward LSTM-RNN architecture

5.4 LSTM Network

A LSTM recurrent neural network architecture considers here to implement the proposed QA model (shows in Fig. 1). The basic LSTM recurrent neural network architecture shows in Fig. 4. To train the proposed QA model, we used the TensorFlow[5] platform using the Keras[6] library.

At first, the word-embedding layer uses for both questions and answer encoding. Then LSTM recurrent neural network layer is used for passing the encoded question where the number of dense layers is considered 512, 256, 128, and 64 for each LSTM memory network block. To increase the accuracy of the proposed architecture dropout layer is used. For training the LSTM memory network, 0.2 dropouts are used after the word embedding layer to avoid the over-fitting problem. The LSTM memory network was trained for 100 epochs with Adam optimizer and identifies the highest training and testing accuracy around 83%

---

[5] https://www.tensorflow.org/.
[6] https://keras.io/.

and 81%, respectively. Finally, the activation layer considers for passing the true-positive information and predicting the final fixed-length output.

6   Performance Analysis

In this section, we experimented proposed QA model on the Wiki-QA data set to observe the performance and identify the effectiveness.The experiment has conducted through training the LSTM recurrent neural network with an attention mechanism and optimization algorithm. Besides, multiple layers of LSTM architecture considered increasing the performance of the proposed model. For semantically representing the question-answer pairs, we used the attention mechanism and optimization technique that makes the proposed model convincing.

The proposed QA approach validates through considering the different range of epochs numbers on the Wiki-QA data set. To find a stable result, we gradually increase the epoch number (number of iteration about 100 times) from 0 to 100 to reveal the best accuracy.

Fig. 5 shows that, in the proposed model, changes in the epoch numbers (10 to 100) can increase or decrease the loss function and overall accuracy. However, in both training and testing, as the number of iteration has increased, the loss in training and testing has decreased (Fig. 5(a–c)). Therefore, a certain number of iteration or epochs could reduce the loss percentages and increase the performance.

Here, Table 1 shows the overall accuracy of the Wiki-QA data set for both training and testing phases. It shows that in 30 epochs for the training and testing phase, the proposed model achieves 78.56% and 76.20% accuracy, respectively. Similarly, the accuracy of both training and testing has increased by around 2% in 50 epochs. In contrast, the proposed model shows significant improvement in both training and testing accuracy of 83.62% and 81.90% in the 100 epoch range, which is the maximum for the Wiki-QA data set. It depicts that a certain number of iteration could able to increase accuracy and improve the final results.

Table 1. Overall accuracy of the proposed QA model considering different range of epoch number

No of epoch	Accuracy (Train)	Accuracy (Test)
30	78.56%	76.20%
50	80.79%	78.12%
100	83.62%	81.90%

To evaluate the proposed approach's effectiveness, we conclude the compared result of the proposed QA model with existing popular QA models. According to the multi-layer prototype, the proposed model compared with Li et al. [26] which proposed a multi-layer LSTM-based QA model without using attention

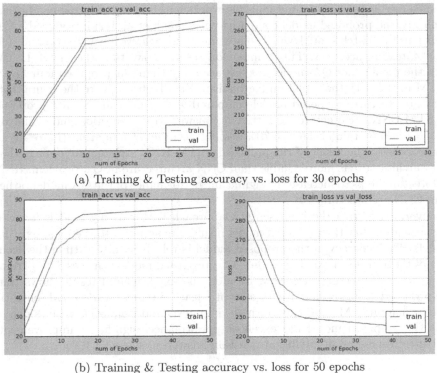

(a) Training & Testing accuracy vs. loss for 30 epochs

(b) Training & Testing accuracy vs. loss for 50 epochs

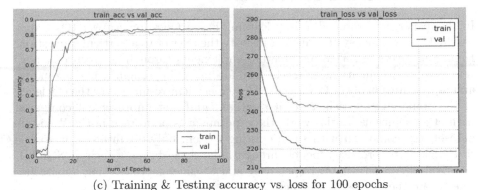

(c) Training & Testing accuracy vs. loss for 100 epochs

Fig. 5. Training and testing loss with accuracy according to epoch size 100 (a–c)

mechanism and achieve 37.11% training and 37.67% testing accuracy. Comparing this result with proposed model result, it emphasizes that multi-layer LSTM architecture with attention mechanism produces a better result than without attention mechanism. Through an attention mechanism, each word of the QA pair gets more weight. Besides, an attention mechanism represents the semantic meaning or relation among multiple words without any external conditions. Similarly, the proposed model compared with Haigh et al. [12] introduced a QA model using match-LSTM with pointer net and achieve 60.30% training and 45.00% testing accuracy. This comparison result shows that the match-LSTM with the pointer net QA model has less accuracy than our proposed four layers LSTM model with the attention mechanism. Finally, to evaluate the neural network architecture, proposed model compared with Tan et al. [8] that proposed a LSTM-CNN architecture for retrieving answer to the question and achieving overall 67.20% training and 65.70% testing accuracy. Comparing the result of this work with the result of the proposed LSTM-RNN architecture depicts that RNN architecture is more reliable than CNN architecture in QA problem. RNN architecture helps to increase the significant performance of the QA model rather than CNN architecture.

However, comparing the proposed QA approach with existing models it could be speculated that the presented model results are superior to the most existing models. The accuracy of the proposed model is relatively higher than the accuracy of the addressed existing models. Our proposed model achieves 83% (Training) and 81% (Testing) accuracy. Besides, the experimental result shows that the proposed model is effective for semantic feature representation and semantic feature mapping. Moreover, performance analysis and evaluation results depict that the proposed model has excellent performance, significant improvement, and effectiveness in an intelligent question answering system.

7   Conclusion and Future Work

In this paper, we proposed a deep learning-based Question Answer (QA) approach. The proposed system aims to identify the answer to a particular question. Here, four layers of LSTM architecture implements through the recurrent neural network with an attention mechanism for improving semantic understanding and representation of the QA tokens. The experiment result proves the efficiency of the proposed approach. Also, it shows a significant improvement compared with other solutions. In this work, the experiment is performed in the Wiki-QA dataset and compared with some existing system only. In future, our goal is to evaluate our proposed model through large-scale datasets such as SQuAD, SemEval-cQA, and SciQA, respectively and consider more existing approaches to evaluate the proposed approach widely.

References

1. Diefenbach, D., Both, A., Singh, K., Maret, P.: Towards a question answering system over the semantic web. Seman. Web 11(3), 421–439 (2018). https://doi.org/10.3233/SW-190343
2. Yu, W., et al.: A technical question answering system with transfer learning. In: Proceedings of the 2020 Conference on Empirical Methods in Natural Language Processing: System Demonstrations, pp. 92–99 (2020). https://doi.org/10.18653/v1/2020.emnlp-demos.13

3. Bhuiyan, H., Oh, K.J., Hong, M.D., Jo, G.S.: An effective approach to generate Wikipedia infobox of movie domain using semi-structured data. J. Internet Comput. Serv. **18**(3), 49–61 (2017)
4. Ara, J., Bhuiyan, H.: Semantic recognition of web structure to retrieve relevant documents from Google by formulating index term. Adv. Comput. Commun. Comput. Sci. (2019). https://doi.org/10.1007/978-981-13-6861-5_43
5. Bouziane, A., Bouchiha, D., Doumi, N., Malki, M.: Question answering systems: survey and trends. Procedia Comput. Sci. **73**, 366–375 (2015). https://doi.org/10.1016/j.procs.2015.12.005
6. Saxena, A., Tripathi, A., Talukdar, P.: Improving multi-hop question answering over knowledge graphs using knowledge base embeddings. In: Proceedings of the 58th Annual Meeting of the Association for Computational Linguistics, pp. 4498–4507 (2020). https://doi.org/10.18653/v1/2020.acl-main.412
7. Banerjee, P., Baral, C.: Knowledge fusion and semantic knowledge ranking for open domain question answering. arXiv preprint arXiv:2004.03101 (2020)
8. Ara, J., Bhuiyan, H.: Upgrading YouTube video search by generating tags through semantic analysis of contextual data. In: Uddin, M.S., Bansal, J.C. (eds.) Proceedings of International Joint Conference on Computational Intelligence. AIS, pp. 425–437. Springer, Singapore (2020). https://doi.org/10.1007/978-981-15-3607-6_34
9. Pradel, C., Sileo, D., Rodrigo, Á., Peñas, A., Agirre, E.: Question answering when knowledge bases are incomplete. In: Arampatzis, A., et al. (eds.) CLEF 2020. LNCS, vol. 12260, pp. 43–54. Springer, Cham (2020). https://doi.org/10.1007/978-3-030-58219-7_4
10. Sadhuram, M.V., Soni, A.: Natural language processing based new approach to design factoid question answering system. In: 2020 Second International Conference on Inventive Research in Computing Applications (ICIRCA), pp. 276–281. IEEE (2020). https://doi.org/10.1109/ICIRCA48905.2020.9182972
11. Gao, S., Chen, X., Ren, Z., Zhao, D., Yan, R., et al.: Meaningful answer generation of e-commerce question-answering. arXiv preprint arXiv:2011.07307 (2021)
12. Haigh, A., van de Graaf, C., Rachleff, J.: Extractive question answering using match-LSTM and answer pointer
13. Minaee, S., Kalchbrenner, N., Cambria, E., Nikzad, N., Chenaghlu, M., Gao, J. et al.: Deep learning based text classification: a comprehensive review. arXiv preprint arXiv:2004.03705 (2021)
14. Abbasiyantaeb, Z., Momtazi, S.: Text-based question answering from information retrieval and deep neural network perspectives: a survey. arXiv preprint arXiv:2002.06612 (2021)
15. Feng, M., Xiang, B., Glass, M.R., Wang, L., Zhou, B., et al.: Applying deep learning to answer selection: a study and an open task. In: 2015 IEEE Workshop on Automatic Speech Recognition and Understanding (ASRU), pp. 813–820. IEEE (2015). https://doi.org/10.1109/ASRU.2015.7404872
16. Wang, D., Nyberg, E.: A long short-term memory model for answer sentence selection in question answering. In: Proceedings of the 53rd Annual Meeting of the Association for Computational Linguistics and the 7th International Joint Conference on Natural Language Processing, pp. 707–712 (2015)
17. Tan, M., Santos, C.D., Xiang, B., Zhou, B.: LSTM-based deep learning models for non-factoid answer selection. arXiv preprint arXiv:1511.04108 (2015)
18. Yin, W., Yu, M., Xiang, B., Zhou, B., Schütze, H., et al.: Simple question answering by attentive convolutional neural network. arXiv preprint arXiv:1606.03391 (2016)

19. Zhang, L., Ma, L.: Coattention based bilstm for answer selection. In: 2017 IEEE International Conference on Information and Automation (ICIA), pp. 1005–1011. IEEE (2017). https://doi.org/10.1109/ICInfA.2017.8079049

20. Larson, T., Gong, J.H., Daniel, J.: Providing a simple question answering system by mapping questions to questions

21. Sharma, Y., Gupta, S.: Deep learning approaches for question answering system. Procedia Comput. Sci. **132**, 785–794 (2018). https://doi.org/10.1016/j.procs.2018.05.090

22. Cai, L., Zhou, S., Yan, X., Yuan, R.: A stacked BiLSTM neural network based on coattention mechanism for question answering. Comput. Intell. Neurosci. (2019). https://doi.org/10.1155/2019/9543490

23. Rohit, G., Dharamshi, E.G., Subramanyam, N.: Approaches to question answering using LSTM and memory networks. In: Bansal, J.C., Das, K.N., Nagar, A., Deep, K., Ojha, A.K. (eds.) Soft Computing for Problem Solving. AISC, vol. 816, pp. 199–209. Springer, Singapore (2019). https://doi.org/10.1007/978-981-13-1592-3_15

24. Yildirim, Ö.: A novel wavelet sequence based on deep bidirectional LSTM network model for ECG signal classification. Comput. Biol. Med. **96**, 189–202 (2018). https://doi.org/10.1016/j.compbiomed.2018.03.016

25. Zhao, Q., Lyu, S., Zhang, B., Feng, W.: Multiactivation pooling method in convolutional neural networks for image recognition. Wireless Commun. Mobile Comput. (2018). https://doi.org/10.1155/2018/8196906

26. Li, Y.: Two layers LSTM with attention for multi-choice question answering in exams. In: IOP Conference Series: Materials Science and Engineering, vol. 323, no. 1, p. 012023. IOP Publishing (2018)

A Machine-Learning-Based Framework for Supporting Malware Detection and Analysis

Alfredo Cuzzocrea[1,2]([✉]), Francesco Mercaldo[3], and Fabio Martinelli[3]

[1] University of Calabria, Rende, Italy
`alfredo.cuzzocrea@unical.it`
[2] LORIA, Nancy, France
[3] IIT-CNR, Pisa, Italy
`{francesco.mercaldo,fabio.martinelli}@iit.cnr.it`

Abstract. Malware is one of the most significant threats in today's computing world since the number of websites distributing malware is increasing at a rapid rate. The relevance of features of unpacked malicious and benign executables like mnemonics, instruction opcodes, API to identify a feature that classifies the executables is investigated in this paper. By applying Analysis of Variance and Minimum Redundancy Maximum Relevance to a sizeable feature space, prominent features are extracted. By creating feature vectors using individual and combined features (mnemonic), we conducted the experiments. By means of experiments we observe that Multimodal framework achieves better accuracy than the Unimodal one.

Keywords: Malware · Machine learning · Opcode · Classification · Static analysis · Dynamic analysis · Hybrid analysis · Security

1 Introduction

Malware or malicious code is harmful code injected into legitimate programs to perpetrate illicit intentions. The primary source of infection is represented by the Internet. It also enables malcode to enter the system without users' knowledge. The primary sources used by the malicious users are free compilers, games, web browsers, etc. Most transactions performed using Internet attracted malware writers to gain unauthorized access to the machines to extract valuable information concerning the user and system. The stealthy malware exploits computer systems' vulnerabilities in order to open the computer systems' to remote monitoring. In particular, modern malware are classified in metamorphic or polymorphic [3]. Hence, it is required to develop a non-signature-based anti-malware detector for the effective detection of modern-day malware.

A. Cuzzocrea—This research has been made in the context of the Excellence Chair in Computer Engineering – Big Data Management and Analytics at LORIA, Nancy, France.

O. Gervasi et al. (Eds.): ICCSA 2021, LNCS 12951, pp. 353–365, 2021.
https://doi.org/10.1007/978-3-030-86970-0_25

Several classes distinguish the malwares by considering the breach policy and its introduction in the system. For example, metamorphic malwares have the ability to change as they propagate [26]. This is in contrast to cloned software, where all instances of a piece of software are identical. Metamorphism is usually considered by malware writers.

From a theoretical point of view, metamorphism enable malware to become undetectable To make viruses more resistant to emulation, virus writers have developed numerous advanced metamorphic techniques. For example, both decryptor and virus body are changed by metamorphic virus.

Furthermore, they change generation after generation. That is, a metamorphic virus changes its "shape" but not its behaviour. Instruction permutation, register re-assignment, code permutation using conditional instructions, no-operation insertion and many other are techniques implemented by malware writers.

The study of the malware is called malware analysis [17]. Malware analysis is an essential part of extracting run-time behaviour such as the API calls, import and export functions, etc. of the program. Hiding the malware execution from the malware analyzer is the primary battle. The counterpart that analyses the malware are projects tools such that Ether [13].

The motivation behind Ether is that it must not induce any side-effects to the host environment that are detectable by malware. The application of hardware virtualization extension such as Intel VT [16] are the building blocks of Ether. The target OS environment is the location where the system resides. In order to avoid this analysis modern malware implements anti-debugging, anti-instrumentation, and anti-VM techniques to thwart the run time analysis. Hence, there are no guest software modules answerable to detection, and there is no deprivation that arises from incomplete system emulation. Apart from providing anti-debugging facilities, de-armoring dynamically is a possible for using Ether.

In this paper, we present an innovative technique for malware detection. The experiments conducted on malware and benign executables are in Portable Executable (PE) format. By seeing the submissions received over the Virus Total [29], which performs offline scanning of malicious samples is the motivation for using Win32 PE executable. In particular, we define a set of features (as mnemonics, instruction opcodes, and API calls) obtained from benign and malicious executable.

2   Related Work

API calls have been used in the past research for modeling program behaviour [27,38] and for detecting malware [2,39]. The behaviour of the malicious programs in a specific malware class differs considerably from programs in other malware classes and benign programs is the main idea exploited in this paper.

In [25], the authors extracted prominent API calls statically using IDA-Pro. Thus, all the late-bounded API calls that are made using GetProcAddress, LoadLibraryEx, etc. are not taken into account. For packed malware the proposed approach is not working. Instead, the proposed approach is shown to be effective against the aforementioned obfuscation techniques, encryption or decryption, and compression. It is also effective against new metamorphic versions of the same program as the dynamic behaviour remains the same, and only a single signature prepared for an entire malware family.

In [4], a mining algorithm for performing dynamic analysis and for detecting obfuscated malicious code is used by the authors. Despite the good performance, the computational time is of several seconds for each single program.

In [23] authors uses static analysis to detect system call locations and run-time monitoring to check all system calls made from a location identified during static analysis. In [27], dynamic monitoring for detecting worms and other exploits is used. The main limit of the proposed work is its applicability only for detecting worms and exploits that use hard-coded addresses of API calls. MetaAware [39] identifies patterns of system or library functions called from a malware sample statically to detect its metamorphic version.

In [12,36], the statistical properties of a set of metamorphic virus variants are represented by using a Hidden Markov Models (HMMs). The authors generate a database by using: Mass Code Generator (MPCGEN), Virus Creation Lab for Win32 (VCL32), Next Generation Virus Construction Kit (NGVCK), and Second Generation virus generator (G2). The HMM was trained on a family of metamorphic viruses and used to determine whether a given program is similar to the viruses the HMM represents.

In [19] a method to find the metamorphism in malware constructors like NGVCK, G2, IL_SMG, MPCGEN is proposed. The method executes each malware samples in a controlled environment like QEMU and it monitors API calls using STraceNTX.

Furthermore, they used the proposed method for estimating the similarity between malware constructors used in the study by computing the proximity indices amongst malware constructors. They found that the NGVCK generated malware were least similar to other constructors. The fact that the NGVCK generated malware invariants are metamorphic compared to all other malware constructors is highlighted by the authors.

In the presented literature, the main lack is the consideration of metamorphic viruses. The unimodal framework used earlier had many drawbacks as classification done based on a single modality. Only a single attribute type is used by the framework. Instead, in the present work, an innovative method for detecting obfuscated variants or mutants of malware dynamically is presented. Here we use a multimodal framework which involves learning based on multiple modalities. In particular, we extract prominent features using Minimum Redundancy and Maximum Relevance (mRMR) and Analysis of Variance (ANOVA). We also concatenate these prominent features, we developed the ensemble classifier from the various feature set which classifies into malware or benign.

3   Malware Detection and Analysis: A Machine Learning Methodology

3.1   Software Armoring

Software armoring or Executable Packing is the process of compressing/encrypting an executable file and prepending a stub that is responsible for decompressing/decrypting the executable and initiating its execution. When execution starts, the original executable code is unpacked and transferred by the stub for controlling it. Nowadays, the vast majority of the malware authors use packed executables to hide from detection. It is also due to software armoring that more and more false detection of malware comes up.

Whether the malware is armored or not is the first necessary knowledge needed for starting with its analysis. In the proposed methodology, Ether is considered as a tool for de-armoring. This is due to its characteristic to be not signature-based and transparent to malware. Signature-based unpacking tools like VMPacker [31], GUnPacker [14] requires that the executable is packed with standard packers. In order to extract the signature of a program we use PEiD [22]. In general, the authors of the malware write their stub and prepend it to the executable [5], moreover stymie's the signature-based detection. The two methods used commonly to unpack the actual malicious code are: (1) to reverse engineering the unpack algorithm, then to use a scripting language to re-implement it and finally to run the script on the packed file as in OllyDbg [21] or IDAPro [15]; (2) at run time, use a breakpoint just previously the malicious code which requires manual intervention [28].

Ether detects all the writes to memory, a program does and dumps the program back in the binary executable form. A hash-table for all the memory maps is created. Whenever a write to a slot in the hash table is available then a Original Entry Point is reported. This is the starting point of execution of a packed executable.

3.2   Feature Extraction

In our approach, we used API calls to measure the dynamic malware analysis and mnemonic or opcode, instruction opcode, 4-gram mnemonic as a measure of static malware analysis. The features are extracted by using the following methods: (1) Usage of VeraTrace, the API Call tracer [30]. Veratrace is an Intel PIN-based API call tracing program for Windows. API calls can be traced by this program of de-armored programs obtained from Ether. A Frequency Vector Table (FVT) is obtained by parsing the output of Veratrace for final API traces. (2) Mnemonics and instruction opcode features are extracted using an open source tool namedscriptsizeObjDump [20]. The FVT of mnemonics and instruction opcode are obtained by parsing both these features. (3) The instruction opcode and mnemonics features are extracted by running scriptsizeObjDump [20] (a freely available open-source tool). A parser for both these features has been developed for obtaining the FVT of mnemonics and instruction opcode.

In particular, only a subset of the features mentioned above is used for our analysis. The selection is made by using feature selection methods. Thus, the selected features are the ones that most represent variation in the data.

3.3   Normalization

In statistics, a standard score indicates the number of standard deviations an observation or datum is above or below the observed mean. By subtracting the feature mean from an individual feature raw score and dividing this difference by the feature standard deviation the score is obtained. This procedure is usually called standardizing or normalizing. The frequency vector table is a table populating feature as the column and their frequencies in a particular sample as a row. The majority of the classifier and feature selection methods need normalization and discretization into states of the frequency vector table. In the proposed approach, the Z-Score is chosen for normalization.

Z–Score. Standard scores are also called z-values, z-scores, normal scores, and standardized variables, the use of "Z" is because the normal distribution is also known as the "Z distribution". The comparison of a sample to a standard normal deviate (standard normal distribution, with $\mu = 0$) is their main usage. Nevertheless, as shown in Eq. 1, they can be defined without the assumption of normality. In statistics, both the mean and the standard deviation are used for standardizing a general feature X.

$$Z_j = \sum_{i=1}^{n} \frac{X_i^j - \mu_j}{\sigma_j},$$

$\qquad(1)$

where μ_j is the mean of all the frequencies obtained from samples of the jth feature, $\sigma_j = \sqrt{\text{Var}(X)}$ is the standard deviation of the jth feature, Z_j is the Z–Score of the jth feature, X_i^j is the frequency of jth feature in i^{th} sample and n is the total number of samples. The following equation defines the variance:

$$Var(X_j) = \sum_{i=1}^{n} \frac{(X_i^j - \mu_j)^2}{n}.$$

$\qquad(2)$

3.4   Feature Selection

Given the financially-motivated nature of these threats, methods of recovery mandate more than just remediation. In general, understanding the runtime behaviour of modern malware is the main need. It includes basic artifacts or features of any Portable Executable like the mnemonics, API Call, instruction opcode, etc. As Fig. 1 shows, the proposed methodology consider mnemonics, instruction opcode, API Calls [19] and 4-gram mnemonic [24] for Feature Fusion [18] to analyze a malware.

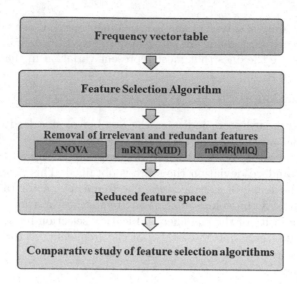

Fig. 1. Feature selection

Today's malware is increasing at a tremendous rate, resulting in the proliferation of high dimensional data getting accumulated, thus challenging state-of-the-art data mining techniques. The importance of feature selection in data mining is becoming widely recognise [35]. This is due to the high dimension of the data extracted. For example, the mnemonics features includes x86 instructions (such as push, xor, pop, test etc.) instruction opcodes, API Call (GetProcAddress, LoadLibrary etc.). Feature selection effectively reduces the data dimensionality by removing redundant and irrelevant features.

Many feature selection algorithms have been developed for different purposes. Moreover, each algorithm is characterized by the usage of its criterion for feature selection that will be explained later. These techniques are of paramount importance if several irrelevant features are considered otherwise learning models cannot generalize in identifying a new sample. Besides the selection of features, we have also presented a comparative study of the algorithms used for feature selection.

3.5 Classification

Databases are rich with hidden information that can be used for intelligent decision making. In order to extract to extract models describing classes, classification is used. It has been successfully applied in several settings (e.g., [9]). This kind of data analysis enable us to better understand the pattern in the data. Classification predicts categorical (discrete, unordered) labels. The Support Vector Machine Classifier [35] (SVM) is used to classify unknown samples. SVM belongs to the class of supervised learning methods. In the proposed application it has been trained by using 2000 different software (1000 malware and 1000 benign samples).

4   Malware Detection and Analysis: Framework Architecture

To classify unknown samples, we used malwares such as viruses, trojans, backdoor, and worms obtained from the VX Heavens repository, freely available for research purpose [32]. The benign programs are considered from the Windows library. The computation of the of features like Mnemonic, instruction opcode, API calls [1], and 4-gram mnemonic needs the samples to be unpacked.

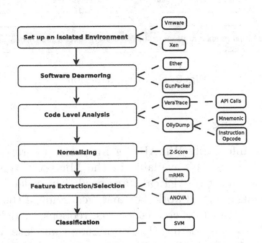

Fig. 2. Architecture

So, the executable files were made to run on a hardware virtualized machine such as Xen [37]. The actual host machine is not affected by the viruses during the dynamic API tracing step is the main advantage of using an emulator.

In Fig. 2, we show the complete architecture of the proposed methodology. A total of 1000 malware samples and 1000 benign samples are considered. Then, they are executed on the hardware virtualized machine using Veratrace to capture the API calls. The emulator moves all the text files with the information of the API calls to the actual host OS. The frequency of each API call is found. The following sections discuss the concepts used here.

5   Feature Extraction Methods

Once the samples unpacked, we carry out the code-level analysis of the samples. Features are extracted using parsers for each feature, as described in Fig. 3.

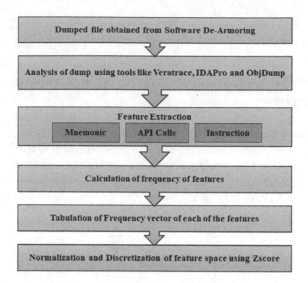

Fig. 3. Feature extraction

5.1 API Calls Tracing

The Windows API, informally WinAPI, is Microsoft's core set of application programming interfaces (APIs) available in the Microsoft Windows operating systems. In Windows, a set of API calls are needed in order for an executable program to perform its assigned work. For instance, some of the API calls for file management are OpenFile that creates, opens, reopens, or deletes a given file; DeleteFile that deletes an existing file; FindClose that closes a file search handle that is opened by FindFirstFile, FindFirstFileEx, or FindFirstStreamW function; FindFirstFile that searches a file inside a directory or look for a sub-directory name that matches a specified name;GetFileSize that retrieves, in bytes, the size of a specified file. Thus, no executable program can run without the API calls. Hence, a good measure to record the behaviour of an executable is by using API calls made by an executable.

Veratrace is a Windows API Call Tracer. It can trace all the calls made by a process to the imported functions from a DLL. Veratrace traces the API calls by mandating the unpacking of the samples. Whenever packed, the import table would be populated with GetProcAddress() and LoadLibrary()-like API calls. Trace files contain information about all the API Calls made by these executables. For the purposes of parsing all traces, a parser was designed. API Traces are passed to the parser as input yielding the following output: (1) All API traces of malware and benign programs are kept in malware_api and benign_api directory, respectively. (2) API Trace parser takes these input directories and creates a file named api.csv. FVT contains a grid representing the frequencies of all the features.

5.2   Mnemonic, Instruction Opcode and 4-Gram Mnemonic Trace

Following mnemonics, instruction opcode, 4-gram mnemonic form the structural aspect of the program. In a matter of tracing them, a static analysis of the sample is required. Alongside performing Static Analysis using the open-source ObjDump tool, we developed a customized parser to filter mnemonics and create a feature vector table with the occurrence of mnemonics as elements of the vectors. Likewise, we carry out the vector representation for opcodes.

6   Classification Procedure

Data classification is a two step process.

1. In the first step, a classifier describing a predetermined set of data classes or concepts is built. The first step being the learning step (or training phase), a classification algorithm, in that step, builds the classifier by analyzing or learning from a training set made up of database tuples and their associated class labels. A tuple or feature, X, is represented by an n-dimensional attribute vector, $X = (A_1 \& A_2 \& ... \& A_n)$, where, n depicts measurements made on the tuple from n database attributes, respectively, A_1, A_2, ..., A_n. Each tuple, X, is assumed to belong to a predefined class determined by another database attribute called the class label. The class label attribute is represented by discrete unordered values. The individual tuples making up the training set namely the training tuples are selected from the database under analysis. The class label of each training tuple is known as supervised learning as the classifier already told, to which class a particular tuple belongs.
2. In the second step, the classification model is used. Test data were created through test tuples and their class. This data was not used in the classifier's training. The percentage of test set tuples that are correctly classified by the classifier is identified as the accuracy of a classifier on a given test set. The associated class label of every test tuple compared with the learned classifiers class prediction for the concerned tuple.

The SVM model is exploited in our approach. Linear as well as nonlinear data are classified by the SVM classification method. It employs a nonlinear mapping to turn the original training data into a higher dimensional data. Within this new dimension, it searches for the linear optimal separating hyperplane (a decision boundary separating the tuples of one class from another). With an appropriate nonlinear mapping to a sufficiently high dimension, a hyperplan can always separate data from two classes. The SVM defines this hyperplane through support vectors (essential training tuples) and margins (defined by the support vectors).

Although the training time of even the fastest SVMs prolonge, they are highly accurate due to their ability to model complex nonlinear decision boundaries. A short description of the learned model is provided by the support vectors.

Our approach is a two-class problem as we have to classify the samples as either benign or malware samples.

We obtain a nonlinear SVM by extending the approach for linear SVMs as follows. In the first step, original input data is transformed into a higher dimensional space using a nonlinear mapping. Once the data has transformed into the new space, a linear separating hyperplane in the new space is searched by the next step. We end up with a quadratic optimization problem that is solved using the linear SVM formulation [33]. This way, a nonlinear, separating surface in the original space is obtained, which is represented by the maximal marginal hyperplane found in the new space. In addition, a nonlinear mapping function is used to transform the training tuples. It turns out that it is mathematically equivalent to applying a kernel function to the original input data. In this way, in the original input space, potentially in a much lower dimensionality the performance of the calculation is done. *Gaussian Radial Basis Function kernel* (RBF) was used.

Other than the SVM Classifier, our proposed method classifies unseen samples using Random Forest, which is an ensemble classifier and also AdaBoost, which uses out–of–bag(oob) data set. We present a comparative study of SVM, Random Forest and AdaBoost. We apply classification using WEKA [34].

7 Experimental Assessment and Analysis

In this section, we discuss the experiment's setup, results obtained and the analysis of the result. We collected a total of 2000 Portable Executables which consists of 1000 malware samples gathered from sources VxHeaven (650) [32], User Agency (250), Offensive Computing (100) [30], besides that we collected benign samples from Windows XP System32 Folder (450), Windows7 System32 Folder (100), MikTex/Matlab Library (400) and Games (50).

We extracted mnemonics or opcodes from 2000 samples. In Fig. 4, we show the experimental results obtained from feature reduction using mRMR (MID and MIQ) and ANOVA. We derived these results after sample classification using SVM, AdaBoost, Random Forest, J48. Five mnemonic based models were constructed at a variable length, starting from 40 to 120 at an interval of 20. Among these five models, the best result with a strong positive likelihood ratio of 16.38 for the feature-length 120 using Random Forest was provided by ANOVA. Low error rate and speed are the main advantages of this model. Nevertheless, mnemonic based features can be effortlessly modified using code obfuscation techniques. Generally, metamorphic and polymorphic malware employ these obfuscation techniques to bypass the detectors provided by current anti-malware free and commercial software.

8 Conclusion and Future Work

In this paper, we proposed a new hybrid approach for malware detection, integrating the prominent attributes realized from a set of unimodal classifiers, which

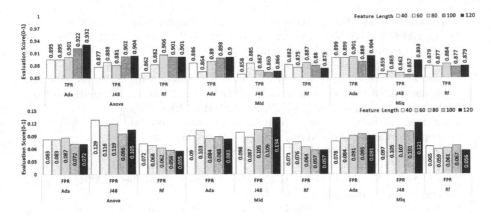

Fig. 4. Performance of mnemonic features

employ both static and dynamic model for malware analysis. Here, a substantial performance gain over the set of unimodal classifiers, was derived from the multimodal feature fusion technique we followed and from which it derived its features. Our proposed solution consistently yielded superior outputs in comparison with the unimodal classifier set we employed, with an accuracy rate exceeding 95% on average.

In future, our method can be extended by analyzing the malware and benign samples using ensemble features and classifying the unseen samples using ensemble classifiers by using different voting schemes. The unseen samples are voted by each classifier from the pool as malware or benign, and accordingly, the sample with the maximum number of votes would be classified. Currently our data set consists of bi-class, malware or benign, a plausible extension to the classification scheme in the future is toward a multi-class classification. Another line of future research predicates the integration of our proposed framework with novel features dictated by the emerging big data trend (e.g., [6–8,10,11]).

Acknowledgements. This research has been partially supported by the French PIA project "Lorraine Université d'Excellence", reference ANR-15-IDEX-04-LUE.

References

1. Alkhateeb, E.M., Stamp, M.: A dynamic heuristic method for detecting packed malware using naive bayes. In: International Conference on Electrical and Computing Technologies and Applications (ICECTA), pp. 1–6. IEEE (2019)
2. Bergeron, J., Debbabi, M., Erhioui, M.M., Ktari, B.: Static analysis of binary code to isolate malicious behaviors. In: WETICE 1999: Proceedings of the 8th Workshop on Enabling Technologies on Infrastructure for Collaborative Enterprises, Washington, DC, USA, pp. 184–189. IEEE Computer Society (1999)
3. Bulazel, A., Yener, B.: A survey on automated dynamic malware analysis evasion and counter-evasion: pc, mobile, and web. In: Proceedings of the 1st Reversing and Offensive-oriented Trends Symposium, pp. 1–21 (2017)

4. Christodorescu, M., Jha, S., Kruegel, C.: Mining specifications of malicious behavior. In: ESEC-FSE 2007: Proceedings of the the 6th joint meeting of the European Software Engineering Conference and the ACM SIGSOFT Symposium on The Foundations of Software Engineering, pp. 5–14, New York, NY, USA. ACM (2007)
5. Chuan, L.L., Yee, C.L., Ismail, M., Jumari, K.: Automating uncompressing and static analysis of conficker worm. In: 2009 IEEE 9th Malaysia International Conference on Communications (MICC), pp. 193–198. IEEE (2009)
6. Cuzzocrea, A.: Improving range-sum query evaluation on data cubes via polynomial approximation. Data Knowl. Eng. **56**(2), 85–121 (2006)
7. Cuzzocrea, A., Matrangolo, U.: Analytical synopses for approximate query answering in OLAP environments. In: Galindo, F., Takizawa, M., Traunmüller, R. (eds.) DEXA 2004. LNCS, vol. 3180, pp. 359–370. Springer, Heidelberg (2004). https://doi.org/10.1007/978-3-540-30075-5_35
8. Cuzzocrea, A., Moussa, R., Xu, G.: OLAP*: effectively and efficiently supporting parallel OLAP over big data. In: Cuzzocrea, A., Maabout, S. (eds.) MEDI 2013. LNCS, vol. 8216, pp. 38–49. Springer, Heidelberg (2013). https://doi.org/10.1007/978-3-642-41366-7_4
9. Cuzzocrea, A., Mumolo, E., Fadda, E., Tessarotto, M.: A novel big data analytics approach for supporting cyber attack detection via non-linear analytic prediction of IP addresses. In: Gervasi, O., et al. (eds.) ICCSA 2020. LNCS, vol. 12249, pp. 978–991. Springer, Cham (2020). https://doi.org/10.1007/978-3-030-58799-4_70
10. Cuzzocrea, A., Saccà, D., Serafino, P.: A hierarchy-driven compression technique for advanced OLAP visualization of multidimensional data cubes. In: Tjoa, A.M., Trujillo, J. (eds.) DaWaK 2006. LNCS, vol. 4081, pp. 106–119. Springer, Heidelberg (2006). https://doi.org/10.1007/11823728_11
11. Cuzzocrea, A., Serafino, P.: LCS-Hist: taming massive high-dimensional data cube compression. In: Proceedings of the 12th International Conference on Extending Database Technology: Advances in Database Technology, pp. 768–779 (2009)
12. Damodaran, A., Di Troia, F., Visaggio, C.A., Austin, T.H., Stamp, M.: A comparison of static, dynamic, and hybrid analysis for malware detection. J. Comput. Virol. Hacking Tech. **13**(1), 1–12 (2017)
13. Ether: http://ether.gtisc.gatech.edu/
14. Gunpacker. http://www.woodmann.com/collabarative/tools/
15. Ida Pro: http://www.hex-rays.com/idapro/
16. Intel: http://www.intel.com/
17. Mandiant: http://www.mandiant.com/
18. Masud, M.M., Khan, L., Thuraisingham, B.: A hybrid model to detect malicious executables. In: Proceedings of IEEE International Conference on Communications, ICC 2007, pp. 1443–1448. IEEE (2007)
19. Nair, V.P., Jain, H., Golecha, Y.K., Gaur, M.S., Laxmi, V.: Medusa: metamorphic malware dynamic analysis using signature from api. In: Proceedings of the 3rd International Conference on Security of Information and Networks, SIN 2010, pp. 263–269, New York, NY, USA. ACM (2010)
20. Objdump. https://ubuntu.pkgs.org/16.04/ubuntu-universe-amd64/dissy_9-3.1_all.deb.html
21. Ollydbg. http://www.ollydbg.de
22. Peid: http://www.peid.info
23. Rabek, J.C., Khazan, R.I., Lewandowski, S.M., Cunningham, R.K.: Detection of injected, dynamically generated, and obfuscated malicious code. In: WORM 2003: Proceedings of the 2003 ACM workshop on Rapid malcode, pp. 76–82, New York, NY, USA. ACM (2003)

24. Santos, I., Penya, Y.K., Devesa, J., Bringas, P.G.: N-grams-based file signatures for malware detection. ICEIS (2), **9**, 317–320 (2009)
25. Sathyanarayan, V.S., Kohli, P., Bruhadeshwar, B.: Signature generation and detection of malware families. In: Mu, Y., Susilo, W., Seberry, J. (eds.) ACISP 2008. LNCS, vol. 5107, pp. 336–349. Springer, Heidelberg (2008). https://doi.org/10.1007/978-3-540-70500-0_25
26. Sharma, A., Sahay, S.K.: Evolution and detection of polymorphic and metamorphic malwares: a survey. arXiv preprint arXiv:1406.7061 (2014)
27. Sun, H.-M., Lin, Y.-H., Wu, M.-F.: API monitoring system for defeating worms and exploits in MS-windows system. In: Batten, L.M., Safavi-Naini, R. (eds.) ACISP 2006. LNCS, vol. 4058, pp. 159–170. Springer, Heidelberg (2006). https://doi.org/10.1007/11780656_14
28. Vilkeliskis, T.: Automated unpacking of executables using dynamic binary instrumentation (2009)
29. Virus Total. http://www.virustotal.com/stats.html
30. Veratrace: http://www.offensivecomputing.net/
31. Vmpacker. http://www.leechermods.com/
32. VX heavens. http://vxheaven.0l.wtf/
33. Wadkar, M., Di Troia, F., Stamp, M.: Detecting malware evolution using support vector machines. Expert Syst. Appl. **143**, 113022 (2020)
34. Open source Machine Learning Software Weka. http://www.cs.waikato.ac.nz/ml/weka/
35. Witten, I.H.: Frank, and E. Morgan Kaufmann, Practical Machine Learning Tools and Techniques with Java Implementation (1999)
36. Wong, W., Stamp, M.: Hunting for metamorphic engines. J. Comput. Virol. **2**(3), 211–229 (2006)
37. Xen: http://www.xen.org
38. Zhang, B., Yin, J., Hao, J.: Using fuzzy pattern recognition to detect unknown malicious executables code. In: Wang, L., Jin, Y. (eds.) FSKD 2005. LNCS (LNAI), vol. 3613, pp. 629–634. Springer, Heidelberg (2005). https://doi.org/10.1007/11539506_78
39. Zhang, Q., Reeves, D.S.: Metaaware: identifying metamorphic malware. In: Twenty-Third Annual Computer Security Applications Conference (ACSAC 2007), pp. 411–420. IEEE (2007)

Finite Set Algebra in Secondary School Using Raspberry Pi with Mathematica

Robert Ipanaqué-Chero[✉][iD], Ricardo Velezmoro-León[iD],
Felícita M. Velásquez-Fernández[iD], and Daniel A. Flores-Córdova[iD]

Departamento de Matemática, Universidad Nacional de Piura,
Urb. Miraflores s/n Castilla, Piura, Peru
{ripanaquec,rvelezmorol,fvelasquezf,dflores}@unp.edu.pe

Abstract. The objective of this research is to automate the graphical representation of Finite Set Algebra, in order to provide: an alternative as a learning tool or a means to teach the association of mathematical concepts with the programming language, determine the construction rules and the Command programming in the subject of Basic Education Mathematics. This document describes a new Mathematica package, DiscreteSets.m programmed in PiWolfram. Since the Mathematica license is expensive, a Raspberry Pi computer is used since, thanks to its low cost and a free version of Mathematica that runs on it, it makes the new Mathematica package accessible to a greater number of users. To show the performance of the new package, a full set of illustrative examples is provided. This package admits as input a certain operation between two or three discrete and finite sets, whose elements can be numbers, letters or certain figures. And it returns as output the Venn diagram corresponding to the given operation, with the location of the elements of the sets similar to how they would be located manually. Allowing students to interactively learn and dabble in the PiWolfram programming language. Programming has been the cause of the great advances in technology. For this reason, it is important to incorporate technological education into the learning of Mathematics so that students appreciate the real importance of mathematics today.

Through mathematics we can understand the world around us. It is present in various human activities: daily life, social and cultural aspects, in such a way that it is an essential key to understanding and transforming culture. Having a mathematical development allows us to participate in the world in any of the aforementioned aspects.

Keywords: Automate · Algebra · Finite set · Raspberry Pi

1   Introduction

Through mathematics we can understand the world around us. It is present in various human activities: daily life, social and cultural aspects, in such a way that it is an essential key to understand and transform culture. Having a

© Springer Nature Switzerland AG 2021
O. Gervasi et al. (Eds.): ICCSA 2021, LNCS 12951, pp. 366–379, 2021.
https://doi.org/10.1007/978-3-030-86970-0_26

mathematical development allows us to participate in the world in any of the aforementioned aspects [2].

For this, it is important that the student understands that the competencies in the area of mathematics are linked with others, integrating in such a way that it allows them to understand, analyze, the variables that intervene when solving problems.

Some of the competences allow the student to characterize equivalences, generalize regularities and changes of one magnitude with respect to another through rules that allow finding values, restrictions and making predictions about the behavior of a phenomenon [3].

One of the relevant topics for the achievement of the aforementioned competence is the Algebra of Sets. Why should school students study, Algebra as a whole? Because they are present in everyday life through certain verbs that relate elements of two groups that are part of social, economic and cultural life, also in scientific matters. It is important to understand the nature of finite set Algebra through the type of association that exists between physical or mental objects that belong to certain sets.

According to many previous studies, learning mathematics through programming is a great advantage today. Well, it gives students a greater capacity for logical reasoning through structured thinking. The Wolfram Mathematica programming language has been used as a tool to address the Algebra of finite sets, which can be worked through an algorithmic design. We will define a program as a set of commands executed in a computer language that allow a machine to perform a series of operations automatically [4]. Programming is one of the few mental skills that allows students to explore the dynamics of their own thoughts, increase their capacity for self-criticism by correcting their mistakes, and also establish their logical reasoning through systematization [5].

Currently, software design and programming languages are in high demand in work environments, however, they are not addressed in the national curriculum of Peru. In other countries such as Spain, the US, Canada, EUROPE, the Middle East, Africa, China, New Zealand and Australia, they introduced the programming language as a means of curricular learning that influences all subjects [6].

The programming language should be incorporated in the last years of basic education or in the courses of the first year of university studies of general training. For this, mathematical knowledge must be structured in a similar way to the structure of a program in such a way that it is related to the concepts of computer science. For this, mathematical knowledge must be structured in a similar way to the structure of a program in such a way that it is related to the concepts of computer science. For students to acquire this ability according to [7], they achieve this under the guidance of the teacher, receiving the appropriate guidance on how to proceed so that they acquire the skill and then the student begins the exercise process in the necessary amount and with the appropriate frequency, so that reproduction is easy and whether to debug.

It is highlighted that the subject of finite set algebra allows us to analyze, classify and order the knowledge acquired by developing the complex conceptual network in which we store our learning. Therefore, given the importance of the subject, we seek to introduce programming; where the teacher guides the students to contemplate the presence of the association, determines the construction rules and programming commands.

In the work for research purposes, the theoretical concepts of the Algebra of finite sets were elaborated and taking into consideration the statement of UNESCO (2015): "the performance of Latin American students is affected by socioeconomic factors". In addition, considering the introduction of ICTs in all social areas, it is necessary to acquire digital skills through the use of the computer, this situation exacerbates learning gaps. An alternative to reverse this situation is to introduce the use of the Raspberry Pi microcomputer to education due to its low cost. Therefore, a finite set Algebra programming package was designed in the PiWolfram language. The objectives of this research were: Design a programming package in the PiWolfram language for learning Finite Set Algebra. Strengthen the computational thinking of high school students so that they have the ability to create algorithms and use different levels of abstraction to solve problems.

The structure of this article is as follows: Sect. 2 asks the research question: Why should school students study Algebra as a whole? and the importance of ensemble algebra is highlighted. In the Sect. 3 introduces the concepts of set, Venn diagram, cardinal of a set and algebra of set. It also talks about the Raspberry Pi Minicomputer and its Raspbian operating system, which is designed to interact with the PiWolfram language. In this language a mathematics package has been designed to represent the algebra of finite sets. Then in Sect. 4 Various studies related to our research that have been applied in secondary education are presented. Finally, the Sect. 5 closes with the main conclusions of this article where it is highlighted that the learning of Finite Set Algebra by students, through PiWolfram programming packages, strengthens reasoning and programming skills using a low-cost minicomputer.

2   Fundamentals

Currently there are advances in the teaching of mathematics in basic education, in this advance the use of technology as a tool has played an important role, it is an indicator of educational quality the use of ICT in teaching processes, technology at students' reach, such as the computer, which has made it possible for the explanation of certain mathematical arguments not to be so monotonous and boring, but rather to be perceived as a game when interacting with programming codes [8].

In basic education, the problems of finite set algebra, are currently solved manually and the only help some software provides is that they allow you to choose the figure you want to represent a set, offering no solution to the graphical representation with diagrams of Venn for operations between finite sets [9].

Students will be taught the automation and graphical representation of finite set algebra in order to provide an alternative to the traditional teaching methods of this subject in Basic Education Mathematics that incorporate the use of technological tools such as, in our case, commands and programs generated with Mathematica.

The contribution of this work consists of the development of a new Mathematica package, DiscreteSets.m, which automates the solution of problems related to operations in the algebra of finite sets. This package admits as input a certain operation between two or three discrete and finite cardinal sets, whose elements can be numbers, letters or certain figures. And it returns as output the Venn diagram corresponding to the given operation, with the location of the elements of the sets similar to how they would be located manually. In such a way that the user is in the possibility of obtaining an automatic solution of this type of operations by entering the respective entries.

There are very positive educational experiences with Wolfram Mathematica. An example is the teachers who incorporated it as an educational program, in their teaching processes together with an appropriate methodology such as Tpack and the creation of dynamic activities [10].

To encode the DiscreteSets.m package a Raspberry Pi 4 Model B computer is used, because it is possible to buy these computers for a price from 35 dollars and also because there is a free version 12 of Mathematica that runs in Raspbian OS for Raspberry Pi. This facilitates access to the package by a greater number of users [11–13].

3   Definitions and Basic Concepts

3.1   Finite Set Algebra

For programming the Algebra of Finite Sets, we will use definitions and examples that appear in this subsection have been taken of [2].

Set. A set is a well defined collection of objects. Objects in the set are called elements. Like other mathematical concepts, sets have their own notation and terminology. In general, we name sets with italicized capital letters and, when possible, list specific elements between braces: {}. For example, $A = \{1, 2, 3\}$ means that set A contains the elements 1, 2 and 3. To describe specific objects in the set, we can use the symbol $\in$ to replace the words "is an element of" and $\notin$ to replace the words "is not an element of". For instance, we can write $1 \in A$ and $4 \notin A$.

Representing Sets. To this point, we have already seen two represent sets: rosters and set-builder notation. A roster is simply a list of the elements in the set. For example: $A = \{1, 3, 5, 7\}$ $B = \{blue, white, red\}$ Another way to represent sets is to use set-builder notation: $\{x \mid x \text{ satisfies some condition}\}$ The vertical bar ($\mid$) replaces the words "such that". So, we read set-builder notation as "The set of all x such that x satisfies some condition".

Empty Set. The empty set, or null set, written {} or ∅, is the set that contains no elements.

Universal Set. The universal set, or universe, written $\mathbb{U}$, is the set of all possible elements in a given discussion.

Venn Diagrams. Venn diagrams are a third way to represent sets that use the notion of universal sets. The standard convention is to represent the universe with a rectangle and sets within the universe as circles. Let's look at Fig. 1, set A (the circle) is in universe $\mathbb{U}$ (the rectangle). As we will see, Venn diagrams are a powerful way to represent sets and the relationships among them.

Fig. 1. Basic Venn diagram.

Subset and Proper Subset. Set A is a subset of set B, written A ⊆B, if and only if, every element of A is also an element of B. If B has an element that is not in A, then A is a proper subset of B, written A ⊂ B.

Cardinal Number. The cardinal number of a set A, written n(A), is the number of elements in A.

Finite Set. A set is finite if and only if:

1. It is empty, or
2. There exists a 1–1 correspondence between the set and a set of the form {1,2,3,...,n}, where n is a natural number.

Union of Two Sets. The union of the two sets A and B, written A ⊎ B, is the set of all elements in A or in B or in both, let's look at Fig. 2(a). In set-builder notation,

A ∪B = {x|x∈A∨x∈B}

Intersection of Two Sets. The intersection of two sets A and B, written A ∩ B, is the set of all elements common to A and B, let's look at Fig. 2(b). In set-builder notation,

A ∩B = {x|x∈A∧x∈B}

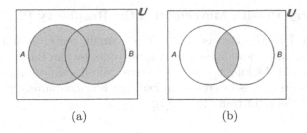

(a)   (b)

Fig. 2. (a) Venn diagram of A ∪ B. (b) Venn diagram of A ∩ B.

Set Difference. The set difference of set B from set A, written A\B, is the set of all elements in A that are not in B (Fig. 3). In set-builder notation,
    A\B = {x|x∈A∧x∉B}

Complement of a Set. The complement of a set A, written AC, is the set of all elements in the universe U that are not in A (Fig. 3). In set-builder notation,
A^c={x|x∈ 𝕌∧x∉A}

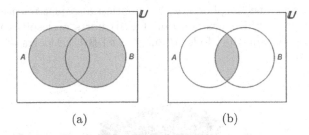

(a)   (b)

Fig. 3. (a) Venn diagram of A \ B. (b) Venn diagram of A^C.

Symmetric Difference of Two Sets. The symmetric difference is equivalent to the union of both relative complements (Fig. 4), that is:
    A/$triangle$B = (A\B)∪(B\A)

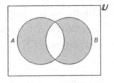

Fig. 4. Venn diagram of A△B.

3.2   Raspberry Pi with Mathematica the Raspberry Pi

La Raspberry Pi. The Raspberry Pi. It is a small, powerful and inexpensive single board computer developed over several years by the Raspberry Pi Foundation. If someone is looking for an inexpensive, small, easy-to-use computer for an upcoming project, or is interested in learning how computers work, then the Raspberry Pi is the right thing to do [1].

El Hardware Raspberry Pi. The Raspberry Pi's BCM2835 CPU runs at 700 MHz and its performance is roughly equivalent to that of a 300 MHz Pentium II computer that was available in 1999 [3].

The Raspberry Pi Foundation. It is a non-profit charity, founded in 2006 by Eben Upton, Rob Mullins, Jack Lang, and Alan Mycroft. The aim of this charity is to promote the study of computing among a generation that did not grow up with BBC Micro or Commodore 64 [1].

Raspberry Pi Models. It has several different variants. Model A is a low cost version and sadly omits the USB hub chip. This chip also works as a USB to Ethernet converter. The Raspberry Pi Foundation also just released the Raspberry Pi Model B + which has additional USB ports and solves many of the power issues surrounding the Model B and Model B USB ports [12], let's look at Fig. 5.

Fig. 5. Raspberry Pi 4 model B.

Operating System: Raspbian. Raspbian is the most popular Linux-based operating system based on Debian, which has been specifically modified for the Raspberry Pi (hence the name Raspbian). Raspbian includes customizations that are designed to make the Raspberry Pi easier to use and includes many different out-of-the-box software packages [12].

Wolfram Language. Makes cutting-edge computing accessible to everyone. Unique in its approach to building vast knowledge and automation, the Wolfram Language scales from a single line of easy-to-understand interactive code to production systems of one million lines [8].

Mathematica. Es un producto original y más conocido de Wolfram, destinado principalmente a la informática técnica para la investigación, el desarrollo y la educación. Basado en Wolfram Language, Mathematica es 100% compatible con otros productos [8].

Raspberry Pi Con Mathematica. Una versión completa de Wolfram Language está disponible para la computadora Raspberry Pi y viene incluida con el sistema operativo Rasp bian. Los programas se pueden ejecutar desde una línea de comandos de Pi o como proceso kground, asícomo a través de una interfaz de computadora portátil en Pi o en una computadora remota. En Pi, Wolfram Language admite acceso programático directo a puertos y dispositivos Pi estándar [8] (Fig. 6).

3.3   Free Wolfram Language and Mathematica on Every Raspberry Pi

The Wolfram Language and Mathematica are included in the Raspberry Pi NOOBS (New Out of Box Software) installation system.

Fig. 6. Mathematica and every Raspberry Pi.

In the Raspberry Pi 4 version, Mathematica Version 12 and the Wolfram Language are available. With a wide range of features, including a significant expansion of numerical, mathematical and geometric computing, audio and signal processing, text and language processing, machine learning, neural networks and others. Version 12 brings Mathematica users new levels of power and effectiveness, with updates to its functions and areas. Mathematica 12 runs significantly faster on the Raspberry Pi 4 than it did in previous versions. Where you can carry out projects from the creation of weather panels to the creation of tools that use machine learning and the execution of command line scripts [17].

3.4   A Mathematica Package for the Graphic Representation of the Finite Set Algebra

This section describes the new Mathematica package, DiscreteSets.m, we developed for the graphic representation of the finite set algebra. We will begin by indicating that Mathematica does not incorporate any native function to obtain the Venn diagram of the operations between two or three given sets, and even less to obtain a Venn diagram in which the elements of the assemblies are placed, according to the operations that are indicated. To encode the new package we use

Mathematica's ability to manipulate and convert strings of characters. Thanks to this it is possible to enter the operations between sets affected by quotes, which is the standard way to enter strings in Mathematica. Mathematica's ability to manipulate lists allowed us to enter the elements of all the sets that participate in the operations. Mathematica's ability to represent circumferences and regions of the plane is also used. This capacity allows to represent the sets by means of circles and to shade those parts that correspond to the solution of the given operations. Finally, the algorithms with which the polygons that make up the different regions into which the plane is divided (according to the number of sets involved in the given operation) are used to take some vertices, of such polygons, on whose coordinates are locate the elements of the sets.

Ahora, comenzamos nuestra discusión del nuevo paquete accediendo a Mathematica en Raspberry Pi, let's look at Fig. 7.

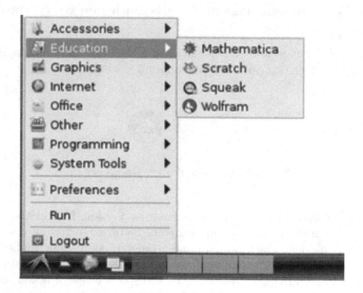

Fig. 7. Accessing mathematica in Raspberry Pi.

Then, we loading the package:
<<DiscreteSets.m

The main function, VennDiagram, supports a maximum of five arguments and at least two, namely:

- an operation between sets as a character string.
- the sets /(one to three sets7) and, optionally, the universe according to the syntax.

$VennDiagram["operation", A \rightarrow \{a1, ..., an\}, ..., Universe \rightarrow \{u1, ..., um\}]$
Below are some examples that show the functionality of the VennDiagram function. The first example is arguably one that students should know how to deal with:
VennDiagram["A", A $\rightarrow$ {1, 2, 3, 4, 5, 7}. Let's look at Fig. 8.

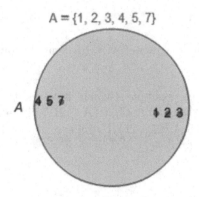

Fig. 8. Venn diagram of set A.

The same comment applies here (the reader is kindly invited to try manually these four examples by him/herself):
$VennDiagram["A \cup B", A \rightarrow \{1,2,3,4,5,7\}, B \rightarrow \{3,4,8,9,11\}, Universe \rightarrow \{0,1,2,3,4,5,6,7,8,9,10,11\}]$. Let's look at Fig. 9(a).
$VennDiagram["A \cap B", A \rightarrow \{1,2,3,4,5,7\}, B \rightarrow \{3,4,8,9,11\}, Universe \rightarrow \{0,1,2,3,4,5,6,7,8,9,10,11\}]$. Let's look at Fig. 9(b).
$VennDiagram["A \setminus B", A \rightarrow \{1,2,3,4,5,7\}, B \rightarrow \{3,4,8,9,11\}, Universe \rightarrow \{0,1,2,3,4,5,6,7,8,9,10,11\}]$. Let's look at Fig. 10(a).
$VennDiagram["A \triangle B", A \rightarrow \{1,2,3,4,5,7\}, B \rightarrow \{3,4,8,9,11\}, Universe \rightarrow \{0,1,2,3,4,5,6,7,8,9,10,11\}]$. Let's look at Fig. 10(b).

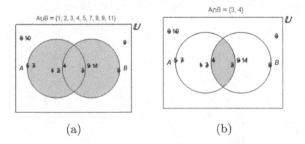

(a)   (b)

Fig. 9. (a) Venn diagram of A $\cup$ B. (b) Venn diagram of A $\cap$ B.

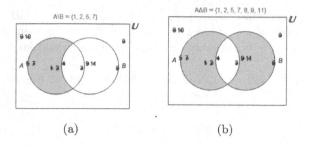

Fig. 10. (c) Venn diagram of A\B. (d) Venn diagram of A △ B.

On the contrary, this example (which involves three sets) is not that easy: $VennDiagram[``((A \cup B) \cap C) \setminus (A \cap B)", A \rightarrow \{1, 2, 3, 4, 5, 7\}, B \rightarrow \{3, 4, 8, 9, 11\}, C \rightarrow \{2, 3, 4, 6, 10, 11\}, Universe \rightarrow \{0, 1, 2, 3, 4, 5, 6, 7, 8, 9, 10, 11, 12\}]$. Let's look at Fig. 11.

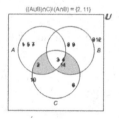

Fig. 11. Venn diagram of $((A \cup B) \cap C) \setminus (A \cap B)$.

This other example is not quite simple either: $VennDiagram[``(A \triangle B) \cap (C \setminus B)", A \rightarrow \{1, 2, 3, 4, 5, 7\}, B \rightarrow \{3, 4, 8, 9, 11\}, C \rightarrow \{2, 3, 4, 6, 10, 11\}, Universe \rightarrow \{0, 1, 2, 3, 4, 5, 6, 7, 8, 9, 10, 11\}]$. Let's look at Fig. 12(a).

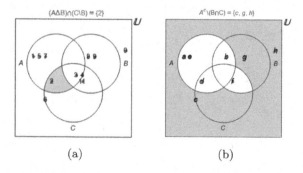

Fig. 12. (a). Venn diagram of $(A \triangle B) \cap (C \setminus B)$. (b) Venn diagram of Ac \ (B ∩ C).

In the following example sets' elements are letters of the alphabet: $VennDiagram["A^C \setminus (B \cap C)", A \rightarrow \{a, b, d, e\}, B \rightarrow \{b, f, g\}, C \rightarrow \{c, d, f\}, Universe \rightarrow \{a, b, c, d, e, f, g, h\}]$. Let's look at Fig. 12(b).

In this last example sets' elements are symbols: $VennDiagram["(A \cup B) \cap (C \triangle B)^{C}", A \rightarrow \{\diamond, \triangledown, \square, \blacksquare\}, B \rightarrow \{\blacklozenge, \heartsuit, \square\}, C \rightarrow \{\star, \heartsuit, \nabla, \blacklozenge\}, Universe \rightarrow \{\nabla, \square, \lozenge, \blacklozenge, \blacksquare, \blacktriangledown, \star, \heartsuit, \spadesuit\}]$. Let's look at Fig. 13.

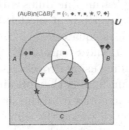

Fig. 13. Venn diagram of $A^C \setminus (B \cap C)$

4 Related Research

Much has been said and written about the potential applications of computer algebra systems to mathematics education. In fact, this is a hot topic of many conferences and workshops related to educational topics, where they make arguments in favor of the inclusion of computer algebra systems in the mathematics curriculum from the middle school level to the upper level as a middle school. aid in problem solving Its use in education should always be subject to feedback according to the teacher's experiences and research in education [14].

By contrast, less has been said on the specific subject of teaching and learning the graphic representation of finite set algebra [15]. Among those rare papers on this issue, the authors found a very nice work a question and answer site for users of Wolfram Mathematica in which a user showed how Mathematica can be adequately used to create a Venn Diagram for three sets [16]. However, the focus in such a document was basically the location of the elements in the diagram, according to the set to which they belonged, without taking into account any type of operation. Because of this limited goal, [16] did not introduce any new command; standard Mathematica commands seem to be good enough to make pictures, as opposed to our approach, for which new symbolic commands are actually required. In this sense, this paper goes far beyond [16]. We have introduced a computer tool that can be used for Mathematics of Basic Education. Even teachers and instructors can benefit from this tool. For instance, we can propose to our students to verify if the result of an operation between two or three sets is mathematically correct.

Among the various jobs related to research, we highlight the work done by Handley, N. and Meacham, S. [18]. They were raised by objective to create solutions to help elementary students learn, in mastery of counting using visual representations. Using a Raspberry Pi they developed prototypes of low-cost interactive solutions that can help learning in this area.

The research by Musyrifah, E., Rabbani, H., Sobiruddin, D. and Khairunnisa K. [19] that sought to incorporate Mathematica software into the learning of Mathematics as an interactive tool. They emphasize that the available texts seem monotonous and do not allow students to understand derivative concepts. They developed didactic materials in learning modules with the Wolfram Mathematica application using the ADDIE development model, being evaluated by teachers and students, who considered it adequate.

The study by Kaufmann, O. and Stenseth, B [20], a group of high school students who applied programming using Processing to solve mathematical problems was observed. An important finding they obtained is that the students progressed in the argumentation categories from the simplest to the most elaborate.

In each of the aforementioned works, the aim is to improve the quality of mathematics learning using specialized mathematics software with an interactive tool that strengthens cognitive abilities, to achieve the students' competencies. Developing in them skills, abilities and attitudes to understand the world around them and be able to create algorithms that help automate solutions.

5   Conclusions

This article introduces a new Mathematica package that enables you to place the elements that satisfy operations between two or three sets automatically in the graphical representation of such operations in a Venn diagram. The package includes commands for graphical output that allow users to grasp the ideas behind this concept. Additionally, a Raspberry Pi computer is used since, thanks to its low cost and a free version of Mathematica that runs on it, it makes the Mathematica package accessible to a greater number of users. It should be noted that there is a slight variation between the code executed on a traditional laptop and that executed on a Raspberry Pi, this occurs specifically in the construction of graphics using primitives. To show the performance of the main commands, a full set of illustrative examples is provided. Teachers will be able to use the Mathematica package to make their classes very interactive using the appropriate strategies or to teach programming using the Mathematica programming language after training in the Software.

References

1. Harrington, W.: Learning Raspbian. Packt Publishing Ltd. (2015)
2. MINEDU, Qué y cómo aprenden nuestro niños. http://www.minedu.gob.pe/DeInteres/pdf/documentos-secundaria-matematica-vii.pdf. Accessed 24 Feb 2021
3. MINEDU. http://www.minedu.gob.pe/curriculo/pdf/03062016-programa-nivel-secundaria-ebr.pdf. Accessed 24 Feb 2021
4. Wilkes, M.: Automatic digital computers Location, 1st edn. Methuen & Co., Ltd (1956)

5. Briz, A., Serrano, A.: Aprendizaje de las matemáticas a través del lenguaje de programación R en Educación Secundaria. SciELO, vol. 30(1) (2018)
6. Yábar, J.M.: Informática y matemáticas. 'Quién apoya a quién?. Uno-Revista de Didáctica de las Matemáticas. (Núm)006
7. López, M.: 'Sabes enseñar a describir, definir, argumentar.- Revista de Didáctica de las Matemáticas La Habana: Pueblo y Educación. 1990
8. Pelgrum, W.J., Law, N.: ICT in Education Around the World: Trends, Problems and Prospects, vol. 77. Unesco, International Institute for Educational Planning (2003)
9. Collins, S.O.: Raspberry Pi 4 User Guide: The New Updated Guide to Master the New Raspberry Pi 4 and Make, Build. Independently Published, Or Hack a Variety of Amazing Projects (2019)
10. Castañeda, F.: Actividades de aprendizaje para la enseñanza de la Matemática utilizando Geogebra y Wolfram Mathematica para estudiantes de décimo año de educación general básica de la Unidad Educativa Capitán Edmundo Chiriboga. Universidad Nacional de Chimborazo (2020). http://dspace.unach.edu.ec/handle/51000/6516. Accessed 24 Feb 2021
11. Raspbian OS Programming with the Raspberry Pi. Apress, Berkeley, CA (2019). https://doi.org/10.1007/978-1-4842-4212-4
12. Wolfram: Wolfram Language y Mathematica gratis en cada Raspberry Pi. https://www.wolfram.com/raspberry-pi/index.php.es?source=footer. Accessed 24 Feb 2021
13. Xataka: Raspberry Pi 4 es oficial: una completa actualización con procesador Cortex-A72, hasta 4 GB de RAM y desde 35 dólares. https://www.xataka.com/ordenadores/raspberry-pi-4-caracteristicas-precio-fichatecnica. Accessed 4 Feb 2020
14. Salat, R.: El potencial de los sistemas de álgebra computacional.Revista de Didáctica de las Matemáticas, vol (81) (2012)
15. Zevallos, A: Como hacer Diagramas de Ven Online- Representación Gráfica de Conjuntos en línea. https://wolframalpha0.blogspot.com/2014/01/como-hacer-diagramas-de-venn-online.html. Accessed 4 Feb 2020
16. StackExchange: Create a Venn Diagram. https://mathematica.stackexchange.com/questions/134014/create-a-venn-diagram. Accessed 4 Feb 2020
17. Wolfram: Mathematica 12 disponible en la nueva Raspberry Pi 4. https://blog.wolfram.com/2019/07/11/mathematica-12-available-on-the-new-raspberry-pi-4/. Accessed 4 Feb 2020
18. Handley, N., Meacham, S.: Raspberry Pi based solution for Primary Schools Mathematics Education. In: Conference: BCS SQM/Inspire Conference 2016
19. Musyrifah, E. et al.: Development of wolfram mathematica application-assisted learning module on derivative in high school. J. Phys. Conf. Ser. **1836**(1). https://doi.org/10.1088/1742-6596/1836/1/012076
20. Kaufmann, O., Stenseth, B.: Programning in mathematics education. Int. J. Math. Educ. Sci. Technol. (52) (2021). https://doi.org/10.1080/0020739X.2020.1736349

Using Visual Analytics to Reduce Churn

Priscilla A. Karolczak and Isabel H. Manssour$^{(\boxtimes)}$ (iD)

Pontifical Catholic University of Rio Grande do Sul, PUCRS. School of Technology.
Av. Ipiranga, 6681, Porto Alegre, RS 90619-900, Brazil
Priscilla.Karolczak@edu.pucrs.br, isabel.manssour@pucrs.br

Abstract. The technological advance and the use of data for decision-making drive business and increase competitiveness among companies. They must understand the environment in which they operate and quickly respond to the needs of their customers, preventing them from canceling their services, maximizing profit, and benefit their own organizations. The main goal of this work is to present a visual analytics approach to deal with a large amount of unstructured, complex, and dynamic data and improve companies' ability to detect the probability of losing a client at an early stage. We processed the probability of subscription cancellation using the machine learning algorithm Random Forest, and we allowed similar customers comparison using the k-nearest neighbor's algorithm. Then, we developed two main visualizations: a general dashboard that displays the probability of subscription cancellation of each client and the variables that can influence this decision; and a radar chart that displays many quantitative variables and allows comparison with other similar customers. To validate our approach, we present a case study with a data set from a hosting services company that uses the Platform as a Service model. Through the application of informal interviews, we concluded that the provided visualizations helped teams in the process of reducing churn rate and therefore maintaining a growing customer base's company. Our visual analytics solution allowed the analysis and information understanding to create strategies and make assertive decisions for professionals involved in the retention process.

Keywords: Churn · Visual analytics · Machine learning

1 Introduction

The advancement of technology and the use of data for decision-making boost business and increase competitiveness among companies. In this context, companies must understand the scenario in which they operate and respond quickly to the needs of their clients. Thus, the focus becomes the customers, through the analysis of their behaviors, to prevent them from canceling their services [10]. The loss of loyal clients is a threat to companies in several sectors, so efforts should be directed more towards retaining existing customers than attracting new ones [7].

This research was financed in part by PUCRS. Isabel Harb Manssour also would like to thank the financial support of the CNPq Scholarship - Brazil (308456/2020-3).

© Springer Nature Switzerland AG 2021
O. Gervasi et al. (Eds.): ICCSA 2021, LNCS 12951, pp. 380–393, 2021.
https://doi.org/10.1007/978-3-030-86970-0_27

For this, companies have teams responsible for the Customer Relationship Management (CRM) process, which aims to manage customer information to maximize their loyalty [11]. Its main objective is to satisfy customers to prevent them from canceling their services. For the Platform as a Service (PaaS) market, which provides infrastructure and technology for the development and hosting of applications, the turnover rate, or churn rate, is one of the main metrics to assess business performance. Churn is a marketing term in which the customer is interested in another organization or product [2].

One challenge is to reduce cancellation requests and prevent the customer from reaching that decision. Therefore, it is essential to understand the factors that motivate you to request cancellation. The ability to detect the probability of losing a customer at an early stage is something that every company wants to achieve [3]. However, within a large customer base and with a significant number of variables, it is complex to identify patterns and factors that influence cancellation requests. Analyzing data from various sources of information promises to address answers to these questions. However, this analysis requires data integration, which is difficult for users without software development skills [4].

According to Keim et al. [8], visual analytics is an interdisciplinary research area that combines, among others, visualization, data mining, and statistics, making the way of processing data and information understandable for an analytic discourse. Therefore, the use of visual analytics [3] can be an alternative to assist in the analysis of a large amount of unstructured, complex, and dynamic data, supporting the retention function. In this context, tools and techniques are created to allow people to synthesize and obtain information from this kind of data, detecting the expected and discovering the unexpected.

Thus, the main goal of this work is to present a visual analytics approach to support CRM teams to create strategies and act at the right time, that is, before a customer request a service cancellation. Our prototype implementation allows analyzing different data sources and identifying the variables that affect customer cancellation through interactive graphs and statistical data. Our main contributions are:

- The provided visualizations that allow analyzing the probability of subscription cancellation of each client, as well as to compare the similar ones;
- A high accuracy of the used data mining algorithms;
- The possibility of using the approach with your own data sets;
- The validation of the proposed approach through a case study that shows how it helps to reduce the churn rate.

The remainder of the paper is organized as follows. Section 2 presents some related work. The proposed approach is described in Sect. 3. Section 4 details the obtained results through the presentation of a real case study and the feedback obtained with an informal interview with domain experts. A discussion about the future work directions is also presented. In the last section, we outline our conclusions.

2 Related Work

There are several works already developed involving the study of prediction models for churn and client loyalty based on ML (Machine Learning) methods. Noyan and Simsek [12] and Zakaria et al. [17] conducted studies to understand customer loyalty. The first [12] revealed that clients' loyalty depended on price, discount, product quality, service quality, perceived value, and satisfaction, with customer satisfaction being the most important factor influencing their loyalty. The second [17] concluded that the store's partnership program, premiums, insurance, discount, and price had a significant impact on customer satisfaction and loyalty.

An approach to better understand clients' behaviors and classify them based on the RFM (Recency, Frequency, Monetary Value) and LTV (Life Time Value) model using genetic algorithm for the analysis was presented by Chan [1]. Thus, the customers with the highest value could be identified for the recommendation programs. Qureshi et al. [15] used ML techniques, such as regression, RNA (Artificial Neural Network), and decision trees, to predict churn for telecommunications customers. The results showed that decision trees had the best performance to identify potential churners. In the study by Khodabandehlou and Rahman [9], five variables were added to the RFM model, with emphasis on discount, delivery time, and price. An accuracy of 97.92 was obtained compared to the specific use of RFM, using the RNA method with the highest performance.

Table 1 presents the comparison between studies involving churn and ML techniques. We note that the variables of the RFM model are widely used in retail, and the price variable is used in almost all situations. In telecommunications, with a more focused aspect to services, customer behavior is used. Decision trees are used in different scenarios, but studies show that it is necessary to analyze each need and test more than one technique to find the best results.

Table 1. Comparison between churn studies and ML techniques.

Ref.	Variables	Techniques	Sector
[12]	Price, discount, service quality, perceived value, satisfaction	SEM (Structural Equation Modeling)	Commerce (supermarkets)
[17]	Store's partnership program, premiums, insurance, price	Decision tree, boosting, logistic regression	B2B/retail
[1]	RFM	Genetic algorithm	Retail resale
[15]	Customer behavior, such as call time	Regression, RNA, decision tree	Telecommunications
[9]	RFM, discount, delivery time, price	RNA	Commerce

The study by Dintakurthi [3] is the closest to the objective of this work. Through the analysis for the manufacturing industry, they identified the variables that affected churn and developed a model to support the after-sales functions. As a result, they made available geolocation graphs indicating places where there is more churn and helping in decision making, for example, identifying the regions where should be prioritized attention to churn. Bar and line graphs are also available to show the distribution of transactions and their relationship to churn and the impact of the price on churn.

Grammel [4] provided support to end-users when analyzing multiple data sources. They used Tableau to explore a widget-centric approach to data analysis and integration. Oelke et al. [13] presented techniques for analyzing positive and negative opinions expressed by customers. They developed visual analytics solutions to support analysts. An example is the interactive circular correlation map that offers a detailed view of the data and allows you to find correlations between the different aspects of the data set. Park [14] presented a visual analytics tool that reads a variety of textual data sources and extracts important keywords, relationships, and events using ontology and natural language processing methods. It provides users with an integrated and interactive search interface to facilitate data investigation. The keywords are aggregated according to the date and the news where they were published.

Table 2 presents a comparison of the works that support users' functions. This table shows that there are several charts available for each need and that the combination of one or more interactive graphics is an important practice for churn analysis.

Table 2. Summary of visual analytics studies.

Ref.	Objective	Chart/Visual model
[3]	Predict churn and support after-sales functions	Geolocation, bar, and line chart
[4]	Support end users in data integration	Cognitive support structure widget-based
[13]	Support CRM analysts	Interactive circular correlation map
[14]	Provide an integrated and interactive research interface for investigation	Bubble chart/keyword cloud

Among the nine works covered, only four present visualization techniques that support end-users. In five works, studies were carried out with different types of techniques, using multiple variables. However, none of them present a tool that supports the daily challenge of retaining customers that is constantly updated and interactive. Besides, none of them were made for the PaaS (Platform as a Service) or similar market. More than one study used variables based on the RFM [1,9] model, but none analyzed these variables based on customer interactions by text, such as chat and call systems. Therefore, we identified the opportunity to combine predictive techniques with visual analytics to support customer relationship teams, such as retention, which are updated daily according to client's behavior and the market.

3   Visual Analytics Approach

The following subsections describe our visual analytics solution to support and improve the productivity of CRM teams that work with retention and aim to reduce churn. Through interactive graphics and statistical data, it allows the analysis of customer, product (CRM), and attendance data to identify the variables that affect cancellation. The developed prototype can be used by all companies that adopt a business model based on services, mainly PaaS, Infrastructure as a Service (IaaS), and Software as a Service (SaaS).

To perform some tests and to facilitate the definition of the classification algorithms to be adopted, we initially used Weka[1]. After that, we selected the following tools to automate, create, and integrate these algorithms with other systems: Language Python[2] for the collection, preparation, and pre-processing of data; and Language R[3] and Shiny[4] for the implementation of the visualizations techniques and the web application. CSV (Comma-Separated Values) files are accepted as data entry and the MySQL database[5] was used for persistence and maintenance of the data warehouse.

3.1   Research Methodology

We started developing our approach with data entry definition, including source and variables identification and data preparation. Then we plan the data processing by identifying the algorithms that could be used to create the statistical and predictive models. Finally, we specify the interactive visualization techniques that would be implemented. Figure 1 presents an overview of the development methodology.

3.2   Data Input

Data preparation can be divided into three stages: source identification, collection, and preparation. The identification of data sources was made by analyzing the systems and databases used by the team responsible for the CRM process in a PaaS company, whose main product is the website hosting service. Thus, seven main data sources were identified to obtain the variables used for model creation. Table 3 shows the list of identified sources. A set of 42 variables was identified,

---

[1] https://www.cs.waikato.ac.nz/ml/weka/.
[2] https://www.python.org/.
[3] https://www.r-project.org/.
[4] https://shiny.rstudio.com/.
[5] https://www.mysql.com/.

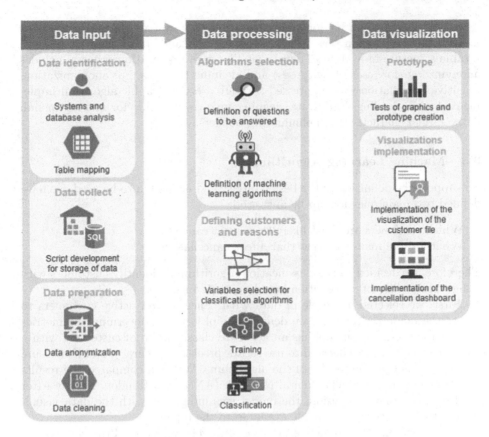

Fig. 1. Work development methodology.

Table 3. Table of data sources.

Source	System	Technology
Customers	CRM	MySQL
Products	CRM	MySQL
Calls	Backoffice	MySQL
Phone calls	Asterisk	MySQL
Websites	Backoffice	MySQL
Financial	Senior	SQL Server
Cancellation	Backoffice	MySQL

such as Lifetime Value, Client Status (active, blocked, debtor), Contracted Plan, Total Contacts and Average Response Time, and access to the control panel of the contracted service in the last 30 days.

Data collection can be done daily from different sources and made available in CSV files. A Python script reads the CSV files and stores their contents in a table structure in a MySQL database. The data preparation process includes anonymizing (for research purposes) and cleaning the data. For anonymization, sensitive information was suppressed through a cryptographic algorithm implemented while reading data. Data cleaning involves checking for missing or null data, special characters, and eliminating outliers.

3.3   Machine Learning Algorithms

To support the definition of the best strategy and algorithm to be used, we have defined two main questions to be answered:

- Which customers are most likely to request cancellation?
- What are the main variables that affect cancellation?

Therefore, we decided to use classification algorithms to identify customers with a high probability of cancellation based on their variables.

First, we selected a sample of data with canceled and active customers to create the model. After that, we defined the objectives concerning the metrics so that the prediction model has an accurate classification of customers with a tendency to cancel. We decided to use recall, precision, accuracy, and F-measure metrics to analyze the results of the algorithms. We then compared the results of two algorithms: Artificial Neural Network (ANN) and Random Forest. After reaching the desired precision, the model was integrated with the data visualization tool proposed to be made available to the teams.

Before training the predictive model definitively, we selected the variables to maximize its performance. Then, we configured the algorithms and performed some tests looking for accuracy of at least 80%. After reaching an accuracy of 81% with the Random Forest algorithm in Weka, we implemented it in Python using the library Scikit-learn[6]. The first step to obtaining a good result was to treat the categorical data by transforming them from string to numeric. The next step was to assemble the training and test data set. For a better result in the model creation, we defined the same number of canceled and active members, 2,000 each. The data set was then divided into training data (85%) and test data (15%). In the first run, all variables were used. For better accuracy, we included a code listing the importance of each variable and defined which ones should remain for the final model. With the optimization of the variables, we obtained an accuracy of 87%.

After we created the model, it was integrated with the data visualization tool. Thus, we were able to consult the data set in the MySQL database to be classified. Analyzing the obtained results, we verified a great variety of values in some variables, making it difficult to visualize their correlation. Therefore, we implemented a binning algorithm to treat the data and organize it on a scale from 0 to 100. To determine the cutoff values, we used the percentile

---

[6] https://scikit-learn.org/.

logic. For variables with a lot of variation, percentiles every 10% were used, and for variables with less variation, they were divided into 10%, 30%, 60%, and 90%. The results were then stored in the database for later consumption by the application responsible for data visualization.

3.4   Interactive Visualizations

Considering the questions to be answered and the objectives of the proposed work, we chose two main visualizations to help teams avoid cancellations. The first one is the visualization of a dashboard that displays the probability of cancellation for each customer and the variables that can influence this decision. The challenge in this visualization was to support a large amount of information easily and on a single screen. For this, we rely on the table lens technique [16], which allows an adequate visualization for the amount of data. In this visualization, illustrated in Fig. 2, each row represents a single record, and the columns represent a specific indicator. For its implementation, we use the LineUp.js[7] library together with R and Shiny.

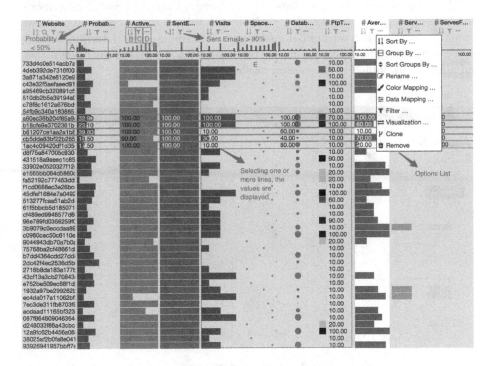

Fig. 2. Retention dashboard.

The first column of Fig. 2 contains the customer's identifier, and the second contains the probability of cancellation. The other columns correspond to the

---

[7] https://lineup.js.org/.

several variables used by the machine learning algorithm to perform the classification. The numeric data is shown as a progress bar, and the strings are shown
as text. Above each column (Fig. 2A), a histogram is displayed to provide an
overview of how the data set is distributed. When selecting one or more lines,
the numerical values are displayed within each bar.

Below the title of each column, three options are displayed. The first one
(Fig. 2B) allows sorting in ascending and descending order. The second (Fig. 2C)
offers the possibility to apply a filter through a free typing box. The third
(Fig. 2D) opens the menu shown at the top right of Fig. 2. The first option
of this menu allows each column to be sorted in descending, ascending, or random order. The second option enables the grouping of rows, for example, to
divide the data into two groups, one with values above 50 % probability and the
other below 50 %. The other options can be used to rename the column; change
the colors of the bars; filter a range of data; change the display of the bars to
alternatives such as absolute value, a symbol of proportion, or points; clone a
column; and remove a column, allowing to focus on the data that most matters
in the analysis.

The second visualization consists of a radar chart that displays the customer's
data considering the quantitative variables of use of the services and allows
comparison with other two similar customers. The similarity is determined using
the k-nearest neighbors [6] data mining technique. Each variable has its own axis,
and all axes are joined in the center, as shown in Fig. 3.

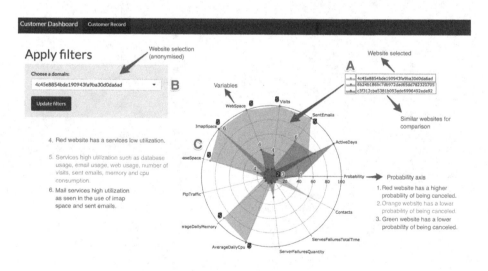

Fig. 3. Customer dashboard.

Each customer on the radar chart is represented by a different color, which
is identified by a legend, as shown in Fig. 3A. Moreover, each customer can have
their data visualization deactivated by just clicking on the legend line that has

its identifier number. It is also possible at any time to select a new customer through a selection list, as shown in Fig. 3B. Finally, it is possible to move the axes by clicking on a variable and dragging as desired, changing the positions of the variables as well as changing the original scale from 0 to 100 (see Fig. 3C).

In the example in Fig. 3, three customers with different cancellation probabilities are being compared. It is possible to observe that the customer presented with red color is who least uses the services and, consequently, is most likely to be canceled.

4   Case Study

In this section, we present some use cases to exemplify the use and potential of the developed approach. Also, to assess its functionality, the approach was made available for five weeks on an experimental basis to three industry analysts responsible for retaining the website hosting product within a PaaS company, the same one that made the data available to develop this work. The main contribution identified by the analysts was the possibility to quickly and easily perceive the probability of a cancellation and to analyze the values of the variables in a unique and interactive view. The goal was to act proactively before a customer request a cancellation. But, in practice, it also helped the team when trying to reverse a cancellation request.

4.1   Data Set

The data used in the case studies were collected and prepared as described in Subsect. 3.2. The selected data set contained information from 85,000 websites, but 1,338 website records were removed with the cleaning process. In this use case, it is considered that there was a cancellation whenever a website is canceled. However, a customer may have contracted one or more plans, and each plan will have an associated website.

4.2   Use and Availability of Services

A practical example occurred in a cancellation request in which the employee realized that the customer had little or no visit to his website, as shown in Fig. 4A. The client had hired a professional to develop a new website. However, it was unsuccessful and, therefore, did not use the service. Thus, an additional product, which allows people without technical knowledge to create a website, was offered, thus maintaining the customer. We noticed that the algorithm had an excellent assertiveness since the probability was indicated as 83.4%.

In general, they noticed a relationship between the use of services and the probability of cancellation. Analysts have identified situations where the more services are used, the less likely the customer will cancel. This finding had already been illustrated in Fig. 3, where the website represented with orange color has a low probability of cancellation and high use of services.

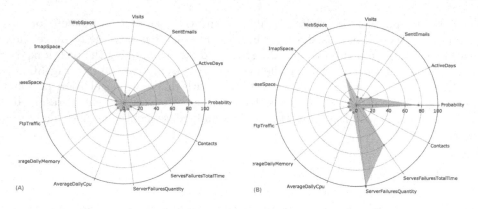

Fig. 4. Chart for reversed (A) and canceled (B) customer.

In another analysis, the sending of e-mails was compared with the cancellation probability. First, a filter was applied to the e-mail sending variable, selecting records with a value greater than 50. Then, only those records with a probability of cancellation, which is greater than 50%, were filtered. The result was the return of only 15 records, and the highest probability is 57%.

In another case, despite the failure to avoid cancellation, analysts were soon able to understand the motivation for the cancellation request through the use of our approach. As can be seen in Fig. 4B, there were many service availability problems as shown by the variables *ServerFailuresQuantity* and *ServerFailuresTotalTime*. The probability assertiveness was 78.1%.

In general, the retention dashboard allows the quick identification of the customers most likely to cancel, and the customer dashboard allows a more detailed analysis for each client or a group of similar clients. Through the case study, the visualizations available in our approach showed that the variables related to the availability of services, such as the number of server failures and the total time of server failure, can lead to the cancellation classification. In other words, it is necessary to invest in the quality of services, as its unavailability is directly linked to possible cancellations.

4.3   Net Promoter Score

The analysts identified that despite being a metric that measures how much the customer is the brand promoter and indirectly satisfied with the contracted services, the Net Promoter Score (NPS) did not influence the increase in the probability of cancellation for most customers. By analyzing the data using the retention dashboard, they observed that out of a universe of 83,662 records, only 600 were identified as retractors with a greater probability of 50% cancellation. When the filter is raised to a probability greater than 80%, only 3 records are displayed.

This result brought a certain surprise to the team that tested our visual analytics approach. They will continue to observe the behavior, but no strategy has been defined for the moment. A situation that still requires analysis is that many records that do not have the NPS survey performed.

4.4 Discussion

We provide all the codes of our visual analytics approach at the GitHub[8]. Thus, it is possible to use it with other data and in different contexts, needing only to make an adjustment and new training of the machine learning algorithms according to the variables that will be used.

As lessons learned from the development of our approach and the case study carried out, we can list:

- The necessity to understand the difficulties of customers in using the service to encourage and retain them, which is facilitated by the use of the customer dashboard (Fig. 3);
- The importance to know the variables that most influence the probability of cancellation, which is facilitated by the use of the retention dashboard (Fig. 2);
- The strategy to proactively monitoring new customers through the proposed visualizations allowing to identify the need to assist in the use of the services because the more you use the services, the lower the chance of cancellation;
- Constant monitoring using the dashboards (Figs. 3 and 2) allows us to identify service failure situations, which can then be adjusted with a focus on recovering customer confidence.

One limitation we faced during this research was obtaining different data sets to perform other case studies. This difficulty is associated with public and private companies with information security contracts, preventing the availability of these data for the general public. Another limitation is the difficulty of providing a generic machine learning solution to work with other data sets associated with different contexts.

As future work, we want to test our visual analytics approach with different data sets. We also aim to validate the cancellation prediction. For this, we intend to create a comparative analysis between the probability indicated by the solution with the reality at each time period. This analysis will serve as input for possible improvements in the prediction algorithm. Another improvement is the development of new interactive visualizations to allow the analysis of the correlation of the cancellation probability with each variable, using, for example, a chord diagram [5]. It is also possible to explore service data and create interactive charts for team analysis in search of discoveries.

Finally, we can say that our visual analytics approach allows CRM teams to make a quick analysis of customer data, even if they are bulky, thus facilitating the creation of strategies to reduce the churn rate. Besides, the algorithms and visualizations used made it possible to answer the research questions presented at the beginning of Subsect. 3.3 and opened space for further research.

---

[8] https://github.com/p-karol/VA-to-Reduce-Churn.

5   Conclusion

The work carried out reinforced the importance of using visual analytics to analyze large volumes of data. It is often complex to keep track of these data due to its dynamism, and it is not easy to relate them because they are spread over several different data sources. It would be more difficult to analyze data with customer information without visual analytics, both for information technology professionals and for an audience considered laymen in the area, such as retention analysts.

Therefore, the main contribution of this work was the dashboards as the resources that improve the ability of visual perception concerning the professionals involved in the customer retention process. They allow the analysis and understanding of information to create strategies and make decisions assertive. Moreover, the central issue in the approach development was the probability of cancellation based on the collected variables, and this result needed to be reliable. Thus, one of its advantages was the excellent performance of the Random Forest data mining algorithm, which obtained an accuracy of 87%.

The use case demonstrates that the final result was gratifying considering that our visual analytics approach brought new skills to a team, allowing them to experience this with interesting analysis, which brings motivation and a roadmap for the evolution of the work to be developed in the future.

References

1. Chan, C.C.H.: Intelligent value-based customer segmentation method for campaign management: a case study of automobile retailer. Expert Syst. Appl. **34**(4), 2754–2762 (2008)
2. Chen, S.H.: The gamma cusum chart method for online customer churn prediction. Electron. Commer. Res. Appl. **17**, 99–111 (2016)
3. Dintakurthi, R.M.R., Venkatraman, B., Mahendran, P., Siddappa, S.: Decision support system for identifying customer churn based on buying patterns in a discrete manufacturing industry. In: 2016 IEEE Annual India Conference (INDICON), pp. 1–5 (2016)
4. Grammel, L.: Supporting end users in analyzing multiple data sources. In: 2009 IEEE Symposium on Visual Languages and Human-Centric Computing (VL/HCC), pp. 246–247 (2009)
5. Humayoun, S.R., Bhambri, K., AlTarawneh, R.: Bid-chord: an extended chord diagram for showing relations between bi-categorical dimensional data. In: Proceedings of the 2018 International Conference on Advanced Visual Interfaces. AVI 2018, Association for Computing Machinery, New York, NY, USA (2018). https://doi.org/10.1145/3206505.3206570
6. binti Jaafar, H., binti Mukahar, N., Ramli, D.A.B.: A methodology of nearest neighbor: design and comparison of biometric image database. In: 2016 IEEE Student Conference on Research and Development (SCOReD), pp. 1–6 (2016). https://doi.org/10.1109/SCORED.2016.7810073
7. Jahromi, A.T., Stakhovych, S., Ewing, M.: Managing b2b customer churn, retention and profitability. Ind. Mark. Manage. **43**(7), 1258–1268 (2014)

8. Keim, D., Andrienko, G., Fekete, J.-D., Görg, C., Kohlhammer, J., Melançon, G.: Visual analytics: definition, process, and challenges. In: Kerren, A., Stasko, J.T., Fekete, J.-D., North, C. (eds.) Information Visualization. LNCS, vol. 4950, pp. 154–175. Springer, Heidelberg (2008). https://doi.org/10.1007/978-3-540-70956-5_7

9. Khodabandehlou, S., Rahman, M.: Comparison of supervised machine learning techniques for customer churn prediction based on analysis of customer behavior. J. Syst. Inf. Technol. **19**, 65–93 (2017). https://doi.org/10.1108/JSIT-10-2016-0061

10. Lai, X.: Segmentation study on enterprise customers based on data mining technology. In: 2009 First International Workshop on Database Technology and Applications, pp. 247–250 (2009)

11. Ngai, E., Xiu, L., Chau, D.: Application of data mining techniques in customer relationship management: A literature review and classification. Expert Syst. Appl. **36**(2, Part 2), 2592–2602 (2009). https://doi.org/10.1016/j.eswa.2008.02.021. http://www.sciencedirect.com/science/article/pii/S0957417408001243

12. Noyan, F., Şimşek, G.G.: The antecedents of customer loyalty. Procedia - Soc. Behav. Sci. **109**, 1220–1224 (2014). https://doi.org/10.1016/j.sbspro.2013.12.615. http://www.sciencedirect.com/science/article/pii/S1877042813052543. 2nd World Conference on Business, Economics and Management

13. Oelke, D., et al.: Visual opinion analysis of customer feedback data. In: 2009 IEEE Symposium on Visual Analytics Science and Technology, pp. 187–194 (2009)

14. Park, J.: Integrated visual analytics tool for heterogeneous text data. In: 2014 IEEE Conference on Visual Analytics Science and Technology (VAST), pp. 325–326 (2014)

15. Qureshi, S.A., Rehman, A.S., Qamar, A.M., Kamal, A., Rehman, A.: Telecommunication subscribers' churn prediction model using machine learning. In: Eighth International Conference on Digital Information Management (ICDIM 2013), pp. 131–136 (2013)

16. Rao, R., Card, S.K.: The table lens: merging graphical and symbolic representations in an interactive focus + context visualization for tabular information. In: Proceedings of the SIGCHI Conference on Human Factors in Computing Systems, CHI 1994, pp. 318–322. Association for Computing Machinery, New York (1994). https://doi.org/10.1145/191666.191776

17. Zakaria, I., Abdul-Rahman, B., Othman, A., Azlina, N., Dzulkipli, M.R., Osman, M.A.: The relationship between loyalty program, customer satisfaction and customer loyalty in retail industry: a case study. Procedia - Soc. Behav. Sci. **129**, 23–30 (2014). https://doi.org/10.1016/j.sbspro.2014.03.643

Generating Formal Software Architecture Descriptions from Semi-Formal SysML-Based Models: A Model-Driven Approach

Camila Araújo[1,2]([✉]), Thais Batista[2], Everton Cavalcante[2],
and Flavio Oquendo[3]

[1] State University of Rio Grande do Norte, Natal, Brazil
camilaaraujo@uern.br
[2] Federal University of Rio Grande do Norte, Natal, Brazil
{thais,everton}@dimap.ufrn.br
[3] IRISA-UMR CNRS/Université Bretagne Sud, Vannes, France
flavio.oquendo@irisa.fr

Abstract. The critical nature of many complex software-intensive systems requires formal architecture descriptions for supporting automated architectural analysis regarding correctness properties. Due to the challenges of adopting formal approaches, many architects have preferred using notations such as UML, SysML, and their derivatives to describe the structure and behavior of software architectures. However, these semi-formal notations have limitations regarding the sought support for architectural analysis. This paper presents an approach to bridge the rigor of formal architecture descriptions and the ease of use of SysML-based notations widely used elsewhere. The main concern is providing formal semantics to SysADL, a SysML-based language to describe software-intensive system architectures. The formal semantics is provided by π-ADL, a formal architecture description language. A model-to-model transformation was defined and implemented to concretize the mapping between the elements of these languages and hence automatically generate formal architecture descriptions in π-ADL from SysADL. This paper describes a proof-of-concept to illustrate the mapping between SysADL and π-ADL and an exploratory study on the transformation performance.

Keywords: Model-driven development · Model transformation · Architecture description language · Formal verification · SysML

1 Introduction

The representation of software architectures is one of the main activities of an architecture-driven software development process as it allows anticipating important decisions regarding the system design. This activity results in architecture

© Springer Nature Switzerland AG 2021
O. Gervasi et al. (Eds.): ICCSA 2021, LNCS 12951, pp. 394–410, 2021.
https://doi.org/10.1007/978-3-030-86970-0_28

descriptions, which can be used to communicate the software architecture among stakeholders and support its maintenance, evaluation, and evolution. Due to the critical nature of many software-intensive systems, architecture descriptions need to be expressed in formal notations as means of better supporting automated architectural analysis [2,6]. The architectural analysis aims at precisely verifying if their designed architectures can meet important properties, e.g., correctness.

As a consequence of the significant learning curve, lack of knowledge among stakeholders, and weak support concerning tooling and guidance on the use of formal approaches, software architects have primarily preferred to use semi-formal notations such as UML, SysML or their derivatives as means of describing structure and behavior of software architectures. This choice has been mainly motivated by their lower learning curve, visuality, and general-purpose scope [13, 18,19]. However, this type of notation has known limitations regarding the sought support for automatically verifying architectural properties.

This paper presents a model-driven approach to bridge the rigor of formal architecture descriptions and the ease of use of SysML-based notations widely used elsewhere. More specifically, the proposed approach intends to provide a formal semantics to SysADL [17], a SysML-based architectural language that combines typical constructs of architectural languages with the use of the popular diagrammatic notation based on the SysML Standard for modeling software-intensive systems. The formal semantics is provided by π-ADL [16], a language based on π-calculus to formally describe software architectures. On the one hand, SysADL was chosen for this work as it is aligned with the ISO/IEC/IEEE 42010 International Standard for architectural descriptions [10]. On the other hand, π-ADL was chosen since it allows formally describing both structure and behavior of dynamic software architectures, unlike most existing architectural languages [4].

The transformation of an architecture description in SysADL to a corresponding one in π-ADL allows for its further formal verification, which is already available for π-ADL architecture descriptions [5]. A mapping process was conceived to establish the relationship between SysADL and π-ADL and their elements. The mapping was concretized by a model-to-model transformation implemented in the ATLAS Transformation Language (ATL) [1,11] upon Eclipse-based frameworks, which are also used in the SysADL [12] and π-ADL [5] tools. In this paper, the proposed approach is validated with an Automated Guided Vehicle (AVG) System [9]. This paper also reports the results of computational experiments to analyze the performance and scalability of the transformation.

The remainder of this paper is structured as follows. Section 2 provides the background of this work. Section 3 details the transformation from SysADL to π-ADL, illustrated with samples of the AVG system. Section 4 reports the exploratory study to analyze the performance of the proposed transformation. Section 5 discusses related work. Section 6 contains some concluding remarks.

2   Background

2.1   SysADL

SysADL [17] is a SysML-based language designed to support multiple view modeling, cross-view checking, validation, and execution of software architectures. This language reconciles the rigorous semantic of architectural languages with the use of popular diagrammatic notation based on the SysML Standard for modeling software-intensive systems. Moreover, SysADL is supported by the SysADL Studio tool [12], which enables software architects to describe software architectures either visually and/or textually.

SysADL defines three software architecture viewpoints for a system, namely (i) *structural*, (ii) *behavioral*, and (iii) *executable*. The structural viewpoint defines the architectural elements composing the structure of a system (*components, ports, connectors*) and relationships among them. SysADL requires declaring all elements before creating their instances in Block Definition Diagrams (BDDs). The Internal Block Diagram (IBD) is used to specify how instances of components and connectors form the configuration of architectures.

The behavioral viewpoint details the behavior of (i) components and connectors through *activities, actions, constraints* and (ii) ports through *protocols*. Activity instances are described in the Activity Diagram by instantiating actions and flows. Activities or actions may have validation constraints specified through expressions in the OMG Action Language for Foundational UML (ALF). Constraints can be also expressed using the Parametric Diagram.

The executable viewpoint represents the concretization of both structural and behavioral viewpoints by simulating the architecture behavior at runtime. The primary purpose of the simulation is to validate the behavior logic regarding the satisfaction of requirements and analysis of architecture functionalities. In the executable viewpoint, it is possible to specify details of each action by using ALF statements and define and instantiate elements. An ALF engine interprets the executable instances to execute the architecture.

Figure 1 illustrates an excerpt from an architectural description in SysADL. Figure 1(a) specifies the *ClientPT* port as a composite of the input port *aC* and the *qC* output port, as well as the *ServerPT* port as a composite of the *aS* input port and the *qS* output port. In SysADL, ports must be defined before being used in the definition of components and connectors.

Figure 1(b) depicts the definition of the *ClientServerCN* connector that uses instances of the previously defined *ClientPT* and *ServerPT* ports. The corresponding IBD specifies the internal configuration of this connector as using the *ClientServerQueryCN* and *ClientServerAnswerCN* connectors to link the *cP* and *sP* composite ports. Figure 1(c) shows the definition of the *ClientServer-ARCH* architecture as a composition of instances of the previously defined *ClientCP* and *ServerCP* components and *ClientServerCN* connector.

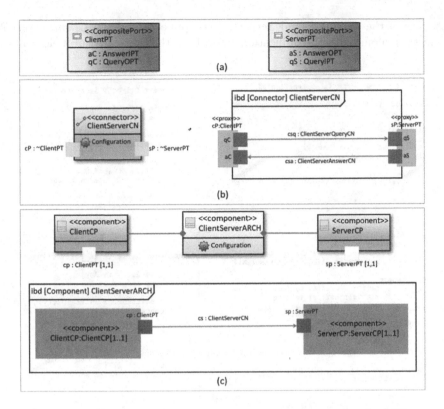

Fig. 1. Excerpt of a client-server architecture as described in SysADL.

2.2 π-ADL

π-ADL [16] is a formal architectural language based on the π-calculus process algebra to describe software architectures under both structural and behavioral viewpoints. The structural viewpoint describes a software architecture in π-ADL, through *components*, *connectors*, and their *composition* to form an architecture. In terms of behavioral viewpoint, both components and connectors comprise a behavior that expresses an architectural element's interaction and internal calculation and uses connections to connect and transmit values.

Figure 2 shows a part of the specification of an architectural description in π-ADL. Figure 2(a) specifies the *ClientServerCN* connector with the *aC* and *qS* input connections and the *qC* and *aS* output connections, followed by their protocol declaration. The behavior of the connector is defined as a composition that instantiates the *ClientServerQueryCN* and *ClientServerAnswerCN* connectors along with unifications that specify the information flow through connections.

Figure 2(b) shows the specifications of the *ClientCP* and *ServerCP* components. The behaviors of *ClientCP* and *ServerCP* components were specified as unobservable, which expresses an internal behavior not externally observable. In Fig. 2(c), the *ClientServerARCH* architecture is shown as a composition of instances of the *ClientCP* and *ServerCP* components and the *ClientServerCN* connectors, along with the unifications that bind these architectural elements.

```
connector ClientServerCN is abstraction(){

    type Query is Any
    type Answer is Any

    connection aC is in (Answer)
    connection qC is out (Query)
    connection qS is in (Query)
    connection aS is out (Answer)

    protocol is{
            (via aC receive Answer
            via qC send Query
            via qS receive Query
            via aS send Answer)*
    }

    behavior is {
        compose{
                csq is ClientServerQueryCN()
                and csa is ClientServerAnswerCN()
            }
            where{
                self::qC unifies csq::qOut
                csq::qIn unifies self::qS
                self::aS unifies csa::aOut
                csa::aIn unifies self::aC
            }
        }
    }
}
```
(a)

```
architecture ClientServerARCH is abstraction(){
        behavior is {
            compose{
                    client is ClientCP()
                    and server is ServerCP()
                    and con is ClientServerCN()
            }
            where{
                client::qC unifies con::qC
                con::qS unifies server::qS
                server::aS unifies con::aS
                con::aC unifies client::aC
            }
        }
}
```
(c)

```
component ClientCP is abstraction(){

    type Query is Any
    type Answer is Any

    connection aC is in (Answer)
    connection qC is out (Query)

    protocol is{
            (via qC send Query
            via aC receive Answer)*
    }
    behavior is {
        unobservable
    }
}
```

```
component ServerCP is abstraction(){

    type Query is Any
    type Answer is Any

    connection qS is in (Query)
    connection aS is out (Answer)

    protocol is{
            (via qS receive Query
            via aS send Answer)*
    }
    behavior is {
        unobservable
    }
}
```
(b)

Fig. 2. Excerpt of a client-server architecture as described in π-ADL.

2.3   Model-to-Model Transformation Process

The Model-Driven Engineering (MDE) approach relies on models as central elements to software development. One of the keys to this approach is performing model-to-model (M2M) transformations [15], which consist in specifying transformation rules that make it possible to produce target models from source models. Figure 3 illustrates a model transformation process. Transformation rules are responsible for define how to map elements of the source model to corresponding ones in the target model.

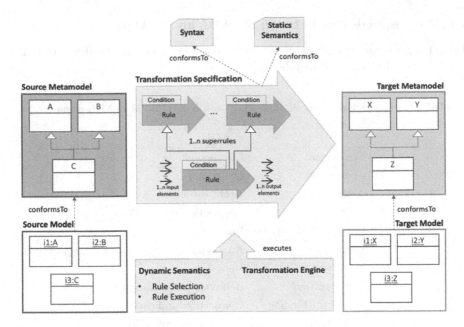

Fig. 3. A model-to-model transformation pattern (adapted from Wimmer et al. [22]).

Both source and target models must conform to their respective metamodels since the transformation rules are based on the abstract syntax of the metamodels. The transformation specification must also comply with the syntax and semantics of the language used for implementing the transformation. This work uses the SysADL and π-ADL metamodels as source and target metamodels, respectively. The goal is transforming an input architecture description in SysADL (conformed to its metamodel) and producing an architecture description in π-ADL (conformed to its metamodel) as output. The mapping process from SysADL to π-ADL and transformation algorithms are shown in Sect. 3.

3 Mapping SysADL Architecture Descriptions to π-ADL

This section describes how to generate π-ADL architecture descriptions from the description of SysADL structural elements. Section 3.1 describes the correspondences between the elements of SysADL and π-ADL. Section 3.2 describes some algorithms to transform an architecture description in SysADL into another one in π-ADL. Section 3.3 presents a process used to perform the translation between the models. Section 3.4 illustrates the proposal by specifying part of the architecture of the Automated Guided Vehicle (AVG) system in SysADL and showing how to translate it to π-ADL.

3.1   Correspondences Between SysADL and π-ADL

Table 1 summarizes the relationships between the main elements of SysADL and π-ADL, each one detailed as follows.

Table 1. Correspondences summary between SysADL and π-ADL elements.

SysADL	π-ADL
Component	Component
Connector	Connector
Architecture	Architecture
Port	Connection declarations
Primitive value types	Basic types
Value type	Any type
Data type	View type
Configuration	Behavior (composition)
Connector binding	Connection unification
Delegation	Connection unification

Component. In SysADL, components are structural elements that represent system functionalities and can be instantiated by other components or an architecture. A simple component performs computation using data available in its ports. A composite component performs computation through the constituent components in their internal structure. In π-ADL, components are created as an abstraction that can be instantiated within the architecture's specification.

Connector. Connectors are responsible for the interaction among components by defining which ports can be connected and how the interaction between connected components occurs. A connector can be either a simple element connecting two or more ports or a composite element embedding other connectors and behavior on its own. In π-ADL, connectors manage interactions among components with constituent elements and/or an internal behavior.

Architecture. The main element of an architecture description is the configuration that specifies the architecture itself. In SysADL, an architecture has a configuration that indicates its structural organization in terms of components and connectors. In π-ADL, an architecture is specified as a composition of component and connector instances.

Port. A SysADL port is an interaction point between a component and other architectural elements. It represents how data flow out from an output port or flow in to an input port. A composite port is composed of multiple port instances. The concept of port is not directly provided in the π-ADL version we have used in this work, but it is supported through the concept of connection representing interaction points as well as binding between interaction points. In this way, data flow by typed connections representing the interaction points of communication channels used to transmit (send/receive) values. Protocols ensure that value types to be transmitted via connections comply with their respective declarations and the order of operations to perform.

Data Types. Both SysADL and π-ADL have four primitive data types, namely Integer, Real, String, and Boolean. These primitive data types are mapped exactly to their counterparts. Composite data types from SysADL are mapped to the View constructed type provided by π-ADL, which is composed of labeled field data. A value type from SysADL is mapped to π-ADL according to its intensity type, and is generally typed as the union *Any* type.

Configuration. The SysADL configuration refers to the structural organization of the architecture on how component and connector instances are linked to each other. A configuration can represent either the structure of a composite component or the overall software architecture. In π-ADL, a composition behavior expresses how independent sub-behaviors in parallel can form a composite behavior associated with the architecture or the composite element. It is possible to instantiate components and connectors to represent either the structure of architectural elements or the general architecture.

Connector Binding. Binding component instances does not happen directly in the sense that there must be a connector instance (linking component–connector ports) between two components. In π-ADL, a component's connection can be attached to a connector's connection to enable these elements to communicate, expressed through the unification of connections.

Delegation. In a SysADL configuration, ports of internal components are not visible to the external composite component. When port data must be visible, the port is linked to a proxy port via a binding connector expressed by delegation. Each port must be of the same data type and direction. In the transformation to π-ADL, SysADL delegations are converted to connection unifications by using the `self` clause to refer to the current component.

3.2   Transformation Rules from SysADL to π-ADL

An M2M transformation can be accomplished by implementing the mapping rules in a transformation language. This section presents a transformation algorithm in pseudocode, independently from the transformation language used to implement it.

Algorithm 1 shows the starting point of the transformation, i.e., the translation of the root element of a SysADL architecture description (*Model*) into the root element of a π-ADL architecture description (*ArchitectureDescription*). The *RuleModel2Architecture* procedure receives *SysADL!Model* as input and produces *PIADL!ArchitecturalDescription* as a result. The *CONS*, *COMPS*, and *ARCHS* sets (lines 3–5) respectively hold all connector, component, and architecture instances from the *SysADL!Model*. The *archElements* attribute of the

---

Algorithm 1. SysADL to π-ADL main transformation rule

1: **procedure** RuleModel2Architecture(*SysADL!Model*)
2:    *ad* : *PiADL!ArchitectureDescription*
3:    *CONS* ← ⋃ *CN* ∈ *SysADL!ConnectorDef*
4:    *COMPS* ← ⋃ *CP* ∈ *SysADL!ComponentDef*
5:    *ARCHS* ← ⋃ *AC* ∈ *SysADL!ArchitectureDef*
6:    *ad.archElements* ← ∅
7:    **for** each *CN* ∈ *CONS* **do**
8:        *conn* ← *RuleConnectorDef2Connector*(*CN*)
9:        *ad.archElements* ← *ad.archElements* ∪ *conn*
10:    **for** each *CP* ∈ *COMPS* **do**
11:        *comp* ← *RuleComponentDef2Component*(*CP*)
12:        *ad.archElements* ← *ad.archElements* ∪ *comp*
13:    *ad.archs* ← ∅
14:    **for** each *AC* ∈ *ARCHS* **do**
15:        *arch* ← *RuleArchitectureDef2Architecture*(*AC*)
16:        *ad.archs* ← *ad.archs* ∪ *arch*
17:    *ad.behavior* ← *UNOBSERVABLE*
18:    **return** *ad*

π-ADL architecture description element must contain the definition of all architectural elements, namely connectors and components. The procedure iterates over the sets of existing components and connectors from the SysADL model and respectively transform them into component and connectors in π-ADL according to the *RuleComponentDef2Component* and *RuleConnectorDef2Connector* auxiliar procedures (lines 7–12). The *archs* attribute of the π-ADL architecture description element must contain the definition of architectures. The last loop (lines 14–16) iterate over the set of architecture definitions from the SysADL model and transform them into the corresponding π-ADL element through the *RuleArchitecturerDef2Architecture* auxiliar procedure. The definition of these auxiliar procedures and an implementation of them are available at https://bit.ly/3huzjTe.

3.3 Mapping Process

Figure 4 illustrates the technical process to generate an architecture description in π-ADL from a SysADL model. The proposed transformation is supported by tools and artifacts complying with the Xtext Eclipse framework, namely (i) SysADL Studio, the tool that supports architectural modeling and execution of SysADL models [12], and (ii) the π-ADL toolchain, which performs syntactic and semantic analysis of architecture descriptions in π-ADL [5]. This infrastructure allows verifying if an architecture description in SysADL is correct according to the syntactic rules of the language and automatically generates a π-ADL file according to the ATL transformation rules. ATL has been chosen as a transformation language due to its maturity, tool support, and community support.

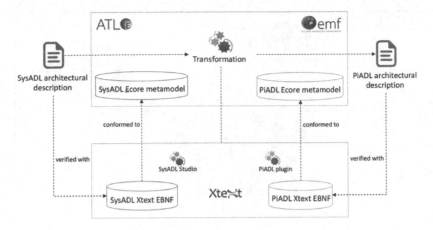

Fig. 4. Process for generating a π-ADL architecture description from a SysADL model.

The SysADL and π-ADL Ecore metamodels are parameters to the ATL transformation. The ATL Module checks for semantic errors in the transformation while the ATL compiler is automatically called for each ATL file, thus compiling transformations to the ASM assembly language [1]. ASM bytecode is then executed by the ATL Virtual Machine, which is specialized in handling models and provides a set of instructions for model manipulation.

3.4 An Illustrative Example: The AVG System

In this section, our proposed approach is illustrated through the Automated Guided Vehicle (AVG) system [9], showing how its architecture description in SysADL can be automatically translated to a corresponding specification in π-ADL by following the correspondences drawn in Sect. 3.1. AGV is a real-time system composed of a *motor*, which is commanded to start/stop moving and sends started/stopped responses, and an *arrival sensor* to detect when AVG has arrived at a station. If the station is the destination, then the AGV stops; otherwise, it continues to the next station. Moreover, there is a *robot arm* for loading and unloading a part onto and off of the AGV. The AGV system interacts with two other existing systems by receiving commands from the *Supervisory System* and sending vehicle acknowledgments to it, as well as sending the vehicle status to an external *Display System*.

Figure 5 shows an IBD corresponding to the configuration the *IAGVSystemCP* component in SysADL. This configuration is defined as the connection of the instances of *RobotArm*, *ArrivalSensor*, *VehicleControl*, and *Motor* components via instances of *commandArm*, *locationVehicle*, *commandMotor*, and *notificationMotor* connectors. The *in_outData* composite port and the *sendStatus* output port are connected to the *VehicleControl* component instance.

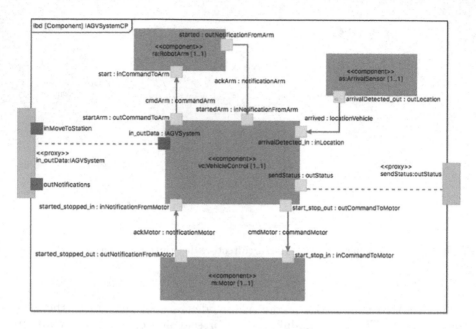

Fig. 5. Configuration of the *IAGVSystemCP* component in SysADL.

Listing 1.1 shows an ATL implementation of the transformation of component definitions in SysADL to component declarations in π-ADL, through the *lazyComponentDef2Component* rule. The attribute mapping of these elements is shown in lines 6–19. Except for the *name* attribute, all attributes are mapped by calling other ATL rules.

The result of the transformation is an architecture description in π-ADL, such as the excerpt shown in Listing 1.2. The *IAGVSystemCP* component ports have been transformed into connections with their respective protocol (lines 2–10). The configuration was transformed into a composition behavior where components (lines 13–16) and connectors (lines 17–21) were instantiated with their bindings (lines 23–32). Finally, connections involving the component ports (delegations in SysADL) were transformed into unifications (lines 33–35) with the use of the *self* clause. Once the transformation to π-ADL has been realized, it is possible to formally verify software architecture properties through the π-ADL toolchain [5,20]. The complete architecture description of the AVG system in SysADL and the resulting one in π-ADL are available at https://bit.ly/3huzjTe.

4   Evaluation

The effective adoption of formal verification approaches by the software industry faces significant barriers, such as the scalability of solutions and the high learning curve of the mathematical formalisms inherent to the techniques that are not part of stakeholders' routines. In particular, scalability is critical when

```
1   lazy rule lazyComponentDef2Component {
2     from
3       sSC: SYSADL!ComponentDef (not sSC.oclIsTypeOf(SYSADL!
            ArchitectureDef))
4     to
5       pC: PIADL!Component (
6         name <- (if(sSC.isBoundary=false) then sSC.name else
              'boundary' + sSC.name endif),
7         typeDecl <- typeDecs -> iterate(tdecl; tdecls:
              OrderedSet(PIADL!TypeDeclaration) = OrderedSet{}
              | tdecls.including(thisModule.lazyTuple2TypeDec
              (tdecl))),
8         connections <- sSC.ports -> iterate(con; cons:
              Sequence(PIADL!ConnectionDeclaration) = Sequence
              {} | cons.including(
9           if (con.definition.oclIsTypeOf(SYSADL!
              CompositePortDef)) then
10            con.definition.ports -> iterate(port; ports:
              Sequence(PIADL!ConnectionDeclaration) =
              Sequence{} | ports.including(thisModule.
              lazyCPortUse2ConDec(port)))
11          else
12            thisModule.lazySPortUse2ConDec(con)
13          endif)),
14        protDecl <- thisModule.lazyComp2ProtocolDec(sSC),
15        behavior <- if (sSC.isBoundary = true or sSC.composite
              .oclIsUndefined()) then
16            thisModule.lazyUnobservable2BehavDec(PIADL!
              Unobservable)
17          else
18            thisModule.lazyCDefConfig2CBehavDec(sSC)
19          endif
20      )
21  }
```

Listing 1.1. An ATL rule that implements the transformation of component definitions in SysADL into component declarations in π-ADL.

analyzing software architectures in industrial contexts since software systems are becoming even more complex with an increasing number of components and connectors. Therefore, evaluating the proposed transformation of SysADL architecture descriptions into π-ADL descriptions concerning scalability is a key point for supporting industrial applications.

The evaluation considered a set with six software architectures descriptions in SysADL. The smallest description had 10 elements and the largest one had 85 elements (see Table 2). In the transformation from SysADL to π-ADL, there is a significant increase in the rule's complexity when dealing with composite elements, such as ports, connectors, and components.

The ATL transformation was performed ten times for each test architecture and the observed execution time was registered. The experiments were performed on a computer with an Intel Core i5 1.6 GHz processor, 8 GB of RAM, and macOS High Sierra version 10.136 as the operating system. Table 3 shows the descriptive statistics of the experiment results.

```
1   component IAGVSystemCP is abstraction() {
2      connection sendStatus is out Status
3      connection inMoveToStation is in VehicleData
4      connection outNotifications is out
          NotificationToSupervisory
5
6      protocol is {
7        (via sendStatus send Status
8         via inMoveToStation receive VehicleData
9         via outNotifications send NotificationToSupervisory)*
10     }
11     behavior is {
12       compose {
13         m is Motor()
14         and aS is ArrivalSensor()
15         and ra is RobotArm()
16         and vc is VehicleControl()
17         and arrived is locationVehicle()
18         and ackArm is notificationArm()
19         and cmdArm is commandArm()
20         and ackMotor is notificationMotor()
21         and cmdMotor is commandMotor()
22       } where {
23         aS::arrivalDetected_out unifies arrived::1OPT
24         arrived::1IPT unifies vc::arrivalDetected_in
25         ra::started unifies ackArm::naOPT
26         ackArm::naIPT unifies vc::startedArm
27         vc::startArm unifies cmdArm::caOPT
28         cmdArm::caIPT unifies ra::start
29         m::started_stopped_out unifies ackMotor::nmOPT
30         ackMotor::nmIPT unifies vc::started_stopped_in
31         vc::start_stop_out unifies cmdMotor::cmOPT
32         cmdMotor::cmIPT unifies m::start_stop_in
33         vc::sendStatus unifies self::sendStatus
34         self::inMoveToStation unifies vc::inMoveToStation
35         vc::outNotifications unifies self::outNotifications
36       }
37     }
38  }
```

Listing 1.2. Description of the *IAGVSystemCP* component in π-ADL.

The Shapiro-Wilk test was applied to verify if data followed a normal distribution at a significance level $\alpha = 0.05$. After confirming the sample normality (p-values $> \alpha$), the Pearson correlation coefficient r was calculated to statistically measure the degree of correlation between the amount of simple/composite structural elements and the execution of the ATL transformation from SysADL to π-ADL. The obtained values were $r = 0.9627$ for simple elements and $r = 0.9383$ for composite elements, thus indicating a strong correlation between the number of elements and the execution time in both cases (see Table 3).

Another important observation was that the growth trend of the execution time is polynomial. Equation 1 and Eq. 2 respectively represent the obtained polynomials for simple and composite elements, which were also used to estimate the execution times shown in Table 4. It is possible to observe that the trans-

Table 2. Architecture descriptions considered in the experiments.

Architecture description	Types	Ports		Connectors		Components		Architectures	Total
		S	C	S	C	S	C		
A1	5	12	2	6	1	6	2	1	35
A2	13	16	2	8	1	11	2	1	54
A3	12	8	0	4	0	7	3	1	35
A4	5	2	0	0	0	2	0	1	10
A5	5	4	0	2	0	2	0	1	14
A6	20	24	6	11	3	17	3	1	85

S = simple element, C = composite element

Table 3. Descriptive statistics and correlation analysis of execution time measures (in seconds).

Architecture description	Simple elements	Composite elements	Median	Standard deviation
A1	24	6	0.0365	0.0116
A2	35	6	0.1015	0.0610
A3	19	2	0.0505	0.0088
A4	4	1	0.0045	0.0010
A5	8	1	0.0065	0.0025
A6	52	13	0.2110	0.0504
p-value (Shapiro-Wilk)	0.7699	0.2439		
Pearson coefficient r	0.9627	0.9383		

formation of an architecture containing only composite elements takes twice the execution time of architecture with only simple elements. When the architecture has thousands of elements, the execution time increases up to approximately 10 h, even considering an architecture with only simple elements.

$$y = 7E{-}05x^2 + 0.0003x + 0.0024, R^2 = 0.9841 \tag{1}$$

$$y = 0.0007x^2 + 0.007x + 0.0075, R^2 = 0.9011 \tag{2}$$

Besides scalability, the approach proposed in this paper aimed to tackle the inherent complexity of adopting formal architectural languages. The use of a SysML-based language such as SysADL and the automated transformation to a formal notation (in this case, π-ADL) have the potential of alleviating the learning curve typically required by a formal verification strategy. In this perspective, software architects could benefit from the proposed approach in that they can use a well-known, standardized notation to produce architecture descriptions that can be further formally verified with respect to architectural properties.

5   Related Work

Enoiu et al. [8] introduced a way of integrating architectural models with verification techniques implemented through the ViTAL tool. The approach allows expressing functional behavior in the EAST-ADL architectural language for

Table 4. Estimated execution time in terms of number of simple/composite elements.

Elements	Execution time	
	Simple	Composite
10^1	0.04 s	0.15 s
10^2	3.53 s	7.71 s
10^3	5.84 min	11.78 min
10^4	9.72 h	19.46 h

automotive embedded systems as timed automata models, which have precise semantics and can be formally verified. The proposal uses model-based techniques to allow for the automated transformation between different design models. In this perspective, the ViTAL tool allows transforming functional EAST-ADL models into the UPPAAL PORT tool to support model verification.

Taoufik et al. [21] proposed using model transformations to translate UML 2.0 architecture descriptions into Wright, a formal CSP-based architectural language, towards verifying the behavioral consistency of software architectures. The Wright descriptions are automatically translated into a CSP specification acceptable to the FDR2 model checker through the Wr2fdr tool.

Maraoui and Cariou [14] proposed a strategy for the formal verification of the composition of Web services through the integration of a mediation protocol. The approach provides an MDE process that guides developers through a series of transformations between models, including SysML, to obtain a formal code in the ACME architectural language (with Armani) towards verification.

Similar to this work, the studies previously mentioned define their approaches of assigning formal semantics to an architectural language by applying MDE techniques, but their concerns are limited to behavioral aspects only or are specific to an application domain. Oppositely, the approach presented in this paper deals with a general-purpose SysML-based architectural language that allows describing software architectures under both structural and behavioral viewpoints, independently from the application domain or architectural style.

6   Concluding Remarks

Formally expressing architecture descriptions in critical domains allows for automated verification of architectural properties of software-intensive systems. This paper presented the definition of a formal semantics for a SysML-based ADL, SysADL, through model transformations to a formal language based on the π-calculus process algebra, π-ADL. As π-ADL supports the formal verification of architectural properties, SysADL specifications transformed to π-ADL can be verified with tools available for this formal language. Note that the transformation models from SysADL to π-ADL are based on the denotational semantics of SysADL in terms of π-ADL.

The transformation from SysADL to π-ADL was implemented using ATL to validate the proposed approach. Furthermore, some experiments were performed to analyze the performance of the transformation and estimate its scalability. Obtained results showed a strong correlation between the number of simple and composite elements and the time spent to perform the transformation.

Ongoing work aims to extend the transformation from a behavioral viewpoint and compare the performance results of the ATL implementation with other transformation languages, such as QVTo [3] and Kermeta [7], to check for significant performance differences. Furthermore, the SysADL Studio tool is being integrated with the π-ADL toolchain to include, in addition to the automatic generation of π-ADL architecture descriptions, also their automatic formal verification of architectural properties, in particular to guarantee correctness.

References

1. ATL Developer Guide. https://wiki.eclipse.org/ATL/Developer_Guide
2. Araujo, C., Cavalcante, E., Batista, T., Oliveira, M., Oquendo, F.: A research landscape on formal verification of software architecture description. IEEE Access **7**, 171752–171764 (2019)
3. Barendrecht, P.: Modeling transformations using QVT Operational Mappings. Eindhoven University of Technology, The Netherlands (Apr), Tech. rep. (2010)
4. Cavalcante, E., Batista, T., Oquendo, F.: Supporting dynamic software architectures: from architectural description to implementation. In: 12th Working IEEE/IFIP Conference on Software Architecture, pp. 31–40. IEEE, USA (2015)
5. Cavalcante, E., Quilbeuf, J., Traonouez, L.-M., Oquendo, F., Batista, T., Legay, A.: Statistical model checking of dynamic software architectures. In: Tekinerdogan, B., Zdun, U., Babar, A. (eds.) ECSA 2016. LNCS, vol. 9839, pp. 185–200. Springer, Cham (2016). https://doi.org/10.1007/978-3-319-48992-6_14
6. Dias, F., et al.: Empowering SysML-based software architecture description with formal verification: from SysADL to CSP. In: Jansen, A., et al. (eds.) ECSA 2020, LNCS, vol. 12292, pp. 101–117. Springer, Switzerland (2020)
7. Drey, Z., Faucher, C., Fleurey, F., Mahé, V., Vojtisek, D.: Kermeta language - Reference Manual (2009). http://www.kermeta.org/docs/KerMeta-Manual.pdf
8. Enoiu, E.P., Marinescu, R., Seceleanu, C., Pettersson, P.: ViTAL: a verification tool for EAST-ADL models using UPPAAL PORT. In: 17th International Conference on Engineering of Complex Computer Systems, pp. 328–337. IEEE, USA (2012)
9. Gomaa, H.: Software modeling and design: UML, use cases, patterns, and software architectures. Cambridge University Press (2011)
10. ISO/IEC/IEEE 42010: Systems and Software Engineering - Architecture Description. ISO, Switzerland (2011)
11. Jouault, F., Kurtev, I.: Transforming models with ATL. In: Bruel, J.M. (ed.) Satellite Events at the MoDELS 2005 Conference, LNCS, vol. 3844, pp. 128–138. Springer, Berlin (2006)
12. Leite, J., Batista, T., Oquendo, F., Silva, E., Santos, L., Cortez, V.: Designing and executing software architectures models using SysADL studio. In: 2018 IEEE International Conference on Software Architecture Companion, pp. 81–84. IEEE, USA (2018)

13. Malavolta, I., Lago, P., Muccini, H., Pelliccione, P., Tang, A.: What industry needs from architectural languages: a survey. IEEE Trans. Softw. Eng. **39**(6), 869–891 (2013)
14. Maraoui, R., Cariou, E.: A mediation based approach for formal verification of web services composition. In: 2017 International Conference on Engineering & MIS. IEEE, USA (2017)
15. Mens, T.: Model transformation: a survey of the state of the art. In: Babau, J.P., et al. (eds.) Model-Driven Engineering for distributed real-time systems, pp. 1–19. ISTE Ltd/John Wiley & Sons Ltd, United Kingdom/USA (2013)
16. Oquendo, F.: π-ADL: an architecture description language based on the higher-order typed π-calculus for specifying dynamic and mobile software architectures. ACM SIGSOFT Software Engineering Notes **29**(3), 1–14 (2004)
17. Oquendo, F., Leite, J., Batista, T.: Software Architecture in Action: Designing and executing architectural models with SysADL grounded on the OMG SysML Standard. UTCS, Springer, Cham (2016). https://doi.org/10.1007/978-3-319-44339-3
18. Ozkaya, M.: The analysis of architectural languages for the needs of practitioners. Softw. Pract. Experience **48**(5), 985–1018 (2018)
19. Ozkaya, M.: Do the informal and formal software modeling notations satisfy practitioners for software architecture modeling? Inf. Softw. Technol. **95**, 15–33 (2018)
20. Quilbeuf, J., Cavalcante, E., Traonouez, L.-M., Oquendo, F., Batista, T., Legay, A.: A Logic for the Statistical Model Checking of Dynamic Software Architectures. In: Margaria, T., Steffen, B. (eds.) ISoLA 2016. LNCS, vol. 9952, pp. 806–820. Springer, Cham (2016). https://doi.org/10.1007/978-3-319-47166-2_56
21. Taoufik, S.R., Tahar, B.M., Mourad, K.: Behavioral verification of UML2.0 software architecture. In: 12th International Conference on Semantics. Knowledge and Grids, pp. 115–120. IEEE, USA (2016)
22. Wimmer, M., et al.: Surveying rule inheritance in model-to-model transformation languages. J. Object Technol. **11**(2), 1–46 (2012)

Lean Rehabilitative-Predictive Recommender Games for Seniors with Mild Cognitive Impairment: A Case Study

Chien-Sing Lee[(✉)] and Wesly Yii

Department of Computing and Information Systems, School of Engineering and Technology,
Sunway University, Petaling Jaya, Malaysia
chiensingl@sunway.edu.my, 18066233@imail.sunway.edu.my

Abstract. The number of dementia seniors is increasing. It is important to maintain their quality of life as much as possible and reach the masses. However, there are not that many personalized recommender systems for dementia seniors. Hence, this study aims to develop simple fun games for seniors with mild cognitive impairment (the stage before dementia) and their caregivers, which would be simple, fun, and rehabilitative. Designed based on the Montreal Cognitive Assessment test (MOCA) and the Mini Mental State Examination (MMSE), our games assess users' game plays, and recommend suitable difficulty levels based on two initial calibrations. The first calibration is for the memory game and the second calibration is for the fishing practice concentration game. The caregiver can also adjust the difficulty level manually. In the future, based on gameplay data, the system will be able to predict seniors' condition, and follow-up with advice from practitioners. User testing highlight interesting correlations between/among intra-Technology Acceptance Model (TAM), intra-User Experience (UX) and inter-TAM-UX constructs.

Keywords: Rehabilitative · Games · Lean · Assessment · Recommender · Mild cognitive impairment

1 Introduction

Throughout the world, millions of people suffer from dementia. According to the World Health Organization (WHO), as of 2019, there are 50 million dementia patients [1]. Dementia affects not only the individual but also their families and the society. The cost of handling dementia globally is estimated to be $818 billion per year [2]. Although there is encouraging progress in the medical field, the intricacies of the disease make it difficult to pinpoint the causes [3].

1.1 Problem to Be Addressed

While dementia is generally not curable at this point of time [4], many research and professional bodies have tried to possibly delay the progress of dementia in a patient.

© Springer Nature Switzerland AG 2021
O. Gervasi et al. (Eds.): ICCSA 2021, LNCS 12951, pp. 411–428, 2021.
https://doi.org/10.1007/978-3-030-86970-0_29

We note that there is a lack of recommender-enhanced mobile-based games for dementia patients in the market, which automatically customize the level of difficulty to the player, or would enable the senior/the caregiver, to choose the level of difficulty.

1.2 Objectives

We aim to train and improve the cognitive abilities of individuals with mild cognitive impairment, to delay progress into dementia. We hope to do so, through simple, yet fun rehabilitative games. Designed based on the Mini Mental State Examination (MMSE) [5] test, and to a certain extent, the Montreal Cognitive Assessment (MOCA) [6] test, *MindRegen*, consists of two mini-games, which are the Matching game (addressing attention and memory) and the Fishing game (addressing attention and motor skills). Rehabilitation and calibration are enabled by tracking the seniors' game play, and assessing and adjusting the game's level of difficulty to the seniors' current level of ability. We hope the gameplay data can also be used to predict seniors' eventual ability.

2 Related Work

2.1 Dementia

Dementia is defined as a disease that can cause progressive decline in a person's cognitive abilities. It affects both learning and reasoning abilities to the point that it disrupts the person's daily activities. There are several types of dementia, including Alzheimer's disease, Lewy dementia, Cerebro-vascular dementia, Fronto-temporal dementia, Parkinson's disease, Hippocampal sclerosis and mixed dementia. Among all the types of dementia, Alzheimer's disease is by far the most common (60–70%) followed by Vascular and Lewy body dementia [2, 7, 8].

2.2 Mini Mental State Examination (MMSE) and Montreal Cognitive Assessment (MOCA)

There are many versions to the Mini Mental State Examination (MMSE) [5] test. However, the common assessments include spatial orientation, object recognition/recall, attention and language. Similar to the MMSE, the MOCA [6] focuses on visual-spatial, recall, attention/concentration, and abstract reasoning. The difference lies in the addition of time and place orientation.

[9] have experimented with affordance design, for cognitive access, with seniors with the Alzheimer's Disease Foundation of Malaysia (ADFM). Design is based on the four MMSE constructs and prior discoveries/insights from the creative industries on semiotics, lean sustainable design, and three types of cognition [10]. Among the activities, seniors piece together shapes, as part of a leisure activity. The placement of the shapes reflect object recognition and visual orientation abilities. The heart shape symbolizes the seniors' main concern. Other shapes are interpreted in relation to it.

The use of semiotics in [9] is supported by [11]. [11] recognize the use of models and semiotics-based artifacts for improving communication and for identifying and modelling elicited requirements. Their study uses semiotics to elicit non-functional requirements, i.e., for privacy. It is a novel bridge between Art and Science.

Pursuant from [9, 12]'s meta-analysis, abstracts from prior inclusive design systems. These are designed based on the MMSE. However, user testing involves active aging seniors, who are hand-phone users, not MCI seniors. Findings indicate that germane load in [13]'s Cognitive Load Theory, is confirmed as a generative design factor, which can motivate self-regulation and tradeoff intrinsic and extraneous cognitive loads. [9, 12] are rehabilitative, and predictive.

2.3 Games for Dementia

2.3.1 Big Brain Academy

Big Brain Academy (BBA) contains a variety of puzzle games. In the "Think" category, players are challenged to solve *logical* questions. For the "Memorize" category, players have to utilize their *recall* and *memorization* abilities. The "Analyze" category provides the players with puzzles that require *reasoning* whereas the "Compute" category involves *Math* tests. The last category "Identify" aims to train *visualization*.

All of them allow 60 s for a player to complete the game. When a player gets the wrong answer, BBA will lower the difficulty level and deduct points and the timer for the player. BBA can be played in three different modes: Test Mode, Practice mode and versus mode. At the end of the game, BBA will show the player's performance in terms of their weaknesses and strengths [14]. A study [15] indicates that seniors playing BBA report a decrease in stress levels and is calmer in mind.

2.3.2 CogniFit

CogniFit, an interactive mental game application that contains more than 20 games, aids cognitive skills stimulation through personalized daily training. It trains cognitive abilities in different aspects, such as *memory, concentration, attention, mental arithmetic, executive functions, reasoning, planning, mental agility, coordination*. It also shows detailed statistics to track the user's progress and performance [16]. The technology used to develop the game is the Cognitive Assessment and Individualized Training System (ITS). ITS is a real-time application that configures every user's training experience, by using advanced algorithms on information provided by assessments on the user. ITS constantly monitors the cognitive level of the user in real time to develop proper adjustments. It then provides each user with a distinct assessment with suitable number of tasks and levels of difficulty. This will provide the most optimal training experience to every user. Apart from that, the Cognitive Assessment test is used to determine the cognitive ability of a user by comparing it with performance from demographic peers with age and gender as the primary variables.

2.3.3 MindMate

MindMate (MM) is a mobile application, which caters to people with Alzheimer's disease or dementia. It aims to reduce the deterioration of the disease by encouraging seniors with dementia to have a holistic lifestyle. MM also promotes reminiscence therapy in improving cognitive abilities and quality of life for the elderly [17]. It provides a digital platform for brain games, nutrition, exercise advisories and social interaction. The games available in MM are classified based on four core cognitive areas, which include Problem Solving, Speed, Memory and Attention.

A Finnish intervention [18] found that aside from improvements in cognitive functions, participants' overall health improves. The research [19] also finds that the MIND (Mediterranean-DASH Intervention for Neurodegenerative Delay) diet helps to slow down dementia by 53%.

2.3.4 WiiFit

WiiFit, is designed by Hiroshi Matsunaga for the Nintendo Wii console. It offers a variety of exercising choices, such as Yoga, Strength Training, Aerobics and Balance Games. The Wii Balance Board allows the user to stand on top of it, while doing the exercises. The user needs to imitate the postures on-screen while a virtual personal trainer provides feedback. The *WiiFit* is able to improve balance and gait, important in preventing falls [20], as well as balance and walking speed [21].

3   Methodology

We have applied the waterfall methodology. The requirements for the game are gathered from literature review on games aimed at improving the condition of patients with dementia. The review covers how seniors with dementia would interact with a certain game and how recommendations are developed for them. These requirements are then included in a survey form. We next ask the training coordinator in a Malaysian Alzheimer's Association for her opinion on the survey (second part of requirements gathering), to obtain her consent, prior to distributing the game and survey online.

The game engine, *Godot,* simplifies all the physics of the character and the platforms. The programming language GDScript, is a high-level, programming language. It uses a syntax like Python (blocks are indent-based, and many keywords are similar). It is optimized for, and tightly integrated with the *Godot* Engine, allowing great flexibility for content creation and integration.

User testing is via Google survey, with the Alzheimers' Disease Foundation of Malaysia's (ADFM) caregivers, MCI seniors. It is the same foundation as that in [9]'s and initially, [12]'s. The differences are different caregivers and seniors (some seniors have passed away), the past study's commentator/vettor is the head nurse but for this study, the training coordinator, past is physical materials but for this, totally digital.

4   Systems Design and Development

4.1   Mechanisms

There are three mechanisms: a) assessment, b) rehabilitation and c) prediction. The predictive mechanism is to predict the user's next level of ability based on the current gameplay. These three mechanisms are also used by [22–24]. Furthermore, users can choose the level of difficulty for any game. By default, a recommender system will calibrate to enable seniors and their caregivers to play the game at a level which they wish. The difficulty level can be manually adjusted if the caregiver thinks it is not suitable, by selecting other difficulty levels in the difficulty selection screen or by skipping the calibration phase.

Other design considerations are the device (tablet/Web-based/executable file), comfort while playing (whether seated/standing), separation of motor and cognitive skills training, suitable duration [25], short and simple instructions, different levels of difficulty (adjusted to error rate of less than 5% and a reaction time of less than 5 s [26] or successful completion of three consecutive tasks without mistake, or successful completion of 80% of the game over six sessions [27]), and feedback [28].

4.2   Game Flow Overview and User Interface

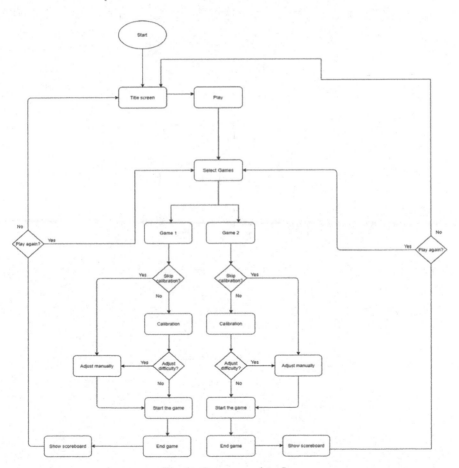

Fig. 1. Game overview flow

An overview of the game flow is presented in Fig. 1. The user interface and human-computer interaction are designed partly based on [29]'s design principles, partly based on non-functional requirements, i.e., portability, effectiveness, efficiency and reliability, and ultimately, fun.

4.3   Game 1's Flowchart and Functional Requirements (to Train Memory Skills)

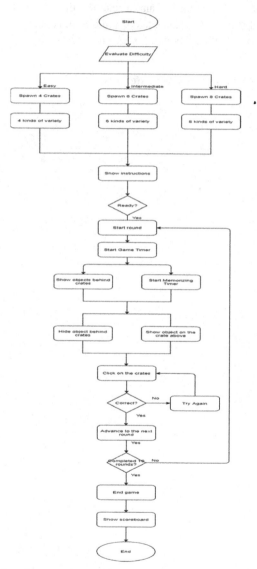

Fig. 2. Game system flow for game 1: matching cards (calibration 1), followed by recommender, based on score and time taken thresholds

The flowchart, and user interfaces are presented in Fig. 2. There are various non-functional requirements, but to us, the most important are convenience, portability, and reliability. Some of Game 1's functional requirements are shown in Table 1.

Table 1. Some functional requirements for game 1 (memory game)

The features inside the game shall be distinguishable and recognizable	The game shall implement large objects with contrasting colours
The difficulty of the game shall have the ability to be manually adjusted	The difficulty of the game can be adjusted by the caregivers if needed
The game shall not take a long time to play	Playing one round of the game shall not exceed 4 min
The game shall be diverse in terms of difficulty (Game 1)	Different difficulty levels will trigger different number of platforms and object variety, e.g. 4 Easy, 9 Intermediate, 16 Difficult or 2 Easy, 4 Intermediate, 6 Hard

4.4 Confirmation of Design (User Requirements)

A survey consisting of five questions are distributed to caregivers of the Alzheimer's Foundation via an online Google Form, to gather and confirm requirements identified from the above literature review and our proposed design. All questions are measured on a Likert scale of 1 to 5, with 1 being very unlikely and 5 very likely.

The first question (Fig. 3) inquires to what extent users would like to set the difficulty level by themselves, and to have the option of choosing which game they would like to play. The results indicate that majority of the users like this function.

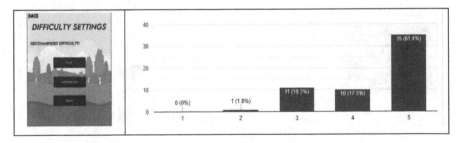

Fig. 3. Degree of preference for availability of difficulty settings and options

The second question (Fig. 4 aims to determine whether the seniors would like the system, to automatically recommend a difficulty level for the user, based on his/her performance. The results indicate that it is positively-perceived.

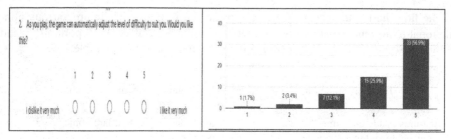

Fig. 4. Preference for automatic adjustment/recommendation of digitadifficulty level

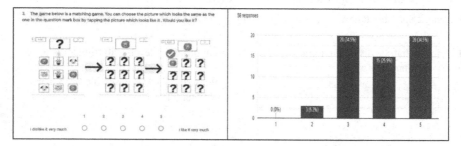

Fig. 5. Degree of preference for the memory game

The third question (Fig. 5 aims to determine whether the users like the gameplay of the first game. This game aims to improve the user's memory skills. More than 80% of the users like this game.

The fourth question (Fig. 6 aims to determine whether users like the gameplay of the second game (Fishing Practice). This game aims to improve the user's concentration and reaction speed. Findings indicate that this game is quite well-perceived, though it is not as popular as the first game.

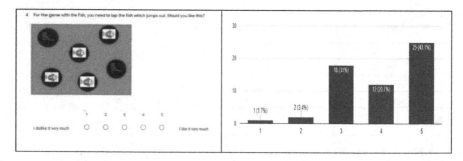

Fig. 6. Degree of preference for the fishing practice (attention and reaction speed)

The final question (Fig. 7) aims to determine whether users would like a scoreboard, which displays their performance after they end the game. The results are mixed, with

4 out of 30 respondents disliking it. This may probably be due to the worry of losing motivation, if they get poor results or lose privacy. However, most like the scoreboard.

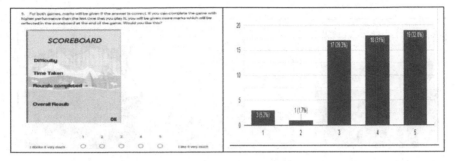

Fig. 7. Degree of preference for scoreboard

4.5 Refinements Based on Confirmation of User Requirements

Based on the confirmation of requirements above, the games, with the following interfaces (Figs. 8, 9 and 10) are finalized for user testing.

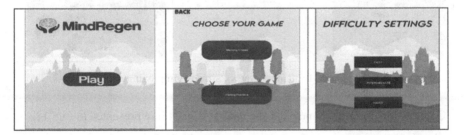

Fig. 8: Main page (left), game selection (middle), manual difficulty selection (right)

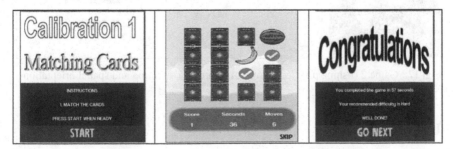

Fig. 9. Calibration test 1: matching cards (memory)

The games begin with 2 calibrations. The first calibration (Fig. 9) will recommend the level of difficulty for game 1, i.e., memory crates, and the second calibration (Fig. 10) will recommend the level of difficulty for game 2, i.e., fishing practice.

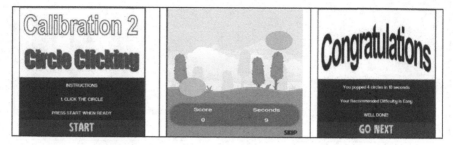

Fig. 10. Calibration test 2: popping the circles (attention and spatial ability)

Game 1's instructions page and the game itself are presented in Fig. 11.

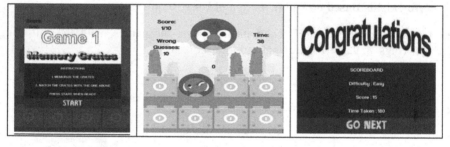

Fig. 11. Game 1 (memory crates), aimed at improving memory

Game 2's instructions game page and the game itself, are presented in Fig. 12.

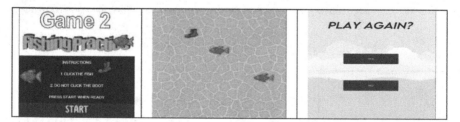

Fig. 12. Game 2 (fishing practice), aimed at improving attention and spatial orientation

5   User Testing

User testing is carried out end of May 2021. It is totally virtual, due to stringent movement control order procedures in the country. A questionnaire, consisting of 21 questions, is developed, integrating constructs from [30, 31]'s Technological Acceptance Models (TAM) 1, 2, and 3 and [32]'s User Experience (UX) questionnaire.

Questions 1–3 evaluates ease of use, questions 4–12 usefulness, and questions 13–14 the excitement and motivation the games create. Perceptions towards the recommender and difficulty levels fall under usefulness. Questions 15–16 evaluates the dependability/reliability of the games in terms of the ability of the game to run without problems. Question 17 assesses whether the games could capture the interest of the users, question 18 the efficiency of the games and question 19 the attractiveness of the games. Lastly, Question 20–21 assess whether the users are interested to continue playing the game in the future. Questions 1–12 and 20–21 are based on the TAM 1, 2, and 3, while questions 13–19 are based on the UX questionnaire.

We receive thirty caregivers'/seniors' responses to our Google survey. Figure 13 illustrates the cumulative score for each question. Figure 14 shows the average scores by TAM and UX grouping. The graph shows a high score of 4.4 for ease-of-use, indicating that the game can be easily managed by the seniors. Similarly, usefulness and intention to use in the future score an average of 4.2 each. Although attractiveness scores 3.8, the respondents find the games stimulating (4.05). We need to improve with more visually appealing designs.

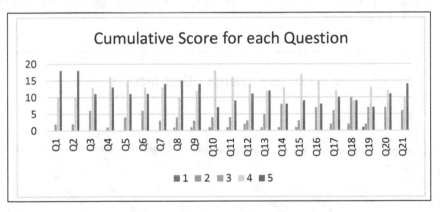

Fig. 13. Cumulative score for each question

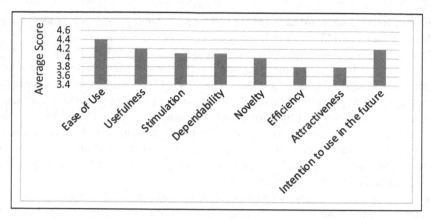

Fig. 14. Average scores by TAM and UX grouping

Next, Pearson correlation for intra-TAM, intra-UX and inter-TAM-UX constructs are investigated. Some of the correlations investigated are Q1 (ease-of-use)-Q7 (usefulness), Q13 (stimulation)-Q15 (dependability), Q17 (novelty)-Q18 (efficiency), Q19 (attractiveness)-Q20 (intention to use in the future). Outcomes are presented in Figs. 15, 16, 17, 18, 19, 20 below.

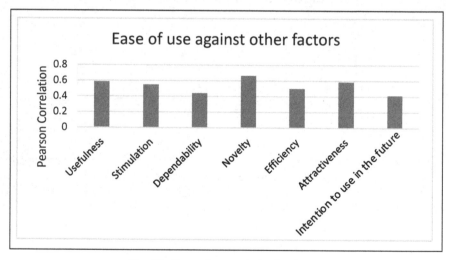

Fig. 15. Ease of use against other factors

These figures indicate that stimulation, followed by novelty and attractiveness (Fig. 22) are the three most important design factors to these seniors. Furthermore, attractiveness is most highly correlated with novelty (Fig. 21). Novelty in turn is equally highly correlated with stimulation and attractiveness, followed by ease of use and intention to use in the future (Figs. 17 and 19).

Moreover, perception towards the recommender's dynamic adjustment of the difficulty level receives an average score of 4.5 (Q4) and the reasonableness of the generated difficulty level for each game (Q5), receives an average score of 4.2. Degree of skewness for Q4 and Q5 are −0.19, and −0.30 respectively.

Surprisingly, the averages for the questions on clarity with regards to the functions of buttons, the Fishing game for improving concentration, the Fishing game for improving recognition and the scoreboard at the end of the game, are 4.5, 4.3, 4.3 and 4.1 respectively, and are skewed −0.96, −0.95, −0.98, −0.90 respectively. These are encouraging findings. Surprisingly, they also confirm the findings in [9, 12], though with different modality and over time.

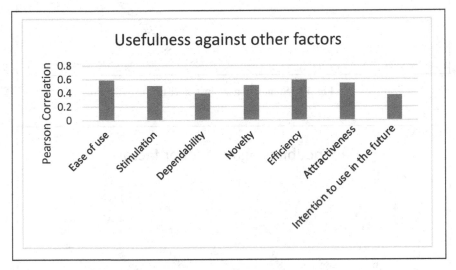

Fig. 16. Usefulness against other factors

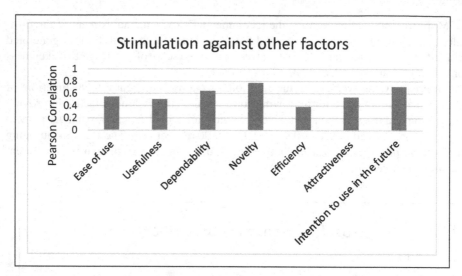

Fig. 17. Stimulation against other factors

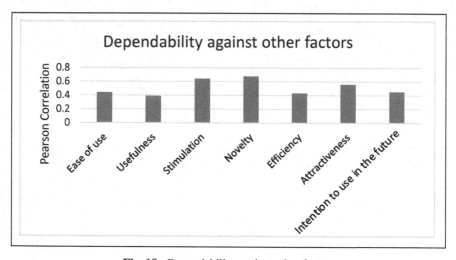

Fig. 18. Dependability against other factors

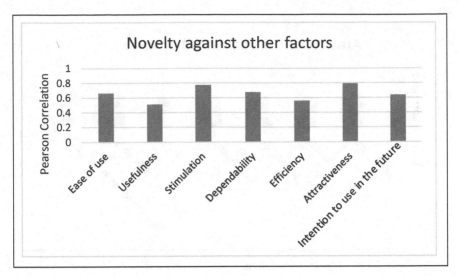

Fig. 19. Efficiency against other factors

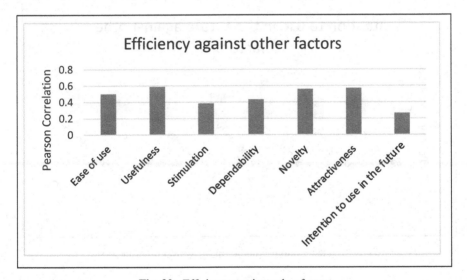

Fig. 20. Efficiency against other factors

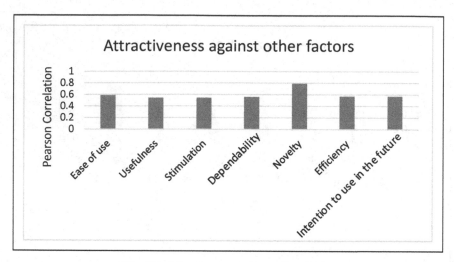

Fig. 21. Attractiveness against other factors

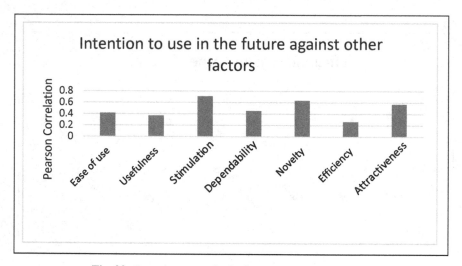

Fig. 22. Intention to use in the future against other factors

6   Conclusions

We have developed games, each with its own calibration cycles. Findings indicate that it is possible to extend these games to MCI seniors at home, where they can play with their caregivers. Due to the small sample size, findings are not generalizable. In the future, if we have funding, data may be captured and shared with doctors, (with family consent), for further interventions. We hope there would be positive improvements, similar to medically-grounded brain training randomized trials in the US, UK, EU.

Acknowledgement. This extended capstone paper builds on a Malaysian Grant Scheme ref. FRGS/ 2016/ ICT04/SYUC/01/1, ethics approval SUREC 2016/067, and [9, 12] (Fulbright). We thank the ADFM, caregivers, seniors, Ms. Jenny Ho. For this paper, the first author thanks UTAR, Dr. K. Daniel Wong (Daniel Wireless Software Pte. Ltd., SG), Dr. John H. Hughes (USUHS), Maryland, USA; Prof. Suat-Cheng Peh, Sunway University, Dr. Foong Chee Choong (Sunway Medical Centre), for the hand-cognitive rehabilitation home concept; enriching the FRGS' subsequent diagnostic-recommender proposals in 2018/2019. Registration fee is via the past Fulbright Visiting Scholar Fellowship grant.

References

1. Dementia: Who.int (2020). https://www.who.int/news-room/fact-sheets/detail/dementia. Accessed 15 Nov 2020
2. Alzheimer's Disease International: World Alzheimer Report 2015, Alzheimer's Disease International (2015)
3. Sims, R., Hill, M., Williams, J.: The multiplex model of the genetics of Alzheimer's disease. Nat. Neurosci. **23**(3), 311–322 (2020). https://doi.org/10.1038/s41593-020-0599-5
4. Tom, S.: Characterization of dementia and Alzheimer's disease in an older population: updated incidence and life expectancy with and without Dementia. Am. J. Publ. Health **105**(2), 408–413 (2015). https://doi.org/10.2105/ajph.2014.301935
5. Pangman, V.C., Sloan, J., Guse, L.: An examination of psychometric properties of the minimental status examination and the standardized mini-mental status examination: implications for clinical practice. Appl. Nurs. Res. **13**(4), 209–213 (2000). https://doi.org/10.1053/apnr.2000.9231.PMID11078787
6. Nasreddine, Z.S., et al.: The montreal cognitive assessment, MoCA: a brief screening tool for mild cognitive impairment. J. Am. Geriatric Soc. **53**(4), 695–699 (2005). https://doi.org/10.1111/j.1532-5415.2005.53221.x.PMID15817019
7. Husband, A., Worsley, A.: Different types of dementia. Pharma. J. **277** (2006)
8. Sonnen, J., et al.: Pathologic correlates of dementia in individuals with Lewy body disease. Brain Pathol. **20**(3), 654–659 (2010)
9. Lee, C.S., Wong, K.D.: Design thinking and semiotics to increase socio-cognitive-affective engagement: an inclusive design human factors case study. In: IEEE International Conference on Industrial Engineering and Engineering Management, pp. 264–268 (2017)
10. Lee, C.S.: Exploring possibilities for synergizing embodied, embedded and extended cognition: implications to STEM Education. In: ICCE, pp. 425–434 (2017)
11. Mendes, L.M., de Franco Rosa, F., Bonacin, R.: Uma revisão sobre o uso da semiótica na análise eespecificação de requisitos de privacidade (review of the state of art about the use of semiotics in data privacy requirements elicitation). Anais do WCF, vol. 6, pp. 31–36 (2019). XV WCF 23–24 set 2019
12. Lee, C.S., Hughes, J.H.: Refocusing on cognitive load design through a meta-analysis on learnability, goal-based intentions and extensibility: towards personalized cognitive-social-affective engagement among seniors. In: International Conference on Intelligent Software, Methodologies, Tools and Techniques, pp. 456–469 (2019)
13. Sweller, J.: Cognitive load during problem solving: effects on learning. Cogn. Sci. **12**(2), 257–285 (1988)
14. Big Brain Academy: IGN. https://www.ign.com/articles/2006/06/02/big-brain-academy. Accessed 19 Nov 2020

15. Fernández-Calvo, B., Rodríguez-Pérez, R., Contador, I., Rubio-Santorum, A., Ramos, F.: Eficacia del entrenamiento cognitivo basado en nuevas tecnologías en pacientes con demencia tipo Alzheimer [Efficacy of cognitive training programs based on new software technologies in patients with Alzheimer-type dementia]. Psicothema 23(1), 44–50 (2011)
16. CogniFit: Brain Training, Brain Games, Memory Games, and Brain Fitness with CogniFit. https://www.cognifit.com/. Accessed 19 Nov 2020
17. Alzheimer's app: MindMate. https://www.mindmate-app.com/. Accessed 19 Nov 2020
18. Neurology, T.L.: Pointing the way to primary prevention of dementia. Lancet Neurol. 16(9), 677 (2017). https://doi.org/10.1016/s1474-4422(17)30256-9
19. Hagerty, J.R., Morris, M.C.: Devised a diet to reduce risks of dementia. Wall Street J. (2020). https://www.wsj.com/articles/martha-clare-morris-devised-a-diet-to-reduce-risks-of-dementia-11582903800
20. Padala, K.P., et al.: Wii-Fit for improving gait and balance in an assisted living facility: a pilot study. J. Aging Res. 2012, 1–6 (2012). https://doi.org/10.1155/2012/597573
21. Dougherty, J., Kancel, A., Ramar, C., Meacham, C., Derrington, S.: The effects of a multi-axis balance board intervention program in an elderly population. Mo. Med. 108(2), 128–132 (2011)
22. Sayed, K.: Interactive digital serious games for dementia. Int. J. Comput. Games Technol. 1 (2014). https://doi.org/10.1155/2014/701565
23. van Adel, J.M., Taler, V., Leiva, R.: The montreal cognitive assessment (MOCA) in geriatric rehabilitation. Psychom. Prop. Rehabil. Outcomes Int. Psychog. 23(10), 1582–1591 (2011)
24. Okochi, J., Yamaguchi, H.: Rehabilitation to live better with dementia: rehabilitation for dementia. Geriatr. Gerontol. Int. 18, 1529–1536 (2018)
25. Dietlein, C., Bock, B.: Recommendations on the design of serious games for people with dementia. EAI Endorsed Trans. Game-Based Learn. 5(17), 159528 (2019). https://doi.org/10.4108/eai.11-7-2019.159528
26. Fenney, A., Lee, T.: Exploring spared capacity in persons with dementia: what Wii can learn. Act. Adapt. Aging 34(4), 303–313 (2010). https://doi.org/10.1080/01924788.2010.525736
27. Tarraga, L.: A randomized pilot study to assess the efficacy of an interactive, multimedia tool of cognitive stimulation in Alzheimer's disease. J. Neurol. Neurosurg. Psychiat. 77(10), 1116–1121 (2006). https://doi.org/10.1136/jnnp.2005.086074
28. Lee, G., Yip, C., Yu, E., Man, D.: Evaluation of a computer-assisted errorless learning-based memory training program for patients with early Alzheimer's disease in Hong Kong: a pilot study. Clin. Interv. Aging 8, 623–633 (2013). https://doi.org/10.2147/CIA.S45726
29. Schneiderman, B., Plaisant, C., Cohen, M. Jacobs, S., Elmqvist, N., Diakopoulos, N.: Designing the User Interface: Strategies for Effective Human-Computer Interaction. 6th edn (Global). Pearson, London (2017)
30. Davis, F.D., Bagozzi, R.P., Warshaw, P.R.: User acceptance of computer technology: a comparison of two theoretical models. Manag. Sci. 35(8), 982–1003 (1989)
31. Ventakesh, V., Bala, H.: Technology acceptance model 3 and a research agenda on interventions. Decis. Sci. 39(2), 273–315 (2008)
32. Schrepp, M., Hinderks, A., Thomaschewski, J.: Design and evaluation of a short version of the user experience questionnaire (UEQ-S). Int. J. Interact. Multimedia Artif. Intell. 4(6), 103–108 (2017)

Early Fault Detection with Multi-target Neural Networks

Angela Meyer[✉] [iD]

Bern University of Applied Sciences, 2501 Biel, Switzerland
angela.meyer@bfh.ch

Abstract. Wind power is seeing a strong growth around the world. At the same time, shrinking profit margins in the energy markets let wind farm managers explore options for cost reductions in the turbine operation and maintenance. Sensor-based condition monitoring facilitates remote diagnostics of turbine subsystems, enabling faster responses when unforeseen maintenance is required. Condition monitoring with data from the turbines' supervisory control and data acquisition (SCADA) systems was proposed and SCADA-based fault detection and diagnosis approaches introduced based on single-task normal operation models of turbine state variables. As the number of SCADA channels has grown strongly, thousands of independent single-target models are in place today for monitoring a single turbine. Multi-target learning was recently proposed to limit the number of models. This study applied multi-target neural networks to the task of early fault detection in drive-train components. The accuracy and delay of detecting gear bearing faults were compared to state-of-the-art single-target approaches. We found that multi-target multi-layer perceptrons (MLPs) detected faults at least as early and in many cases earlier than single-target MLPs. The multi-target MLPs could detect faults up to several days earlier than the single-target models. This can deliver a significant advantage in the planning and performance of maintenance work. At the same time, the multi-target MLPs achieved the same level of prediction stability.

Keywords: Condition monitoring · Fault detection · Multi-target neural networks · Normal behaviour models · Wind turbines

1 Introduction

The global wind power capacity is growing strongly with a total installed volume of 651 GW in 2019 and an increase of 76 GW in 2020 [1]. The newly installed wind turbines are getting larger and increasingly more complex. At the same time, the operating cost of wind farms still makes up a major fraction, approximately 30%, of their lifetime cost [2]. Major faults can result in days and even weeks of downtime [3, 4]. Therefore, they can substantially reduce the owner's return on investment and pose a considerable economic risk. As a result, many operators want to closely monitor the health state of their turbines in order to be alerted as early as possible of any developing technical problems and to prevent any major damage and downtime. To this end, an automated condition

© Springer Nature Switzerland AG 2021
O. Gervasi et al. (Eds.): ICCSA 2021, LNCS 12951, pp. 429–437, 2021.
https://doi.org/10.1007/978-3-030-86970-0_30

monitoring of wind turbine subsystems provides an essential prerequisite for informed operational decision making and fast responses in case of unforeseen maintenance needs [5, 6].

Data-driven automated monitoring methods have been proposed, amongst others, based on sensor data logged in the turbines' supervisory control and data acquisition (SCADA) systems [7–10]. Temperature can be an important indicator of different types of developing machine problems such as mechanical faults which can give rise to excessive friction generating heat. Therefore, a major focus of the proposed SCADA-based condition monitoring approaches is the temperature-based detection of developing faults in the wind turbine subsystems based on models of the turbine's normal operation behaviour in the absence of operational faults [11–22]. The present study focusses on the gear bearing temperature as an indicator of developing gearbox faults.

The goal of this study is to assess the potential of multi-target regression models for the automated SCADA-based fault detection. Specifically, this work has investigated and compared the delays in detecting gear bearing faults using single-target versus multi-target models of the turbines' normal operation. Moreover, the stability of the alarm signal after the first detection of a developing fault is being assessed.

The remainder of this paper is structured as follows. Section 2 provides a brief overview of previous work in this field. Section 3 describes the data sources and the training and testing of the multi- and the single target regression models. The analysis and results are discussed in Sect. 4. Conclusions and possible future work are proposed in Sect. 5.

2 Related Work

Normal behaviour modelling has become an established technique in wind turbine condition monitoring and fault detection [7, 8]. Normal behaviour models characterize the machine state during normal operation in the absence of faults. They have been in use for monitoring the health state of turbine subsystems such as the gearbox [23, 24] and the generator [25]. Normal behaviour models have also successfully been employed for monitoring the active power generation [15, 26, 27]. We refer to [8] for a comprehensive review of SCADA-based condition monitoring and normal behaviour models of wind turbines.

Multi-target machine learning models [28–32] are regression or classification models which predict multiple target variables simultaneously. It has been demonstrated in other fields that multi-target models hold the potential to enable an increased prediction accuracy compared to single-target models and are less susceptible to overfitting the training data [29, 33–34]. In the field of wind turbine monitoring, we have recently introduced multi-target regression models for simultaneously monitoring the growing number of SCADA channels, and we demonstrated that they can reduce the effort of SCADA-based normal behaviour monitoring in wind turbine condition monitoring [35].

3 Data and Methods

Condition monitoring data from the SCADA system of three commercial onshore turbines was analyzed in this work. The turbines are variable-speed three-bladed horizontal axis systems with pitch regulation from an onshore wind farm. Their rated power was 3.3 MW, and they operated with a 3-stage planetary/helical gearbox. The turbines' rotors were 112 m in diameter with the hub located at 84 m height above ground. The turbines' cut-in, rated and cut-out wind speeds were specified at 3 m/s, 13 m/s and 25 m/s, respectively.

In this study, fourteen months of ten-minute mean SCADA signals served to train and test the models specified below. The data were anonymized to maintain the privacy of the wind farm operator. We report the results for one of the wind turbines. It was randomly selected and the results were not affected by the choice of turbine. We focus on monitoring the gear bearing condition based on the temperature of the gear bearing. The temperature is an important SCADA-based indicator of incipient fault processes in gearbox components [8, 19, 20, 22, 23]. In the present study, the condition of the gear bearing has been monitored based on two normal operation models of the bearing temperature. Wind speed v_{wind}, wind direction α_{wind} and air temperature T_{air} constitute the models' input variables which were provided as ten-minute averages of measurements from nacelle-mounted anemometers and thermometers. The input variables were selected due to their relevance for explaining and predicting the behaviour of the target variables.

A multi-target fully connected feedforward neural network was designed to predict the gear bearing temperature T_{gear} along with the hydraulic oil temperature T_{oil} and the transformer winding temperature T_{tr} from the input variables at high accuracy, T_{gear}, T_{oil}, $T_{tr} \sim v_{wind} + \alpha_{wind} + T_{air}$. The single-target model estimates the gear bearing temperature only, $T_{gear} \sim v_{wind} + \alpha_{wind} + T_{air}$. In addition, the two fully connected feedforward neural networks (multi-layer perceptrons, MLP) were trained and tested to assess the normal operating behaviour of the gear bearing based on the provided SCADA data. The model architectures were developed to obtain a high predictive accuracy on the training set without overfitting the training data. In this process, the number of neurons and weights to be trained was increased only if this resulted in higher predictive accuracy. The resulting model architectures are detailed in Table 1.

In this work, our goal is to systematically assess the ability of multi-target neural networks to detect developing faults that result in rising component temperatures. We demonstrate this approach by the example of the gear bearing temperature. However, it is equally applicable to faults in other subsystems and other components that result in elevated SCADA-logged temperatures.

A major challenge in data-driven fault detection and isolation is the scarcity of actually observed fault instances. We addressed this point by combining gear bearing temperature measurements with a multitude of synthetic temperature trends in order to mimic the bearing temperature rise induced by a developing fault. To this end, a synthetic temperature trend was overlaid on the normalized gear bearing temperature signal. One of ten different linear temperature trends was added to the normalized bearing temperature. Temperature trends with integer slopes in the range of 1 to 10 were used to simulate slowly and fast evolving fault processes. The temperature trend onset time was randomly sampled from a two-week time window in months 12 and 13 of the 14-months observation period. Fifty different onset times have been randomly sampled from the two-week window for each of the ten temperature slopes. This ensured that the results did not depend on the choice of the trend onset time.

Table 1. The architectures of the multi-layer perceptrons.

Model	Architecture
Multi-layer perceptron (MLP) with three target variables	Two dense hidden layers with 4 neurons in the first layer and 19 neurons in the second layer, batch normalization, and a 3-neuron output layer. Dropout was applied at a rate of 10% to avoid overfitting
Single-target multi-layer perceptron	Three dense hidden layers with 4 neurons each in the first and second layers and 5 neurons in the third layer, batch normalization and a single node in the output layer. Dropout rate of 10%

4   Results and Discussion

Two common alarm criteria [7] were applied using the residuals of the gear bearing temperature. The residuals were computed as the difference of the observed temperature of the gear bearing and its temperature predicted by the normal operation behaviour models of Table 1. Figure 1 illustrates the residuals and alarms. According to the first criterion, an alarm was raised when the 99.9^{th} percentile of the residuals distribution was exceeded for more than 8 h in the past 24 h. On the other hand, the second criterion triggered an

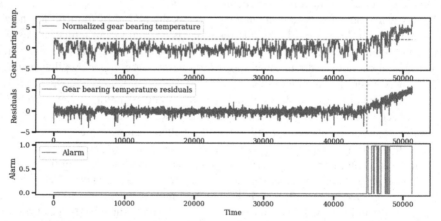

Fig. 1. Normalized gear bearing temperature and residuals. An alarm was raised when the 99.9^{th} percentile of the residuals distribution was exceeded for more than 8 h during the past 24 h (alarm criterion 1). The dashed line indicates the first alarm event. The temperature units are dimensionless due to normalization.

alarm whenever the rolling mean of the residuals computed over the past 8 h exceeded the 99.9th percentile of the residuals distribution. The fault detection capabilities of the multi-target and single-target normal behaviour models were compared with regard to the delay of detecting the induced faults. Moreover, the stability of the triggered alarms was assessed following the detection. The stability is computed as the fraction of true positive alarms after the first detection.

Figure 2 shows the resulting alarms for a fault instance with fast rising bearing temperature and a randomly sampled onset time. The false positive detection rate is zero for both criteria but there are false negatives after the first alarm detection by both alarm criteria. Following the first detection, the alarms are initially unstable and stabilize within three to seven days after the first alarm was triggered, as shown in Fig. 2. We found that the alarm signal stabilized faster in the case of the first criterion, while the second criterion generated a less stable signal that required a week to stabilize in the case shown in Fig. 2. This alarm instability was caused by the signal-to-noise ratio which was relatively small immediately after the first detection and increased over time along with the overlaid increasing temperature.

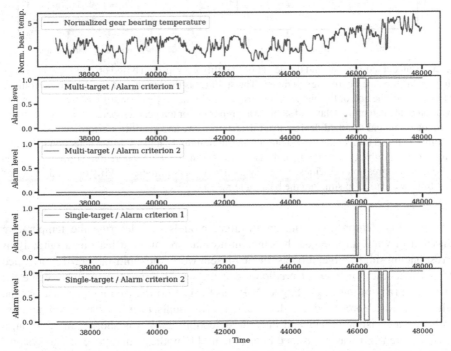

Fig. 2. A fast evolving gear bearing fault with trend slope 10 is shown at one of the 50 randomly sampled onset times. The gear bearing temperature starts to rise near time step 44000 in this case. The alarm criteria are compared based on the multi-target and the single-target normal behaviour models of the gear bearing temperature. Alarm criterion 2 detects the trend earlier but is less stable than alarm criterion 1.

The detection delays and detection stabilities are reported in Figs. 3 and 4 in terms of the means and the standard deviations computed over the 50 randomly sampled onset times. The fault detection delays could be quantified precisely as the faults were induced at known times in terms of temperature trend onset times.

We found that the multi-target model could detect the induced faults at least as fast as the single-target model. In the majority of the fault instances in this case study, the multi-target MLP even enabled a somewhat faster detection of the gear bearing fault than the single-target network. As shown in Figs. 3 and 4, the multi-target MLP facilitated shorter detection delays for both slow and fast trends and regardless of the chosen alarm criterion.

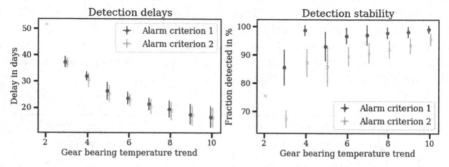

Fig. 3. Fault detection delay and the detection stability are reported based on *multi-target* normal behavior MLP of the bearing temperature. The stability is the fraction of true positives from the first alarm until the end of the observation period, i.e. till the end of month 14. Mean and standard deviation of the detection delay and stability are reported for ten gear temperature trends. For each trend, the mean and standard deviations of detection delay and stability were computed from 50 randomly sampled temperature trend onset times in months 12 and 13. Mean detection delay and mean detection stability were computed based on randomized temperature trend onsets times and for 10 trend progression velocities, assuming a linear temperature trend. Fifty randomly selected onset times were used to simulate the bearing temperature increase.

This study demonstrates that multi-target models can describe the temperature behaviour of wind turbine gear bearings in normal operation at least as accurately as corresponding single-target models. In the present work, the multi-target MLP provided a test-set accuracy of 0.49 for predictions of the gear bearing temperature, while the test-set accuracy of the single-target MLP was 0.51 in normalized temperature. As the multi-target MLP produced smaller prediction residuals in many cases, temperature trends became visible earlier and thus the detection delay was shorter. In addition, a paired sample t-test was performed to test the null hypothesis that there are no systematic differences in the detection delays based on the normal behaviour descriptions of the multi-target MLP versus the single-target MLP. The null hypothesis was clearly rejected $(p < 10^{-21})$ in favor of the alternative hypothesis of shorter detection delays based on the multi-target MLP. The test result was confirmed for both alarm criteria.

Similarly, a paired sample t-test was performed to test for systematic differences of the detection stability. This provided a more ambiguous picture. Based on alarm criterion 2, the multi-target MLP enabled a more stable alarm after the first detection $(p$

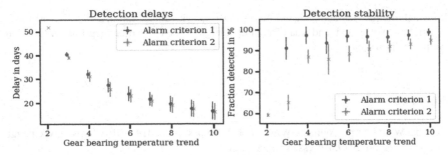

Fig. 4. Fault detection delay and stability as in Fig. 3 but based on the *single-target* normal operation MLP of the gear bearing temperature.

$< 10^{-14}$). However, no systematic difference in the detection stability was found in the case of alarm criterion 1 ($p = 0.22$) Comparing both alarm criteria, it is also found that the second criterion detects faults significantly earlier than the first criterion, but this comes at the cost of significantly reduced detection stabilities (Figs. 3 and 4).

In summary, this study demonstrated that multi-target MLPs can detect faults as fast as and in some cases even earlier than single-target MLPs, and at the same time achieve the same level of detection stability. Comparing Figs. 3 and 4, the detection speed-up observed in this study ranged from several hours to several days. The earlier detection enabled by the proposed multi-target approach can deliver a significant advantage in the planning and performance of maintenance activities. If wind farm operators learn about a developing fault hours or up to several days ahead, they have more time to respond and schedule inspections and adjustment work that may prevent more serious damage and component replacement.

5   Conclusions

This study investigated how multi-target neural networks compare to single-target models with regard to the speed and accuracy of detecting incipient faults in wind turbines from SCADA data. We analyzed the delays in detecting gear bearing faults in onshore wind turbines. Gear bearing faults can result in anomalous temperature increases that are detectable from SCADA data. In this work, synthetic temperature trends were overlaid on the gear bearing temperature SCADA signals in order to facilitate a systematic analysis despite the scarcity of fault observations. We compared the detection delay and the detection stability of the alarm signals among the multi- and single-target models based on different alarm criteria. In this study, we found that multi-target neural networks can meet and even go below the detection delays achieved with single-target MLPs of the turbine normal operation. At the same time, the multi-target MLPs achieved the same level of prediction stability. We demonstrated that the detection of temperature-related faults could be accelerated by up to several days compared to the state-of-the-art fault detection with single-target models. With regard to future studies, we propose to also investigate the potential of multi-target models for normal behaviour modelling and fault detection tasks based on high-frequency data, in particular from vibration measurements in the drive train, which did not form part of this paper.

Acknowledgments. The author thanks Bernhard Brodbeck, Janine Maron, Dimitrios Anagnostos of WinJi AG, Switzerland, and Kaan Duran of Energie Baden-Wuerttemberg EnBW, Germany, for valuable discussions.

References

1. Global Wind Energy Council GWEC: Global Wind Report 2019 (2019). https://gwec.net/glo bal-wind-report-2019. Accessed 5 June 2021
2. International Renewable Energy Agency: Renewable power generation costs in 2019 (2019). https://www.irena.org. Accessed 5 June 2021
3. Faulstich, S., Hahn, B., Tavner, P.: Wind turbine downtime and its importance for offshore deployment (2011). https://doi.org/10.1002/we.421
4. Pfaffel, S., Faulstich, S., Rohrig, K.: Performance and reliability of wind turbines: a review. Energies (2017). https://doi.org/10.3390/en10111904
5. Garcia Marquez, F., Tobias, A., Pinar Perez, J., Papaelias, M.: Condition monitoring of wind turbines: techniques and methods. Renew. Energy **46**, 169–178 (2012). https://doi.org/10. 1016/j.renene.2012.03.003
6. Fischer, K., Coronado, D.: Condition monitoring of wind turbines: state of the art, user experience and recommendations. VGB PowerTech **7**, 51–56 (2015)
7. Stetco, A., et al.: Machine learning methods for wind turbine condition monitoring: a review. Renew. Energy **133**, 620–635 (2019). https://doi.org/10.1016/j.renene.2018.10.047
8. Tautz-Weinert, J., Watson, S.: Using SCADA data for wind turbine condition monitoring - a review. IET Renew. Pow. Gener. **11**(4), 382–394 (2017). https://doi.org/10.1049/iet-rpg.2016. 0248
9. Helbing, G., Ritter, M.: Deep learning for fault detection in wind turbines. Renew. Sustain. Energy Rev. **98**, 189–198 (2018). https://doi.org/10.1016/j.rser.2018.09.012
10. Qiao, W., Lu, D.: A survey on wind turbine condition monitoring and fault diagnosis—part II: signals and signal processing methods. IEEE Trans. Industr. Electron. **62**, 6546–6557 (2015). https://doi.org/10.1109/TIE.2015.2422394
11. Zaher, A., McArthur, S., Infield, D., Patel, Y.: Online wind turbine fault detection through automated SCADA data analysis. Wind Energy **12**, 574–593 (2009). https://doi.org/10.1002/ we.319
12. Kusiak, A., Verma, A.: Analyzing bearing faults in wind turbines: a data-mining approach. Renew. Energy **48**, 110–116 (2012). https://doi.org/10.1016/j.renene.2012.04.020
13. Salameh, J., Cauet, S., Etien, E., Sakout, A., Rambault, L.: Gearbox condition monitoring in wind turbines: a review. Mech. Syst. Sig. Process. **111**, 251–264 (2018). https://doi.org/10. 1016/j.ymssp.2018.03.052
14. McKinnon, C., Turnbull, A., Koukoura, S., Carroll, J., McDonald, A.: Effect of time history on normal behaviour modelling using SCADA data to predict wind turbine failures. Energies (2020). https://doi.org/10.3390/en13184745
15. Meyer, A., Brodbeck, B.: Data-driven performance fault detection in commercial wind turbines. In: Proceedings of the 5th European Conference of the Prognostics and Health Management Society (PHME20) (2020). ISBN 978-1-93-626332-5
16. Liu, Y., Wu, Z., Wang, X.: Research on fault diagnosis of wind turbine based on SCADA data. IEEE Access (2020). https://doi.org/10.1109/ACCESS.2020.3029435
17. Zhang, S., Lang, Z.: SCADA-data-based wind turbine fault detection: a dynamic model sensor method. Control. Eng. Pract. (2020). https://doi.org/10.1016/j.conengprac.2020.104546

18. Liu, X., Du, J., Ye, Z.: A condition monitoring and fault isolation system for wind turbine based on SCADA data. IEEE Trans. Industr. Inf. (2021). https://doi.org/10.1109/TII.2021. 3075239
19. Yang, Y., Liu, A., Wang, J.: Fault early warning of wind turbine gearbox based on multi-input support vector regression and improved ant lion optimization. Wind Energy (2021). https:// doi.org/10.1002/we.2604
20. Astolfi, D., Scappaticci, L., Terzi, L.: Fault diagnosis of wind turbine gearboxes through temperature and vibration data. Int. J. Renew. Energy Res. (2017)
21. Leahy, K., Hu, R., O'Sullivan, D.: Diagnosing and predicting wind turbine faults from SCADA data using support vector machines. Int. J. Prognost. Health Manag. (2018)
22. Zeng, X., Yang, M., Bo, Y.: Gearbox oil temperature anomaly detection for wind turbine based on sparse Bayesian probability estimation. Int. J. Electr. Pow. Energy Syst. (2020)
23. Wang, Y., Infield, D.: Supervisory control and data acquisition data-based non-linear state estimation technique for wind turbine gearbox condition monitoring. IET Renew. Pow. Gener. 7, 350–358 (2013). https://doi.org/10.1049/iet-rpg.2012.0215
24. Wang, L., Zhang, Z., Long, H., Xu, J., Liu, R.: Wind turbine gearbox failure identification with deep neural networks. IEEE Trans. Industr. Inf. 13, 1360–1368 (2017). https://doi.org/ 10.1109/TII.2016.2607179
25. Guo, P., Infield, D.: Wind turbine generator condition monitoring using temperature trend analysis (2012). https://doi.org/10.1109/TSTE.2011.2163430
26. Kusiak, A., Zheng, H., Song, Z.: Online monitoring of power curves. Renew. Energy 34, 1487–1493 (2009). https://doi.org/10.1016/j.renene.2008.10.022
27. Schlechtingen, M., Santos, F., Achiche, S.: Using data-mining approaches for wind turbine power curve monitoring: a comparative study. IEEE Trans. Sustain. Energy 4, 671–679 (2013). https://doi.org/10.1109/TSTE.2013.2241797
28. Caruana, R.: Multitask learning. Mach. Learn. 28, 41–75 (1997). https://doi.org/10.1023/A: 1007379606734
29. Piccart, B.: Algorithms for multi-target learning. Doctoral thesis, KU Leuven (2012)
30. Borchani, H., Varando, G., Bielza, C., Larranaga, P.: A survey on multi-output regression. Wiley Interdiscipl. Rev. Data Min. Knowl. Discov. (2015). https://doi.org/10.1002/widm. 1157
31. Waegeman, W., Dembczyński, K., Hüllermeier, E.: Multi-target prediction: a unifying view on problems and methods. Data Min. Knowl. Disc. 33(2), 293–324 (2018). https://doi.org/ 10.1007/s10618-018-0595-5
32. Xu, D., Shi, Y., Tsang, I., Ong, Y., Gong, C., Shen, X.: Survey on multi-output learning. IEEE Trans. Neural Netw. Learn. Syst. 31, 2409–2429 (2020). https://doi.org/10.1109/TNNLS. 2019.2945133
33. Kocev, D., Dzeroski, S., White, M., Newell, G., Griffioen, P.: Using single- and multi-target regression trees and ensembles to model a compound index of vegetation condition. Ecol. Model. 220(8), 1159–1168 (2009)
34. Segal, M., Xiao, Y.: Multivariate random forests. Wiley Interdiscipl. Rev. Data Min. Knowl. Discov. (2011). https://doi.org/10.1002/widm.12
35. Maron, J., Anagnostos, D., Brodbeck, B., Meyer, A.: Gear bearing fault detection using multi-target neural networks. In: Wind Energy Science Conference WESC 2021. Hanover, Germany, May 2021

Challenges Regarding the Compliance with the General Data Protection Law by Brazilian Organizations: A Survey

Edna Dias Canedo[2]([✉])[iD], Vanessa Coelho Ribeiro[1,3][iD],
Ana Paula de Aguiar Alarcão[1][iD], Lucas Alexandre Carvalho Chaves[1],
Johann Nicholas Reed[1], Fábio Lúcio Lopes Mendonça[1][iD],
and Rafael T. de Sousa Jr[1][iD]

[1] National Science and Technology Institute on Cyber Security, Electrical
Engineering Department (ENE), University of Brasília (UnB), Brasília, DF, Brazil
{vanessa.ribeiro,Johann.reed,fabio.mendonca}@redes.unb.br,
160001986@aluno.unb.br, desousa@unb.br
[2] Department of Computer Science, University of Brasília (UnB), Brasília, DF, Brazil
ednacanedo@unb.br
[3] University Center Uniprojeção, Brasília, DF, Brazil

Abstract. Brazilian organizations must comply with the Brazilian General Data Protection Law (LGPD) and this need must be carried out in harmony with legacy systems and in the new systems developed and used by organizations. In this article we present an overview of the LGPD implementation process by public and private organizations in Brazil. We conducted a literature review and a survey with Information and Communication Technology (ICT) professionals to investigate and understand how organizations are adapting to LGPD. The results show that more than 46% of the organizations have a Data Protection Officer (DPO) and only 54% of the data holders have free access to the duration and form that their data is being treated, being able to consult this information for free and facilitated. However, 59% of the participants stated that the sharing of personal data stored by the organization is carried out only with partners of the organization, in accordance with the LGPD and when strictly necessary and 51% stated that the organization performs the logging of all accesses to the personal data. In addition, 96.7% of organizations have already suffered some sanction / notification from the National Data Protection Agency (ANPD). According to our findings, we can conclude that Brazilian organizations are not yet in full compliance with the LGPD.

Keywords: Brazilian General Data Protection Law – LGPD · LGPD compliance · Personal data processing · Data privacy

1 Introduction

The presence of Information and Communication Technology (ICT) in society brought several benefits, such as access to information in an easier way and

O. Gervasi et al. (Eds.): ICCSA 2021, LNCS 12951, pp. 438–453, 2021.
https://doi.org/10.1007/978-3-030-86970-0_31

the possibility of communicating with people who are geographically distant. However, this strong presence often requires the use of personal data both in services, in the availability of resources and in business/commercial transactions. This process leads to an exponential growth in the use of personal data, making it necessary to review the regulation of how this data is treated for the protection of individuals [1].

Based on this assumption, several laws have emerged with the aim of protecting personal data, such as the General Data Protection Regulation (GDPR) [2] in the European Union, Australian Privacy Principles (APP) in Australia [3] and California Consumer Privacy Act (CCPA) in the United States [4]. With the creation of these laws, the way in which countries manipulate and store information has changed. After the creation of the GDPR, for example, several countries in the world had to modify or even create laws that regulate the protection of personal data in order to comply with the guidelines defined in the European regulation [5].

Brazil was one of the countries that needed to adapt to the GDPR guidelines, since the nations belonging to the European Union make up the country's main economic partners [6,7]. Thus, in September 2020, Law No. 13,709 / 18, called the Brazilian General Data Protection Law (LGPD), came into force [8], [9]. With the LGPD, companies that have personal information stored in their systems must pay attention and follow the guidelines and principles established so that the data is kept safe.

It is possible to observe the movement of many companies to adapt to the LGPD and to create tools that facilitate this implementation in the daily lives of the institutions. An example is the LGPD Educational platform, created by the Data Processing Service - (SERPRO), which offers professional training and certification for the public and private sectors in disciplines related to privacy and protection of personal data [10]. It should be noted that by the end of 2020, surveys pointed out that around 64% of Brazilian companies were not in compliance with the new legislation [11].

Given the above, the objective of this article is to identify the current situation of the implementation of LGPD in Brazil, in public and private organizations, identifying the level of knowledge and compliance with the legislation. The main findings of this article were: more than 75% of the survey participants informed their organization that they treat personal information for legal purposes (according to the permissions granted by legislation); more than 62% stated that the organization controls personal data considered sensitive by the LGPD; more than 46% stated that The organization maintains an inventory of all sensitive information (including personal data) stored, processed or transmitted by the Institution's IT systems, at physical locations and remote service providers and more than 57% stated that the organization has a specific privacy policy related to privacy and data protection.

2   Background

Law No. 13,709 / 2018, called the Brazilian General Data Protection Law (LGPD) [12], of August 14, 2018, is the law that regulates activities involving the processing of personal data, including digital media, with the purpose of protecting the fundamental rights of the citizen to freedom and privacy, as well as the free development of the personality of the natural person, as natural persons are called within the law. Although these fundamental rights are already protected by the Federal Constitution of 1988, it was only after the enactment of the LGPD that the privacy of personal data acquired a similar relevance [13]. It is noteworthy that this regulation did not revoke the previous laws that were used to deal with data privacy (Consumer Protection Code, Brazilian Civil Code and Civil Framework of the Internet), which still remain in force, since they do not conflict with the new law, but are complemented by it, as there were many gaps that needed to be filled around this issue.

Article 1 of the LGPD gives a general definition of what will be covered in the Law, a general scope and an objective. Article 2, on the other hand, deals with the fundamentals of data protection that the Law will be based on, such as respect for privacy, freedom of expression, free initiative and human rights [12].

Article 3 establishes that the law applies to any processing operation carried out by legal or natural personnel, under public or private law, regardless of the means, provided that the data processing has been carried out in national territory, or has the objective of the provision of goods and services, or that data has been collected in Brazil. Whereas Article 4 deals with cases to which the Law does not apply: treatment carried out by natural persons for private, journalistic and artistic purposes, academic or for public security, national defense, State security and investigative activities [12].

Article 5 of the Law establishes the definitions for terms used in the law, such as the definition of titleholder, personal data, sensitive personal data, database, consent, shared use of data and others that can be found in the law [12]. Article 6, on the other hand, defines the principles to be observed in the data processing process as the purpose for this, adequacy, data quality, transparency, accountability and accountability, among others.

Articles 7 and 8 concern the possibility of processing personal data. In order for this process of handling personal information to be legitimate, the legislation lists an exhaustive list of circumstances that allow this treatment. Article 9 of the law deals with the data subject's rights in relation to facilitated access to information on data processing. Article 10, in turn, deals with the duties of the data controller in relation to the legitimate purposes of carrying out the processing of the data.

The LGPD covers any transaction involving the processing of data by individuals or legal entities, under public or private law, that is carried out in Brazilian territory, regardless of where the data is initially located. Personal data is classified as information related to a natural person, such as Individual Taxpayer Registration (CPF), Number of the General Registry (RG), Name and Education Level. The processing of personal data is defined by law as any operation carried out with personal data, including any activity carried out that uses personal information, such as storage, transmission, modification, use and access. In addition, the law requires companies to collect the minimum possible amount of data they need and process it only on legal grounds. The Law also encompasses scenarios where the data are from individuals located or whose data are collected in the national territory. Thus, in order to comply with the LGPD, Brazilian companies must take, among others, the following measures:

1. Analyze and verify which data falls within the scope of the LGPD as this information is processed;
2. Correct incomplete, inaccurate or outdated data;
3. Obtain information about public and private entities with which the controller shared the data;
4. Obtain information about the possibility of the information holder denying consent and the consequences of such denial.

Although the LGPD was based on the GDPR and has many aspects in common, the Brazilian privacy regime has some particularities, which must be taken into account by those who are subject to this law. LGPD does not apply only to companies located in Brazilian territory: In addition to being employed by any company that operates in Brazil or that sells its goods and services there, it also applies to global businesses that are processing information in Brazil or that use personal data of individuals located in Brazil, regardless of the means used for processing and whether this is done directly or indirectly.

The qualification of a legal basis for data processing is one of the main differences between LGPD and GDPR. According to the LGPD, the protection offered is limited to data that qualifies as personal data, offering two levels of protection, depending on whether the data is classified as personal data or sensitive personal data. However, the vast majority of data subject to the application of the LGPD fall within the general definition of personal data, which is considered "information about an identified or identifiable natural person" (Article 5, I) [12]. When compared to the European regulation, this definition applies not only to information directly linked to an individual, but also to any element that can be used to identify or potentially contribute to the identification of a person [13]. Although the GDPR has six legal bases for processing and the data controller company must choose one of them as a justification for using personal information, the LGPD has ten, which are defined in its article 7:

1. Have the consent of the data subject;
2. To aim to fulfill a legal or regulatory obligation of the controller;
3. Execute a contract or preliminary procedures related to a contract to which the data subject is a party, at the request of the data subject;
4. Execute public policies provided for in laws or regulations, or based on contracts, agreements or similar instruments;
5. Conduct studies by research entities that guarantee, whenever possible, the anonymity of personal data;

6. Exercise rights in judicial, administrative or arbitration proceedings;
7. To aim to protect the life or physical security of the data subject or third parties;
8. Protect health, in a procedure carried out by health professionals or by health entities;
9. Meet the legitimate interests of the controller or of third parties, except when the fundamental rights and freedoms of the data subject, which require the protection of personal data, prevail;
10. Protect credit (referring to a credit score).

As for sensitive personal data, defined as information that relates to "racial or ethnic origin, religious belief, political, union or religious opinion, philosophical or political", the LGPD offers even broader protection. One of the fundamentals for the processing of this type of Personal Information is the prevention of fraud. According to this requirement, data processing must be legitimate, legal, non-fraudulent and transparent. They also require data from the controllers in order to guarantee the accuracy, completeness and security of the data and to provide the holders with access and portability rights. In addition, it is also possible for data controllers to report violations and dispense with the transfer of data to countries that do not have adequate protection conditions.

The protection of children's data has also been strengthened after the LGPD regulation. Although the protection of children and adolescents is guaranteed through Law No. 8,069 / 1990 (Statute of Children and Adolescents) and CF/88 itself, with the new personal data protection law, a child protection structure has been implemented. in which it is an absolute priority for all parties involved in the child's development [13]. In order for data to be processed for children and adolescents, controllers must have the consent of the child's legal guardian even before collecting the personal information of children under 16 years of age. In addition, any Information addressed to this audience must be clear and simple.

Both the GDPR and the LGPD determine that the data controller must provide the data holder, when requested, with a copy of all their personal information, which can be delivered electronically or in print within up to thirty days, with the extension of the deadline being permitted. However, the LGPD is somewhat more rigid in the way that information holders must respond to their respective owners. According to Brazilian law, in its article 19, the data controller company has up to fifteen (non-extendable) days to respond to the request of the holder, and must provide a clear and complete statement regarding the information it has. This statement must provide clarification of the source of the data, the lack of registration, the criteria used and the purpose of the treatment. However, the LGPD still allows the parent company to respond to the holder in a simplified format, making it possible to report fewer elements and less details.

Another similarity that can be observed between the LGPD and the GDPR is the requirement to hire a Data Protection Officer (DPO) by the data controllers. This professional should serve as a communication channel between the controllers, the information holders and the national data inspection body, the

National Data Protection Authority (ANPD) – which is responsible for inspecting and imposing sanctions. This obligation is given to oversee compliance and governance, in addition to including training, management and data governance. In addition, the DPO will be responsible for collecting complaints and communications from data subjects and ANPD, guiding employees on good practices and fulfilling other duties determined by the controller or by complementary rules [13].

The LGPD encourages institutions to comply with their requirements by imposing penalties, which may be warnings, suspensions, financial penalties or a ban on the exercise of their activities, for companies that fail to comply with the rules. To impose punishment or for non-compliance, LGPD regulatory and supervisory authorities have administrative powers to impose fines. In cases of more severe penalties, the fine imposed can reach 2% of the company's revenues or a maximum of R$ 50,000,000.00 for an infraction [13].

2.1 Related Works

Ribeiro and Canedo [7] carried out a study on the implementation of LGPD at the University of Brasilia (UnB), in order to minimize the challenges and identify the criteria for the implementation of security of personal data, and the main difficulties faced for the application, under this Law, how to identify all databases that store personal information, check what types of personal data are collected in the University's systems and forms, check for potential vulnerabilities that may cause loss, leakage or impair information access. Through the Analytical Hierarchy Process (AHP) method, the main criteria for the implementation of LGPD at UnB were determined, which are: Data Protection Level, Security Risk, Incident Severity and Data Privacy Risk.

Alencar [14] carried out a study comparing the Brazilian and Chilean Data Protection Laws in relation to the decision criteria on the adequacy of third countries provided for by the General Data Protection Regulation of the European Union (GDPR). Alencar made an analysis of the necessary requirements for adequacy granted by the European Commission, which are: fundamental principles, conducting the processing of sensitive data, automated decisions, supervisory authorities and mechanisms to protect data subjects. Thus, it was concluded that both countries have outdated legislation in relation to the European Regulation. This shows how the interest of Brazil and Chile in fulfilling the requirements is for economic purposes and not in actually protecting the rights of data subjects. The study also showed that the main flaw in Brazilian legislation is the lack of consolidation of a supervisory authority, which in this case would be due to the implementation of the National Data Protection Authority (ANPD). And finally, it was concluded that Chilean law is completely in line with European regulations, while Brazilian law is only partially due to the absence of effective supervisory authority and there is no express right of opposition on the part of the data subject regarding automated decisions.

Cantelle [15] held a debate on the importance of the existence of a legal framework on data protection in Brazil that establishes tools for the exercise of data subjects' rights, minimum standards of security and privacy, legitimate use of data and consent of the holder. The author concludes that Marco Civil is very positive for the future development of the Brazilian General Data Protection Law, as the Brazilian population is in need of legal information about the limits of web users. In addition, it establishes that informing and educating citizens regarding new legislation will be essential.

Machado et al. [16] carried out an analysis based on the increase in security reports on sensitive data leaks and how it has affected companies and governments. It identified the socioeconomic impacts of significant leaks and how the latest data protection laws can help to combat the lack of investment in information security. In this article, the authors argue that recent reports show that 50% of security incidents are caused by malicious employees or former employees, and that a possible solution is to decrease access to data in plain text, using encrypted databases. In the case of Brazil, the LGPD provides for a fine of up to 2% of the annual turnover of legal entities in case of leakage of sensitive data, or up to R$50 million.

Souza et al. [17] presented the development of a tool that automates the service to users regarding LGPD issues. This way, the bot is able to answer questions instantly about information related to the Brazilian General Data Protection Law. In this study, an entire infrastructure using the LUIS service (Language Understanding Intelligent Service) and Natural Language Processing so that the communication between the user and the bot occurs in a more natural way, without seeming to be a conversation with the computer. With this study, a knowledge bank for making inquiries about LGPD and identifying the user's intention is used to provide this service.

Lehfeld et al. [18] studied how the Brazilian General Data Protection Law (LGPD) was applied and effective in protecting consumers and its relationship with the occurrence of cyber-crimes, mainly related to consumer protection in the face of abusive practices by companies with fundamental data. This research showed that, although the Internet is constantly growing, the ways of combating digital crimes are still elementary, with few activities effectively typified as crimes. The study also deals with the vulnerability of cyber space under the optimum of digital consumption and how the Brazilian General Data Protection Law reaffirms the need for individuals to be protected as virtual consumers and owners of their data.

Renner et al. [19] carried out a study in Brazilian organizations regarding the adequacy to LGPD based on a survey carried out with IT workers in each company in the public and private sphere. A survey was applied with the objective of carrying out an assessment and in relation to the adequacy and perception of organizations in relation to the LGPD, collecting the individual perception of the IT employees of each company. From there, the level of compliance that these companies have with the Law can also be analyzed.

The survey respondents consisted of several IT professionals from the public and private sectors from different areas and profiles who are involved in the processing of personal data, such as developers, systems analysts, requirements analysts and project managers. Out of a total of 105 IT professionals, 64% of employees worked in companies with more than 400 employees, while 17.1% work in companies with less than 20 employees. Of the total, 37.1% work for the private sector and the remaining 62.9% work for the public sector. As a result, the authors identified that only 25% of the companies, according to the professionals, adhered to the personal data processing from LGPD, while 11% did not. These and other results showed that the principles of LGPD are still in the initial phase of implementation in companies and a large proportion of correspondents stated that their company had not yet started a plan or initiative to implement LGPD [19]. This study indicated that it is possible to observe the lack of maturity in relation to governance and data management, privacy and information security on the part of most Brazilian organizations and that a possible solution to this lack of maturity will be the establishment of the National Data Protection Agency (ANPD) to monitor the adherence of organizations to the LGPD.

Canedo et al. [20] conducted a survey with ICT professionals from different organizations to gain an insight into how these people understand the LGPD guidelines and how these guidelines are being addressed in the software development process in companies. With the completion of the study, the authors observed that the majority of professionals who responded to the survey are aware of the changes that LGPD would bring to company processes. In addition, these professionals declared that they do not have the necessary knowledge to implement the principles and guidelines of the LGPD, in addition to the organizational environment itself interfering in privacy practices. The authors observed a certain deficiency in the disclosure of information and the promotion of training on LGPD by companies to train their members, as well as in their own privacy policies (which were often unknown to employees). In addition, a literature review was also carried out in order to identify proposed methodologies for obtaining software privacy and privacy requirements. According to the authors, several studies have been found in the literature that propose some techniques for software privacy, but no reports have been found of the use of these methodologies by the industries, being restricted only to theory.

3   Study Settings

The aim of this study is to understand and understand the guidelines of the Brazilian General Data Protection Law and how its principles are being treated by Brazilian organizations. Thus, we conducted a literature review, in order to identify related works related to LGPD. In addition, to investigate and understand how the Law is doing, what aspects were addressed by the organizations that implemented the LGPD and the current scenario in the adequacy of these organizations in relation to the LGPD, we conducted a survey. The questionnaire was prepared considering the principles of the LGPD and the related works identified in the literature that addressed the implementation of the legislation and the implementation guides.

The survey developed has 40 questions, 38 of which are closed and 02 are open, using mainly the Likert scale [21]. The survey was designed along three axes: demographic issues, general issues unrelated to the LGPD, and questions about the principles and compliance with the LGPD. We made the survey available through google forms and the response time of the participants was approximately ten minutes. In addition, the survey was available for 30 days. In total, 61 practitioners responded to the survey.

4   Results

The survey was answered by 61 professionals in the area of Information and Communication Technology (ICT) from public and private Brazilian organizations. Figure 1 shows the distribution of the nature of the organizations that the survey participants work with. 57.4% of participants work in private organizations and 42.6% in public organizations. With this number, we can see that there is a good balance in the nature of the organizations that the participants work with, helping to measure the adequacy of Brazilian companies in general.

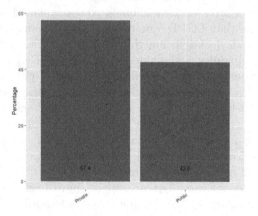

Fig. 1. Nature of the organization

Regarding the role of the ICT practitioner in the organization, 24.6% have the role of responsible for Infrastructure and Networks, 18% are professionals in the area of Information Security, 14.8% are developers, 13.1% are professionals in Database and Business Intelligence, 9.8% work in ICT Management or Governance and 8.2% are Software Engineers. 77% of the participants are male and only 23% are female. 16.4% of the participants stated that they have less than one year of experience, 11.5% stated that they were between 1 and 3 years old, 13.1% stated that they were between 4 and 6 years old, 4.9% between 7 and 9 years old, 6.6% between 10 and 13 years old, 11.5% between 14 and 15 years old and 36.1% stated that they have more than 16 years of experience, as shown in

Fig. 2. With these results we can conclude that most participants have extensive experience in the area of ICT.

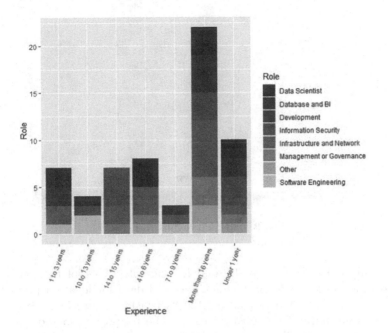

Fig. 2. Role in organization and experience

92% of the participants reported that they are aware of the Brazilian General Data Protection Law (LGPD), as shown in Q5 of Fig. 3. 61% of the participants reported that their organization is in compliance with the LGPD, 16% did not know how to inform, and 23% of the participants informed that their organization is not in conformity with the LGPD, as presented in Q6 of Fig. 3. 56% of the participants informed that the organization has a responsible sector to verify the conformity of its activities with to LGPD, 16% were unable to inform and 28% reported that the organization does not have a responsible sector to verify compliance (Fig. 3 Q7). 46% of the participants reported that the organization has a specific professional who is responsible for data protection (DPO - Data Protection Officer) (Fig. 3 Q8). This finding is worrying because less than 50% of organizations have a DPO and one of the obligations determined by the LGPD is that the organization has a DPO. 54% of respondents stated that organizations provide training and lectures to raise employee awareness and guidance on practices that must be adopted to comply with data protection principles (Fig. 3 Q9). 33% of organizations use a specific methodology or technique to elicit privacy requirements. This is also far below what is expected, as a data privacy policy needs to be defined so that organizations can comply with the LGPD (Fig. 3 Q10).

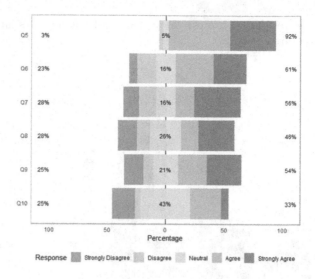

Fig. 3. Question results 5 to 10

Regarding issues related to the LGPD principles, 75% of the participants reported that the organization handles personal data for legal purposes (Fig. 4 Q13), according to the permissions granted by legislation and 79% reported that the organization only uses personal data for the purpose proposed and agreed with the user, limiting their processing to the minimum possible to carry out their activities (Fig. 4 Q14). 69% strongly agree and agree that the process used by the organization is transparent to users, providing them with clear and accurate information about what is being done with their personal data and about the treatment agents (Fig. 4 Q15). 77% of the participants reported that the organization is concerned with the security of the personal data it has control over, adopting effective technical and administrative measures to provide protection against unauthorized access to this information and against accidental or unlawful situations of destruction, loss, alteration or dissemination of personal data (Fig. 4 Q16). 54% stated that the holders have free access to the form and duration that their data is being treated, being able to consult this information for free and facilitated, as shown in 4. Finally, 67% reported that the stored personal data they have accuracy, clarity, relevance and are constantly being updated to fulfill the purpose of their treatment (Fig. 4 Q18).

In matters related to LGPD compliance, 62% of organizations control personal data deemed sensitive by LGPD (Fig. 5 Q19). In 57% of organizations there is a different treatment for sensitive information (Fig. 5 Q20); 46% of organizations maintain an inventory of all sensitive information (including personal data) stored, processed or transmitted by the Institution's IT systems, at physical locations and remote service providers (Fig. 5 Q21). 80% of organizations allow access to personal data only by authorized persons (Fig. 5 Q22). This

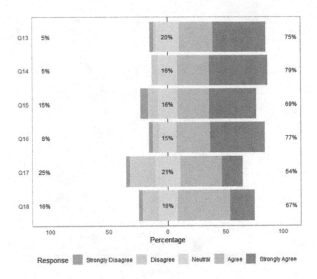

Fig. 4. Principles of LGPD

result allows us to infer that the organizations participating in the survey are in compliance with the principles of the LGPD.

Only 25% of the organizations treat personal data of children and adolescents (Fig. 5 Q23) and only 22% of them carry out the different treatment of the personal data of children and adolescents (Fig. 5 Q24). According to this result, we concluded that more than 70% of the organizations participating in the survey do not deal with data from children and teenagers.

41% of suppliers, contractors and service providers linked to the organizations participating in the survey are in compliance with the LGPD guidelines (Fig. 5 Q25) and share personal data stored only with partners of the organization, in accordance with the LGPD and when strictly necessary (Fig. 5 Q26). 77% of organizations periodically back up data controlled by them (Fig. 5 Q27) and risks related to the security of personal data are identified and documented frequently by 62% of organizations (Fig. 5 Q28). Finally, 51% of organizations log all access to the personal data they maintain, as shown in Q29 in Fig. 5.

49% of organizations regularly review log records in order to identify anomalies or abnormal events (Fig. 6 Q30) and 56% of organizations have an incident response and remediation plan for data leakage situations (Fig. 6 Q31). 31% of participants reported that organizations have performance indicators (Key Performance Indicator - PKI) to check for gaps in data governance and privacy policy (Fig. 6 Q32) and 30% of organizations continually discard unnecessary and excessive data (Fig. 6 Q33). 49% of organizations constantly assess whether there is a legal need to keep personal data, as well as how long it will be kept (Fig. 6 Q34). Regarding physical documents (on paper, such as resumes), 48% of participants reported that the organizations they work with perforate the documents before they are discarded (Fig. 6 Q35). In 57% of organizations there is

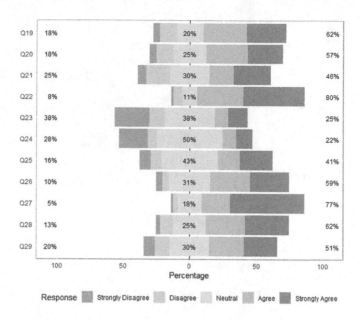

Fig. 5. Compliance LGPD Q19 to Q29

a specific privacy policy related to privacy and data protection (Fig. 6 Q36) and 62% have information security policies accessible to all employees in the organization (Fig. 6 Q37). 46% of organizations use encryption techniques to anonymize controlled data (Fig. 6 Q38) and only 33% of organizations anonymize all user data, as shown in Q39 in Fig. 6. This result is quite worrying, since that one of the principles of LGPD is data anonymization and few organizations carry out this activity.

96.7% of the professionals who participated in the survey, reported that their organizations have already suffered some sanction/notification from the National Data Protection Agency (ANPD), as shown in Fig. 7. This result allows us to conclude that Brazilian organizations are not yet in compliance with the LGPD, since the number of sanctions by the body responsible for controlling the implementation of the LGPD is very high.

In relation to open questions, participants were asked in relation to which methodology and/or technique the organization uses to elicit privacy requirements. Some responses were: "In the process of implementing the LGPD, we use risk management to carry out the controls" and "We use the Trusted Development Guide, developed by the organization itself". Regarding which techniques or methodologies practitioners know and use to elicit privacy requirements in the organization, some of the answers were: Lean Inception, Questionnaires, Personas, Use Cases, Prototyping, Brainstorming and Interviews.

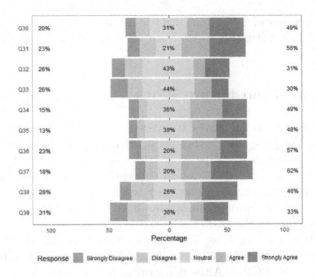

Fig. 6. Compliance LGPD Q30 to Q39

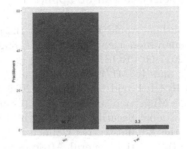

Fig. 7. Organizations that have suffered sanctions/ANPD notifications

5 Threats to Validity

As with any survey that investigates users' perceptions of a given scenario, we have some threats to validity. Regarding the fidelity of the practitioners' responses, we cannot guarantee that they all responded according to the real scenario of the organizations and whether the information really represents what the organization does on a daily basis. To mitigate this threat, we made it clear to all participants that the name of the professional and the organization in which they work would not be disclosed and that the responses would not be disclosed to expose the organization's possible weaknesses.

Another threat is related to the number of responses. In Brazil, we understand that there are a large number of public and private organizations and we understand that 61 participants are a small sample in relation to the number of organizations existing in Brazil. As the participants are from different public and private organizations, we believe that this sample is quite significant, as it allows us to have an overview of the current situation, but of course, we also understand that we cannot generalize this result from the sample that participated in the survey.

6   Conclusions and Future Work

In this article, we conduct a survey to investigate how companies are adapting to the LGPD principles and whether they are in compliance with them. Most of the practitioners who participated in the survey have more than 16 years of experience in ICT. The results reveal that practitioners know the principles of LGPD and that the organizations they work with are adapting to what is required by legislation. Most organizations have a DPO and are training their employees in relation to the LGPD.

According to our findings, we can conclude that Brazilian organizations are adopting sensitive data control mechanisms and are committed to achieving compliance with the LGPD. Although, concerns about the anonymization of user data are not yet a concern of organizations and several organizations have received sanctions/notifications that can be applied by the ANPD.

Acknowledgments. The authors would like to thank the support of the Brazilian research, development and innovation agencies CAPES (grants 23038.007604/2014-69 FORTE and 88887.144009/2017-00 PROBRAL), CNPq (grants 312180/2019-5 PQ-2, BRICS2017-591 LargEWiN, and 465741/2014-2 INCT in Cybersecurity) and FAP-DF (grants 0193.001366/2016 UIoT and 0193.001365/2016 SSDDC), as well as the cooperation projects with the Ministry of the Economy (grants DIPLA 005/2016 and ENAP 083/2016), the Institutional Security Office of the Presidency of the Republic (grant ABIN 002/2017), the Administrative Council for Economic Defense (grant CADE 08700.000047/2019-14), and the General Attorney of the Union (grant AGU 697.935/2019).

References

1. Erickson, A.: Comparative analysis of the eu's gdpr and brazil's lgpd: enforcement challenges with the lgpd. Brook. J. Int. L. **44**, 859 (2018)
2. European Union. General data protection regulation (gdpr). Intersoft Consult. **1**(1), 1–100 (2018). Accessed 24 Oct 2020
3. Australian Government Federal Register of Legislation. Privacy act (1988) https://www.legislation.gov.au/details/c2021c00139, Accessed 4 Apr 2021
4. Code Section Group: California consumer privacy act of 2018. Accessed 14 2021
5. Ribeiro, R.C., Canedo, E.D.: Using mcda for selecting criteria of lgpd compliant personal data security. In: The 21st Annual International Conference on Digital Government Research, dg.o '20, New York, NY, USA, pp. 175–184. Association for Computing Machinery (2020). https://doi.org/10.1145/3396956.3398252
6. Ministério da Economia. Comercio exterior brasileiro, https://www.gov.br/produtividade-e-comercio-exterior/pt-br/assuntos/comercio-exterior/estatisticas, Accessed 30 Mar 2021

7. Ribeiro, R.C., Canedo, E.D.: Using MCDA for selecting criteria of LGPD compliant personal data security. In: Eom, S.J., Lee, J. (eds.) dg.o '20: The 21st Annual International Conference on Digital Government Research, Seoul, Republic of Korea, 15–19 June 2020, pp. 175–184. ACM (2020)
8. Presidência da República. Lei Geral de Proteção de Dados (2018). http://www.planalto.gov.br/ccivil_03/_ato2015-2018/2018/lei/L13709.htm, Accessed 22 Mar 2021
9. Governo Federal. Guia de boas práticas para implementação na administração pública federal (2020). https://www.gov.br/governodigital/pt-br/governanca-de-dados/guia-lgpd.pdf, Accessed 20 Mar 2021
10. Serpro. Governo federal lança plataforma lgpd educacional. Serpro (2021)
11. Serpro. Pesquisa indica que 64% das empresas não estão em conformidade com a lgpd. InfoMoney (2020)
12. Pereira Neto Macedo. Brazilian general data protection law (lgpd). Braz. Natl. 1(1), 1–16 (2018). Accessed 18 Oct 2020
13. Khyara Passos. Compliance with brazil's new data privacy legislation: What u.s. companies need to know. SSRN (2021)
14. Alencar, A.D.S.: ProteÇÃo de dados pessoais no brasil e no chile: Uma anÁlise comparativa sob a perspectiva da decisÃo de adequaÇÃo da comissÃo europeia. Observatório da LGPD, pp. 1 (2020)
15. Cantelle, A.: Marco civil da internet e proteção de dados pessoais 2017. conteúdo Jurídico, p. 1 (2020)
16. Machado, R., Kreutz, D., Paz, G., Rodrigues, G.: Vazamentos de dados: Histórico, impacto socioeconômico e as novas leis de proteçãoo de dados. Anais da Escola Regional de Redes de Computadores (ERRC) (2019)
17. Ferreira De Jesus, A.P., et al.: Robôde conversação baseado em inteligência artificial para treinamento na lei geral de proteção de dados pessoais. Unisanta Science and Technology, pp. 1–10 (2020)
18. de Souza Lehfeld, L., Celiot, A., Siqueira, O.N., Barufi, R.B.: A (hiper)vulnerabilidade do consumidor no ciberespaço e as perspectivas da lgpd. In: Revista Eletrônica Pesquiseduca, p. 10 (2021)
19. Ferrão, S.R., Carvalho, A., Canedo, E.D., Costa, M.P., Cerqueira, A.: Diagnostic of data processing by brazilian organizations-a low compliance issue. Academic Editor Willy Susilo (2021)
20. Canedo, E.D., Calazans, A., Masson, E., Costa, P., Lima, F.: Perceptions of ict practitioners regarding software privacy. Entropy **22**, 429 (2020)
21. Allen, I.E., Seaman, C.A.: Likert scales and data analyses (2007)

A Novel Approach to End-to-End Facial Recognition Framework with Virtual Search Engine ElasticSearch

Dat Nguyen Van[2,3]([✉]), Son Nguyen Trung[1], Anh Pham Thi Hong[1], Thao Thu Hoang[1], and Ta Minh Thanh[1,4]

[1] Research and Development Department, Sun Asterisk, Hanoi, Vietnam
{nguyen.trung.son,pham.thi.hong.anh,hoang.thu.thao}@sun-asterisk.com
[2] VinAI Research, Hanoi, Vietnam
v.datnv21@vinai.io
[3] University of Engineering and Technology, Hanoi, Vietnam
[4] Le Quy Don Technical University, 236 Hoang Quoc Viet, Hanoi, Vietnam
thanhtm@mta.edu.vn

Abstract. Facial recognition has been one of the most intriguing, interesting research topics over years. It involves some specific face-based algorithms such as facial detection, facial alignment, facial representation, and facial recognition as well; however, all of these algorithms are derived from heavy deep learning architectures, which leads to limitations on development, scalability, flawed accuracy, and deployment into publicity with mere CPU servers. It also requires large datasets containing hundreds of thousands of records for training purposes. In this paper, we propose a full pipeline for an effective face recognition application which only uses a small Vietnamese-celebrity datasets and CPU for training that can solve the leakage of data and the need for GPU devices. It is based on a face vector-to-string tokens algorithm then saves face's properties into Elasticsearch for future retrieval, so the problem of online learning in Facial Recognition is also tackled. In comparison with another popular algorithms on the dataset, our proposed pipeline achieves not only higher accuracy, but also faster inference time for real-time face recognition applications.

Keywords: Facial recognition · Visual search engine · End-to-end applications · Online learning · ElasticSearch

1 Introduction

1.1 Overview

Thanks to the rapid development of technologies, facial recognition is increasingly better and evolving. Many companies all over the world are paying great attention to facial recognition technology for their authentication systems

© Springer Nature Switzerland AG 2021
O. Gervasi et al. (Eds.): ICCSA 2021, LNCS 12951, pp. 454–470, 2021.
https://doi.org/10.1007/978-3-030-86970-0_32

instead of using traditional verification methods such as fingerprints or iris. It thus has been developed in the future as well as making it practical in many other fields. More specifically, the workable applications of the facial recognition technology in every aspect of life are diverse [30,39], such as security [29], internet of things and mobile systems [2], real-time identification systems [5], bio-metric systems based on motion detection and facial features [32].

It is no wonder that a full pipeline for the facial recognition applications requires some underlying, complex algorithms consisting of: extracting frontal human faces in given images, called Face Detection [16,24,26,50], optional face alignments for aligning face's positions [47], the necessity of representing faces in the form of numeric vectors [17,43], and distinguishing faces [46] relied on these continuous vectors from previous steps. Critical reviews of such algorithms are listed below:

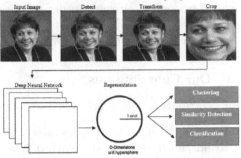

Fig. 1. An overview of full pipeline for face recognition applications

- Face detection, since the 2000s, had a lot of different approaches, but all were either slow or delivered low accuracy, or both. A big innovation came in 2001 when Viola and Jones invented the Haar-based cascade classifier [41], and in 2002, it was continuously improved by Lienhart and Maydt [25]. As a result, the algorithm has become faster and could be run in real-time with 95% accuracy on a difficult dataset. Until 2010, along with the explosion of deep learning [14,38], the research area has reached the state-of-the-art accuracy.
- Attaining facial feature vectors after performing face detection is greatly essential for the recognition step. Algorithms [13,35] are examples to convert these cropped face's images into vectors of specified dimensions that denote the most crucial face's characteristics.
- For identity recognition, well-known approaches based on deep learning architecture like [1,12,35], with the concept of spare representation [46], basing on clustering [36], or Cosine Loss proposed by Hao Wang in the paper [42].

A common full pipeline of facial recognition can be described in Fig. 1.

1.2 Challenging Issues

Each of the algorithms mentioned above has proved its strength and made significant contributions to the research area of face recognition application; however, there are still some issues that need to be worked on. First of all, it can be easily seen that these methods mostly offer only one part amongst the full pipeline of face recognition. The adoption of deep models demonstrate more impressive accuracy than the other approaches, but those models need to be trained on an extremely large, diverse dataset with the size up to hundreds of thousands or millions of images. As a result, it seems to be too slow for face application's development, deployment, management or comes at the expense of high-priced physical devices. The increasing demand for GPU servers for training and deployment has been even more ubiquitous than ever. Another challenge is that the existing online learning problems [33] make these models subject to the requirement of periodic training in order to preserve the system's accuracy whenever new faces are added. Furthermore, a scarcity of body research for the full face recognition pipeline, it need to be researched and widen.

1.3   Our Contributions

Our contributions are summarized as follows:

1. With all these drawbacks discussed above, we have proposed a new approach to an end-to-end facial recognition application in the form of a complete pipeline consisting of: development, deployment, and model version management.
2. In comparison to other well-known methods, our pipeline not only acquires an impressive prediction accuracy with a very challenging dataset that solves the dearth of collection of face data, but it has also gained a very speedy time response for real-time applications.
3. Moreover, the cost for necessary physical devices is reduced exponentially by applying a vector-to-string token algorithm, so that we can train, release the model directly in CPU servers.
4. Last but not least, instead of using a deep learning model for identifying faces, ES is leveraged for storing, creation, retrieval of face identity, thus the online learning problems in face recognition apps are also tackled.

1.4   Roadmap

The rest of the paper is organized as follows. Section 2 presents reviews of related works. Data pre-processing and re-balancing methods are shown in Sect. 3. Section 4 discusses the proposed methods of facial recognition system and experimental results are presented in Sect. 5. Our conclusions and future works are described in Sect. 6.

2   Related Works

2.1   Face Detection

Face detection, which means determining the location and size of a human face in a digital image, is a fundamental step for many face-related technologies. There is a variety of face-based algorithms that take face detection as the foundation to acquire adequate accuracy, such as face verification [13], face recognition [48], face anti-spoofing [22]. In the other works, the purpose of the face detection algorithm is to improve the accuracy of these algorithms by eliciting only the frontal faces as the algorithm's inputs. To date, some adoption of deep learning architectures using the concept of convolutional neural network (CNN) for this step like MTCNN [50], Cascade CNN [24], R-CNN [16], SSD [26], *etc* have achieved remarkable progresses.

2.2 Face Representation

Face representation (in other words, facial-features extraction) is the process of encoding raw facial images into continuous vector representation in high-dimensional feature space [43]. Traditionally, facial features are extracted by manually design patterns as edges, lines, four-rectangle features in Viola Jones algorithm [40] or grids of Histograms of Oriented Gradient (HOG) descriptors [11].

Nowadays, statistical facial-features extraction has been performed automatically and more efficiently (in terms of both time and feature quality [43]), through CNN, a class of deep neural networks [17]. Allowing spatial features preservation, CNNs are suitable for learning feature embedding for 2-D topology data type, including images and facial pictures [17]. Using facial embedding learned by CNNs out-performs most traditional methods in several downstream tasks including Face Recognition, Face Verification [43].

2.3 Principal Component Analysis (PCA)

PCA [15] is a dimensionality reduction technique that was created in 1901 by Karl Pearson. It uses recognition of statistical design to shrink dimensionality and extract features. This approach has been used in various applications since its appearance, such as handwritten recognition, neuroscience, quantitative finance, and image compression.

2.4 FaceNet Architecture in OpenFace

FaceNet [35] was developed in 2015 by Google researchers for the face recognition task. Essentially, a CNN is responsible for extracting features of the face image. The CNN training was performed on large datasets (vggface, MS-Celeb-1M). The key feature of FaceNet is that it uses the Triplet loss function to minimize the distance between similar faces and maximize the distance to dissimilar faces:

$$loss(\mathbf{A}, \mathbf{P}, \mathbf{N}) = max(\|\mathbf{f}(\mathbf{A}) - \mathbf{f}(\mathbf{P})\|^2 - \|\mathbf{f}(\mathbf{A}) - \mathbf{f}(\mathbf{N})\|^2 + \alpha, 0),$$

where $\mathbf{A}$ is anchor input, $\mathbf{P}$ is a positive input, $\mathbf{N}$ is a negative input, α is a margin between positive and negative pairs.

3   Dataset

Our proposed model is implemented on the VN-Celeb[1] dataset that is a collection of Vietnamese celebrities's images collected from google image search. The dataset contains 24.125 images belonging to 1020 famous Vietnamese people (1020 classes). More specifically, the aver-

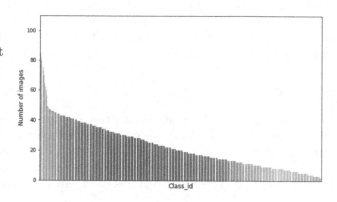

Fig. 2. The number of images in each class

age number of photos in each class is around 23; 7 classes have only 2 shots, and the class with the most images has up to 105. It can be said that the dataset is very challenging both in term of size and of each class' proportion. To elaborate,

- The numbers of images of each class are severely imbalanced.
- The size of dataset is small in comparison to other face recognition dataset.
- Some problems affecting differences recognition such as: lighting condition, posing, and picture quality (Fig. 3).

Fig. 3. Some noise images in the VN-Celeb dataset like black and white, blurred, low resolution, masked.

Firstly, the image's proportion between classes is severely imbalanced. Table 1 illustrates the percentage of the classes classified by the range of the number of photos. We can see that the proportions of classes in two ranges: [2, 5) and [50, 105] are only 5%, 3% respectively; meanwhile, the classes belonging to scope [30, 50) account for 27%. Especially, classes having [5, 30) elements show the highest percentage of 65%. As you can see in Fig. 2, the distribution of images in each class is extremely disproportionate, because the number of photos stretch widely from 2 to 105 per class.

---

[1] https://drive.google.com/drive/folders/1I3KXcGpmm6zpw_y07p-7wIKt5K08iOgc.

Secondly, the size of the VN-Celeb dataset consists of 24.125 images, while the number of classes is 1020. In other words,

Table 1. The image proportion

No shots	[2, 5]	[5, 30]	[30, 50]	[50, 105]
Percentage	5%	65%	27%	3%

this task would fall into the problem of few-shot learning [44] - generalizing from a few training examples.

Another issue is the variation of the dataset images in lighting conditions, pose, and image quality... Many photos show only one part of the face, have the head upside-down, or show no face. Moreover, there are both grayscale and RGB color photos included.

4 Proposed Model

In this section, the proposed pipeline would be described step by step and illustrations will be provided in order to demonstrate the idea in more detail.

As mentioned in the previous section, the dataset is challenging. The size is only 24.1k images containing 1020 classes, which is of much smaller size and much more imbalanced comparing to other datasets such as FaceNet's [35] private dataset with 100M-200M, DeepFace [37] using a private dataset with 4.4M images, OpenFace [2] training on combined dataset CASIA-WebFace [49] and FaceScrub [28], and so on. Therefore, before the dataset can be passed into the Face Embedding model to get representing vectors for training purposes in the encoding phase, data cleaning, re-balancing, and some other data preprocessing techniques must be performed. Our proposed pipeline application is shown in Fig. 4.

4.1 Face Detection

This phase plays an integral role in the accuracy of the whole application, and is the foundation for subsequent steps. After experimenting some tools like Dlib [21], Haar Cascades [3], MTCCN [50], *etc* to extract faces from given images, we decide to take 2 methods for this phase, these methods are interchangeable depending on particular spec. Firstly, we train SSD [26] with based network MobileNet on Labeled Faces in the Wild dataset [23] to take advantage of the fast inference time of MobileNet architecture [20]. Another choice is using a pre-trained library, MTCNN [50]. Although detecting faces with MobileNet-SSD gives better performance, the bounding boxes results are not as good as those of MTCNN counterparts. So, hinging on the purposes, we can switch between two methods.

4.2 Data Pre-processing

The dataset implemented is really challenging due to its size, its properties, and the proportion of the image in each class that needs to be preprocessed. To mitigate adverse effects and enhancing the system's accuracy, we generated more data to classes in which the number of images lower than 20 by applying some non-geometric preprocessing techniques including Blurring, Sharpening, Smoothing, Histogram Equalization, Gamma Correction $\Longrightarrow$ DOG Filtering $\Longrightarrow$ Contrast Equalization [34]. The dataset obtained after preprocessing is passed through the next step to get face representation for training in the encoding phase.

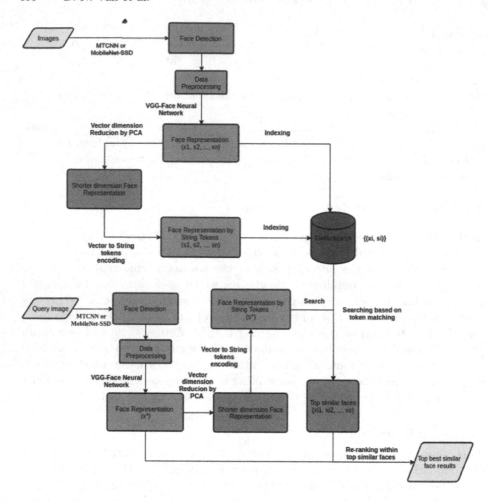

Fig. 4. Workflow of the application

4.3   Face Embeddings with VGG-Face

At this step, the embedding vectors containing the most informative face elements represented as a form of the numeric array is extracted from a DNN. Based on experiments and evaluation on some deep learning architectures like FaceNet [35], VGG-Face [4], ArcFace [13], VGG-Face was chosen because of both its accuracy and performance running on our framework. The based model is ResNet50 [18], all Fully Connected Layers are removed for converting face image into numeric representing vectors that can be used for recognizing purpose.

4.4   Reducing Face Vector Dimensions by PCA

The default dimension output of VGG-Face is 2048. As far as we know, it is very difficult for real-time applications to run on the server without GPU, since the computational cost is very high and may hurt the performance of our whole framework detrimentally. So some data dimension reduction algorithms need to be implemented. An attempt of training and evaluating a list of array dimensions in the range [256, 512, 1024] was conducted through a combination between this phase and the encoding phase to find the best dimension for production deployment (which will be explained in the next subsection). Finally, after assessing both accuracy and performance deliberately, we decided to apply PCA [45] to compress from 2048 to 512 dimensions as the best dimension to represent face data.

4.5   Encoding Face Embeddings to String Tokens

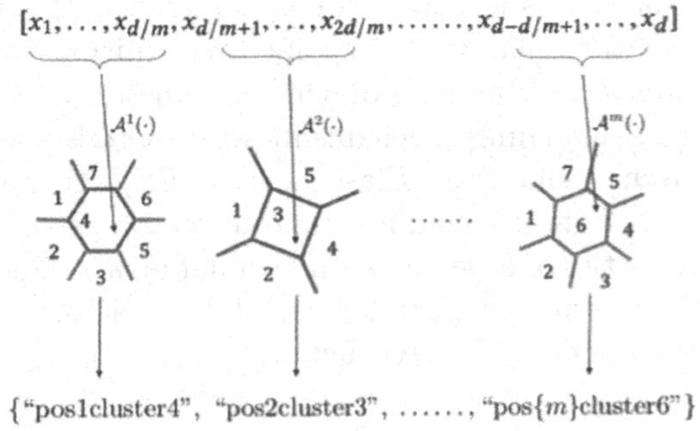

In this section, we explain the reason why we decided to change the data-type for face representation. Actually, it is possible to recognize which name of a person after getting numeric face embedding vectors extracted from VGG-Face by using well-known similarity methods like Euclidean, Cosin, or a deep learning classification model. However, using these methods to calculate similarity within all face vectors of high dimensions to find the best similar faces

$$[x_1, \ldots, x_{d/m}, x_{d/m+1}, \ldots, x_{2d/m}, \ldots, x_{d-d/m+1}, \ldots, x_d]$$

$A^1(\cdot)$ $A^2(\cdot)$ $A^m(\cdot)$

{"pos1cluster4", "pos2cluster3", ..., "pos{m}cluster6"}

Fig. 5.   Illustration   of   the   subvector-wise clustering

is computationally expensive and time-consuming. Besides, another option of using deep learning algorithms for this task is proposed, but it has long inference time and high latency due to the complexity of deep model architectures for recognition, not to mention inefficiency regarding the online learning problems. The problems mentioned above have proved that the application is challenging and difficult to put into practice as well as to apply in real-life. In this paper, we utilize an encoding algorithm from [27] to convert numeric face vectors into collections of string tokens that can be retrieved faster with Elasticsearch, which is beneficial for real-time applications. The encoding algorithm is inspired by the idea of subvector-wise clustering. More specifically, with any numeric face vector $\mathbf{x} \in \mathbb{R}^d$, we divide it into m position as below:

$$[x_1, \ldots, x_{d/m}, x_{d/m+1}, \ldots, x_{2d/m}, \ldots, x_{d-d/m+1}, \ldots, x_d]$$

Considering m position as m subvectors $= \{x^1, x^2, \ldots, x^m\}$, and $P^i := \{x_1^i, x_2^i, \ldots, x_n^i\}$ where $i = \{1, \ldots, n\}$ as the collection of the ith subvectors merged from each position in the whole dataset. We apply k-means algorithm to cluster each P^i into k clusters ($k > 1$). By doing that, we are able to encode the original numeric face vector into string tokens. For example:

$$[\text{"pos1cluster2"}, \text{"pos2cluster\{k\}"}, \ldots, \text{"pos\{m\}cluster1"}]$$

is an example vector encoded from an origin numeric face vector applying the encoding algorithm above with m positions and k clusters. The illustration of the encoding algorithm is in Fig. 5.

4.6   Data Indexing and Searching Method

We provide insights about the data indexing progress and the method for retrieving face identity. We elaborate in two parts: Data Indexing and Searching Method as follows.

Data Indexing. After getting string tokens from the previous encoding phase, we need to store them for face retrieval afterward. Adding to the fact that we are aware that retrieving face recognition from string representation is much faster than from high dimension numeric vectors, in order to achieve an optimal string matching mechanism, we leverage the concept of inverted-index-based search engine from Elasticsearch (ES). To index data into ES, we construct a JSON format including numeric vectors, string tokens together with some other properties of a person such as a name, age, address, image path, and so on. Such JSON format depicts for each face in a total of 1020 classes. The format of the JSON is demonstrated below:

```
1  body = {
2        "user_id": user_id ,
3        "user_name": user_name ,
4        "image_url": image_url ,
5        "embed_vector": embed_vector ,
6        "string_token": string_tokens ,
7        ...
8  }
```

In the end, we have an array of JSON with the length equals to that of the training dataset. Then, we simply use available ES API (Application Programming Interface) functions to index data into Elasticsearch Server.

Searching Method. In this step, we follow the steps of implementation Cun (Matthew) Mu [27] did in his paper, but for face data.

With any query image, we apply the same steps as the training steps explained above to attain a string token $\hat{s}$, then use it for searching. Top r similar faces are obtained relied on overlap between string tokens set $\hat{s}$ and the ones stored in ES server $\{s_1, s_2, ..., s_n\}$

$$i_1, i_2, ..., i_r = \operatorname*{argmax}_{i \in \{1,2,...,n\}} | \hat{s} \cap s_i | \tag{1}$$

ES provides us with RESTful API for searching conveniently, all we need to do is to build a JSON-encoded request body which would instruct the ES server to compute and then return the visual search results.

In the JSON format, we establish 2 prime parts inside for getting sorted results from ES. The first part taking function score query [8] that is responsible for finding top r faces that share the most common in string tokens with $\hat{s}$, then using a custom rescore API function [9] provided by ES to re-sort the collection of top r vectors above. The JSON request body plays an indispensable role in our end-to-end applications.

4.7   Model Serving and Management with Tensorflow

In any AI-related application, model management is necessary in order to operate the system more easily, smoothly, conveniently. At this point, we have had several trained models from Face Detection, Face Embedding to vector-to-string encoding model that needs to served for inference time whenever receiving requests to face recognition server. With these two first deep models, instead of storing the physical file and loading it directly in the server, we convert these models into TensorFlow PB format then using TensorFlow serving API [10] to serve and manage in a separate server. Regarding the last one, the vector-to-string encoding model, Data Version Control [7] is leveraged.

4.8   Django Framework for Development and Deployment

To this point, in order to make our application available to the community by providing open API functions, a framework is necessary. Our application was developed and deployed on the Django Framework [6], one of the most popular framework using Python.

5   Experiments

In this section, to prove the validity of our proposed pipeline, we compare our proposed pipeline to another well-known framework, OpenFace [2] on the VN-celeb dataset. In addition, we also illustrate the impact of the number of positions as well as the number of clusters in the vector-to-string encoding algorithm.

5.1   Experimental Environment

Our proposed pipeline experiments are conducted on a computer with IntelCore i5-4460 CPU @3.2 GHz, 16 GB of RAM, and 256 GB SSD hard disk. The models are implemented with the python 3.6.8 environment.

5.2   Evaluation Method

Accuracy. Accuracy is the most important metric which is used to evaluate the efficiency and generality of almost every model. It describes how well the model performs by providing a ratio between the number of correct predictions and the total elements of testing set. The formula is shown below:

$$accuracy = \frac{\sum_{i=0}^{c-1} \sum_{j=0}^{n_{ci}} E(y_j^*, \hat{y}_j)}{N}, \tag{2}$$

where N is the length of test set, c is the number of classes that need to predict, n_{ci} provide how many items belonging to class ci with $i = \{0, 1, ..., c-1\}$, and $E(y_j^*, \hat{y}_j)$ is a boolean method to compare y_j^* and $\hat{y}_j$ which return "1" if $y_j^* = \hat{y}_j$ and "0" if otherwise.

Recall. Recall is the fraction of the number of positive prediction of classes to it's actual positive. Recall is defined as below:

$$recall = \frac{1}{c} \sum_{i=0}^{c-1} \frac{\sum_{j=0}^{n_{pi}} (E(y_j^*, \hat{y}_j) == 1)}{n_{ci}} \tag{3}$$

As you can see in the recall formula: c is the number of classes that need to predict, n_{pi} illustrates the count of positive prediction, n_{ci} show the actual positive of each class and $E(y_j^*, \hat{y}_j) == 1$ indicates one correct prediction.

5.3 Experimental Analysis

In this section, to prove the efficiency of our proposed pipeline, we have re-implemented and trained OpenFace [2] on the above dataset to compare with our proposed algorithm.

OpenFace with VN-celeb Dataset. In order to allow a fair comparision, we re-implement OpenFace [2] following the same steps as we do with our algorithm. It is noteworthy that the dataset only contains frontal, portrait images of Vietnamese celebrities, so the face detection step for training can be ignored. In the first place, we do data preprocessing as in Sect. 4.2 before feeding into FaceNet's triplet loss [35] to train for a total of 150 epochs based on this empirical experiments. However, since its weights have been trained in 500k images, instead of initializing for the whole deep architecture, we do fine-tuning techniques by resetting weights of some last FaceNet's layers and freezing all the remaining layers, then warming up the model in 30 epochs. After that, we unfreeze and train the whole model in the last 120 epochs.

After training completed, we have a collection of numeric vectors in 128 dimensions from the trained model generating above that characterize the face's properties. As mentioned in [2], the final step is to put these vectors through the Support Vector Machine (SVM) [19] from Scikit-learn [31] as the classifier for the distinction between each individual.

Our Proposed Pipeline. To gain our full pipeline, we follow the steps described in Sect. 4. We also ignore face detection step for the reason mentioned above. In the next step, data preprocessing techniques in Sect. 4.2 is applied, the output then passed through VGG-Face [4] with all fully connected layers eliminated to get face representation vectors of 2048 dimensions as default. It is clear that the dimension of 2048 is too long for a real-time face recognition application, so we decide to apply PCA [45] to reduce the number of dimensions from 2048 to 512 to select the most useful face principal components. At this point, it is to encode numeric face vectors to string tokens, following Sect. 4.5, we divide all face vectors of 512 dimensions in the training dataset into m positions: $P^i := \{x_1^i, x_2^i, ..., x_n^i\}$, where $i = \{1, ..., n\}$, n is the length of the training dataset. Then, we apply separate k-means algorithm from Scikit-learn [31] library to each ith collection vector P^i. The array of trained k-mean's models are saved for future inference.

Using k-mean's models which have been trained to get string tokens, we combine these tokens with some other personal properties such as name, address, phone number, division, nationality, email, numeric face vectors, *etc* to build a JSON object as in Sect. 4.6 for indexing data into ES server.

For inference, to get the individual identity, we just need to build a JSON request body as in Sect. 4.6 and make use of ES's searching API for face retrieval. Finally, the top best 5 similar faces would be returned. The individual identity is the name field of a record with the highest score calculated by ES score function.

Particularly, all deep learning, as well as k-means models, are protected and managed by TensorFlow serving and DVC respectively as we describe in Sect. 4.7. Finally, we develop and deploy our full pipeline application with the Django framework[2].

5.4 Experimental Results and Comparison

In this section, we compare the results of our proposed method to that of Open-Face [2], which we implemented in Sect. 5.3.

According to result's statistic and comparison, it can be concluded that our proposed application bring about many benefits to face recognition applications. Let's see some statistic tables and chart below:

Table 2, Table 3, Table 4 represent statistic tables we built for the purpose of comparing the accuracy, recall and inference time between OpenFace [2] and ours pipeline using numeric face vectors of different dimensions in range of [256, 512, 2048] for vector-to-string tokens algorithms. They also demonstrate the impact of the number of positions, clusters in the encoding algorithm on the evaluation metrics. It is important to note that our accuracy completely outperforms that of the OpenFace counterparts throughout all row records in the three tables with relatively equivalent in two recall columns, some of ours are higher, especially in Table 3 and Table 4. Moreover, by encoding high-dimensional vectors into string tokens, we have the ability to gain the same performance as OpenFace did. The OpenFace lib just takes face vectors of 128 dimensions, meanwhile, we got a comparable inference time with far higher dimensional vectors that obviously comprise more facial features.

---

[2] https://www.djangoproject.com/.

Table 2. Ours evaluation metrics affected by *Npositions* and *Nclusters* with our face vector of *256* dimensions and OpenFace

N positions	N clusters	Accuracy		Recall		Inference time	
		Ours	OpenFace	Ours	OpenFace	Ours	OpenFace
64	19	**90.93**	87.92	**88.40**	87.90	0.078 s	**0.0445 s**
64	20	**90.74**	87.92	**88.33**	87.90	0.097 s	**0.0445 s**
64	21	**91.03**	87.92	87.66	**87.90**	0.106 s	**0.0445 s**
64	22	**91.52**	87.92	87.84	**87.90**	0.083 s	**0.0445 s**
32	19	**90.97**	87.92	86.76	**87.90**	0.064 s	**0.0445 s**
32	20	**91.20**	87.92	86.84	**87.90**	0.063 s	**0.0445 s**
32	21	**91.47**	87.92	86.89	**87.90**	0.06 s	**0.0445 s**
32	22	**91.39**	87.92	87.18	**87.90**	0.055 s	**0.0445 s**
16	19	**92.51**	87.92	81.21	**87.90**	**0.042 s**	0.0445 s
16	20	**92.97**	87.92	81.24	**87.90**	**0.039 s**	0.0445 s
16	21	**91.95**	87.92	81.84	**87.90**	**0.041 s**	0.0445 s
16	22	**92.05**	87.92	81.32	**87.90**	**0.03 s**	0.0445 s

Table 3. Ours evaluation metrics affected by *Npositions* and *Nclusters* with our face vector of *512* dimensions and OpenFace

N positions	N clusters	Accuracy		Recall		Inference time	
		Ours	OpenFace	Ours	OpenFace	Ours	OpenFace
64	19	**90.03**	87.92	**89.67**	87.90	0.147 s	**0.0445 s**
64	20	**90.88**	87.92	**90.07**	87.90	0.087 s	**0.0445 s**
64	21	**91.03**	87.92	**89.01**	87.90	0.078 s	**0.0445 s**
64	22	**90.89**	87.92	**89.81**	87.90	0.095 s	**0.0445 s**
64	23	**91.78**	87.92	**89.48**	87.90	0.095 s	**0.0445 s**
32	19	**92.62**	87.92	84.95	**87.90**	**0.044 s**	0.0445 s
32	20	**92.50**	87.92	83.78	**87.90**	0.052 s	**0.0445 s**
32	21	**92.77**	87.92	85.07	**87.90**	0.054 s	**0.0445 s**
32	22	**92.21**	87.92	84.42	**87.90**	0.067 s	**0.0445 s**
32	23	**92.52**	87.92	84.36	**87.90**	0.055 s	**0.0445 s**
16	19	**93.94**	87.92	79.02	**87.90**	0.05 s	**0.0445 s**
16	20	**93.89**	87.92	79.78	**87.90**	0.049 s	**0.0445 s**
16	21	**93.47**	87.92	80.50	**87.90**	**0.044 s**	0.0445 s
16	22	**93.57**	87.92	80.83	**87.90**	**0.039 s**	0.0445 s
16	23	**93.68**	87.92	80.49	**87.90**	0.046 s	**0.0445 s**

Table 4. Ours evaluation metrics affected by *Npositions* and *N clusters* with our face vector of *2048* dimensions and OpenFace

N positions	N clusters	Accuracy		Recall		Inference time	
		Ours	OpenFace	Ours	OpenFace	Ours	OpenFace
128	32	**93.52**	87.92	**91.59**	87.90	0.175 s	**0.0445 s**
64	32	**93.40**	87.92	**90.70**	87.90	0.131 s	**0.0445 s**
32	32	**94.02**	87.92	**88.56**	87.90	0.1 s	**0.0445 s**

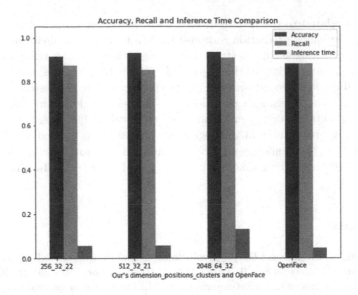

Fig. 6. Accuracy, recall and inference time comparison

More specifically, in Table 4 both our accuracy and recall metrics are far better than the OpenFace, but the time for face retrieval is not as good for real-time application. Besides, in Table 2 and Table 3, we partition it into 3 main parts to demonstrate our performance that includes: 64 positions, 32 positions, 16 positions with a range of [19–22] clusters, an addition cluster of 23 in Table 3. With group of 64 positions, there is every likelihood that ours overshadows the other with greater accuracy and recall, but the search time gets an average of 0.09 s per query. More balanced in the second group, our pipeline acts more efficiently with a significant improvement of accuracy and marginally lower in the other metrics. Last but not least, the last group come faster search time, much more precise and reach a peak of an approximate accuracy of 94%; however, the recall metric seems to be quite modest.

Taking account into Table 2 and Table 3, all in all, the second one definitely showcases more effectively with higher accuracy and recall, but the response time is a little bit slower than the other one. Referring to Fig. 6, we construct a bar chart of accuracy, recall and searching time that chooses the best numbers of positions, clusters for the encoding algorithm of different dimensional face representation vectors.

6   Conclusion and Future Works

In this paper, we have proposed a new approach for an end-to-end facial recognition application with full pipeline for both development and deployment. As shown in evaluation analysis above, our pipeline acquires an impressive prediction accuracy when facing with a very challenging dataset, which help solve the problem related to the dearth of face data. Also, the proposed pipeline has resulted in very quick prediction response time in real-time application. Furthermore, by applying a vector-to-string token algorithm, we can train the model directly in computers without the need of GPU, which means the cost for expensive physical devices needed for training purpose could be reduced.

Finally, instead of using a deep learning model for identifying faces, ES is leveraged for better storing, creation, and retrieval of face identity, thus the online learning problems in face recognition apps are also tackled.

In the future, our tendency research is finding a solution for enhancing accuracy of the vector-to-string tokens algorithm to get even better face recognition results.

References

1. Almabdy, S., Elrefaei, L.: Deep convolutional neural network-based approaches for face recognition. Appl. Sci. **9**, 4397 (2019). https://doi.org/10.3390/app9204397
2. Amos, B., Ludwiczuk, B., Satyanarayanan, M.: OpenFace: a general-purpose face recognition library with mobile applications. Technical report CMU-CS-16-118, CMU School of Computer Science (2016)
3. Bradski, G.: The OpenCV library. Dr. Dobb's J. Softw. Tools (2000)
4. Cao, Q., et al.: VGGFace2: A dataset for recognising faces across pose and age (2018). arXiv: 1710.08092 [cs.CV]
5. Chowdhry, D.A., et al.: Smart security system for sensitive area using face recognition. In: 2013 IEEE Conference on Sustainable Utilization and Development in Engineering and Technology (CSUDET), pp. 11–14 (2013)
6. Django Contributors. Django 3.1 (2020). https://www.djangoproject.com/
7. DVC Contributors. Iterative, DVC: Data Version Control - Git for Data & Models (2020). https://doi.org/10.5281/zenodo.012345
8. Elasticsearch Contributors. Function Score query 6.8 (2019). https://www.elastic.co/guide/en/elasticsearch/reference/6.8/query-dslfunction-score-query.html
9. Elasticsearch Contributors. Rescoring 6.8 (2019). https://www.elastic.co/guide/en/elasticsearch/reference/6.8/search-request-rescore.html
10. Tensorflow Contributors. Tensorflow Serving 6.8 (2019). https://github.com/tensorflow/serving
11. Dalal, N., Triggs, B.: Histograms of oriented gradients for human detection. In: 2005 IEEE Computer Society Conference on Computer Vision and Pattern Recognition (CVPR 2005), vol. 1, pp. 886–893 (2005)
12. Deb, D., Nain, N., Jain, A.K.: longitudinal study of child face recognition (2017). arXiv: 1711.03990 [cs.CV]
13. Deng, J., et al.: ArcFace: additive angular margin loss for deep face recognition (2019). arXiv: 1801.07698 [cs.CV]
14. Deng, J., et al.: RetinaFace: single-stage dense face localisation in the wild (2019). arXiv: 1905.00641 [cs.CV]
15. Karl Pearson, F.R.S.: LIII. On lines and planes of closest fit to systems of points in space. London Edinburgh Dublin Philos. Mag. J. Sci. **2**(11), 559–572 (1901). https://doi.org/10.1080/14786440109462720
16. Girshick, R., et al.: Rich feature hierarchies for accurate object detection and semantic segmentation (2014). arXiv: 1311.2524 [cs.CV]

17. Goodfellow, I., Bengio, Y., Courville, A.: Deep Learning. MIT Press (2016). http://www.deeplearningbook.org
18. He, K., et al.: Deep residual learning for image recognition (2015). arXiv: 1512.03385 [cs.CV]
19. Hearst, M.A., et al.: Support vector machines. IEEE Intell. Syst. Their Appl. **13**(4), 18–28 (1998)
20. Howard, A.G., et al.: MobileNets: efficient convolutional neural networks for mobile vision applications (2017). arXiv: 1704.04861 [cs.CV]
21. King, D.E.: Dlib-Ml: a machine learning toolkit. J. Mach. Learn. Res. **10**, 1755–1758 (2009). ISSN: 1532-4435
22. Komulainen, J., Hadid, A., Pietikainen, M.: Context based face anti-spoofing, pp. 1–8, September 2013. https://doi.org/10.1109/BTAS.2013.6712690
23. Huang, G.B., Learned-Miller, E.: Labeled faces in the wild: updates and new reporting procedures. Technical report UM-CS-2014-003. University of Massachusetts, Amherst, May 2014
24. Li, H., et al.: A convolutional neural network cascade for face detection. In: 2015 IEEE Conference on Computer Vision and Pattern Recognition (CVPR), pp. 5325–5334 (2015)
25. Lienhart, R., Maydt, J.: An extended set of Haar-like features for rapid object detection. In: Proceedings of International Conference on Image Processing, vol. 1, p. I (2002)
26. Liu, W., et al.: SSD: single shot multibox detector. In: Leibe, B., Matas, J., Sebe, N., Welling, M. (eds.) ECCV 2016. LNCS, vol. 9905, pp. 21–37. Springer, Cham (2016). https://doi.org/10.1007/978-3-319-46448-0_2. ISBN 978-3-319-46447-3
27. Mu, C., et al.: Towards practical visual search engine within elasticsearch (2019). arXiv: 1806.08896 [cs.CV]
28. Ng, H.-W., Winkler, S.: A data-driven approach to cleaning large face datasets. In: 2014 IEEE International Conference on Image Processing, ICIP 2014, pp. 343–347, January 2015. https://doi.org/10.1109/ICIP.2014.7025068
29. Owayjan, M., et al.: Face recognition security system, December 2013
30. Parmar, D., Mehta, B.: Face recognition methods & applications. Int. J. Comput. Technol. Appl. **4**, 84–86 (2014)
31. Pedregosa, F., et al.: Scikit-learn: machine learning in Python. J. Mach. Learn. Res. **12**, 2825–2830 (2011)
32. Rima, S., et al.: Smart security surveillance using IoT, pp. 659–663, August 2018. https://doi.org/10.1109/ICRITO.2018.8748703
33. Sahoo, D., et al.: Online deep learning: learning deep neural networks on the fly (2017). arXiv: 1711.03705 [cs.LG]
34. Satish, A., Devarajan, N.: Preprocessing technique for face recognition applications under varying illumination conditions. Glob. J. Comput. Sci. Technol. Graph. Vis. **12**, 13–18 (2012)
35. Schroff, F., Kalenichenko, D., Philbin, J.: FaceNet: a unified embedding for face recognition and clustering. In: 2015 IEEE Conference on Computer Vision and Pattern Recognition (CVPR), June 2015. https://doi.org/10.1109/cvpr.2015.7298682. http://dx.doi.org/10.1109/CVPR.2015
36. Shi, Y., Otto, C., Jain, A.K.: Face clustering: representation and pairwise constraints. IEEE Trans. Inform. Forensics Secur. 13(7), 1626–1640 (2018). https://doi.org/10.1109/tifs.2018.2796999. https://dx.doi.org/10.1109/TIFS.2018
37. Taigman, Y., et al.: DeepFace: closing the gap to human-level performance in face verification, September 2014. https://doi.org/10.1109/CVPR.2014.220

38. Tang, X., et al.: PyramidBox: a context-assisted single shot face detector (2018). arXiv: 1803.07737 [cs.CV]

39. Tolba, A., El-Baz, A., El-Harby, A.: Face recognition: a literature review. Int. J. Signal Process. **2**, 88–103 (2005)

40. Vikram, K., Padmavathi, S.: Facial parts detection using Viola Jones algorithm. In: 2017 4th International Conference on Advanced Computing and Communication Systems (ICACCS), pp. 1–4 (2017)

41. Viola, P., Jones, M.: Rapid object detection using a boosted cascade of simple features, vol. 1, p. I-511, February 2001. ISBN: 0-7695-1272-0. https://doi.org/10.1109/CVPR.2001.990517

42. Wang, H., et al.: CosFace: large margin cosine loss for deep face recognition (2018). arXiv: 1801.09414 [cs.CV]

43. Wang, M., Deng, W.: Deep face recognition: a survey (2018). arXiv: 1804.06655 [cs.CV]

44. Wang, Y., Yao, Q.: Few-shot learning: a survey. CoRR abs/1904.05046 (2019). arXiv: 1904.05046. http://arxiv.org/abs/1904.05046

45. Wold, S., Esbensen, K., Geladi, P.: Principal component analysis. Chemometr. Intell. Lab. Syst. **2**(1), 37–52 (1987). Proceedings of the Multivariate Statistical Workshop for Geologists and Geochemists. ISSN 0169-7439. https://doi.org/10.1016/0169-7439(87)80084-9. http://www.sciencedirect.com/science/article/pii/0169743987800849

46. Wright, J., et al.: Robust face recognition via sparse representation. IEEE Trans. Pattern Anal. Mach. Intell. **31**, 210–227 (2009). https://doi.org/10.1109/TPAMI.2008.79

47. Yang, H., et al.: An empirical study of recent face alignment methods, November 2015. https://doi.org/10.13140/RG.2.1.4603.8484

48. Yang, J., et al.: Nuclear norm based matrix regression with applications to face recognition with occlusion and illumination changes. IEEE Trans. Pattern Anal. Mach. Intell. **39**(1), 156–171 (2017)

49. Yi, D., et al.: Learning face representation from scratch (2014). arXiv: 1411.7923 [cs.CV]

50. Zhang, K., et al.: Joint face detection and alignment using multitask cascaded convolutional networks. IEEE Signal Process. Lett. **23**, 1499–1503 (2016). https://doi.org/10.1109/LSP.2016.2603342

Distributed Novelty Detection
at the Edge for IoT Network Security

Luís Puhl(✉)[iD], Guilherme Weigert Cassales[iD], Helio Crestana Guardia[iD],
and Hermes Senger[iD]

Universidade Federal de São Carlos, São Carlos, Brazil
https://www2.ufscar.br/

Abstract. The ongoing implementation of the Internet of Things (IoT)
is sharply increasing the number and variety of small devices on edge networks. Likewise, the attack opportunities for hostile agents also increases,
requiring more effort from network administrators and strategies to
detect and react to those threats. For a network security system to operate in the context of edge and IoT, it has to comply with processing,
storage, and energy requirements alongside traditional requirements for
stream and network analysis like accuracy and scalability. Using a previously defined architecture (IDSA-IoT), we address the construction and
evaluation of a support mechanism for distributed Network Intrusion
Detection Systems based on the MINAS Data Stream Novelty Detection
algorithm. We discuss the algorithm steps, how it can be deployed in a
distributed environment, the impacts on the accuracy and evaluate performance and scalability using a cluster of constrained devices commonly
found in IoT scenarios. The obtained results show a negligible accuracy
loss in the distributed version but also a small reduction in the execution
time using low profile devices. Although not efficient, the parallel version showed to be viable as the proposed granularity provides equivalent
accuracy and viable response times.

Keywords: Novelty detection · Intrusion detection · Data streams ·
Distributed system · Edge computing · Internet of Things

1 Introduction

The Internet of Things (IoT) brings together a wide variety of devices, including mobile, wearable, consumer electronics, automotive and sensors of various
types. Such devices can either be accessed by users through the Internet or connect to other devices, servers and applications, with little human intervention
or supervision [1,8,13,15]. Security and privacy is a major concern in the IoT,
especially regarding devices having access to user personal data like location,
health and many other sensitive data [12]. Furthermore, if compromised, such

The authors would like to thank Brazilian funding agencies FAPESP and CNPq for
the financial support.

devices can also be used to attack other devices and systems, steal information, cause immediate physical damage or perform various other malicious acts [9]. As an additional concern, IoT devices likely have a long lifespan, less frequent software patches, growing diversity of technologies combined with lack of control over the software and hardware of such devices by the host organization (where they are deployed), which considerably increases the attack surface.

Because most IoT devices have limited resources (i.e., battery, processing, memory and bandwidth), configurable and expensive algorithm-based security techniques are not usual, giving way to network based approaches [17]. Machine Learning (ML) techniques, for instance, have been studied for years to detect attacks from known patterns or to discover new attacks at an early stage [2,11]. A recent survey [15] shows that ML based methods are a promising alternative which can provide potential security tools for the IoT network making them more reliable and accessible than before.

Despite the promising use of ML to secure IoT systems, studies found in the literature [2,11,15] are limited to traditional ML methods that use static models of traffic behavior. Most existing ML solutions for network-based intrusion detection cannot maintain their reliability over time when facing evolving attacks [10,16]. Unlike traditional methods, stream mining algorithms can be applied to intrusion detection with several advantages, such as: *(i)* processing traffic data with a single read; *(ii)* working with limited memory (allowing the implementation in small devices commonly employed in edge services); *(iii)* producing real-time response; and *(iv)* detecting novelty and changes in concepts already learned.

Given the recent [4,10,16] use of Data Stream Novelty Detection (DSND) in network data streams, this paper shows the effects of adapting these mechanisms to edge services for use in IoT environments. Our proposal, called *MFOG*, adapted the IDSA-IoT architecture [3] using the DSND algorithm MINAS [5,7], making it suitable to run on a distributed system composed of small devices with limited resources on the edge of the network. Using our newer version of the MINAS algorithm, we have experimentally evaluated how the distribution affects the capability to detect changes (novelty) in traffic patterns and its impact on the computational efficiency. Finally, some distribution strategies and policies for the data stream novelty detection system are discussed.

This paper is organized as follows: Sect. 2 reviews the chosen DSND algorithm MINAS. A distributed extension of MINAS, including its implementation and evaluation are presented in Sect. 3 and in Sect. 4 we show how we evaluated *MFOG* and the discuss results we found. Finally, Sect. 5 summarizes the main findings and presents possible future work.

2   MINAS

MINAS [5,7] is an offline-online DSND algorithm, meaning it has two distinct phases. The first phase (offline) creates an initial model set with several clusters based on a clustering algorithm with a labeled training set. Each cluster can be associated with only one class of the problem, but each class can have many clusters.

During its online phase, which is the main focus of our work, MINAS performs three tasks in (near) real-time over a potentially infinite data stream: classification, novelty detection and model update, as shown in Algorithm 1.

MINAS attempts to classify each incoming unlabeled instance according to the current decision model. Instances not explained by the current model receive the *"unknown"* label and are stored in the unknowns-buffer. When the unknowns-buffer size reaches a preset threshold, MINAS executes the Novelty Detection function. After a set interval, samples in the unknowns-buffer are considered to be noise or outliers and removed. The algorithm also has a mechanism to forget clusters that became obsolete and unrepresentative of the current data stream distribution, removing them from the Model and storing in a Sleep Model for recurring patterns detection [7].

Input: ModelSet, inputStream
Output: outputStream
Parameters: cleaningWindow, noveltyDetectionTrigger
1 **Function** MinasOnline(*Model, inputStream*):
2 | UnknownsBuffer $\leftarrow \emptyset$; SleepModel $\leftarrow \emptyset$;
3 | lastCleanup $\leftarrow 0$; noveltyIndex $\leftarrow 0$;
4 | **foreach** *sample$_i$* $\in$ *inputStream* **do**
5 | | nearest $\leftarrow$ nearestCluster (sample, Model);
6 | | **if** *nearest.distance* $\leq$ *nearest.cluster.radius* **then**
7 | | | sample.label $\leftarrow$ nearest.cluster.label;
8 | | | nearest.cluster.lastUsed $\leftarrow i$;
9 | | **else**
10 | | | sample.label $\leftarrow$ unknown;
11 | | | UnknownsBuffer $\leftarrow$ UnknownsBuffer $\cup$ sample;
12 | | | **if** | *UnknownsBuffer* | $\geq$ noveltyDetectionTrigger **then**
13 | | | | novelties $\leftarrow$ NoveltyDetection (Model $\cup$ SleepModel, *UnknownsBuffer);
14 | | | | Model $\leftarrow$ Model $\cup$ novelties;
15 | | | **if** $i >$ (*lastCleanup* + cleaningWindow) **then**
16 | | | | Model $\leftarrow$ moveToSleep (Model, *SleepModel, lastCleanup);
17 | | | | UnknownsBuffer $\leftarrow$ removeOldSamples (UnknownsBuffer, lastCleanup);
18 | | | | lastCleanup $\leftarrow i$;
19 | | outputStream.append(sample);

Algorithm 1: Our interpretation of MINAS [7].

The Novelty Detection function, illustrated in Algorithm 2, groups the instances to form new clusters, and each new cluster is validated to discard the non-cohesive or unrepresentative ones. Valid clusters are analyzed to decide if they represent an extension of a known pattern or a completely new pattern. In both cases, the model absorbs the valid clusters and starts using them to classify new instances.

Parameters: minExamplesPerCluster, noveltyFactor

```
1  Function NoveltyDetection(Model, Unknowns):
2  |  newModelSet ← ∅;
3  |  foreach new in clustering (Unknowns) do
4  |  |  if ( |new.sampleSet| ≥ minExamplesPerCluster ) ∧ (new.silhouette > 0) then
5  |  |  |  nearest ← nearestCluster (new, Model);
6  |  |  |  if nearest.distance < (nearest.cluster.radius × noveltyFactor) then
7  |  |  |  |  new.label ← nearest.cluster.label;
8  |  |  |  |  new.type ← "extension";
9  |  |  |  else
10 |  |  |  |  new.label ← noveltyIndex;
11 |  |  |  |  noveltyIndex ← noveltyIndex +1;
12 |  |  |  |  new.type ← "novelty";
13 |  |  |  Unknowns ← Unknowns − new.sampleSet;
14 |  |  |  newModelSet ← newModelSet ∪ new;
15 |  return newModelSet;
```

Algorithm 2: MINAS [7] Novelty Detection task.

3 Proposal

In this work we investigate an appropriate architecture for performing DSND at the edge, allowing small IoT devices to detect undesirable network behavior. To that end, we propose and evaluate *MFOG*, a distributed DSND system following the IDSA-IoT architecture [3] and based on a distributed version of the algorithm MINAS [7]. Our approach explores distributed computing and a trivial load balancer to enable low profile devices to classify and detect unwanted traffic in a scalable, edge focused way.

However, given the distributed nature and the typical use of small computing devices in IoT scenarios as well as the need handle network speeds, new challenges arise: *(i)* the classification phase of the algorithm must occur in parallel at different nodes; *(ii)* the novelty detection phase, which provides the model evolution, must also be asynchronous; *(iii)* the algorithm complexity (time and space) must allow it to be processed by modest computing devices (i.e., small memory and low processor performance).

NIDS monitor network traffic, and analyze the characteristics of each flow to identify any intrusion or misbehavior. However, this problem requires both fast and accurate response [4]: fast response is needed to have a proper reaction before harm can be cast to the network and to cope with the traffic without imposing loss or delay in the NIDS or observed network; accurate response is required as not to misidentify, especially the case of false positive that leads to false alarms. To achieve those goals, we leverage fog computing.

In common IoT scenarios, data is captured by small devices and sent to the cloud for any compute or storage tasks, but this is not feasible in a NIDS scenario. Fog computing infrastructure aims to offload processing from the cloud providers by placing edge devices closer to end-users and/or data sources.

In our proposal, fog and cloud computing resources are combined to minimize the time elapsed between a flow descriptor ingestion and intrusion alarm, performing the classification step of MINAS running multiple classifier instances. After the initial classification, the resulting label can be used immediately, but if the sample is labeled as *unknown*, this sample must be stored and the novelty detection step will be triggered.

To have a better overview of our proposal and how it integrates with existing IoT environments, Fig. 1 depicts such scenario showing from bottom to top: IoT devices directly connected to a (local) gateway network; this gateway network could be as simple as a single Internet router or be more complex by connecting to private clouds or containing more devices providing fog computing capabilities; lastly, available over the internet, the traditional public cloud provides inexpensive computing and storage on demand. In this scenario, the further apart resources are, the more network resources need to be employed, and, as with any networked system, the higher is the latency.

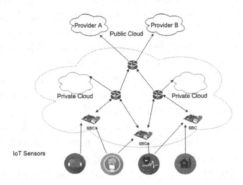

Fig. 1. IDSA-IoT [3] physical architecture and deployment scenario overview.

The overall *MFOG* architecture has two main modules, Classification and Novelty Detection, which implement the MINAS main tasks. The Classification Module performs the same task of the MINAS Online phase and is the focal point for parallelism and distribution in our proposal. It is replicated in the fog and runs on each cluster node, using a configurable number of threads (limited to the node CPU core count).

The Novelty Detection Module can also be replicated, the choice being one instance per local network, one global cloud instance, or both. This module also handles the homonymous task of MINAS Online phase, receiving all the samples labeled with *unknown*, storing them in an internal *unknown-buffer*, and, when this buffer is full, performing the MINAS Novelty Detection task (clustering followed by validation).

3.1   Policies

The design of our distributed DSND architecture includes partitioning the functionalities of MINAS and establishing the appropriate data flows between different actors. Changes to placement and behavior can have different impacts and should be chosen with care. The decisions following these discussions can be organized in several policies, some of them were recurring during our implementation discussions and are:

- Regarding the allocation of the Novelty Detection Module:
  - At each fog node: patterns will be only detected if sufficient samples of them occur in the local observed network, use of the local node processing power, and a model synchronization mechanism between networks must be added;
  - In the cloud: detect patterns even when scattered on each local network, each sample with *unknown* label must be sent from edge to cloud implying increased internet link usage and increased delay between the appearance of a pattern, its detection and propagation to fog classifiers;
  - On both: local *unknown* buffer is maintained and novelty detection is local as well, once a sample is considered as noise or outlier it shall be sent to the cloud where the process repeats but with global data. This choice needs an even more complex model synchronization mechanism.
- Regarding the model cleanup (forget mechanism): Even when a global novelty detection is used, local models can be optimized for faster classification using the local model statistics by sorting by (or removing) least used clusters;
- Lastly, reclassification of *unknowns*: In the novelty detection task in MINAS, the *unknown-buffer* is effectively classified using the new set of clusters. In Algorithm 2 line 13, the new valid cluster (novelty or extension) includes the set of samples composing that cluster, thus, if this new label assignment was put forth to the system output it would introduce delayed outputs, more recent and perhaps more accurate. Also, it would change the system data stream behavior from a *map* (meaning each input has one output) to a *flatMap* (each input can have many outputs).

3.2   Implementation

The original MINAS algorithm has a companion implementation[1] (*Ref*) written in Java using MOA library base algorithms such as K-means and CluStream, but our implementation only used K-means. Another difference between *Ref* and

---

[1] Available at http://www.facom.ufu.br/~elaine/MINAS.

$MFOG$ is the definition of cluster radius derived from the distances of elements forming the cluster and the cluster's center. Ref uses the maximum distance while $MFOG$ uses the standard deviation of all distances as described in [7].

The stream formats for input and output are also of note. As input, the algorithm takes samples ($\vec{v}$), which are a sequence of numbers with dimension d. In addition to $\vec{v}$, for both training and evaluation, the class identifier is provided as a single character, along with a unique item identifier (uid), which can otherwise be determined from the sample index in the input stream.

As its output, the algorithm returns the original sample $\vec{v}$ followed by the assigned label. Adjustments can easily be made to provide the output results as a tuple containing uid and the assigned label.

For evaluation purposes, an $MFOG$ implementation[2] was made using MPI (*Open MPI 4.0.4*). The program is organized in a single program multiple data (SPMD) programming model, so a single version of the $MFOG$ program was initiated on all nodes, being that one of them would perform the root role, while the others ran as leaves, the program entry point is illustrated on Algorithm 3 and the overall sequence of interactions is shown in Fig. 2.

Each leaf node runs a model adjustment thread and multiple (up to the number of cores) classifier threads. The leaf tasks are illustrated in Algorithm 4.

On the root process, illustrated in Algorithm 5, a sampler thread is responsible for distributing the sampled flow information ($\vec{v}$) to the classifier nodes, using a round-robin load balancing scheme. The other thread on the root process is responsible for receiving the classification results and for processing the unknown samples in the search for novelties.

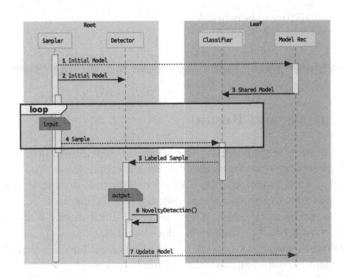

Fig. 2. $MFOG$ life line overview.

[2] Available at https://github.com/luis-puhl/minas-flink.

Parameters: mpiRank
Input: inputStream
Output: outputStream

```
1  Function Mfog(inputStream, outputStream):
2  |  Model ← ∅; ModelLock ← new Lock ();
3  |  if mpiRank = 0 then root
4  |  |  new Thread (Detector, [outputStream, Model, ModelLock]);
5  |  |  Sampler (inputStream, Model, ModelLock);
6  |  else leaf
7  |  |  new Thread (modelReceiver, [Model, ModelLock]);
8  |  |  Classifier (Model, ModelLock);
```

Algorithm 3: *MFOG*: main MPI entry-point.

```
1   Function Classifier(Model, ModelLock):
2   |  while True do
3   |  |  sampe ← receive (SampleType, root);
4   |  |  if sample = EndOfStream then break;
5   |  |  sample.label ← "unknown";
6   |  |  with readLock (ModelLock)
7   |  |  |  nearest ← nearestCluster (sample, Model);
8   |  |  if nearest.distance ≤ nearest.cluster.radius then
9   |  |  |  sample.label ← nearest.cluster.label;
10  |  |  send (root, SampleType, sample);
11  Function modelReceiver(Model, ModelLock):
12  |  while True do
13  |  |  cl ← receive (ClusterType, root);
14  |  |  if cl = EndOfStream then break;
15  |  |  with writeLock(ModelLock)
16  |  |  |  Model ← Model ∪ cl;
```

Algorithm 4: *MFOG* Leaf Tasks: Model Receiver and Classifier.

4   Experiments and Results

Aiming to evaluate our proposal for the effects of distributed novelty detection in a IoT NIDS scenario, we implemented an experimental setup, composed of three Raspberry Pi 3 model B single board computers connected via Ethernet Switch. The idea was to create a simple cluster simulating an IoT network with constrained resources at the edge of the network. This cluster stored all source code, binaries (compiled and linked in place) and data sets. In our setup, the data set is stored in the root's node SD card and is read for each experiment. All experiments were executed in this cluster for isolation of otherwise unforeseen variations and for safe software comparison with constant hardware.

The data set used is the December 2015 segment of Kyoto 2006+ data set[3] (Traffic Data from Kyoto University's Honeypots) [14] containing 7 865 245

---

[3] Available at http://www.takakura.com/Kyoto_data/.

```
    Parameters: mpiSize
 1  Function Sampler(inputStream, Model, ModelLock):
 2  │  dest ← 1;
 3  │  foreach sample ∈ inputStream do
 4  │  │  if typeOf (sample) is Cluster then
 5  │  │  │  broadcast (ClusterType, sample, root);
 6  │  │  │  with writeLock (ModelLock)
 7  │  │  │  │ Model ← Model ∪ sample;
 8  │  │  │  continue;
 9  │  │  send (dest, SampleType, sample);
10  │  │  dest ← dest +1;
11  │  │  if dest > mpiSize then dest ← 1;
    Parameters: cleaningWindow, noveltyDetectionTrigger
12  Function Detector(outputStream, Model, ModelLock):
13  │  UnknownSet ← ∅; lastCleanup ← 0;
14  │  while True do
15  │  │  sample ← receive (SampleType, any);
16  │  │  if sample = EndOfStream then break;
17  │  │  outputStream.append(sample);
18  │  │  if sample.label = "unknown" then
19  │  │  │  UnknownSet ← UnknownSet ∪ sample;
20  │  │  │  if | UnknownSet | ≥ noveltyDetectionTrigger then
21  │  │  │  │  novelties ← NoveltyDetection (Model, *UnknownSet);
22  │  │  │  │  with writeLock (ModelLock)
23  │  │  │  │  │ Model ← Model ∪ novelties;
24  │  │  │  │  foreach cluster ∈ novelties do
25  │  │  │  │  │ broadcast (ClusterType, cluster, root);
26  │  │  │  if  sample.uid > ( lastCleanup + cleaningWindow) then
27  │  │  │  │  UnknownSet ← removeOldSamples (UnknownSet, lastCleanup);
28  │  │  │  │  lastCleanup ← sample.uid;
```

Algorithm 5: *MFOG* Root Tasks: Sampler and Detector.

samples. From the original data set, we filtered only samples associated with normal traffic or known attack types identified by existing NIDS, and attack types with more than 10 000 samples for significance, as previously done by [3]. The remaining samples then were normalized so each feature value space (e.g., IP Address, Duration, Service) is translated to the Real interval $[0, 1]$.

The resulting derived data set is then stored in two sets, training set and test set, using the holdout technique. However, for the training set we filter in only normal class resulting in 72 000 instances. For the test set we use 653 457 instances with 206 278 instances with "N" (normal) class and 447 179 instances with "A" (attack) class. Note that this choice results in overfitting for the normal class and, under-fitting for the attack class as the system first needs to detect a novel class and then add it to the model.

4.1 Measurements and Visualizations

We have used two types of evaluation measurements for each experiment: a measure of the full experiment execution time and, a set of qualitative measurements extracted by a Python script.

Our evaluation script was build following reference techniques like multi-class confusion matrix with label-class association [7] to extract classification quality measurements. This script takes two inputs, the test data set and the captured output stream, and gives as outputs the confusion matrix, label-class association, final quality summary with: *Hits* (true positive), *Misses* (Err), *Unknowns* (UnkR); and stream visualization chart with per example instance summary with novelty label markers.

In the confusion matrix $M = m_{ij} \in \mathbb{N}^{c \times l}$, computed by our evaluation script, each row denotes the actual class c and each column denotes the predicted label l present in the captured output stream. Thus, each cell $M_{c,l}$ contains the count of examples from the test data set of class c, found in the output stream with the label l assigned by the experiment under evaluation.

For the data set under use, original classes are $c \in \{N, A\}$, and for the labels we have the training class *"N"*, *unknown* label *"-"* and the novelties $i \in \mathbb{N}$.

Added to the original confusion matrix M are the rows *Assigned* and *Hits*. *Assigned* row represents which original class c (or if *unknown*, *"-"*) the label l is assigned to, this is computed by using the original class if $c = l$ or by associated novelty label to original class as described in [6] Sect. 4.1 (class from where the most samples came from). *Hits* row shows the true positive count for each label l with assigned class c, being the same value as cell $M_{c,l}$. The *Hits* row is also used to compute the overall true positive in the summary table and stream visualization chart. One complete matrix is shown in Table 1a.

For the measurements summary table, six measurements from two sources are displayed. Three measures *Hits*, *Unknowns* and *Misses* represented as ratio of the captured output stream, extracted from the evaluation python program, computed as follows: *Hits* (true positive rate) is the sum of the *Hits* row in the extended confusion matrix; *Unknowns* is the count of examples in the captured output stream marked with the *unknown* label (*"-"*); *Misses* is the count of all examples in the captured output stream marked with a label distinct from the *Assigned* original class and are not marked as unknown.

Furthermore in the measurement summary table, *Time*, *System* and *Elapsed* represented in seconds, are extracted from *GNU Time 1.9*. *Time* is the amount of CPU seconds expended in user-mode (indicates time used doing CPU intensive computing, e.g., math). *System* is the amount of CPU seconds expended in kernel-mode (for our case, it indicates time doing input or output). *Elapsed* is the real-world (wall clock) elapsed time and indicates how long the program took to complete. Our four main experiments are shown in Table 2.

Lastly, the stream visualization chart shows the summary quality measurement (*Hits, Unknowns, Misses*) computed for each example in the captured output stream. The Horizontal axis (x, domain) plots the index of the example

Table 1. Confusion matrices and qualitative measurements

(a) Reference implementation

Labels	-	N	1	2	3	4	5	6	7	8	9	10	11	12
Classes														
A	3774	438750	123	145	368	8	52	165	1	1046	161	2489	71	26
N	8206	193030	0	79	44	0	0	0	229	181	154	4066	289	0
Assigned	-	N	A	A	A	A	A	A	N	A	A	N	N	A
Hits	0	193030	123	145	368	8	52	165	229	1046	161	4066	289	26

(b) Sequential implementation

Labels	-	N	0	1	2	4	5	6	7	8	10
Classes											
A	16086	429765	94	995	104	0	23	3	29	46	34
N	12481	193642	3	94	0	47	0	0	0	11	0
Assigned	-	N	A	A	A	N	A	A	A	A	A
Hits	0	193642	94	995	104	47	23	3	29	46	34

(c) Parallel single-node

Labels	-	N	0	1	2	3	4
Classes							
A	12282	433797	147	952	0	0	1
N	3088	203019	40	99	27	5	0
Assigned	-	N	A	A	N	N	A
Hits	0	203019	147	952	27	5	1

(d) Parallel multi-node

Labels	-	N	0	1	2	3	4
Classes							
A	12378	433631	117	886	0	162	5
N	3121	202916	40	96	105	0	0
Assigned	-	N	A	A	N	A	A
Hits	0	202916	117	886	105	162	5

and the vertical axis (y, image) shows the measurement computed until that example index on the captured output stream.

Adding to the stream visualization chart, novelty label markers are represented as vertical lines indicating *when* in the captured output stream a new label first appeared. Some of the novelty label markers include the label itself ($l \in \mathbb{N}$) for reference (showing every label would turn this feature unreadable due to overlapping). Figure 3 shows complete stream visualization charts.

4.2   Discussion

Four main experiments are presented for discussion: (a) reference implementation of MINAS (Ref) [7]; (b) new implementation in sequential mode; (c) new implementation in single-node, multi-process mode (1×4) and (d) new implementation in multi-node, multi-process mode (3×4). Each experiment uses the adequate binary executable, initial model (or training set for Ref) and test set to compute a resulting output stream which is stored for qualitative evaluation. The summary of all four experiments is shown in Table 2.

The comparison of the first two experiments (a and b) provides a validation for our implementation, while the latter three (b, c and d) serve as showcase for the effects of distribution.

As stated, to validate our implementation we have compared it to Ref (the original MINAS companion implementation), so we extracted the same measurements using same process for both a and b, which can be viewed in Tables 1a,

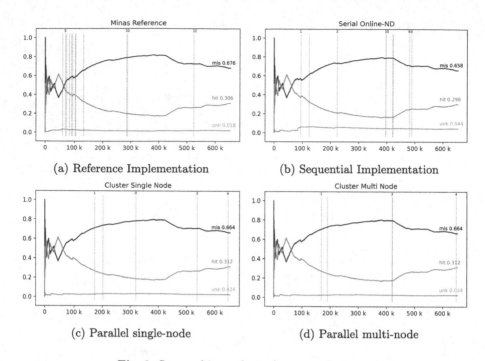

(a) Reference Implementation   (b) Sequential Implementation

(c) Parallel single-node   (d) Parallel multi-node

Fig. 3. Stream hits and novelties visualization.

1b and for ease of comparison in Table 2 the summary can be compared side by side.

In general, the observed classification quality measurements are very similar, and only diverge slightly where *a* has more *Hits* and *Misses* whereas *b* shifted those to *Unknowns*. This phenomenon was watched very closely during development and we found that it was due to small changes to MINAS parameters, MINAS internals like K-means ordering, cluster edge inclusion and cluster radius formula as stated in Subsect. 3.2.

As for the time measurements in Table 2 our implementation used less time to analyze the test data set. This is mostly due to the stop condition on the internal K-means algorithm; while *Ref* uses a fixed iteration limit of 100, our implementations adds the "no improvement" check and stops earlier in most cases, which in turn reduces the time taken on the *NoveltyDetection* function. There are also small optimizations on the *nearestCluster* function (minimal distance from sample to cluster center in the set) affecting the *classifier* task and *NoveltyDetection* function. One can also note that *Ref* time in *a* includes the Offline phase while our implementation runs it once and reuses the initial model for *b*, *c* and *d*. In the table the offline time this is shown as a separate column.

As for the effects of running the classification processes on the small devices as MPI nodes with our implementation, we observe an increase of time when we go from 1 to 4 instances in a single node (*b* and *c* respectively), hinting that our choice of load distribution is not as effective as we expected. Further experiments

Table 2. Collected measures summary.

Experiment metric	Ref (a)	Offline	Sequential (b)	Single node (c)	Multi node (d)
unk	11980		28567	15370	15499
	0.018333		0.043717	0.023521	0.023718
hit	199708		195017	204151	204191
	0.305618		0.298438	0.312416	0.312478
err	441769		429873	433936	433767
	0.676049		0.657843	0.664061	0.663802
Time (s)	2761.83	194.12	80.79	522.10	207.14
System (s)	7.15	0.075	11.51	47.77	157.61
Elapsed (s)	2772.07	194.27	93.03	145.04	95.38
Latency (s)	$4.24 \cdot 10^{-3}$		$1.42 \cdot 10^{-4}$	$2.22 \cdot 10^{-4}$	$1.46 \cdot 10^{-4}$
Processors	1	1	1	4	12
Speedup				0.6414092	0.9753617
Efficiency				0.1603523	0.0812801

were conducted with the number of instances varying from 1 (sequential) to 12 (3 nodes with 4 CPUs each), but that caused no impact on the true positive rate (*Hits*) and elapsed time. More detailed time measurements can be seen in Fig. 4, where we observe near constant time for *elapsed* (near 100 s), the *system* increases gradually while *user* decreases at the same rate. We interpret this behavior as a display of potential for gains using a better load balancing than our choice of round-robin such as micro-batching for better *compute-to-communication ratio* (CCR). In general, Fig. 4 shows no speedup but also no penalty for scaling to more than 4 instances.

Nevertheless, we can also show the effects of delay in the Classify, Novelty Detection, Model Update and Classify feedback loop. Comparing *b* and *c* we observe a reduction in Novelty labels on the Confusion Matrix (Tables 1b and 1c) from 9 to 5. The same effect is observed on the stream visualization (Figs. 3b and 3c) where our sequential implementation has fewer novelty markers, and they appear later, but the measures keep the same "shape". Comparing *c* and *d* the difference is even smaller, (Figs. 3b and 3c) as they both suffer the expected delay in the feedback loop due to asynchronous task execution.

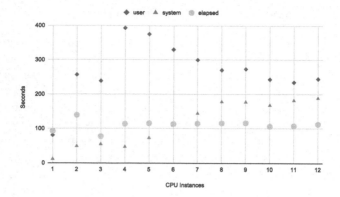

Fig. 4. Time measurements per added instance.

5   Conclusion

Data Stream Novelty Detection (DSND) can be a useful mechanism for Network Intrusion Detection (NIDS) in IoT environments. It can also serve other related applications of DSND using continuous network or system behavior monitoring and analysis. Regarding the tremendous amount of data that must be processed in the flow analysis for DSND, it is relevant that this processing takes place at the edge of the network. However, one relevant shortcoming of the IoT, in this case, is the reduced processing capacity of such edge devices.

In this sense, we have put together and evaluated a distributed architecture for performing DSND applied at network flow descriptors at the edge. Our proposal, *MFOG*, is a distributed DSND implementation based on the MINAS algorithm and the main goal of this work is to observe the effects of our approach to a previously sequential only algorithm, especially in regards to time and quality metrics.

While there is some impact on the predictive metrics, this is not reflected on overall classification quality metrics indicating that distribution of MINAS has a negligible loss of accuracy. In regards to time and scale, our distributed executions was faster than the previous sequential implementation of MINAS, but efficient data distribution was not achieved as the observed time with each added node remained near constant.

Overall, *MFOG* and the idea of using distributed flow classification and novelty detection while minimizing memory usage to fit in smaller devices at the edge of the network is a viable and promising solution. Further work include the investigation of other DSND algorithms, other clustering algorithms in MINAS and analysis of varying load balancing strategies.

Acknowledgment. This study was financed in part by the Coordenação de Aperfeiçoamento de Pessoal de Nível Superior - Brasil (CAPES) - Finance Code 001, and Programa Institucional de Internacionalização – CAPES-PrInt UFSCar (Contract 88887.373234/2019-00). Authors also thank Stic AMSUD (project 20-STIC-09),

FAPESP (contract numbers 2018/22979-2, and 2015/24461-2) and CNPq (Contract 167345/2018-4) for their support.

References

1. Abane, A., Muhlethaler, P., Bouzefrane, S., Battou, A.: Modeling and improving named data networking over IEEE 802.15.4. In: 2019 8th International Conference on Performance Evaluation and Modeling in Wired and Wireless Networks (PEMWN), pp. 1–6 (2019). https://doi.org/10.23919/PEMWN47208.2019. 8986906
2. Buczak, A.L., Guven, E.: A survey of data mining and machine learning methods for cyber security intrusion detection. IEEE Commun. Surv. Tutor. **18**(2), 1153–1176 (2016)
3. Cassales, G.W., Senger, H., De Faria, E.R., Bifet, A.: IDSA-IoT: an intrusion detection system architecture for IoT networks. In: 2019 IEEE Symposium on Computers and Communications (ISCC), pp. 1–7, June 2019. https://doi.org/10. 1109/ISCC47284.2019.8969609. https://ieeexplore.ieee.org/document/8969609/
4. da Costa, K.A., Papa, J.P., Lisboa, C.O., Munoz, R., de Albuquerque, V.H.C.: Internet of things: a survey on machine learning-based intrusion detection approaches. Comput. Netw. **151**, 147–157 (2019). https://doi.org/10.1016/j. comnet.2019.01.023
5. Faria, E.R., Gama, J.A., Carvalho, A.C.P.L.F.: Novelty detection algorithm for data streams multi-class problems. In: Proceedings of the 28th Annual ACM Symposium on Applied Computing, SAC 2013, pp. 795–800. Association for Computing Machinery, New York (2013). https://doi.org/10.1145/2480362.2480515. https:// doi.org/10.1145/2480362.2480515
6. de Faria, E.R., Gonçaalves, I.R., Gama, J., Carvalho, A.C.P.D.L.F.: Evaluation of multiclass novelty detection algorithms for data streams. IEEE Trans. Knowl. Data Eng. **27**(11), 2961–2973 (2015). https://doi.org/10.1109/TKDE.2015. 2441713. http://ieeexplore.ieee.org/document/7118190/
7. de Faria, E.R., de Leon Ferreira Carvalho, A.C.P., Gama, J.: MINAS: multiclass learning algorithm for novelty detection in data streams. Data Min. Knowl. Discov. **30**(3), 640–680 (2016). https://doi.org/10.1007/s10618-015-0433-y
8. HaddadPajouh, H., Dehghantanha, A., Parizi, R.M., Aledhari, M., Karimipour, H.: A survey on internet of things security: requirements, challenges, and solutions. Internet Things 100129 (2019)
9. Kolias, C., Kambourakis, G., Stavrou, A., Voas, J.: DDoS in the IoT: Mirai and other botnets. Computer **50**(7), 80–84 (2017)
10. Lopez, M.A., Duarte, O.C.M.B., Pujolle, G.: A monitoring and threat detection system using stream processing as a virtual function for big data. In: Anais Estendidos do XXXVII Simpósio Brasileiro de Redes de Computadores e Sistemas Distribuódos, pp. 209–216. SBC, Porto Alegre (2019). https://sol.sbc.org.br/index. php/sbrc_estendido/article/view/7789
11. Mitchell, R., Chen, I.R.: A survey of intrusion detection techniques for cyber-physical systems. ACM Comput. Surv. (CSUR) **46**(4), 55 (2014)
12. Sengupta, J., Ruj, S., Bit, S.D.: A comprehensive survey on attacks, security issues and blockchain solutions for IoT and IIoT. J. Netw. Comput. Appl. **149**, 102481 (2020)

13. Shanbhag, R., Shankarmani, R.: Architecture for internet of things to minimize human intervention. In: 2015 International Conference on Advances in Computing, Communications and Informatics, ICACCI 2015, pp. 2348–2353 (2015). https://doi.org/10.1109/ICACCI.2015.7275969

14. Song, J., Takakura, H., Okabe, Y., Eto, M., Inoue, D., Nakao, K.: Statistical analysis of honeypot data and building of Kyoto 2006+ dataset for NIDS evaluation. In: Proceedings of the 1st Workshop on Building Analysis Datasets and Gathering Experience Returns for Security, BADGERS 2011, pp. 29–36 (2011). https://doi.org/10.1145/1978672.1978676

15. Tahsien, S.M., Karimipour, H., Spachos, P.: Machine learning based solutions for security of internet of things (IoT): a survey. J. Netw. Comput. Appl. **161**(November 2019) (2020). https://doi.org/10.1016/j.jnca.2020.102630

16. Viegas, E., Santin, A., Bessani, A., Neves, N.: BigFlow: real-time and reliable anomaly-based intrusion detection for high-speed networks. Future Gener. Comput. Syst. **93**, 473–485 (2019)

17. Zhou, J., Cao, Z., Dong, X., Vasilakos, A.V.: Security and privacy for cloud-based IoT: challenges. IEEE Commun. Mag. **55**(1), 26–33 (2017). https://doi.org/10.1109/MCOM.2017.1600363CM

Artificial Intelligence Application in Automated Odometer Mileage Recognition of Freight Vehicles

Pornpimol Chaiwuttisak[(✉)]

Department of Statistics, Faculty of Science, King Mongkut's Institute of Technology
Ladkrabang, Bangkok, Thailand
pornpimol.ch@kmitl.ac.th

Abstract. Transportation cost management is necessary for entrepreneurs in industry and business. One way to do this is to report the daily mileage numbers read by the employees of a company, but still encountering errors in human mileage reading, resulting in incorrect information received and difficulties to plan effective revenue management. As well as, increasing workload and creating complications for employees in checking mileage information. Therefore, the objective of this research is to create a machine learning model for detecting and reading the mileage numbers 2 types of freight vehicles: Analog and Digital. It can be divided into 2 parts: 1) detect mileage is used to identify the position of the mileage in the image and cut only the mileage by removing the unrelated background from the image 2) detect numbers and reads miles. Both use object detection with the Faster-RCNN. The results show that to detect the position of the miles and cut only the number of miles to read the numbers correctly, 187 images from 220 test images, which is correct for the model, representing 85%. The results of the study achieve satisfactory performance that meet the requirements needed for real-life applications in the transportation and logistics industry.

Keywords: Object detection · Mileage reading · Faster-RCNN · Freight vehicles

1 Introduction

The advancement of Artificial Intelligence (AI) technology has played an important role in the digital era. It has been extensively used to facilitate convenience and enhance operational efficiency in various businesses and industries. Moreover, it helps to create business opportunities in competitiveness for the industry. Applications of AI have been applied widely in transportation to drive the new perspectives in innovative solutions and to support performance transportation operations.

Freight transportation is one of the backbones of industrial systems and it is also a major component of the supply chain. Energy consumption has significant effect on operating costs of freight vehicle. It can be estimated from the number of miles/kilometres traveled on an odometer which is an instrument for indicating the distance traveled by

© Springer Nature Switzerland AG 2021
O. Gervasi et al. (Eds.): ICCSA 2021, LNCS 12951, pp. 487–496, 2021.
https://doi.org/10.1007/978-3-030-86970-0_34

a vehicle. Auditing daily mileage is an approach to monitor the length of journeys and keep controls for avoiding fuel fraud.

Therefore, the objective of the research is to develop the learning model based on AI to automated mileage detection and reading from an odometer on freight vehicles. We applied to the company operating in the industries of beverages consisting of several distribution centers. In which each way of transporting the products, the mileage will be manually read before leaving the distribution center and the mileage will be recorded when freight vehicles arrive at the point of delivery and after the freight vehicles return to the distribution center. The aforementioned data was used to calculate and verify the amount of oil used in each trip to use for managing logistic costs. However, the problem was found that the inaccuracy and completeness of the data was the result of human record errors.

2 Literature Reviews

2.1 Artificial Neural Network

The Artificial Neural Network (ANN) is based on the simulation concept of the human brain's activities with creating computers to be intelligent in learning as well as humans. Moreover, it can practice and deploy knowledge to solve various problems. Therefore, many researchers have applied the concept of neural networks to several applications ranging from easy to complex tasks.

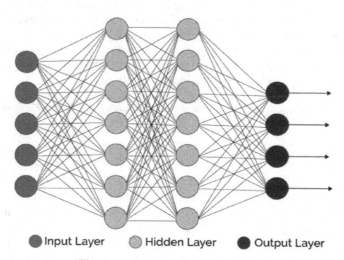

● Input Layer   ◐ Hidden Layer   ● Output Layer

Fig. 1. Neural network architecture [6]

The architecture of the neural network is showed in Fig. 1. In principle, it can be divided into 3 layers as follows:

1. The input layer is the first layer of the neural network. It has a neuron that acts like a human neuron to receive and transmit information. The number of neurons in this

layer is equal to the number of input data. This layer is responsible for receiving data into the neural network model.

2. The hidden layer is an intermediate layer of the neural network, which has an effect on the model's learning efficiency. It serves the functions to combine the multiplication of the input data and the weight, and then convert to the result in the output layer.
3. The output layer is the last layer in the structure of the neural network model. It consists of output neurons. The number of neurons in this layer is equal to the number of output variables that represent the model's output of the model.

2.2 Convolutional Neural Network

Convolutional Neural Network (CNN) is very similar to the neural network. The difference is that the convolutional neural network is designed specifically for the task that uses images as an input. The layer of a convolutional neural network is different from that of the artificial neural network. The convolutional neural network has neurons arranged in three dimensions consisting of width, height and depth. It is used in object detection, identifying locations objects in images.

The most common layers of the convolutional neural network consisted of 5 layers, each of which is detailed as follows:

1. An input layer is often used to express visual data as a matrix. For a color (RGB) image, there will be 3 matrices associated with the image. For example, the color image is 32 × 32 (width × height), so the input layer will have an input dimension of 32 × 32 × 3.
2. A convolution layer consists of a filter or kernel to calculate the weight of the area which the filter has moved through. The filter can have multiple layers to increase the number of feature maps. Each feature map learns different information about the image, such as border or dot pitch.
3. An activation layer is composed of a function that is in between the input feeding and the output connected to the layers. A popular activation function in convolution neural networks is the Rectified Linear Unit (ReLU) function, where all the negative value is turned to 0, while the positive value is still retained and the size does not change. In addition, one popular function, the softmax function which is a classification function, is used in the last layer of the network to classify inputs into multiple categories based on considering the maximum probability as in Eq. 1.

$$f(x)_j = \frac{e^{x_j}}{\sum_{i=1}^{k} e^{x_i}} \tag{1}$$

4. The Pooling layer is a layer that reduces the width and height of the image but retains the depth of the image. The popular pooling techniques consist of 2 types: Max pooling and Average pooling.
5. The fully connected layer is the last layer for classification. Namely, the neurons in this layer connect to the activation of the neurons in the previous layer.

The concept of R-CNN is to use a Selective Search (SS) approach to propose around 2000 Regions-Of-Interest (ROI), which are then fed into a convolutional neural network to extract features. These features were used to classify the images and their object boundaries using Support Vector Machines (SVM) and regression methods.

[1] studied the mileage extraction from odometer images for the automating auto insurance process. The company provides customers with an easy and flexible way in order to provide all the information they need when reporting a claim or request for a quote. It also reduces human errors and speeds up the process of data collection, where accurate mileage readings are essential to apply to car insurance quotes and process claims in sequence. Therefore, the study objective was to extract the mileage from the odometer image using object detection which will be divided into two parts to locate the miles in the image by using the structure of Faster R-CNN and Single Shot Detector (SSD), As a result, the Faster R-CNN structure was more accurate than the model built with the Single Shot Detector (SSD) structure.

[2] and [5] improve accuracy on automatic license plate recognition (ALPR) systems.by applying the Convolutional Neural Networks (CNNs) to use of synthetically generated images. [3]. introduced the Faster R-CNN which use selective search for region proposals and convolutional feature maps as input. It can improve the speed of the object detection. [7] presented text detection method based on multi-region proposal network (Multi-RPN) to detect texts of various sizes simultaneously. As a result, the proposed Multi-RPN method improved detection scores and kept almost the same detection speed as compared to the original Faster R-CNN.

[8] studied Indian road license plate detection and reading using Faster-RCNN to detect and read Indian license plates. It is challenging due to the diverse nature of license plates such as font, size, and number of lines. It applied the structure of the Faster R-CNN and extracted the features with Resnet50 andVGG16. As a result, Resnet50 provided higher mAP than VGG16. The accuracy in detecting and reading car license plates was accounting for 88.5%.

[9] studied traffic sign detection system using a convolutional neural network to train for extraction of the characteristics of traffic signs using You Only Look Once (YOLO), There were 9 types of mandatory traffic signs, including no overtaking signs, no u-turn signs to the right, no u-turn sign to the left, no right turn sign, no left turn sign, signs do not turn right or make a u-turn, signs do not turn left or make a u-turn, the sign prohibits changing the lane to the right and finally, the sign prohibits changing the lane to the left. The design of this traffic sign detection system consists of 4 parts: preparation of training data sets, building a model, testing the model and performance evaluation. The model was tested with a total of 108 traffic sign data sets and had a precision value of 0.87%, a recall value of 0.89% and F-Measure equals to 0.88%.

3   Research Methodology

Data used in this research are RGB images saved as a JPEG file format with a variety of resolutions starting from the resolution of 720×720 pixels to 2730×1536 pixels. The different types of freight vehicles are found in the company. However, there are two types of odometers in those various vehicles are used: analog and digital. Analog odometer can be divided into two subtypes: blue and black (see in Fig. 2), while the digital odometer can be divided into 2 subtypes (see in Fig. 3).

Fig. 2. Analog odometer (Color figure online)

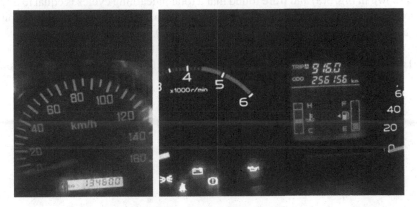

Fig. 3. Digital odometer

There are a total of 635 images used in the study, divided into 390 analog images and 245 digital images. A simple random sampling (SRS) technique is used to divide images based on the type of odometers into two dataset: training dataset accounting for approximately 80% of the total of data and testing dataset accounting for approximately 20% of the total of data. Moreover, data augmentation is applied to build the various images such as number of pixels in images to make images lighter or darker, noise enhancement, image quality degradation, and image rotation to increase a size of training dataset for improving the model.

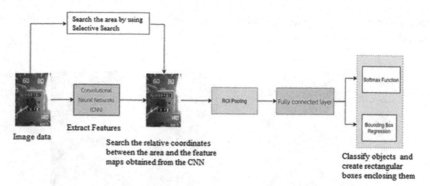

Fig. 4. The process of object detection

Figure 4 shows the process to detect the area of images containing the mileage based on the Faster-RCNN algorithms [4].

4 Experimental Results

The learning rate for detecting mileage is defined in a range of 0.0005 and 0.0002. The results shown in Fig. 5(a) and b are found that a total of learning cycles is equal to 24,046 iterations and take a time about 1 h 39 min 57 s with mAP (mean Average Precision) of 0.733 and mAP@iou0.5 equal to 0.996. Total loss is 0.0476.

From Fig. 6(a) and (b), the total time used to train the model for mileage reading is 3 h 30 min 6 s with mAP (mean Average Precision) of 0.835 and mAP@iou0.5 equal to 0.989. Total loss is 0.2898.

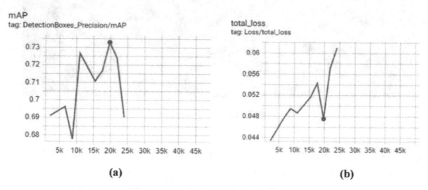

(a)   (b)

Fig. 5. The detection object model

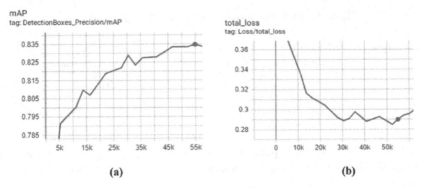

Fig. 6. The mileage reading model

Figure 7 shows an example of testing the object detection model (a) the mileage detection and (b) cutting specific area of mileage numbers. Figure 8 shows the rectangular frame to predict mileage numbers by the model.

Fig. 7. An example of the result while training the model for mileage detection

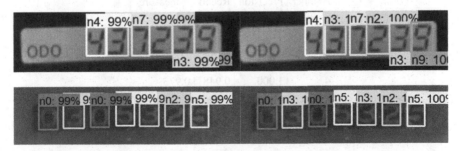

Fig. 8. An example of the result while training the model for reading mileage

The totals of 220 images are used to test the accuracy of object detection and reading the 6-digit of mileage in both the analog and digital odometer. The experimental results are shown in Table 1.

Table 1. Results of object detection and mileage reading

Type of odometer	# Images of testing dataset	Accuracy of object detection	%	Accuracy of mileage reading	%
Analog	68	67	98.53	57	83.82
Digital	152	148	97.37	130	85.53

As Table 2 of Chi-square hypothesis testing, it was found that the accuracy ratio in detecting and reading the miles on both analog and digital odometer is equal to 85:15 at the significant level of 0.05.

Table 2. Chi-square hypothesis testing

Test statistics	
Chi-square	0.074
df	1
Asymp. sig.	0.786

The Precision, Recall and F-Measure of reading mileages on analog and digital odometers are shown in Tables 3 and 4 respectively. The result shows the precisions of numbers 2, 4 and 8 in the analog odometer are equal to 1. It can say that the model is accurately and precisely detected and classified in every time it is predicted to be the numbers 2, 4 and 8 of the analog numbers as well as the digital numbers shown in

Table 3. Result of the precision, recall, F-measure for analog mileage

Number	Precision	Recall	F-measure
1	0.941	1.000	0.970
2	1.000	1.000	1.000
3	0.931	0.982	0.956
4	1.000	0.948	0.973
5	0.983	1.000	0.991
6	0.978	0.978	0.978
7	0.976	1.000	0.988
8	1.000	0.933	0.965
9	0.955	0.977	0.966
0	0.946	0.981	0.963
Average	**0.971**	0.98	0.975

Table 4, the 6 and 9 with precision of 1, while digital numbers 1 and 0 have less precision than other numbers. There is an error in the classification of numbers, for example the number 1 or 0 cannot be detected or classified properly in some images. Recall shows that the model can predict the numbers correctly with the solution relative to the total number of solutions of that number. From Table 1, it is found that the recall of the analog numbers 1, 2, 5 and 7 are equal to 1. Namely, the model is able to correctly detect analog numbers that contain numbers 1, 2, 5 and 7 as well as digital numbers consisting of 0 2 3 6 and 7 shown in Table 2, while recall values for analog numbers 4 and 8 are equal to 0.948 and 0.933 which are less than other numbers. Moreover, F-Measure is preferred to evaluate the performance of the learning model because it is the mean between precision and recall.

Table 4. Result of the precision, recall, F-measure for digital mileage

Number	Precision	Recall	F-measure
1	0.917	0.989	0.952
2	0.993	1.000	0.996
3	0.989	1.000	0.994
4	0.976	0.976	0.976
5	0.987	0.987	0.987
6	1.000	1.000	1.000
7	0.963	1.000	0.981
8	0.990	0.970	0.980
9	1.000	0.969	0.984
0	0.921	1.000	0.959
Average	0.974	0.989	0.981

5   Discussion and Conclusion

When comparing to the model of the company applied on the same dataset, it is found that our model for mileage detection and mileage read with the Faster-RCNN framework provides considerable accuracy. The accuracy of the model for detecting and reading analog and digital mileage was 87.5% and 93% respectively, while the accuracy of the company's model based on the commercial tools was 50% and 87% respectively. Thus, the developed machine learning model could replace the relatively expensive commercial software. This enables the company to detect errors in the mileage reading by drivers. The mileage figures are also used to provide time frame for fleet maintenance on time, and to plan efficient logistics management strategies. It can also be applied in conjunction with GPS data analysis to prevent fuel theft.

However, it is found that 33 images were incorrect mileage detection and reading the miles. Mistakes can be divided 3 types: errors in detecting mileage, errors to detect numbers, and errors in reading miles. It is found that some images have interference in detecting mileage and reading mileage such as reflection on an odometer, scratches on an odometer. This is some external factors that cannot be controlled. Consequently, it causes an error in the mileage detection and reading of the mileage.

References

1. Acharya, S., Fung, G.: Mileage Extraction from Odometer Pictures for Automating Auto Insurance Processes. https://doi.org/10.3389/fams.2019.00061. Accessed 05 May 2020
2. Bulan, O., Kozitsky, V., Ramesh, P., Shreve, M.: Segmentation- and annotation-free license plate recognition with deep localization and failure identification. IEEE Trans. Intell. Transp. Syst. **18**(9), 2351–2363 (2017)
3. Girshick, R: Fast R-CNN. In the IEEE International Conference on Computer Vision (ICCV), Beijing (2015)
4. Liu, L., et al.: Deep Learning for Generic Object Detection: A Survey. https://doi.org/10.1007/s11263-019-01247-4. Accessed 05 May 2020
5. Masood, S.Z., Shu, G., Dehghan, A., Ortiz, E.G.: License plate detection and recognition using deeply learned convolutional neural networks. CoRR abs/1703.07330 (2017)
6. McDonald, C.: Neural networks. https://towardsdatascience.com/machine-learning-fundam entals-ii-neural-networks-f1e7b2cb3eef. Accessed 22 Apr 2020
7. Nagaoka, Y., Miyazaki, T., Sugaya, Y., Omachi, S.: Text detection by faster R-CNN with multiple region proposal networks. In: 14th IAPR International Conference on Document Analysis and Recognition (ICDAR), Kyoto, pp. 15–20 (2017)
8. Ravirathinam, P., Patawari, A.: Automatic License Plate Recognition for Indian Roads Using Faster-RCNN. https://www.researchgate.net/publication/338116738_Automatic_Lic ense_Plate_Recognition_for_Indian_Roads_Using_Faster-RCNN. Accessed 22 Apr 2020
9. Sirikong, N.: Traffic Signs Detection System by Using Deep Learning Graduate School Conference, pp. 952–959 (2019)

A Three-Fold Machine Learning Approach for Detection of COVID-19 from Audio Data

Nikhil Kumar, Vishal Mittal, and Yashvardhan Sharma[✉]

Birla Institute of Technology and Science, Pilani, RJ 333031, India
{f20170658,f20170080,yash}@pilani.bits-pilani.ac.in

Abstract. Most work on leveraging machine learning techniques has been focused on using chest CT scans or X-ray images. However, this approach requires special machinery, and is not very scalable. Using audio data to perform this task is still relatively nascent and there is much room for exploration. In this paper, we explore using breath and cough audio samples as a means of detecting the presence of COVID-19, in an attempt to reduce the need for close contact required by current techniques. We apply a three-fold approach of using traditional machine learning models using handcrafted features, convolutional neural networks on spectrograms and recurrent neural networks on instantaneous audio features, to perform a binary classification of whether a person is COVID-positive or not. We provide a description of the preprocessing techniques, feature extraction pipeline, model building and a summary of the performance of each of the three approaches. The traditional machine learning model approaches state-of-the-art metrics using fewer features as compared to similar work in this domain.

Keywords: Audio processing · Machine learning · Classification · Convolutional neural networks · Recurrent neural networks · COVID-19

1 Introduction

Our aim is to detect the presence of COVID-19, given an audio sample of a person either breathing or coughing. This is essentially a binary classification task (normal: 0, covid: 1). There are several features that can be extracted from an audio sample and fed into machine learning (ML) models to perform the classification task. This has been the traditional method of feeding audio data into ML models, i.e., by extracting features by hand.

Alternatively, neural networks have been shown to perform well on classification tasks [7], while also not requiring hand-crafted features. Convolutional neural networks (CNNs) have been around since the 1990s [6], and have been used for face and object detection tasks. The biggest advantage of using CNNs is that filters need not be handcrafted, as they are learnt by the network. Interest was rekindled upon seeing their success at the ImageNet Large Scale Visual Recognition Challenge (ILSVRC) 2012 [5].

© Springer Nature Switzerland AG 2021
O. Gervasi et al. (Eds.): ICCSA 2021, LNCS 12951, pp. 497–511, 2021.
https://doi.org/10.1007/978-3-030-86970-0_35

Another class of neural networks is the recurrent neural network (RNN), which has seen success with sequence data, such as text processing, speech processing and time-series forecasting [7]. Long short-term memory networks (LSTMs) are a special kind of RNN, containing memory cells to keep track of information over a sequence [3]. We explore the use of LSTMs to process breath and cough audio sequences in this project, owing to their success on sequence data.

2   Related Work

2.1   COVID-19 Sounds Project at the University of Cambridge

Brown et al. developed an application to crowdsource audio samples of COVID-positive and COVID-negative individuals [1]. Two modalities were used – breath and cough audio samples. The data was divided broadly into individuals with asthma (but not COVID-positive), COVID-positive individuals and healthy individuals. The COVID-negative class was subdivided into individuals who reported having cough as a symptom and those who did not.

Two kinds of models were applied – traditional machine learning models on handcrafted features and a convolutional model based on VGG-16, made for audio data called VGGish. 477 features were extracted using a combination of global and instantaneous features, along with their statistics, and fed into traditional ML models. Spectrograms of the audio samples were extracted and fed into VGGish.

Three separate binary classification tasks were performed, aimed at differentiating between different combinations of COVID-positive and COVID-negative individuals, taking other meta-information (such as whether the patient reported having a cough or asthma) into account. However, our work will focus on the plain vanilla task of distinguishing COVID-positive and normal samples from the raw audio data itself without using the metadata.

2.2   Coswara Project at the Indian Institute of Science

Researchers at the Indian Institute of Science worked on collecting a database of cough, breath and voice samples called Coswara [15]. Several standard traditional audio processing features were used, and traditional machine learning models were fitted. Their paper discusses the results they obtained on a random forest classifier on a 9-way classification problem.

2.3   Approach Using MFCCs and CNNs at the University of Manchester

Dunne et al. built a CNN classifier using the Mel-cepstral frequency coefficients, which gave an accuracy of 97.5% on a small dataset and a 2-way classification problem of COVID-positive vs. normal [2].

However, we will not be considering accuracy as the only metric in our project, given the highly unbalanced nature of the dataset we are using. Precision and recall are considered more reliable metrics in disease-detection tasks, as compared to raw accuracy scores.

2.4 What Is New in This Project?

Given the sequential nature of the audio data, we leveraged recurrent neural networks. To the best of our knowledge, there has been very little to no work done in using RNNs for this task (one of the few works is [13], which uses a combination of transformers and RNNs).

3 Dataset Used

We obtained all the data used in this project from the University of Cambridge's COVID-19 Sounds Project [1] under an academic license. The dataset contained 1134 breath samples and 1135 cough samples. Each audio type was further divided into:

- Asthma samples (COVID-negative), all samples reporting having a cough
- COVID-positive samples, reporting having a cough
- COVID-positive samples, reporting not having a cough
- Healthy samples, reporting having a cough
- Healthy samples, reporting not having a cough

The samples were collected by [1] from two sources – an Android application and a web application. We merged the two sources together using some simple Python scripts. We also combined the COVID with cough and COVID without cough samples together, and the healthy with cough and healthy without cough together, since our goal is binary classification of covid vs. normal, and not obtaining the predictions given whether we know someone reports having cough or not. The task at hand (i.e., binary classification into covid and normal) does not require the class asthma, so we discarded those samples.

The distribution of asthma:covid:normal samples in both breath and cough data was roughly 15 : 15 : 70. We maintained a similar distribution in the dataset split into train, validation and test sets (more on this in Sect. 5.1).

4 Proposed Technique

We use a three-fold approach to detect COVID-19 using:

- Traditional ML model using support vector machine (SVM)
- CNN on mel-spectrograms
- LSTM on sequential audio features

We used 108 features for the machine learning models. All features for these models were handcrafted, and are the typical set of features that are used in audio processing. We used the Librosa [9] library in Python for the signal processing tasks. A detailed description of the features used is given in Sect. 6.2.

For the CNN model, we brought all audio samples to the same length and then generated the log-scaled-spectrograms and mel-spectrograms for the audio samples. More information about this procedure is given in Sect. 5.2.

For the LSTM model, we split the audio sample into frames and extracted features across each frame, giving us an array of feature sequences, with shape [num_timesteps, num_features].

5   Data Preprocessing

5.1   Partitioning the Data

For the ML model, we performed an 80 : 20 split of the data into train and test. We did not keep a hold-out validation set, but performed a 10-fold cross validation, owing to the small size of the dataset. For the CNN and LSTM models, we performed an 70 : 15 : 15 split of the data into train, validation and test. In both cases, we ensured that the original distribution of each class (asthma, covid and normal) was maintained while performing the split (Tables 1, 2).

Table 1. Train-test split (80 : 20) for ML model)

Class	Breath		Cough	
	Train	Test	Train	Test
asthma	133	34	134	34
covid	137	35	142	36
normal	636	159	631	158

Table 2. Train-validation-test split (70 : 15 : 15) for CNN and LSTM models

Class	Breath			Cough		
	Train	Validation	Test	Train	Validation	Test
asthma	133	17	17	134	17	17
covid	137	17	18	142	18	18
normal	636	79	80	631	79	79

Note that in each column, asthma:covid:normal is approximately 15 : 15 : 70, which is the same as for the original distribution.

5.2 Loopback and Clipping

The audio samples all have varying lengths (as shown in Fig. 1). This does not cause a problem for the traditional machine learning models because we consider aggregate features for the whole audio sample, so there is no time sensitivity. However, it causes problems for the convolutional and recurrent models as:

- All the spectrograms have the same dimension in pixels. A spectrogram has time along the x-axis. However, we cannot have different scales on the x-axis for different images (If the audio samples have different lengths, and the length of the x-axis is the same for all images, the scale converting from x-axis units to time in seconds will be different for each image).
- Each audio sequence must have the same length as LSTMs require input of shape [num_samples, num_timesteps, num_features].

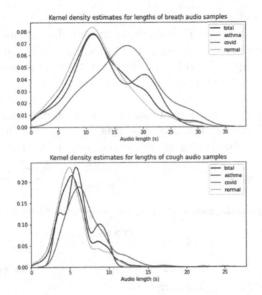

Fig. 1. Distribution of lengths of audio samples

6 Feature Extraction

6.1 Types of Features

An audio signal is an analog signal. However, to store an audio signal as a file on a computer, it needs to be converted into a digital signal through sampling and quantization. Sampling is the discretization of time and quantization is the discretization of the audio intensity.

– The number of samples taken in a unit interval of time is the sampling rate. Alternatively, the duration of each sample is the reciprocal of the sampling rate. A common value of sampling rate is 44100Hz. Human hearing has an upper bound at around 22050Hz. The Nyquist-Shannon theorem [14] tells us that to extract reliably a maximum frequency f from the signal and to avoid aliasing, we need to sample it with a minimum frequency of $2f$.
– An n-bit audio clip uses 8 bits used for each audio level. So an 8-bit audio clip has $2^8 = 256$ levels. Higher the value of n, higher the resolution of audio levels.

We used a sampling rate of 16000 Hz ($= 2 \times 8000$ Hz), because the exploratory data analysis revealed that most of the significant frequencies were found up to 8000 Hz. Researchers working on [1] used the same sampling rate, on the same dataset. There are two kinds of features that are used in audio signal processing [12] – instantaneous and global.

Instantaneous Features. The audio sample is divided into frames. Frames can overlap with each other. Frame size is the number of samples within each frame. The distance between consecutive frames is called the hop length. The ends of the audio clip are padded to allow for framing over the edges. Audio features are then calculated over these frames, giving a sequence of features over the timesteps.

The relation between number of samples in the audio (audio_len), number of samples in the frame (frame_len), padding used on each side (pad_len), hop length (hop_len) and the length of the output vector (output_len) is

$$\text{output\_len} = \left\lfloor \frac{\text{audio\_len} + 2 \times \text{pad\_len} - \text{frame\_len}}{\text{hop\_len}} \right\rfloor + 1 \qquad (1)$$

Librosa [9] by default uses:

$$\text{pad\_len} = \left\lfloor \frac{\text{frame\_len}}{2} \right\rfloor \qquad (2)$$

Values of the above parameters we used in our work are frame_len = 256, pad_len = $\lfloor \frac{256}{2} \rfloor = 128$ and hop_len = 64.

Instantaneous features are perfect for LSTMs. CNNs too can use these features in the form of spectrograms, which are calculated using an algorithm called the short-time Fourier transform (STFT). STFT uses the same idea of calculating over frames.

We did not use instantaneous features for the traditional machine learning models for two reasons:

– The mean audio length is 13.04s for the breath samples and 5.82s for the cough samples. Approximating the mean audio length as 10s, using sampling rate as 16000Hz and the parameters defined in Sect. 6.1, a simple back-of-the-envelope calculation using Eq. 2 yields an output sequence of length 2000+ timesteps. These are too many features to pass into a traditional ML model (curse of dimensionality).

– Passing timestep features to a traditional ML model is not useful as it cannot pick up on the sequencing.

Global Features. Global features are features that are calculated over the complete audio sample, i.e., no framing. These are well-suited for traditional machine learning models. We have not used global features for our models.

Aggregate Instantaneous Features. Since we cannot pass the raw instantaneous feature sequences extracted from the audio samples into the traditional ML models, we calculated statistics across the sequence and used these as input (since we get a single value for each audio sample that summarizes the sequence for us).

We used 18 instantaneous features summarized using mean, median, root mean square value, maximum, minimum and RMS energy weighted mean, giving us $18 \times 6 = 108$ features.

6.2   Features Used

We shortlisted commonly used features in audio processing tasks ands extracted them from the data:

Root Mean Squared Energy (RMSE). It gives an idea of how much energy is contained in the signal [9]. RMSE for a frame is given by

$$\text{RMSE}_{\text{frame}} = \sqrt{\frac{1}{n} \sum_{i=1}^{n} y_i^2} \tag{3}$$

where i iterates over the samples in the frame, n is the number of samples in the frame and y_i is the signal value for the ith sample in the frame.

Zero Crossing Rate (ZCR). It is the number of times a signal crosses the y-axis in a time interval. A high ZCR indicates noisy data [12]. ZCR for a frame is given by

$$\text{ZCR}_{\text{frame}} = \frac{1}{2} \sum_{i=0}^{n-1} |\,\text{sgn}(y_{i+1}) - \text{sgn}(y_i)\,| \tag{4}$$

where the indices are the same as for Eq. 3 and sgn is the signum function.

Spectral Centroid (SC). It gives an idea about what value of frequency f most of the frequencies are distributed [12]. One can think of it as a measure of central tendency or the centre of mass of the frequencies present in the Fourier transform of the signal. SC for a frame is given by

$$SC_{frame} = \frac{\sum\limits_{j=1}^{m} M_j f_j}{\sum\limits_{j=1}^{m} M_j} \tag{5}$$

where j iterates over the frequencies in the frame, m is the number of frequencies in the frame, M_j is the magnitude corresponding to the jth frequency f_j in the frame.

Spectral Bandwidth (SB). It gives an idea about the spread of the frequencies around a mean value [9], like a measure of dispersion such as variance or standard deviation. We used the 2nd order spectral bandwidth $(p = 2)$, which is the full width at half-maximum of the plot of power vs. frequency. SB of order 2 for a frame is given by

$$SB2_{frame} = \frac{1}{2} \sum\limits_{j=1}^{m} M_i (f_i - SC_{frame})^2 \tag{6}$$

Spectral Rolloff (SR). It is the frequency below which $p\%$ of energy of the spectrum for a frame lies [1]. p is a threshold, typical values for p are 85 and 95 [12]. We used $p = 85$ for our work.

Mel-Frequency Cepstral Coefficients (MFCCs). MFCCs are obtained by taking the discrete cosine transform (DCT) of the mel-spectrogram. MFCCs are one of the most popular features in audio processing [1]. Typically, the first 13 coefficients are used.

6.3 Spectrograms

The audio signal $y(t)$ in the time domain does not directly give us information about its frequency content. To obtain the frequency and their magnitudes, one has to take the Fourier transform of $y(t)$. This is represented by $y(\omega) = \mathcal{F}[y(t)]$. However, in this process, we lose the time information as the Fourier transform is calculated over the complete audio signal.

A solution to maintain both information related to time and frequency is to calculate the Fourier transform over frames of the audio signal (as described in Sect. 6.1).

This gives us the magnitude of the frequencies as a function of time $M(f, t)$. A spectrogram is a heatmap of function M, usually with time along the x-axis and frequency along the y-axis.

The Fourier transform assumes that the signal tapers off and goes to zero at the edges. However, this is not true for the individual frames, so we apply a windowing function (Hann window) that makes this condition true. Keeping a hop length less than frame length (i.e., having overlapping frames) helps to avoid the loss of information at the edges of a frame due to windowing.

After performing the STFT on the audio signal, we changed:

- The frequency scale to the mel scale
- The magnitude of the frequencies into a logarithmic scale (decibels).

This is because a lot of the information is concentrated at the lower frequencies, so the magnitudes of the higher frequencies appear almost 0. The scaling helps visualize the magnitudes of the lower frequencies clearly.

The mel scale attempts to model human pitch perception, as the variation between perceived pitch and frequency is not linear. Rather, it is logarithmic. One of many empirical formulae to convert frequency from Hz to mels is as follows [10]:

$$m = 2595 \log_{10} \left(1 + \frac{f}{700} \right) \tag{7}$$

6.4   Input Vectors to LSTMs

Inputs to LSTMs have the shape [num_samples, num_timesteps, num_features]. We have considered only instantaneous features as inputs to the LSTM, because we want the time-ordering to be preserved.

7   Experiments and Results

The models[1] we used are described in detail below, organized by the audio type (breath or cough).

7.1   SVM Model

We used Scikit-learn [11] to train the SVM.

Procedure:

1. Applied principal component analysis (PCA) on the data to reduce the dimensionality of the data, using 20 components (which explained 84% of the variance).
2. Passed the output from the PCA into a support vector machine classifier with a radial basis function (RBF) kernel.
3. Performed grid search with 10-fold cross validation using the hyperparameters below:
   - C: L2 regularization parameter for SVM. Lower values of C give higher regularization.
   - γ: Coefficient for the kernel.
4. Refitted model on all 10 folds using the best hyperparameters found.

Results on test set are detailed below. We have taken covid as the positive class to calculate metrics (Table 3).

---

[1] Source code can be found at www.github.com/nikhilkmr300/covid-audio and www.github.com/vismit2000/covid-audio.

Table 3. Performance (SVM)

		P					P	
		normal (0)	covid (1)				normal (0)	covid (1)
T	normal (0)	157	2		T	normal (0)	153	4
	covid (1)	6	29			covid (1)	1	34
	(a) Breath					(b) Cough		

Metric	Breath	Cough
Precision	0.935	0.895
Recall	0.828	0.971
F1	0.879	0.931
Accuracy	0.959	0.976

(c) Metrics

T = True   P = Predicted

7.2   CNN Model

We were able to obtain excellent results using an SVM by directly using the imbalanced dataset, however the CNN model was highly sensitive to the class imbalance. Deep neural networks are also prone to overfitting on small datasets, as is in this case, so we used a relatively simple architecture.

The model, however, learned to predict all samples as `normal` because of the heavy class imbalance, leading to an accuracy equal to the percentage of `normal` samples in the dataset and a recall of 0. We tried the following approaches to remedy the situation:

- Giving a higher weight to the minority class.
- Undersampling the majority class.

neither of which caused any significant improvement.

The poor performance of the CNN can be attributed to:

- Tiny dataset consisting of 1000 samples. Augmentation can help only to a certain extent.
- Large class imbalance between `covid` and `normal` samples.

both of which point to more data being required.

Another reason could also be the fact that the mel spectrograms of the `covid` and `normal` classes, at least to the human eye, look quite similar (refer Fig. 2). While it could be argued that the model can capture hidden features that the human eye misses, after some trials it does not seem that the CNN is capable of extracting from the raw spectrograms significant information to discern between the classes.

The poor performance of the CNN can be understood in the context of [1]. Brown et al. [1] used the pretrained VGGish model to extract 256 features from a spectrogram. They also mention that since VGGish takes a spectrogram as input, there might be a loss of temporal information, and hence they use the VGGish features along with 477 handcrafted features. In our case, with only a raw spectrogram as input and a CNN with a straightforward architecture, post factum, it almost seems unreasonable to expect it to perform well. It appears that until more data is available, tweaking the CNN model is going to yield marginal results at best.

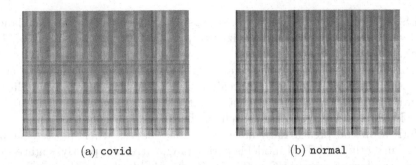

(a) covid   (b) normal

Fig. 2. Mel spectrograms of a COVID-positive vs. a normal breath audio sample

The procedure we followed is detailed below:

1. Resized images to $(128, 128)$ using bilinear interpolation to reduce computational cost.
2. Performed augmentation on already generated mel spectrograms, shifting it upto a range of 0.2 times the width. This only affects the time axis, which preserves semantics, because we can shift the audio forward and backward without affecting the class of the sample or meaning of the spectrogram. However, if we shift the y-axis, we would be changing the frequency values on the y-axis, and then the spectrograms would not be comparable. Rotating, zooming and cropping the spectrogram are not meaningful transformations.

Table 4. CNN architecture

Layer	Activation	Layer information
Convolution 2D	relu	$f_c = 16, f_s = (5, 5)$
Max pooling 2D	None	$f_s = (2, 2)$
Convolution 2D	relu	$f_c = 32, f_s = (5, 5)$
Max pooling 2D	None	$f_s = (2, 2)$
Convolution 2D	relu	$f_c = 64, f_s = (3, 3)$
Max pooling 2D	None	$f_s = (2, 2)$
Flatten	None	None
Fully connected	relu	$n = 32$
Dropout	None	$p = 0.5$
Fully connected	sigmoid	$n = 1$

$f_c =$ Filter count, $f_s =$ Filter/pool size
$p =$ Dropout probability, $n =$ Number of neurons

3. Trained the model using the architecture detailed below with early stopping, with the Adam optimizer [4] with an initial learning rate of 1×10^{-4}, decreasing it by a factor of 0.05 on plateauing of validation loss (Table 4).

## 7.3	LSTM Model

The LSTM model, like the CNN model, was highly sensitive to the imbalanced nature of the data, and was incentivized to predict all samples as being of class `normal`.

Considering that deep networks are highly prone to overfitting with small datasets, we used relatively simple architectures. However, the downsampling of the majority class caused a performance hit, which is possibly the reason why the LSTM is unable to perform as well as the SVM (Tables 5, 6).

Metrics on the breath data have very noisy curves when plotted against epoch. Currently the LSTM model is performing, extremely poorly, slightly better than random chance (AUC = 0.569). It is probably due to the small size of the dataset, rendered even smaller because of downsampling.

The LSTM model performs significantly better on the cough data than on the breath data.

Table 5. LSTM architecture

Layer	Activation		Layer	Activation
LSTM ($u = 32$)	tanh		LSTM ($u = 64$)	tanh
LSTM ($u = 32$)	tanh		LSTM ($u = 64$)	tanh
FC ($n = 32$)	relu		FC ($n = 64$)	relu
FC ($n = 1$)	sigmoid		FC ($n = 1$)	sigmoid
(a) Breath			(b) Cough	

u = Number of hidden units, n = Number of neurons FC = Fully connected layer

Table 6. Test metrics (LSTM)

Metric	Breath	Cough
Precision	0.600	0.750
Recall	0.667	0.833
AUC	0.569	0.802
Accuracy	0.611	0.778

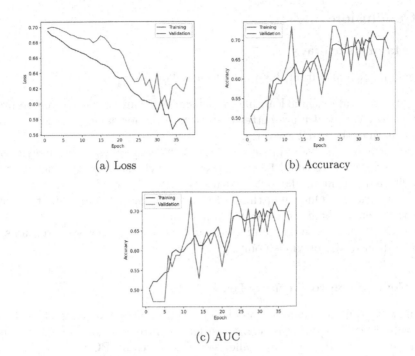

(a) Loss (b) Accuracy

(c) AUC

Fig. 3. Training and validation metrics for breath data

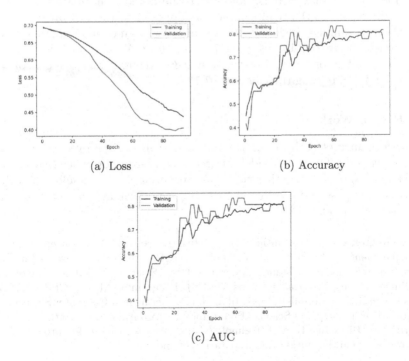

(a) Loss (b) Accuracy

(c) AUC

Fig. 4. Training and validation metrics for breath data

8   Conclusion

8.1   Major Takeaways

The major takeaways from this work are as follows:

- The traditional machine learning model shows promise, scoring well on precision and recall, which are better indicators of performance in medical applications than the usual accuracy.
- Feeding models partially processed data in the form of aggregate metrics over raw time sequences could be more useful, as evidenced by the fact that the traditional ML model far outperforms the LSTM models using raw instantaneous features. One could think of this as the model being able to process partially *metabolized* data over the raw data.
- The CNN is unable to differentiate the classes from the raw spectrograms, an attestation to the previous point.

8.2   Comparison to Work by Brown et al.

Brown et al. [1] used a combination of 256 features extracted using VGGish, 477 handcrafted features and other combined features to get a total of 773 features, which were then reduced to a smaller set of features by PCA. Task 1 in [1] is in line with the work done in this paper. They obtained a precision of 0.72 and a recall of 0.69 on task 1. It is, however, to be noted that they used only 298 breath samples and 141 cough samples, as opposed to the 773 breath samples and 773 cough samples used by us for training. We used a larger amount of data, with fewer features (our 108 vs. their 773 before PCA), and obtained improved a precision metric of 0.935 (breath) and 0.895 (cough) and a recall of 0.828 (breath) and 0.971 (cough) (refer Sect. 7.1).

8.3   Future Work

The performance of each of the three models can be improved as more data becomes available. Once the models approach similar performance to each other, they can be combined together to create an ensemble. Ensembling has shown promise in the past and can be used to obtain performance better than any of the constituent models [8].

Acknowledgements. The authors would like to convey their sincere thanks to the Department of Science and Technology (ICPS Division), New Delhi, India, for providing financial assistance under the Data Science (DS) Research of Interdisciplinary Cyber Physical Systems (ICPS) Programme [DST/ICPS/CLUSTER DataScience/2018/Proposal-16:(T-856)] at the Department of Computer Science, Birla Institute of Technology and Science, Pilani, India. The authors are also thankful to the authorities of Birla Institute of Technology and Science, Pilani, for providing basic infrastructure facilities during the preparation of the paper.

References

1. Brown, C., et al.: Exploring automatic diagnosis of COVID-19 from crowdsourced respiratory sound data. CoRR abs/2006.05919 (2020). https://arxiv.org/abs/2006.05919

2. Dunne, R., Morris, T., Harper, S.: High accuracy classification of COVID-19 coughs using mel-frequency cepstral coefficients and a convolutional neural network with a use case for smart home devices (August 2020). https://doi.org/10.21203/rs.3.rs-63796/v1

3. Hochreiter, S., Schmidhuber, J.: Long short-term memory. Neural Comput. **9**(8), 1735–1780 (1997)

4. Kingma, D.P., Ba, J.: Adam: a method for stochastic optimization. In: Bengio, Y., LeCun, Y. (eds.) 3rd International Conference on Learning Representations, ICLR 2015, San Diego, CA, USA, 7–9 May 2015, Conference Track Proceedings (2015). http://arxiv.org/abs/1412.6980

5. Krizhevsky, A., Sutskever, I., Hinton, G.E.: ImageNet classification with deep convolutional neural networks. Commun. ACM **60**(6), 84–90 (2017)

6. LeCun, Y., et al.: Backpropagation applied to handwritten zip code recognition. Neural Comput. **1**(4), 541–551 (1989). https://doi.org/10.1162/neco.1989.1.4.541

7. LeCun, Y., Bengio, Y., Hinton, G.: Deep learning. Nature **521**(7553), 436–444 (2015)

8. Maclin, R., Opitz, D.W.: Popular ensemble methods: an empirical study. CoRR abs/1106.0257 (2011). http://arxiv.org/abs/1106.0257

9. McFee, B., et al.: librosa: Audio and music signal analysis in python. In: Proceedings of the 14th Python in Science Conference, vol. 8, pp. 18–25 (2015)

10. O'Shaughnessy, D.: Speech Communication: Human and Machine. Addison-Wesley series in electrical engineering, Addison-Wesley Publishing Company, Boston (1987). https://books.google.co.in/books?id=mHFQAAAAMAAJ

11. Pedregosa, F., et al.: Scikit-learn: machine learning in python. J. Mach. Learn. Res. **12**(85), 2825–2830 (2011). http://jmlr.org/papers/v12/pedregosa11a.html

12. Peeters, G.: A large set of audio features for sound description (similarity and classification) in the cuidado project. CUIDADO IST Proj. Rep. **54**, 1–25 (2004)

13. Pinkas, G., Karny, Y., Malachi, A., Barkai, G., Bachar, G., Aharonson, V.: Sars-cov-2 detection from voice. IEEE Open J. Eng. Med. Biol. **1**, 268–274 (2020). https://doi.org/10.1109/OJEMB.2020.3026468

14. Shannon, C.E.: Communication in the presence of noise. Proc. IRE **37**(1), 10–21 (1949). https://doi.org/10.1109/JRPROC.1949.232969

15. Sharma, N., et al.: Coswara - a database of breathing, cough, and voice sounds for COVID-19 diagnosis. Interspeech 2020 (October 2020). https://doi.org/10.21437/interspeech.2020-2768. http://dx.doi.org/10.21437/Interspeech.2020-2768

Hydrogen Economy and Its Production Impact on Automobile Industry Forecasting in Ecuador Using Principal Component Analysis

Joffre Espin[1] , Eduardo Estevez[2] ,
and Saravana Prakash Thirumuruganandham[3]([✉])

[1] Facultad de Ingeniería en Sistemas, Electrónica e Industrial,
Universidad Técnica de Ambato, Ambato, Ecuador
[2] Universidade da Coruña, Grupo de Polímeros, Departamento de Física y Ciencias
de la Tierra, Escuela Universitaria Politécnica, Serantes, Avda. 19 de Febrero s/n,
15471 Ferrol, Spain
[3] Faculty of Industrial Engineering, Centro de Investigación en Mecatrónica y
Sistemas Interactivos - MIST del Instituto de Investigación, Desarrollo e Innovación,
Universidad Tecnológica Indoamérica, Ambato, Ecuador
saravanaprakash@uti.edu.ec

Abstract. Hydrogen vehicles are operating in many parts of the world. However, in South America these cars are not on the streets. Even there are merchant hydrogen plants in the continent, hydrogen stations are not available yet. Doing a PCA analysis with H_2 Vehicle Simulator Framework, Principal component 1 is related with amount of storage of hydrogen. Second principal component is related with autonomy and the last principal component with raw distance. Nowadays hydrogen price is around 12 USD per kilo in Europe and it is expensive for South America. The goal is to achieve a price of 2.15 USD in future with different renewable energies like: nuclear energy, hydro electrical, biomass and photo voltaic. Other goal is to reduce the vehicle price because the cost of a hydrogen car is 2.62 more expensive on average than a combustion car in Ecuador. One alternative to incentive the introduction of hydrogen vehicles is with hydrogen taxis vehicles like Paris is doing. Photo voltaic energy available in Tungurahua province is enough to produce hydrogen for all the taxis of Ambato city.

Keywords: PCA · Hydrogen · Vehicle · Simulation · Cost

1 Introduction

Hydrogen (H_2) is the first element in the Periodic Table and the main constituent of matter in the Universe with approximately 75%. On Earth, H_2 is also found in abundance, but adhered to organic components such as methane or oxygen, forming water [1]. Currently, H_2 is considered an Energetic Vector like cells, batteries and all fossil fuels because it is a substance that stores energy that

O. Gervasi et al. (Eds.): ICCSA 2021, LNCS 12951, pp. 512–526, 2021.
https://doi.org/10.1007/978-3-030-86970-0_36

can be released in a controlled way later [2]. H_2 as an energy vector serves as "fuel" and it must be manufactured through the separation of the other elements because it is not found in its pure state on the planet. Because the emission result is water and electricity, H_2 is considered the clean fuel of the present and future [3]. To replace fossil fuels due to the pollution to the environment that their use generates, we have chosen to use renewable energy to boost our means of transport. Such is the case that today you have electric or battery-powered motorcycles, cars, trucks, trains, boats and even aeroplanes. However, these means of transport are not found in abundance or on par with H_2 as an energy source. Although H_2 is commonly recognized as the fuel of the future, it does not burn like gasoline, LPG, or diesel. In fact, H_2 generates energy to move an electric motor and thus obtain the movement of the vehicle. So, talk about an hydrogen vehicle is talk about an electric vehicle, whose difference from commonly known electric vehicles is in how electricity is generated. While an H_2 vehicle uses this gas to generate electricity through a fuel cell integrated with it and thus drive the engine, an electric vehicle receives the electrical charge generated in a Power Plant (solar, wind, hydro or combustion) to insert it into a battery and finally propel the vehicle. There are some processes to manufacture H_2 at an industrial level like: grey hydrogen, blue hydrogen and green hydrogen. Gray hydrogen consists of obtaining H_2 through natural gas, oil or coal, generating carbon dioxide (CO_2) and carbon monoxide (CO). Steam Methane Reforming (SMR) is a pressure swing adsorption purification technology. An example of grey hydrogen is compressing methane with steam and heat, thus obtaining H_2. Blue hydrogen is generated by electrolysis, which consists of separating hydrogen from oxygen in water by applying electrical energy. If in the electrolysis process the electricity comes from renewable energies such as solar, wind, water; then it is called green hydrogen. The electrolysis process loses efficiency between 20 to 30% of the energy. A feasible technology to implement in Ecuador is through residual biomass suitable for the production of bio-ethanol and thus obtain hydrogen and acetaldehyde [4]. Another potential source of hydrogen production is hydroelectricity. Just as a study has been developed on the production of H_2 for an economy in Colombia, it can be developed for Ecuador since it has 27 hydroelectric plants and where an approximate waste of 50% is assumed [5,6]. According to the Hydrogen Tools organization, in 2021 10068 Hydrogen Fueled Vehicles are operating in the US. The vehicles are subjected to the cycle mentioned during their operation. However, in South America, that kind of vehicles are not on the streets. Just a few projects or demo hydrogen cars are available in few cities. To apply hydrogen economy in the country, Ecuador has to overcome the following barriers: low and disperse investigation activity; low academic formation needed to operate and innovate on hydrogen technologies; lack of legal framework and regulations that incentive SESH penetration and incipient formation of collaborative networks in research and promotion [7]. Principal component analysis (PCA) is a technique of dimensional reduction. It means turn high dimensional data in a lower dimensional form using sophisticated underlying mathematical principles. Possible correlated variables are into smaller variables called principal components. PCA can be applied for all topics like health: analysis of Munich

Functional Developmental Diagnosis to Industry and technology: relationship between important properties of bio diesel and its chemical composition, reduction of different features of products and customers or to describe the correlation and necessity of dependent and independent variables on a research work and mathematical model [8–11]. The aim of this paper is to present hydrogen economy and its production for hydrogen operating vehicles in Ecuador.

– We show the merchant and captive hydrogen plants available in south America and their production capacity.
– We mention three alternatives to produce hydrogen in Ecuador based on energies: nuclear, hydro, biomass and photo voltaic.
– We run H_2 Vehicle Simulation Framework with 700 bar compressed gas system as a hydrogen storage system for four different standard drive cycles.
– We apply Principal Component Analysis (PCA) to the results of H_2 Vehicle Simulation Framework.
– We show a comparative of best-selling commercial hydrogen vehicles in USA and best-selling commercial combustion vehicles in Ecuador according to three main principal component analysis: storage of H_2, autonomy and raw distance.
– We present an alternative to introduce hydrogen vehicles and overcome the egg hydrogen dilemma with a demand of 2055 hydrogen taxis operating in Ambato city thanks photo voltaic hydrogen production of the province.

The remainder of this paper is organized as follows: Sect. 2 mentions the materials; Sect. 3 describes methods applied; Sect. 4 presents the results and discussion, while Sect. 5 contains the conclusions.

2 Materials

2.1 Literature

Academic articles, international organisms web pages, Ecuadorian government web pages, reviews, reports and news mentioned on references were used. Free data source excel sheets were downloaded from Hydrogen Tools organization. https://h2tools.org/. Excel sheets used are: "Asia merchant hydrogen plants Jan 2016.xlsx"; "Worldwide refinery hydrogen production capacities by country Jan 2017.xlsx"; "Worldwide refinery hydrogen production capacities by country Jan 2017.xlsx" and "US Hydrogen-Fueled Vehicles - FY21Q2.xlsx"

2.2 H_2 Vehicle Simulation Framework Configured at 700 Bar Storage System

H_2 Vehicle Simulation Framework is a Matlab tool that simulates a light commercial vehicle with a fuel hydrogen cell (PEM) and its hydrogen storage system. Test cases are played on the vehicle simulation framework related to the storage system. Driving conditions are associated with standard drive cycles on a specific test case. Fist case is the Ambient drive cycle or Fuel economy test. Cycles involved in it are UDDS which means low speeds in stop-and-go urban traffic and HWFET or Free-flow traffic at highway speeds. The second case is Aggressive drive using the cycle US06 for higher speeds; harder acceleration and breaking. The third case is Cold drive with FTP-45 cycle at colder ambient temperature ($-20\,°C$). Finally, Hot drive cycle related to SC03 standard is used for hot ambient ($35\,°C$) conditions [12]. A 700 bar compressed gas system is used as a hydrogen storage system. Because hydrogen is very light, at atmosphere pressure, transport 1 Kg of H_2 would need an 11 000 Lt tank for more than 100 Km. That's a problem. The solution, compress H_2 at 700 bar to transport 4–5 Kg of it given 500 Km of autonomy.

2.3   Information Management Software: Microsoft Excel and Minitab 18

Microsoft excel is used to organizing numbers and data with functions and formulas. Info organized is represented on histograms as figures of this paper. Minitab 18 is a software used in high education and industry because his statistics analysis, graphics capacity, quality evaluation and experiment designer.

3   Methods

One of the most important tools on industry to represent qualitative data or quantitative data of a discrete type of variable are histograms [13]. Variables are on X axis while frequency values are on Y axis. The graphic with a surface bar is proportional to the frequency of the values presented. Figure 1 to Fig. 3 are histograms built with free data source excel sheets from Hydrogen Tools organization once information was leaked and organized with Microsoft excel software. H_2 Vehicle Simulation Framework was downloaded from the Hydrogen Materials Advanced Research Consortium organization (HyMARC) and run on Matlab R2014a [12]. A compressed 700 bar was selected on Storage System. The fuel economy test was chosen for the first test case. The second case was an aggressive cycle. The third case was the cold cycle and finally, the hot cycle. All the tests were input with 0.7 auxiliary loads and the run. Each drive cycle has his standard and configuration. Variables result of each simulation at the end are: H_2 delivered [kg], H_2 used [kg], Usable H_2 [kg], Storage system mass [kg], Storage system volume [L], Gravimetric capacity [%], Volumetric capacity [g/L], Temperature [°C], Pressure [bar], Raw distance [miles], On-board efficiency [%], Calculated fuel economy [mpgge] and Calculated range [miles]. Storage system mass, Store system volume and On-board efficiency were excluded for PCA. PCA is the first step to analyses large data sets. This technique uses a transform vector to reduce dimensional and project the data in the directions of large variance [14]. PCA plays an important role in efficiently reducing dimensional of data, but preserving the variation present in the data set. It can be represented and applied like:

$$1st\ principal\ axis = argmax_{||a||=1} Var\left[a^T X\right] \tag{1}$$

$$Var\left[a^T X\right] = \frac{1}{n}\sum_{i=1}^{n}\left\{a^T\left(X_i - \frac{1}{n}\sum_{j=1}^{n}X_j\right)\right\}^2 = a^T V_{XX} a \qquad (2)$$

$$V_{XX} = \frac{1}{n}\sum_{i=1}^{n}\left(X_i - \frac{1}{n}\sum_{j=1}^{n}X_j\right)\left(X_i - \frac{1}{n}\sum_{j=1}^{n}X_j\right)^T \qquad (3)$$

4   Results

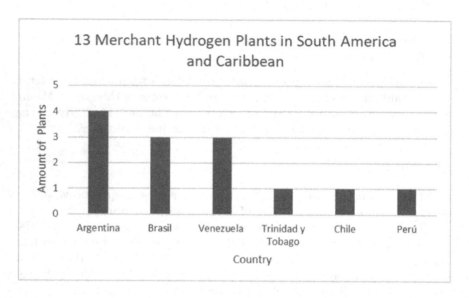

Fig. 1. Merchant hydrogen plants in South America and the Caribbean (2016) [15]

In 2016 Argentina had 4 commercial H_2 production plants, three of these run by Air Liquide and one by Galileo. Two Air Liquide production companies have as their source of H_2 gas production, while the other plant of the company does it for SMR. In Brazil, Air Liquide owns two Plants, one of which is for SMR. The other Plant in that country corresponds to Air Products, producing H and CO. In Venezuela, two plants correspond to Hyundai-Wison, while the other is owned by Linde (BOC) by SMR. The production plant in Trinidad and Tobago is Air Liquide, in Chile, it is Linde (BOC) and in Peru, it corresponds to Praxair, all the latter by SMR.

According to Fig. 1, Argentina, Brazil and Venezuela cover 76.92% of the number of commercial H_2 production plants in South America and the Caribbean. Ecuador does not appear on the list due to the following barriers: scant and scattered research activity; minimum training offer of the human talent necessary for the operation and innovation in H_2 technologies; lack of a legal and regulatory framework that encourages the penetration of the SESH and the incipient formation of collaborative networks in research and promotion [7].

To achieve an H_2 economy in Ecuador, it is important to produce it in a massive, clean, safe and efficient way. An alternative is implementing a Nuclear Hydrogen Production Plant. VHTR, a Very High-Temperature Reactor, is a helium gas-cooled graphite-moderated thermal neutron spectrum reactor that can provide electricity and process heat for a wide range of applications, including hydrogen production [16]. The characteristics of said Plant and the approximate costs to it are shown in Table 1 [17]. The cost of production of 1 kg of H_2 would be approximately \$2.15 in the Nuclear Hydrogen Production Plant [17].

Table 1. (a) Specification for nuclear hydrogen production plant [17], (b) Summary of costs for nuclear hydrogen plant (USD M/year) [17]

Design Parameters	Specifications
Reactor Plant	VHTR
Hydrogen Plant	SI-Based Plant
Thermal Output	4 x 600 MWth
Operating Life	60 years
Plant Availability	90%
Capacity Factor	90%
Fuel Cycle	Open

(a)

Account Description	Amount ($M/year)
Reactor Plant Capital Cost	115.7
Reactor Plant O&M Cost	73.4
Nuclear Fuel Cost	81.8
Decommissioning Cost	1.1
Total Annual Cost	272.0
SI Plant Capital Cost	85.1
SI Plant O&M Cost	34.8
Energy Cost	222.8
Total Annual Cost	342.7

(b)

The Fig. 2(a) shows the different producers and their capacity for: the volume of non-condensate H_2 at 1 atm pressure and 0 °C or Normal cubic meter per hour (Nm^3/hr), thousand standard cubic feet per day (MSCF/day) and kilograms per day (Kg/day). Hyundai-Wison located in Venezuela is the Company that has the most capacity in all the exposed aspects. In the second place, regarding the capacity of Normal cubic meter per hour, there is Air Liquide, then there is Linde and finally Praxair. In second place in terms of capacity in standard cubic feet per day is Linde, followed by Air Liquide and finally Praxair. Finally, the second place regarding capacity in Kg per day is for Linde, then there is Air Liquide and finally Praxair. It should be noted that Air Liquide opened in 2002, Hyundai-Wison opened in 2014, Linde in 1996 and Praxair in 2016.

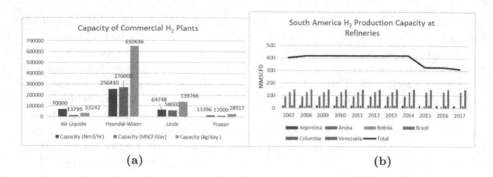

(a)   (b)

Fig. 2. (a) Capacity of commercial H_2 Plants in South America and Caribbean (2016) [15], (b) South America H_2 production capacity at refineries (2017) [18]

Table 2. South America captive H_2 production capacity at refineries [18]

Country	2007	2008	2009	2010	2011	2012	2013	2014	2015	2016	2017
Argentina	19	19	19	19	19	19	19	19	19	19	19
Aruba	93	93	93	93	93	93	93	93	–	–	–
Bolivia	0	14	14	14	14	14	14	14	14	14	–
Brazil	126	126	126	126	126	126	126	126	126	126	126
Columbia	18	18	18	18	18	18	18	18	18	18	18
Venezuela	148	148	148	148	148	148	148	148	148	148	148
Total	**404**	**418**	**418**	**418**	**418**	**418**	**418**	**418**	**325**	**325**	**311**

The South American capacity for captive hydrogen production in refineries is shown in Fig. 2(b) and Table 2. Captive hydrogen is the amount of H_2 that the Plant produces for its consumption in million standard cubic feet per day. As with commercial H_2, Venezuela ranks first in terms of captive H_2 production until 2017. In second place is Brazil and in third place until 2014 Aruba. As of 2015, third place is occupied by Argentina.

H_2 vehicles operating or planned to operate in the United States with an updated date of March 31, 2021, are shown in Fig. 3(a) [19]. The Toyota Mirai and Honda Clarity account for more than 80% of hydrogen-powered vehicles in the United States according to Fig. 3(a).

Even though there are hydrogen-producing plants in South America, observe these cars on streets is not frequent. Today, different automotive companies already have a commercial hydrogen product such as Toyota with its Mirai sedan, Hyundai with the Nexo jeep or Honda with the Clarity sedan, but these are not currently available in the South American continent. The 9 existing units are demos or projects [20]. The base cost of Toyota Mirai is $49500 in California [21], Hyundai Nexo is $72850 or $82000 more or less [22] and Honda Clarity $58490 [23]. In Ecuador, the best-selling vehicles in 2017 and which remain until 2020 are the Kia Sportage R jeep, the Chevrolet D-Max pick-up and the Chevrolet Sail

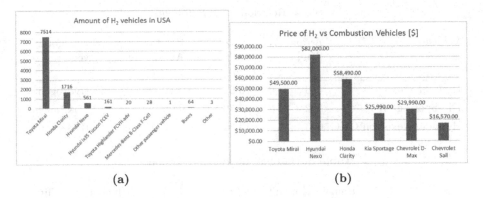

Fig. 3. (a) Amount of H_2 vehicles operating in USA (2021) [19]. (b) Price of H_2 vehicles vs Combustion [$] (2020–2021)

sedan [24,25]. The price for these cars is around $25,990, $ 29,990 and $16,570 respectively [26,28].

Table 3. Detail of price of H_2 and combustion vehicle

Vehicle	Price USD	Energetic vector	Avarage Price USD
Toyota Mirai	49,500.00	Hydrogen	63,330.00
Hyundai Nexo	82,000.00	Hydrogen	
Honda Clarity	58,490.00	Hydrogen	
Kia Sportage	25,990.00	Gasoline	24,183.33
Chevrolet D-Max	29,990.00	Diesel	
Chevrolet Sail	16,570.00	Gasoline	
		Relation	**2.62**

According to Fig. 3(b) and Table 3, commercial hydrogen cars not yet available in Ecuador are priced 2.62 times higher than the average price of the best-selling vehicles in the country in 2020. Comparing the autonomy and the price of filling the tank of the Hyundai Nexo and the Kia Sportage R, the following results are obtained: Both, the Hyundai Nexo and the Kia Sportage are 5-passenger Jeep-type vehicles. On the one hand, taking into account the information from the manufacturer of Nexo of 1 Kg of H_2 per 100 Km and the price of the Kilo of H_2 at $ 12, the Price/Autonomy ratio is 0.12 $/Km [29]. Taking advantage of the Green Hydrogen Catapult (GHC) project, the price of 1 Kg of H_2 would be $ 2 by 2026 and with it, the ratio drops to 0.02 $/Km in the Hyundai Nexo [30,31]. On the other hand, as of April 17, 2021, and with data from the manufacturer, the Kia Sportage R, with an extra gasoline price of $ 1.91 per gallon, requires $ 27.79 to cover the range of 552 km, obtaining a

Table 4. Relation price autonomy

Vehicle	Autonomy [Km]	Price fuel tank [$]	Relation Price/ Autonomy [$/Km]	Energetic vector
Hyundai Nexo	666	79.92	0.12	Hydrogen
Kia Sportage R	552	27.79	0.05	Gas
Hyundai Nexo con GHC	666	13.32	0.02	Hydrogen

Table 5. Gasoline vs. H_2 taxi economy

Vehicle	Range [Km]	Amount energetic vector	Price energetic vector	Cost energetic vector [USD]	Incomes [USD]	Utility [USD]
Gasoline Taxi (gal)	300	12	1.91	22.92	45	22.08
H2 GHC Taxi (Kg)	300	3	2.15	6.45	45	38.55

ratio of 0.05 \$/Km [26,32]. Paris in 2015 already had 5 hydrogen taxis. In 2019 the number of taxis amounted to 100, while by 2020 600 units were expected [33,34]. In 2018 the number of taxis registered in Ambato was 2055 units [35]. A taxi travels up to 300 km a day in 20 races consuming 10 to 12 gallons of extra gasoline. In this way, the driver of the unit has an income of \$ 40 to \$ 45 [36]. A taxi driver in Ambato if he used a hydrogen unit with GHC would obtain a daily profit of approximately \$ 38.55 versus \$ 22.08 that he obtains for using a gasoline vehicle. Amount of hydrogen required for the circulation of 2055 taxis during a year: 300 km $\frac{1\,Kg}{100\,Km}$ * 365 days * 2055 taxis = 2.2502 * 10^6 Kg (Tables 4, 5).

Tungurahua has a potential for hydrogen production from Solar Photo voltaic Energy, which in one year would be capable of producing 3.29 * 10^6 Kg, which is enough to cover the demand of the 2055 taxis [37]. H_2 vehicles need a safe "hydro generation" and continuous supply as infrastructure for refuelling this energy vector. A viable deployment of such a refuelling infrastructure requires an initial market for vehicles, such as city taxis. Using the city's fleet of taxis as consumers, the hydrogen chicken and egg dilemma can be overcome: "What comes first: Assemble the infrastructure to lower the price and promote consumption or Consumption to accelerate the implementation of the infrastructure". In addition to meeting the objectives of ensuring the supply of hydrogen, having a city refuel the energy vector and promote the use of H_2 [38] (Tables 6, 7 and 8).

The first principal component represents 75.8% of the total variation. The variables that are more correlated with Principal Component 1 (PC1) are: H_2 delivered (0.362), H_2 used (0.362), Usable H_2 (0.362), Pressure (−0.362), Gravi-metric capacity [%] (0.360) and Volumetric capacity [g/L] (0.360). PC1 has a positive correlation with the first three variables and the last two. However, it

Table 6. Summary simulation storage system 700 bar

N°	Detail	H_2 delivered [kg]	H_2 used [kg]	Usable H_2 [kg]	Gravimetric cap.[%]	Volumetric cap.[g/L]	Temperature[°C]	Pressure[bar]	Raw distance[miles]	Calculated fuel economy[mpgge]	Calculated range[miles]
1	Fuel Economy test (UDDS+HWY, 24C)	5.67	5.67	5.67	4.8	25.3	7.5	5	464	55	311
2	Aggressive cycle (US06, 24C)	5.66	5.66	5.66	4.8	25.3	−6.4	6	328	58	328
3	Cold cycle (FTP-75, −20C)	5.4	5.4	5.4	4.5	24.1	−29.1	25	403	75	403
4	Hot Cycle (SC03, 35C)	5.66	5.66	5.66	4.8	25.3	19.3	6	409	72	409

Table 7. Eigenanalysis of the correlation matrix

Principal component (PC)	1	2	3	4	5	6	7	8	9	10
Eigenvalue	7.5763	1.3769	1.0467	0.0000	0.0000	0.0000	0.0000	0.0000	0.0000	0.0000
Proportion	0.7580	0.1380	0.1050	0.0000	0.0000	0.0000	0.0000	0.0000	0.0000	0.0000
Cumulative	0.7580	0.8950	1.0000	1.0000	1.0000	1.0000	1.0000	1.0000	1.0000	1.0000

has a negative correlation with Pressure. So, increase the values of H_2 delivered, H_2 used, Usable H_2, Gravimetric capacity and Volumetric capacity will increase the value of the first principal component. But, increase the value of Pressure will decrease the value of PC1. The first three components have an Eigenvalue of more than 1. According to Kaiser criterion, components with a value of more than 1 are used as Principal Components. The first three principal components explain 100% of the variation of the data. So, PC1, PC2, PC3 are used. Also, the Scee plot shows that Eigenvalues form a straight line after the third principal component.Results show that principal component 1 has a positive association with H_2 delivered, H_2 used, Usable H_2, Gravimetric capacity and Volumetric capacity, but a negative association with Pressure. Five positive PC1 are the amount of H_2 before, while and after a hydrogen vehicle function. The second principal component is associated with Calculated fuel economy and calculated range. So, this component measures the autonomy of the vehicle. Finally, the third principal component is Raw distance. The loading plot shows the variables that are heavier on each component. Influences approaching −1 or 1 indicate that the variable significantly affects the component. Influences close to 0 means

Table 8. Eigenvectors

Variable	PC1	PC2	PC3	PC4	PC5	PC6	PC7	PC8	PC9	PC10
H_2 delivered [kg]	0,362	0,064	−0,062	0,398	0,172	−0,333	0,707	0,236	−0,081	0,009
H_2 used [kg]	0,362	0,064	−0,062	0,398	0,172	−0,333	−0,707	0,236	−0,081	0,009
Usable H_2 [kg]	0,362	0,064	−0,062	0,398	0,172	0,667	0,000	−0,471	−0,081	0,009
Gravimetric capacity [%]	0,360	0,073	−0,095	−0,476	0,187	0,408	0,000	0,577	−0,226	−0,208
Volumetric capacity [g/L]	0,360	0,073	−0,095	−0,476	0,187	−0,408	0,000	−0,577	−0,226	−0,208
Temperature [C]	0,302	0,466	0,091	−0,079	−0,700	0,000	0,000	0,000	−0,181	0,392
Pressure [bar]	−0,362	−0,061	0,050	0,165	−0,008	0,000	0,000	0,000	−0,912	−0,064
Raw distance [miles]	0,015	0,272	0,925	−0,000	0,224	0,000	0,000	0,000	0,034	−0,135
Calculated fuel economy [mpgge]	−0,272	0,536	−0,202	−0,113	0,533	0,000	0,000	0,000	−0,002	0,549
Calculated range [miles]	−0,227	0,629	−0,250	0,160	−0,129	0,000	0,000	0,000	0,111	−0,660

that the variable has a low influence on the component. In Fig. 4(b), Volumetric capacity, H_2 used, Usable H_2, Gravimetric capacity and H_2 delivered have a positive influence on component one. Pressure has a negative influence on PC 1. Calculated fuel economy and Calculated range have a strong positive influence on principal component 2. It means autonomy of the vehicle, an important factor now to decide which car to buy.

Table 9. Eigenvectors for PC

Variable	PC1	PC2	PC3
H_2 delivered [kg]	0.362	0.064	−0.062
H_2 used [kg]	0.362	0.064	−0.062
Usable H_2 [kg]	0.362	0.064	−0.062
Gravimetric capacity [%]	0.360	0.073	−0.095
Volumetric capacity [g/L]	0.360	0.073	−0.095
Temperature [$°C$]	0.302	0.466	0.091
Pressure [bar]	−0.362	−0.061	0.050
Raw distance [miles]	0.015	0.272	0.925
Calculated fuel economy [mpgge]	−0.272	0.536	−0.202
Calculated range [miles]	−0.227	0.629	−0.250

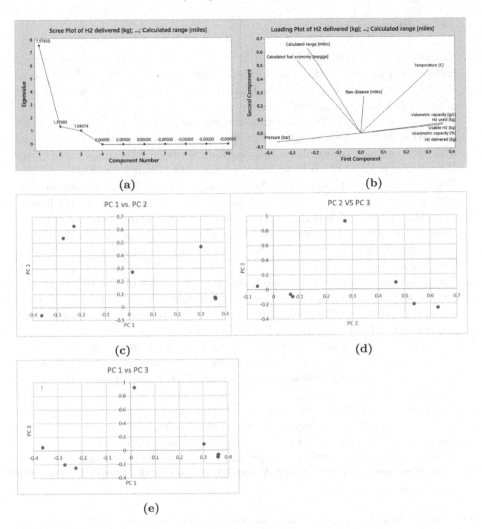

Fig. 4. (a)Scree plot, (b)Loading plot, (c) PC 1 vs PC 2, (d) PC 2 vs PC 3, (e) PC 1 vs PC

Information from Table 9. can be used with other methodology such as Factor Analysis to improve knowledge. Principal Components become the new axes that provide the best angle for viewing and evaluating the data, so that the differences between the observations are more visible. Figure 4(c), (d), and (e) represent a contrast the three principal components to each other. Values close to 1 are strong correlated. A point on the corners of the graphs is expected for a strong correlation between both principal components. Points close to X axis and 1 value means that a strong relation only with that principal component. The same happened with Y axis.

5   Conclusion

1. Ecuador has the potential to produce hydrogen through biomass and photo voltaic solar energy despite the different current barriers. Similarly, the possibility of implementing a Nuclear Plant to produce H_2. The goal is the cost of $ 2.15 per kilogram of hydrogen.
2. The hydrogen vehicle, Toyota Mirai, is the most accessible car for the implementation of this technology in the country due to its price. Commercial hydrogen cars not yet available in Ecuador are priced 2.62 times higher than the average price of the best-selling vehicles in the country in 2020.
3. The comparison of the Hyundai Nexo to H2 with the gasoline-powered Kia Sportage shows that if the first car circulated in Ecuador, it would have a Price-Range ratio of $ 0.12/km compared to $ 0.05/km of the second car. However, if one of the advantages of the Green Hydrogen Catapult project were applied, the ratio would be more favourable for the Hyundai Nexo with 0.02 $/Km.
4. The profit that a taxi driver would have after a workday of 8 to 10 h travelling 300 km approximately is $ 22.08 using a gasoline vehicle at $ 38.55 if using a hydrogen vehicle and with the advantage of the Green Hydrogen Catapult project or the Implementation of the Nuclear Plant.
5. The photo voltaic solar energy that is generated in the Province of Tungurahua would cover the demand of 2,055 city taxis in the city of Ambato. Furthermore, with the hydrogen-powered taxi fleet, the hydrogen chicken and Egg dilemma can be broken.
6. H2 Vehicle Simulation Framework is a three-dimensional scale that assesses the H_2 amount of the vehicle, the autonomy of the car and raw distance. This dimension will be important at the moment to buy an H_2 vehicle.
7. A strong correlation between two principal components can be appreciate on a graph where a point of X and Y axis should be in the corner because the values will be close to 1.

Acknowledgements. Authors have no conflict of interest. Mr. Joffre Espin, Mr .Eduardo Estevez thank for research support of the seed grant "Computational modelling of biomaterials and applications to bioengineering and infectious disease, Universidad Technologica Indoamérica, Ecuador", awarded to S.P.T.

Source. https://github.com/joffreespin/PaperH2

References

1. Ramírez-Samperio, J.A.: Las cinco formas más elegantes de morir en el Universo, vol. 1, no. 1 (2019). https://doi.org/10.29057/h.v1i1.4777
2. Aguado, R., Casteleiro-Roca, J.-L., Jove, E., Zayas-Gato, F., Quintián, H., Calvo-Rolle, J.L.: Hidrógeno y su almacenamiento: el futuro de la energía eléctrica (2021). https://doi.org/10.17979/spudc.9788497497985

3. Fúnez Guerra, C., Almasa Rodríguez, E., Fuentes Ferrara, D.: El hidrógeno: vector energético del futuro (2010)
4. Donoso Quimbita, C., Ortiz Bustamante, V., Amón De La Guerra, B., Herrera Albarracín, R.: Diseño de un reactor continuo para la producción de hidrógeno y acetaldehído a partir de etanol en Ecuador. UTCiencia **5**(1), 30–40 (2018)
5. Carvajal Osorio, H., Babativa, J., Alonso, J.: Estudio sobre producción de H con hidroelectricidad 2 para una economía de hidrógeno en Colombia. Ing. Compet. **12**(1), 31–42 (2010)
6. García Nieto, A.: Centrales hidroeléctricas en Ecuador (2018)
7. Posso Rivera, F., Sánchez Quezada, J.: La Economía del Hidrógeno en el Ecuador: oportunidades y barreras. Av. Cienc. Ing. **6**(2) (2014). https://doi.org/10.18272/aci.v6i2.187
8. Pazera, G., Młodawska, M., Młodawski, J., Klimowska, K.: Principal component analysis of munich functional developmental diagnosis. Pediatr. Rep. **13**, 227–233 (2021). https://doi.org/10.3390/pediatric13020031
9. Jahirul, M.I., et al.: Investigation of correlation between chemical composition and properties of biodiesel using principal component analysis (PCA) and artificial neural network (ANN). Renew. Energy **168**, 632–646 (2021). https://doi.org/10.1016/j.renene.2020.12.078
10. Bandyopadhyay, S., Thakur, S.S., Mandal, J.K.: Product recommendation for e-commerce business by applying principal component analysis (PCA) and K-means clustering: benefit for the society. Innov. Syst. Softw. Eng. **17**, 45–52 (2021). https://doi.org/10.1007/s11334-020-00372-5
11. Patil, H., Dwivedi, A.: Prediction of properties of the cement incorporated with nanoparticles by principal component analysis (PCA) and response surface regression (RSR). Mater. Today Proceedings **43**(2), 1358–1367 (2021). https://doi.org/10.1016/j.matpr.2020.09.170
12. HyMARC: Hydrogen Storage Systems Modeling. https://www.hymarc.org/models.html. Accessed 17 Apr 2021
13. Kalaivani, S., Shantharajah, S.P., Padma, T.: Agricultural leaf blight disease segmentation using indices based histogram intensity segmentation approach, pp. 9145–9159 (2020). https://doi.org/10.1007/s11042-018-7126-7
14. Gray, V.: Principal Component Analysis: Methods, Applications and Technology (2017)
15. HydrogenTools: Merchant Hydrogen Plants (2016). https://h2tools.org/hyarc/hydrogen-data/merchant-hydrogen-plant-capacities-asia. Accessed 17 Apr 2021
16. Deokattey, S., Bhanumurthy, K., Vijayan, P.K., Dulera, I.V.: Hydrogen production using high temperature reactors: an overview. Techno Press **1**(1), 013–033 (2013). https://doi.org/10.12989/eri.2013.1.1.013
17. Kim, J., Lee, K., Kim, M.: Calculation of LUEC using HEEP software for nuclear hydrogen production plant. Trans. Korean Nucl. Soc. Spring Meet. Jeju Korea **47**(8) (2015)
18. HydrogenTools: Worldwide Refinery Hydrogen Production Capacities (2017). https://h2tools.org/hyarc/hydrogen-data/refinery-hydrogen-production-capacities-country. Accessed 17 Apr 2021
19. HydrogenTools: US Hydrogen-Fueled Vehicles (2021). https://h2tools.org/hyarc/hydrogen-data/inventory-us-over-road-hydrogen-powered-vehicles. Accessed 17 Apr 2021
20. HydrogenTools: International Hydrogen Fueled Vehicles (2021). https://h2tools.org/hyarc/hydrogen-data/inventory-international-over-road-hydrogen-powered-vehicles. Accessed 17 Apr 2021

21. Toyota: Mirai 2021 (2021). https://www.toyota.com/espanol/mirai/. Accessed 17 Apr 2021
22. Hyundai: Nexo (2021). https://www.hyundai.com/es/modelos/nexo.html. Accessed 17 Apr 2021
23. Edmunds: 2021 Honda Clarity Electric (fuel Cell) (2021). https://www.edmunds.com/honda/clarity/2021/electric-fuel-cell/. Accessed 17 Apr 2021
24. Ulloa Masache, J.L., Velasco Vicuña, A.F.: Evaluación del consumo de combustible en vehículos, utilizando diferentes estrategias cambios de marcha. Universidad del Azuay (2018)
25. El Universo: Cuáles han sido los 10 carros preferidos en Ecuador en 2020, Motores (2020). https://www.eluniverso.com/entretenimiento/2020/11/24/nota/8060893/cuales-han-sido-10-carros-preferidos-ecuador-2020/. Accessed 17 Apr 2021
26. Chevrolet: Sail (2021). https://www.chevrolet.com.ec/autos/sail-sedan. Accessed 17 Apr 2021
27. Kia: Sportage R (2021). https://www.kia.com/ec/showroom/sportage-r.html. Accessed 17 Apr 2021
28. Chevrolet: D-Max (2021). https://www.chevrolet.com.ec/pick-ups/dmax-hi-ride-pick-up. Accessed 17 Apr 2021
29. Motor.es: El hidrógeno en los coches: ventajas e inconvenientes (2021). https://www.motor.es/que-es/hidrogeno::text=Celkilode hidrógeno,%2C9 kg%2F100 km. Accessed 30 Apr 2021
30. Hyundai: Todo sobre Hyundai NEXO (2021). https://www.hyundai.com/es/zonaeco/eco-drive/modelos/todo-sobre-hyundai-nexo-primer-coche-hidrogeno-espana::text=ElHyundaiNEXOdisponede,%2C2itroscadauno. Accessed 17 Apr 2021
31. United Nations Framework Convention on Climate Change: Green Hydrogen Catapult (2020). https://racetozero.unfccc.int/green-hydrogen-catapult/. Accessed 17 Apr 2021
32. Petroecuador: Precios de venta de combustibles (2021). https://www.eppetroecuador.ec/?p=8062. Accessed 17 Apr 2021
33. La Vanguardia: El hidrógeno alimenta nuevos taxis en París (2015). https://www.lavanguardia.com/motor/tendencias/20151210/30714240786/hyundai-ix35-fuel-cell-taxi-hidrogeno-paris.html. Accessed 01 May 2021
34. FuelCellWorks: Taking a Taxi in Paris: Now with Hydrogen! (2019). https://fuelcellsworks.com/news/taking-a-taxi-in-paris-now-with-hydrogen/. Accessed 01 May 2021
35. El Heraldo: Matriculados 24.063 vehículos en Ambato (2018). https://www.elheraldo.com.ec/matriculados-24-063-vehiculos-en-ambato/. Accessed 01 May 2021
36. El Comercio: Costos operativos de taxis se ajustarán con el subsidio (2019). https://www.elcomercio.com/actualidad/taxis-subsidio-costos-operativos-gasolina.html : :text=Mera calcula que un taxi, una utilidad de hasta 25. Accessed 01 May 2021
37. Posso, F.R., Sánchez, J.P., Siguencia, J.: Estimación del Potencial de Producción de Hidrógeno a partir de Energía Solar Fotovoltaica en Ecuador. Rev. Técnica "energía" (12), 373–378 (2016)
38. Campíñez Romero, S., Colmenar Santos, A., Pérez Molina, C., Mur Pérez, F.: A hydrogen refuelling stations infrastructure deployment for cities supported on fuel cell taxi roll-out. Energy **148**, 1018–1031 (2018). https://doi.org/10.1016/j.energy.2018.02.009

A Neural Network Approach to High Cost Patients Detection

Franklin Messias Barbosa$^{(\boxtimes)}$ and Renato Porfirio Ishii$^{(\boxtimes)}$

Federal University of Mato Grosso do Sul, Campo Grande, MS, Brazil
franklin.barbosa@aluno.ufms.br, renato.ishii@ufms.br

Abstract. The growing aging of the world's population along with several environmental, social and economic factors, end up posing major challenges for public health. One challenge is the detection and treatment of high cost patients, i.e., a small percentage of patients associated with majority of expenditures in healthcare. The early detection of patients who may become high cost in the future can be used to better target interventions focusing on preventing their transition or, in the case of those who are already in such condition, to allow appropriate approaches, rather than generic ones. In both cases, the detection of such patients can be beneficial, reducing avoidable costs and improving patients' condition. In order to make such detection, this work has focused on using deep learning techniques, specifically, Neural Networks, along with a dataset composed of survey answers applied by the United States government, called Medical Expenditure Panel Survey (MEPS) and attributes gathered from the literature. For the purposes of this work, 11 years of the MEPS dataset were considered, including the years from 2006 to 2016. The models created have shown results ranging between 83% to 90% on metrics such as accuracy, precision, recall, specificity and f1-score. This work also aimed to make the creation and testing of such networks easier, by providing the tools developed during its evolution on GitHub.

Keywords: Machine learning · Deep learning · Big data · Classification · Smart cities

1 Introduction

Public health is an area of great importance both globally and nationally, which also makes it important to efficiently manage the resources available in this area. One of the challenges in this regard, which can be observed in both public and private contexts, are high cost patients, i.e. a small portion of users who represent a large share of the total expenditure. Such a phenomenon can be observed not only in Brazil [10], but also in several other countries, such as Canada [14] and USA [11].

In a report published in 2013, the Kaiser Family Foundation, an American non-profit organization that works on public health issues in the USA, points out

© Springer Nature Switzerland AG 2021
O. Gervasi et al. (Eds.): ICCSA 2021, LNCS 12951, pp. 527–540, 2021.
https://doi.org/10.1007/978-3-030-86970-0_37

that in Medicaid [8], public health plan in the United States, 5% of the highest cost beneficiaries at the time accounted for 54% of total expenses [11].

Thus, a possible approach would be to identify such patients while they still have average costs, that is, before they become high cost, as pointed out by Chechulin [6]. Such identification may be useful as a guide to targeted interventions, in order to prevent the transition of the patient, preventing the condition complications already present and preventing the emergence of new problems, when possible, thus improving not only the efficiency of the agents responsible for healthcare, but also preventing avoidable expenses.

Similarly, the identification of patients who are already high cost is also beneficial, as stated by Blumenthal [5], high cost and high need patients are usually people that even receiving substantial services, have critical health needs that are not attended to. This portion of the population often receives inefficient care, such as unnecessary hospitalizations. According to Blumenthal [5], by giving high priority to the care of these individuals, the resources can be used where they are likely to produce better results, with lower costs.

1.1   Problem Definition

Given the current economic scenario, in addition to factors such as population growth and aging, adequate resource management is becoming increasingly important, especially in critical areas such as health. In order to assist decision making and, consequently, optimize efforts and resources, adequate data processing technologies are essential.

In this scenario, the use of machine learning algorithms provides an improvement in the detection of high cost patients compared to the traditional methods used previously. A possible alternative is to use Deep Learning approaches to deal with this problem. As this technique ends up discovering previously unknown interrelationships in the data sets in which they are applied, our hypothesis is that its use may present an improvement in the results found in previous works.

1.2   Objectives

The objectives of this work include the implementation of deep learning models for the high cost patient prediction problem, as well as to compare the results with those obtained in other previous studies, the development of the necessary tools for creating such models in order to simplify future projects in this area. And, as a final objective, to validate the set of attributes defined by Shenas [15] in a larger and more current dataset.

In order to achieve these objectives, different datasets were initially considered and after analyzing their contents, availability, ease of access and amount of data, the MEPS dataset was chosen for this work, more specifically, 11 years were considered, including the years of 2006 to 2016.

This paper is organized as follows: Sect. 2 presents brief theoretical reference, Sect. 2.1 expands on how the MEPS dataset is structured and what information can be obtained from it. Section 2.2 explains the preprocessing steps taken to build the final dataset. Section 2.3 presents an explanation on how the preprocessing steps were implemented. Finally, on Sect. 2.4, the metrics used to evaluate the constructed model are presented and explained. Section 3 summarises the results obtained from the constructed models, and Sect. 4 the results and contributions that were derived from this work.

2   Materials and Methods

2.1   Dataset

The Medical Expenditure Panel Survey, or MEPS, "is a set of large-scale surveys of families and individuals, their medical providers, and employers across the United States. MEPS is the most complete data source on the cost and the use of health care and health insurance coverage" [2]. The results derived from these surveys are periodically released as public domain data, and aim to present information on which health services are used more frequently, how frequently they are used, what are the associated costs, how they are paid for. In addition, to several other information that may be important for research in this area, including, mainly, information at the individual level. Concerning patient privacy, according to AHRQ (Agency for Healthcare Research and Quality), all information is collected with assurances of confidentiality, and are to be used for research purposes only, with unauthorized disclosure of survey data being subject to substantial fines [3].

The MEPS dataset is divided by years, and further divided by panels, with each year, a new panel is started, that panel will be concluded in the following year, meaning that a full panel contains data for each individual during a two years period. These data include information on several topics defined by the AHRQ as follows [4]:

- Unique person identifiers and survey administration variables, which contain information related to conducting the interview, the household and family composition and person level status code;
- Geographic Variables, which indicate the census region for the individual;
- Demographic Variables, which include information about the demographic characteristics of each person, such as age, sex, race, ethnicity, marital status, educational attainment and military service;
- Income and Tax Filing Variables provide income and tax-related information, such as poverty status and number of dependents;
- Person-Level Condition Variables, which refers to the individual's physical and mental health status, including several disease diagnostics, such as angina, bronchitis, high cholesterol, among others. In addition, it includes information on how old the individual was when the condition was diagnosed, if the condition persists or if it was recently treated.
Information on the diagnostics of rare diseases, more specifically, for types of cancer that appear on the list of rare diseases of the National Institutes of

Health, such as cancer of the esophagus, pancreas, larynx, leukemia, among others are presented in the same way, in order to protect the anonymity of the interviewees;

- Health Status Variables include information about the individual's physical conditioning, addressing issues such as the difficulty of lifting 10 pounds, climbing 10 steps, walking 3 blocks, walking 1 mile, standing up for 20 min, and other activities that are common in everyday life, such as crouching, or picking up elevated objects, in addition to other physical limitations, such as low vision or hearing. It also indicates if there are any social or cognitive limitations, as well as behavioral factors. Finally, information about the clinical history is included in this class of variables, indicating the time since certain procedures were performed, including mammography, colonoscopy and hysterectomy, in addition to others;
- Disability Days Variables, which contains questions about time lost from work or school because of a physical illness or injury, or a mental or emotional problem or days lost treating other people with such conditions;
- Access to Care Variables carry information on basic access to healthcare, what is the provider and if they find any difficulty when needing such services;
- Employment Variables consist of person-level indicators such as employment status, and hourly wage and occupation area;
- Health Insurance Variables indicate which health insurance the interviewee has, when applicable, and information on the history of plans used;
- Finally, utilization, expenditure, and source of payment variables contain information such as the number of times that each health service was used, such as outpatient visits, dental services, prescription drugs, emergency care, among others. The total cost of the procedures associated with the interviewee and how they were paid for. For the problem of high cost patients, these variables can be used to define the classes of each individual in the training set.

2.2   Data Preprocessing

The data preprocessing phase consists of a set of techniques and activities on the chosen dataset. This process is intended to prevent a series of problems that may arise in the future due to the state in which the data can be received at first, in addition to ensuring that a greater amount of relevant information can be extracted from the original dataset, contributing to the quality of the proposed models.

For the purposes of this work, the processes described by Shenas [15] were followed, in a work that defined the minimal set of attributes on the MEPS dataset. More specifically, only individuals older than 17 were considered in his work, those that have non negative record for their expenditure (values for neither years are missing) and those with a positive person-level weight, since records with a zero person-level weight do not represent any part of the population and should not be considered in studies conducted on the US household population [15]. In total, 39 attributes were chosen, divided in 5 categories:

1. Demographics
   - Age, Sex, Race, HIDEG (highest degree of education), Region, Marry and POVCAT (family income as percent of poverty line);
2. Health Status
   - RTHLTH (perceived health status), MNHLTH (perceived mental health status), ANYLIM (any limitation on daily activities), and BMINDX (body mass index);
3. Preventive Care
   - Check (routine checkup), BPCHEK (blood pressure check), CHOLCK (cholesterol check), NOFAT (if a doctor advised a lower cholesterol diet), EXRCIS (if a doctor has advised more exercise), ASPRIN (frequent aspirin use), BOWEL (sigmoidoscopy or colonoscopy), STOOL (blood stool test), and DENTCK (frequency of dental checkup);
4. Priority Conditions
   - HIBPDX (high blood pressure), CHDDX (coronary heart disease), ANGIDX (angina), MIDX (myocardial infarction), OHRTDX (other heart disease), STRKDX (stroke), ASTHDX (asthma), EMPHDX (emphysema), CANCERDX (cancer), DIABDX (diabetes), ARTHDX (arthritis), CHOLDX (high cholesterol), PC (presence of any priority condition), and PCCOUNT (number of conditions present);
5. Visits Counts
   - OBTOT (office based visits), OPTOT (outpatient visits), ERTOT (emergency room visits), IPDIS (inpatient hospital stays), and RXTOT (number of prescription medicines).

These attributes were also preprocessed according to Shenas' [15] results, mainly, several reductions in the number of possible categories were proposed by them, these reductions include joining similar categories, time periods or normalizing values, in order to reduce the number of possible values for each attribute, making their interpretation easier [15], those changes are detailed below.

"Age" was divided in 4 categories, ages in the 18–49 range were replaced by 1, 50–65 by 2 and those older than 65 by 3. In addition, all records with an age value lesser than 18 are removed, since many of the other attributes are not applicable in these cases. "Race", originally a categorical variable with 6 possible values was reduced to 3 categories, with values of 1, 2 or 3 for whites, blacks and others, respectively. "HIDEG" was changed from a categorical variable that represented the Highest Degree of Education (high school, bachelor's, master's, doctorate, etc.) to a binary flag, with 1 for having a higher education, and 0 for lacking it.

The Marry variable was reduced from its original 10 possible values to a 0–2 range, with 0 being "never married", 1 for "married" and 2 for "widowed", "separated" or "divorced", this was done by excluding the category for underage people, since all minors were excluded from the final dataset, by joining the widowed, separated and divorced classes. The Marry attribute also had classes for people who married, widowed or separated in the current round, those were also converted to the new value.

Finally, POVCAT, represents the family's income as percent of poverty line. Originally, this attribute had 5 possible values, ranging from 1–5 and representing "poor/negative", "near poor", "low income", "middle income" and "high income", respectively. In the final dataset, these values were reduced to a 1–3 range, with 1 representing "poor/negative/near poor", 2 representing "low/middle" and 3 "high income". The remaining attributes from the demographics category were left unchanged.

From the Health Status category, aside from BMINDX, the body mass indicator, all attributes are left unchanged. BMINDX originally contained a continuous value and was changed to an ordinal attribute on the range of 1–4, representing "underweight", "normal weight", "overweight" and "obese". RTHLTH and MNHLTH represent the individual's self perceived physical and mental health status, in a 1–5 range, representing "excellent", "very good", "good", "fair" and "poor" health. Finally, ANYLIM is a binary attribute indicating if the individual has any limitations in activities of daily living, such as eating or dressing or limitations in instrumental activities of daily living, such as housework or shopping. The preventive care category contains attributes that indicate the time since the individual's last routine checkup (Check), blood pressure test (BPCHEK), cholesterol check (CHOLCK), if a doctor has advised the person to eat fewer high cholesterol foods (NOFAT), if a doctor has advised the individual to exercise more (EXRCIS), if the person takes aspirin frequently (ASPRIN), if the person ever had a sigmoidoscopy or colonoscopy (BOWEL), if a blood stool test was ever necessary (STOOL) and frequency of dental check-up (DENTCK). All attributes on this category are either binary, in which case were left unchanged, or follow a similar pattern for values, ranging from 1–6, where each value represents the number of years since said event or exam occurred, exceptions being 5, which represents 5 or more years and 6, representing never. These values were reduced to a 0–2 range, where 0 represents never, 1 representing said event occurred within the past year and 2 for 2 years or more. Additionally, DENTCK was changed to a 0–2 range, where 0 represents never, 1 being twice a year or more and 2 less than twice a year.

The priority conditions category contains information on diagnostics of several health conditions for each individual, such as high blood pressure (HIBPDX), coronary heart disease (CHDDX), angina or angina pectoris (ANGIDX), heart attack, or myocardial infarction (MIDX), other heart diseases (OHRTDX), stroke (STRKDX), asthma (ASTHDX), emphysema (EMPHDX), cancer (CANCERDX), diabetes (DIABDX), arthritis (ARTHDX) and high cholesterol (CHOLDX). Since all attributes are binary, all were left unchanged. The PC and PCCOUNT attributes are not present in the original MEPS dataset, being derived from the previous attributes, PC is a binary attribute that indicates if the individual has any of the listed priority conditions and PCCOUNT accepts values in a 0–4 range, where 0–3 is the number of conditions and 4 represents 4 or more conditions.

Finally, the visits counts category indicate the number of visits the individual has had to several medical providers, such as office based (OBTOT), hospital outpatient (OPTOT), emergency room (ERTOT), inpatient hospital stays (IPDIS) and the number of prescription medicines (RXTOT). All attributes in this category were normalized using minmax scaling, which changes every value to a 0–1 range using the equation described in 1, where x' is the new value for the attribute, x_{min} is the minimal value found for that attribute, and similarly, x_{max} is its maximum value.

$$x' = \frac{x - x_{min}}{x_{max} - x_{min}} \tag{1}$$

2.3 Preprocessing Implementation

As one of this work's objectives, the whole preprocessing and data preparation steps defined by Shenas [15] and explained on the previous section were automated for future studies using the MEPS dataset, this process was divided in several steps, detailed below, allowing easier understanding, accessibility and eventual modification of the source code. The only prerequisites for using this implementation are a working Python 3 environment and the required MEPS data files, in csv format.

1. The first step consists on applying the conditions mentioned before to the original full year dataset. This step excludes records with a personal level weight of 0, those with an age lesser than 18 and those with a missing value for total expenditure. Since at this point we have the full dataset, adding or removing restrictions can easily be done by editing this file;
2. The second step is used only for the 2006 and 2007 full year files. Prior to 2008, the cancer diagnostic attribute was not present in the full year file, being instead found in a separate medical conditions file. This step searches the medical condition file and appends the cancer diagnostic attribute to the main file. Files from 2008 onwards can skip this step;
3. The third step creates the PC and PCCOUNT attributes, by counting how many priority conditions are present on each record. Priority conditions can be easily changed on this step, by simply adding or removing their names from the list;
4. The fourth step retrieves the second year expenditure for each individual in the main file. Records without a second year expenditure value are excluded;
5. The fifth step creates a HIGHCOST attribute for each individual, at this point the user can choose whether to consider the first or second year expenses to determine the high cost population, in addition to the percentage of the dataset that will be considered high cost. In other words, the user can choose to consider the top 5% as the high cost population considering their second year expenses, in this scenario, 5% of the dataset, those with the highest values for their total expenditure on year 2, will have the HIGHCOST attribute set to true, while the remaining will have it as false;
6. The sixth step applies changes to the values of the records, applying minmax normalization, or changing value intervals, according to the specifications

listed above. Over the years, several fields in the MEPS dataset have had their names, or even values, changed. HIDEG, for instance, the field representing the highest degree of education an individual has achieved, is not present in the 2014 and 2013 full year files, instead, we use a different field, called EDUYRDG (years of education or highest degree) which provides similar information. In this step we take such changes into account, ensuring that each of the 39 variables described above are present;

7. The seventh step deals with missing or invalid values in each of the relevant fields in the dataset. By default, MEPS uses certain values for questions the individual has failed to provide an answer for, for instance, if they do not know the answer to a question, the value is set as -8, if they refused to answer, the value is set as -7, and so on for various other reasons. For our purposes, such differences are not relevant, as such, for every field where such notation is used, every value under 0 is replaced by -1, this is done so that all invalid values are equal, preventing invalid assumptions to be made based on the original MEPS value notation. Furthermore, this also allows for an easier identification of invalid values, in case the user wishes to exclude them from the final dataset or impute them by using other methods;

8. The eighth step can be further divided into 4 parts, in short, it allows the user to add fields that were not included in the original 39, and, depending on the format the data is in, either normalize it, reduce its range (for the variables that refer to time since last event) and remove negative values as mentioned above. This allows the user to easily change the fields in the final dataset, while applying the necessary data preprocessing. At the end of this step all fields not used are removed from the dataset;

9. The ninth step handles the final dataset, in it, we divide the dataset into high cost and low cost, based on the variable created previously, in order to keep the dataset balanced, we count the number of records in the high cost dataset, and randomly pick the same number of records in the low cost dataset. Since the low cost and high cost datasets are not deleted in the process, this can be repeated several times, each time creating a new low cost dataset with randomized records.

2.4   Model Evaluation

Assuming any binary classifier that takes an example and return as a hypothesis one between two possible classes, "A" and "B", for example, any prediction returned by that classifier will fit into one of four possible scenarios: the model predicted class correctly predicted "A" for the example; the model predicted class "A", but the example belonged to class "B"; the model returned "B" as a prediction and the example was class "A"; or, finally, the model returned "B" and the example really belonged to the "B" class. Viewing the classes "A" and "B" as positive and negative, the cases can be represented with matrix 1, where the first line represents the prediction returned by the classifier, and the first column represents the true class of the example. This table is called the confusion matrix [9].

Table 1. Example of a confusion matrix

	A	B
A	True Positive	False Negative
B	False Positive	True Negative

After constructing a model, at the end of the testing process, the matrix can be filled, according to the results found, and from it some metrics are calculated that can be useful when assessing the quality of the constructed model. The first metric is accuracy, which represents the rate at which the model was able to correctly predict the class of any given example, and is defined in 2.

$$Accuracy = \frac{True\ Positive + True\ Negative}{Total\ Number\ of\ Examples} \qquad (2)$$

Accuracy presents a valid metric for the analysis of models, however, it is not reliable in certain cases. More specifically, it does not represent well the predicting capacity of a model whose dataset is unbalanced [16]. In other words, if most of the examples in the test set are from the same class, we can have a model that classifies all entries as being in the majority class. As most examples will be predicted correctly, the accuracy of this model will be high, even if it is not actually making any kind of prediction itself. Thus, there are other metrics, similar to accuracy, that are less sensitive to this type of scenario and can be used in conjunction with it to better represent the capacity of the evaluated model.

One of these metrics is the recall, also referred to as sensitivity, which measures the hit rate of the positive class. In other words, of all the examples that belong to the positive class, the recall indicates how many were classified correctly. It can be calculated using Eq. 3. A model that does not produce false negatives has a value of 1. A third metric that can be considered is precision, defined by Eq. 4, which indicates the proportion of predictions for the positive class that were correct. A model that does not produce false positives has a precision of 1. Finally, the rate of false positives is similar to recall, and is defined by Eq. 5 [9,13].

$$Recall = \frac{True\ Positive}{True\ Positive + False\ Negative} \qquad (3)$$

$$Precision = \frac{True\ Positive}{True\ Positive + False\ Positive} \qquad (4)$$

$$False\ Positive\ Rate = \frac{False\ Positive}{False\ Positive + True\ Negative} \qquad (5)$$

From these three metrics, a fourth can be derived, the f1 score, which is defined as the harmonic mean between precision and the recall and is calculated according to Eq. 6. It gives equal weight to both measures, so the f1 score of a model is high if both recall and accuracy are high. On the other hand, if either one decreases, the final f1 score will also be lower. This metric is important for the evaluation of models where both metrics have similar importance, that is, in problems where it is not advantageous to increase either of them at the expense of the other [13].

$$f1 = 2 \times \frac{Precision \times Recall}{Precision + Recall} \tag{6}$$

3 Results

The resulting dataset was used to train several neural networks using Keras [7], 11 years (2006–2016) of data from the MEPS dataset were used, resulting in a dataset with $125,457$ total records. We have considered the top 5% of the population as high cost, as such, the dataset can be divided into high cost, containing $6,280$ records, and low cost, containing $119,177$. Since the classes are clearly unbalanced, we follow the steps detailed in the previous section, keeping the $6,280$ original high cost individuals and randomly selecting the same quantity of low cost records, keeping the dataset balanced. During training, the resulting dataset is divided randomly between training and validation sets, where the training set contains 80% of the records, while the validation set contains the remaining 20%.

Each model was trained separately 100 times, each using different low cost records, the results for each network were logged, and an average result for each metric was calculated using the results found in each iteration. This process was done twice, once using the same year expenditure as a parameter to define high cost individuals, and then using the second year's expenses.

The models created were all fully connected, sequential networks. Different networks created had a different number of layers, number of nodes in each layer, different activation functions, number of dropout layers and dropout probability.

The model used to generate the results shown on Tables 2 and 3 contained 3 hidden layers, with 32, 64 and 32 nodes respectively, with a dropout layer between layers 1 and 2, and another, between layers 2 and 3, the dropout rate for both of those layers was set to 0.5. A final dropout layer was added between the third and output layers, with a dropout rate of 0.1. The activation function used on the layers was the Rectified Linear Unit (relu), for the hidden layers, and the hyperbolic tangent function (tanh) for the output layer. When compiling the model, the RMSprop algorithm was used as the optimizer, the loss function used was binary cross entropy and the metric accuracy. Finally, the model was trained for 100 epochs, with a batch size of 32.

For the same year prediction, the metrics for the best performing model are detailed on Table 2. For models trained to predict high cost individuals of the following year, the best results were detailed on Table 3.

All results are similar to the ones obtained by neural networks constructed by Shenas [15], shown on Table 4 except for accuracy, which was slightly higher. This corroborates the minimal set of attributes defined in that work, in addition to

Table 2. Metrics for our model when using the top 5% expenditures as target for the same year

	Value	Standard deviation
Sensitivity	90%	0.018
Specificity	83%	0.02
Accuracy	86%	0.007
Precision	84%	0.01
F1 score	87%	0.006

Table 3. Metrics for our model when using the top 5% expenditures as target for the following year

	Value	Standard deviation
Sensitivity	91%	0.02
Specificity	84%	0.03
Accuracy	88%	0.009
Precision	86%	0.02
F1 score	88%	0.008

suggesting that similar or better results can be achieved with a greater quantity or more recent data.

Table 4. Metrics presented by Shenas [15] when using the top 5% expenditures as target using decision trees (C5.0 and CHAID) and Neural Networks (NN)

	C 5.0	CHAID	NN
Sensitivity	56%	90%	87%
Specificity	96%	86%	86%
Accuracy	96%	86%	86%

As a second point of comparison we can point the results presented by Meehan [12], shown on Table 5, who used traditional machine learning algorithms to generate models for the same problem.

A final point of comparison can be made by observing the confusion matrices generated by both our work and the ones reported by Shenas, shown on Tables 6, 7 and 8. On all three Tables, we compare the total number of predictions for each case, similar as what can be observed on Table 1. On the top headers, "Predicted HC" means the model predicted an individual as a high cost patient, while "Predicted LC" means the model predicted the individual as not being high cost, similarly, the headers on the left show the real class for the

Table 5. Results presented by Meehan [12]

	Logistic regression	Naive Bayes	J48
Sensitivity	74%	**74.10%**	67.20%
Specificity	77.10%	76.90%	**82.20%**
Accuracy	**75.55%**	75.49%	74.73%
F-Measure	**75.50%**	**75.50%**	74.60%

predictions, with "Actual HC" meaning the individual was a high cost patient, and "Actual LC" the opposite. This way, it is possible to grasp how many wrong classifications each model has made for each of the possible error types (classifying a high cost as a non high cost or classifying a non high cost as high cost), as well as how many predictions were correctly made.

Table 6. Confusion matrix presented by Shenas [15] when using the top 5% expenditures as target (Neural Network)

	Predicted HC	Predicted LC
Actual HC	405	56
Actual LC	891	6884

Table 7. Confusion matrix presented by Shenas [15] when using the top 5% expenditures as target (CHAID)

	Predicted HC	Predicted LC
Actual HC	460	52
Actual LC	1262	7790

Table 8. Confusion matrix generated by one of our tests, when using the top 5% expenditures as target

	Predicted HC	Predicted LC
Actual HC	1160	229
Actual LC	100	1023

The matrices show that, although the previously mentioned metrics were similar, the proportion of cases where the model returned a high cost prediction for a low cost patient was higher on Shenas' model. For the neural network model this puts precision at around 31%, with the CHAID model at 26%, while the average precision for the models generated in this work is over 80%.

4   Conclusion and Contributions

The main conclusions drawn from this work include the validation of the previously defined set of attributes using more data from a wider period and the results obtained in the high cost patient prediction problem using neural networks, both conclusions indicate that neural networks are a valid approach to this problem, especially if more data is available in the future.

With proper data collection from multiple sources, both geographical, as in data collected from several countries, as well as technological, with the use of sensors and other technologies, the scope of this work could certainly be expanded to other countries and scenarios. Since the minimal set of attributes defined by Shenas [15] has shown good results even when expanding the time period from three to eleven years, these attributes are a strong starting point for this expansion, which could motivate relevant healthcare agents to start their own datasets, allowing for richer dataset to be analyzed, which, in turn could lead to both better predictions for high cost individuals, along with the associated benefits mentioned previously, as well as an increase in resource efficiency.

Our contributions include the creation of the necessary tools to preprocess MEPS dataset files, tested in 11 years of data, from 2006 to 2016. These tools allow the user to create models without the need to handle the raw data and year to year differences in the MEPS dataset, such as attributes or value changes, and are available on github [1]. As well as the previously mentioned conclusions, which could aid in future research concerning the high cost patient prediction problem.

Acknowledgments. The InterSCity project is sponsored by the INCT of the Future Internet for Smart Cities funded by CNPq, proc. 465446/2014-0, Coordenação de Aperfeiçoamento de Pessoal de Nível Superior – Brasil (CAPES) – Finance Code 001, and FAPESP, proc. 2014/50937-1 and 2015/24485-9. This paper is based upon work supported by Fundect, Programa de Educação Tutorial (PET/MEC/FNDE), and Federal University of Mato Grosso do Sul (UFMS).

References

1. Github repository. https://github.com/franklin-ll/Highcost-MEPS. Accessed 6 Apr 2021
2. AHRQ: Agency for healthcare research and quality, about MEPS. https://www.meps.ahrq.gov/mepsweb/. Accessed 6 Apr 2021
3. AHRQ: Agency for healthcare research and quality, confidentiality of MEPS data. https://meps.ahrq.gov/communication/participants/confidentialitymcp.shtml. Accessed 28 June 2021
4. AHRQ: Agency for healthcare research and quality, MEPS data file types. https://bit.ly/3AaBXow. Accessed 6 Apr 2021
5. Blumenthal, D., Anderson, G., Burke, S., Fulmer, T., Jha, A.K., Long, P.: Tailoring complex-care management, coordination, and integration for high-need, high-cost patients a vital direction for health and health care. Documentos/National Academy of Medicine, Washington, DC, vol. 9, pp. 1–11 (2016). https://bit.ly/2UkH5FR

6. Chechulin, Y., Nazerian, A., Rais, S., Malikov, K.: Predicting patients with high risk of becoming high-cost healthcare users in Ontario (Canada). Healthc. Policy **9**, 68–81 (2014). www.longwoods.com/content/23710

7. Chollet, F., et al.: Keras (2015). https://keras.io/. Accessed 6 Apr 2021

8. CMCS: The center for medicaid and CHIP services. www.medicaid.gov/about-us/organization/index.html

9. Google: Machine learning crash course. https://developers.google.com/machine-learning. Accessed 6 Apr 2021

10. Kanamura, A.H., Viana, A.L.D.: Gastos elevados em plano privado de saúde: com quem e em quê. Documentos/Scielo **41**, 814–820 (2007). https://www.scielosp.org/article/rsp/2007.v41n5/814-820/

11. Kanamura, A.H., Viana, A.L.D.: The Kaiser commission on medicaid and the uninsured. Documentos/Kaiser Family Foundation **1**, 27 (2013). https://kaiserfamilyfoundation.files.wordpress.com/2010/06/7334-05.pdf

12. Meehan, J., Chou, C.A., Khasawneh, M.T.: Predictive modeling and analysis of high-cost patients. In: Industrial and Systems Engineering Research Conference (2015)

13. Powers, D.M.W.: Evaluation: from precision, recall and F-factor to ROC, informedness, markedness & correlation. School of Informatics and Engineering, Flinders University of South Australia (2007). https://bit.ly/3w3XcoK

14. Rashidi, B., Kobewka, D.M., Campbell, D.J.T., Forster, A.J., Ronksley, P.E.: Clinical factors contributing to high cost hospitalizations in a Canadian tertiary care centre. BMC Health Serv. Res. **1**, 27 (2017). https://bmchealthservres.biomedcentral.com/track/pdf/10.1186/s12913-017-2746-6

15. Shenas, S.A.I.: Predicting high-cost patients in general population using data mining techniques (2012). https://doi.org/10.20381/ruor-6153

16. Sokolova, M., Lapalme, G.: A systematic analysis of performance measures for classification tasks. Inf. Process. Manag. **45**(4), 427–437 (2009). https://doi.org/10.1016/j.ipm.2009.03.002. https://www.sciencedirect.com/science/article/pii/S0306457309000259

Prediction of the Vigor and Health of Peach Tree Orchard

João Cunha[1], Pedro D. Gaspar[1,2(✉)] ⓘ, Eduardo Assunção[1] ⓘ, and Ricardo Mesquita[1] ⓘ

[1] Electromechanical Engineering Department, University of Beira Interior, Rua Marquês d'Ávila e Bolama, 6200 Covilhã, Portugal
{joao.tomas.mota.cunha,dinis,eduardo.assuncao,Ricardo.mesquita}@ubi.pt
[2] C-MAST—Center for Mechanical and Aerospace Science and Technologies, 6201-001 Covilhã, Portugal

Abstract. New technologies are a great support for decision-making for agricultural producers. An example is the analysis of orchards by way of digital image processing. The processing of multispectral images captured by drones allows the evaluation of the health or vigor of the fruit trees. This work presents a proposal to evaluate the vigor and health of trees in a peach orchard using multispectral images, an algorithm for segmentation of trees canopy, and application of vegetable indexes. For canopy segmentation, the Faster R-CNN convolutional neural network model was used. To predict the health of the peach trees, the vegetable indexes NDVI, GNDVI, NDRE, and REGNDVI were calculated. The values of the NDVI, GNDVI, NDRE, REGNDVI indexes obtained for the healthiest tree were 0.94, 0.86, 0.58, and 0.57, respectively. With the application of this method, it was possible to conclude that the use of multispectral images together with image processing algorithms, artificial intelligence, and plant indexes, allows providing relevant information about the vigor or health of the cultures serving to support the decision making in agricultural activities, helping the optimization of resources, reduction of time and cost, maximizing production, facilitating the work of agricultural explorers.

Keywords: Image processing · Health · Vigor · Canopy segmentation · Vegetation indexes · Faster R-CNN · Multispectral images

1 Introduction

Modern agriculture is increasingly dependent on systems based on new technologies, which involve robotics, image processing, remote monitoring, and even artificial intelligence in the development of methods to support decision-making and use of information relevant to the evolution of agriculture [1].

© Springer Nature Switzerland AG 2021
O. Gervasi et al. (Eds.): ICCSA 2021, LNCS 12951, pp. 541–551, 2021.
https://doi.org/10.1007/978-3-030-86970-0_38

There have been many efforts to move beyond species assessment to functional aspects of vegetation condition. Vegetation condition is the measure of vegetation response to stress. Good condition is associated with green, photosynthetically active vegetation, while stressors such as water and nutrient deficiencies or pest infestations result in low or poor condition. Physiologically, plants respond to stress by reducing chlorophyll activity and subsequently producing other pigments. These responses can be measured in the remote sensing signal in both visible and near infrared (NIR) [2]. There is a strong correlation between vegetation composition, structure, and function and signatures observed by remote sensing instruments. These remote sensing systems can be used to identify and map a variety of phenomena, including small to large variations in the characteristics, health, and condition of vegetation across a landscape [3].

Remote sensing data from satellites and airborne sensors typically have large coverage areas and are highly dependent on the constellation of viewing and illumination angles and atmospheric conditions [4]. To address these issues, the application of UAV-based multispectral image data is used to assess vegetation condition in-situ, rather than using remote sensing data. The vegetation index is generated from combinations of two or three spectral bands (red and near-infrared being the most used), whose values are summed, divided, or multiplied to produce a single value (index) that can be used as an indicator of the amount of vigor of vegetation [5].

To calculate vegetation indexes it is necessary to delineate the area of interest, i.e., canopy areas, which can be supported by a computer vision method.

There are several traditional methods of segmentation of manually performed hearts. Unlike those commonly used, this job requires an artificial intelligence method of image segmentation called Faster R-CNN (Faster Region-based Convolutional Neural Network) [6]. This model, contrary to the traditional and commonly used, has as main difference and advantage relating to the obtaining of vectors, based on the characteristics of images automatically and optimized. In addition, a vegetation index, such as the Normalized Difference Vegetation Index (NDVI), is used to evaluate peach trees.

2   Related Work

Several studies have been conducted in this sense. Underwood et al. [7] developed a mobile terrestrial "scanning" system for almond orchards, capable of mapping the distribution of flowers and fruits and predicting the productivity of individual trees. By stifling the canopy volume and studying the images taken by the terrestrial mobile vehicle that examines either the orchard and registered the data of the LiDAR (Light Detection and Ranging) sensor and camera sensors, it is possible to estimate the density of flowers and fruits. They evaluated 580 fruit trees at peak flowering, fruiting, and just before the rabbit hutch for two subsequent years where the canopy volume had the strongest linear relationship with the production with R2 $1/4 = 0:77$ for 39 tree samples in two years. Hunt Jr. et al. [8] evaluated the Triangular Greenness Index (TGI), developed to be sensitive to the chlorophyll content of the leaf, applying nitrogen during vegetative growth to avoid yield losses, but only the portions with severe nitrogen deficiency (very low chlorophyll content) were discoverable. In this case, they concluded that the TGI may be the spectral index by which digital devices mounted on low altitude aerial platforms can be used for a low-cost assessment of fertilizer needs. In order to contribute to this research topic, this paper evaluates the vigor and health of peach trees using multispectral images in a Faster R-CNN convolutional neural network algorithm for segmentation of trees canopy, and the calculus of vegetable indexes (NDVI, GNDVI, NDRE, and REGNDVI) to predict the health of the peach trees.

3   Materials and Methods

3.1   Multispectral Camera

The Micasense RedEdge-MX multispectral camera shown in Fig. 1 was used to acquire the images. This camera simultaneously captures five discrete spectral bands Blue (B), Green (G), Red (R), Red Edge (RE), and near-infrared (NIR). In general, these images (5 bands) are used to generate accurate and quantitative information on cultures [9,10].

Fig. 1. Micasense RedEdge-MX multispectral camera.

This camera weight around 230 g and has dimensions 8.7cm×5.9cm×4.54cm, being suited to be mounted on drones. The sensor acquires the spectral bands in the following Wavelength (nm): Blue: 475 nm center, 32 nm bandwidth; Green: 560 nm center, 27 nm bandwidth; Red: 668 nm center, 14 nm bandwidth; Red Edge: 717 nm center, 12 nm bandwidth; Near-IR: 842 nm center, 57 nm bandwidth. The Ground Sample Distance (GSD) is 8 cm per pixel (per band) at 120 m and it able to capture a 12-bit RAW image (all bands) per second in a field of view of 47.2° HFOV.

3.2   Image Database

The tree canopy image database was obtained with the MicaSense RedEdge-MX camera attached to a drone. The images were captured in the peach orchard located on the Quinta Nova, Tortosendo (Portugal).

For this study, 20 multispectral images of the five channels Red (R), Green (G), Blue (B), Near IR (NIR), and Red Edge (RE) channels were used. In each image, there are approximately 16 tree peach canopy. Therefore, the database includes a total of approximately 320 trees canopy.

The R, G, B, NIR, and RE image channels are (slightly) spatially misaligned due to the physical distance between each sensor/lens set. This phenomenon is known as parallax. To resolve this issue, the camera manufacturer provides an alignment software [11]. This software was used to align the 5 bands of the respective images. Figure 2 shows the 5 bands of an image used in this work.

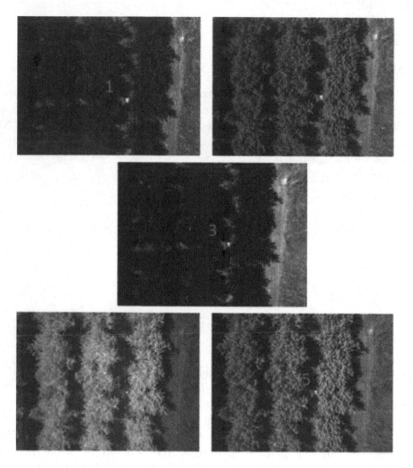

Fig. 2. 1) Channel B, 2) Channel G, 3) Channel R, 4) Channel NIR, 5) Channel RE.

3.3 Vegetation Indexes

Red light is strongly absorbed by the photosynthetic pigments found in green leaves, while near-infrared light crosses or is reflected by living leaf tissues, regardless of their color, acting in the solar reflective spectral range between 390 nm and 1040 nm. Each vegetation index is a number generated by some combination of remote sensor bands (images obtained by the camera) and has some relation to the amount of vegetation (i.e., dense or sparse) in a given image pixel.

Several vegetation indexes are examined and the concept of a formula library is introduced. In this study, a series of equations for different vegetation indexes were used [12]. NDVI is probably the most commonly used index. The difference between the NIR and R reflectance is divided by the sum of the two reflectances, as shown in Eq. 1.

$$NDVI = \frac{NIR - Red}{NIR + Red} \tag{1}$$

NDVI values range from +1.0 to −1.0. Healthy vegetation has low red light reflectance and high near-infrared reflectance, resulting in high NDVI values [13]. At the same time, sparse vegetation such as shrubs and pastures or old plants can give moderate NDVI values (between 0.2 and 0.5) [14]. Thus, increasing positive NDVI values indicate an increase in green vegetation. Conversely, NDVI values near zero and decreasing negative values indicate non-vegetated features, such as barren surfaces (rock and soil), water, snow, ice, and clouds [13,14].

3.4   Tree Canopy Segmentation

There are several methods of detaining or segmenting objects into digital images, such as the Faster R-CNN method [6] and Mask R-CNN (Mask Region-based Convolutional Neural Network) [15]. These models are just some of the state-of-the-art methods for location and segmentation. In this study, the Faster R-CNN model was used to detect the trees canopy. This model is based on a convolutional neuronal network, which learns how to automatically extract the necessary characteristics of a digital image to solve a computational vision task. The Faster R-CNN network receives a digital image as input and produces the output of the various location positions of the objects found in the input image.

Using machine learning TensorFlow and open source image processing software OpenCV, the proposed model is performed as follows:

- Align the 5-band images (channels) using the Micassense framework.
- Annotation of images: aims to manually locate the objects on which the network is to be trained. For this work, the annotation tool "LabelImg" [16] was used to annotate the tree canopy regions in the training images.
- Training model: the Faster R-CNN model is available in the Github repository Tensorflow Model Garden [17]. We split the annotated tree dataset into training and validation data and performed the training in the model.

3.5   Proposed Method

The proposed method for checking the vitality of a tree is to evaluate the various vegetation indices of a region where a tree canopy is located in a digital image, since these vegetation indices may reflect the health of a tree. For example, healthy vegetation results in high NDVI values and old or not healthy plants result in moderate NDVI values [13,14].

The pipeline for the proposed method is as follows:

1. Perform canopy segmentation using the Faster R-CNN model. The results are rectangular bounding boxes of the detected tree canopies.
2. Each rectangular region of interest detected by the Faster R-CNN model (step 1) is approximated to an ellipse using the OpenCV library. Since the tree canopies are relatively circular.
3. Calculate the indexes within the ellipse.

4   Results and Discussion

This study shows a method based on convolutional neuronal networks for detection of tree canopies, as well as the application of vegetation indexes as a way to evaluate the vegetation condition of the trees in the peach orchard. Figure 3 shows the visual result obtained in a test image of the Quinta Nova orchard (Covilhã, Portugal).

The rectangles represent the trees made by the Faster R-CNN model. Ellipses are the regions of interest, obtained from the rectangles found, in which the vegetation indexes corresponding to each tree are obtained.

It can be verified that the model performed detection on all trees. In addition, the detection are well centered in the canopy. Thus, evidence that the model is robust and efficient for the design of trees canopy.

Table 1 shows the results of the NDVI vegetation index analysis for each tree in the test image.

According to Table 1, of the 15 trees in the test image, one (tree 3) had an average NDVI value of 0.88, indicating that this was the tree with the lowest vigor compared to the other trees. Visual analysis of this tree (Fig. 3) shows that its branches are sparser. That is, this tree is slightly less dense. This observation justifies its lower average NDVI value compared to the other trees.

It can also be seen that there are two trees (2 and 12) with an average NDVI of 0.95. According to the index, these two trees are the healthiest or developed compared to the others. Most trees (1, 4, 5, 6, 9, 11, and 14) have an average NDVI between 0.93 and 0,94. These trees are also very healthy and with great vigor.

It can also be seen that there are two trees (2 and 12) with an average NDVI of 0.95. According to the index, these two trees are the healthiest and best developed compared to the others. Most trees (1, 4, 5, 6, 9, 11, and 14) have an average NDVI between 0.93 and 0.94. These trees are also very healthy and with great vigor.

Table 2 shows the results of four vegetal indexes, whose values are normalized so that a direct comparison between results can be performed. The values of the GNDVI (Green Normalized Difference Vegetation Index), NDRE (Normalized

Fig. 3. Visual result of the proposed method.

Table 1. Analysis of NDVI vegetation index for the tree canopies in the test image. The green color is the highest value, and the red is the lowest value.

Tree ID	Average	Maximum	Minimum
1	0.94	0.99	0.32
2	0.95	0.99	0.47
3	0.88	0.99	0.25
4	0.94	0.99	0.35
5	0.94	0.99	0.51
6	0.94	0.99	0.34
7	0.90	0.99	0.32
8	0.91	0.98	0.36
9	0.93	0.99	0.26
10	0.92	1.00	0.37
11	0.94	0.99	0.33
12	0.95	1.00	0.44
13	0.91	0.99	0.32
14	0.93	1.00	0.33
15	0.90	0.99	0.32

Difference Red Edge Index), REGNDVI (Red Edge Green Normalized Difference Vegetation Index) confirm the analysis previously done with the NDVI index.

Tree 3 is the least developed, as its vegetation index is lower than that of the other trees. Tree 12, on the other hand, is the one with the highest values in most of the indices, so it has the highest vigor in comparison.

Table 2. Average values of five vegetation indexes for the tree canopies in the test image. The green color is the highest value and the red is the lowest value.

Tree ID	NDVI	GNDVI	NDRE	REGNDVI
1	0.94	0.86	0.58	0.57
2	0.95	0.84	0.54	0.57
3	0.88	0.79	0.48	0.51
4	0.94	0.86	0.58	0.56
5	0.94	0.85	0.57	0.57
6	0.94	0.84	0.54	0.57
7	0.90	0.81	0.51	0.52
8	0.91	0.80	0.49	0.53
9	0.93	0.85	0.57	0.54
10	0.92	0.82	0.50	0.55
11	0.94	0.84	0.53	0.57
12	0.95	0.88	0.59	0.56
13	0.91	0.8	0.48	0.52
14	0.93	0.85	0.58	0.56
15	0.90	0.8	0.50	0.52

Figure 4 shows the training loss curve for the 320 canopy training images and 10000 iterations. Figure 5 shows the Mean Average Precision (mAP) for validation during training.

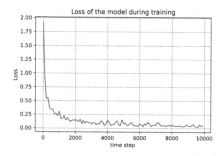

Fig. 4. Training loss of canopy detection.

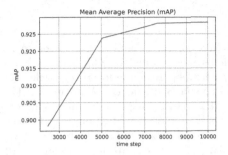

Fig. 5. Mean average precision of validation.

5   Conclusions

The technologies of analysis of multispectral images, captured by drones, manage to be an asset for effective evaluation of vegetation condition of fruit trees. Combining technologies with computational image processing models, makes possible the precise calculation of vegetation indexes, through the use of several bands of combined light, which leads to the prediction of the vigor of the trees. This work has applied a method based on convolutional neural networks for the detection of canopy trees, as well as the application of plant indices as a way to infer the vegetation condition of fruit trees in a peach orchard. The model performed the canopy detection of all trees, which provides a result of its robustness and efficiency. Various vegetation indexes were calculated to infer the vigor of the trees. The values of the GNDVI, NDRE, and REGNDVI indexes confirm the analysis performed with the NDVI index. In the test image, trees are distinguished whose vigor is less, while others have higher than average values. These results allow the fruit grower to analyze in detail and on the ground what are the potential causes of these variations and eventually correct them. Examples like the one used in this work are relevant to support decision-making in agricultural activities, making it possible to optimize resources, reduce time and cost, maximize production, and can be used by farmers, agronomists, and re-searchers. The scope of this study was limited in terms of data (i.e., images) to evaluate non-healthy trees and further study could assess the proposed model with this data.

Acknowledgement. This research work is funded by the PrunusBot project - Autonomous con-trolled spraying aerial robotic system and fruit production forecast, Operation No. PDR2020-101-031358 (leader), Consortium No. 340, Initiative No. 140, promoted by PDR2020 and co-financed by the EAFRD and the European Union under the Portugal 2020 program. The authors thank the opportunity and financial support to carry on this project to Fundação para a Ciência e Tecnologia (FCT) and R&D Unit "Centre for Mechanical and Aerospace Science and Technologies" (C-MAST), under project UIDB/00151/2020.

References

1. Shanmugam, S., Assunção, E., Mesquita, R., Veiros, A., D. Gaspar, P.: Automated weed detection systems: a review. KnE Eng. **5**(6), 271–284 (2020). https://knepublishing.com/index.php/KnE-Engineering/article/view/7046

2. Schrodt, F., de la Barreda Bautista, B., Williams, C., Boyd, D.S., Schaepman-Strub, G., Santos, M.J.: Integrating biodiversity, remote sensing, and auxiliary information for the study of ecosystem functioning and conservation at large spatial scales. In: Cavender-Bares, J., Gamon, J.A., Townsend, P.A. (eds.) Remote Sensing of Plant Biodiversity, pp. 449–484. Springer, Cham (2020). https://doi.org/10.1007/978-3-030-33157-3_17

3. Serbin, S.P., Townsend, P.A.: Scaling functional traits from leaves to canopies. In: Cavender-Bares, J., Gamon, J.A., Townsend, P.A. (eds.) Remote Sensing of Plant Biodiversity, pp. 43–82. Springer, Cham (2020). https://doi.org/10.1007/978-3-030-33157-3_3

4. Dorigo, W., Zurita-Milla, R., de Wit, A., Brazile, J., Singh, R., Schaepman, M.: A review on reflective remote sensing and data assimilation techniques for enhanced agroecosystem modeling. Int. J. Appl. Earth Obs. Geoinf. **9**(2), 165–193 (2007). https://www.sciencedirect.com/science/article/pii/S0303243406000201. advances in airborne electromagnetics and remote sensing of agro-ecosystems

5. Multispectral Sensors: Drone-based Data Capture and Processing (April 2021). https://www.precisionhawk.com/agriculture/multispectral

6. Ren, S., He, K., Girshick, R., Sun, J.: Faster R-CNN: towards real-time object detection with region proposal networks. IEEE Trans. Pattern Anal. Mach. Intell. **39**(6), 1137–1149 (2017)

7. Underwood, J.P., Whelan, B., Hung, C., Sukkarieh, S.: Mapping almond orchard canopy volume, flowers, fruit and yield using lidar and vision sensors. Comput. Electron. Agric. **130**, 83—96 (2016). https://doi.org/10.1016/j.compag.2016.09.014

8. Hunt, E.R., Doraiswamy, P.C., McMurtrey, J.E., Daughtry, C.S., Perry, E.M., Akhmedov, B.: A visible band index for remote sensing leaf chlorophyll content at the canopy scale. Int. J. Appl. Earth Obs. Geoinf. **21**, 103–112 (2013). https://www.sciencedirect.com/science/article/pii/S0303243412001791

9. User Guide for MicaSense Sensors (April 2021). https://support.micasense.com/hc/en-us/articles/360039671254-User-Guide-for-MicaSense-Sensors-Altum-RedEdge-M-MX-Dual-Camera-System-

10. RedEdge User Manual (PDF Download) – MicaSense Knowledge Base (April 2021). https://support.micasense.com/hc/en-us/articles/215261448-RedEdge-User-Manual-PDF-Download-

11. Active Image Alignment (April 2021). https://micasense.github.io/imageprocessing/Alignment.html

12. Hunt, E.R., Doraiswamy, P.C., McMurtrey, J.E., Daughtry, C.S., Perry, E.M., Akhmedov, B.: A visible band index for remote sensing leaf chlorophyll content at the canopy scale. Int. J. Appl. Earth Obs. Geoinf. **21**, 103–112 (2013)

13. Saravanan, S., Jegankumar, R., Selvaraj, A., Jacinth Jennifer, J., Parthasarathy, K.: Chapter 20 - utility of landsat data for assessing mangrove degradation in muthupet lagoon, South India. In: Ramkumar, M., James, R.A., Menier, D., Kumaraswamy, K. (eds.) Coastal Zone Management, pp. 471–484. Elsevier (2019). https://www.sciencedirect.com/science/article/pii/B9780128143506000203

14. NDVI, the Foundation for Remote Sensing Phenology (April 2021). https://www.usgs.gov/core-science-systems/eros/phenology/science/ndvi-foundation-remote-sensing-phenology?qt-science_center_objects=0#qt-science_center_objects
15. He, K., Gkioxari, G., Dollár, P., Girshick, R.: Mask R-CNN. In: 2017 IEEE International Conference on Computer Vision (ICCV), pp. 2980–2988 (2017)
16. LabelImg is a graphical image annotation tool (April 2021). https://github.com/tzutalin/labelImg
17. TensorFlow Model Garden (April 2021). https://github.com/tensorflow/models

Multimodal Emotion Recognition Using Transfer Learning on Audio and Text Data

James J. Deng[1]([☒]), Clement H. C. Leung[2]([☒]), and Yuanxi Li[3]

[1] MindSense Technologies, Hong Kong, People's Republic of China
james@mindsense.ai
[2] School of Science and Engineering, The Chinese University of Hong Kong,
Shenzhen, People's Republic of China
clementleung@cuhk.edu.cn
[3] Hong Kong Baptist University, Hong Kong, People's Republic of China
csyxli@comp.hkbu.edu.hk

Abstract. Emotion recognition has been extensively studied in a single modality in the last decade. However, humans express their emotions usually through multiple modalities like voice, facial expressions, or text. In this paper, we propose a new method to find a unified emotion representation for multimodal emotion recognition through speech audio, and text. Emotion-based feature representation from speech audio is learned by an unsupervised triplet-loss objective, and a text-to-text transformer network is constructed to extract latent emotional meaning. As deep neural network models trained by huge datasets exhaust a lot of unaffordable resources, transfer learning provides a powerful and reusable technique to help fine-tune emotion recognition models trained on mega audio and text datasets respectively. Automatic multimodal fusion of emotion-based features from speech audio and text is conducted by a new transformer. Both the accuracy and robustness of proposed method are evaluated, and we show that our method for multimodal fusion using transfer learning in emotion recognition achieves good results.

Keywords: Multimodal emotion recognition · Multimodal fusion · Transformer network

1 Introduction

In the last decade, deep learning like convolution neural network (CNN), recurrent neural network (RNN), and other deep models have proven useful in many domains, including computer vision, speech and audio processing, and natural language processing. Many applications like face recognition, speech recognition, and machine translation have achieved great success. Research on emotion recognition makes quick improvements as well. However, as emotion is complex and determined by a joint function of pharmacological, cognitive, and environmental variables, emotion recognized by different people may be ambiguous or even

© Springer Nature Switzerland AG 2021
O. Gervasi et al. (Eds.): ICCSA 2021, LNCS 12951, pp. 552–563, 2021.
https://doi.org/10.1007/978-3-030-86970-0_39

opposite. The single modality for emotion recognition is insufficient and incomplete. For example, in a conversation, people's voices, text of speech content, facial expressions, body language, or gestures all convey emotional meanings. Thus, it is inappropriate to recognize people's emotions through a single modality. Multimodal modeling is a natural and reasonable process for emotion recognition. Although multimodal analysis has been extensively studied and some have achieved remarkable results in specific constraints, it cannot simply apply these methods in different environments. Therefore, building a universal representation for emotional information from multiple modalities is necessary and useful in the task of multimodal emotion recognition. In this paper, to simplify, we propose a method to find a good emotional information representation using both speech audio and text information for emotion recognition.

Nowadays training models using mega dataset consumes huge resource and have become more and more difficult and less affordable for most of researchers, institutes or companies. For example, training GPT-3 would cost at least $4.6 million, because training deep learning models is not a clean, one-shot process. There are a lot of trial and error and hyperparameter tuning that would probably increase the cost sharply. Transfer learning provides a powerful and reusable technique to help us solve resource shortages and save model training costs. This paper adopts this strategy and employs the excellent fruits of pre-trained models by mega datasets of audio and text as the basis of our work. The pre-trained model of speech audio is trained on a Youtube-8M dataset named AudioSet by Google. This is an extremely large-scale dataset of manually annotated audio clips. A model named VGGish is an audio feature embedding produced by training a modified VGGNet model to predict video-level tags from this dataset, which is widely used for audio classification. In addition, another model TRIpLet Loss network (TRILL) is trained from AudioSet again and achieves good results for several audio tasks like speaker identification, and emotion recognition. We use the fine-tuned TRILL model to extract the speech audio features as the representation for the modality of speech audio. As for text representation, we adopt fine-tuned Text-To-Text Transfer Transformer (T5) [23] model trained by a common crawl (C4) dataset to extract text embeddings. Operations of transfer learning have greatly accelerated model training and applied in specific domains. To reuse the fruits of transfer learning especially for learning embeddings from speech audio and text, we adopts the strategy of transfer learning to construct emotional information representation.

We expect the fused data is more informative and synthetic. To retain more hidden emotional information for multimodal fusion, we concatenate speech audio features and text embeddings extracted from fine-tuned pertained models. The concatenated feature representation is passed to a transformer-based encoder-decoder network with an attention mechanism. This means that the final predictor module can balance the influence of information transmission. A quantitative evaluation shows that our models outperform the existing methods in emotion recognition. The main contribution of this paper is summarized as follows: (1) we propose an effective method to fuse multiple modalities on the basis of transfer learning; (2) we evaluate the performance of the proposed method and make an attempt at a general semantic framework. The rest of the paper is structured as follows: Sect. 2 describes the literature review, Sect. 3 discusses the proposed methodologies and overall architecture, Sect. 4 explains the experimental setup and results, and Sect. 5 summarizes our work and discusses the future work.

2   Literature Review

In psychology and philosophy, emotion is a highly complex subjective, and conscious experience. There is no clear definition of what an emotion is. Consequently, different emotion theories or models have been proposed by different researchers in the last century. These models are usually rooted in two emotion theories: discrete emotion theory and dimensional emotion theory. Discrete emotion theory employs a finite number of emotional descriptors or adjectives [11] to express basic human emotions (e.g., joy, sadness, anger, contempt, happiness). For example, Ortony et al. [19] proposed an emotion cognition model commonly known as the OCC model to hierarchically describe 22 emotion descriptors. For the sake of simplicity, some researchers [21] used coarser-grained partition such as happy, neutral, and sad. Dimensional emotion theory states that emotion should be depicted in a psychological dimensional space, which can resolve the shortcomings of discrete emotion theory. Two or three dimensional emotion models such as arousal-valence [24], resonance-arousal-valence [3], and arousal-valence-dominance [16] are widely used in different application domains for example music emotion recognition. Dimensional emotion theory is more likely to be used in computational emotion systems or emotion synthesis. The advantage of discrete emotion theory is that it is easy to understand and use in practice, especially in emotion recognition, In this paper, for simplicity, we shall adopt discrete emotion theory to represent emotion.

In recent years many research works of emotion recognition have been done in a single modality setting. A survey [5] of methods is summarized to address three important aspects of the design of a speech emotion recognition system. The first one is the choice of suitable features for speech representation. The second issue is the design of an appropriate classification scheme and the third issue is the proper preparation of an emotional speech database for evaluating system performance. Domain expert knowledge makes significant for manually constructing high-quality features [18]. Recent research has mostly focused on deep representation learning methods, either supervised, semi-supervised, or unsupervised. Successful representations improve the sample efficiency of ML algorithms by extracting most information out of the raw signal from the new data before any task-specific learning takes place. This strategy has been used successfully in many application domains. Many deep learning methods like CNN [10], Deep Belief Network (DBN) [9], Long Short-Term Memory (LSTM) [28], Autoencoder [2,13] have been used to construct various deep neural networks, achieving a good performance in speech emotion recognition. Sentiment analysis

[15] is the task of automatically determining from the text the attitude, classified by positive, negative, and neutral attitude. Knowledge-based and statistical methods are usually used to extract text features like word embedding (e.g., Word2Vec), pair-wise correlation of words, and parts-of-speech (POS) tag of the sentence. Recently, transformer architecture [26] is rather popular in dealing with natural language processing like General Language Understanding Evaluation (GLUE) [27]. The model of Bidirectional Encoder Representations from Transformers (BERT) [4] is constructed by 12 Encoders with bidirectional self-attention mechanism. A unified text-to-text format, in contrast to BERT-style models that can only output either a class label or a span of the input, can flexibly be used on the same model, loss function, and hyperparameters on any NLP tasks. This unified framework provides valuable insight for performing semantic classification like recognizing emotion from the text. This paper inherits this text-to-text transformer architecture.

Transfer learning [20] and domain adaptation have been extensively practiced in machine learning. In specific domains, there is often only a few data available, and it is difficult to train an accurate model by using these small datasets. However, many modalities (e.g., speech, text) have the same essence and low-level elements. Thus, transfer learning can well overcome the small dataset limitation. A sharing learned latent representation [6] or deep models like VGGish or BERT is transferred to be used on another learning task, usually achieving an exciting performance. [14] uses DBN of transfer learning to build sparse autoencoder for speech emotion recognition. [12] introduces a sent2affect framework, a tailored form of transfer learning to recognize emotions from the text. Another Universal Language Model Fine-tuning (ULMFiT) [8] is carried out to evaluate several text classification and outperforms state-of-the-art models. Therefore, to make full use of pre-trained models by mega dataset, we adopt transfer learning of multiple modalities. Though there exists some work on multimodal emotion recognition, for example, canonical correlational analysis [17], joint feature representation [22,25], or generative adversarial network (GAN) for multimodal modeling, there is less work on the aspect of transfer learning on multimodal. This inspires us to use transfer learning and multimodal fusion to build a unified emotional representation of speech audio and text.

3   Methodology

This paper concentrates on dual modalities: speech and text. The aim is to find a unified and reusable representation of emotional information from these dual modalities. This section will first introduce the emotional features of speech audio, and text, respectively. Then the proposed method for fusing speech and text for multimodal emotion recognition will be described.

3.1   Emotion-Based Feature Representation for Audio Data

As speech recognition has been extensively researched and achieved great success in the last decade, many speech emotion recognition tasks also adopt acoustic

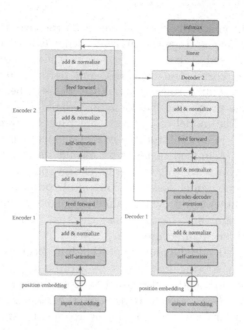

Fig. 1. An encoder-decoder Transformer architecture with residual skip connections.

features like Mel-frequency Cepstral Coefficient (MFCC), deep auto-encoders in Deep Neural Network (DNN), or end-to-end with attention-based strategy. However, the same content of speech usually expresses the different emotions corresponding to different voice attributes like pitch, rhythm, timbre, or context. Emotional Aspects of the speech signal generally change more slowly than the phonetic and lexical aspects used to explicitly convey meaning. Therefore, we need to find a good representation for emotion-related tasks to be considerably more stable in time than what is usually adopted in speech recognition.

To consider the temporal characteristic of speech audio, we represent a large and unlabeled speech collection as a sequence of spectrogram context windows $X = x_1, x_2, \ldots, x_N$, where each $x_i \in^{F \times T}$. We aim to learn a embedding $g :^{F \times T} \to^d$ from spectrogram context windows to a $d-$dimensional embedding space such that $\|g(x_i) - g(x_j)\| \leq \|g(x_i) - g(x_k)\|$ when $i - j \leq i - k$. We can express this embedding formulated by learning a triplet loss function. Suppose a large collection of example triplets is represented by $z = (x_i, x_j, x_k)$, where $i - j \leq \tau$ and $i - k\rangle\tau$ for some suitably chosen time scale τ. The τ represent the specific duration of each given audio clip. The whole loss function $\Theta(z)$ is expressed as follows:

$$\Theta(z) = \sum_{i=1}^{N} [\|g(x_i) - g(x_j)\|_2^2 + \|g(x_i) - g(x_k)\|_2^2 + \delta] \tag{1}$$

where $\| \bullet \|$ is the L_2 norm, $[\bullet]$ represents standard hinge loss and δ is non-negative margin hyperparameter.

3.2   Text-to-Text Transformer Representation

Since the transformer architecture has achieved high performance in a number of NLP tasks like The General Language Understanding Evaluation (GLUE) benchmark, Sentiment analysis (SST-2), SQuAD question answering, we adopt the transformer architecture as well. Given a sequence of text obtained from speech, we first map the tokens of the initial input sequence to an embedding space, and then input the embedded sequence to the encoder layer. The encoder layer is composed of a stack of blocks, and each block consists of a self-attention layer followed by a small feed-forward network. Layer normalization in the self-attention layer and feed-forward network is considered, where the activations are only re-scaled and no additive bias is applied. In addition, a residual skip is connected from input to output of the self-attention layer and feed-forward network, respectively. After that, the dropout is calculated within the feed-forward network, on the skip connection, on the attention weights, and at the input and output of the entire stack. As for the decoder layer, except for the similar block in the encoding layer, it contains a standard attention operation after each self-attention layer. The self-attention mechanism in the decoder also uses a form of auto-regressive or causal self-attention, which only allows the model to attend to past outputs. The output of the final decoder block is feed into a dense layer with a softmax output, whose weights are shared with the input embedding matrix. All attention mechanisms in the transformer are split up into independent "heads" whose outputs are concatenated before being further processed. The whole text-to-text transformer architecture is illustrated in Fig. 2. In a transformer, instead of using a fixed embedding for each position, relative position embeddings produce a different learned embedding according to the offset between the "key" and "query" being compared in the self-attention mechanism. We use a simplified form of position embeddings where each "embedding" is simply a scalar that is added to the corresponding logit used for computing the attention weights.

3.3   Multimodal Fusion of Speech Audio and Text

Considering the success of transformer architecture, we also adopt this attention-enhanced architecture in the fusion of speech audio features and text embeddings extracted along the operations described in previous subsections. As the speech consists of both audio and corresponding text, to retain more emotional information used for multimodal fusion, we directly concatenate the speech audio features and text embeddings. The combined features are processed by a transformer model, and the output of the model is regarded as the fusion of multimodal. The overall fusion of speech audio and text is illustrated in Fig. 2. In the process of fusion, we want to minimize the loss of both global and regional segments that convey emotional information. Speech audio features and texts are treated as sequences to be encoded into the learning procedure, while on the decoder side the target fused emotional features are generated. Suppose concatenated features of emotional representation of speech audio and text are denoted

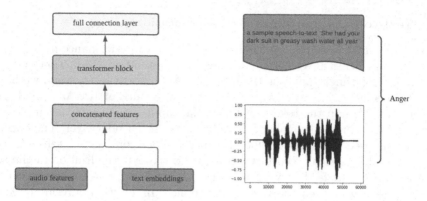

Fig. 2. Left part shows multimodal fusion of speech audio and text on embeddings extracted from transfer learning; right part shows an sample speech with anger emotion.

by x, the reconstruction x' of the same shape as x is achieved by decoder, the simplest loss function is calculated by

$$\Phi = ||x - x'||^2 \tag{2}$$

We add an explicit regularizer in the above loss function that forces the model to learn an encoding robust to slight variations of input features extracted from transfer learning. The regularization loss term is regarded as the squared Frobenius norm $\| A \|_F$ of the Jacobian matrix J for the hidden layer activations with respect to the input features. The final loss function is represented by

$$\Phi = ||x - x'||^2 + \lambda \sum_i \| \bigtriangledown_x h_i \|^2 \tag{3}$$

where $\bigtriangledown_x h_i$ refers to the gradient field of hidden layer activations with respect to the input x, summed over all i training samples. The Frobenius norm and Jacobian matrix J are given as follows:

$$\| A \|_F = \sqrt{\sum_{i=1}^{m} \sum_{j=1}^{k} \| a_{ij} \|^2} \tag{4}$$

$$J = \begin{pmatrix} \frac{\delta a_1^{(h)}(x)}{\delta x_1} & \cdots & \frac{\delta a_1^{(h)}(x)}{\delta x_m} \\ \frac{\delta a_2^{(h)}(x)}{\delta x_1} & \cdots & \frac{\delta a_2^{(h)}(x)}{\delta x_m} \\ \vdots & \ddots & \vdots \\ \frac{\delta a_n^{(h)}(x)}{\delta x_1} & \cdots & \frac{\delta a_n^{(h)}(x)}{\delta x_m} \end{pmatrix} \tag{5}$$

Finally, we consider the emotional information representation as multimodal fusion that satisfies the minimization of the loss function in Eq. 3.

Table 1. Comparison of multimodal fusion performance with single modality through the different embeddings on different emotion datasets.

Models	IEMOCAP	SAVEE	Mean
VGGis FC1	65.3%	57.7%	61.5%
VGGish Finetuned	61.4%	59.3%	60.4%
YAMNet layer 10	63.2%	62.3%	62.8%
YAMNet Finetuned	67.6%	62.7%	65.2%
TRILL distilled	70.5%	67.8%	69.2%
TRILL Finetuned	73.8%	68.6%	71.2%
Text-to-Text Transformer (T5)	75.7%	72.3%	75.5%
TRILL-T5 Multimodal Fusion	81.3%	77.4%	79.4%

4   Experiments and Results

Before presenting the results from our empirical study, we briefly introduce the necessary background topics required to understand the experiments and results. Youtube-8M dataset is a benchmark dataset that expands ontology of 632 audio event classes and a collection of 2,084,320 human-labeled 10-second sound clips drawn from YouTube videos. This dataset have been used in a number of audio tasks like speaker identification, emotion recognition. Several famous models like VGGish, Yamnet, and TRILL are trained by this benchmark dataset. VGGish is audio embedding generated by training a modified VGGNet model, and Yamnet employs Mobilenet_v1 depthwise-separable convolution architecture to predicts 521 audio event classes. TRILL is trained to find a good non-semantic representation of speech and exceeds state-of-the-art performance on a number of transfer learning tasks. Another new open-source pre-training dataset, called the Colossal Clean Crawled Corpus (C4) is used in many NLP tasks. The text-to-text transfer transformer, named T5 model, also achieves state-of-the-art results on many NLP benchmarks. Therefore, in our experiments, we reuse these pre-trained models in single modality, respectively.

To evaluate the results of multimodal fusion through transfer learning, we choose two emotion datasets. Emotional Dyadic Motion Capture (IEMOCAP) [1] is a multimodal and multispeaker database, containing approximately 12 h of audiovisual data, including video, speech, motion capture of face, text transcriptions. Another dataset is SAVEE [7] database recorded from four native English male speakers, supporting 7 emotion categories: anger, disgust, fear, happiness, neutral, sadness, and surprise.

In the experiment of transfer learning, we select both the final output and intermediate representations of the given pre-trained models. As for speech audio, in the TRILL model, we use the final 512-dimensional embedding layer and the pre-ReLU output of the first 19-depth convolutional layer. For Vggish, we use the final layer and the first fully connected layer. For YAMNet, we use the final pre-logit layer and the 5 depth-separable convolutional layer outputs. We use the emotion dataset of IEMOCAP and SAVEE for fine-tune training. As for text processing, we use the Text-to-Text transformer model with a maximum sequence length of 512 and a batch size of 128 sequences. During fine-tuning, we continue using batches with 128 length-512 sequences, and a constant learning rate of 0.001. The number of embedding dimensionality is set to 512.

We used different pre-trained models like VGGish, YAMNet, TRILL, and T5 to fine-tune for emotion recognition in a single modality. The obtained emotion representation of speech audio and text embedding are concatenated to pass through a transformer network. Multimodal fusion of emotional information through a transformer network generates a unified representation for emotion recognition. Table 1 shows the comparison of several single modalities and multimodal fusion results for emotion recognition. We can see that fine-tuning the final embedding of pre-trained models gives a clear boost to emotion recognition. In addition, TRILL shows better performance than VGGish and YAMNet. Text-to-Text Transformer (T5) model shows better performance in emotion recognition from the text. The average of Text-to-Text Transformer accuracy achieves up to 75.5% in the emotion dataset. Multimodal fusion shows better results than single modality for emotion recognition. This explains that multiple modalities like speech and text convey more emotional information than a single modality. The multimodal fusion can well employ complementary information.

Table 2. Average performance of the different embeddings on four selected emotions.

Models + Dataset	Happy	Anger	Sad	Natural
TRILL (IEMOCAP_Audio)	77.2%	81.9%	72.3%	66.8%
TRILL (SAVEE_Audio)	73.3%	84.3%	73.3%	66.7%
T5 (IEMOCAP_Text)	79.6%	82.7%	72.2%	64.9%
T5 (SAVEE_Text)	75.0%	81.7%	71.7%	66.7%
Multimodal Fusion (IEMOCAP)	83.3%	85.2%	77.5%	71.8%
Multimodal Fusion (SAVEE)	85.0%	86.7%	78.3%	71.7%
Mean	78.9%	83.8%	74.2%	68.1%

As speech audio samples of IEMOCAP datasets in some emotion categories are rather small, we only used four emotion categories (e.g., happy, sad, anger, and natural) for analysis of differences in both two emotion datasets. Table 2 shows the comparison of single and multiple modalities on different emotion categories. We can see that the emotion category happy and anger achieves the best recognition results than that of sad and neutral. The mean accuracy of anger emotion recognition is up to 83.8%, while the mean accuracy of neutral emotion recognition is only 68.1%. Through studies of false-positive results, we find that in the above-mentioned emotion datasets, it is not obvious and easy for a human to recognize samples that convey neutral or sad emotional meaning. Table 2 also shows the same conclusion that the performance of multimodal fusion through a transformer network exceeds single modality in Table 1. We also investigated the robustness of multimodal features' representation. The RAVDESS dataset is a multimodal validated English dataset that contains speech, song, audio, and video files that represent 8 emotions. As the speech audio and text are rather short, we used them to test the robustness of emotion representation through multimodal fusion. The results of the selected four emotion categories are shown in Table 3. The robustness is evaluated by mean squared error (MSE) criteria. The MSE of prediction on RAVDESS dataset is calculated and the values are limited to a small range, and we think transfer learning achieves good results on a different dataset, the proposed method on multimodal emotion feature fusion indeed contains rich emotional information and can be used in other zero-shot or one-shot emotion learning problems.

Table 3. Evaluation of robustness of our multimodal emotion recognition method on different datasets.

Dataset	Happy	Anger	Sad	Natural
IEMOCAP & SAVEE	84.2%	86.0%	77.9%	71.8%
RAVDESS	82.2%	87.8%	77.8%	76.1%

5 Conclusion

In this paper, we proposed a new method for multimodal emotion recognition. As deep neural network models trained by huge datasets exhaust a lot of unaffordable resources, we made full use of transfer learning. Large excellent pre-trained models like TRILL and Text-to-Text Transformer from a single modality are used and fine-tuned on the emotion dataset. The extracted emotional information from speech audio and text embedding are concatenated and passed to a transformer network to fuse. The experiments compared the performance of single modality and multiple modalities (speech and text) for emotion recognition which showed noticeable advantages of the latter over the former. Furthermore, the robustness of transfer learning and multimodal fusion is also tested by an extra dataset. We showed that our proposed method for multimodal emotion recognition achieves good results and can be used in other zero-shot or one-shot emotion learning tasks.

References

1. Busso, C., et al.: IEMOCAP: interactive emotional dyadic motion capture database. Lang. Resour. Eval. **42**(4), 335–359 (2008)

2. Cibau, N.E., Albornoz, E.M., Rufiner, H.L.: Speech emotion recognition using a deep autoencoder. Anales de la XV Reunion de Procesamiento de la Informacion y Control **16**, 934–939 (2013)
3. Deng, J.J., Leung, C.H., Milani, A., Chen, L.: Emotional states associated with music: classification, prediction of changes, and consideration in recommendation. ACM Trans. Interact. Intell. Syst. (TiiS) **5**(1), 4 (2015)
4. Devlin, J., Chang, M.W., Lee, K., Toutanova, K.: Bert: Pre-training of deep bidirectional transformers for language understanding. arXiv preprint arXiv:1810.04805 (2018)
5. El Ayadi, M., Kamel, M.S., Karray, F.: Survey on speech emotion recognition: features, classification schemes, and databases. Pattern Recogn. **44**(3), 572–587 (2011)
6. Hamel, P., Davies, M.E., Yoshii, K., Goto, M.: Transfer learning in mir: sharing learned latent representations for music audio classification and similarity (2013)
7. Haq, S., Jackson, P.J., Edge, J.: Speaker-dependent audio-visual emotion recognition. In: AVSP, pp. 53–58 (2009)
8. Howard, J., Ruder, S.: Universal language model fine-tuning for text classification. arXiv preprint arXiv:1801.06146 (2018)
9. Huang, C., Gong, W., Fu, W., Feng, D.: A research of speech emotion recognition based on deep belief network and SVM. Mathematical Problems in Engineering, 2014 (2014)
10. Huang, Z., Dong, M., Mao, Q., Zhan, Y.: Speech emotion recognition using CNN. In: Proceedings of the 22nd ACM international conference on Multimedia, pp. 801–804 (2014)
11. Izard, C.E., Malatesta, C.Z.: Perspectives on emotional development i: differential emotions theory of early emotional development. In: The First Draft of this Paper was Based on an Invited Address to the Eastern Psychological Association, 1 April 1983. John Wiley & Sons (1987)
12. Kratzwald, B., Ilić, S., Kraus, M., Feuerriegel, S., Prendinger, H.: Deep learning for affective computing: text-based emotion recognition in decision support. Decis. Support Syst. **115**, 24–35 (2018)
13. Latif, S., Rana, R., Khalifa, S., Jurdak, R., Epps, J., Schuller, B.W.: Multi-task semi-supervised adversarial autoencoding for speech emotion recognition. IEEE Trans. Affect. Comput. **abs/1907.06078** (2019)
14. Latif, S., Rana, R., Younis, S., Qadir, J., Epps, J.: Transfer learning for improving speech emotion classification accuracy. arXiv preprint arXiv:1801.06353 (2018)
15. Medhat, W., Hassan, A., Korashy, H.: Sentiment analysis algorithms and applications: a survey. Ain Shams Eng. J. **5**(4), 1093–1113 (2014)
16. Mehrabian, A.: Basic dimensions for a general psychological theory implications for personality, social, environmental, and developmental studies (1980)
17. Mittal, T., Bhattacharya, U., Chandra, R., Bera, A., Manocha, D.: M3er: multiplicative multimodal emotion recognition using facial, textual, and speech cues. In: Proceedings of the AAAI Conference on Artificial Intelligence, vol. 34, pp. 1359–1367 (2020)
18. Nwe, T.L., Foo, S.W., De Silva, L.C.: Speech emotion recognition using hidden markov models. Speech Commun. **41**(4), 603–623 (2003)
19. Ortony, A., Clore, G.L., Collins, A.: The Cognitive Structure of Emotions. Cambridge University Press, Cambridge (1990)
20. Pan, S.J., Yang, Q.: A survey on transfer learning. IEEE Trans. Knowl. Data Eng. **22**(10), 1345–1359 (2009)

21. Pohle, T., Pampalk, E., Widmer, G.: Evaluation of frequently used audio features for classification of music into perceptual categories. In: Proceedings of the Fourth International Workshop on Content-Based Multimedia Indexing, vol. 162 (2005)
22. Poria, S., Chaturvedi, I., Cambria, E., Hussain, A.: Convolutional MKL based multimodal emotion recognition and sentiment analysis. In: 2016 IEEE 16th international conference on data mining (ICDM), pp. 439–448. IEEE (2016)
23. Roberts, A., Raffel, C.: Exploring transfer learning with t5: the text-to-text transfer transformer. Accessed on, pp. 23–07 (2020)
24. Thayer, R.E.: The Biopsychology of Mood and Arousal. Oxford University Press, Oxford (1990)
25. Tzirakis, P., Trigeorgis, G., Nicolaou, M.A., Schuller, B.W., Zafeiriou, S.: End-to-end multimodal emotion recognition using deep neural networks. IEEE J. Selected Topics Signal Process. **11**(8), 1301–1309 (2017)
26. Vaswani, A., et al.: Attention is all you need. arXiv preprint arXiv:1706.03762 (2017)
27. Wang, A., Singh, A., Michael, J., Hill, F., Levy, O., Bowman, S.R.: GLUE: a multi-task benchmark and analysis platform for natural language understanding. arXiv preprint arXiv:1804.07461 (2018)
28. Zhao, J., Mao, X., Chen, L.: Speech emotion recognition using deep 1d & 2d cnn lstm networks. Biomed. Signal Process. Control **47**, 312–323 (2019)

International Workshop on Automatic Landform Classification: Spatial Methods and Applications (ALCSMA 2021)

Molodensky Seven Parameter Transformation for Precise Urban Mapping

Javier Urquizo[1](✉) and Clifford Mugnier[2]

[1] Escuela Superior Politécnica del Litoral, FIEC, Campus Gustavo Galindo,
Km. 30.5 vía perimetral, Guayaquil, Ecuador
`jurquizo@espol.edu.ec`
[2] Department of Civil and Environmental Engineering, Louisiana State University,
Baton Rouge, LA 70803, USA
`cjmce@lsu.edu`
`http://www.espol.edu.ec, https://www.lsu.edu`

Abstract. The existing coordinate system for the entire country of
Ecuador is the Provisional South American Datum of 1956 (PSAD56)
which originated in La Canoa, Venezuela. This coordinate system is considered one of the major datums of the world and is generally associated
with the countries of South America that comprise the Andes Moun-
tain Range. The system is referenced to the International Ellipsoid and
is rigorously constrained by gravimetric observations. Shortly after the
fall of the Iron Curtain, the traditionally classified gravity system of
the U.S. Department of Defense was offered to private geodesists on a
case-by-case basis. The Guayaquil project was one of those offered access
to the WGS84 geoid. The availability of cartographic data in different
datums made it necessary to create a new application software. This software needs to look at the input and output of all the components and
consolidate all the information into a final depository a block based Geographical Information System (adjusted to the World Geodetic System
1984 using Normal Mercator projection). This paper shows a Molodensky
seven parameter transformation shift from a PSAD56 classical geodetic
system to WGS84 Word Geodetic System 1984 for precise urban mapping applications. All future work within the city of Guayaquil will be
able to use the GPS technology with no systematic bias. Such facility
using the PSAD56 system is not possible.

Keywords: Spatial interpolation · Record imputation · Distributed
target scenarios · Domestic energy consumption · Structure detection ·
Urban energy modelling

1 Introduction

The PSAD56 is a modern, classical geodetic datum, and has been considered
a model for other continental regions within the technological constraints of

© Springer Nature Switzerland AG 2021
O. Gervasi et al. (Eds.): ICCSA 2021, LNCS 12951, pp. 567–587, 2021.
https://doi.org/10.1007/978-3-030-86970-0_40

the late 1950's to the middle of the 1960's. In 1982, a branch of the Ecuadorian Army the 'Instituto Geográfico Militar' (IGM) started a project called "Plano de Guayaquil", a series of maps at 1: 1000 scale were compiled by photogrammetric means in 1984 using a 1983 basic horizontal control, the projection used was a modified version from the Transverse Mercator projection. These maps were referenced to the PSAD56 data. As a Horizontal Datum, PSAD56 provides a frame of reference for measuring locations on the earth's surface, this defines the origin and orientation of latitude and longitude lines. All data is based on an ellipsoid, which is approximate to the shape of the earth. Since the beginning of the 18th century, scientists have estimated the largest and smallest axis on the earth's ellipsoid. The goal in all of this was to find the most convenient way for mapping a particular region or city. At the beginning (before 1960) the estimated ellipsoid was used in conjunction with an initial point on the earth's surface to produce a local horizontal datum. The starting point in PSAD56 is La Canoa – Venezuela. This point is assigned a latitude, longitude and elevation on the ellipsoid and related to a known point. All ground control measurements are then relative to this local datum (ellipsoid and starting point).

The Transverse Mercator or Gauss-Krüger projection is a conformal mapping of the earth ellipsoid where a central meridian is mapped into a straight line at constant scale. Because it cannot be expressed in terms of elementary functions, the mapping is usually computed by means of a truncated series [1]. The resulting mapping approximates the true map. Lee [2] implemented the formulas for the exact mapping and credits E. H. Thompson (1945) for their development. If the mapping is needed at greater distances from the central meridian, the use of an algorithm based on the exact mapping, an accuracy of 9 nm (1 nm = one billionth of a meter) is needed. Instead, the Krüger series, truncating equations to order n^6 with double precision gives an accuracy of 5 nm for distances up to 3,900 km from the central meridian [3]. The initial Grid established by IGM was based on a Gauss-Krüger Transverse Mercator projection. The scale factor at origin, $m_o = 1.0$, the central meridian of the belt (C.M.), $\lambda_o = 79°$ 53' 05.8232" West of Greenwich, the False Easting at C.M. = 624 km, and the False Northing = 10,000,051.000 m.

As technology progressed in the field of global geodesy, the satellite datums that began to emerge in the late 1960's through the 1970's rapidly began to render classical geodetic datums obsolete. Those with virtually unlimited budgets experienced the luxury of applying a comparative "perfect" ruler to the old classical systems, including the PSAD56.

With the declassification of the World Geodetic Systems of 1972 by Thomas Seppelin [4] in 1976; the world was allowed a tantalizing view of the possibilities of the future. The WGS72 datum (NWL9D) was the major impact in global geodesy, but the Cold War necessitated the continued restricted access of detailed gravity data on a worldwide basis known as the Geoid [5]. The late 1980's allowed civilian use of the new NAVSTAR satellite system otherwise known as the Global Positioning System (GPS). The reference frame of GPS is the World Geodetic System 1984 (WGS84) [6]. WGS84 is the standard of the Department of Defense.

The origin is the centre of mass of the earth. The WGS-84 reference ellipsoid is a geocentric ellipsoid of revolution.

Datum shift techniques and parameters are needed to relate positions on local datums [7] as the PSAD56. The most accurate approach for obtaining WGS 84 coordinates is to acquire satellite tracking data at the site of interest and position it directly in WGS 84 using the Satellite Point Positioning technique. Section 2 presents a detailed mathematical modelling of the transformation formulae.

2   Mathematical Modelling

Let us start with the definition of the terms involved in the datum shift transformation. The US Army Corps of Engineers [8] defines the following terms. Figure 1 shows the geometric relationship between Geodetic, Geocentric and Cartesian coordinates, and Fig. 2 shows the relationship between the reference ellipsoid, the geoid and the surface of the earth.

Datum is any numerical o geometrical quantity or set of such quantities specifying the reference coordinate system used for geodetic control in the calculation of coordinates of points on the earth. **Horizontal datum** specifies the coordinate system in which latitude and longitude of points are located. **Vertical datum** is the surface to which elevation are referred. **Elevation** is the vertical distance measured along the local plumb line from a vertical datum, usually mean sea level or the geoid, to a point on the earth. **Ellipsoid** is the surface generated by an ellipse rotating about in its axes. Also called ellipsoid of revolution. **Equator** is the line of zero geodetic latitude; the great circle described by the semi-major axis of the reference ellipsoid as it is rotated about the semi-minor axis. **Geocentric Cartesian coordinates** are the coordinates (X, Y, Z) that define the position of a point with respect to the centre of mass of the earth. **Geodetic coordinates** (geodetic position) are the quantities of latitude, longitude, and geodetic height (ϕ, λ, h) that define the position of a point on the surface of the earth with respect to the reference ellipsoid. **Geodetic height** (ellipsoidal height, h) is the height above the reference ellipsoid, measured along the ellipsoidal normal through the point in question. **Geodetic latitude** (ϕ) is the angle between the plane of the Equator and the normal to the ellipsoid through the computation point. Geodetic latitude is positive north of the equator and negative south of Ecuador. **Geodetic longitude** (λ) is the angle between the plane of a meridian and the plane of the prime meridian. **Geographic coordinates** are the quantities of latitude and longitude that define the position of a point on the surface of the earth. **Geoid** is the equipotential surface of the earth's gravity field approximated by undisturbed mean sea level of the oceans. **Grid reference system** is a plane-rectangular coordinate system usually based on, and mathematically adjusted to, a map projection in order that geodetic positions (latitudes and longitudes) may be readily transformed into plane coordinates.

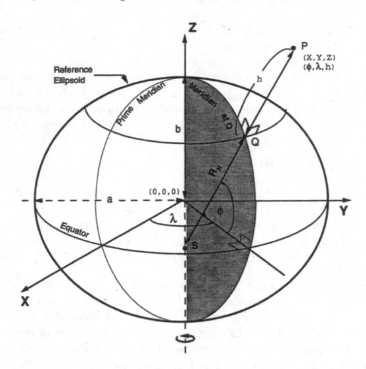

Fig. 1. The geometric relationship between Geodetic and Cartesian coordinates.

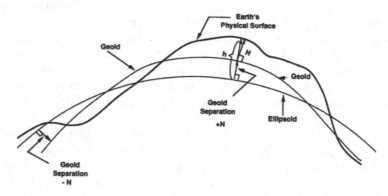

Fig. 2. Relationship between the reference ellipsoid, the geoid and the surface of the earth.

Figure 2 shows that the Z axis, the semi-minor axis is nearly parallel to the axis of rotation of the earth. The equatorial radius a is equal to 6,378,137 m at the equator, flattening f equals to 1/298.257223563, the first eccentricity squared is equal to 0.00669437999013 and the value of b equals to 6,356,752.3142 m.

2.1   Geodetic to Cartesian Coordinate Conversion

If the geodetic coordinates (ϕ, λ, h) are known, then the Cartesian coordinates (X, Y, Z) are given in Eq. 1, Eq. 2, and Eq. 3.

$$X = (R_N + h)cos\phi cos\lambda, \tag{1}$$

$$Y = (R_N + h)cos\phi sin\lambda, \tag{2}$$

$$Z = (R_N b^2/a^2 + h)sin\phi, \tag{3}$$

Where R_N, the radius of curvature in the prime vertical is given by Eq. 4.

$$R_N = a^2/\sqrt{a^2 cos^2\phi + b^2 sin^2\phi} \tag{4}$$

2.2   Cartesian to Geodetic Coordinate Conversion

In the reverse case, If the Cartesian coordinates (X, Y, Z) are known, the geodetic coordinates (ϕ, λ, h) are given in Eq. 5, Eq. 7, and Eq. 11. Computing λ for values of X different from zero, all values of Y is in Eq. 5.

$$\lambda = \arctan Y/X \tag{5}$$

Computing ϕ is an iterative procedure of Bowring [14]. Following Rapp [15] an initial approximation to a variable β is given by Eq. 6

$$\tan \beta_0 = aZ/b\sqrt{X^2 + Y^2} \tag{6}$$

Once β_0 is obtained, use Eq. 7 to substitute it of for β.

$$\tan \phi = (Z + \epsilon'^2 b \sin^3 \beta)/\sqrt{X^2 + Y^2} - a\epsilon^2 cos^3\beta \tag{7}$$

Where the values of ϵ^2 and ϵ'^2 are given by Eqs. 8 and 9.

$$\epsilon^2 = (a^2 - b^2)/a^2 = 2f - f^2 \tag{8}$$

$$\epsilon'^2 = (a^2 - b^2)/b^2 = \epsilon^2/(1 - \epsilon^2) \tag{9}$$

This approximation of ϕ is substituted into Eq. 10

$$\tan \beta = (1 - f)\tan \phi \tag{10}$$

Equation 10 gives an updated approximation to β. This procedure of using the latest approximation of β to produce an updated approximation of ϕ (Eq. 7) can be continued until the updated value of ϕ is close enough to the previous value.

Once ϕ is computed, the value of h is in Eq. 11

$$h = (\sqrt{X^2 + Y^2}/ \cos \phi) - R_N \tag{11}$$

Where R_N is in Eq. 4.

2.3   Seven Parameter Geometric Transformation

The seven parameter transformation is the most general transformation between local and global Cartesian coordinates. The assumption is that the local (PSAD56) and WGS84 Cartesian axes differ by seven parameters: three rotational parameters (ε, φ, ω), a scale change δS, and three origin shift parameters (δX, δY, δZ). Their relations are in Fig. 3, and Eqs. 12, 13 and 14.

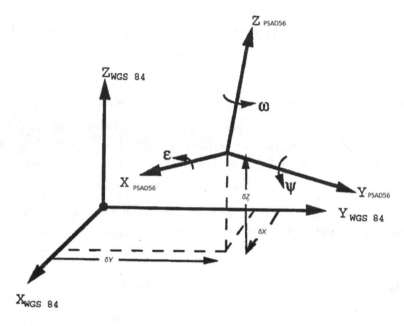

Fig. 3. Relationship between WGS84 and PSAD56 coordinate axis in a seven parameter geometric transformation.

$$X_{WGS84} = X_{PSAD56} + \delta X + \omega Y_{PSAD56} - \varphi Z_{PSAD56} + \delta S X_{PSAD56} \quad (12)$$

$$Y_{WGS84} = Y_{PSAD56} + \delta Y - \omega X_{PSAD56} + \varepsilon Z_{PSAD56} + \delta S Y_{PSAD56} \quad (13)$$

$$Z_{WGS84} = Z_{PSAD56} + \delta Z + \varphi X_{PSAD56} - \varepsilon Y_{PSAD56} + \delta S Z_{PSAD56} \quad (14)$$

2.4   Three Parameter Geometric Transformation

In certain cases, the three parameter geometric transformation can be applied directy to convert local geocentric coordinates to WGS84 geocentric coordinates.

The resulting Eqs. 15 are a subset of Eqs. 12, 13 and 14 with ε, φ, ω, and δS equal to zero.

$$X_{WGS84} = X_{PSAD56} + \delta X; Y_{WGS84} = Y_{PSAD56} + \delta Y; Z_{WGS84} = Z_{PSAD56} + \delta Z \tag{15}$$

2.5   Seven Parameter Bursa Wolf Model

We are given a set of rectangular coordinates, (X_{PSAD56}, Y_{PSAD56}, Z_{PSAD56}), in an "PSAD56" system and we want to transform these coordinates into the "WGS84" system to obtain (X_{WGS84}, Y_{WGS84}, Z_{WGS84}) $\hat{X}_{WGS84}$. The general geometry of the transformation has the translation parameters δX, δY, δZ, which will be designated T in vector form. The three rotation angles ε, φ, ω. A positive rotation is a counter-clockwise rotation about an axis when viewed from the end of the positive axis in right-handed coordinate systems. The mathematical relationship between coordinates in the seven parameter Bursa Wolf model can be written in the form of a vector Eq. 16 assuming the rotation angles are small (a few seconds of arc) [9]. Malys [10] has studied the numerical impact of the small angle approximation in obtaining Eq. 16. He found that the disagreement between an element was at the level of 0.5×10^{-11} when the rotation angles were on the order of 1"; on the order of 0.5×10^{-10} when the angles were on the order of 3"; and on the order of 0.5×10^{-9} when the angles were on the order of 9". An error of 0.5×10^{-9} propagates into a coordinate error on the order of 3 mm. We should note that the order of rotation is not important when the angles are small, as Eq. 16 is independent of the order of rotation.

$$\hat{X}_{WGS84} = (1 + \delta S) \begin{bmatrix} 1 & \omega & -\varphi \\ -\omega & 1 & \varepsilon \\ \varphi & -\varepsilon & 1 \end{bmatrix} \hat{X}_{PSAD56} + \begin{bmatrix} \delta X \\ \delta Y \\ \delta Z \end{bmatrix} \tag{16}$$

2.6   Seven Parameter Molodensky Model

Various interpretations and application of the Molodensky transformation have been given in Badekas [11], Leick and van Gelder [12], Soler [13] and others. Our research follows the Molodensky-Badekas transformation: a seven-parameter conformal transformation (or similarity transformation) linking rotations and translations between the X,Y,Z Cartesian axes and a scale factor to changes in the Cartesian coordinates.

Figure 4 shows the geometry of the Molodensky transformation that makes use of a centroid. The X, Y, Z axes of system 1 are rotated by very small angles ε, φ, ω, from the X, Y, Z axes of system 2, and the origins of the two systems are displaced. The $\overline{X_1}$, $\overline{Y_1}$, $\overline{Z_1}$ system is a centroidal system whose origin is at a centroid G of a set of points in system 1 and whose axes are parallel to the X,Y,Z axes of system 1. The mathematical relationship between coordinates in the seven parameter Bursa Wolf model can be written in the form of a vector Eq. 17 assuming the rotation angles are small.

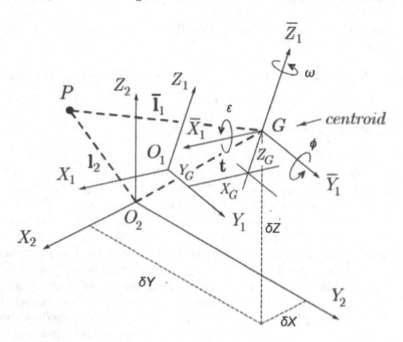

Fig. 4. Relationship between WGS84 and PSAD56 coordinate axis in a seven parameter Molodensky geometric transformation.

$$\hat{X}_{WGS84} = (1 + \delta S) \begin{bmatrix} 1 & \omega & -\varphi \\ -\omega & 1 & \varepsilon \\ \varphi & -\varepsilon & 1 \end{bmatrix} \; (\widehat{X - X_G})_{PSAD56} + \begin{bmatrix} \delta X \\ \delta Y \\ \delta Z \end{bmatrix} \qquad (17)$$

In summary, the magnitude of the shift parameters is minimized in the Molodensky model by the scalar because the geocentric coordinates of the origin point, La Canoa, is subtracted from each geocentric coordinate of the data points. That is the primary difference between Bursa-Wolf and Molodensky. The scalar reduces the translations. We prefer to use the Molodensky model to the Bursa-Wolf model in cases like this when the local area is a great distance from the datum origin. Note that the net computational results (of shift transformations) are identical, but the shift parameters in this case are less correlated and thus they appear more meaningful with the Molodensky model.

2.7   The Mercator Projection

As a projection falls into the cylindrical group (it is the cylindrical orthomorphic), it has several characteristics which are common to all cylindrical projections. First, meridians and parallels are straight lines set at right angles. Secondly, the equator is the correct length, but as all other parallels are the same length as the equator, they are progressively exaggerated in length pole wards. Clearly, the pole cannot be shown on this projection. On the globe, the pole is a point, but on any cylindrical projection, it would be represented as a straight line of the same length as the equator and would, therefore, be infinitely exaggerated. To maintain the property of orthomorphism, the scale along the meridian at the pole would also have to be infinitely exaggerated and so the pole would be infinitely distant from the equator.

The new Normal Mercator parameters of the City of Guayaquil are COUNTRY: Republic of Ecuador, GRID: Guayaquil/ESPOL La Rotonda, DATUM: WGS84, ELLIPSOID: WGS84, Semi-major axis = 6,378,137.000, 1/flattening = 298.25722356300, Units: 1 m = 1.000000000 Meters, Projection type = 41 Normal Mercator, Central Meridian = −79° 52' 45.160", Latitude at origin = −2° 11' 33.0900", Scale factor at origin = 0.999272829, False easting = 500,000.0000 m and False northing = 2,242,320.5100 m.

3   Data Available from Field Survey

The object control for the photos was obtained from two sources. The Horizontal Control was from the GPS survey made by Offshore Navigation Inc (see Fig. 5) in Guayaquil and the Vertical Control was from the Monographs from the IGM. For a full breakdown see Table 1 and Table 2 below. Table 1 shows the grid coordinates (Easting and Northing) and the a priori standard deviation and Table 2 shows the elevation and a priori standard deviation. The geodetic survey of Guayaquil produced thirty points plus fifteen kinematic points. From those the points Jordan (GPS6), Antena C-10 (GPS15) and Faro (GPS11) were sent to Defense Mapping Agency for its adjustment to the ephemeris and referencing to WGS84.

There were three objectives to the field survey.

1. The first objective was to establish absolute WGS84 control points on IGM second order monuments. The data for these points were collected for four hours. During this time, both code and carrier data on L1 and L2 were collected in accordance to DMA specifications. Data from these stations were processed by DMA using precise ephemeris and orbit smoothing to obtain absolute WGS84 coordinates.

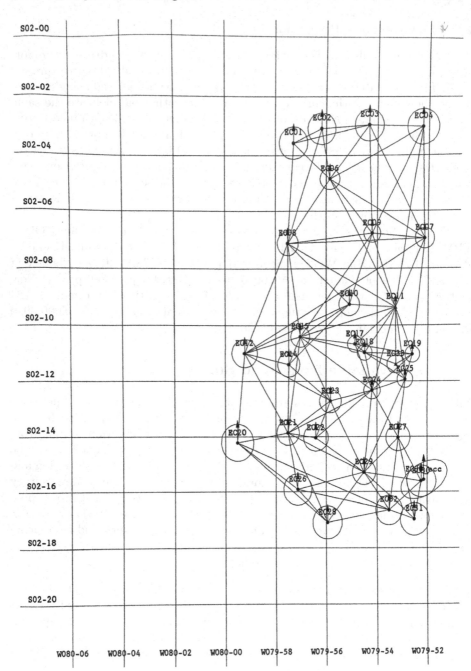

Fig. 5. GPS Ecuador (EC) data available from field survey.

Table 1. Horizontal Control Points (ONI), all units are space units (meters).

Control point	Easting	Northing	Elevation	Easting	Northing	Elevation
ID	Position			A priori standard deviation		
GPS1	491,599.246	2,014,685.053	23.278	0.007	0.007	0.010
GPS2	493,654.311	2,015,585.286	5.542	0.006	0.006	0.010
GPS3	497,076.149	2,015,812.419	14.360	0.007	0.007	0.010
GPS4	500,967.861	2,015,705.493	7.233	0.007	0.007	0.010
GPS6	494,254.413	2,012,380.494	94.438	0.005	0.005	0.008
GPS7	501,136.259	2,008,458.029	4.733	0.004	0.004	0.008
GPS8	491,183.733	2,008,145.422	52.971	0.005	0.005	0.008
GPS9	497,318.375	2,008,785.593	5.290	0.004	0.004	0.008
GPS10	495,660.659	2,004,257.312	100.301	0.005	0.005	0.008
GPS12	487,980.567	2,001,021.131	6.802	0.006	0.006	0.009
GPS14	491,246.484	2,000,275.383	11.910	0.005	0.005	0.008
GPS15	492,093.995	2,002,121.963	316.442	0.004	0.004	0.007
GPS17	496,091.295	2,001,637.492	76.487	0.004	0.004	0.007
GPS18	496,788.643	2,001,077.778	27.077	0.003	0.003	0.007
GPS19	500,278.543	2,000,948.621	3.781	0.003	0.004	0.007
GPS21	491,182.747	1,995,750.284	3.599	0.005	0.005	0.008
GPS23	494,306.867	1,997,848.204	34.653	0.005	0.005	0.008
GPS26	491,896.507	1,991,959.302	2.066	0.007	0.007	0.009
GPS27	499,239.497	1,995430.020	42.698	0.006	0.006	0.008
GPS28	494,078.448	1,989,778.300	3.509	0.007	0.007	0.009
GPS29	496,751.453	1,993,113.360	26.717	0.005	0.005	0.008
GPS31	500,460.263	1,989,991.929	2.101	0.007	0.007	0.009
GPS32	498,574.265	1,990,580.463	3.060	0.006	0.006	0.009
4N	493,592,961	2,012,306.257	11.030	0.036	0.043	0.094
6N	495,137.181	2,013,490.837	10.141	0.050	0.048	0.120
7N	494,823.537	2,014,589.045	8.369	0.051	0.056	0.123
2S	492,268.085	2,006,812.702	42.235	0.027	0.028	0.066
3S	492,581.441	2,005,593.927	44.126	0.032	0.027	0.078
5S	491,946.455	2,000,275.306	13.255	0.059	0.066	0.174
6S	491,483.872	1,998,369.640	4.467	0.069	0.076	0.213
7S	491,399.715	1,996,350.465	5.165	0.080	0.088	0.249

2. The second objective was to survey approximately thirty points that were identified in the new aerial photos. Some of these points were also second order stations established in Guayaquil by the IGM. For these points GPS receivers were configured for synchronous data acquisition on five survey points for each session. There were three sessions each day and each session lasted forty-five minutes. For each session data was collected from a five or six satellite constellation.

3. The third objective was to run a kinematic survey where possible. Lines were run alongside the free-way that runs around Guayaquil. The control station for this phase was EC-08.

Table 2. Vertical Control Points (ONI), all units are space units (meters).

Control point	Elevation	Elevation
ID	Position	A priori standard deviation
PV-2	10.305	0.005
PV-3	6.537	0.005
PV-4	4.072	0.005
PV-14	5.306	0.005
PV-15	6.057	0.005
PV-27	13.531	0.005
PV-29	8.559	0.005
PV11-A	3.974	0.005
PV12-A	3.501	0.005
PV-16A	25.039	0.005
PV22-A	4.051	0.005
PV25-A	3.037	0.005
PV26-A	2.538	0.005
PB-B2	3.326	0.005
PV-B4	3.584	0.005
PV-B7L3	3.220	0.005
PV-B7L4	3.220	0.005
PV-B10	3.206	0.005
PV-B11	3.808	0.005
PV-B12	3.734	0.005
PV-B13	3.121	0.005
PV-B15	2.985	0.005
PV-B16	3.244	0.005
PV-C4	3.210	0.005
PV-C7	1.604	0.005
PV-C9	3.180	0.005
PV-C10	3.413	0.005

4 The Datum Shift Parameter Transformation Results

Local PSAD56 geodetic coordinates can be shifted to WGS84 coordinates in three steps. Fist, convert local geodetic PSAD56 to local geocentric Cartesian coordinates using Eqs. 1, 2 and 3. Second, shift the local geocentric Cartesian coordinates to WGS84 geocentric Cartesian coordinates using Eqs. 12, 13, and 14. Finally, convert WGS84 geocentric coordinates to WGS84 geodetic coordinates using Eqs. 5, 7, and 11.

In order to transform between datums several methods have been offered [16]. From those, the methods based upon Molodensky and Bursa-Wolfe transformations, referred to as three- and seven-parameter transformation, have received great acceptance [17,18]. The three-parameter transformation accounts for shifts in the origin of the X, Y, and Z axes at the centre of the earth. This in essence, is like sliding the latitude/longitude graticule across the surface of the earth. The seven-parameter transformation accounts for the shifts in origin, as well as rotation about each for the three axes and a difference in scale. Section 4.1 uses the three-parameter model. Data from one or more positions are required to derive a three-parameter geometric transformation. Section 4.2 uses the generalized geometric transformation model. This model assumes the origins of the two coordinate systems are offset from the other, the axes are not parallel and there is a scale difference between the two datums. This seven-parameter model needs at least three well-spaced positions. Section 2.5 and Sect. 2.6 show the Bursa Wolf and Molodensky seven parameter transformation formulae, respectively.

4.1   Provisional Three Transformation Parameters

To confirm the common points in the two datums, the geodetic coordinates were transformed to geocentric coordinates independently, in each datum. These coordinates were used to calculate three transformation parameters from WGS84 to PSAD56, which are the translations in X, Y, Z. Note that these parameters are provisional parameters because it is an initial transformation of WGS84 to PSAD56, just to confirm the common points. The three parameters were used to transform the geocentric coordinates of the GPS points in WGS844 to PSAD56. The geocentric coordinates of the GPS points, already in PSAD56, were transformed to geodetic coordinates. The WGS84 data is shown in Table 3 and the PSAD56 data is shown in Table 4.

Table 3. Geodetic coordinates of GPS (WGS84) – data provided by Offshore Navigation.

	Latitude	Longitude	Geodetic height
JORDAN (GPS 6)	$-2°$ 04' 50.97029"	$-79°$ 57' 17.03262"	94.438 m
FARO (GPS 11)	$-2°$ 09' 20.74185"	$-79°$ 53' 17.29297"	22.220 m
ANTENA C-10 (GPS 15)	$-2°$ 10' 24.00518"	$-79°$ 57' 01.02111"	316.442 m
SUBURBIO (GPS 23)	$-2°$ 12' 43.14517"	$-79°$ 55' 49.40620"	34.653 m
PANORAMICO (GPS 25)	$-2°$ 11' 55.83892"	$-79°$ 52' 53.39967"	68.560 m
FERTISA (GPS 29)	$-2°$ 15' 17.29233"	$-79°$ 54' 30.29234"	26.717 m

Table 4. Geodetic coordinates of PSAD-56 – data provided by IGM.

	Latitude	Longitude	Geodetic height
JORDAN (GPS 6)	−2° 04' 38.07738"	−79° 55' 43.14658"	92.740 m
FARO (GPS 11)	−2° 09' 08.81595"	−79° 53' 09.35307"	21.765 m
ANTENA C-10 (GPS 15)	−2° 10' 12.07286"	−79° 56' 53.07339"	315.856 m
SUBURBIO (GPS 23)	−2° 12' 31.21684"	−79° 55' 41.45763"	35.241 m
PANORAMICO (GPS 25)	−2° 11' 43.90930"	−79° 52' 45.46013"	68.614 m
FERTISA (GPS 29)	−2° 15' 05.36320"	−79° 54' 22.34870"	27.209 m

The geodetic coordinates of WGS-84 points are transformed to geocentric coordinates as shown in Table 5 and the geodetic coordinates of PSAD-56 points area transformed to geocentric coordinates as shown in Table 6.

Table 5. Geocentric coordinates – converted from Geodetic WGS-84.

	X	Y	Z
JORDAN (GPS 6)	1,114,418.813	−6,275,878.449	−230,009.979
FARO (GPS 11)	1,119,029.765	−6,274,672.056	−238,317.507
ANTENA C-10 (GPS 15)	1,112,261.692	6,276,099.370	−240,270.451
SUBURBIO (GPS 23)	1,114,362.856	−6,275,274.572	−244,530.417
PANORAMICO (GPS 25)	1,119,732.941	−6,274,409.799	−243,079.750
FERTISA (GPS 29)	1,116,735.965	−6,274,657.224	−249,261.248

Table 6. Geocentric coordinates – converted from Geodetic PSAD-56.

	X	Y	Z
JORDAN (GPS 6)	1,114,704.833	−6,276,093.956	−229,646.278
FARO (GPS 11)	1,119,317.707	−6,274,889.135	−237,954.029
ANTENA C-10 (GPS 15)	1,112,549.642	6,276,316.706	−239,906.779
SUBURBIO (GPS 23)	1,114,651.138	−6,275,493.197	−244,166.985
PANORAMICO (GPS 25)	1,120,021.030	−6,274,627.621	−242,716.236
FERTISA (GPS 29)	1,117,024.205	−6,274,875.941	−248,897.851

The initial (approximate) three parameter transformation are in Table 7.

Table 7. Initial three parameters geometric transformation.

	δX	δY	δZ
Three parameter transformation	288.21935 m	−217.08152 m	363.68780 m

The PSAD56 geodetic coordinates using the initial parameter transformation are shown in Table 8 and the modified Transverse Mercator map coordinates are shown in Table 9.

Table 8. Geodetic coordinates of PSAD-56 – using initial three parameter transformation.

	Latitude	Longitude
GPS 1	−2° 03' 23.0419"	−79° 57' 09.0744"
GPS 2	−2° 02' 53.7308"	−79° 56' 02.5694"
GPS 3	−2° 02' 46.3353"	−79° 54' 11.8334"
GPS 4	−2° 02' 49.8166"	−79° 52' 05.8916"
GPS 6 (JORDAN)	−2° 04' 38.0762"	−79° 55' 43.1490"
GPS 7	−2° 06' 45.7854"	−79° 52' 00.4417"
GPS 8	−2° 06' 55.9638"	−79° 57' 22.5207"
GPS 9	−2° 06' 35.1207"	−79° 54' 03.9943"
GPS 10	−2° 09' 02.5515"	−79° 54' 57.6403"
GPS 11 (FARO)	−2° 09' 08.8094"	−79° 53' 09.3442"
GPS 12	−2° 00' 47.9121"	−79° 59' 06.1798"
GPS 14	−2° 11' 12.1910"	−79° 57' 20.4896"
GPS 15 (ANTENA C-10)	−2° 10' 12.0724"	−79° 56' 53.0630"
GPS 17	−2° 10' 27.8451"	−79° 54' 43.7042"
GPS 18	−2° 10' 46.0675"	−79° 54' 21.1369"
GPS 19	−2° 10' 50.2722"	−79° 52' 28.1983"
GPS 20	−2° 14' 01.0303"	−79° 59' 23.8659"
GPS 21	−2° 13' 39.5105"	−79° 57' 22.5521"
GPS 22	−2° 13' 50.4848"	−79° 56' 16.0648"
GPS 23 (SUBURBIO)	−2° 12' 31.2108"	−79° 55' 41.4508"
GPS 24	−2° 12' 07.8808"	−79° 54' 03.3481"
GPS 25 (PANORAMICO)	−2° 11' 43.9048"	−79° 52' 45.4517"
GPS 26	−2° 15' 42.9268"	−79° 56' 59.4535"
GPS 27	−2° 13' 49.9365"	−79° 53' 01.8232"
GPS 28	−2° 16' 53.9284"	−79° 55' 48.8424"
GPS 29 (FERTISA)	−2° 15' 05.3562"	−79° 54' 22.3400"
GPS 30	−2° 15' 25.0890"	−79° 52' 06.0540"
GPS 30-	−2° 15' 21.5880"	−79° 52' 00.1016"
GPS 31	−2° 16' 46.9735"	−79° 52' 22.3171"
GPS 32	−2° 16' 27.8141"	−79° 53' 23.3509"
GPS 33	−2° 11' 11.2206"	−79° 53' 07.7171"

Table 9. Modified Transverse Mercator map coordinates of PSAD-56 – using initial three parameter transformation.

	Easting (meters)	Northing (meters)
GPS 1	616,482.681	9,772,662.071
GPS 2	618,537.894	9,773,562.462
GPS 3	621,960.039	9,773,789.692
GPS 4	625,892.107	9,773,682.764
GPS 6	619,138.143	9,770,357.426
GPS 7	626,020.448	9,766,434.786
GPS 8	616,067.437	9,766,121.978
GPS 9	622,202.369	9,766,762.364
GPS 10	620,544.665	9,762,233.879
GPS 11	623,891.196	9,762,041.695
GPS 12	612,864.591	9,758,997.301
GPS 14	616,130.568	9,758,233.879
GPS 15	616,978.001	9,760,098.373
GPS 17	620,975.363	9,759,614.021
GPS 18	621,672.726	9,759,054.315
GPS 19	625,162.648	9,758,925.174
GPS 20	612,318.489	9,753,065.461
GPS 21	616,067.054	9,753,726.686
GPS 22	618,121.530	9,753,389.688
GPS 23	619,191.032	9,755,824.693
GPS 24	622,222.446	9,756,541.356
GPS 25	624,629.496	9,757,277.807
GPS 26	616,780.971	9,749,935.877
GPS 27	624,123.600	9,753,406.635
GPS 28	618,962.877	9,747,755.081
GPS 29	621,635.662	9,751,090.035
GPS 30	625,846.836	9,750,483.930
GPS 30-	626,030.764	9,750,591.465
GPS 31	625,344.296	9,747,968.781
GPS 32	623,458.410	9,748,557.281
GPS 33	623,941.476	9,758,281.732

After this Section the GPS 10 might be Mapasingue Auxiliar, which is a third order station in the IGM monographs. Therefore, our research has an additional point (a total of seven points) to compute the datum shift seven parameter transformation.

4.2   Final Seven Transformation Parameters

Dr Munendra Kumar from the Defense Mapping Agency established the parameters of the Bursa-Wolfe Model and the parameters of the Molodensky Model. The parameters of the Molodensky model are shown in Table 10.

Table 10. Parameters of the Molodensky model.

δ x	δ y	δ z	Rx	Ry	Rz	δs
–263.91 m	–25.05 m	–285.81 m	+3.54"	–3.42"	–36.88"	–0.00000361

A computational test point for instance, is station Panoramico where the PSAD 56 coordinates are: ϕ = 02° 11' 43.9093" South, λ = 79° 52' 45.4601" West, and h = 68.614 m. The Panoramico WGS 84 coordinates are: ϕ = 02° 11' 55.8406" South, λ = 79° 52' 53.4010" West, and h = 68.530 m.

Each PSAD56 city block were shifted from the PSAD56 classical system to the new WGS84 datum using custom software development. As the city grew mainly to the west and north, it was necessary to have additional sources of data for these new developments in the city. This research also build a block based Geographic Information System from compilation out of stereoscopic models produced from new controlled photos directly in WGS1984 for the new developments. Figure 6 shows the map4439 in the Guayaquil Block based Geographical Information System.

Figure 6 shows the coincidence of the digitized Mylar map information shifted from PSAD-56 Modified Transverse Mercator to WGS84 Normal Mercator Projection, and the compiled photo information directly in WGS84 Normal Mercator Projection.

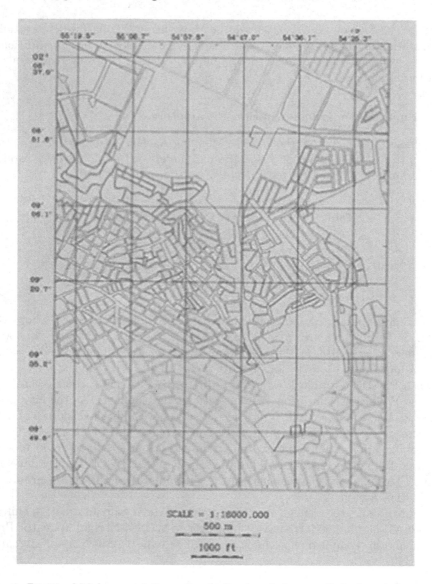

Fig. 6. Digitized Mylar map information (yellow) and the complied photo information (red). (Color figure online)

5   Discussion

The datum shift approach done for Guayaquil, although a proper approach, was also constrained by budget. The solution for a least square fit to the Geocentric Reference System for the Americas (SIRGAS) is substantially better for the country as a whole, but it must be realized that for a larger area there must be a compromise that will result in some error and compromise for any given region.

Dr. Kumar was involved in the Guayaquil project because the geoid was still a U.S. military classified Secret thing back then. Since then, the Earth Gravity Model 1996 (EGM96) is now completely unclassified and is available over the internet. That will allow one to convert between WGS84 ellipsoid heights and Mean Sea Level elevations to a world-wide accuracy of about one meter – for elevations. The best for Ecuador would be to obtain a least square fit for all (100%) of the IGM points in and surrounding the Guayaquil new developments if the fit to IGM existing control is the objective. With 100% of the points, however many that might be, that will give the best compromise specifically related to the project at hand. Looking ahead for Guayaquil, the mathematical model for the Datum Shift chosen depends on just how close one wish to find a fit to existing IGM control. Visiting 100% of the points in specific area might be expensive for 4–8 h' observation at each point, that is something that might be done in order to match existing IGM maps and IGM control.

The vertical datums currently used in Latin America refer to different tide gauges and, therefore, at different sea levels and at different times. These do not take into account the variations of the heights and the reference level with time, and in general, the extension of the vertical control by means of levelling nets does not include the reductions due to the effects of gravity. Consequently, the heights associated with them present considerable discrepancies between neighbouring countries, do not allow the exchange of vertical information either on a continental or global scale and are not able to support the practical determination of heights from the Global navigation satellite system (GNSS) technique. SIRGAS is working on a new vertical reference system defined in terms of potential quantities (W_0 as the reference level and geopotential numbers or heights as fundamental coordinates). In accordance with this, each country will be able to introduce the type of physical heights that it prefers, along with the corresponding reference surface: geoid for orthometric heights or quasi-geoid for normal heights. The definition of the new vertical reference system for SIRGAS is identical to the definition of the International Height Reference System (IHRS) described in Resolution No. 1, 2015, of the International Geodesy Association (IAG). The realization of the new vertical reference system for SIRGAS should be a regional densification of the International Height Reference Frame (IHRF).

Current technology has placed an emphasis on Continuously Operating Reference Stations (CORS) for GPS-derived national control. That is what it has done in the state of Louisiana. Louisiana State University (LSU) now has 50 CORS sites throughout the entire state. That is 50 GPS dual frequency geodetic receivers permanently bolted to buildings in Louisiana, and all simultaneously working as a Real Time Network. LSU is now selling subscriptions to Surveyors so that they can connect to its GPS RTN and achieve zero-length baseline positions via cell phone data links. The new realization of the North American Datum of 1983 for the United States is now based exclusively on CORS - nationwide. As time goes on, each country will eventually do the same, including Ecuador. CORS networks are a reality in Guatemala, Honduras, Argentina, and Brazil.

The earthquake that occurred in April 2016 on the Northwest coast of Ecuador severely affected the Geodetic Reference Frame of the country, physically reflected in the damage produced in the stations of (GNSS) Network and landmarks of the GPS National Network; and in a non-tangible way in the deformation of its coordinates, invalidating the georeferencing of any cartographic product in the country. Before 2016, Ecuador maintained the GNSS Network for Continuous Monitoring of Ecuador (REGME) and the Nation GPS Network of Ecuador (RENAGE) referring to the International Terrestrial Reference (ITRF) Frame. The earthquake damage of the Geodetic Frame leaving the country without having a guaranteed and quality reference.

6   Conclusions

A new Grid was devised for the City of Guayaquil based on the Normal Mercator projection, WGS 84 Datum. The origin is at the Rotonda of Simón Bolívar in downtown Guayaquil. We occupied a number of existing IGM triangulation stations as well as new photo-identifiable points based on recent IGM aerial photography. Co-located fiducial points were observed according to DMA/NIMA specifications, and NIMA performed the subsequent fiducial point adjustment to the precise ephemeris. The solution from PSAD56 to WGS 84 for a 7–Parameter Molodensky model (using the PSAD56 origin at La Canoa with Northern latitude), yielded: DX $= -263.91$ m, DY$= -25.05$ m, DZ $= -285.81$ m, scale $= -3.61 \times 10^{-6}$, Rz $= -36.88$", Ry $= -3.42$", Rx $= +3.54$". This research solve for the 7-Parameter Moledensky model using rigorous least squares techniques.

The current Geographical Information System being established is based on the WGS84 Datum using the U.S. Department of Defense satellites. This basic geometrical foundation will serve all practical applications of municipal works for decades.

References

1. Krüger, L.: Konforme Abbildung des Erdellipsoids in der Ebene (Conformal mapping of the ellipsoidal earth to the plane). Royal Prussian Geodetic Institute 1912, New Series, vol. 52, pp. 1–172 (1912)
2. Lee, T.: Conformal projections based on Jacobian elliptic functions. Cartographica **13**, 67–101 (1976)
3. Karney, C.F.: Transverse Mercator with an accuracy of a few nanometers. J. Geod. **8513**, 475–485 (2011)
4. Seppelin, T.: The department of defense world geodetic system 1972. Can. Surv. **28**, 496–506 (1974)
5. Fisher, I.: The role of the geoid in datum transformation. Can. Surv. **28**, 507–513 (1974)
6. Decker, B.L.: World geodetic system 1984. In: Proceedings of the Fourth International Geodetic Sysmposium on Satellite Positioning, The University of Texas at Austin, United States, 28 April–2 May 1986

7. Defense Mapping Agency: Department of Defense World Geodetic System 1984. Its definition and relationships with local Geodetic Systems. In DMA TR 8350.2; Publishing House: Fairfax, USA (1991)
8. U.S. Army Corps of Engineers: Topographic Engineering Center. Handbook for Transformation of Datums, Projections, Grids, and Common Coordinate Systems. In TEC-SR-7; Publishing House: Alexandria, USA (1966)
9. Rapp, R.: Geometric Geodesy Part II. Ohio State University Department of Geodetic Science and Surveying, Columbus (1993)
10. Malys, S.: Dispersion and Correlation Among Transformation Parameters Relating Two Satellite Reference Frames. Ohio State University Department of Geodetic Science and Surveying, Columbus (1988)
11. Badekas, J.: Investigations Related to the Establishment of a World Geodetic System Ohio State University Department of Geodetic Science and Surveying (1969)
12. Leick, A., van Gelder, B.: On Similarity Transformations and Geodetic Network Distortions Based on Doppler Satellite Ohio State University Department of Geodetic Science and Surveying (1975)
13. Soler, T.: On Differential Transformations Between Cartesian and Curvilinear (Geodetic) Coordinates Ohio State University Department of Geodetic Science and Surveying (1975)
14. Bowring, B.: Transformation from spatial to geographical coordinates. Surv. Rev. **181**, 323–327 (2011)
15. Rapp, R.: Geometric geodesy part I. Ohio State University Department of Geodetic Science and Surveying, vol. 28, pp. 123–124 (1984)
16. Mueller, I.: Review of problems associated with conventional geodetic datums. Can. Surv. **28**, 514–523 (1974)
17. Janssen, V.: Understanding coordinate reference systems, datums and transformations. Int. J. Geoinform. **5**, 41–53 (2009)
18. Abbey, D.A., Featherstone, W.E.: Comparative review of Molodensky-Badekas and Burša-Wolf methods for coordinate transformation. J. Surv. Eng. **146**(3), 04020010 (2020)

Deep and Ensemble Learning Based Land Use and Land Cover Classification

Hicham Benbriqa[1,2], Ibtissam Abnane[1], Ali Idri[1,2(✉)], and Khouloud Tabiti[2]

[1] Software Project Management Research Team, ENSIAS, Mohammed V University,
Rabat, Morocco
{ibtissam.abnane,ali.idri}@um5.ac.ma
[2] Digital 4 Research-MSDA, Mohammed VI Polytechnic University, Ben Guerir, Morocco
{benbriqa.hicham,khouloud.tabiti}@um6p.ma

Abstract. Monitoring of Land use and Land cover (LULC) changes is a highly encumbering task for humans. Therefore, machine learning based classification systems can help to deal with this challenge. In this context, this study evaluates and compares the performance of two Single Learning (SL) techniques and one Ensemble Learning (EL) technique. All the empirical evaluations were over the open source LULC dataset proposed by the German Center for Artificial Intelligence (EuroSAT), and used the performance criteria -accuracy, precision, recall, F1 score and change in accuracy for the EL classifiers-. We firstly evaluate the performance of SL techniques: Building and optimizing a Convolutional Neural Network architecture, implementing Transfer learning, and training Machine learning algorithms on visual features extracted by Deep Feature Extractors. Second, we assess EL techniques and compare them with SL classifiers. Finally, we compare the capability of EL and hyperparameter tuning to improve the performance of the Deep Learning models we built. These experiments showed that Transfer learning is the SL technique that achieves the highest accuracy and that EL can indeed outperform the SL classifiers.

Keywords: Land use · Land cover · Machine learning · Deep learning · Deep feature extraction · Ensemble learning · Hyperparameter optimization

1 Introduction

Land use and Land cover (LULC) classification refers to the arrangement into groups of the "human activities on and in relation to the land" and the "biophysical cover of the Earth's surface" [12, 16, 23]. LULC classification has several applications such as spanning forestry, agriculture, urban planning, and water-resources management [22]. And thanks to the availability of images of the Earth's surface (remote sensing) and to image processing techniques, researchers can construct their own LULC dataset depending on the LULC classification problem they are dealing with. For instance, the community has already produced several open source LULC datasets (e.g. Merced Land Use [29], UC HistAerial [21], EuroSAT [9]).

© Springer Nature Switzerland AG 2021
O. Gervasi et al. (Eds.): ICCSA 2021, LNCS 12951, pp. 588–604, 2021.
https://doi.org/10.1007/978-3-030-86970-0_41

To solve LULC classification problems, the literature suggests different approaches, and Convolutional Neural Networks (ConvNets) are still the most popular one [5], as they integrate the tedious feature extraction from the input images [19]. Besides, ConvNets have recently achieved major successes in many Computer Vision (CV) applications [15]. However, building ConvNets is challenging as it requires setting the values of several hyperparameters, such as the learning rate, the number of hidden layers and the batch size [17]. This drawback has motivated the use of another approach, that is Transfer learning [1, 21]. It consists of using ConvNets that were trained on an image-related task, and adapting the knowledge they had acquired to the new problem [11]. This approach reduces the time required to design and develop the models since it provides a ready to use ConvNet architecture. Nonetheless, it still leaves a set of hyperparameters to be fine-tuned by the user. Hence, the third approach we found in the previous works: Deep Feature extraction. It tries first to extract the visual features from images, and then feed them to a Machine Learning (ML) classification technique [2]. This approach achieves relatively good performance with little fine tuning. But it does not generally outperform the previously mentioned approaches [3].

The use of ML in LULC classification problems has lately proved very promising [25]. However, we noticed some limitations in the previous works, namely few pretrained models are employed and the performance of the resulting classifiers depend highly on the context. Given the cost of building new architectures, most previous works tend to use Transfer learning instead. Nevertheless, the number of Deep Learning (DL) pretrained architectures they use is still limited [1, 26]. Additionally, an important limitation when using ML/DL techniques is that different techniques give different performances depending on the context. In fact, there is no general best/worst technique in all contexts. Thus, we propose the use of Ensemble Learning (EL) techniques to remedy this limitation. The use of EL to address LULC problems is not thoroughly studied in the literature as most works focus on implementing and optimizing Transfer learning single models, and those who did employ EL techniques, have only combined ML models and did not investigate the use of DL models as members [8, 13].

This study intends to overcome the aforementioned limitations by means of: 1) thoroughly implementing single DL/ML techniques with hyperparameters optimization to find the best single model; and 2) proposing an EL approach to aggregate the performance of the best single DL/ML models.

Indeed, we firstly evaluate the three state of the art ML techniques. Specifically: we build and optimize a ConvNet architecture referred to us LULC-Net, we use several pretrained ConvNet architectures and fine tune their hyperparameters and finally we use the ConvNet part of two of these pretrained models as Feature extractors, and train ML algorithms on the resulting features. Secondly, we build different EL models: Machine Learning Ensembles (MLEnsembles), Deep Ensembles (DLEnsembles) and Hybrid Ensembles (HybridEnsembles) to solve the LULC classification problem. MLEnsembles are ensembles where the members are the ML classifiers we developed thanks to the third state of art approach (i.e. Deep Feature extraction) (see Fig. 1). DLEnsembles on the other hand are composed of DL classifiers (i.e. Transfer learning). Finally, HybridEnsembles are ensembles combining all the classifiers we built (i.e. ML and DL). Further, we study their results and investigate the use of EL as an alternative to hyperparameter tuning when it comes to improving the performance of well-performing models.

The main research contributions this paper are:

- Evaluating three state of art single ML/DL techniques and approaches in LULC classification problems using hyperparameter optimization (i.e. new ConvNet architecture, End to End Transfer learning and Deep Feature extraction).
- Proposing and assessing the use of different EL combinations for LULC classification.
- Investigating the use of EL to further improve the performance of single DL models compared with hyperparameter optimization.

Thus, this study discusses the following research questions:

- **RQ1**: Which one of the SL techniques (i.e. ConvNet, Transfer learning and Deep Feature extraction) performs the best on the EuroSAT dataset?
- **RQ2**: Does any of the proposed EL combinations (i.e. MLEnsembles, DLEnsembles and HybridEnsembles) outperform the best SL models?
- **RQ3**: Is EL a better alternative to hyperparameter tuning when it comes to increasing the accuracy of a well performing DL model?

The rest of this paper is organized as follows: Sect. 2 presents the dataset, optimization technique and evaluation metrics we chose to use in this empirical study, and discusses the hyperparameter tuning of the different models we built. Then, the performances of the SL and EL classifiers are presented and compared in Sect. 3. Section 4 provides a summary of the findings of this study and summarizes ongoing works.

2   Materials and Methods

2.1   Materials

This section presents the materials we used in this work, the experimental design and its implementation.

Dataset Description. As part of a research project that aimed at proposing a novel dataset and DL benchmark for LULC classification, the German Center for Artificial Intelligence (DFKI)[1] constructed and made open source a new LULC dataset: EuroSAT [9]. It is composed of 27000 64 by 64-pixel images: 2000 to 3000 images per class. The team chose 10 classes of LULC that were frequent and visible in the Sentinel-2A images: Industrial Buildings, Residential Buildings, Sea and Lake, Herbaceous Vegetation, Annual Crop, Permanent Crop, River, Highway, Pasture, Forest.

Hyperparameter Optimization. To fine tune the hyperparameters of our classifiers, we set accuracy to be the objective function and Random Search (RS) as the optimization algorithm. RS lets the hyperparameters be chosen at random within a range of values and

---

[1] https://www.dfki.de/en/web.

decides the best configuration that maximizes the objective function (i.e. accuracy) [4]. It is computationally cheap, and outperforms Grid Search when some hyperparameters are far more important than others [4]. We implemented this phase using Keras Tuner [18]; a Keras hyperparameter optimization library.

Metrics. In order to evaluate the performance of our classifiers, and since EuroSAT is a balanced dataset, this study uses the following benchmark metrics: Accuracy, Precision, Recall, and F1 score. These popular criteria are defined by Eqs. 1–4.

$$Accuracy = \frac{TP + TN}{TP + TN + FP + FN} \tag{1}$$

$$Precision = \frac{TP}{TP + FP} \tag{2}$$

$$Recall = \frac{TP}{TP + FN} \tag{3}$$

$$F1score = 2 \times \frac{Precision \times Recal}{Precision + Recall} \tag{4}$$

where TP, FP, TN and FN stand for True Positive. False Positive, True Negative and False Negative respectively. To evaluate the improvement of EL classifiers over the best single classifier, we use another metric as well, and that is Change In Accuracy. This metric compares the Accuracy of the Ensemble (AoE) to that of the Best Single Classifier (AoBSC) in the ensemble. Equation 5 defines AoBSC.

$$AoBSC = max\,(AoBaseLearner\,1\,,\,AoBaseLearner\,2,\dots,\,AoBaseLearner\,N\,-\,1,$$
$$AoBaseLearner\,N) \tag{5}$$

where $AoBaseLearner_i$ is Accuracy of the ith Base Learner, and N is Number of Base Learners used.

Lastly, Eq. 6 defines the Change In Accuracy metric.

$$ChangeInAccuracy(e) = AoE(e) - AoBSC(e) \tag{6}$$

where AoE(e) is Accuracy of the ensemble e, and AoBSC(e) is Accuracy of the Best Single Classifier in the ensemble e. A negative value indicates that the best single classifier outperformed the ensemble.

2.2 Experimental Design

This study focuses on providing an evaluation of the performance of diverse single ML/DL techniques in LULC classification problems and proposing a new technique based on ensemble learning as shown in Fig. 1. We used the images and labels in the EuroSAT dataset as input data, and implemented four ML approaches: build a new ConvNet architecture (LULC-Net), End to End Transfer learning, DL Feature Extraction and Ensemble Learning.

During the process of building the different classifiers, we used the following configuration: categorical cross entropy as loss function, Softmax as an activation function of the last layer and a 80–20% data split. Next, we will present the implementation details of the building process of the different classifiers.

LULC-Net: To propose a new ConvNet architecture for LULC classification tasks, we began with a shallow ConvNet randomly configured, and while seeking to maximize the accuracy, we tuned a list of hyperparameters that according to [17] affect both the structure and the training process of ConvNets. Table 1 shows the hyperparameters we chose to tune, their respective ranges as well as the optimal values returned by the RS technique we implemented using the Keras Tuner library.

Transfer Learning. A Transfer learning model is a combination of a DL Feature extractor (i.e. ConvNet layers) and a set of dense layers (i.e. fully connected layers). Training the Transfer learning models was a two steps process: (1) we trained the fully connected layers (initially composed of two layers, 512 and 10 neurons respectively), then fine-tuned the weights of the pretrained models. The first step trained the dense layers part of the model, and did not permit change to occur in the ConvNet part (i.e. the pretrained layers were frozen). And (2) let the learning process change the weights of the top 30% layers of the pretrained part (i.e. unfreeze the top 30% layers), and thus adapt to the new task [7] (Table 2).

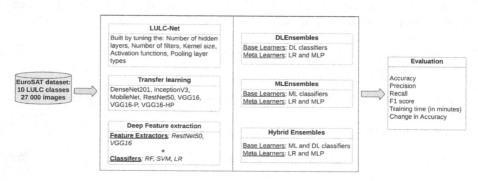

Fig. 1. Experimental design.

Deep Feature Extraction. As it was previously pointed out, the first part of every ConvNet architecture is one that performs feature extraction [19]. Therefore, we used the convolutional part of the VGG16 and RestNet50 pretrained models to extract visual features from the images, and fed them to ML algorithms. These features are predictions of the ConvNet parts of VGG16 and RestNet50 on the EuroSAT images stored in a CSV file. And the ML models that we trained were mainly: Support Vector Machines (SVM), Random Forest (RF) and Logistic Regression (LR). Each with their default configuration implemented in the Sikit-learn library [20] (Table 3).

Table 1. Fine tuning of LULC-Net: Hyperparameters, range and optimal values.

Hyperparameter	Range	Optimal value
Optimizer	Adam, RMSprop, Stochastic Gradient Descent	Adam
Batch size	20, 32, 64, 128	32
Number of ConvNet blocks	min = 3, max = 25, step = 1	7
Number of filters	min = 32, max = 256, step = 32	64, 224, 192
Kernel size	3, 5 and 7	5
Activation function	Tanh, ReLu, Leaky Relu	ReLu
Learning rate	0.01, 0.0001, 0.00001	0.0001

Table 2. Fine tuning of the fully connected layers on top of VGG16.

Hyperparameter	Range	Optimal value
Optimizer	Adam, RMSprop, Stochastic Gradient Descent	Adam
Batch size	20, 32, 64, 128	32
Number of hidden layers	min = 1, max = 3, step = 1	2
Number of neurons	min = 32, max = 1024, step = 32	960
Dropout rate	min = 0, max = 0.6, step = 0.1	0.2
Activation function	Tanh, ReLu	ReLu
Learning rate	0.01, 0.0001, 0.00001	0.0001

Table 3. Fine tuning of the Machine Learning classifiers trained on VGG16's extracted features.

Algorithm	Hyperparameter	Range	Optimal value
SVM	C	1, 10, 50 and 100	10
	Kernel	Linear, RBF, Sigmoid	Linear
RF	Number of estimators	min = 100 max = 1500, step = 100	1000
	Max depth	min = 3 max = 100, step = 1 and None	10
	Max feature	auto, sqrt and log2	auto
LR	C	100, 50, 20, 10, 1, 0.1, 0.01	100
	Solver	newton-cg, lbfgs, liblinear	liblinear

EL: Stacked Generalization. We chose to develop the EL models using the stacked generalization technique, which combines learners and finds the best combination of their predictions using another model (i.e. meta learner) [27]. Figure 2 shows the implementation process we followed, we put the classifiers we built so far into three groups of base learners. The first set was composed of Deep Learners namely: VGG16, RestNet50, DenseNet201, VGG16 Places365 (VGG16-P) and VGG16HybridPlaces (VGG16-HP)

(DLEnsembles) [14]. The second was made of ML models: SVM, RF, K Nearest Neigh-
bour (KNN) and Gaussian Native Bayes (Gaussian NB) trained on the features extracted
by VGG16 and RestNet50 (MLEnsembles). Finally, the last set of base learners was a
combination of both approaches and sets of base learners (HybridEnsembles). The meta
learners we experimented with were LR or Multilayer perceptron (MLP) and the fine
tuning of this approach focused on finding the best combination of base learners and
the best meta learner. By comparing the increase in accuracy introduced by EL and
fine-tuning DL models we aim at investigating the ability of EL methods to improve the
accuracy of well performing single models.

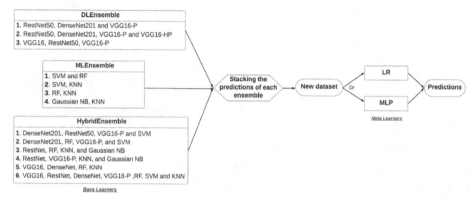

Fig. 2. Ensemble learning technique.

3   Results and Final Discussion

This last section will attempt to answer the RQs of our work by presenting and discussing
the findings of our work.

3.1   Performance Comparison Between ML Models

The first single classifier we built was a new ConvNet architecture: LULC-Net. Its
accuracy, F1 score and training times are respectively: 94.24%, 94.36% and 7 min (12
epochs using the GPU support of Kaggle kernels). However, LULC-Net does suffer
slightly from overfitting. Indeed, among the LULC classifiers we developed, the one
that overfits the data the most is this newly proposed ConvNet architecture. The Transfer
learning classifiers on the other hand achieved state of art performance (96% and 97%
accuracy) [5, 28]. Table 4 summarizes the performance achieved by our single Transfer
learning classifiers after 25 epochs of training. The DenseNet201 model achieves the
best accuracy and F1 score (96.88% and 96.90% respectively). However, because of its
depth [10], it takes the longest to train (30 min). On the other hand, VGG16 manages
to achieve 96.81% accuracy, after just 11 min of training. Nonetheless, the performance

of the VGG16 architecture decreases slightly if it was trained on the Places dataset [30] instead of ImageNet [15] (e.g. VGG16-P). One other notable finding is that InceptionV3's performance is poor on EuroSAT compared to that of the other models we used, and this is even if InceptionV3 outperforms all of them on the ImageNet dataset [24]. Lastly, MobileNet is the pretrained model that performed the poorest on EuroSAT.

Overall the Deep Feature extraction classifiers underperformed the Transfer learning ones (see Table 5). Additionally the Deep Feature extraction model that performs the best is RestNet50 + Logistic Regression (93.86% accuracy). However, it takes RestNet50 several minutes to extract the features, which in turn elongates the classification process. Further, all metrics considered, Logistic Regression outperforms all other algorithms regardless of the Feature extractor used. In fact, it trains in fewer minutes, and it achieves the best accuracy with RestNet50's features and second best with VGG16's. Finally, although it takes longer to extract the features, accuracy-wise RestNet50 is a better feature extractor than VGG16. For instance, training Random Forest on RestNet50's features improved its accuracy significantly, and that is the case for Logistic Regression as well.

In terms of accuracy and training time, the findings of the SL experiments indicate that Transfer learning performs the best, followed by LULC-Net and Deep Feature extraction classifiers. In fact, Tables 4 and 5 show that all the Transfer learning classifiers produced a better accuracy than the one ML models returned. Moreover, the experiments also demonstrate that the training time of the ML models is longer than that of the Transfer learning models: most DL models train in less than 20 min, whereas most of the ML models take more than 20 min to converge.

Table 4. Performance of the fine tuned Transfer learning models

Architecture	Dataset	Accuracy (%)	Precision (%)	Recall (%)	F1 score (%)	Training time (minutes)
DenseNet201	ImageNet	96.88	96.97	96.83	96.90	30
InceptionV3	ImageNet	95.44	95.75	95.44	95.59	18
MobileNet	ImageNet	94.70	94.69	94.52	94.60	10
RestNet50	ImageNet	96.77	96.79	96.69	96.74	15
VGG16	ImageNet	96.81	96.87	96.80	96.83	11
VGG16-HP	Places/ ImageNet	96.37	96.42	96.36	96.39	15
VGG16-P	Places	96.33	96.28	96.25	96.25	14

Table 5. Performance of the Deep Feature Extractors and fine tuned ML models.

Feature extractor	Algorithm	Accuracy %	Precision %	Recall %	F1 score %	Training time (minutes)a
RestNet50	LR	92.60	92.56	92.60	92.56	10
	RF	85.86	85.97	85.86	85.91	20
	SVM	92.98	92.95	92.87	92.96	35
VGG16	LR	93.86	93.87	93.86	93.84	24
	RF	87.26	87.68	87.26	87.01	27
	SVM	93.53	93.53	93.53	93.51	41

[a]Including 7.6 min and 21 min of Feature extraction with VGG16 and RestNet50 respectively

3.2 Performance Comparison Between EL and SL Models

The results of the experiments we conducted to answer this question show that the accuracies of the DLEnsembles and HybridEnsembles are far better than those of the MLEnsembles. In fact, the HybridEnsemble$_3$ + LR has the worst accuracy among all DL and Hybrid ensembles (95.74%); however, its accuracy is still approximately 6% better than the accuracy of the best performing ML ensemble (i.e. MLEnsemble$_1$ + MLP, 89.92%). Moreover, DL and Hybrid ensembles are also better at improving their Best Single Classifier's accuracy. With the exception of the MLEnsemble$_4$, all the ML ensembles have either slightly increased the accuracy of their Best Single Classifier (less than 0.30% increase), or decreased it. On the other hand, when they use LR as a meta learner, most of the DL and Hybrid ensembles have an accuracy that is approximately 1% better than that of their Best Single Classifier.

More insights can be drawn from Table 6. First, the best classifier this approach managed to build is a Deep EL classifier, namely: DLEnsemble$_1$ which achieved 97.92% accuracy. Second, although DenseNet201 and VGG16 have very close accuracy values, combining DenseNet201 instead of VGG16 with VGG16-P and RestNet50 resulted in a better performing ensemble. In fact, DLEnsemble$_1$ + LR composed of DenseNet201, VGG16-P and RestNet50 produced an accuracy of 97.92%, while DLEnsemble$_3$ + LR where we substitute DenseNet201 with VGG16, manages to classify only 97.11% of the unseen images. Nonetheless, DLEnsemble$_1$ takes longer to train than DLEnsemble$_3$, and that is because DLEnsemble$_1$ combines DenseNet201 whose time-cost is higher than VGG16's. Similarly, ensembling the SVM classifier with KNN performed much better than MLEnsemble$_3$. However, this high performance of ensembles where SVM is a member also comes at a high time-cost, as SVM is the model that takes the longest to train among the ML classifiers we built. Finally, the hybrid experiments show also this trade off as well: the hybrid ensembles that performed the best combined models that have the highest cost (in terms of their training time).

Finally, as a meta learner, LR outperforms MLP when the ensemble members are DL models, whereas MLP produces a better accuracy if the base learners are mainly ML. In fact, the DL ensembles + LR all returned a higher accuracy than that they produced with MLP. Similarly, the hybrid ensembles composed mostly of DL classifiers performed better with LR (accuracy-wise). On the other hand, MLP increases the accuracy of ML ensembles. With the exception of MLEnsemble$_3$, all ML ensembles scored a higher accuracy with MLP than with LR.

3.3   Performance Comparison Between Hyperparameter Tuning and the EL Approach

Hyperparameter optimization was crucial to improving the performance of our single ML/DL models, especially the DL. Prior to fine tuning, DenseNet201 and VGG16 achieved 89% and 87% accuracy respectively. However, optimizing the same models (i.e. unfreezing their top 33% pretrained layers) increased their respective accuracies to 96.88% and 96.81%. Nonetheless, the EuroSAT benchmark model proposed by DFKI has an accuracy of 98.57% [9]. One way to increase the performance of our DL classifiers, is to extend the search space during the hyperparameter optimization. Another way is to use an EL technique and aggregate the knowledge they had previously acquired. Table 7 compares both methods as it shows the increase in accuracy of VGG16, DenseNet201 and MobileNet introduced by hyperparameter optimization and EL. The Current performance column presents the results discussed in Table 4. The third column we find the performances (accuracy and time cost) attained after further-optimizing the models; that is after unfreezing 66% of their top pretrained layers. The last column presents the performance obtained after the aggregation of the classifier in question with others in a stacked generalization ensemble. The time cost in this case is the summation of the cost of each member in the ensemble, including the meta learner.

Optimizing the hyperparameters by letting more layers adapt to the new task slightly increased the accuracy of each classifier. Actually, the model whose accuracy saw improvement the most is the MobileNet classifier, which is also the model that performed the poorest after the first fine tuning phase. More importantly, the use of EL improved all classifiers' accuracy more than fine tuning; however, it does require more time as it trains and combines several models.

Table 6. Performance of the stacked generalization Ensemble

Approach	Ensembles	Members	AoE %		AoBSC[a]	Change In Accuracy %	
			LR	MLP		LR	MLP
DL	DLEnsemble$_1$	DenseNet, RestNet, VGG16-P	97.92	97.18	DenseNet(96.77%)	+1.15	+0.41
	DLEnsemble$_2$	DenseNet, RestNet, VGG16-P, VGG16-HP	97.7	97.22	DenseNet(96.77%)	+0.93	+0.55

(continued)

Table 6. (*continued*)

Approach	Ensembles	Members	AoE %		AoBSC[a]	Change In Accuracy %	
			LR	MLP		LR	MLP
	DLEnsemble$_3$	VGG16, RestNet, VGG16-P	97.11	96.74	VGG16 (96.66%)	+0.45	+0.37
	All DL classifiers	DenseNet, RestNet, VGG16, VGG16-P, VGG16-HP	97.74	97.07	DenseNet(96.77%)	+0.97	+0.30
ML	MLEnsemble$_1$	SVM, RF	89.85	89.92	SVM(89.66%)	+0.19	+0.26
	MLEnsemble$_2$	SVM, KNN	89.85	89.48	SVM(89.66%)	+0.19	−0.18
	MLEnsemble$_3$	RF, KNN	82.77	85.40	RF(85.5%)	−2.73	−0.1
	MLEnsemble$_4$	Gaussian NB, KNN	78.51	80.07	KNN(73.37%)	+5.14	+6.34
	All ML classifiers	RF, SVM, Gaussian NB, KNN	89.29	89.74	SVM(89.66%)	−0.4	+0.8
Hybrid	HybridEnsemble$_1$	DenseNet, RestNet, VGG16-P, SVM	97.55	96.92	DenseNet(96.77%)	+0.74	+0.15
	HybridEnsemble$_2$	DenseNet, RF, VGG16-P, SVM	97.70	97.18	DenseNet(96.77%)	+0.89	+0.41
	HybridEnsemble$_3$	RestNet, RF, KNN, Gaussian NB	95.74	96.07	RestNet(96%)	−0.26	+0.07
	HybridEnsemble$_4$	RestNet, VGG16-P, KNN, Gaussian NB	96.92	96.22	VGG16-P(96.33%)	+0.92	−0.11
	HybridEnsemble$_5$	VGG16, DenseNet, RF, KNN	97.74	97.03	DenseNet(96.77%)	+0.97	+0.26

(*continued*)

Table 6. (*continued*)

Approach	Ensembles	Members	AoE %		AoBSC[a]	Change In Accuracy %	
			LR	MLP		LR	MLP
	HybridEnsemble$_6$	VGG16, RestNet, DenseNet, VGG16-P, RF, SVM, KNN	97.74	97.18	DenseNet(96.77%)	+0.97	+0.41

[a]To conduct these experiments we trained and tested the DL models another time. Hence the differences between the results in this column and the ones in Table 4

Table 7. The increase in accuracy introduced by hyperparameter fine tuning and EL

Model	Current performance		Further Fine tuning		Ensemble learning	
	Accuracy %	Training time	Accuracy %	Training time	Accuracy %	Training time
DenseNet201	96.88	30	97.07	32	97.92[a]	60
VGG16	96.81	11	96.92	14	97.74[b]	56
MobileNet	94.70	10	95.14	11	97.14[c]	40

[a]VGG16, RestNet50 and DenseNet201.
[b]DLEnsemble$_1$.
[c]MobileNet and DenseNet201 ensemble.

3.4 Discussion

To tackle the LULC classification from satellite images problem we used three SL approaches. The first SL approach consisted of optimizing a ConvNet architecture, and resulted in a classifier that achieves a relatively good accuracy (94%) but that overfits the data. This inability to generalize to unseen data can be due to its low expressive power, which has led to an incapacity to capture the target function. Experiments with the second approach (i.e. Transfer learning) helped us build classifiers that perform better (accuracy-wise). They also demonstrated that when it comes to LULC classification, models pretrained on the ImageNet dataset outperform the ones trained on the Places dataset; in fact, VGG16 pretrained on ImageNet achieved 96.81% accuracy and 96.33% when pretrained on Places. The third approach uses VGG16 and RestNet50 as feature extractors, stores the features they extract from EuroSAT and trains several ML algorithms on them. This last SL technique performed the poorest and some of the algorithms it uses are expensive (time-wise). The one that trains the longest is SVM, and that is due to the fact that the implementation we used is based on LIBSVM [6], which makes "the fit time scale at least quadratically with the number of samples" [20]. Overall, the SL classification technique that achieved the best results is Transfer learning. End to end classifiers outperformed the ML approach in terms of accuracy and training time.

Nonetheless, this last approach and as it can use white-box ML algorithms, has the advantage of returning interpretable predictions.

To outperform the results the SL approach produced we propose the use of EL classifiers (i.e. stacked generalization) (MLEnsembles, DLEnsembles and HybridEnsembles) that combine the knowledge the SL models had gained. Overall, the findings suggest that ensembles composed of DL classifiers perform best (accuracy-wise). This should not come as a surprise since DL single learners achieve the highest accuracy among the SL models we built. Similarly, and since they performed the worst as single classifiers, ensembles of ML models also performed the poorest (accuracy-wise). All in all, the DL and ML ensembles achieved an increase in accuracy of the SL classifiers we built.

Using SL techniques, we managed to build well-performing models. In the previous Subsect. 3.3 we presented experiments that would answer our third RQ, which dealt with the viability of EL as an alternative to fine tuning when it comes to improving the performance of optimized models. However, these experiments showed once more the high accuracy and high time-cost trade-off (see Table 7). In fact, the EL approach helped the three SL classifiers increase their accuracy significantly; nonetheless, the cost of building an EL classifier aggregates the cost of building each one of its members. Plus, choosing the models to combine is a challenging task given that a model's performance is directly related to its context. On the other hand, the hyperparameter optimization process has a lower time-cost, but results in smaller improvements.

4 Conclusion and Future Work

The purpose of this work was to use different ML techniques and the freely available satellite data, to build LULC classification systems. In fact, we aimed at first optimizing, evaluating and comparing the performance of single ML/DL techniques. Second, comparing their results to those of the EL classifiers we built using the stacked generalization. Third, investigating the ability of EL and hyperparameter tuning to increase the accuracy of single DL models. The experiments we conducted in this work helped achieve these objectives and answer this work's research questions:

- **RQ1: Which one of the SL techniques (i.e. ConvNet, Transfer learning and Feature extraction) performs the best on the EuroSAT dataset?**

When it comes to SL techniques, the end to end classifiers (i.e. LULC-Net and Transfer learning) perform the best (accuracy-wise). In fact, pretrained models achieved state of art accuracy and trained fast (e.g. VGG16). This performance was mainly obtained thanks to the hyperparameter optimization steps. On the other hand, single ML models produced a lower accuracy, and took longer to train and optimize.

- **RQ2: Does any of the proposed EL combinations outperform the best SL models?**

Several insights can be drawn from the experiments we conducted to answer this question. Firstly, ensembles with DL base learners (i.e. DLEnsembles and HybridEnsembles) tend to have a higher accuracy than that of the ensembles that only have ML base learners (i.e. MLEnsembles). Moreover, the performance of a stacked generalization ensemble increases if the base learners are diverse. Finally, these experiments also proved that in terms of accuracy, the EL approach outperforms SL.

- **RQ3: Is EL a better alternative to hyperparameter tuning when it comes to increasing the accuracy of a well performing DL model?**

Our results showed that EL increased the accuracy of single DL models more than hyperparameter optimization. However, using EL presents several challenges, such as the high time-cost, and difficulty of choosing the members of the ensemble. Hyperparameter optimization on the hand had a lower time-cost, but only introduced a slight increase in accuracy to the DL models.

Ongoing work aims at using the classifiers this work built to construct a Moroccan LULC dataset, which to the best of our knowledge will be the first of its kind. Then, we will use this newly constructed dataset, to develop Land use & Land cover monitoring systems of Moroccan lands.

Appendix A - Comparison of the Performance of Our Models on Moroccan LULC Images

We aim to use the models we built during this work to construct a Moroccan LULC dataset. Indeed, we will choose one of these models to be used as an annotator of Moroccan satellite images. Therefore, we decided to test these classifiers' ability to generalize to images of Moroccan regions. In this appendix we present the classification results of the four classifiers (i.e. LULC-Net, VGG16, VGG16 + LR and DLEnsemble$_1$). The nine images we tested our models on are of the city of Casablanca where Industrial, Residential, Sea & Lake and highway classes are present, and of the North of Morocco where classes such as Forest, Pasture, River and Annual Corp are present.

- **LULC-Net**

Image from: North
Classified as: Annual Corp

Image from: Casablanca-Sea
Classified as: Industrial

Image from: Casablanca
Classified as: Industrial

Image from: Casablanca
Classified as: Industrial

Image from: Casablanca
Classified as: Industrial

Image from: North
Classified as: Industrial

Image from: Casablanca
Classified as: Industrial

Image from: North
Classified as: Industrial

Image from: North
Classified as: Industrial

- **VGG16**

Image from: North
Classified as: Sea&Lake

Image from: Casablanca-Sea
Classified as: Sea&Lake

Image from: Casablanca
Classified as: Permanent crop

Image from: Casablanca
Classified as: Industrial

Image from: Casablanca
Classified as: Highway

Image from: North
Classified as: Industrial

Image from: Casablanca
Classified as: Permanent Crop

Image from: North
Classified as: Sea&Lake

Image from: North
Classified as: Sea&Lake

- **VGG16 + LR**

Image from: North
Classified as: Herbaceous Vegetation

Image from: Casablanca-Sea
Classified as: Residential

Image from: Casablanca
Classified as: Sea&Lake

Image from: Casablanca
Classified as: Industrial

Image from: Casablanca
Classified as: Industrial

Image from: North
Classified as: Herbaceous Vegetation

Image from: Casablanca
Classified as: Industrial

Image from: North
Classified as: Herbaceous Vegetation

Image from: North
Classified as: Residential

- **DLEnsemble₁**

Image from: North
Classified as: Sea&Lake

Image from: Casablanca-Sea
Classified as: Sea&Lake

Image from: Casablanca
Classified as: Permanent crop

Image from: Casablanca
Classified as: Residential

Image from: Casablanca
Classified as: Highway

Image from: North
Classified as: Sea&Lake

Image from: Casablanca
Classified as: Residential

Image from: North
Classified as: Sea&Lake

Image from: North
Classified as: Annual crop

References

1. Alam, A., Bhat, M.S., Maheen, M.: Using Landsat satellite data for assessing the land use and land cover change in Kashmir valley. GeoJournal **85**(6), 1529–1543 (2019). https://doi.org/10.1007/s10708-019-10037-x
2. Athiwaratkun, B., Kang, K.: Feature representation in convolutional neural networks CoRR abs/1507.02313 (2015). http://arxiv.org/abs/1507.02313
3. Aung, S.W.Y., Khaing, S.S., Aung, S.T.: Multi-label land cover indices classification of satellite images using deep learning. In: Zin, T.T., Lin, J.-W. (eds.) ICBDL 2018. AISC, vol. 744, pp. 94–103. Springer, Singapore (2019). https://doi.org/10.1007/978-981-13-0869-7_11
4. Bergstra, J., Bengio, Y.: Random search for hyper-parameter optimization. J. Mach. Learn. Res. **13**, 281–305 (2012)
5. Bernasconi, E., Pugliese, F., Zardetto, D., Scannapieco, M.: Satellite-net: automatic extraction of land cover indicators from satellite imagery by deep learning (2019)
6. Chang, C.C., Lin, C.J.: LIBSVM: a library for support vector machines. ACM Trans. Intell. Syst. Technol. **2**, 27:1–27:27 (2011). http://www.csie.ntu.edu.tw/cjlin/libsvm
7. Chollet, F.: Complete guide to transfer learning fine-tuning in keras (2020)
8. Ghimire, B., Rogan, J., Rodriguez-Galiano, V., Panday, P., Neeti, N.: An evaluation of bagging, boosting, and random forests for land-cover classification in cape cod, Massachusetts, USA. GIScience Remote Sens. **49**, 623–643 (2012). https://doi.org/10.2747/1548-1603.49.5.623
9. Helber, P., Bischke, B., Dengel, A., Borth, D.: Eurosat: A novel dataset and deep learning benchmark for land use and land cover classification. IEEE J. Sel. Top. Appl. Earth Obs. Remote Sens. **12**(7), 2217–2226 (2019)
10. Huang, G., Liu, Z., Weinberger, K.Q.: Densely connected Convolutional networks. CoRR abs/1608.06993 (2016). http://arxiv.org/abs/1608.06993
11. Hussain, M., Bird, J.J., Faria, D.R.: A study on cnn transfer learning for image classification. In: Lotfi, A., Bouchachia, H., Gegov, A., Langensiepen, C., McGinnity, M. (eds.) UKCI 2018. AISC, vol. 840, pp. 191–202. Springer, Cham (2019). https://doi.org/10.1007/978-3-319-97982-3_16
12. Jansen, L., Gregorio, A.: Land Cover Classification System (LCCS): Classification Concepts and User Manual (2000)

13. Jozdani, S.E., Johnson, B.A., Chen, D.: Comparing deep neural networks, ensemble classifiers, and support vector machine algorithms for object-based urban land use/land cover classification. Remote Sens. **11**(14), 1713 (2019). https://doi.org/10.3390/rs11141713

14. Kalliatakis, G.: Keras VGG16 places365 github repository (2017). https://github.com/GKalliatakis/Keras-VGG16-places365

15. Krizhevsky, A., Sutskever, I., Hinton, G.: Imagenet classification with deep convolutional neural networks. Neural Inf. Process. Syst. 25 (2012). https://doi.org/10.1145/3065386

16. Lo, C.: Applied remote sensing. Geocarto Int. **1**(4), 60 (1986). https://doi.org/10.1080/10106048609354071

17. Mohd Aszemi, N., Panneer Selvam, D.D.D.: Hyperparameter optimization in convolutional neural network using genetic algorithms. Int. J. Adv. Comput. Sci. Appl. **10**, 269–278 (2019). https://doi.org/10.14569/IJACSA.2019.0100638

18. O'Malley, T., Bursztein, E., Long, J., Chollet, F., Jin, H., Invernizzi, L., et al.: Keras Tuner (2019). https://github.com/keras-team/keras-tuner

19. O'Shea, K., Nash, R.: An introduction to convolutional neural networks. CoRRabs/1511.08458 (2015). http://arxiv.org/abs/1511.08458

20. Pedregosa, F., et al.: Scikit-learn: machine learning in python. J. Mach. Learn. Res. **12**, 2825–2830 (2011)

21. Ratajczak, R., Crispim-Junior, C.F., Faure, E., Fervers, B., Tougne, L.: Automatic land cover reconstruction from historical aerial images: an evaluation of features extraction and classification algorithms. IEEE Trans. Image Process. **28**(7), 3357–3371 (2019)

22. Smith, G.: Remote sensing of land use and land cover: principles and applications, by C. P. Giri, Boca Raton. Int. J. Remote Sens. **35**(6), 2392–2393 (2014). https://doi.org/10.1080/01431161.2014.891560

23. Sokal, R.R.: classification: purposes, principles, progress, prospects. Science **185**(4157), 1115–1123 (1974). https://doi.org/10.1126/science.185.4157.1115

24. Szegedy, C., Vanhoucke, V., Ioffe, S., Shlens, J., Wojna, Z.: Rethinking the inception architecture for computer vision (2016). https://doi.org/10.1109/CVPR.2016.308

25. Talukdar, S., Singha, P., Mahato, S., Shahfahad, S.P., Liou, Y.-A., Rahman, A.: Land-use land-cover classification by machine learning classifiers for satellite observations—a review. Remote Sens. **12**(7), 1135 (2020). https://doi.org/10.3390/rs12071135

26. Tong, X.Y., et al.: Land-cover classification with high-resolution remote sensing images using transferable deep models (2018)

27. Wolpert, D.H.: Stacked generalization. Neural Netw. **5**(2), 241–259 (1992). https://doi.org/10.1016/S0893-6080(05)80023-1

28. Cheng, X., Namjoshi, N., Rodriguez, R.: Temporal analysis of regional sustainability using CNNs and satellite data (2018)

29. Yang, Y., Newsam, S.: Bag-of-visual-words and spatial extensions for land-use classification, GIS 2010, pp. 270–279. Association for Computing Machinery, New York (2010). https://doi.org/10.1145/1869790.1869829

30. Zhou, B., Lapedriza, A., Khosla, A., Oliva, A., Torralba, A.: Places: A 10 million image database for scene recognition. IEEE Trans. Pattern Anal. Mach. Intell. **40**(6), 1452–1464 (2018). https://doi.org/10.1109/TPAMI.2017.2723009

Integrated Methods for Cultural Heritage Risk Assessment: Google Earth Engine, Spatial Analysis, Machine Learning

Maria Danese(✉) [iD], Dario Gioia [iD], and Marilisa Biscione [iD]

ISPC-CNR, Tito Scalo, 85100 Potenza, Italy
maria.danese@cnr.it

Abstract. Cultural heritage risk assessment is an important task. Usually cultural heritage inside the consolidated city is more protected than cultural heritage spread over the territory. In this paper a method is proposed that integrates different technologies and platforms: from Google Earth Engine (GEE) and machine learning to desktop GIS and spatial analysis. The aim is to map the different vulnerability and hazard layers that constitute the cultural heritage risk map of the rural area in the Matera Municipality.

Keywords: Natural hazard · Urban sprawl · Google Earth Engine · Machine learning · Matera · Cultural heritage

1 Introduction

The integration of the new Information and Computer Technologies (ICT) is important not only for knowledge and analysis of cultural heritage, but also for the extraction of information on the basis of their protection. This is the case of cultural heritage risk maps, that are useful to assess future natural or human transformations, that can affect cultural heritage [1].

Analysis and mapping of natural hazards or anthropic impact, are crucial to investigate the vulnerability of a landscape and plan possible mitigation or protection actions of cultural heritage. Such analyses can be done in a more effective way thanks to the integration of spatial and remote sensing technologies, by using GIS and the new GEE platform launched in 2001. With these instruments also a protocol for cultural heritage risk assessment could be defined, based on the definition of risk [2], Eq. 1)

$$R = H \times V \times E \qquad (1)$$

In this paper, we considered this definition in particular for the cultural heritage spread over the territory, because it is more exposed to natural and anthropic risks, while cultural heritage inside the urban areas is generally more known and protected.

In Eq. 1, risk R is expressed as the product of the following components:

© Springer Nature Switzerland AG 2021
O. Gervasi et al. (Eds.): ICCSA 2021, LNCS 12951, pp. 605–619, 2021.
https://doi.org/10.1007/978-3-030-86970-0_42

- H is hazard. For the Matera municipality, different type of hazards were considered: the urban growth as anthropic hazard, because it could interfere in a negative way in the presence of elements of archeological interest; landslides, floods, and erosion as natural hazards, for the impact that they have over the cultural heritage in general, but also, more in particular, over the spread archaeological findings.
- V is vulnerability. There are many ways to assess vulnerability maps for cultural heritage. A basic method is represented by the mapping of cultural heritage in a territory [3]; other methods use predictive models to consider also the archaeological potential of the studied area [4–6].
- E is exposure and indicates the value of the elements at risk. In this work, we consider all the cultural heritage elements as exposed to the same level of risk.

2   Datasets and Methods

2.1   Google Earth Engine©

GEE is a free, powerful platform that allows the analysis and monitoring of remote sensing big data at the planetary scale.

The first GEE version went out in 2001 [7], but it was publicly released in 2010. 2001 could be the internal release date of the platform. and today it is still used very much in many application fields [8] in general for the global –scale earth observation data extraction and analysis [9], but more in particular, just to cite some examples, for land use mapping [10] and monitoring and natural risk assessment ([11] and reference therein), for population mapping [12], climate data analysis [13] and large scale management of cultural heritage [14].

The use of GEE offers many advantages. It gives the possibility to work with large dimension data (satellite data but also vector GIS data), without downloading them on pc, even if the download is possible, also because many of these data are freely available. At the same time, the calculation time is very fast, thanks to the cloud elaboration, more than it would be possible on a personal computer. It also gives the possibility to upload the user's data, if needed. Moreover, the GEE platform has an efficient versioning system and offers a flexible access to its code through APIs. Finally, even if the code developed by the user is online, the ownership remains to its developer, even if it could decide to share it.

2.2   Machine Learning: Random Forest Method

Google Earth Engine offers many classification methods inside its platform. Between these, we choose an ensemble method taken from machine learning: the random forest classifier, a method introduced by [15, 16].

Ensemble methods 1) run the same algorithm many times or 2) put different machine learning algorithms, in order to execute a more effective analysis. The random forest classifier belongs to the first category because it uses recursively the decision trees for classification, according to the following simplified steps:

1. A training set is chosen inside the population dataset;
2. The decision tree classification is executed;
3. Then the process is repeated in order to improve the learning process and to return the best classification.

In satellite data analysis random forest recently started to be very used in many applications (see for example [17, 18] and reference therein).

In this paper, the random forest method was used to extract from Landsat data the land uses changes over 30 years and in particular to extract the urban growth [19] of the Matera case study.

2.3 Spatial Analysis

In this paper, map algebra is used as spatial analysis method useful to calculate some of the hazard parameters and then to combine hazard and vulnerability to obtain the risk map.

The classic definition of Map Algebra [20] expresses it as a "high level spatial modeling language, including base elements (operators), complex elements (functions) and formal language (instructions), together with elements needed to program and develop complex models".

Usually map algebra functions are classified in local, focal and zonal [21]. In local functions cell-by-cell operations are done; in focal functions operations are executed inside a defined neighborhood of the input raster; finally in zonal functions, operations are done inside another zonal raster.

3   The Matera Case Study

3.1   The Matera Landscape

The study area includes the administrative territory of Matera, which, from a geological viewpoint, is located between the westernmost sectors of the foredeep of the southern Apennines belt and the carbonate ridges of the Apulia foreland [22, 23]. The landscape is carved in a thick succession of Cretaceous limestones, which exhibit typical features of a karst environment [22, 24, 25]. Starting from the middle-late Pliocene, the subsidence of the western slope of the Apulia foreland promoted the deposition of coarse-grained mixed lithoclastic-bioclastic deposits (i.e. which unconformably overlie the Cretaceous limestone. Calcarenite di Gravina Fm.) passes upward to marine silty clay of the Argille subappennine Fm., Early Pleistocene in age [23]. Starting from the-Middle Pleistocene, this sector is affected by a regional uplift [22, 26], which promoted the progressive emersion of the area and the formation of deep gorges (the so-called Gravine), formed in the Cretaceous bedrock [22].

3.2 The Matera Municipality and its Cultural Heritage

The case study of this research is the territory of Matera, a centuries-old town lo-cated in the Basilicata Region, in Southern Italy. The ancient core of the town, worldwide known as the Sassi and the nearby "Murgia Materana" Park were classified as a UNESCO World Heritage Site since 1993 (http://whc.unesco.org/en/list/670, accessed 12 April 2021). The protected area comprises about 80 small rock-hewn villages that are still almost intact due to the lack of urbanization in the whole area, albeit with significant conservation problems. The park is important for the cultural heritage it mainly preserves architectonic and archaeological heritage in the rupestrian landscape, with many rock-cut settlements (rock churches, farms, dry stone walls, ancient calc-tuff quarry).

We chose the municipality of Matera for three reasons:

1) The Sassi and the Murgia Materana Park represent an extraordinary kind of settle-ment which have been continuously inhabited by man for two millennia. The cultural heritage is huge and partially not yet discovered and preserved;
2) The UNESCO and 2019 European Capital of Culture acknowledgments on the one hand require an effort for conservation and risk assessment, and on the other, they give a greater tourist, economic and urban impetus to the modern town;
3) The particular history of the territorial planning between the 50s and 70s cer-tainly influenced the growth of the modern city (for example the laws for the re-loca-tion of inhabitants in new residential and rural districts designed by sociologists and urban planners - laws 619/1952, 299/1958, 126/1967, 1043/1971).

4 Vulnerability Analysis

In addition to studies on the loss of cultural heritage [27] and on the appropriate actions to prevent its loss [28], it is important to evaluate previously the effective vulnerability of the cultural heritage at a territorial scale in order to safeguard it and provide precise guidelines for territorial planning and landscape protection [29, 30].

Vulnerability is the aptitude of an artefact to suffer damage, and it is generally as-sociated to the state of conservation. It is usually evaluated with a multidisciplinary study: specific parameters are applied to the artefact characteristics, such as the appearance of the surface, the constructive and structural data, the use and the safety.

In this study, the assessment of vulnerability is on a territorial scale, and it can be classified as a first level vulnerability, based only on the presence of the cultural heritage over the territory. In particular, the consolidated built area was not considered, while all the cultural heritage in the periurban and rural areas were considered to assess vulnerable areas [Fig. 1] in a buffer of 150 m around each element.

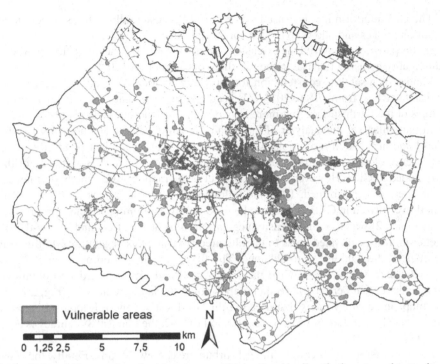

Fig. 1. Vulnerability map based on the presence of the cultural heritage in the extra-urban territory.

5   Hazard Analysis

In this work, we evaluate the following natural and anthropic hazards for the Matera municipality: for what concerns anthropic hazard, the urban growth was considered; whereas natural hazard are represented by soil erosion, landslides, and floods. In the next paragraphs, we explained how each of these hazard layers were evaluated in order to assess a cultural heritage risk map.

5.1   Urban Growth

Land use changes and in particular the urban growth could affect both cultural heritage and archaeological findings distribution over the territory. In the last years, remote sensing and the increasing calculation capability of computers helped very much to monitor this phenomenon, however some limitations remain.

Among them, it is worth to note:

- the monitoring of land use changes is not an easy task, because a robust estimation of the parameter needs landcover maps always updated and with high resolution;
- remote sensing data elaboration, for large dataset, is time consuming and requires large hardware resources.

The GEE platform in this sense is very useful because it allows to use freely high-resolution images and allows to reduce computing time.

In this paper, we analyzed the extension and the variation of the urban area in Matera Municipality by using Landsat in the last three decades (1990–2020).

Landsat is a satellite constellation for the Earth remote observation. There were 8 Landsat launches, from '70s to 2013. It is used for studying many natural and anthropic changes of the Earth surface, but more in particular for land-use changes.

In particular, for the analysis of the urban growth in Matera Municipality Landsat 5 and Landsat 8 (30 m resolution) were: 1) filtered in order to clear pixel affected by the presence of clouds 2) analyzed with the random forest method. Both the algorithm are included in GEE. However, for what concerns the random forest method, it needs a training set, to be performed. Thus, the following four types of areas were chosen and partially vectorized in GEE, in order to be used before as training set and further as test set: 1) urban, 2) water, 3) high density vegetation (indicated as tree) and 4) low-density vegetation, from cultivated areas to uncultivated (indicated as vegetation). In this way, two maps of the land use were obtained. The first is for the period between 1988 and 1991 (Fig. 2a), where we let GEE produce a mediated image in order to obtain the best product in that period; the second map is about the land use in 2020 (Fig. 2b). For each map, an accuracy index that uses the digitized test set and a Kappa statistic was calculated [31]. The Landsat 5 accuracy is equal to 0.92, the Landsat 8 accuracy is equal to 0.89.

As it is possible to observe in Table 1, urban areas grew 24 km^2. Finally, the urban areas were extracted and compared in order to obtain areas where the city is growing (Fig. 3).

5.2 Landslides and Floods

Maps of the hydrogeological (i.e. landslides and floods) hazards were derived through a revision of literature data and available official cartography (see for example [32]). More specifically, the basic information about the spatial distribution of landslide phenomena and flood areas has been extracted from the data stored in the hydrogeological plan of the Basin Authority of Basilicata with local revision based on photo-interpretation. Such data are summarized in Fig. 4.

5.3 Soil Erosion Map

Spatial distribution of erosion and deposition was delineated through the application of the Unit Stream Power Erosion Deposition (USPED) model [33]. The model estimates the average soil loss (A, annual average soil loss [Mg ha^{-1}yr^{-1}]) using the structure of the well-known RUSLE empirical equation [34]:

$$A = R \cdot K \cdot LS \cdot C \cdot P \tag{2}$$

where R [MJ mm h^{-1}ha^{-1}yr^{-1}] is the rainfall intensity factor, which has been estimated using the following equation:

$$EI_{30} = 0.1087 \cdot (P_{24})^{1.86} \tag{3}$$

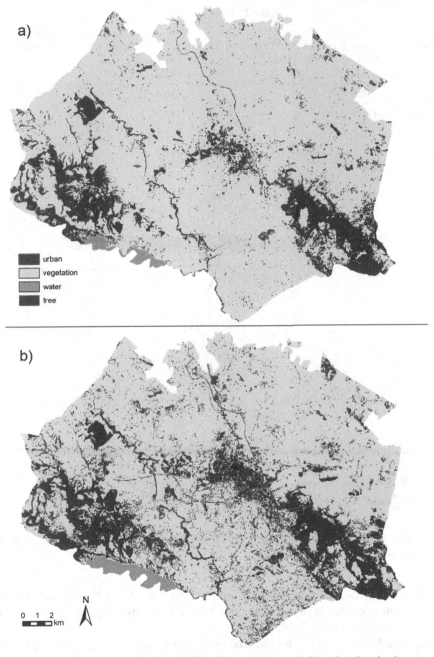

Fig. 2. Random forest classification of the land cover extracted from Landsat in the years a) 1988–1991, b) 2020

Table 1. Land use areas.

Land use	Areas of classes (km²) extracted from Landsat 5 (1988–1991)	Areas of classes (km²) extracted from Landsat 8 (2020)
Urban	9.6	34.0
Tree	425.4	385.7
Water	6.0	7.8
Vegetation	71.7	85.3

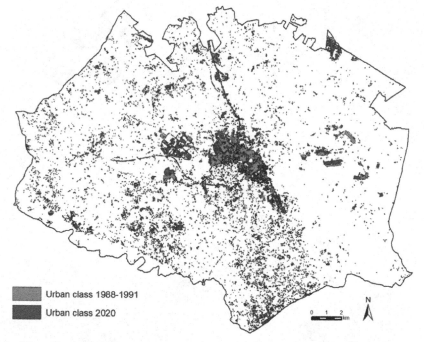

Fig. 3. The map of the urban growth extracted with GEE and random forest classification for the Matera municipality

with, EI30 is the rainfall erosivity in MJ mm h⁻¹ ha⁻¹ yr⁻¹ and P_{24} the daily rainfall amount in mm [35]. Mean annual R-factor value has been estimated for the period from January 2010 to December 2020.

K [Mg h MJ⁻¹ mm⁻¹] is the soil erodibility factor, which was estimated using the following equation that includes soil texture and permeability:

$$K = [2.1 \cdot 10^{-4} \cdot (12 - M) \cdot [(Si + fS)(100 - C)] \cdot 1.14 + 3.25(A - 2) + 2.5(P - 3)]/100 \tag{4}$$

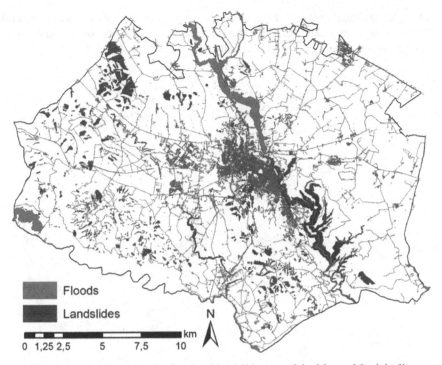

Fig. 4. Natural hazards: the flood and landslide map of the Matera Municipality

where M is the organic matter content (%), Si is the silt content (%), 2 to 50 μm, fS is the fine sand content (%) 50 μ to100 μm, C is the clay content (%) less than 2 μm, S is the sand content (%) 50 μm to 2 mm, A is the structure and P is the permeability class (within the top 0.60 m). K-factor map was drawn by combining lithological and land-use maps with values ranging from 0 to 0.1.

According to the procedures proposed by [33], the topographic factor LS is computed as the divergence of sediment flow (change in sediment transport capacity) using a 5 m DEM.

C [dimensionless] is the land cover factor and was derived starting from a land use map of the study area, provided by the Basilicata Regional Authority (http://rsdi.reg ione.basilicata.it). Using the basic structure of the Corine Land Cover project [36], the C-factor [dimensionless] has been calculated according to literature data [37].

The results of the application of the USPED model are shown in Fig. 5. Each pixel has been classified using the following three classes of erosion/deposition:

1 - Moderate to extreme erosion (> -5 Mg ha^{-1} yr^{-1}).
2 - Stable ($-5/5$ Mg ha^{-1} yr^{-1}).
3 - Moderate to extreme deposition (>5 Mg ha^{-1} yr^{-1}).

A simple statistical analysis of the frequency distribution of erosion and deposition areas allowed us to evaluate the following data: erosion areas occupy 6.9 km^2, stable areas are the majority (390 km^2) and deposition areas are equal to 7.7 km^2.

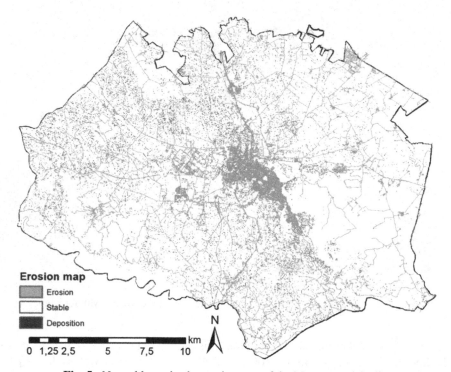

Fig. 5. Natural hazards: the erosion map of the Matera municipality

6   Results and Final Discussion

The different layers obtained in the previous paragraphs were combined with map algebra techniques, in particular with cell-by-cell functions in order to obtain: 1) the hazard map (Fig. 6); 2) the final risk map as a result of the overlay between vulnerability and hazards (Fig. 7).

Quantitative results show that hazard areas cover a total surface of 63.37 km^2 where about 30.7 km^2 are occupied by floods and landslides, 6.9 km^2 by erosion and 25.7 by the urban growth. Moreover, a relevant surface amount is affected by the overlay of two or three hazard factors (Table 2).

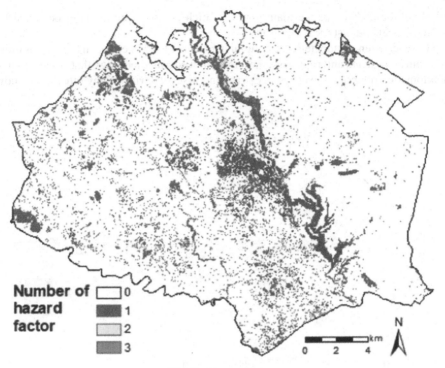

Fig. 6. Hazard factors

Table 2. Areas and percent over the study are occupied by 0, 1, 2 or 3 hazard factors and vulnerable areas

	Area (km^2)	%
Number of hazard factors		
0	330.1	85.3
1	54.6	14.1
2	2.3	0.6
3	0.04	0.01
Vulnerability areas	10.1	2.6

The risk map represents the synoptic view of the overlay of vulnerability and hazard (Main map Fig. 7, Table 3), shows that very few areas are really at risk, even if there are many areas characterized by a complex risk, where different types of hazard coexist (inset maps, Fig. 7). Of course, areas characterized by a more complex and heterogeneous risk should have a priority in the planning strategies of protection. Our results also highlight that natural hazards are mainly located in sectors featured by high-relief, steep slopes, and along the thalweg of the main gorges, which are not preferentially occupied by archaeological and historical elements.

However, even, fortunately, the cultural heritage of the Matera open territory is quite safe from natural and anthropic risk, the paper showed a methodology that could become an automatic protocol to investigate, extract and assess the risk of any other case study.

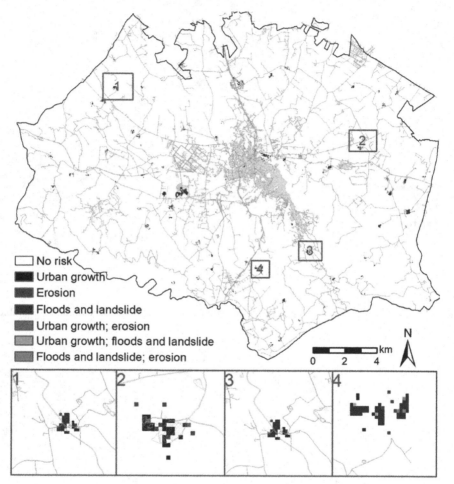

Fig. 7. The final risk map. Inset maps show some areas where different types of risks coexist, in overlay and in proximity.

However, to improve the performance of the model, it would be appropriate to go deepen also in exposure and the state of conservation of cultural heritage, according to the different types of heritage and its intrinsic features.

Table 3. Risk areas and percentage found

	Area (m^2)	% of the vulnerability areas
Type of risk		
Urban growth	1000503	10
Erosion	124356.3	1
Floods and landslides	185827.9	2
Urban growth & erosion	8478.84	0.1
Urban growth & floods and landslides	4945.99	0.1
Floods and landslide & erosion	10598.55	0.1

Acknowledgement. Authors want to thank Guido Ceccarini of the European Commission Joint Research Centre, who freely shared the script which GEE elaboration for urban growth analysis are inspired.

References

1. Serra, M., D'Agostino, S.: Archeologia Preventiva. Agenzia Magna Grecia, Albanella (SA) (2010)
2. Cruden, D.M., Varnes, D.J.: Landslide types and processes. In: Turner, A.K., Shuster, R.L. (eds.) Landslides: Investigation and Mitigation, pp. 36–75. National Academies Press, Washington DC (1996)
3. Biscione, M., Danese, M., Masini, N.: A framework for cultural heritage management and research: the cancellara case study. J. Maps **14**, 576–582 (2018)
4. Danese, M., Gioia, D., Biscione, M., Masini, N.: Spatial methods for archaeological flood risk: the case study of the neolithic sites in the Apulia Region (Southern Italy). In: Murgante, B., et al. (eds.) ICCSA 2014. LNCS, vol. 8579, pp. 423–439. Springer, Cham (2014). https://doi.org/10.1007/978-3-319-09144-0_29
5. Danese, M., Masini, N., Biscione, M., Lasaponara, R.: Predictive modeling for preventive archaeology: overview and case study. Cent. Eur. J. Geosci. **6**(1), 42–55 (2014). https://doi.org/10.2478/s13533-012-0160-5
6. Yan, L., et al.: Towards an operative predictive model for the songshan area during the yangshao period. ISPRS Int. J. Geo-Inf. **10**, 217 (2021)
7. https://blog.google/outreach-initiatives/sustainability/introducing-google-earth-engine/
8. Tamiminia, H., Salehi, B., Mahdianpari, M., Quackenbush, L., Adeli, S., Brisco, B.: Google Earth Engine for geo-big data applications: a meta-analysis and systematic review. ISPRS J. Photogramm. Remote. Sens. **164**, 152–170 (2020)
9. Moore, R.T., Hansen, M.C.: Google Earth Engine: a new cloud-computing platform for global-scale earth observation data and analysis. In: AGU Fall Meeting Abstracts (2011)
10. Ravanelli, R., et al.: Monitoring the impact of land cover change on surface urban heat island through google earth engine: proposal of a global methodology, first applications and problems. Remote Sens. **10**, 1488 (2018)
11. Mutanga, O., Kumar, L.: Google Earth Engine applications. Remote Sens. **11** (2019)
12. Patel, N.N., et al.: Multitemporal settlement and population mapping from Landsat using Google Earth Engine. Int. J. Appl. Earth Obs. Geoinf. **35**, 199–208 (2015)

13. Huntington, J.L., et al.: Climate engine: cloud computing and visualization of climate and remote sensing data for advanced natural resource monitoring and process understanding. Bull. Am. Meteor. Soc. **98**, 2397–2410 (2017)
14. Agapiou, A.: Remote sensing heritage in a petabyte-scale: satellite data and heritage Earth Engine© applications. Int. J. Digit. Earth **10**, 85–102 (2017)
15. Kleinberg, E.M.: On the algorithmic implementation of stochastic discrimination. IEEE Trans. Pattern Anal. Mach. Intell. **22**, 473–490 (2000)
16. Kleinberg, E.M.: An overtraining-resistant stochastic modeling method for pattern recognition. Ann. Stat. **24**, 2319–2349 (1996)
17. Belgiu, M., Drăguţ, L.: Random forest in remote sensing: a review of applications and future directions. ISPRS J. Photogramm. Remote. Sens. **114**, 24–31 (2016)
18. Sheykhmousa, M., Mahdianpari, M., Ghanbari, H., Mohammadimanesh, F., Ghamisi, P., Homayouni, S.: Support vector machine versus random forest for remote sensing image classification: a meta-analysis and systematic review. IEEE J. Selected Topics Appl. Earth Obs. Remote Sens. **13**, 6308–6325 (2020)
19. Cao, G.Y., Chen, G., Pang, L.H., Zheng, X.Y., Nilsson, S.: Urban growth in China: past, prospect, and its impacts. Popul. Environ. **33**, 137–160 (2012)
20. Demers, M.: GIS modeling in raster (2001)
21. Tomlin, C.D.: GIS and Cartographic Modeling. Esri Press, Redlands, California (2013)
22. Beneduce, P., Festa, V., Francioso, R., Schiattarella, M., Tropeano, M.: Conflicting drainage patterns in the Matera Horst Area, Southern Italy. Phys. Chem. Earth **29**, 717–724 (2004)
23. Tropeano, M., Sabato, L., Pieri, P.: Filling and cannibalization of a foredeep: the Bradanic Trough Southern Italy. Geol. Soc. Spec. Publ. **191**, 55–79 (2002)
24. Gioia, D., Sabato, L., Spalluto, L., Tropeano, M.: Fluvial landforms in relation to the geological setting in the "Murge Basse" karst of Apulia (Bari Metropolitan Area, Southern Italy). J. Maps **7**, 148–155 (2011)
25. Teofilo, G., Gioia, D., Spalluto, L.: Integrated geomorphological and geospatial analysis for mapping fluvial landforms in Murge basse karst of Apulia (Southern Italy). Geosciences (Switzerland) **9**, 418 (2019)
26. Gioia, D., Schiattarella, M., Giano, S.: Right-angle pattern of minor fluvial networks from the ionian terraced belt, Southern Italy: passive structural control or foreland bending? Geosciences **8**, 331 (2018)
27. Pérez-Hernández, E., Peña-Alonso, C., Hernández-Calvento, L.: Assessing lost cultural heritage. a case study of the eastern coast of Las Palmas de Gran Canaria city (Spain). L. Use Policy **96**, 104697 (2020)
28. de Noronha Vaz, E., Cabral, P., Caetano, M., Nijkamp, P.: Urban heritage endangerment at the interface of future cities and past heritage: a spatial vulnerability assessment. Serie Research Memoranda 0036, VU University Amsterdam, Faculty of Economics, Business Administration and Econometrics (2011)
29. Fry, G.L.A., Skar, B., Jerpåsen, G., Bakkestuen, V., Erikstad, L.: Locating archaeological sites in the landscape: a hierarchical approach based on landscape indicators. Landsc. Urban Plan. **67**, 97–107 (2004)
30. Agapiou, A., et al.: Impact of Urban sprawl to cultural heritage monuments: the case study of paphos area in Cyprus. J. Cult. Herit. **16**, 671–680 (2015)
31. Congalton, R.: Putting the Map Back in Map Accuracy Assessment, pp. 1–11 (2004)
32. Lazzari, M., Gioia, D., Anzidei, B.: Landslide inventory of the Basilicata region (Southern Italy). J. Maps **14**, 348–356 (2018)
33. Mitasova, H., Hofierka, J., Zlocha, M., Iverson, L.R.: Modelling topographic potential for erosion and deposition using GIS. Int. J. Geogr. Inf. Syst. **10**, 629–641 (1996)
34. Renard, K.G.: Predicting soil erosion by water: a guide to conservation planning with the Revised Universal Soil Loss Equation (RUSLE). United States Government Printing (1997)

35. Capolongo, D., Diodato, N., Mannaerts, C.M., Piccarreta, M., Strobl, R.O.: Analyzing temporal changes in climate erosivity using a simplified rainfall erosivity model in Basilicata (Southern Italy). J. Hydrol. **356**, 119–130 (2008)
36. Büttner, G.: CORINE land cover and land cover change products. In: Manakos, I., Braun, M. (eds.) Land Use and Land Cover Mapping in Europe. RSDIP, vol. 18, pp. 55–74. Springer, Dordrecht (2014). https://doi.org/10.1007/978-94-007-7969-3_5
37. Renard, K.G., Foster, G.R., Weesies, G.A., Porter, J.P.: RUSLE: revised universal soil loss equation. J. Soil Water Conserv. **46**, 30–33 (1991)

A Spatial Method for the Geodiversity Fragmentation Assessment of Basilicata Region, Southern Italy

Maria Danese[1] , Dario Gioia[1](✉) , Antonio Minervino Amodio[1],
Giuseppe Corrado[2] , and Marcello Schiattarella[2]

[1] Istituto di Scienze del Patrimonio Culturale, Consiglio Nazionale delle Ricerche (ISPC-CNR),
Contrada Santa Loja, 85050 Tito Scalo (Potenza), Italy
`dario.gioia@cnr.it, antonio.minervinoamodio@ispc.cnr.it`
[2] Dipartimento delle Culture Europee e del Mediterraneo (DiCEM), Basilicata University,
75100 Matera, Italy

Abstract. The heterogeneity of the abiotic factor, or geodiversity, is an important issue for the landscape genetics. In this paper, a spatial method for the assessment of the geodiversity of Basilicata region in southern Italy is proposed through the use of fragmentation indexes. The fragmentation indices use the values derived from the overlay of three different maps: i) landform map; ii) climate map; iii) geological map. Three indexes have been applied to obtain the geodiversity of the study area: i) Shannon's diversity index (SHDI); ii) Shannon's Evenness Index (SHEI); iii) Simpson's Diversity Index (SIDI). The results show how the three indices offer similar results. High values of geodiversity can be observed: i) in the frontal sector of the chain; ii) where the complexity of the landscape forms is high. Low values of geodiversity are observed in the Ionian coast belt. These results can represent a basic but valuable tool for the delineation of the main macro-areas with different geodiversity values.

Keywords: Geodiversity · Spatial analysis · Landform classification · Climatic classes · Fragmentation indexes · Basilicata region (southern Italy)

1 Introduction

Geodiversity is defined as the "diversity of the earth features and systems" [1]. Its conservation or "Geoconservation" is important: 1) by its own right; 2) for the appearance of a landscape; 3) for the planning and the management of many territorial aspects; 4) for contributing to the conservation of the ecological systems and consequently for sustainability. Quantitative assessment of the regional geodiversity of a landscape can represent a valuable parameter for the identification of homogeneous sectors of wide and complex geological and morpho-climate landscapes. In literature there many attempts to calculate a geodiversity map, in different geodynamics and morpho-structural contexts (see for example [2, 3]).

O. Gervasi et al. (Eds.): ICCSA 2021, LNCS 12951, pp. 620–631, 2021.
https://doi.org/10.1007/978-3-030-86970-0_43

Even if geodiversity maps are diffused and calculated also by different national or local Agencies, however, nowadays, a unique method for the geodiversity calculation does not exist.

In this paper the geodiversity is estimated by spatial analysis methods as a combination of three physical aspects of Basilicata region, a 10,000 km²-wide region of southern Italy: 1) the landform classes, 2) the climatic zones and 3) the geological features. The study area includes a relatively narrow transect of the southern Apennine chain, where poly-deformed litho-structural units belonging to different domains of the chain, clastic deposits of Pliocene to Quaternary tectonic basins and foredeep areas, and carbonate of the Apulia foreland coexist. Such a wide richness in terms of morpho-tectonic domains, litho-structural features, and landform variability can represent a valuable peculiarity for the identification of areas prone to the development of sustainable geotourism. The geodiversity fragmentation is assessed through the use of well-known fragmentation indexes and was summarized in regional scale maps.

2 Case Study

The southern Apennines are a northeast-verging fold-and thrust belt mainly composed of shallow-water and deep-sea sedimentary covers, deriving from Mesozoic–Cenozoic circum-Tethyan domains, covered by Neogene-Pleistocene foredeep and satellite-basin deposits [4] (Fig. 1). These units formed the backbone of the chain and overthrust on the Apulia carbonate of the foreland area.

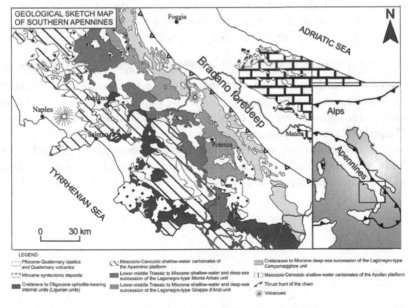

Fig. 1. Study area location

Contractional structure of the belt was dismember by high-angle faults, which are responsible for i) the formation of longitudinal and transversal fault-bounded basins [5–7], ii) the displacement of several generations of planation surfaces [5, 8], and iii) the re-organization and control of many hydrographic networks [5, 9, 10].

3 Parameters and Methods

The method used to assess geodiversity fragmentation consisted of the combination of three main elements: the map of landforms of Basilicata region, the map of climatic areas obtained by using climate data, and the map with the main lithological/geological classes of the study area.

Results are summarized in three fragmentation index map in order to assess the fragmentation degree of the Basilicata region landscape. In the following paragraphs, the methods and the parameters used to calculate each of the three elements are explained.

3.1 Landform Calculation

The first step consisted of the automatic landform classification. Many methods exist in literature, and each method uses different criteria, resolution of the input data, number of variables, and algorithms. Some of these were also compared in selected sectors of the study area to understand which one is more effective [11, 12] and test that a reliable method is linked to the concept of "geomorphon", developed by [13].

In such a method, a pattern-based classification that self-adapts to the local topography is used, through the construction of Local Ternary Patterns (LTM), also called geomorphons. These are the basic micro-structures that form any type of morphological classes. Consequently, landform classes are extracted from LTP and their combination [13].

In the method of geomorphons, the main parameters are: 1) the input DEM; 2) the flatness threshold that defines which slope should have areas to consider flat; 3) the inner and outer search radius. For Basilicata region DEM with a resolution of 5 m was used. The flatness threshold was imposed equal to 5°, while the inner and outer radii were imposed equal to 25 m and 500 m, respectively.

The resulting landform map consists of ten classes, as shown in Fig. 2. Table 1 shows the area and percentage distribution data from each of the ten classes in the study area, the main landform is the slope one, which covers 31% of the study region (Table 1).

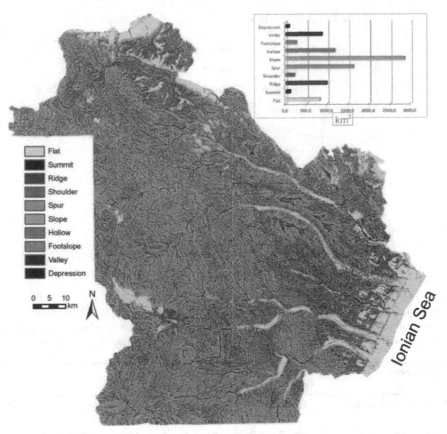

Fig. 2. Landform classes calculated with the method of geomorphons.

3.2 Climate Classification

The second step is represented by the delineation of the main climatic zones of the study area. To this aim, the criteria of Koppen classification was used (see [14] for more details). It allows to define climatic zones by using mean precipitation and temperature data (Table 2).

Data on rainfall and temperatures were taken from the site of the Functional Centre of Basilicata region and then they were interpolated using universal kriging. The resulting 5 m cell-sized maps were combined in order to apply the Koppen classification, which revealed 3 climatic zones (Fig. 3) in the general cluster of the temperate climate group: a mild temperate humid zone, a mild temperate dry winter zone, and a hot-summer Mediterranean climate.

Table 1. Landform classes distribution in Basilicata region.

Landform code	Landform	Area (m^2 * 10^6)	% of landscape
0	Flat	819.0	9.1
1	Summit	108.0	1.2
2	Ridge	989.7	11.0
3	Shoulder	201.8	2.2
4	Spur	1623.9	18.1
5	Slope	2854.2	31.8
6	Hollow	1177.2	13.1
7	Footslope	250.1	2.8
8	Valley	874.8	9.7
9	Depression	86.8	1.0

Table 2. Koppen climate classes (T_i = average temperatures of each month from January to December; P = precipitation). Our study region showed only three classes: Cwa (Mild temperate dry winter), Cfa (Mild temperate humid) and Csa (Hot-summer Mediterranean climate).

Type/Subtype	Criteria Rainfall/temperature regime
A - Tropical climates	$T_{min} \geq 18°C$
B – Dry climates	70% P_{ann} falls in summer months and $P_{mean\ ann} < 20T + 280$; 70% P_{ann} falls in winter and $P_{mean\ ann} < 20T$; 70% P_{ann} falls in summer months and $P_{mean\ ann} < 20T + 140$;
C - Temperate climates	$0 < T_{i\ min} < 18°C$ & $T_{i\ max} > 10°C$
w	$P_{driest\ winter} < P_{wettest\ summer}/10$
s	$P_{driest\ summer} < 30mm$ & $P_{driest\ summer} < P_{wettest\ winter}/10$
f	Precipitation distributed throughout year. Neither w or s
a	$T_{warmest\ month} \geq 22°C$
b	Four of the $T_{warmest\ month} \geq 10°C$ & $T_{warmest\ month} < 22°C$
c	Three of the $T_{warmest\ month} \geq 10°C$ & $T_{warmest\ month} < 22°C$
D – Continental climates	$T_{i\ min} < 0°C$ & $T_{i\ max} > 10°C$
E – Polar and alpine climates	$T_i \leq 10$ $\forall$ $1 < i < 12$

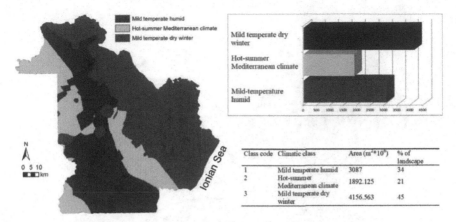

Class code	Climatic class	Area (m²*10⁸)	% of landscape
1	Mild temperate humid	3087	34
2	Hot-summer Mediterranean climate	1892.125	21
3	Mild temperate dry winter	4156.563	45

Fig. 3. Climate classes of the study area.

3.3 Geological Data

In the third step, the geological map was delineated through a revision of the official geological cartography (Fig. 4). The map includes data from different geological maps at different scales and the definition of the main lithological complexes classes was carried out taking into account three different characteristics: 1) type of rock; 2) formation/sedimentation environment; 3) location in the frame of the main domains of the chain. A code was assigned and the area occupied by each class was estimated (Table 3).

3.4 Geodiversity Map

The landform, climatic and geological maps were combined with the help of map algebra. The resulting raster map was estimated as follows:

$$C \cdot 10^3 + L \cdot 10^2 + G \cdot 10^0 \tag{1}$$

where C is the climatic class, L is the landform, G is the geology. The three rasters were summed in order to obtain a geodiversity map characterized by numerical categories, where the first left digit represents the climatic class, the second one represents the landform type, and finally the last two digits represent the geology. In this way, even if 281 classes were obtained, that are not clearly distinguishable by the human eyes, the raster is an information system, where in each pixel it is possible to query and know the different elements of the landscape.

For example, in the class 1312, we have the morphogenetic zone 1, the landform class 2 and the geologic category 12.

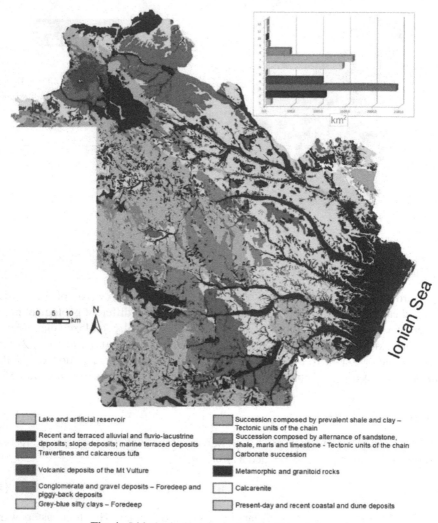

Fig. 4. Lithological/geological units of the study area.

Table 3. Distribution of geological classes in Basilicata.

Geology code	Geology	Area (m^2 * 10^{-6})	% of landscape
0	Lake and artificial reservoir	94.4	1.1
1	Recent and terraced alluvial and fluvio-lacustrine deposits; slope deposits; marine terraced deposits	1095.7	13.3
2	Travertines and calcareous tufa	2420.2	29.4
3	Volcanic deposits of the Mt Vulture	1039.9	12.6
4	Conglomerate and gravel deposits – Foredeep and piggy-back deposits	3.1	0.04
5	Grey-blue silty clays – Foredeep	1419.2	17.02
6	Succession composed by prevalent shale and clay – Tectonic units of the chain	1610.9	19.5
7	Succession composed by alternance of sandstone, shale, marls and limestone - Tectonic units of the chain	449.8	5.5
8	Carbonate succession	48.5	0.6
9	Metamorphic and granitoid rocks	22.5	0.3
10	Calcarenite	19.3	0.2
12	Present-day and recent coastal and dune deposits	20.0	0.2

.

3.5 Fragmentation Indexes

For the fragmentation analysis and the geodiversity assessment three indexes were used: the Shannon's Diversity Index [15] (SHDI), Eq. 2; the Shannon's Evenness Index (SHEI), Eq. 3; the Simpson's Diversity Index (SIDI), Eq. 4 ([16] and references therein).

$$SHDI = - \sum_{i=1}^{n} (P_i ln P_i) \tag{2}$$

$$SHEI = - \sum_{i=1}^{n} \frac{(P_i ln P_i)}{ln n} \tag{3}$$

$$SIDI = 1 - \sum_{i=1}^{n} P_i^2 \tag{4}$$

In the Eqs. 2, 3 and 4, n is the number of classes, also called, in this type of analysis, patch units. P_i is, instead, the proportion occupied by the i class over the study area.

Each index has a different range:

- the SHDI index is, without an upper limit, greater or equal to 0;
- the SHEI ranges between 0 and 1;

- the SIDI ranges between 0 and 1.

The SHDI and the SIDI express the richness of diversity (also called entropy degree in [15] and how it is quantitatively composed and distributed, by considering the number of different classes in a specific area and the surface by them occupied. These indexes can be calculated globally or locally. In this paper, a local version of the indexes was chosen and calculated. The shape of the local area of calculation is a circle with a radius of 1 km. Even if both SHDI and SIDI assess richness, SIDI highlights mainly the most common classes present in the distribution.

The SHEI index, instead, is useful to quantify how much the classes are even between them and where even classes are distributed. Also for this index, the local version was used, again by choosing a circular shape and a radius of 1 km.

4   Results and Final Considerations

The geodiversity obtained for the investigated region using the Shannon diversity index (SHDI) shows an interval between 0 and 3.25 (Fig. 5). High diversity values are observed: 1) in the frontal sectors of the chain; 2) along the ridges that divide the catchment areas

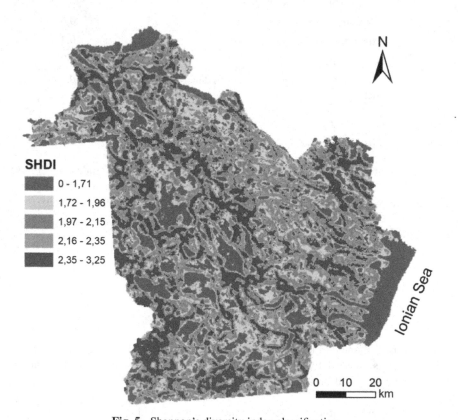

Fig. 5. Shannon's diversity index classification.

of the most important rivers; 3) where the landscape is dominated by the high-relief of morphostructural ridges made of shallow-water carbonates and deep-sea hard rocks. Lower values of geodiversity (0 > SHDI < 1.71) are observed mainly along the Ionian coastal belt (i.e. easternmost sectors of the study area).

The Shannon's Evenness index (SHEI) shows values between 0 and 1 (Fig. 6). Higher values (>2.35) of geodiversity are observed above all in areas characterized by a high variability of landscape shapes. This index also shows the lower values (<1.71) along the Ionian coast and in the central part of the map.

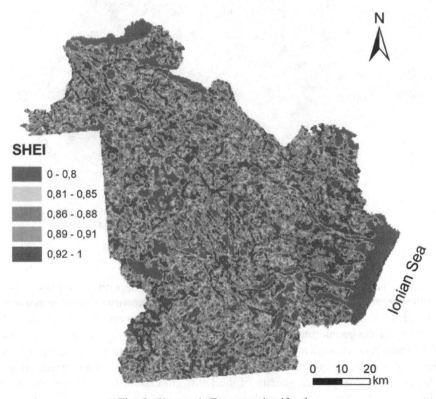

Fig. 6. Shannon's Evenness classification.

The Simpson index (SIDI) shows a distribution of geodiversity very similar to that shown by the SHDI index (Fig. 7). In fact, higher values of geodiversity are observed along the passage between the chain and the foredeep, on the border between the different hydrographic basins. Low values are observed along the Ionian coast.

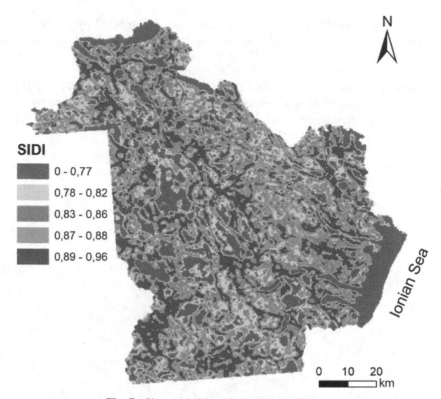

Fig. 7. Simpson's Diversity index classification.

For SHDI and SIDI index the high values of geodiversity depend on the passage from one lithology to another, this means that there is a high diversity of geomorphological forms for the same climate. Another factor of high geodiversity for all indices is the complexity of the geomorphological system, in fact the landscape sectors featured by well-hierarchized drainage networks correspond to sectors with higher values of geodiversity. Probably, in this case, the deep incision promoted the high fragmentation of the landscape a strong heterogeneity of the landform classes (Fig. 2). Instead, as all the indices show, low values of geodiversity are found along the Ionian coast, where characteristics of landscape shapes, climate, and geological information are almost the same. However, the geodiversity values shown by the three indices are similar in many cases, the SHDI and SIDI indices show almost the same results.

The results obtained allow us to identify the areas with different geodiversity and allow a macro subdivision into areas with different geological/geomorphological characteristics. In the future, the implementation of other parameters can allow the improvement of the final quality of the geodiversity maps.

References

1. Sharples, C.: Concepts and principles of geoconservation. Tasmanian Parks & Wildlife Service Website, 79 p. (2002)
2. Benito-Calvo, A., Pérez-González, A., Magri, O., et al.: Assessing regional geodiversity: the Iberian Peninsula. Earth Surf. Proc. Land. **34**(10), 1433–1445 (2009)
3. Pereira, D.I., Pereira, P., Brilha, J., et al.: Geodiversity assessment of Paraná State (Brazil): an innovative approach. Environ. Manag. **52**(3), 541–552 (2013)
4. Pescatore, T., Renda, P., Schiattarella, M., et al.: Stratigraphic and structural relationships between Meso-Cenozoic Lagonegro basin and coeval carbonate platforms in southern Apennines, Italy. Tectonophysics **315**(1–4), 269–286 (1999)
5. Schiattarella, M., Giano, S.I., Gioia, D.: Long-term geomorphological evolution of the axial zone of the Campania-Lucania Apennine, southern Italy: a review. Geol. Carpath. **68**(1), 57–67 (2017)
6. Giano, S.I., Gioia, D., Schiattarella, M.: Morphotectonic evolution of connected intermontane basins from the southern Apennines, Italy: the legacy of the pre-existing structurally controlled landscape. Rendiconti Lincei **25**, 241–252 (2014)
7. Gioia, D., Schiattarella, M., Mattei, M., et al.: Quantitative morphotectonics of the Pliocene to Quaternary Auletta basin, southern Italy. Geomorphology **134**(3–4), 326–343 (2011)
8. Schiattarella, M., Giano, S.I., Gioia, D., et al.: Age and statistical properties of the summit palaeosurface of southern Italy. Geogr. Fis. Din. Quat. **36**(2), 289–302 (2013)
9. Capolongo, D., Cecaro, G., Giano, S.I., et al.: Structural control on drainage network of the south-western side of the Agri River upper valley (Southern Apennines, Italy). Geogr. Fis. Din. Quat. **28**(2), 169–180 (2005)
10. Beneduce, P., Festa, V., Francioso, R., et al.: Conflicting drainage patterns in the Matera Horst Area, southern Italy. Phys. Chem. Earth **29**(10), 717–724 (2004)
11. Gioia, D., Danese, M., Bentivenga, M., Pescatore, E., Siervo, V., Giano, S.I.: Comparison of different methods of automated landform classification at the drainage basin scale: examples from the southern Italy. In: Gervasi, O., et al. (eds.) ICCSA 2020. LNCS, vol. 12250, pp. 696–708. Springer, Cham (2020). https://doi.org/10.1007/978-3-030-58802-1_50
12. Gioia, D., Bavusi, M., Di Leo, P., et al.: A geoarchaeological study of the Metaponto coastal belt, southern Italy, based on geomorphological mapping and GIS-supported classification of landforms. Geografia Fisica e Dinamica Quaternaria **39**(2), 137–148 (2016)
13. Jasiewicz, J., Stepinski, T.F.: Geomorphons—a pattern recognition approach to classification and mapping of landforms. Geomorphology **182**, 147–156 (2013)
14. Beck, H.E., Zimmermann, N.E., McVicar, T.R., et al.: Present and future Köppen-Geiger climate classification maps at 1-km resolution. Sci. Data **5**(1), 180214 (2018)
15. Shannon, C.E., Weave, W.: The Mathematical Theory of Communication, 125 p. University of Illinois Press, Urbana (1975)
16. Ciaian, P., Guri, F., Rajcaniova, M., et al.: Land fragmentation and production diversification: a case study from rural Albania. Land Use Policy **76**, 589–599 (2018)

International Workshop on Application of Numerical Analysis to Imaging Science (ANAIS 2021)

A Forward-Backward Strategy for Handling Non-linearity in Electrical Impedance Tomography

Martin Huska[✉], Damiana Lazzaro, and Serena Morigi

Department of Mathematics, University of Bologna, Bologna, Italy
{martin.huska,damiana.lazzaro,serena.morigi}@unibo.it

Abstract. Electrical Impedance Tomography (EIT) is known to be a nonlinear and ill-posed inverse problem. Conventional penalty-based regularization methods rely on the linearized model of the nonlinear forward operator. However, the linearized problem is only a rough approximation of the real situation, where the measurements can further contain unavoidable noise. The proposed reconstruction variational framework allows to turn the complete nonlinear ill-posed EIT problem into a sequence of regularized linear least squares optimization problems via a forward-backward splitting strategy, thus converting the ill-posed problem to a well-posed one. The framework can easily integrate suitable penalties to enforce smooth or piecewise-constant conductivity reconstructions depending on prior information. Numerical experiments validate the effectiveness and feasibility of the proposed approach.

Keywords: EIT inverse problem · Forward-backward algorithm · Nonlinear optimization · Regularization

1 Introduction

Electrical impedance tomography (EIT) is the problem of determining the electrical conductivity distribution of an unknown medium by making voltage and current measurements at the boundary of the object. As a radiation-free non-invasive monitoring tool, EIT has attracted much attention in the last two decades in a variety of biomedical applications. EIT is a typical ill-posed inverse problem, in the sense that the solution to the inverse problem can fail to exist or be extremely sensitive to small errors in the observations [4]. Since in practice all measurements in EIT experimental setup incorporate noise, a naive solution of this inverse problem with noisy data is typically a meaningless solution. Moreover, differently from the well-assessed tomographic imaging modalities, e.g., computed tomography (CT) and magnetic resonance imaging (MRI), the image reconstruction problem of EIT is characterized by a highly nonlinear forward operator. The forward EIT operator allows calculation of EIT measurement data from a given conductivity distribution.

© Springer Nature Switzerland AG 2021
O. Gervasi et al. (Eds.): ICCSA 2021, LNCS 12951, pp. 635–651, 2021.
https://doi.org/10.1007/978-3-030-86970-0_44

To reduce the ill-posedness, a regularization term is commonly introduced in the mathematical model to use the prior information to constrain the conductivity reconstruction problem. The prior information represents the distribution characteristics of the target conductivity, e.g., smooth, blocky, or sparse, and is modeled by a specially designed regularization term, e.g., Tikhonov, total variation (TV), or ℓ_1-norm [9]. To simplify the computational difficulty of the EIT inverse problem, classical regularization methods rely on the linearized model of the nonlinear forward operator. However, the linearization of the EIT formulation is only a rough approximation of the real measurement system, and this affects the reconstruction results.

We propose a variational framework for EIT reconstructions, which allows to solve the nonlinear ill-posed EIT problem via a forward-backward splitting iterative strategy where a sequence of regularized linear optimization problems are solved, thus turning the ill-posed problem to a sequence of well-posed ones.

The proximal Forward–Backward algorithm is a standard first-order optimization method to minimize a function which is defined as the sum of a differentiable function f and an easily proximable proper lower semicontinuous function g, and it consists in alternating a gradient step on f and a proximal step on g. In the proposed variational framework the proximal step involves a penalty which, according to the a priori knowledge on the conductivity distribution, can be either Tiknonov-like or Total-Variation-like. In the former case we enforce smooth reconstructions, while in the latter we assume piecewise-constant conductivity distribution as the prior information.

The paper is organized as follows. The remaining of this section is devoted to a brief overview on variational approaches to the EIT problem. In Sect. 2 notations and the inverse EIT formulation are introduced. Section 3 formulates the regularized Gauss-Newton method for the EIT problem. The proposed variational approach and the forward-backward iterative strategy are described in Sect. 4. Numerical experiments validate the effectiveness and feasibility of the proposed approach in Sect. 5. Conclusions are drawn in Sect. 6.

1.1 Related Works

Variational approaches to the solution of the EIT problem are currently the most successful and widely used both for linearized and non-linearized EIT reconstruction methods. A well-known variational formulation for linearized EIT problems relies on a least squares formulation with Tikhonov regularizer to favour smooth conductivity reconstructions [12]. Generalized Tikhonov approaches allow for a more general smoothing penalty term, which includes approximations of differential operators or structural priors [17]. To promote the reconstruction of piece-wise constant conductivity regions, many variational approaches included another common regularizer which relies on the ℓ_1 norm since it maintains the discontinuities in the reconstructed profiles. This regularization method is generally referred to as Total Variation (TV) [3]. This feature makes it particularly suited for medical and industrial applications (e.g. definition of inter-organ and inter-phase boundaries). However, the use of the TV functional for regularization leads to the minimization of a non-differentiable objective function in the inverse formulation.

The numerical solution of the optimization problems derived from the linear EIT variational approaches with linear regularizers reduces to the use of an efficient solver for linear systems, while the presence of a non-differentiable regularizer, such as the TV one, leads to more sophisticated optimization strategies such as Alternating Directions Method of Multipliers [9], or Primal Dual - Interior Point Method (PD-IPM) [3].

For what concerns the optimization methods for the nonlinear EIT inverse formulations, the literature is based on the Regularized Gauss-Newton (RGN) method. The Newton One-Step Algorithm (NOSER [6]), which performs only the first step of the Gauss-Newton's method with linear regularizer, represents a significant milestone in the investigation of EIT solvers. In [2] the GN method is enriched with an Iterative Reweighted strategy to deal with the non-differentiability of the TV regularizer. A relaxed inexact proximal Gauss–Newton method is proposed in [11] which deals with non-smooth TV regularization to better track the sharp edges in the conductivity.

We propose a reconstruction variational framework for the solution of the non-linear least squares EIT inverse problem which allows to include both linear and non-linear regularization and leads to the solution of a sequence of simple regularized linear least squares optimization problems via a forward-backward splitting strategy, thus converting the ill-posed problem into a well-posed one.

2   Inverse EIT Formulation

In inverse EIT, small alternating currents are applied to conducting surface electrodes attached at the boundary of the object Ω. The measured voltages V_m on the electrodes are used to reconstruct electrical conductivity distribution σ of the internal part of the object.

In the corresponding forward EIT problem one wants to find the electric potential u in the interior of the object Ω and at the electrodes, given some applied current and inner conductivity σ. Following the accurate Complete Electrode Model (CEM), introduced in [5], the forward EIT problem can be formulated as follows:

$$\begin{cases} \nabla \cdot (\sigma(x)\nabla u(x)) = 0 \text{ in } \quad \Omega, \\[2mm] u + z_l\sigma\frac{\partial u}{\partial n} = V_l \qquad \text{on} \quad E_l, \, l = 1,..,L, \\[2mm] \int_{E_l} \sigma\frac{\partial u}{\partial n}\,\mathrm{d}s = I_l \qquad \text{on} \quad \Gamma, \\[2mm] \sigma\frac{\partial u}{\partial n} = 0 \qquad\qquad \text{on} \quad \tilde{\Gamma}, \end{cases} \tag{1}$$

where Γ ($\tilde{\Gamma}$) is the boundary $\partial\Omega$ with (without) electrodes, V_l is the unknown voltage to be measured by l-th electrode E_l when the currents I_l are applied, z_l

are the contact impedances. The solution of forward EIT problem amounts to solving the boundary problem (1).

The *Forward Operator* $\tilde{F}$, which operates between the Hilbert spaces X and Y, maps the conductivity σ to the solution of the forward problem:

$$\tilde{F} \colon \mathfrak{S} \subset X \to Y \tag{2}$$
$$\sigma \mapsto (u, V)$$

where $\mathfrak{S} = \{\sigma \in L^\infty(\Omega) \mid \sigma\nabla u = 0\}$, denotes the domain of definition of $\tilde{F}$.

Proposition 1: The operator $\tilde{F}$ that maps $\sigma \in \mathfrak{S}$ to the solution of the Forward Problem with current vector I is Fréchet differentiable, meaning that

$$\lim_{\|\delta\sigma\|_\infty \to 0} \frac{\|\tilde{F}(\sigma + \delta\sigma) - \tilde{F}(\sigma) - \tilde{F}'(\sigma)\delta\sigma\|_Y}{\|\delta\sigma\|_\infty} = 0. \tag{3}$$

The proof of Proposition (1) is provided in [14], where it is also proved that $\tilde{F}'$ is Lipschitz continuous.

From now on we restrict the conductivities σ to a finite dimensional space of piecewise polynomials. We consider the object domain Ω discretized into n_T subdomains $\{\tau_j\}_{j=1}^{n_T}$ and σ constant over each of them. Given a Finite Elements Model (FEM) of an EIT medium, we calculate the vector of voltages, V_m, for each FEM degree of freedom. For a given stimulation pattern a vector of n_M measurements is acquired, obtained by injecting current through an electrodes pair and then measuring the corresponding voltage V_m induced on another pair of electrodes. Then $F : \mathbb{R}^{n_T} \to \mathbb{R}^{n_M}$ represents the discrete version of the Forward Operator (2) as a nonlinear vector map.

Since $\tilde{F}$ is Fréchet differentiable, F' is a matrix, called the Jacobian of F and denoted by J; each element of $J \in \mathbb{R}^{n_M \times n_T}$ is defined as

$$\{J(u_d, u_m)\}_{i,j} = \int_{\tau_j} \nabla u_d \cdot \nabla u_m d\Omega , \tag{4}$$

where the row index i corresponds to the ith measurement, associated with the dth driving potential u_d and mth measurement potential u_m, while the column index j corresponds to the subdomain τ_j.

Considering measured data corrupted by additive noise, we can assume the following noisy non-linear observation model

$$V_m = F(\sigma) + \eta, \tag{5}$$

where $V_m \in \mathbb{R}^{n_M}$ represents the vector of all the measured electrode potentials whose dimension n_M depends on the choice of a measurement protocol, and $\eta \in \mathbb{R}^{n_M}$ is a zero-mean Gaussian distributed measurement noise vector.

Assuming the non-linear degradation model (5) and the given measurements V_m, the so called absolute imaging problem aims to estimate the (static) conductivity σ by solving the following **non-linear** least squares problem

$$\sigma^* = \arg\min_{\sigma} f(\sigma), \qquad f(\sigma) = \int_{\Omega} (F(\sigma) - V_m)^2 \, d\Omega. \qquad \text{(EITNL)}$$

The underlying optimization problem is hard to solve, as the boundary currents depend non-linearly on the conductivity. This means that the optimization problem is nonconvex.

For the reconstruction of small conductivity changes, conventional approaches rely on the simplest linearized model of the non-linear forward operator F,

$$F(\sigma) \approx F(\sigma_0) + J\delta\sigma = F(\sigma_0) + J(\sigma - \sigma_0), \qquad (6)$$

where J represents the Jacobian matrix defined in (4), and calculated at the initial conductivity estimate σ_0. The reconstruction is thus obtained by solving the following **linear** least squares problem

$$\delta\sigma^* = \arg\min_{\delta\sigma} f(\delta\sigma), \qquad f(\delta\sigma) = \int_{\Omega} (J\,\delta\sigma - \delta V_m)^2 \, d\Omega, \qquad \text{(EITL)}$$

where $\delta\sigma = \sigma - \sigma_0$ and $\delta V_m = V_m - F(\sigma_0)$.

A benefit of the linear approach is that it leads to computationally fast reconstruction, however, the linearization leads to a very ill-conditioned undetermined linear system to solve, and the solution is only valid for sufficiently small deviations from the conductivity σ_0 at which the Jacobian is initially calculated.

In difference imaging model for EIT, the goal is to reconstruct the change in conductivity between two states measured at different times. The conductivity reconstruction in difference imaging is conventionally carried out using a linear approach as in (EITL), based on the difference of the two data sets and the linearization of the forward model (6). In absolute imaging mode the conductivity distribution is reconstructed even in case only the data after the conductivity change is available (background material conductivity σ_0 is not known). However, absolute reconstructions seem to be less robust with respect to modelling errors caused by inaccurately known boundary shape and electrode positions. To account for modelling errors and, at the same time, to avoid the linear approximation of the forward model, one approach is to compute absolute reconstructions of the conductivities σ_0 and σ based on each of the data sets separately, and then to subtract σ_0 from σ to obtain $\delta\sigma$.

The conductivity change $\delta\sigma$ is reconstructed clearly better by the nonlinear observation model over the conventional approach of using difference imaging with a globally linearized observation model [13].

This motivated us to propose an efficient solution to the non-linear optimization problem (EITNL).

3   Regularized Gauss-Newton for EITNL

The Gauss-Newton method is the most commonly used one for minimizing non-linear least squares (NLLS) problems such as the EITNL and performs a line search strategy with a specific choice of a descent direction. It simplifies the Newton-Raphson method which relies on the second-order Taylor's expansion approximation of the function $f(\sigma)$. Specifically, the Newton's method approximates

$$f(\sigma + p) \approx f(\sigma) + \nabla f(\sigma)^T p + \frac{1}{2} p^T \nabla^2 f(\sigma) p, \tag{7}$$

where the gradient and the Hessian of $f(\sigma)$ are given respectively by

$$\nabla f(\sigma) = J(\sigma)^T (F(\sigma) - V_m), \tag{8}$$

and

$$\nabla^2 f(\sigma) = J(\sigma)^T J(\sigma) + \sum_k r_k(\sigma) \nabla^2 r_k(\sigma), \tag{9}$$

with $J(\sigma)$ the Jacobian matrix of $r(\sigma) := F(\sigma) - V_m$.

The search direction p is computed by imposing optimality condition for p to EITNL with $f(\sigma)$ approximated as in (7):

$$\frac{\partial f(\sigma + p)}{\partial p} = \nabla f(\sigma)^T + \nabla^2 f(\sigma) p = 0, \tag{10}$$

which implies that

$$\nabla^2 f(\sigma) p^N = -\nabla f(\sigma). \tag{Newton-Raphson}$$

We can essentially approximate the Hessian matrix by ignoring all the second order terms from $\nabla^2 f(\sigma)$ in (9), so the search direction p in (Newton-Raphson) reduces to the following linear system

$$J^T(\sigma) J(\sigma) p^{GN} = -J^T(\sigma)(F(\sigma) - V_m). \tag{Gauss-Newton}$$

Gauss-Newton method starts from an initial guess σ_0 and performs a line search along the direction p_k^{GN} to obtain the new iterate σ_{k+1} as

$$\sigma_{k+1} = \sigma_k + p_k^{GN}. \tag{11}$$

In (Gauss-Newton), V_m is the measurement vector, the coefficient matrix which involves the Jacobian matrix, is a linear operator, however, due to its compact nature, it has an unbounded (discontinuous) inverse. This causes the solution to be unstable against variations in the data, hence violating Hadamard's third criterion for well posedness. Therefore, applying the Gauss-Newton method to problem (EITNL) yields inaccurate solutions; one could instead employ some form of regularization on the sought solution σ. The regularized Gauss-Newton

(RGN) method, using a generalized Tikhonov regularizer, consists in applying GN to the minimization problem:

$$\sigma^* = \arg\min_{\sigma} \{\mathcal{J}(\sigma;\lambda) = f(\sigma) + \lambda g(\sigma)\}, \quad g(\sigma) = \|L\sigma\|_2^2, \qquad (12)$$

where $L \in \mathbb{R}^{n_T \times n_T}$ is a matrix representing a discrete differential operator.

The gradient and the approximated Hessian matrix of the objective function in (12) are respectively:

$$\nabla\mathcal{J}(\sigma;\lambda) = J(\sigma)^T(F(\sigma) - V_m) + \lambda L^T L\sigma, \qquad (13)$$

$$\nabla^2\mathcal{J}(\sigma;\lambda) = \nabla^2 F(\sigma)(F(\sigma) - V_m) + J(\sigma)^T J(\sigma) + \lambda L^T L.$$

The search direction p_k from the current iterate satisfies the linear system

$$(J(\sigma_k)^T J(\sigma_k) + \lambda L^T L)p_k^{GN} = J(\sigma_k)^T(V_m - F(\sigma_k)). \qquad (14)$$

An efficient minimization of the nonlinear Tikhonov regularized functional (12) is proposed in [16] for small or medium-scale problems, by exploiting the singular value decomposition of the Jacobian matrix J.

In our implementation of the RGN method we consider L to be the second order high-pass filter (Laplace prior) with homogeneous Neumann boundary conditions. Due to the triangular discretization of the domain Ω, each row of the graph Laplacian L has only 4 non-zero elements: it has value -1 for each adjacent triangle, and 3 for the triangle itself.

4   EITNL with Non-linear/Linear Regularization Term

In this section, we present a non-linear reconstruction framework represented by the following NLLS optimization problem

$$\sigma^* = \arg\min_{\sigma}\left\{\mathcal{J}(\sigma;\lambda) = \frac{1}{2}\|F(\sigma) - V_m\|_2^2 + \lambda R(\sigma)\right\}, \qquad (15)$$

where the regularizer term $R(\sigma)$ can assume one of the following forms

(a) $R_{NL}(\sigma) = \phi(\|D\sigma\|_2; a)$;
(b) $R_L(\sigma) = \|L\sigma\|_2^2$.

In (a), $D \in \mathbb{R}^{n_E \times n_T}$ is a matrix representing the discrete gradient operator which vanishes to zero everywhere but the mesh edges, and is defined as follows

$$D_{ij} = \begin{cases} 1 & \text{if } \tau_j \bigcap \tau_k = e_i, \ j < k, \ k \neq j, \ k = 1,\dots,n_T, \\ -1 & \text{if } \tau_j \bigcap \tau_k = e_i, \ j > k, \ k \neq j, \ k = 1,\dots,n_T, \\ 0 & \text{otherwise}, \end{cases} \qquad (16)$$

where the two nonzero entries per row correspond to position of adjacent triangles τ_j, τ_k sharing the edge e_i. Applying homogeneous Neumann boundary conditions, $\nabla \sigma|_{\partial\Omega} = 0$, the rows of D corresponding to the boundary edges can be eliminated from D, thus, n_E will refer to the number of inner mesh edges.

The function $\phi : \mathbb{R} \rightarrow \mathbb{R}^+$ in (a) is a numerically tractable version of the ℓ_0 pseudo-norm promoting sparsity and depending on a parameter a. Among the non-convex sparsity-promoting regularizers characterized by tunable degree of non-convexity, we consider in the proposed FB framework the Minimax Concave Penalty, introduced in [18], which reads as:

$$\phi(t; a) = \begin{cases} -\frac{a}{2}t^2 + \sqrt{2a}t & \text{if } |t| < \sqrt{\frac{2}{a}}, \\ 1 & \text{if } |t| \geq \sqrt{\frac{2}{a}}, \end{cases} \tag{17}$$

where $a > 0$ modulates the concavity of the regularizer.

The non-convex optimization problem (15) is smooth for R_L and non-smooth for R_{NL}.

Motivated by the following result on a generic point $\sigma^{(n+1)}$ of an iterative scheme, the minimization of $\mathcal{J}(\sigma; \lambda)$ in (15) will be carried out by a Forward-Backward (FB) splitting scheme, introduced in [7].

Proposition 2: Let F be a Frechét differentiable operator defined in (2), then

$$\sigma^{(n+1)} \in \text{prox}_{\gamma g}\left(\sigma^{(n)} - \gamma J^T(\sigma^{(n)})(F(\sigma^{(n)}) - V_m)\right) \tag{18}$$

is equivalent to

$$\sigma^{(n+1)} \in \arg\min_{\sigma}\left\{\tilde{f}(\sigma) + g(\sigma)\right\}, \tag{19}$$

where $\tilde{f}(\sigma) = \frac{1}{2}\|F(\sigma^{(n)}) - V_m + J(\sigma^{(n)})(\sigma - \sigma^{(n)})\|_2^2$, and $g(\sigma) = \lambda R(\sigma)$.

Proof: Consider the minimization problem

$$\sigma^* \in \arg\min_{\sigma}\left\{\tilde{f}(\sigma) + g(\sigma)\right\}. \tag{20}$$

The first order necessary conditions for (20), satisfied for σ^*, read as

$$0 \in \partial\tilde{f}(\sigma^*) + \partial g(\sigma^*), \tag{21}$$

where ∂ denotes the subdifferential, and

$$\partial\tilde{f}(\sigma) = J^T(\sigma^{(n)})\left(F(\sigma^{(n)}) - V_m + J(\sigma^{(n)})(\sigma - \sigma^{(n)})\right). \tag{22}$$

For any fixed $\gamma > 0$ the followings are equivalent:

$$0 \in \gamma\partial\tilde{f}(\sigma^*) + \gamma\partial g(\sigma^*) \tag{23}$$
$$0 \in \gamma\partial\tilde{f}(\sigma^*) - \sigma^* + \sigma^* + \gamma\partial g(\sigma^*) \tag{24}$$
$$(I + \gamma\partial g)(\sigma^*) \in (I - \gamma\partial\tilde{f})(\sigma^*) \tag{25}$$

$$\sigma^* \in (I + \gamma \partial g)^{-1}(I - \gamma \partial \tilde{f})(\sigma^*). \tag{26}$$

Equation (26) can be interpreted as fixed point scheme that generates a sequence $\{\sigma^{(n)}\}$ by iterating:

$$\sigma^{(n+1)} \in (I + \gamma \partial g)^{-1}(I - \gamma \partial \tilde{f})(\sigma^{(n)})$$

$$\sigma^{(n+1)} \in (I + \gamma \partial g)^{-1}\left(\sigma^{(n)} - \gamma J^T(\sigma^{(n)})(F(\sigma^{(n)}) - V_m)\right)$$

which can be rewritten as the proximal map (18). ■

Considering that the proximal map (18) can be rewritten as

$$\sigma^{(n+1)} \in \arg\min_{\sigma} \{g(\sigma) \\ + \tfrac{1}{2\gamma}\|\sigma - (\sigma^{(n)} - \gamma J^T(\sigma^{(n)})(F(\sigma^{(n)}) - V_m))\|_2^2\}, \tag{27}$$

and that $F(\sigma^{(n)}) - V_m + J(\sigma^{(n)})(\sigma - \sigma^{(n)})$ represents a good linear approximation model of problem (15), then we can apply a first order optimization method to solve the problem. In particular, given $\sigma^{(0)}$, $\gamma > 0$, $\lambda > 0$, we apply the Forward-Backward scheme which iterates for $n = 1, \dots$ as follows:

Forward Step
$$z^{(n)} = \sigma^{(n)} - \gamma J^T(\sigma^{(n)})(F(\sigma^{(n)}) - V_m)$$
Backward Step
$$\sigma^{(n+1)} \in \arg\min_{\sigma} \left\{ \tfrac{1}{2\gamma} \left\|\sigma - z^{(n)}\right\|_2^2 + \lambda R(\sigma) \right\}.$$

The FB scheme is stopped as soon as the relative residual is less than a given tolerance.

4.1   Using a Linear Regularizer R_L

In this case the Backward step in the FB algorithm is defined as follows

$$\sigma^{(n+1)} \leftarrow \arg\min_{\sigma} \left\{ \frac{1}{2} \left\|\sigma - z^{(n)}\right\|_2^2 + \gamma\lambda\|L\sigma\|_2^2 \right\} \tag{28}$$

with L being the Laplace prior. Being (28) a convex and differentiable minimization problem, an approximate solution $\sigma^{(n+1)}$ is explicitly given by solving the linear system

$$\left(I + \frac{\gamma\lambda}{2}L^T L\right)\sigma = z^{(n)}. \tag{29}$$

The system matrix is symmetric, positive definite, but unstructured, therefore the linear system (29) admits a unique solution obtained very efficiently by a few iterations of a warm-started iterative solver.

4.2   Using a Non-linear Regularizer R_{NL}

The backward step using $\phi(\cdot; a)$ in (17) leads to a nonconvex-nonsmooth optimization problem. Nevertheless, thanks to the properties of the penalty function, we can apply the so called Convex-Non-Convex strategy [1],[15], to ensure the convexity of the overall functional. The convexity conditions are summarized in the following Proposition.

Proposition 3: Let $R_{NL} = \phi(x; a)$, with $\phi(x; a)$ defined in (17), then the sufficient convexity condition for the functional in the Backward step of the FB algorithm is defined as

$$a < \frac{1}{\gamma \lambda e_{MAX}}, \tag{30}$$

where e_{MAX} is the maximum eigenvalue of matrix $D^T D \in \mathbb{R}^{n_T \times n_T}$.

The proof follows from [8].

By introducing the auxiliary variable $t \in \mathbb{R}^{n_E}$, the Backward Step can be rewritten in the following equivalent form

$$\{\sigma^*, t^*\} \leftarrow \arg\min_{\sigma, t} \left\{ \frac{1}{2\gamma} \left\| \sigma - z^{(n)} \right\|_2^2 \right.$$
$$\left. + \lambda \sum_{j=1}^{n_E} \left[\phi\left(|t_j|; a\right) \right] \right\} \tag{31}$$
$$\text{subject to}: \quad t = D\sigma.$$

To solve the constrained optimization problem we introduce the augmented Lagrangian functional

$$\mathcal{L}(\sigma, t; \rho; \lambda, a) = \frac{1}{2\gamma} \left\| \sigma - z^{(n)} \right\|_2^2 + \lambda \sum_{j=1}^{n_E} \phi\left(|t_j|; a\right)$$
$$- \langle \rho, t - D\sigma \rangle + \frac{\beta}{2} \|t - D\sigma\|_2^2$$

where $\beta > 0$ is a scalar penalty parameter and $\rho \in \mathbb{R}^{n_E}$ is the vector of Lagrange multipliers associated with the linear constraint $t = D\sigma$. The following saddle-point problem is then considered:

$$\text{Find } (\sigma^*, t^*; \rho^*) \in \mathbb{R}^{n_T} \times \mathbb{R}^{n_E} \times \mathbb{R}^{n_E}$$
$$\text{s.t.} \quad \mathcal{L}\left(\sigma^*, t^*; \rho; \lambda, a\right) \leq \mathcal{L}\left(\sigma^*, t^*; \rho^*; \lambda, a\right)$$
$$\leq \mathcal{L}\left(\sigma, t; \rho^*; \lambda, a\right)$$
$$\forall(\sigma, t; \rho) \in \mathbb{R}^{n_T} \times \mathbb{R}^{n_E} \times \mathbb{R}^{n_E}$$

Given vectors $\sigma^{(k)}$ and $\rho^{(k)}$ computed at the kth iteration (or initialized if $k = 0$), the $(k+1)$th iteration of the ADMM-based iterative scheme applied to the solution of the saddle-point problem is split into the following three sub-problems:

$$t^{(k+1)} \leftarrow \arg\min_{t \in \mathbb{R}^{n_E}} \mathcal{L}\left(\sigma^{(k)}, t; \rho^{(k)}; \lambda, a\right)$$
$$\sigma^{(k+1)} \leftarrow \arg\min_{\sigma \in \mathbb{R}^{n_T}} \mathcal{L}\left(\sigma, t^{(k+1)}; \rho^{(k)}; \lambda, a\right)$$
$$\rho^{(k+1)} \leftarrow \rho^{(k)} - \beta\left(t^{(k+1)} - D\sigma^{(k+1)}\right).$$

The minimization sub-problem for σ can be rewritten as follows:

$$\sigma^{(k+1)} \leftarrow \arg\min_{\sigma \in \mathbb{R}^{n_T}} \left\{ \frac{1}{2\gamma} \left\| \sigma - z^{(n)} \right\|_2^2 + \left\langle \rho^{(k)}, D\sigma \right\rangle \right.$$
$$\left. + \frac{\beta}{2} \left\| t^{(k+1)} - D\sigma \right\|_2^2 \right\}.$$

The first-order optimality conditions of the quadratic minimization problem lead to the following linear system:

$$\left(\frac{1}{\gamma} I + \beta D^T D \right) \sigma = \frac{1}{\gamma} z^{(n)} + \beta D^T \left(t^{(k+1)} - \frac{1}{\beta} \rho^{(k)} \right).$$

Since $\gamma, \beta > 0$, the system matrix is symmetric, positive definite. However, D represents the discretization of first order derivatives on a unstructured 2D mesh, thus matrix $D^T D$ is unstructured, and the linear system cannot benefit of fast transform solvers. The unique solution of the linear system can be efficiently obtained via a few steps of warm-started iterative solver.

Solving the Sub-problem for t. Recalling the separability property of $\phi(\cdot; a)$, defined in (17), the minimization sub-problem for t can be rewritten in the following element-wise (edge-per-edge) form:

$$t^{(k+1)} \leftarrow \arg\min_{t \in \mathbb{R}^{n_E}} \sum_{j=1}^{n_E} \left\{ \lambda \phi(|t_j|; a) - \rho_j^{(k)} t_j \right.$$
$$\left. + \frac{\beta}{2} \left(t_j - (D\sigma^{(k)})_j \right)^2 \right\}, \tag{32}$$

where constant terms have been omitted. Therefore, solving (32) is equivalent to solving n_E one-dimensional problems for t_j, $j = 1, \ldots, n_E$

$$t_j^{(k+1)} \leftarrow \arg\min_{t \in \mathbb{R}} \left\{ \lambda \phi(|t|; a) - \rho_j^{(k)} t + \frac{\beta}{2} \left(t - (D\sigma^{(k)})_j \right)^2 \right\}. \tag{33}$$

After some algebraic manipulation, we can rewrite (33) in the following form

$$t_j^{(k+1)} \leftarrow \arg\min_{t \in \mathbb{R}} \left\{ \frac{\alpha}{2} \left(t - r_j^{(k+1)} \right)^2 + \phi(|t|; a) \right\}, \tag{34}$$

where

$$\alpha := \frac{\beta}{\lambda}, \qquad r_j^{(k+1)} := (D\sigma^{(k)})_j + \frac{1}{\beta} \rho_j^{(k)}. \tag{35}$$

Based on results in [[8], Prop.7.1], we can ensure strong convexity of the subproblems for the primal variables t_j, $j = 1, \ldots, n_E$, if and only if the following condition holds:

$$\alpha > a \Leftrightarrow \beta > a\lambda. \tag{36}$$

In case that (36) is satisfied, following [[8], Prop.7.1], the unique solutions of the strongly convex problems in (34) can be obtained in closed-form by applying the soft-thresholding operator:

$$t_j^{(k+1)} = \min\left\{\max\left\{\nu - \zeta/|r_j^{(k+1)}|, 0\right\}, 1\right\} r_j^{(k+1)}, \tag{37}$$

where $\nu = \frac{\alpha}{\alpha - a}$, and $\zeta = \frac{\sqrt{2a}}{\alpha - a}$.

5   Numerical Experiments

In this section we evaluate the proposed method on a set of synthetic 2D experiments. All examples but the last one, labeled as "heart and lungs", simulate a circular tank slice of unitary radius with a boundary ring of 16 electrodes; drive current value is set to $0.1\,mA$ and conductivity of the background liquid is set to $\sigma_0 = 0.02\,\Omega m^{-1}$. Measurements are simulated through *opposite injection - adjacent measurement* protocol via EIDORS software using a generic forward mesh of $n_T = 39488$ triangles and $n_V = 19937$ vertices. A coarse backward mesh was employed for inverse EIT solutions, with $n_T = 5056$ triangles without overlapping elements with the ones of the fine forward mesh so that to avoid the so called *inverse crimes* in EIT [3].

The performance is assessed both qualitatively and quantitatively. The quantitative analysis is performed via the following *slice metric*:

$$\epsilon_s = \frac{\|\sigma_{GT} - \sigma^*\|_2^2}{\|\sigma_{GT}\|_2^2}, \tag{38}$$

which measures how well the original conductivity distribution is reconstructed in case a ground truth conductivity distribution σ_{GT} is known. In order to allow for a mesh independent comparison, the conductivity distributions are evaluated on an image structure of dimension 576×576 pixels. In all the experiments and for all the algorithms, we hand-tuned their regularization parameter so as to achieve the lowest possible ϵ_s norm error.

In the last example we applied our framework to a more realistic dataset "heart and lungs" where the measurements are given in 2D circular tank with chest shaped conductivity targets in calibrated phantom at Rensselaer Polytechnic with ACT 3 EIT system, and kindly provided in [10]. This data have been measured with trigonometric injection current pattern over a boundary ring of 32 Electrodes. The tank width is $30\,cm$ in diameter; the electrodes are 2.5 cm long and $1.6\,cm$ tall. The saline conductivity was $424\,mS/m$. The agar targets of the "lungs" had a conductivity of $240\,mS/m$, and the "heart" was $750\,mS/m$. The experimental setting is illustrated in Fig. 6 (first column), thus imposing a piecewise-constant conductivity distribution. Since the conductivity distribution was assumed to be piecewise constant, the FB framework involved the R_{NL} non-linear regularizer, setting the parameter a accordingly to obtain a convex backward step.

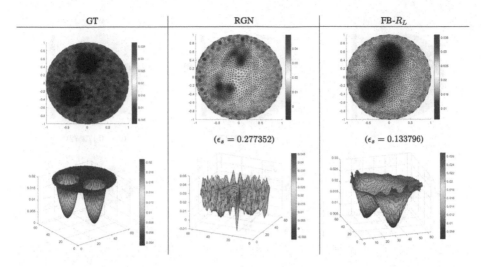

Fig. 1. Reconstructions using smooth R_L regularizer: GT (first column), RGN (second column), our (third column).

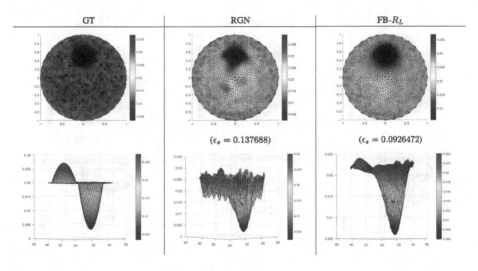

Fig. 2. Reconstructions using smooth R_L regularizer: GT (first column), RGN (second column), our (third column).

In all the examples we compared the Regularized Gauss Newton method (RGN) using Laplace prior with the proposed FB-R_L, or FB-R_{NL} algorithms. The regularization parameters λ for all the experiments have been chosen by trial and error to produce the best reconstructions in terms of ϵ_s. The setup is blind, that is no a priori information about the dimensions or locations of the inclusions is considered, only the kind of conductivity distribution is preliminary given such as *smooth* or *piecewise constant*. This simple a priori knowledge about

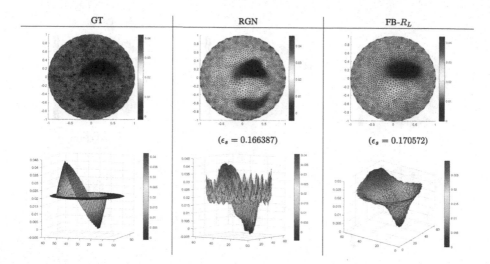

Fig. 3. Reconstructions using smooth R_L regularizer: GT (first column), RGN (second column), our (third column).

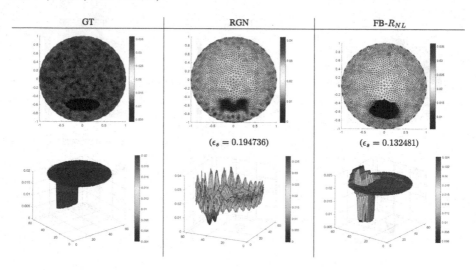

Fig. 4. Reconstructions using non-smooth R_{NL} regularizer: GT (first column), RGN (second column), our (third column).

the feature of the inclusions is used to select the regularizer $R(\sigma)$ in (15) as R_L or R_{NL}, respectively.

First, we compared the results obtained by RGN and FB-R_L algorithms when applied to the reconstruction of conductivity distribution characterized by smooth inclusions. The ground truth conductivity distributions are illustrated in Fig. 1, Fig. 2, Fig. 3 (first columns), and the reconstructed conductivities with RGN and FB-R_L are reported in Fig. 1, Fig. 2, Fig. 3 middle and right columns

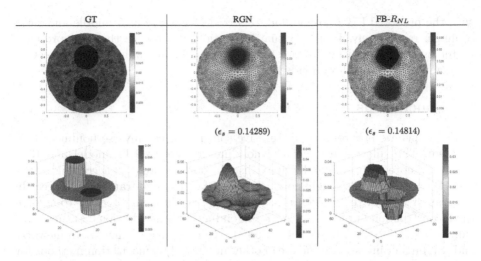

Fig. 5. Reconstructions using non-smooth R_{NL} regularizer: GT (first column), RGN (second column), our (third column).

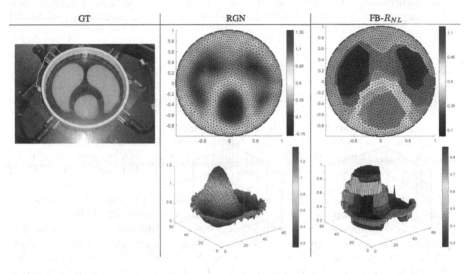

Fig. 6. Reconstructions of the "heart and lungs" example, using non-smooth R_{NL} regularizer: GT (first column), RGN (second column), our (third column).

respectively. Our approach provides always more stable results, which produce better reconstructions. The ϵ_s errors reported in the figures for each reconstruction confirm the improved accuracy of the proposed approach (Fig. 4).

In case the a priori knowledge of the conductivity distribution indicates that the sought anomalies are characterized by piecewise constant shape, the framework FB-R_{NL} is the most suitable to use. In the examples illustrated in Fig. 5, Fig. 6 we compare the reconstruction results obtained by the RGN algorithm with the proposed FB-R_{NL} method, reported in Fig. 5, Fig. 6 middle and right columns, respectively. From a visual inspection we observe that our proposal better preserve the piecewise constant shape of the anomalies with respect to the round-shaped results provided by RGN.

6   Conclusions

In this paper we proposed a FB iterative solution to the inverse nonlinear EIT problem. This allowed to treat the non-linearity of the EIT model and thus to obtain more accurate conductivity distributions. Moreover, according to the prior knowledge on the sought conductivities, the proposed framework can easily integrate a smooth regularizer as well as a sparsity-inducing penalty to enforce the reconstruction of piecewise constant conductivity distributions. Future work will address the theoretical analysis on the convergence of the FB framework, and will further investigate the use of other numerical regularization methods for nonlinear inverse problems to deal with the challenging nonlinear EIT problem.

Acknowledgment. This work was supported in part by the National Group for Scientific Computation (GNCS-INDAM), Research Projects 2020, and in part by MIUR RFO projects.

References

1. Blake, A., Zisserman, A.: Visual Reconstruction. Cambridge MIT Press, Cambridge (1987)
2. Borsic, A.: Regularisation methods for imaging from electrical measurements, Ph.D. thesis, School of Engineering, Oxford Brookes University, UK (2002)
3. Borsic, A., Graham, B.M., Adler, A., Lionheart, W.R.B.: In Vivo impedance imaging with total variation regularization. IEEE Trans. Med. Imaging 29(1), 44–54 (2010)
4. Calderón, A.P.: On an inverse boundary value problem. In: Seminar on Numerical Analysis and its Applications to Continuum Physics, pp. 65–73 (1980)
5. Cheng, K.-S., Isaacson, D., Newell, J.C., Gisser, D.G.: Electrode models for electric current computed tomography. IEEE Trans. Biomed. Eng. 36(9), 918–924 (1989)
6. Cheney, M., Isaacson, D., Newell, J., Simske, S., Goble, J.: NOSER: an algorithm for solving the inverse conductivity problem. Int. J. Imaging Syst. Technol. 2, 66–75 (1990)
7. Combettes, P.L., Wajs, V.R.: Signal recovery by proximal forward-backward splitting. Multiscale Model. Simul. 4, 1168–1200 (2005)
8. Huska, M., Lanza, A., Morigi, S., Selesnick, I.: A convex non-convex variational method for the additive decomposition of functions on surfaces. Inverse Prob. 35(12), 124008–124041 (2019)
9. Huska, M., Lazzaro, D., Morigi, S., Samorè, A., Scrivanti, G.: Spatially-adaptive variational reconstructions for linear inverse electrical impedance tomography. J. Sci. Comput. 84(3), 1–29 (2020). https://doi.org/10.1007/s10915-020-01295-w
10. Isaacson, D., Mueller, J.L., Newell, J.C., Siltanen, S.: Reconstructions of chest phantoms by the d-bar method for electrical impedance tomography. IEEE Trans. Med. Imaging 23(7), 821–828 (2004)

11. Jauhiainen, J., Kuusela, P., Seppanen, A., Valkonen, T.: Relaxed gauss-newton methods with applications to electrical impedance tomography. SIAM J. Imaging Sci. **13**(3), 1415–1445 (2020)
12. Lionheart, W.: EIT reconstruction algorithms: pitfalls, challenges and recent developments. Physiol. Meas. **25**(1), 125 (2004)
13. Liu, D., Kolehmainen, V., Siltanen, S., Seppänen, A.: A nonlinear approach to difference imaging in EIT; assessment of the robustness in the presence of modelling errors. Inverse Prob. **31**(3), 035012:1-035012:25 (2015)
14. Lechleiter, A., Rieder, A.: Newton regularizations for impedance tomography: convergence by local injectivity. Inverse Prob. **24**(6), 065009065009 (2008)
15. Nikolova, M.: Estimation of binary images by minimizing convex criteria. In: Proceedings of the IEEE International Conference on Image Processing, vol. 2, p. 108 (1988)
16. Pes, F., Rodriguez, G.: The minimal-norm Gauss-Newton method and some of its regularized variants. Electron. Trans. Numer. Anal. **53**, 459–480 (2020)
17. Vauhkonen, M., Vadasz, D., Karjalainen, P.A., Somersalo, E., Kaipio, J.P.: Tikhonov regularization and prior information in electrical impedance tomography. IEEE Trans. Med. Imaging **17**(2), 285–293 (1998)
18. Zhang, C.-H.: Nearly unbiased variable selection under minimax concave penalty. Ann. Stat. **38**(2), 894–942 (2010)

International Workshop on Advances in information Systems and Technologies for Emergency management, risk assessment and mitigation based on the Resilience concepts (ASTER 2021)

Seismic Risk Simulations of a Water Distribution Network in Southern Italy

Maurizio Pollino[1]($\boxtimes$) (iD), Antonio Di Pietro[1], Luigi La Porta[1], Grazia Fattoruso[2],
Sonia Giovinazzi[1] (iD), and Antonia Longobardi[3] (iD)

[1] ENEA, TERIN-SEN-APIC Lab, Casaccia Research Centre, Rome, Italy
{maurizio.pollino,antonio.dipietro,luigi.laporta,
sonia.giovinazzi}@enea.it
[2] ENEA, TERIN-FSD-SAFS Lab, Portici Research Centre, Portici (NA), Italy
grazia.fattoruso@enea.it
[3] Department of Civil Engineering, University of Salerno, Fisciano, Italy
antonia.longobardi@unisa.it

Abstract. Critical Infrastructure Protection (CIP) is at the heart of the European and International Agenda. Such issue should be handled also with adequate methodologies and tools to carry out effective prevention and protection of the assets (technological, industrial, strategic) from impacts arising from natural disasters that can harm them and reduce the continuity of services delivered. In this general framework, the RAFAEL Project aims at integrating ad hoc technologies into a comprehensive Decision Support System called CIPCast, which has been implemented to support risk analysis and emergency management in the case of natural hazardous events. A specific module is devoted to the earthquake simulation (CIPCast-ES). The Castel San Giorgio water distribution network (located in Campania Region, Southern Italy) was selected to analyse the seismic risk by simulating four earthquake relevant events, occurred in the past in the area of study. Thus, four different scenarios were produced and, for each of them, two types of results were obtained for each pipe element of the water network, expressed respectively in terms of probability of failure and damage status.

Keywords: Seismic risk · Simulation · Water network · Critical infrastructure protection · CIPCast · DSS

1 Introduction

1.1 The Critical Infrastructure Protection (CIP)

Critical Infrastructure Protection (CIP) is at the heart of the European and International Agenda since several years. Recent data [1] show an increase in the number of disasters affecting Critical Infrastructures (CI), capable of producing huge damage and draining important financial resources. These events, furthermore, can have high costs in terms of casualties. The use of more advanced tools and technologies to promote the preparedness and, consequently, the reduction of global damage (especially in terms of human

© Springer Nature Switzerland AG 2021
O. Gervasi et al. (Eds.): ICCSA 2021, LNCS 12951, pp. 655–664, 2021.
https://doi.org/10.1007/978-3-030-86970-0_45

casualties as well as short- and long-term economic losses) is a strategy that cannot and should not be overlooked.

As far as CIP is concerned, the European Union has since long identified its own line of intervention. The Directive 2008/114/EC of the Council of the EU has identified the importance of protection of a number of infrastructural assets. As stated by the art. 2 of such Directive, a critical infrastructure is defined as *"an element, a system or part of this located in the Member States that is essential for the maintenance of the vital functions of society, health, safety and the economic and social well-being of citizens whose damage or destruction would have a significant impact in a Member State due to the impossibility of maintaining these functions"*. The Directive was subsequently transposed in Italy through the Legislative Decree 65/2011. Finally, the recent proposal (COM/2020/829 final) for a new EU Directive on the "Resilience of Critical Entities" (Directive "RCE", update of Directive 114/2008/EC) will impose on the Member States a stronger governance of the management of the security of CI assets.

Once defined at a legal and institutional level, the CIP problem should also be handled from the technological point of view, by supporting Public Administration and Operators to carry out effective prevention and protection of the assets (technological, industrial, strategic) from impacts arising from natural disasters and other events that can harm them and reduce the continuity of services delivered.

In particular, CI Systems include not only utility networks such as energy, water, telecommunications and transportation systems, but also discrete critical facilities such as hospitals, schools, ports and airports among others. As analysed in several studies [2–4], these infrastructures have become increasingly complex and interdependent.

A typical example of interdependency is that of CI water on CI electricity: there are numerous devices that require the power supply for the relative functioning of the water service. However, the severity and extent of the effects that critical situations in electrical CI can cause in strongly interconnected systems, such as those dedicated to the water service provision, are not yet understood [2]. A methodology to assess the impacts on a water network coming from failures occurred on the power network has been presented by [5, 6]. From such studies it is possible to deduce that the impact assessment is strongly dependent on the characteristics of the water system, on the data availability and – above all – on the CI level of interdependency. These features make the problem hard to be well-defined.

On the other hand, such aspect has induced a high level of vulnerability to a series of natural or non-natural events, which can cause damage or destruction [7, 8]. For an effective protection of the CI, thus, the development of new approaches and methodologies is needed, in order to reduce the vulnerabilities, mitigate the risks and face new threats to which these complex systems, essential for the everyday life, are exposed. There is a large number of studies that illustrates the effects of natural phenomena on CI systems. In particular, the effects that earthquakes have on infrastructures were widely studied in the literature [9–16].

The study described in the present paper focuses on the seismic risk and the impacts of earthquakes on a Water Distribution Network (WDN). To this end, an interactive simulation model was developed to simulate different seismic events, actually occurred in the past within the area of study, assessing the damage levels virtually obtained. Such model was implemented in a Decision Support System (DSS) platform (developed in the framework of the RAFAEL Project, described in Sect. 1.2) and was applied to an actual metropolitan WDN operating in the Southern Italy (Sect. 1.4) to evaluate various damage scenarios in consequence of each simulated earthquake.

1.2 The Context of Application: The RAFAEL Project

The study and the related applicative case presented in this paper have been carried out in the framework of the RAFAEL Project (*"system for Risk Analysis and Forecast for critical infrastructure in ApenninEs dorsaL regions"*, PNR 2015–2020, co-funded by the Italian Ministry of University and Research, Grant no. ARS01_00305, scheduled to end in November 2021, https://www.progetto-rafael.it/). As part of the technological developments undertaken over the last few years, the Project aims at integrating ad hoc technologies, developed within its activities, into a comprehensive DSS platform [17] (namely CIPCast and described in the next Section), which will become the reference platform to provide services to companies and to the Public Administration through the constitution of EISAC.IT programme (www.eisac.it) that will carry out its operational deployment in Italy. RAFAEL deals with the management of numerous CI assessing the relevant damage deriving from natural disastrous events, and evaluating the impact on the services and the consequences on the population and on the infrastructural system. Among its objectives, the RAFAEL project is going to extend the current CIPCast capabilities, by implementing new analytical and simulation tools to further improve the CIP tasks.

1.3 The CIPCast Platform

CIPCast is a DSS platform, which has been conceived for supporting risk analysis and emergency management in the case of natural hazardous events [17]. It includes and manages more than 400 geospatial data layers for spatial analysis, protection of CI and increasing urban/infrastructures resilience. CIPCast can run in operational mode (on a 24/7 basis, as an alert system) and in an off-line mode by simulating events. In the first case, CIPCast gets external data from many different sources (e.g. seismic network, meteo/hydrological models), to establish the current conditions [18]. Then, it produces an expected "damage scenario" for each element of the monitored CI, depending on the type of event expected (and its intensity), considering the specific vulnerabilities of infrastructure elements (pipes, substations, etc.). In the simulating mode CIPCast can be exploited as a stress tester enabling to set-up synthetic natural hazards (e.g. earthquakes) and assessing the resulting chain of events [19, 20]. Moreover, CIPCast can be also exploited as an interactive application for supporting multi-criteria decision processes [21]. All the services are accessible by means of the geographical interface of CIPCast, a WebGIS application that can be used as graphical front-end of data and information provided and/or produced.

   The specific module devoted to the earthquake simulation and risk analysis is called CIPCast-ES (Earthquake Simulator). It has been exploited to carry out the simulations performed during the case study here presented.

1.4 The Case Study

The Castel San Giorgio water distribution network (WDN) was the CI chosen within the RAFEL project in order to define and simulate the seismic risk scenarios. The network (Fig. 1) is located in the Campania Region (Italy) and serves a population of about 14,000 inhabitants, over an area of approximately 13 km². It distributes approximately 60 l/s daily from three wells, which guarantee the entire water supply, through a network of around 60 km of pipelines. More information about the characteristics of the WDN can be found in [6].

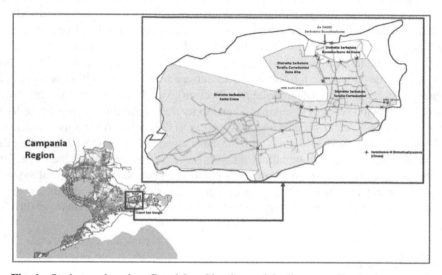

Fig. 1. Study area location: Castel San Giorgio municipality water distribution network

2 Methodology

2.1 Overview

To assess the damage, considering the various approaches analysed and proposed in literature during the last years [16], the following steps were implemented and defined in CIPCast-ES (Fig. 2):

1. Classification of the infrastructure elements according to a defined taxonomy;
2. Simulation set-up: parameters input (e.g. epicentre, magnitude, etc.);
3. Hazard assessment: ground motion intensity measures elaboration;
4. Repair Rate (RR) assessment as a function of the hazard;
5. Probability of failure (P_f) assessment as a function of RR and pipe length;
6. Damage assessment, in order to assign a specific damage level/rate as a function of P_f to each CI element (e.g. a single pipe track).

Then, the CIPCast-ES module (Fig. 3) was exploited to set-up and run different earthquake simulations and, consequently, to produce the relative damage scenarios.

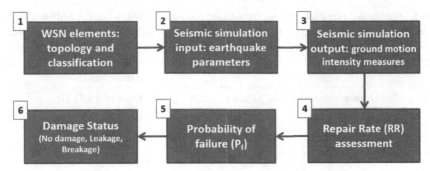

Fig. 2. CIPCast-ES: simulation workflow

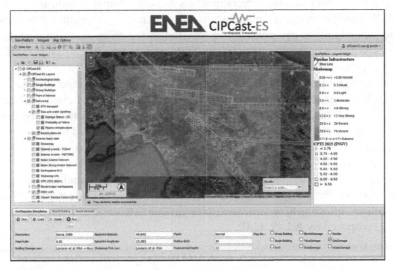

Fig. 3. CIPCast-ES interface: example of visualization of basic data (e.g. seismogenic sources, historical earthquake catalogue) and of the results obtained after a simulation (e.g. Shakemap in terms of peak ground acceleration, PGA)

2.2 Seismic Simulations and Scenarios

In the case-study related to the Castel San Giorgio WDN, once its elements (buried pipes) have been suitably classified and stored into the CIPCast Geodatabase, it has been possible to perform the simulations. The simulations were set-up through the CIPCast-ES user-friendly wizard. The following parameters has been provided as inputs of the application: epicentre coordinates (in latitude and longitude), hypocentre depth (in km) and magnitude. To this end, four past occurred events were selected (their characteristics are reported in Table 1), as significant earthquakes in terms of magnitude and/or proximity to the geographical area where the Castel San Giorgio WDN is located.

Then, it has been selected the area of interest of the simulation, the CI typology to be processed amongst those available (water/gas pipelines, power network, buildings) and finally the GMPE (Ground Motion Prediction Equation) [22] to be used in order

Table 1. Earthquake event selected for the simulations (source: INGV, Italian National Institute for Geophysics and Volcanology, CPTI15 v3.0 - Parametric Catalogue of Italian Earthquakes)

Event	Date	Magnitude	Epicentre coordinates
Penisola Sorrentina	07.31.1561	5.56	40.68 N–14.71 E
Castel San Giorgio	04.27.1930	4.98	40.78 N–14,69 E
Irpinia-Basilicata	11.23.1980	6.81	40.84 N–15.28 E
Sant'Arcangelo T.	09.27.2012	4.2	41.18 N–14.92 E

to describe the ground shaking effects in terms of peak ground motion intensity measures. GMPEs or "attenuation" relationships, allow to model the ground shaking and its associated uncertainty at any given site or location, based on an earthquake magnitude, source-to-site distance, local soil conditions, fault mechanism, etc. Different GMPEs are currently implemented CIPCast-ES.

As proposed by FEMA [23] and by the A.L.A. Guidelines [24] and applied in many similar studies (see [15] for an overview), seismic damages to buried pipes can be expressed by the "Repair Rate" (*RR*) per unit length of pipe, that is the rate between the number of repairs and the length (in kilometres) of a pipe exposed to seismic hazard. This number is a function of pipe material, joint type, soil conditions and diameter size, and of ground shaking, expressed in terms of *PGA* (Peak Ground Acceleration) or *PGV* (Peak Ground Velocity, as used in the present analysis), or else ground failure expressed in terms of *PGD* (Permanent Ground Deformation). Then, to evaluate the *RR* for the WDN pipelines, referring to seismic wave propagation in terms of *PGV*, we have applied the Eq. (1) proposed by A.L.A. [24]:

$$RR = K_1 0.002416 \, PGV \tag{1}$$

where K_1 is a coefficient that considers pipe material, joint type, soil type and conditions and diameter size (determined experimentally and reported in specific tables provided by A.L.A. [24]). *PGV* was evaluated using the GMPE proposed by Bindi et al. [25], which allowed to obtain the spatial distribution of the geometrical mean of the horizontal components of the *PGV*. Subsequently, from the *RR* (representing the number of leaks/breaks per pipe length) the failure probability of a single pipe was calculated according the Eq. (2) proposed (among the others) by [24] and [26]:

$$\overline{P_f} = 1 - e^{-RRL} \tag{2}$$

The failure probability of a single pipe is equal to one minus the probability of zero breaks along the pipe. In addition, a damage status was derived for each single pipe by means of the approach proposed by [27], which allows to classify the pipes according three different levels of damage: "No Damage", "Leakage" and "Breakage".

3 Results

By simulating in CIPCast-ES the four earthquake events selected (Table 1), four different seismic scenarios were obtained, according to the approach described in the previous Section. For each scenario, two types of results were produced for each pipe element of the Castel San Giorgio WDN: i) probability of failure; ii) damage status.

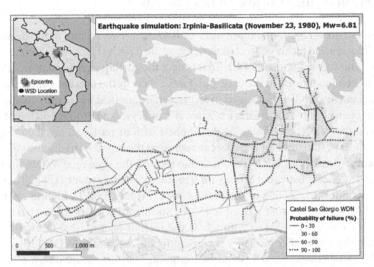

Fig. 4. Earthquake simulation: Irpinia-Basilicata 1980 (Mw = 6.81). Example of scenario produced in terms of probability of failure (in percentage)

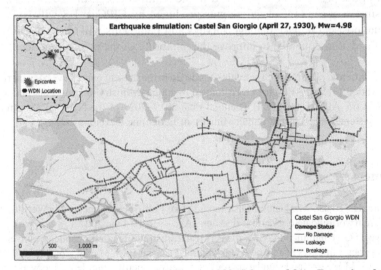

Fig. 5. Earthquake simulation: Irpinia-Basilicata 1980 (Mw = 6.81). Example of scenario produced in terms of damage status ("No Damage", "Leakage" and "Breakage")

As results of the above described simulations, in particular, Fig. 4 shows the probability of failure (expressed in percentage) for each branch of the Castel San Giorgio WDN, in consequence of the ground shaking (expressed in terms of *PGV*) obtained by simulating the Irpinia-Basilicata 1980 earthquake ($M_w = 6.81$, see Table 1). Moreover, still considering the same seismic event, in Fig. 5 is depicted the potential damage status obtained, expressed according to the above-mentioned three levels [27].

4   Conclusions and Future Developments

The activities carried out in the framework of RAFAEL Project aim at the development of methodologies and technological tools (such as the CIPCast platform) for the analysis, monitoring and risk assessment of CI. To this end, the earthquake simulator CIPCast-ES module has been exploited in the case study described in the present paper, focused on the Castel San Giorgio WDN, located in an area characterised by a high seismic hazard. Thus, it has been possible to perform a seismic risk analysis and produce different damage scenarios on the WDN, by simulating the impacts of four pasty occurred earthquakes.

The outputs of the simulation model can represent the inputs of hydraulic simulations, to be used for describing the hydraulic behaviour of the damaged WDN, in order to evaluate and suggest post-event actions for a prompt system recovery [28, 29]. Although this aspect is out of the scope of the present paper, it has been tackled in another task of the RAFAEL Project, as described in [5, 6]. Simulated scenarios can also be used to analyse cascading effects: for example, how the lack of energy (due to a failure on the power network) can further influence the operational level of the WDN [30].

Finally, some considerations about the on-going and the future developments. The CIPCast-ES workflow will be further improved by considering the approaches developed in similar studies, among others [13, 15, 31–34]. Contextually, other earthquakes occurred in the past will be reproduced in the simulator (e.g. the 2016–2017 Central Italy seismic sequence [35]): the outputs produced by CIPCast-ES will be compared with actual post-earthquake damage surveys and data. This will allow to calibrate the simulator and, consequently, to improve the reliability of the results produced.

Acknowledgments. The research activities described in the present paper have been carried out in the framework of: i) RAFAEL project, co-funded by Italian Ministry of University and Research, MUR, Grant no. ARS01_00305; ii) Triennial Plan 2019–2021 of the National Research on the Electrical System, funded by the Italian Ministry of Economic Development.

References

1. Ritchie, H., Roser, M.: Natural Disasters (2014). Published online at OurWorldInData.org. https://ourworldindata.org/natural-disasters. Accessed 19 June 2021
2. Baloye, D.O., Palamuleni, L.G.: Urban critical infrastructure interdependencies in emergency management: findings from Abeokuta, Nigeria. Disaster Prev. Manag. **26**(2), 162–182 (2017)
3. Griot, C.: Modelling and simulation for critical infrastructure interdependency assessment: a meta-review for model characterization. Int. J. Crit. Infrastruct. **6**(4), 363–379 (2010)

4. Chou, C.-C., Tseng, S.-M.: Collection and analysis of critical infrastructure interdependency relationships. J. Comput. Civ. Eng. **24**(6), 539–547 (2010)

5. Longobardi, A., et al.: Water distribution network perspective in RAFAEL, a system for critical infrastructure risk analysis and forecast. In: 21th ICCSA 2021 Conference, RRS2021 Workshop, Cagliari, Italy. LNCS (2021, Accepted paper)

6. Ottobrino, V., Esposito, T., Locoratolo, S.: Towards a smart water distribution network for assessing the effects by critical situations in electric networks. The pilot case of Castel San Giorgio. In: 21th ICCSA 2021 Conference, RRS2021 Workshop, Cagliari, Italy. LNCS (2021, Accepted paper)

7. Luiijf, E., Nieuwenhuijs, A., Klaver, M., van Eeten, M., Cruz, E.: Empirical findings on critical infrastructure dependencies in Europe. In: Setola, R., Geretshuber, S. (eds.) CRITIS 2008. LNCS, vol. 5508, pp. 302–310. Springer, Heidelberg (2009). https://doi.org/10.1007/978-3-642-03552-4_28

8. Rosato, V., et al.: A decision support system for emergency management of critical infrastructures subjected to natural hazards. In: Panayiotou, C.G.G., Ellinas, G., Kyriakides, E., Polycarpou, M.M.M. (eds.) CRITIS 2014, LNCS, vol. 8985, pp. 362–367 (2016). https://doi.org/10.1007/978-3-319-31664-2_37

9. Omidvar, B., Hojjati Malekshah, M., Omidvar, H.: Failure risk assessment of interdependent infrastructures against earthquake, a Petri net approach: case study-power and water distribution networks. Nat. Hazards **71**(3), 1971–1993 (2014)

10. De Risi, R., De Luca, F., Kwon, O.S., Sextos, A.: Scenario-based seismic risk assessment for buried transmission gas pipelines at regional scale. J. Pipeline Syst. Eng. Pract. **9**(4), 1–12 (2018)

11. Eskandari, M., Omidvar, B., Modiri, M., Nekooie, M.A., Alesheikh, A.A.: Geospatial analysis of earthquake damage probability of water pipelines due to multi-hazard failure. Int. J. Geo-Inf. **6**, 169 (2017)

12. Giovinazzi, S., Brown, C., Seville, E., Stevenson, J., Hatton, T., Vargo, J.J.: Criticality of infrastructures for organisations. Int. J. Crit. Infrastruct. **12**(4), 331–363 (2016)

13. Toprak, S., Taskin, F.: Estimation of earthquake damage to buried pipelines caused by ground shaking. Nat. Hazards **40**(1), 1–24 (2007)

14. O'Rourke, T.D., Jeon, S.: Factors affecting the earthquake damage of water distribution systems, optimizing post-earthquake lifeline system reliability. In: Proceedings of the 5th U.S. Conference on Lifeline Earthquake Engineering, Seattle, USA, pp. 379–388 (1999)

15. Makhoul, N, Navarro, Lee, C.J.S., Gueguen, P.: A comparative study of buried pipeline fragilities using the seismic damage to the Byblos wastewater network. Int. J. Disaster Risk Reduct. **51**, 75–88 (2020)

16. Pineda-Porras, O.A., Najafi, M.: Seismic damage estimation for buried pipelines - challenges after three decades of progress. J. Pipeline-Syst.-Eng. Pract. - ASCE **1**, 1–19 (2010)

17. Di Pietro, A., Lavalle, L., La Porta, L., Pollino, M., Tofani, A., Rosato V.: Design of DSS for supporting preparedness to and management of anomalous situations in complex scenario. In: Setola, R., Rosato, V., Kyriakides, E., Rome, E. (eds.) Managing the Complexity of Critical Infrastructures. Studies in Systems, Decision and Control, vol. 90, pp. 195–232. Springer International Publishing, Cham (2016)

18. Taraglio, S., et al.: Decision support system for smart urban management: resilience against natural phenomena and aerial environmental assessment. Int. J. Sustain. Energy Plan. Manag. **24**, 135–146 (2019)

19. Matassoni, L., Giovinazzi, S., Pollino, M., Fiaschi, A., La Porta, L., Rosato, V.: A geospatial decision support tool for seismic risk management: Florence (Italy) case study. In: Gervasi, O., et al. (eds.) ICCSA 2017. LNCS, vol. 10405, pp. 278–293. Springer, Cham (2017). https://doi.org/10.1007/978-3-319-62395-5_20

20. Giovinazzi, S., Di Pietro, A., Mei, M., Pollino, M., Rosato, V.: Protection of critical infrastructure in the event of earthquakes: CIPCast-ES. In: Proceedings of XVII ANIDIS Conference, Pistoia, Italy (2017)
21. Modica, G., et al.: Land suitability evaluation for agro-forestry: definition of a web-based multi-criteria spatial decision support system (MC-SDSS): preliminary results. In: Gervasi, O., et al. (eds.) ICCSA 2016. LNCS, vol 9788, pp. 399–413. Springer, Cham (2016). https://doi.org/10.1007/978-3-319-42111-7_31
22. Douglas, J.: GMPE compendium. http://www.gmpe.org.uk/. Accessed 24 May 2021
23. Federal Emergency Management Agency. HAZUS MH 2.0 Earthquake Technical Manual; Department of Homeland Security, Washington, DC, USA (2011)
24. American Lifelines Alliance. American Lifelines Alliance: seismic fragility formulations for water systems-guideline and appendices (2001)
25. Bindi, D., Pacor, F., Luzi, L., et al.: Ground motion prediction equations derived from the Italian strong motion database. Bull. Earthq. Eng. 9, 1899–1920 (2011)
26. Fragiadakis, M., Christodoulou, S.E.: Seismic reliability assessment of urban water networks. Earthq. Eng. Struct. Dyn. 43, 357–374 (2014)
27. Choi, J., Yoo, D., Kang, D.: Post-earthquake restoration simulation model for water supply networks. Sustainability 10, 3618 (2018)
28. Fattoruso, G., et al.: Valutazione dell'interdipendenza tra rete elettrica e rete di distribuzione idrica nell'analisi del rischio sismico in ambiente ArcGIS. In: Proceedings of Conference ESRI ITALIA 2019 "The Science of Where Envisioning Where Next", Roma, Italy (2019)
29. Fattoruso, G., et al.: Modeling electric and water distribution systems interdependences in urban areas risk analysis. In: Proceedings of 2nd International Conference Citizen Observatories for natural hazards and Water Management Venice, Italy (2018)
30. Rosato, V., Di Pietro, A., Kotzanikolaou, P., Stergiopoulos, G., Smedile, G.: Integrating resilience in time-based dependency analysis: a large-scale case study for urban critical infrastructures. In: Issues on Risk Analysis for Critical Infrastructure Protection. IntechOpen (2021)
31. Gehl, P., Matsushima, S., Masuda, S.: Investigation of damage to the water network of Uki City from the 2016 Kumamoto earthquake: derivation of damage functions and construction of infrastructure loss scenarios. Bull. Earthq. Eng. 19(2), 685–711 (2020). https://doi.org/10.1007/s10518-020-01001-z
32. Lanzano, G., Salzano, E., Santucci de Magistris, F., Fabbrocino, G.: Seismic vulnerability of gas and liquid buried pipelines. J. Loss Prev. Process Ind. 28, 72–78 (2014)
33. Kongar, I., Esposito, S., Giovinazzi, S.: Post-earthquake assessment and management for infrastructure systems: learning from the Canterbury (New Zealand) and L'Aquila (Italy) earthquakes. Bull. Earthq. Eng. 15(2), 589–620 (2015). https://doi.org/10.1007/s10518-015-9761-y
34. Tabucchi, T., Davidson, R., Brink, S.: Simulation of post-earthquake water supply system restoration. Civ. Eng. Environ. Syst. 27(4), 263–279 (2010)
35. INGV Working Group on Central Italy Earthquake. Report on the 2016–2017 Central Italy Seismic Sequence (update 2 February 2017). https://doi.org/10.5281/zenodo.267984.Accessed 20 June 2021

Security and Resilience in Small Centres of Inland Areas

Priscilla Sofia Dastoli[✉] 🆔 and Piergiuseppe Pontrandolfi 🆔

University of Basilicata, DiCEM, CdS Architecture, Via Lanera, 20, 75100 Matera, Italy
{priscilla.dastoli,piergiuseppe.pontrandolfi}@unibas.it

Abstract. Starting from a theoretical reflection on current planning tools and on the concept of urban resilience, the paper supports a new methodological approach that attempts to offer a more structural and not only operational characterization to the Emergency Plans. The analysis of the physical and functional relationships between the various strategic elements of the territory shifts the focus of the emergency from a punctual approach to a more integrated one, capable of directing future decisions relating to the urban planning of the territory in a more appropriate way specific to spatial planning. In fact, although over the years the issues of risk and safety have contributed to orienting the methods of governing the territories, still today, when an adverse event occurs, clear unresolved critical issues emerge that push urban planning to question the new challenges to be faced undertake, especially in terms of emergency management. The focus of the municipality of Moliterno, in the Alta Val d'Agri district (Basilicata, Italy), is shown as a practical example of improvement in seismic emergency management, despite the presence of critical issues, such as the historic centre and the accentuated orography.

Keywords: Security · Emergency management · Inland areas

1 Introduction

Urban planning plays an important role in emergency management and can pursue regeneration and sustainability criteria [1–7], placing them in a resilient perspective of the iland areas, taking into account above all that the dynamic nature of territorial systems leads to awareness that there are risks that cannot be avoided, rather, they must be accepted and managed in the different time phases (ex-ante, in itinere and ex-post) [8, 9]. It is understood that, in order to appropriately address the risks present in the territory, it is first of all necessary to reduce the vulnerability of the reference urban system, and to do so it is also necessary to identify an effective and efficient instrumentation system capable of planning, planning and designing regeneration activities sustainable urban areas, also in terms of emergency management [10–12].

The urban seismic risk assessment is prepared to carry out prevention and mitigation actions for the two main phases following the seismic event: the emergency phase, immediately following the seismic event, and the phase of resumption of activities and reconstruction, in the short-medium term following the event [13, 14].

© Springer Nature Switzerland AG 2021
O. Gervasi et al. (Eds.): ICCSA 2021, LNCS 12951, pp. 665–677, 2021.
https://doi.org/10.1007/978-3-030-86970-0_46

In order to ensure the operation of the system itself after the earthquake, the analysis of the Limit Condition for Emergency (CLE) is introduced in the OPCM 4007/12, which has the purpose of verifying the main physical elements of the management system emergencies defined in the civil protection plan (coordination places, emergency areas and connecting infrastructures).

The study on the Minim urban structure (SUM), on the other hand, represents the modality through which systems of routes, spaces, urban functions and strategic buildings are organized in advance for the response of the settlement to the earthquake, in order to allow a rapid resumption of urban activities after the occurrence of the calamitous event. The SUM raprasents a tool to assess urban resilience, that is the ability of a system to restore its functionality after having suffered an external impact [15].

The main difference is that the SUM overcomes the objective of simple emergency management to also aim at the possibility of ensuring the resumption of the main urban functions [16].

The CLE Analysis and the SUM study come from two different institutional referents, because theoretically they refer to two different moments of post-earthquake intervention; the first comes from the work of the Civil Protection, the second, on the other hand, is a method strictly linked to regional and local planning, therefore possibly provided for in the regional urban planning laws [17].

The innovative contribution that is intended to be illustrated in the following paragraphs, and in particular in the case of experimentation by the municipality of Moliterno, relates to a possible integration between CLE and SUM. To carry out this operation, it is necessary to organize phases that are strictly dependent on each other. It was a question of understanding which elements of CLE and SUM can integrate and allow the achievement of the common goal for the mitigation of seismic risk.

2   The Urban Layout of Moliterno Municipality

To introduce a new methodological approach for emergency management, it is essential to implement a phase of knowledge of the housing fabric, because each inhabited area is unique and therefore reacts subjectively to adverse events, such as an earthquake. The available urban planning tools are analyzed, both in terms of planning and civil protection plans, compared with urban fabric growth [18–20].

The municipalities of the inland areas, such as Moliterno, have a small extension of the housing fabric, since in general the population is less than 5 thousand inhabitants. In the case of the municipality of Moliterno, the main settlement stands on a promontory at an altitude of 880 m above sea level, overlooking the bed of the Sciaura stream. The building fabric is extremely compact in the medieval part, forming a walled curtain, and less so in the more recent area. Access to the town is from the north and west, through the main road axes, and from the south through a road that develops consistently with the main elevations. The State Road 103 connecting the Fondovalle dell'Agri and the A3 Salerno - Reggio Calabria motorway, constitutes the main road network [21].

The historic centre (see Fig. 1) stands in a prominent position with respect to the rest of the town, dominated by the presence of the castle. The edges are compact, consisting of steep ridges on three sides (north, east and south) while to the west the historic centre has no clear physical edges; the historic centre is accessed through paths that are partly accessible to vehicles and partly pedestrian. Due to the presence of buildings of historical architectural value and significant elements, the historic centre, in a good state of conservation, can be considered significant from the point of view of value and quality.

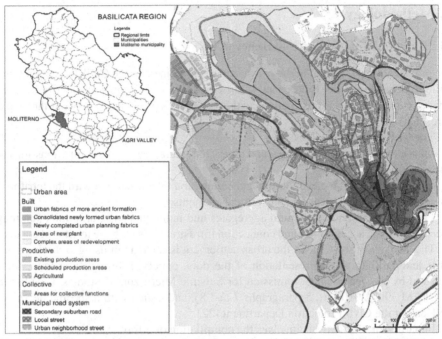

Fig. 1. Framing of the municipality of Moliterno in the Basilicata region *(left)* and identification of the main fabric of the settlement indicated in the urban planning tool *(right)*.

There are no hamlets in the open area and there is a moderate trend towards Parsee settlement. As for the functional characteristics of the settlement system, Moliterno winds between the inhabited centre and the two industrial/artisan areas located one in the terminal part of the town in the western direction (almost completely implemented), the other in the flat area in the north-est direction (only partially implemented). All services and equipment are located outside the historic centre, in the external area to the north, in a hinged area between the historic centre and the more recent expansions. The functions present are located in a compact fabric characterized by an irregular road network in which the original typological and architectural characteristics are recognizable, while the more recent part is characterized by a less regular medium-density fabric developed along the main access road to the inhabited centre. The settlement under construction, on the other hand, stands on the secondary viability in the western direction adjacent to the saturated production/artisan area, has a low-density linear development urbanistic-building fabric which includes areas not implemented in the current PRG intended mainly to the residence.

The periurban and extra-urban area of Moliterno has important elements of morphological-landscape relevance due to the presence of areas intended for the conservation of biodiversity and the production of eco-compatible resources due to the presence of natural and semi-natural ecosystems, as well as for the elements of the network hydrographic.

3 Methodology

3.1 Limit Condition of the Emergency (CLE)

The organizational question of the rescue action is pertinent to the Civil Protection Plan, while the analysis of the Emergency Limit Condition (CLE) aims to identify suitable physical structures and infrastructures within which it is possible to carry out the action of rescue.

The CLE analysis involves:

- the identification of buildings and areas that guarantee strategic functions for the emergency;
- the identification of accessibility and connection infrastructures with the territorial context, buildings and areas and any critical elements;
- the identification of structural aggregates and individual structural units that may interfere with the accessibility and connection infrastructures with the territorial context.

The analysis of the CLE of the urban settlement is carried out using archiving standards and cartographic representation of the data, collected through a specific form prepared by the Technical Commission for Seismic Micro-zoning studies, established by OPCM 3907/2010 (art. 5 paragraphs 7 and 8), and issued with a specific decree of the Head of the Civil Protection Department [22].

In particular, the analysis involves the compilation of 5 forms:

i.   the ES Strategic Building,
ii. AE Emergency Area,
iii. AC Infrastructure Accessibility/Connection,
iv. AS Structural Aggregate,
v.   US Structural Unit.

The main task of the CLE check is to ensure that each settlement has a system of elements capable of supporting the immediate action of the Civil Protection, whose fundamental principles are linked to the speed, efficiency and effectiveness of the intervention.

The CLE verification can represent an element of interest within the preparation of the SUM; the CLE, in fact, although based on the principles of the organization of the emergency, may be able to provide both useful indications for the rescue action and give indications to ordinary planning and specific risk reduction interventions contents.

It therefore emerges that the SUM cannot be reduced to the contents of a civil protection plan, rather it is the way to translate the objectives and contents of a civil protection plan into urban planning terms.

The Moliterno CLE system (see Fig. 2) includes n. 6 Strategic buildings (ES) and n. 7 Emergency Areas (AE). The ES who participate in emergency management are:

1) the Town Hall (headquarters of the COC) - Police Command; 2) the Barracks of the Ca-rabinieri; 3) the Barracks C.F.S.; 4) the 118 station; 5) the alternative headquarters of the COC (ITCG - High School); 6) the Traffic Police Station (Pol-strada). The AE, including Hospitalization areas and Storage areas, which participate in emergency management are: 1) municipal sports field (Hospitalization); 2) square in front of the municipal sports field (Shelter); 3) area adjacent to the municipal sports field (Stacking); 4) area behind the municipal sports field (Stacking); 5) area behind the municipal sports field (Stacking); 6) area behind the municipal sports field (Stacking - covered); 7) area behind the municipal sports field/in front of the gym (Stacking).

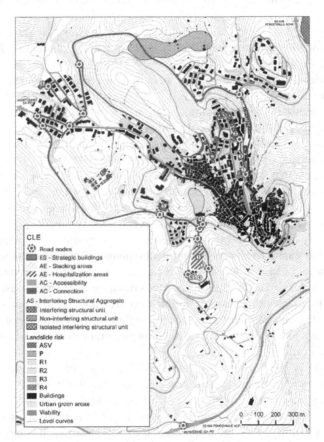

Fig. 2. Map illustrating the configuration of the Limit Condition for the Emergency for the municipality of Moliterno.

In addition to the ES and AE, the accessibility and connection infrastructures (AC) and any critical elements interfering with them belong to the emergency management system, in particular phenomena of landslides/landslides affecting road infrastructures.

Have been identified n. 24 BC. Finally, there are the Structural Aggregates (AS), n. 3, and the Structural Units (US), n. 29, interfering with the connection and accessibility infrastructures.

3.2 Minim Urban Structure (SUM) and Urban Resilience

The strategic elements of the SUM are not only those necessary for the seismic emergency phase - that is, those that respond to the purposes of the Civil Protection Plan - but all those essential for the functioning of the urban structure and for the resumption of ordinary urban activities after the seismic event.

The identification of the SUM is the necessary urban framework for priority interventions and, more generally, to direct public action in the field of seismic prevention [23, 24].

The contents and stages of identifying the SUM can be linked to three points:

- the identification of the strategic components of the urban structure for the urban response to the earthquake;
- the identification of criticalities and system weaknesses of the strategic components of the urban structure;
- the definition of the actions and interventions for the SUM.

All that is strategic from a functional point of view is part of the SUM: the structuring routes, the main road and railway communication routes and related functional nodes, escape routes and safe open spaces are always part of the SUM.

The main economic functions and elements of high symbolic value can also be part of the SUM - as a reason for being rooted in the place - which are a symbol of the social fabric and local culture, the loss of which could compromise the ability to react settlement and recovery following the seismic event.

The criticality of the SUM can be defined as its susceptibility to damage or loss of functionality that can derive from the physical damage of individual elements and/or systems, and which can lead to a loss of systemic functionality, even greater than the sum of the individual damages physical. The criticality assessments useful from an urban planning perspective consist in a quick consideration of the potential criticalities, in qualitative and system terms, of the single elements or component systems; existing specialist assessments, such as building vulnerability, can be incorporated or summarized, if available [25].

The increase in the functionality of the SUM consists of a series of actions and interventions to be implemented both to reduce the criticality of elements and structures, single or system, and to improve the overall behaviour in the event of a seismic event. The reduction or elimination of the criticalities of the SUM is a priority in the provision of interventions to mitigate the urban effects of the earthquake. Such issues may became objectives for the new territorial development policies funded by EU Structural Funds [26–30]. It can consist of interventions for eliminating or reducing specific criticalities but also in interventions for the definition of alternatives, i.e. the creation of elements of redundancy. The overall improvement, as well as through the elimination of specific criticalities or the definition of alternatives, can also derive from new plan forecasts (new strategic buildings, safe spaces and escape routes).

The SUM articulated in its components must be an integral part of the municipal urban planning instrument; therefore, it is necessary that there is a close relationship between the process of defining the SUM and the planning process.

In preparing the cognitive framework, the systemic components of the urban structure for the response to the earthquake must also be identified, i.e. the components of the SUM, they are:

i.   the mobility and accessibility system,
ii. the system of safe open spaces,
iii. the system of buildings and strategic structures,
iv. the system of the main technological networks.

Depending on the specific context conditions, they can also be part of the SUM:

i.   the system of cultural heritage and places of relationship,
ii. the system of economic-productive activities and main urban functions.

In the cognitive phase, the civil protection documents, the relative forecasts and any studies on building vulnerability are taken into consideration.

In the absence of a validated study of the SUM, at this stage they will be at least identify the components of the structure strategic urban areas for recovery (see. Fig. 3), in what has already been assessed the efficiency of those that belong to the system of emergency.

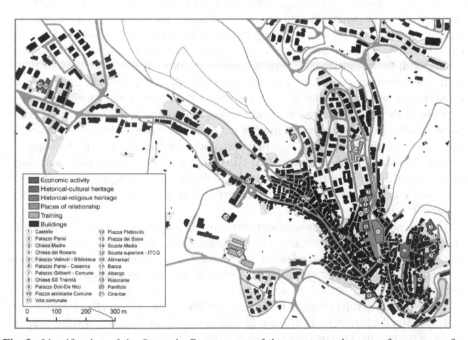

Fig. 3. Identification of the Strategic Components of the structure urban area for recovery for Moliterno municipality.

By doing so, you can have it a well-defined idea of the critical issues of the CLE Analysis and lay the foundations for integrate CLE studies with those for the definition of the SUM.

3.3 Evaluation of the CLE to Supplement the SUM

The CLE analysis serves to identify, in a given settlement or territory, the elements necessary for emergency management, what they are, where they are and what characteristics they possess. In fact, if the Limit Condition for Emergency is reached in a settlement hit by the earthquake, it means that all urban functions are compromised except for most of those necessary for the management of the emergency and its subsequent overcoming.

The evaluation of the CLE is a process that aims to verify whether the emergency management system - in the conditions given at the time of the evaluation - is adequate to ensure that the CLE is verified for that settlement. That is, given the existing elements recognized in the analysis phase, the evaluation serves to verify whether those elements are sufficient to ensure the urban performance necessary for emergency management.

In order to assess the system's resistance in an emergency phase, it is advisable to verify the relationship of this system with the characteristics of the urban structure and the recovery objectives of the settlement following the earthquake. Basically, two objectives must be taken into account in the assessment: one strictly functional and one consistent with the characteristics of the settlement.

To meet the two objectives mentioned, we intend to refer to three criteria for the evaluation of the emergency management system; they are: completeness, efficiency and urban compatibility. The three criteria define increasing urban performance. In fact, an emergency management system can be complete but not efficient, as some elements may not withstand the earthquake or may not be well connected to each other. Similarly, a system can be complete and efficient but not compatible from an urban point of view.

In detail, the completeness criterion is used to verify whether all the essential elements for emergency management are present within the settlement; the efficiency criterion is used to assess whether the set of elements allows to provide an efficient urban response in the emergency phase; finally, the urban compatibility criterion serves to assess whether, in addition to operation in the emergency phase, the system is able to guarantee the maintenance and resumption of strategic urban activities for recovery.

The evaluation of the elements of the CLE must possess such requisites as to meet the objectives described, structured according to the criteria of completeness, efficiency and urban compatibility, and with characteristics that allow its use in ordinary urban planning processes.

For the purposes of the assessment, two objectives were taken into account: one strictly functional and one consistent with the characteristics of the settlement. To address the two objectives mentioned, reference was made to three criteria for the evaluation of the emergency management system: the criterion of completeness, efficiency and urban compatibility.

Following the results obtained for the town of Moliterno, an overall assessment was established on the current CLE analysis. In fact, the verifications and evaluations carried out were put into the system - in a summary table.

3.4 The Improvement of the Urban System for the Management of the Seismic Emergency

A different location of the elements of the emergency management system can generate an overall evaluation of the CLE that is better than the current one.

Basically, for the municipalities of Moliterno and Tramutola - which have insufficient performance - the intention is to improve the criterion of urban compatibility and this is only possible by introducing new elements within the system. The relationship between settlement and components for recovery with the new emergency system leads to the improvement of the assessment of urban compatibility, that is, to a sufficient overall assessment.

The municipality of Moliterno, compared to the CLE configuration, had some criticalities due to the concentration of the AEs in a single area, i.e. near the sports field (AE 1, 2, 3, 4, 5, 6, 7), and to the little consideration of the relationship between the emergency management system and the strategic components for recovery.

To improve the assessment of urban compatibility (see Fig. 4), it was intended to act precisely on a greater distribution of the AEs; in this way it was possible to involve more parts of the settlement and components for the recovery.

In particular, the new AEs are:

8)   area in front of the municipal villa;
9)   Piazza Immacolata;
10) Area in Via Madonna del Vetere, at the ES 6 Traffic Police Barracks;
11) Piazza Kennedy (in Via Parco del Seggio), at the ES 2 Carabinieri Barracks.

For new AEs, it is necessary to identify the AC Connections that can allow accessibility.

In this regard, the road system that is added to that of the emergency management system includes three road sections: Corso Umberto I, Via Amendola and Via Istria. Corso Umberto I retraces part of the Provincial Road which rejoins the AE at the sports field; moreover, this road section is essential because it is the only one that crosses the ancient building fabrics (Zone A identified in the PRG), generating a strong criticality due to the presence of numerous US interfering structural aggregates. These criticalities and interferences will be duly taken into consideration in the subsequent phases of the SUM. Following the identification of new components within the CLE emergency management system, it was possible to carry out a new urban compatibility assessment, in order to verify that the overall assessment is sufficient for the basis of the study on the SUM.

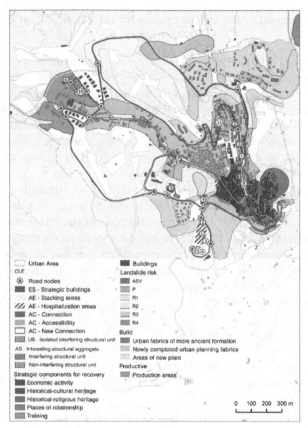

Fig. 4. The map shows the improvement of the urban system for the management of the seismic emergency in Moliterno municipality.

In conclusion, the new elements of the emergency system led to a Medium-high urban compatibility assessment, starting from the partial results (medium-high, high). By comparing this result with the efficiency evaluation, an overall Average (4) evaluation is obtained, which in this case is sufficient to represent the first phase for the identification of the SUM.

4 Conclusion

The improvement of the management system of the emergency was made possible by taking action on the assessment of urban compatibility. The added Stacking Areas (AE) have been located in in order to respond to further needs, related to the resumption of the settlement after the earthquake, which go beyond the initial state of emergency.

Its localization has also considered the presence of areas subjected to hydrogeological risk; however, for an effective definition, it would be useful to fill in the appropriate CLE maps also for these items. In fact, this is mainly a attempt to affirm what - by posing more attention to all fabrics of the residential type settlement and the strategic components for the recovery - it is possible to create a system emergency management that has a significant contribution to the recovery. Following the changes made, the emergency management system for the three municipalities is organized in a manner to have an overall evaluation sufficient; in thus, the CLE can represent the first phase of identification of the Minim Urban Structure (SUM). Generally, in the first phase they are identified precisely the components of the structure strategic urban areas for the response urban to the earthquake. The integration of CLE with SUM, following improvement changes in the organization of the elements of CLE, is only possible when it also meets urban compatibility criteria connect also with smart and sustainable urban development [31–33]. The SUM, therefore, can help the CLE Analysis and bring together the results of this study in the first phase of its global structure. Basically, it comes down to two orders of results; the first is an increase in efficiency of the resources made available for achieve the mitigation objective seismic risk on the urban scale; the second result, is the predisposition of apparently a contact between two realities distant, which come from needs different, where it is however possible establish convergences.

An intersting development perspetive of this research also is to inlcude in the territorial assessment the dimension of ecosystem services as a territorial value under risks [34–36].

References

1. Casas, G.L., Scorza, F.: Sustainable planning: a methodological toolkit. In: Gervasi, O., et al. (eds.) ICCSA 2016. LNCS, vol. 9786, pp. 627–635. Springer, Cham (2016). https://doi.org/10.1007/978-3-319-42085-1_53

2. Scorza, F., Grecu, V.: Assessing sustainability: research directions and relevant issues. In: Gervasi, O., et al. (eds.) ICCSA 2016. LNCS, vol. 9786, pp. 642–647. Springer, Cham (2016). https://doi.org/10.1007/978-3-319-42085-1_55

3. Dvarioniene, J., Grecu, V., Lai, S., Scorza, F.: Four perspectives of applied sustainability: research implications and possible integrations. In: Gervasi, O., et al. (eds.) ICCSA 2017. LNCS, vol. 10409, pp. 554–563. Springer, Cham (2017). https://doi.org/10.1007/978-3-319-62407-5_39

4. Las Casas, G., Scorza, F., Murgante, B.: New urban agenda and open challenges for urban and regional planning. In: Calabrò, F., Della Spina, L., Bevilacqua, C. (eds.) ISHT 2018. SIST, vol. 100, pp. 282–288. Springer, Cham (2019). https://doi.org/10.1007/978-3-319-92099-3_33

5. Las Casas, G., Scorza, F., Murgante, B.: Razionalità a-priori: Una proposta verso una pianificazione antifragile. Sci. Reg. 18, 329–338 (2019). https://doi.org/10.14650/93656

6. Scorza, F., Saganeiti, L., Pilogallo, A., Murgante, B.: Ghost planning: the inefficiency of energy sector policies in a low population density region1. Arch. DI Stud. URBANI E Reg. 34–55 (2020). https://doi.org/10.3280/ASUR2020-127-S1003

7. Las Casas, G., Murgante, B., Scorza, F.: Regional local development strategies benefiting from open data and open tools and an outlook on the renewable energy sources contribution. In: Papa, R., Fistola, R. (eds.) Smart Energy in the Smart City. GET, pp. 275–290. Springer, Cham (2016). https://doi.org/10.1007/978-3-319-31157-9_14

8. Alberti, F.: Il valore territorio. Gestire la complessità per governare le trasformazioni. Alinea Editrice (2009)

9. De Rossi, A.: Riabitare l'Italia: Le aree interne tra abbandoni e riconquiste (curated by). Donzelli editore, Roma (2018)

10. Barca, F.: Disuguaglianze territoriali e bisogno sociale. La sfida delle "Aree Interne". In: Lettura annuale Ermanno Gorrieri, pp. 31–33. Stampa Grafiche TEM, Modena (2016)
11. Francini, M., Palermo, A., Viapiana, M.F.: Integrated territorial approaches for emergency plans. Territorio 2019, 85–90 (2019). https://doi.org/10.3280/TR2019-089011
12. Nigro, R., Lupo, G.: Civiltà Appennino. L'Italia in verticale tra identità e rappresentazioni (2020)
13. Pontrandolfi, P.: Pianificazione fisica e rischi territoriali. Resilienza dei territori tra prevenzione ed emergenza. In: Franco Angeli Editore (ed.) Il piano di emergenza nell'uso e nella gestione del territorio, Milano (2020)
14. Orlando, G., Selicato, F., Torre, C.M.: The use of GIS as tool to support risk assessment. In: van Oosterom, P., Zlatanova, S., Fendel, E.M. (eds.) Geo-information for Disaster Management, pp. 1381–1399. Springer, Heidelberg (2019). https://doi.org/10.1007/3-540-27468-5_95.
15. Bramerini, F., Castenetto, S.: Manuale per l'analisi della Condizione Limite per l'Emergenza (CLE) dell'insediamento urbano, Commissione tecnica per la microzonazione sismica. BetMultimedia, Roma (2014)
16. DPTU - Dipartimento di pianificazione territoriale e urbanistica – Sapienza Università di Roma: Linee guida per l'individuazione della Struttura urbana minima e le valutazioni di vulnerabilità urbana (2010)
17. DATA - Sapienza Università di Roma: Rischio sismico urbano. Indicazioni di metodo e sperimentazioni per l'analisi della condizione limite per l'emergenza e la struttura urbana minima (2013)
18. Saganeiti, L., Favale, A., Pilogallo, A., Scorza, F., Murgante, B.: Assessing urban fragmentation at regional scale using sprinkling indexes. Sustainability 10, 3274 (2018). https://doi.org/10.3390/su10093274
19. Scorza, F., Pilogallo, A., Saganeiti, L., Murgante, B., Pontrandolfi, P.: Comparing the territorial performances of renewable energy sources' plants with an integrated ecosystem services loss assessment: a case study from the Basilicata region (Italy). Sustain. Cities Soc. 56, 102082 (2020). https://doi.org/10.1016/J.SCS.2020.102082
20. Saganeiti, L., Pilogallo, A., Faruolo, G., Scorza, F., Murgante, B.: Territorial Fragmentation and Renewable Energy Source Plants: Which Relationship? Sustainability 12, 1828 (2020). https://doi.org/10.3390/SU12051828
21. Las Casas, G., Scorza, F., Murgante, B.: Conflicts and sustainable planning: peculiar instances coming from Val D'agri structural inter-municipal plan. In: Papa, R., Fistola, R., Gargiulo, C. (eds.) Smart Planning: Sustainability and Mobility in the Age of Change. GET, pp. 163–177. Springer, Cham (2018). https://doi.org/10.1007/978-3-319-77682-8_10
22. Bramerini, F., Fazzio, F., Parotto, R.: La microzonazione sismica e le condizioni limite nella prevenzione urbanistica del rischio sismico. Urban. Doss. 22–28 (2013)
23. La sfida della Resilienza. TRIA Territ. della Ric. su insediamenti e Ambient. 8 (2015)
24. Martinelli, G.: La Sum nel progetto di ricostruzione. Traiettorie preliminari per la rigenerazione urbana nei territori del cratere sismico. In: Atti della XXI Conferenza Nazionale SIU I CONFINI, MOVIMENTI, LUOGHI. Politiche e progetti per città e territori in transizione. Planum Publischer (2019)
25. Cibelli, M.P.: Il miglioramento sismico nel restauro dell'architettura storica. La sperimentazione su di un aggregato in area campana (2016)
26. Scorza, F.: Improving EU Cohesion Policy: the spatial distribution analysis of regional development investments funded by EU structural funds 2007/2013 in Italy. In: Murgante, B., et al. (eds.) ICCSA 2013. LNCS, vol. 7973, pp. 582–593. Springer, Heidelberg (2013). https://doi.org/10.1007/978-3-642-39646-5_42
27. Scorza, F., Casas, G.B.L., Murgante, B.: That's ReDO: ontologies and regional development planning. In: Murgante, B., et al. (eds.) ICCSA 2012. LNCS, vol. 7334, pp. 640–652. Springer, Heidelberg (2012). https://doi.org/10.1007/978-3-642-31075-1_48

28. Scorza, F., Attolico, A.: Innovations in promoting sustainable development: the local implementation plan designed by the Province of Potenza. In: Gervasi, O., et al. (eds.) ICCSA 2015. LNCS, vol. 9156, pp. 756–766. Springer, Cham (2015). https://doi.org/10.1007/978-3-319-21407-8_54
29. Curatella, L., Scorza, F.: Polycentrism and insularity metrics for in-land areas. In: Gervasi, O., et al. (eds.) ICCSA 2020. LNCS, vol. 12255, pp. 253–261. Springer, Cham (2020). https://doi.org/10.1007/978-3-030-58820-5_20
30. Scorza, F.: Training decision-makers: GEODESIGN workshop paving the way for new urban agenda. In: Gervasi, O., et al. (eds.) ICCSA 2020. LNCS, vol. 12252, pp. 310–316. Springer, Cham (2020). https://doi.org/10.1007/978-3-030-58811-3_22
31. Santopietro, L., Scorza, F.: The Italian experience of the covenant of mayors: a territorial evaluation. Sustainability. **13**, 1289 (2021). https://doi.org/10.3390/su13031289
32. Santopietro, L., Scorza, F.: A systemic perspective for the Sustainable Energy and Climate Action Plan (SECAP). Eur. Plan. Stud. (2021, submitting, pending publication)
33. Garau, C., Desogus, G., Zamperlin, P.: Governing technology-based urbanism. In: The Routledge Companion to Smart Cities, pp. 157–174. Routledge (2020). https://doi.org/10.4324/9781315178387-12
34. Pilogallo, A., Saganeiti, L., Scorza, F., Murgante, B.: Assessing the impact of land use changes on ecosystem services value. In: Gervasi, O., et al. (eds.) ICCSA 2020. LNCS, vol. 12253, pp. 606–616. Springer, Cham (2020). https://doi.org/10.1007/978-3-030-58814-4_47
35. Pilogallo, A., Saganeiti, L., Scorza, F., Murgante, B.: Soil ecosystem services and sediment production: the Basilicata region case study. In: Gervasi, O., et al. (eds.) ICCSA 2020. LNCS, vol. 12253, pp. 421–435. Springer, Cham (2020). https://doi.org/10.1007/978-3-030-58814-4_30
36. Pilogallo, A., Scorza, F.: Mapping regulation ecosystem services (ReMES) specialization in Italy. J. Urban Plan. Dev. (2021, submitting, pending publication)

International Workshop on Advances in Web Based Learning (AWBL 2021)

Teaching English Word Order with CorrectWriting Software

Elena Novozhenina⬤, Oleg Sychev⁽✉⁾ ⬤, Olga Toporkova⬤,
and Oksana Evtushenko⬤

Volgograd State Technical University, Volgograd, Russia
o_sychev@vstu.ru

Abstract. The article looks into the problem of using interactive computer technologies in technology-enhanced teaching English at a technical university. An electronic course "English Word Order" using CorrectWriting software for Moodle LMS was developed and used to teach students of various levels of education. The system of formative and summative assessments regarding the word order of English sentences was studied. The students' responses to CorrectWriting questions were analyzed using the Longest Common Subsequence algorithm, allowing determining misplaced, missing and extraneous words. During formative assessments, the students were offered automatic hints on how to fix their mistakes without having to wait for feedback from the teacher. The learning gains and the student survey prove the efficiency of the developed online course which helps to improve students' grammar skills during the classes and unsupervised work. The study also notes a positive role of electronic teaching materials in optimizing the technological and organizational support of the educational process. Further work on enhancing the developed course and software is outlined.

Keywords: E-learning · English as a second language ·
CorrectWriting · Computer-generated feedback · Formative assessments

1 Introduction

The nature of pedagogical activity is changing nowadays along with the rapid development of information technology. It penetrates deeply into our lives, and today one can hardly find any field of activity where a computer is not used to solve professional problems. This process dramatically accelerated during lockdowns caused by COVID-19 pandemic. Education has not remained indifferent to this change which brought profound transformation in the system of teaching and evaluating student progress. This completely applies to the process of teaching foreign languages.

Modern information technologies provide significant opportunities for improving the quality of training, so electronic teaching materials and Internet

The reported study was funded by RFBR, project number 20-07-00764.

technologies are often used in foreign language teaching. The design of electronic educational resources aimed at the formation of relevant students' competencies is a topical issue due to the fact that this is one of the priority tasks in the activities of a foreign language teacher. The ability to quickly modernize and update the components of the educational process as well as to find new efficient ways of teaching is an important characteristic of the methodological literacy of a modern effective teacher [10,15,28]. The Internet offers great opportunities for carrying out such tasks. A big challenge for teachers is learning and growing professionally in order to use digital technologies to enhance students' learning. In spite of the fact that this has become a hot topic today most university students and teachers have a basic level of digital competence. Therefore the institutions of higher education are encouraged to focus on the development of students' and teachers' digital competence, create relevant learning strategies and use appropriate tools to improve the quality of education [32]. Some educators have made an attempt to analyze teachers' digital information skills and measure teachers' digital citizenship highlighting the factors that influence their development and the importance of their acquisition.

Teachers of foreign languages need to master information and communication technologies to be able to implement them in their teaching practice. They are faced not only with the task of choosing the most effective available Internet resources offered but also with the process of designing their own e-learning materials. This helps to optimize the educational process and contributes to the development of teachers' professional and creative activity which improves the quality of training and assessment.

2   Related Work

The use of information and communication technologies in teaching students foreign languages has been studied extensively, with a wide variety of approaches. A number of studies are focused on the use of such means as e-mail, social networks, IP telephony [7,8,19]. There have been research studies conducted on specific areas of online language learning strategies, including mobile learning, with their extensive opportunities to diversify and improve the learning process [13,16,17]. In recent years, a large number of teaching materials have appeared supporting the use of the Internet resources. In addition, many higher educational institutions are actively creating virtual educational environments by introducing programs for electronic support of courses in various disciplines [22,23,29]. Following the majority of experts who are exploring the ways of increasing the efficiency of teaching a foreign language by using multimedia technologies, it should be emphasized that the technologies can be successfully used in the formation of skills and communicative foreign-language competence, as well as information competence [1,6,14]. The main advantage is their enormous methodological and didactic potential, which includes the possibility of creating flexible and adaptive educational programs, quick feedback, a new mode of presentation, extensive training and assessment of the developed skills and abilities.

Knowledge assessment is an important component in the process of achieving desired learning goals. The use of digital technologies in teaching/learning English has given rise to new forms of assessing students' outcomes. Researchers have focused their efforts on analyzing effective assessment techniques which are essential for effective teaching and learning [9,30]. The main purposes of assessments are to facilitate learning, monitor students' success, enhance teaching [4,5,9]. Attention should be paid to a particular aspect: the use of timely feedback. This focus can be justified because of the importance of assessment and especially, of formative assessment which may improve outcomes. An urgent need to assess large numbers of university students makes computer-based assessments (CBAs) a helpful tool to support students' learning in higher education [18,20,21,27]. In spite of observing the comprehensive research in the field of knowledge online assessment [11,12,24,31,33], the relationships between applications of certain forms of knowledge online assessment and their influence on developing students' skills and communicative foreign-language competence are still insufficiently studied.

Motivated by all these ideas we have developed an online teaching assessment instrument for foreign language learning students. The research described in the article is part of the interdisciplinary project of the departments "Foreign languages" and "Software Engineering" in teaching a foreign language at a technical university. The authors created a set of formative and summative assessments for translating sentences from Russian into English to teach "English Word Order". It is based on CorrectWriting plug-in [26] developed at Software Engineering department for the popular learning management system Moodle [2].

3   The Online Course English Word Order

The developed online course "English Word Order" includes assessments for 9 Grammar categories: Statements, Adverbs with verbs in the statement, Negative sentences, General questions, Special questions, Tag questions, Alternative questions, Subject Questions and Reported speech questions. The word order practice is aimed at reinforcing main grammar constructions and vocabulary in the context of English Intermediate level communication. Each category contains 20 formative and 20 summative sentences to be translated from Russian into English. The example is shown in Fig. 1.

This practice is useful for developing grammar and translation skills for different sentence types. A student is given a chance to check the response immediately, to read the mistake messages, and correct the mistakes using the learned rules. Moodle is a modular learning platform for online and blended learning. One of the module types Moodle supports is "question type", allowing third-party developers to produce new kinds of advanced questions to be used by Moodle quizzes. One such module is the CorrectWriting question type, supporting open-answer short-text questions implemented in PHP programming language. It operates by breaking the sentence into lexemes – words and punctuation marks – and detects four kinds of mistakes using Damerau-Levenshtein editing distance and Longest Common Subsequence algorithm:

1. a typo (including a skipped or unnecessary delimiter);
2. a misplaced lexeme (word-order mistake);
3. an extraneous lexeme (it should not be in the answer)
4. an omitted lexeme (it should be in the answer).

The application can determine all the mistakes even if a student has made a few mistakes in one word (e.g. misplacing a word containing a typo).

To stimulate thinking and better grammar rules learning, each lexeme has a description, showing its role in the sentence that determines its place. It allows displaying a mistake message not as "'waited' is misplaced" but as "a predicate is misplaced". This stimulates thinking about the word-order rules, in this case about the place of the predicate. Suppose the sentence "Did it rain or snow last time?" was constructed by a student as "At last time it rain or snow". In this case the following mistakes will be displayed:

1. the auxiliary verb is missing;
2. "At" should not be in response;
3. the adverbial modifier of time misplaced.

The message quality depends on the description of lexical items provided by the question authors. If a student fails to fix the mistake using the mistake message, a number of hints is available to them if allowed by the teacher:

1. a hint "what it is" shows the word in which a mistake is made, for example "The predicate in your sentence is 'waited'";
2. a hint "where to place" gives the right word order, for example "First put the subject, then the predicate, then the adverbial modifier of time";
3. an image "where to place" hint shows where the lexeme should be placed (see Fig. 2).

A teacher can detract points for correcting the answer and using hints that makes sense in summative assessments. At the end of the exercise, if a student didn't provide a correct answer, CorrectWriting generates an image showing how to make the correct answer from student's response (see Fig. 3).

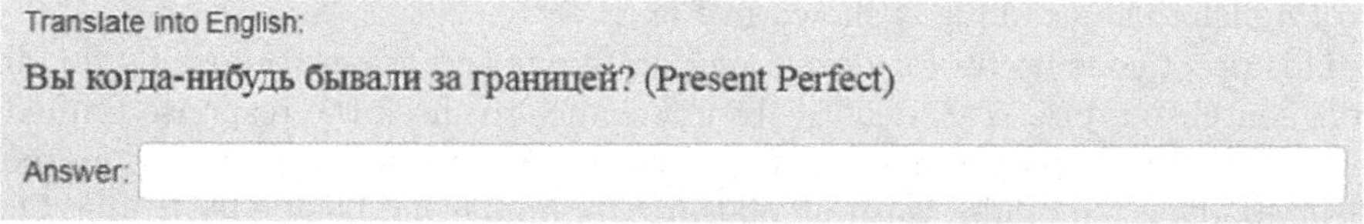

Translate into English:

Вы когда-нибудь бывали за границей? (Present Perfect)

Answer:

Fig. 1. An example of a question

CorrectWriting application provided such advantages as virtually unlimited learning time, the analysis of mistakes, the opportunity of revising the material, and self-study efficiency. It is restricted only by the number of formative

Fig. 2. A hint "where to place": a) when the word order is wrong; b) when the word is missing.

Your response: ~~last t~~me it r~~in or snow~~   ?

Correct answer: [Did]   it rain or snow last time ?

Fig. 3. The final picture of correcting the answer with different types of mistakes.

questions, but more questions can be added by the course authors each year. The training system gives a student the opportunity to not only check the correctness of translation, but also to correct the wrong answer, and to analyze the mistakes using the lexeme descriptions. The formative assessments blend seamlessly into the whole training system, technologically enhancing the way of presentation and revision of learning material. Students acquire knowledge on their own in a more familiar manner. Summative assessments allow conducting a continuous skills and knowledge evaluation. So this way of revising and testing students' knowledge provides an opportunity to change a delivery mode of classroom activities in terms of saving study time and using it for developing more productive skills.

4 Evaluation

We performed two kinds of evaluation for the created course:

1. the students' learning gains;
2. students' perceptions about the course.

4.1 Learning Gains

To evaluate the developed course, 45 students were offered to do a series of quizzes: 27 undergraduate Computer Science (CpS) students and 18 graduate Chemical Engineering (ChE) students. The average knowledge of the students could be evaluated at the levels of "good" and "satisfactory". It should be noted that the classroom hours in English at Volgograd State Technical University are limited to 2 h per week. The testing took place within the 1st term of studies in 2020. At first, the students were given the summative quizzes to measure their baseline performance. After that, the students were able to perform three formative quizzes provided with the hints feedback system during their homework. The number of students' attempts of taking formative quizzes were not limited, with the students being able to improve their grammar skills in word order freely any time. At the end of the term, the second summative quiz (using the same questions) was taken by the students to measure their progress in improving the grammar skills of word order. It is important to note that the time for taking the formative quizzes was not limited whereas the summative quiz timing was limited as well as the number of attempts; during the summative quizzes the students were allowed to answer each question only once.

Table 1. Baseline performance and learning gains (LG)

Student group	Baseline grade		Absolute LG		Relative LG	
	Mean	Stddev	Mean	Stddev	Mean	Stddev
CpS, undergrad	3.52	1.48	2.9	1.77	42.9	23.8
ChE, graduate	3.18	1.42	3.48	1.61	49.34	18.45

Both groups of students had a similar average baseline performance, around 3 out of 10 points. We measured absolute learning gains as the difference between the grades for the second and the first attempts of the summative quiz (SG_2 and SG_1 respectively). However, absolute learning gains are not a good way to compare the progress of the students with different baseline performance as the students with better baseline have less points to acquire before reaching the maximum. So we calculated relative learning gains (RLG), showing the percent of remaining points the student gained (see (1) where 10 is the maximum grade).

$$RLG = \frac{SG_2 - SG_1}{10 - SG_1} \tag{1}$$

The results are shown in Table 1 and Figs. 4, 5 and 6. The two groups of students showed different behavior. Among the undergraduate CpS students, the students with the best baseline performance mostly ignored formative quizzes. They had the worst learning gains (including relative gains, so it cannot be attributed to their high base value), from 1/2 to 2 times worse than the others. One of the worst-performing students made a lot of formative-quiz attempts instead (15, which is about twice more than the other students) and also had the low learning gains. This situation shows that a large number of attempts can be a sign of ineffective usage of formative quizzes (his results are excluded from the plots to make them more readable). Among the other CpS students, the learning gains generally increased with the number of attempts. The dip in absolute learning gains around 4 attempts is caused by the better baseline performance of these students: in Fig. 6 you can see that this dip is significantly lower for relative gains.

Among the graduate ChE students, those who had the worst baseline performance chose to use formative quizzes less or ignore them. They, however, had relatively high learning gains that can be attributed to the other methods

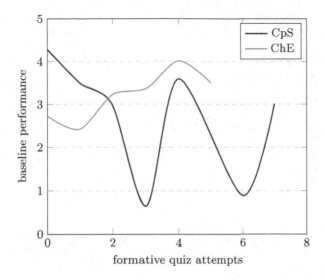

Fig. 4. Average baseline performance of students with different numbers of formative-quiz attempts

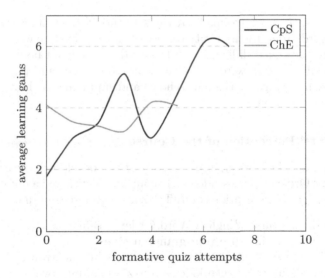

Fig. 5. Average absolute learning gains depending on formative-quiz attempts

of learning: graduate students are older and more proficient in self-education. The students who made less than 4 attempts of formative quizzes had the worst gains; however, the students who made 4 or more attempts had the best learning gains and the highest baseline performance: the best students trained and learned more. This is particularly evident considering the relative gains.

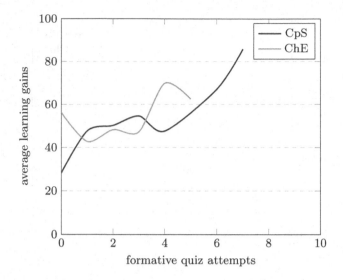

Fig. 6. Average relative learning gains depending on formative-quiz attempts

Nevertheless, the mistakes made by the students in the tests were caused not only by the word order: the necessity to choose the appropriate word and correct tense brought additional difficulties to the students. The authors see the room for improvement of these exercises by entering more words and phrases into the system, giving more hints in the tense choice and supplying the hint system with the theoretical background.

4.2   Students' Perception of the Course

We also evaluated the subjective perception of the developed e-learning course "English Word Order" by the students. The authors developed a questionnaire to find out the students' perceptions of it. The questionnaire contained 6 questions:

1. Do you find the course "English Word Order" useful?
2. Did you manage to improve your grammar skills?
3. How useful were the hints explaining the errors in your translation?
4. What skills do you think such kind of quizzes help improve?
5. Would you like to continue doing such quizzes?
6. Would you like to take part in developing such quizzes?

We asked 45 respondents, including undergraduate and graduate students from various faculties of Volgograd State Technical University who study English as a foreign language (see Table 2). Despite the fact that some researchers note that students may experience difficulties in using ICT for their learning, the results of the survey, in general, indicate the willingness of the respondents (70.5%) to further participate in them and improve their English language skills using the developed application.

Table 2. Survey results

Question number	Percentage of students replying positively
Question 1	91%
Question 2	78.9% (see Table 3 for details)
Question 3	69%
Question 4	See Table 4
Question 5	70.5%
Question 6	51%

Table 3. How much have your grammar skills been improved?

Skills improvement	Percentage of students
10%–39%	16%
40%–69%	33%
70%–89%	27%
90%–100%	2.9%

The results of the survey show that 51% of the respondents expressed the desire to take part in the creating similar tasks when their language proficiency improves. Overall, 91% of all the respondents thought that the developed online course positively: 69% agreed that such applications are useful, 22% thought they were very useful for improving grammar skills, while only 7% considered them not very useful. 78.9% of the participants believed that their grammatical skills improved after these exercises (see Table 3). The respondents were confident that this course allowed them to improve a number of skills, which we will indicate according to the degree of their popularity (see Table 4). The system of hints implemented by the authors of CorrectWriting found a positive response from the respondents: 69% of them noted the hints were useful.

Table 4. Which skills have been improved?

Skill	Percentage of students
Translation skills	75%
Grammar	73%
Independent work skills	50%
Vocabulary	23%
Motivation to learn a foreign language	19%
Computer using	16%
Reading	11%

5   Conclusions and Further Work

The results show that using CorrectWriting questions for formative and summative assessments enhances foreign-language teaching; this form of e-learning is methodologically appropriate. For undergraduate students, the learning gains generally increased with the number of the attempts of the formative quizzes; for graduate students only significant amount of attempts brought good gains. However, too many attempts (significantly more than the other students make) can be an indicator of misusing the quizzes and low learning gains. However, the lack of formative-quiz attempts can indicate learning problems only combined with a low initial performance.

In general, the joined efforts at the interdepartmental level in order to develop and use the electronic course "English Word Order" made it possible to create and analyze the positive learning experience. The survey results indicate the efficiency of this form of training and confirm that the introduction of training based on electronic educational resources can significantly diversify and optimize the educational process. Those findings are supported by the studies of Balci et al. [3], Stefanovic and Klochkova [25], also reporting positive effects of electronic educational resources on educational process. In our study, the training system with hints and mistake analysis allowed teachers to organize efficient self-study of students and save the class time for more sophisticated training (e.g. in communication skills). It also helped to support students during COVID-19 lockdowns.

The results of the study also indicate several areas for further improvement. They include:

- adding options for correct answers, taking into account synonymy and possible abbreviations;
- developing a unified system of descriptions of the word role in a sentence that would be used both in lecture material and in exercises;
- extending the functions of the language-checking software to account for word-formation errors and choosing the appropriate word form in English to make the displayed mistake messages more detailed and understandable.

However, the most difficult and laborious stage of compiling the formative and summative assessment base has already been completed that allows us to hope for their successful further use in the educational process.

References

1. Andreyeva, Y., Fakhrutdinova, A., Korneva, I., Chugunov, A.: The effectiveness of the use of computer technology compared to the traditional methods in the process of a foreign language teaching. Univers. J. Educ. Res. **7**, 21–26 (2019). https://doi.org/10.13189/ujer.2019.071805
2. Arianti, B., Kholisho, Y., Sujatmiko, S.: The development of e-learning use MOODLE as a multimedia learning medium. J. Phys. Conf. Ser. **1539**, 12–33 (2020). https://doi.org/10.1088/1742-6596/1539/1/012033

3. Balcı, B., Çiloğlugil, B., İnceoğlu, M.M.: Open courseware-based logic design course. Croatian J. Educ. **22**(2), 515–562 (2020). https://cje2.ufzg.hr/ojs/index.php/CJOE/article/view/3499
4. Bennett, R.E.: Formative assessment: a critical review. Assess. Educ. Princ. Policy Pract. **18**(1), 5–25 (2011). https://doi.org/10.1080/0969594X.2010.513678
5. Dixson, D.D., Worrell, F.C.: Formative and summative assessment in the classroom. Theory Pract. **55**(2), 153–159 (2016). https://doi.org/10.1080/00405841.2016.1148989
6. Eghtesad, S.: Authentic online resources for learning French. In: 2018 12th Iranian and 6th International Conference on e-Learning and e-Teaching (ICeLeT), pp. 007–012 (2018). https://doi.org/10.1109/ICELET.2018.8586757
7. Eid, M.I., Al-Jabri, I.M.: Social networking, knowledge sharing, and student learning: the case of university students. Comput. Educ. **99**, 14–27 (2016). https://doi.org/10.1016/j.compedu.2016.04.007
8. Gadakchyan, A., Kapitonova, N., Treboukhina, N., Ustinova, N.: Web environment of distance learning. E3S Web Conf. 210, 18015 (2020). https://doi.org/10.1051/e3sconf/202021018015
9. Gaytan, J., McEwen, B.C.: Effective online instructional and assessment strategies. Am. J. Dist. Educ. **21**(3), 117–132 (2007). https://doi.org/10.1080/08923640701341653
10. Greathouse, P., Eisenbach, B.B., Kaywell, J.F.: Preparing teacher candidates to be "effective" in the classroom: lessons learned from national teachers of the year. Clear. House J. Educ. Strateg. Issues Ideas **92**(1–2), 39–47 (2019). https://doi.org/10.1080/00098655.2018.1561405
11. Hettiarachchi, E., Huertas, M.A., Mor, E.: Skill and knowledge e-assessment: a review of the state of the art. IN3 Working Paper Series (2013). https://doi.org/10.7238/in3wps.v0i0.1958
12. Ho, V.W., Harris, P.G., Kumar, R.K., Velan, G.M.: Knowledge maps: a tool for online assessment with automated feedback. Med. Educ. Online **23**(1), 1457394 (2018). https://doi.org/10.1080/10872981.2018.1457394
13. Hoi, V.N., Mu, G.M.: Perceived teacher support and students' acceptance of mobile-assisted language learning: evidence from Vietnamese higher education context. Br. J. Edu. Technol. **52**(2), 879–898 (2021). https://doi.org/10.1111/bjet.13044
14. Kannan, J., Munday, P.: New trends in second language learning and teaching through the lens of ICT, networked learning, and artificial intelligence. Círculo de Lingüística Aplicada a la Comunicación 76 (2018). https://doi.org/10.5209/CLAC.62495
15. Koutrouba, K.: A profile of the effective teacher: Greek secondary education teachers' perceptions. Eur. J. Teach. Educ. **35**(3), 359–374 (2012). https://doi.org/10.1080/02619768.2011.654332
16. Kukulska-Hulme, A.: Will mobile learning change language learning? ReCALL **21**(2), 157–165 (2009). https://doi.org/10.1017/S0958344009000202
17. Ma, G.: The current situations of mobile assisted language learning. In: MacIntyre, J., Zhao, J., Ma, X. (eds.) SPIOT 2020. AISC, vol. 1283, pp. 675–679. Springer, Cham (2021). https://doi.org/10.1007/978-3-030-62746-1_99
18. Malau-Aduli, B.S., Assenheimer, D., Choi-Lundberg, D., Zimitat, C.: Using computer-based technology to improve feedback to staff and students on MCQ assessments. Innov. Educ. Teach. Int. **51**(5), 510–522 (2014). https://doi.org/10.1080/14703297.2013.796711

19. Manca, S., Ranieri, M.: Facebook and the others. Potentials and obstacles of social media for teaching in higher education. Comput. Educ. **95**, 216–230 (2016). https://doi.org/10.1016/j.compedu.2016.01.012

20. Miller, T.: Formative computer-based assessment in higher education: the effectiveness of feedback in supporting student learning. Assess. Eval. High. Educ. **34**(2), 181–192 (2009). https://doi.org/10.1080/02602930801956075

21. Peat, M., Franklin, S.: Supporting student learning: the use of computer-based formative assessment modules. Br. J. Edu. Technol. **33**(5), 515–523 (2002). https://doi.org/10.1111/1467-8535.00288

22. Potkonjak, V., et al.: Virtual laboratories for education in science, technology, and engineering: a review. Comput. Educ. **95**, 309–327 (2016). https://doi.org/10.1016/j.compedu.2016.02.002

23. Shubita, A.F., Issa, G.F.: Using 3d virtual environment as an educational tool in a middle eastern university. J. Comput. Sci. **15**(10), 1498–1509 (2019). https://doi.org/10.3844/jcssp.2019.1498.1509

24. Spivey, M.F., McMillan, J.J.: Classroom versus online assessment. J. Educ. Bus. **89**(8), 450–456 (2014). https://doi.org/10.1080/08832323.2014.937676

25. Stefanovic, S., Klochkova, E.: Digitalisation of teaching and learning as a tool for increasing students' satisfaction and educational efficiency: using smart platforms in EFL. Sustainability **13**(9) (2021). https://doi.org/10.3390/su13094892

26. Sychev, O.A., Mamontov, D.P.: Automatic error detection and hint generation in the teaching of formal languages syntax using correctwriting question type for MOODLE LMS. In: 2018 3rd Russian-Pacific Conference on Computer Technology and Applications (RPC), pp. 1–4 (2018). https://doi.org/10.1109/RPC.2018.8482125

27. Sychev, O., Anikin, A., Prokudin, A.: Automatic grading and hinting in open-ended text questions. Cogn. Syst. Res. **59**, 264–272 (2020). https://doi.org/10.1016/j.cogsys.2019.09.025

28. Toporkova, O., Novozhenina, E., Tchechet, T., Likhacheva, T.: Methodology of a modern foreign language lesson for postgraduate students of technical disciplines. Engl. Lang. Teach. **7**, 35–39 (2014). https://doi.org/10.5539/elt.v7n10p35

29. Ventayen, R.J.M., Orlanda-Ventayen, C.C.: The design of virtual learning environment: a case from Pangasinan state university, open university systems. Int. J. Sci. Technol. Res. **8**(12), 3063–3066 (2019). http://www.ijstr.org/final-print/dec2019/The-Design-Of-Virtual-Learning-Environment-A-Case-From-Pangasinan-State-University-Open-University-Systems.pdf

30. Webber, K.L., Tschepikow, K.: The role of learner-centred assessment in postsecondary organisational change. Assess. Educ. Princ. Policy Pract. **20**(2), 187–204 (2013). https://doi.org/10.1080/0969594X.2012.717064

31. Yeatman, J.D., et al.: Rapid online assessment of reading ability. Sci. Rep. **11**(1), 1–11 (2021). https://doi.org/10.1038/s41598-021-85907-x

32. Zhao, Y., Llorente, A.M.P., Gómez, M.C.S.: Digital competence in higher education research: a systematic literature review. Comput. Educ. 104212 (2021). https://doi.org/10.1016/j.compedu.2021.104212

33. Zlatović, M., Balaban, I.: Adaptivity: a continual adaptive online knowledge assessment system. In: Rocha, Á., Adeli, H., Reis, L.P., Costanzo, S., Orovic, I., Moreira, F. (eds.) WorldCIST 2020. AISC, vol. 1161, pp. 152–161. Springer, Cham (2020). https://doi.org/10.1007/978-3-030-45697-9_15

A Multi-agent Based Adaptive E-Learning System

Birol Ciloglugil[1]($\boxtimes$)(iD), Oylum Alatli[1](iD), Mustafa Murat Inceoglu[2](iD),
and Riza Cenk Erdur[1](iD)

[1] Department of Computer Engineering, Ege University,
35100 Bornova, Izmir, Turkey
{birol.ciloglugil,oylum.alatli,cenk.erdur}@ege.edu.tr
[2] Department of Computer Education and Instructional Technology, Ege University,
35100 Bornova, Izmir, Turkey
mustafa.inceoglu@ege.edu.tr

Abstract. In this paper, a multi-agent based adaptive e-learning system that supports personalization based on learning styles is proposed. Considering that the importance of distance education has increased with the effect of the Covid-19 pandemic, it is aimed to propose an adaptive e-learning system solution that offers more effective learning experiences by taking into account the individual differences in the learning processes of the students. The Felder and Silverman learning style model was used to represent individual differences in students' learning processes. In our system, it is aimed to recommend learning materials that are suitable for learning styles and previous knowledge levels of the students. With the multi-agent based structure, an effective control mechanism is devised to monitor the interaction of students with the system and to observe the learning levels of each student. The purpose of this control mechanism is to provide a higher efficiency in the subjects the students study compared to non-personalized e-learning systems. This study focuses on the proposed architecture and the development of the first prototype of it. In order to test the effectiveness of the system, personalized course materials should be prepared according to the learning styles of the students. In this context, it is planned to use the proposed system in future studies within the scope of a course in which the educational content is personalized.

Keywords: Adaptive E-Learning · Personalization · Multi-agent system · Felder and Silverman learning style model

1 Introduction

E-Learning systems have been used widely for decades; however, providing effective e-learning systems to students has gained significant importance especially after the Covid-19 pandemic [1–3]. Both educators and students have faced problems such as lack of interaction, loss of motivation and less effective learning experiences during this difficult period, which resulted in a significant increase

O. Gervasi et al. (Eds.): ICCSA 2021, LNCS 12951, pp. 693–707, 2021.
https://doi.org/10.1007/978-3-030-86970-0_48

in adaptive e-learning systems research among different kinds of e-learning systems research being conducted.

Adaptive e-learning systems provide personalized solutions by modeling individual differences of students in order to provide them personalized learning experiences [4]. Students may have different learning objectives, preferences, background knowledge levels, individual needs that can affect the way they learn [5,6]. Learning style models are widely used for modeling these individual differences in adaptive e-learning systems [7–10]. In this study, in order to model individual differences in students' learning processes, we used Felder and Silverman learning style model (FSLSM) [11], which is the most-commonly used model in the literature. In addition to FSLSM, previous knowledge levels of the students are also used as a source to provide personalized learning materials to them.

Software agents are a good choice for the implementation of adaptive e-learning systems [12]. Software agents have been one of the widely-used technologies in various fields and industries such as industrial engineering, economics and financial management, medical science, traffic and transportation, defense, business and management [13]. They have also been utilized in the e-learning field to provide multi-agent based e-learning systems [14–23].

Software agents technology is frequently used with Semantic Web technology [14,24]. As an important component of the Semantic Web, ontologies are used to develop data representation models in various domains [25]. In the e-learning field, ontologies are used for modeling students and learning materials [26]. These ontologies can be used in different e-learning systems, and therefore, increase interoperability of them [26]. Learning objects and learning style models have been modeled with ontologies in different studies [4,5,15–17,22,27–30]. E-learning systems that support both multi-agent and Semantic Web technologies include [15,16,22,31,32].

In this paper, we propose a multi-agent based adaptive e-learning system that utilizes ontological representation of Felder and Silverman learning style model as part of the student model. Learning materials provided by the lecturers are also annotated ontologically. Therefore, students can access learning materials personalized according to their learning styles by using the proposed e-learning system. Multi-agent based components of the system provide the interface for both the students and the lecturers using the system. The multi-agent features also support an effective control mechanism to monitor interaction of the students with the system and to observe the learning levels of each student.

The rest of the paper is organized as follows; Sect. 2 briefly introduces the software agent notion and usage of multi-agent systems in different domains. Then, the related work of multi-agent systems in e-learning domain is discussed. Section 3 presents the architecture and the prototype implementation of the proposed multi-agent based adaptive e-learning system. Finally, Sect. 4 concludes the paper with the future work directions intended for the first prototype.

2   Related Work

In this section, first, we will introduce software agents and multi-agent systems briefly. Then, we will examine how multi-agent systems are used in the e-learning domain by discussing various studies.

2.1   Multi-agent Systems

There are different definitions for the term agent. According to the most fundamental definition [33], an agent is a system that perceives its environment using its sensors and interact with its environment using its effectors. In order to classify a program as an agent there are some essential requirements that the program should have [34]: autonomy, social-ability, reactivity and pro-activeness. Autonomy means, programs defined as agents should be able to operate on their own without any intervention. Social ability indicates the necessity of being able to interact with other agents and people via an agent communication language. Reactivity requires the agent to sense its environment and respond to the changes in it when necessary. However, reactivity is not enough. Agents should be able to behave in a goal-directed manner and take initiative when conditions in the environment require so. Another property stated to distinguish agents from programs is temporal continuity [35], which lays the foundation for autonomy, social-ability, reactivity and pro-activeness.

Multi-agent systems is a subfield of distributed artificial intelligence which aims solving complex problems, that can not be solved by single agents, by using cooperating multiple agents. Multi-agent systems aid in the analysis and development of large-scale distributed systems and human-centered systems which are situated in complex, dynamic environments and require parallel processing [36]. That's why these systems are particularly recommended for developing systems composed of loosely coupled entities which can leave or join the system without any restrictions [37]. However, designing and programming multi-agent systems from scratch using plain programming languages is a hard task for engineers. This difficulty led to the proposal of various agent platforms and programming languages. In order to standardize these proposals Foundation of Intelligent Physical Agents of IEEE was formed in 1996. In 2005, FIPA standards were officially accepted.

Although there are many multi-agent systems, platforms and languages [38–40], only JADE, Jadex, JACK, EMERALD are fully FIPA compliant [39]. Among these platforms, JADE, Jadex and EMERALD are open source and free [41]. Since both Jadex and EMERALD are based on JADE [39], a JADE agent can run on both Jadex and EMERALD platforms [42,43]. This enables a prototyped system to be extended for Jadex system to use BDI agent architecture [43] or for EMERALD [42] system to use trust and reputation mechanisms. Among these three platforms, JADE is the most widely used one [44]. The reason for the widespread use of JADE is its ease to learn and develop system prototypes with [45–47] due to its active user community and many examples they have put on the Internet [39,48,49]. This ease of use is one of the reasons why JADE was chosen for prototyping the system introduced in this paper.

JADE is implemented using Java which makes it portable to different systems regardless of operating system or architecture. A JADE agent platform consists of multiple containers which may contain multiple agents. Since each JADE agent has a unique identifier, agents on different computers can communicate with each other transparently [44]. As a result of these features, a JADE platform may be distributed over multiple computers [39]. Additionally, since JADE is a scalable and fully FIPA compliant agent platform [44,50]. It provides the basic services like naming, yellow-pages (provided by the Directory Facilitator in FIPA compliant agent platforms), message-transport and parsing. JADE also has graphical debugging and deployment tools [41], a library of interaction protocols, an extendable kernel and supports the use of ontologies [39,44,50]. Scalability, FIPA compliance and portability of JADE are the other reasons for choosing it for prototyping our system.

2.2   Multi-agent Systems in E-Learning

E-learning systems, due to their inherent nature, are human-oriented, highly complex and distributed systems that require parallel processing [31,51–53]. Multi-agent systems are particularly recommended for modeling and implementing systems with these inherent properties. Therefore, in the literature there are many studies that utilize multi-agent technology for realizing e-learning systems [15–18,20–23,32,51–55]. There are also some recent studies that focus on multi-agent based e-learning system architectures developed with different agent development frameworks [16,22]. Programming environments of agents that contain e-learning materials and student models is another recent application area of multi-agent based e-learning systems [20,21].

There are three main types of agents in agent based e-learning systems: pedagogical, harvester and monitoring agents. Pedagogical agents are the most common type of agents found in e-learning systems [18,31,51–54]. These agents generally guide and motivate students during their studies and are represented as lifelike characters [56]. Pedagogical agents are generally interactive [52], and ask questions or provide solutions to achieve their goals [51]. Harvester agents are learning material collectors and generally run in parallel to harvest data and metadata from remote, heterogeneous repositories [19,22,32,51–53,55,57]. Finally, monitoring agents observe student activity within the system and provide a more personalized and enhanced learning experience by adapting the system via student profiling [15,16,18,54]. Some examples for what monitoring agents observe can be the types of learning materials preferred by the student, how much exercise is done while studying a subject, forum involvement [18], student performance during exercises and persistence on difficult tasks [54]. The educational components of agent based e-learning systems present in the literature can usually be classified based on these three agent types.

AgCAT catalog service [32] is a system that utilizes harvester agents only. There are three types of harvester agents in AgCAT: finder agents, librarian and inter-librarian agents. Finder agents search and retrieve learning objects registered by librarian agents. Librarian agents retrieve metadata from learning object repositories and store them in local catalogs. Inter-librarian agents federate learning object catalogs of remote AgCat systems.

Another adaptive system that utilizes agents is HAPA [53]. HAPA combines pedagogical and harvester agents. It is a system for supporting students in Java programming tasks too. Main focus of the system is code completion tasks. These tasks can be used as tests and exercises. HAPA harvester agents which run in parallel collect Java code examples from the web and give them to the classifier module which matches each code example with suitable lectures present in the system. Teachers using the system choose from these ready to use materials, select the parts to be completed by the students, and adds necessary hints to construct their lessons. These hints are used by the pedagogical agents. These agents interact with the student directly, observe his/her performance and adapt to him/her. They give the hints provided by the teacher and recommend additional learning material when student success decreases.

MagMAS [52] system is an example for adaptive systems that incorporate pedagogical agents which also act as monitoring agents. MagMAS is a system for Java programming tasks that focuses on code completion tasks. It is a re-implementation of the Mag system [58] by means of agents. MagMAS system uses pedagogical agents which also monitor the students [52]. In this system, marks and scores of each learner along with the time he/she spent on each material and past performances and learning history are monitored. The information gathered is used to build a learner model, which in turn is used by the pedagogical agents to recommend suitable learning materials, and give hints where necessary to each learner.

ElectronixTutor [54] is a system where pedagogical agents are used as tutors. This system aids in electronics education utilizing a recommender system. The system observe student behavior using a component called Learning Record Store which uses this data for building student models. Learning Record Store observes several components of the student performance like subject knowledge and grit while studying on topics. The resultant student model is used by the recommender system for material and tutor recommendation later on. There are specialized pedagogical agents for student profiles that can be classified as beginner, intermediate and advanced students. The recommender system decides which pedagogical agent to suggest depending on the student performance.

A system that utilizes pedagogical agents, which was also an inspiration for the system prototyped in this study, is eTeacher [18]. The pedagogical agent "eTeacher" in this system acts as a monitoring agent too. An overview of the "eTeacher" agent's functionality is shown in Fig. 1. The "eTeacher" agent observes students' interactions with the e-learning system inconspicuously. It gathers data about the student like his/her learning style, the number and type of the exercises done, his/her grades, and participation in chat rooms. A Bayesian network is used for finding out the students' learning styles. The student profile is then used by the "eTeacher" to provide guidance when needed regarding the materials to continue with, chats to participate in or questions to answer.

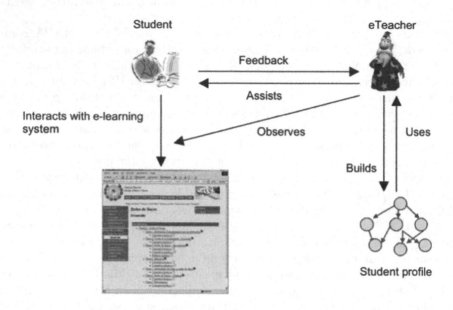

Fig. 1. An overview of the "eTeacher" agent's functionality in [18].

3   Multi-agent Based Adaptive E-Learning System

In this section, architecture of the proposed multi-agent based adaptive e-learning system is detailed based on related studies. After that, a prototype implementation of the proposed architecture is described.

3.1   Architecture of the Proposed E-Learning System

The architectural design of the proposed the multi-agent based adaptive e-learning system is based on related work such as [18,22,59].

As stated in the previous section, functionality that the agents offer in our system is inspired by the "eTeacher" agent in [18]. In a similar manner, we designed "student" agent type in our system that provides personalized learning materials to the students based on their learner models and keeps track of the interaction of the students with the e-learning system to assist them for an enhanced learning experience. Additionally, we designed another agent type called "lecturer" agent in our system that assists the lecturers in providing adaptive learning materials.

The proposed e-learning system is also based on the functionalities offered by intelligent tutoring systems. The architecture of the Protus intelligent tutoring system proposed by [59] is given in Fig. 2. Software agents were not used in [59]; however, the architecture provided in this study contains components such as domain module, learner model, application module, adaption module and session monitor that offer similar functionalities. The learner model in [59] provides similar functionality for building and using a student profile in the "eTeacher" system [18]. Besides, session monitor is also similar as it observes students' interactions with the system during a session. The "student" agent type in our system performs the operations supported by the adaption and application modules autonomously. Finally, the functionality of the domain module is provided by the "lecturer" agents in our system.

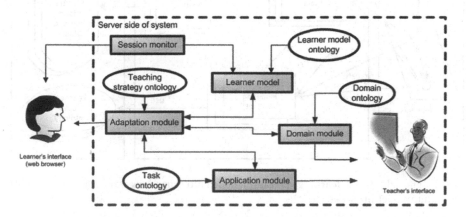

Fig. 2. The architecture of the Protus intelligent tutoring system in [59].

Another study that our system is based on is the multi-agent based architecture proposed by [22]. The architecture of [22] which is based on the Agents and Artifacts Metamodel [60], is demonstrated in Fig. 3. The environment modeling in multi-agent systems was emphasized in [22] and the focus was on the abstraction of the components in the environment of agents such as student models and learning object repositories with artifacts in a similar manner to [20,21]. In this study, we focused fully on how to support the functionalities provided by the agents in an intelligent way, and therefore, decided to skip the environment modeling aspect of multi-agent system programming.

As a result of the literature review, we decided to design an architecture that focuses on functionalities supported by the related studies. Thus, we designed two types of agents named as "student" and "lecturer" and focused on the interaction of them. As the names suggest, these agents represent and are used by students and lecturers, respectively.

"Student" agents monitor the interaction between the student and the system. The behaviors observed include the time spent on certain types of learning materials and mistakes made during learning sessions. "Student" agents are also responsible for choosing the learning materials suitable for the student. As students engage with these materials, further observations are done, gathered data

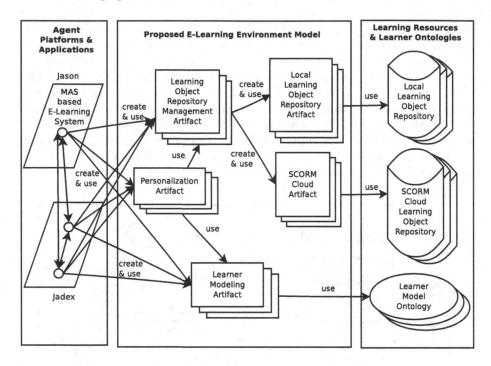

Fig. 3. The adaptive e-learning environment architecture based on Agents and Artifacts Metamodel proposed in [22].

are processed and the resulting data are stored in the student model. If success level of a student an agent is responsible for drops significantly, a notification is sent to the "lecturer" agent for further reasoning about the student.

"Lecturer" agents request metadata from lecturers for the learning materials they supply. "Lecturer" agents work collaboratively with "student" agents, process the student data on notifications they received from "student" agents, and assist the students by means of the information they supply to student agents if they are at risk of dropping the course. When the activities of both agent types are considered, both of them can be considered as pedagogical and monitoring agents.

The architecture of the proposed adaptive e-learning system is demonstrated in Fig. 4. In our architecture, "student" and "lecturer" agents communicate with their users via separate interfaces. All of the agents register themselves to the DF (Directory Facilitator) agent. This enables them to search for other agents via DF and communicate with them using agent messages later on. Student modeling layer includes student models in which student information such as demographic information, learning styles, courses taken and interaction history with the e-learning system are stored. Learning materials layer contains learning contents personalized according to learning styles.

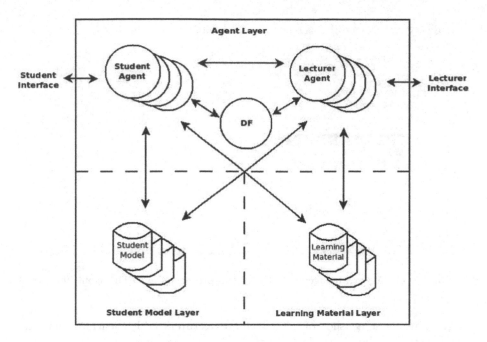

Fig. 4. Architecture of the proposed multi-agent based adaptive e-learning system.

3.2   Prototype Implementation

We implemented a prototype of the proposed adaptive e-learning architecture with JADE agent development platform. JADE is one of the most commonly used agent platforms which provides a GUI based interface. There are e-learning systems such as [16,19] that have been successfully implemented with JADE. Thus, we decided to use JADE for the prototype implementation.

Ontologies serve as a common vocabulary between software agents and enhance system interoperability. This is also true for e-learning systems [61]. In addition to this, ontologies facilitate learning material description, sharing and search on the Web [61]. There are many studies that use ontologies for student and learning material modeling [16,23,27,28,30,59]. Thus, we decided to use ontological representation for student modeling in our prototype. An ontological student model based on Felder and Silverman, Kolb and Honey-Mumford learning style models was used in [23]. In our prototype, we utilized a subset of this model [23] and focused only on the Felder and Silverman learning style model. However, we extended this model subset to store the interaction history of students with the e-learning system.

JADE provides the Remote Agent Management GUI shown in Fig. 5 where agents running on the prototype can be observed and managed. A student agent named "learner1" and a lecturer agent called "provider1" running on JADE can be seen in Fig. 5.

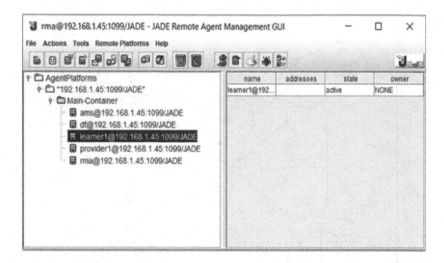

Fig. 5. Remote Agent Management GUI of the prototype in JADE agent platform.

An example screenshot of a GUI where lecturers add a learning material to the system is shown in Fig. 6. In this GUI, the lecturer enters metadata of a learning material such as title, description, keywords and learning style, and then, selects the source file of the learning material.

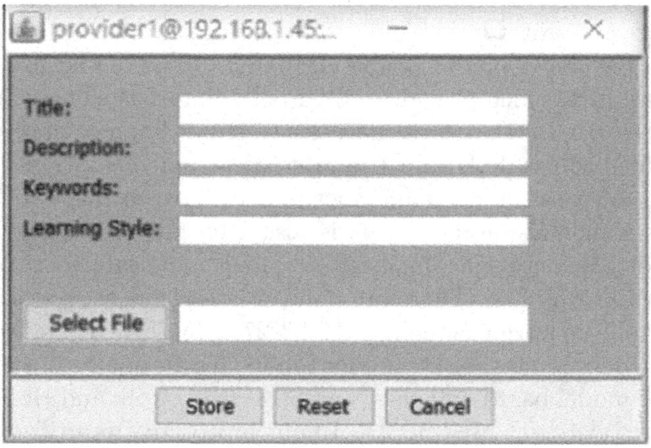

Fig. 6. An example GUI of the prototype for adding learning materials.

The functions provided at this stage of the prototype implementation are adding learning materials to the system, learning material recommendation according to student learning style (which is implemented based on our previous work in [62]), and determining at risk students who interact with the system at a lower rate than the average interaction rate. These functions are still under development and will be enhanced as the implementation goes on.

4 Conclusion

In this paper, we presented a multi-agent based adaptive e-learning system that is prototyped with JADE agent platform. By providing adaptivity based on learning styles of the students, it is aimed to provide higher success levels compared to non-personalized e-learning systems. The multi-agent based nature of the system provides an effective control mechanism to monitor the interaction of students with the system and to observe their learning levels. We used an ontology to model students to increase re-usability of the student model and interoperability of the proposed system.

As future work, a learning material ontology can be developed to provide a modular ontological model that can be used by other e-learning systems, too. In this regard, the prototype implementation with JADE can be extended to utilize other agent platforms such as Jason and Jadex. Additionally, in the following phases of the prototype implementation, we plan to add enhanced reasoning capability to the lecturer agent.

This study focuses on a multi-agent based architecture proposal. Implementation of the prototype still continues. When the prototype is fully implemented, we plan to test the impact of the architecture on student performance. In order to conduct these tests, personalized learning materials should be prepared. In this context, as another future work, it is planned to use the proposed system in a course where the educational content is personalized with Felder and Silverman learning styles model.

Acknowledgments. This study was supported by Ege University Scientific Research Projects Coordination Unit (Project number 18-MUH-035).

References

1. Soni, V.D.: Global Impact of E-learning during COVID 19 (2020). Available at SSRN: https://ssrn.com/abstract=3630073
2. Almaiah, M.A., Al-Khasawneh, A., Althunibat, A.: Exploring the critical challenges and factors influencing the E-learning system usage during COVID-19 pandemic. Educ. Inf. Technol. **25**(6), 5261–5280 (2020). https://doi.org/10.1007/s10639-020-10219-y
3. Elumalai, K.V., et al.: Factors affecting the quality of E-learning during the COVID-19 pandemic from the perspective of higher education students. In: COVID-19 and Education: Learning and Teaching in a Pandemic-Constrained Environment, p. 189 (2021)
4. Sangineto, E., Capuano, N., Gaeta, M., Micarelli, A.: Adaptive course generation through learning styles representation. Univ. Access Inf. Soc. **7**(1–2), 1–23 (2008). https://doi.org/10.1007/s10209-007-0101-0

5. Essalmi, F., Ayed, L.J.B., Jemni, M., Kinshuk, Graf, S.: A fully personalization strategy of e-learning scenarios. Comput. Hum. Behavior **26**(4), 581–591 (2010)
6. Ciloglugil, B., Inceoglu, M.M.: User modeling for adaptive e-learning systems. In: Murgante, B., et al. (eds.) ICCSA 2012. LNCS, vol. 7335, pp. 550–561. Springer, Heidelberg (2012)
7. Akbulut, Y., Cardak, C.S.: Adaptive educational hypermedia accommodating learning styles: a content analysis of publications from 2000 to 2011. Comput. Educ. **58**(2), 835–842 (2012)
8. Truong, H.M.: Integrating learning styles and adaptive e-learning system: current developments, problems and opportunities. Comput. Hum. Behav. **55**, 1185–1193 (2015)
9. Ozyurt, O., Ozyurt, H.: Learning style based individualized adaptive e-learning environments: content analysis of the articles published from 2005 to 2014. Comput. Hum. Behav. **52**, 349–358 (2015)
10. Ciloglugil, B.: Adaptivity based on Felder-Silverman learning styles model in E-learning systems. In: 4th International Symposium on Innovative Technologies in Engineering and Science, ISITES 2016, pp. 1523–1532 (2016)
11. Felder, R.M., Silverman, L.K.: Learning and teaching styles in engineering education. Eng. Educ. **78**(7), 674–681 (1988)
12. Gregg, D.G.: E-learning agents. Learn. Organ. **14**, 300–312 (2007)
13. Khemakhem, F., Ellouzi, H., Ltifi, H., Ayed, M.B.: Agent-based intelligent decision support systems: a systematic review. IEEE Trans. Cogn. Dev. Syst., 1 (2020). https://doi.org/10.1109/TCDS.2020.3030571
14. Ciloglugil, B., Inceoglu, M.M.: Developing adaptive and personalized distributed learning systems with semantic web supported multi agent technology. In: 10th IEEE International Conference on Advanced Learning Technologies, ICALT 2010, Sousse, Tunesia, 5–7 July 2010, pp. 699–700. IEEE Computer Society (2010)
15. Dung, P.Q., Florea, A.M.: An architecture and a domain ontology for personalized multi-agent e-learning systems. In: Third International Conference on Knowledge and Systems Engineering, KSE 2011, pp. 181–185. IEEE. (2011)
16. Rani, M., Nayak, R., Vyas, O.P.: An ontology-based adaptive personalized e-learning system, assisted by software agents on cloud storage. Knowl.-Based Syst. **90**, 33–48 (2015)
17. Sun, S., Joy, M., Griffiths, N.: The use of learning objects and learning styles in a multi-agent education system. J. Interact. Learn. Res. **18**(3), 381–398 (2007)
18. Schiaffino, S., Garcia, P., Amandi, A.: eTeacher: providing personalized assistance to e-learning students. Comput. Educ. **51**(4), 1744–1754 (2008)
19. Sandita, A.V., Popirlan, C.I.: Developing a multi-agent system in JADE for Information management in educational competence domains. Procedia Econ. Finance **23**, 478–486 (2015)
20. Ciloglugil, B., Inceoglu, M.M.: Exploiting agents and artifacts metamodel to provide abstraction of E-learning resources. In: 17th IEEE International Conference on Advanced Learning Technologies, ICALT 2017, Timisoara, Romania, 3–7 July 2017 (2017)
21. Ciloglugil, B., Inceoglu, M.M.: An agents and artifacts metamodel based E-learning model to search learning resources. In: Gervasi, O., et al. (eds.) ICCSA 2017. LNCS, vol. 10404, pp. 553–565. Springer, Cham (2017). https://doi.org/10.1007/978-3-319-62392-4_40
22. Ciloglugil, B., Inceoglu, M.M.: An adaptive E-learning environment architecture based on agents and artifacts metamodel. In: 18th IEEE International Conference on Advanced Learning Technologies, ICALT 2018, Mumbai, India, 9–13 July 2018 (2018)

23. Ciloglugil, B., Inceoglu, M.M.: A learner ontology based on learning style models for adaptive E-learning. In: Gervasi, O., et al. (eds.) ICCSA 2018. LNCS, vol. 10961, pp. 199–212. Springer, Cham (2018). https://doi.org/10.1007/978-3-319-95165-2_14

24. Marik, V., Gorodetsky, V., Skobelev, P.: Multi-agent technology for industrial applications: barriers and trends. In: 2020 IEEE International Conference on Systems, Man, and Cybernetics (SMC), pp. 1980–1987 (2020)

25. Berners-Lee, T., Hendler, J., Lassila, O.: The semantic web. Sci. Am. **284**(5), 34–43 (2001)

26. Ciloglugil, B., Inceoglu, M.M.: Ontology usage in E-learning systems focusing on metadata modeling of learning objects. In: International Conference on New Trends in Education, ICNTE 2016, pp. 80–96 (2016)

27. Essalmi, F., Ayed, L.J.B., Jemni, M., Kinshuk, Graf, S.: Selection of appropriate e-learning personalization strategies from ontological perspectives. Interact. Des. Archit. J. IxD&A **9**(10), 65–84 (2010)

28. Valaski, J., Malucelli, A., Reinehr, S.: Recommending learning materials according to ontology-based learning styles. In: Proceedings of the 7th International Conference on Information Technology and Applications, ICITA 2011, pp. 71–75 (2011)

29. Yarandi, M., Jahankhani, H., Tawil, A.R.H.: A personalized adaptive e-learning approach based on semantic web technology. Webology **10**(2), Art-110 (2013)

30. Kurilovas, E., Kubilinskiene, S., Dagiene, V.: Web 3.0-based personalisation of learning objects in virtual learning environments. Comput. Hum. Behav. **30**, 654–662 (2014)

31. Gago, I.S.B., Werneck, V.M.B., Costa, R.M.: Modeling an educational multi-agent system in MaSE. In: Liu, J., Wu, J., Yao, Y., Nishida, T. (eds.) AMT 2009. LNCS, vol. 5820, pp. 335–346. Springer, Heidelberg (2009). https://doi.org/10.1007/978-3-642-04875-3_36

32. Barcelos, C., Gluz, J., Vicari, R.: An agent-based federated learning object search service. Interdisc. J. E-Learn. Learn. Objects **7**(1), 37–54 (2011)

33. Norvig, P.R., Intelligence, S.A.: A Modern Approach. Prentice Hall, Upper Saddle River (2002)

34. Wooldridge, M.J., Jennings, N.R.: Intelligent agents: theory and practice. Knowl. Eng. Rev. **10**(2), 115–152 (1995)

35. Franklin, S., Graesser, A.: Is It an agent, or just a program?: a taxonomy for autonomous agents. In: Müller, J.P., Wooldridge, M.J., Jennings, N.R. (eds.) ATAL 1996. LNCS, vol. 1193, pp. 21–35. Springer, Heidelberg (1997). https://doi.org/10.1007/BFb0013570

36. Julian, V., Botti, V.: Multi-agent systems. Appl. Sci. **9**(7), 1402 (2019). MDPI AG

37. Mariani, S., Omicini, A.: Special issue "multi-agent systems". Appl. Sci. **9**(5), 954 (2019). MDPI AG

38. Leon, F., Paprzycki, M., Ganzha, M.: A review of agent platforms. In: Multi-Paradigm Modelling for Cyber-Physical Systems (MPM4CPS), ICT COST Action IC1404, pp. 1–15 (2015)

39. Kravari, K., Bassiliades, N.: A survey of agent platforms. J. Artif. Soc. Soc. Simul. **18**(1), 11 (2015)

40. Bordini, R.H., et al.: A survey of programming languages and platforms for multi-agent systems. Informatica **30**(1), 33–44 (2006)

41. Pal, C. V., Leon, F., Paprzycki, M., Ganzha, M.: A review of platforms for the development of agent systems. arXiv preprint arXiv:2007.08961 (2020)

706 B. Ciloglugil et al.

42. Kravari, K., Kontopoulos, E., Bassiliades, N.: EMERALD: a multi-agent system for knowledge-based reasoning interoperability in the semantic web. In: Konstantopoulos, S., Perantonis, S., Karkaletsis, V., Spyropoulos, C.D., Vouros, G. (eds.) SETN 2010. LNCS (LNAI), vol. 6040, pp. 173–182. Springer, Heidelberg (2010). https://doi.org/10.1007/978-3-642-12842-4_21

43. Braubach, L., Lamersdorf, W., Pokahr, A.: Jadex: implementing a BDI-infrastructure for JADE agents. EXP Search Innov. **3**(3), 76–85 (2003)

44. Bergenti, F., Caire, G., Monica, S., Poggi, A.: The first twenty years of agent-based software development with JADE. Auton. Agents Multi-Agent Syst. **34**(2), 1–19 (2020). https://doi.org/10.1007/s10458-020-09460-z

45. Balachandran, B.M., Enkhsaikhan, M.: Developing multi-agent E-commerce applications with JADE. In: Apolloni, B., Howlett, R.J., Jain, L. (eds.) KES 2007. LNCS (LNAI), vol. 4694, pp. 941–949. Springer, Heidelberg (2007). https://doi.org/10.1007/978-3-540-74829-8_115

46. Zhao, Z., Belloum, A., De Laat, C., Adriaans, P., Hertzberger, B.: Using Jade agent framework to prototype an e-Science workflow bus. In: Seventh IEEE International Symposium on Cluster Computing and the Grid (CCGrid 2007), pp. 655–660 (2007)

47. Kularbphettong, K., Clayton, G., Meesad, P.: e-Wedding based on multi-agent system. In: Demazeau, Y., et al. (eds.) Trends in Practical Applications of Agents and Multiagent Systems. Advances in Intelligent and Soft Computing, vol. 71, pp. 285–293. Springer, Heidelberg (2010). https://doi.org/10.1007/978-3-642-12433-4_34

48. van Moergestel, L., Puik, E., Telgen, D., van Rijn, R., Segerius, B., Meyer, J.J.: A multiagent-based agile work distribution system. In: 2013 IEEE/WIC/ACM International Joint Conferences on Web Intelligence (WI) and Intelligent Agent Technologies (IAT), vol. 2, pp. 224–230. IEEE (2013)

49. Scutelnicu, A., Lin, F., Kinshuk, Liu, T., Graf, S., McGreal, R.: Integrating JADE agents into moodle. In: Proceedings of the International Workshop on Intelligent and Adaptive Web-Based Educational Systems, pp. 215–220 (2007)

50. Bellifemine, F.L., Caire, G., Greenwood, D.: Developing Multi-agent Systems with JADE, vol. 7. Wiley, Hoboken (2007)

51. Klašnja-Milićević, A., Vesin, B., Ivanović, M., Budimac, Z., Jain, L.C.: Agents in E-learning environments. In: E-Learning Systems. ISRL, vol. 112, pp. 43–49. Springer, Cham (2017). https://doi.org/10.1007/978-3-319-41163-7_5

52. Ivanović, M., Mitrović, D., Budimac, Z., Vesin, B., Jerinić, L.: Different roles of agents in personalized programming learning environment. In: Chiu, D.K.W., Wang, M., Popescu, E., Li, Q., Lau, R. (eds.) ICWL 2012. LNCS, vol. 7697, pp. 161–170. Springer, Heidelberg (2014). https://doi.org/10.1007/978-3-662-43454-3_17

53. Ivanović, M., Mitrović, D., Budimac, Z., Jerinić, L., Bădică, C.: HAPA: harvester and pedagogical agents in e-learning environments. Int. J. Comput. Commun. Control **10**(2), 200–210 (2015)

54. Graesser, A.C., et al.: ElectronixTutor: an intelligent tutoring system with multiple learning resources for electronics. Int. J. STEM Educ. **5**(1), 15 (2018)

55. Sharma, S., Gupta, J.P.: A novel architecture of agent based crawling for OAI resources. Int. J. Comput. Sci. Eng. **2**(4), 1190–1195 (2010)

56. Heidig, S., Clarebout, G.: Do pedagogical agents make a difference to student motivation and learning? Educ. Res. Rev. **6**(1), 27–54 (2011)

57. De la Prieta, F., Gil, A.B.: A multi-agent system that searches for learning objects in heterogeneous repositories. In: Demazeau, Y., et al. (eds.) Trends in Practical Applications of Agents and Multiagent Systems. Advances in Intelligent and Soft Computing, vol. 71, pp. 355–362. Springer, Heidelberg (2010). https://doi.org/10.1007/978-3-642-12433-4_42
58. Ivanović, M., Pribela, I., Vesin, B., Budimac, Z.: Multifunctional environment for e-learning purposes. Novi Sad J. Math. **38**(2), 153–170 (2008)
59. Klašnja-Milićević, A., Vesin, B., Ivanović, M., Budimac, Z.: E-Learning personalization based on hybrid recommendation strategy and learning style identification. Comput. Educ. **56**(3), 885–899 (2011)
60. Ricci, A., Piunti, M., Viroli, M.: Environment programming in multi-agent systems: an artifact-based perspective. Auton. Agent. Multi-Agent Syst. **23**(2), 158–192 (2011). https://doi.org/10.1007/s10458-010-9140-7
61. Dalipi, F., Idrizi, F., Rufati, E., Asani, F.: On integration of ontologies into e-learning systems. In: 2014 Sixth International Conference on Computational Intelligence, Communication Systems and Networks, pp. 149–152. IEEE (2014)
62. Ciloglugil, B., Inceoglu, M.M.: A Felder and Silverman learning styles model based personalization approach to recommend learning objects. In: Gervasi, O., et al. (eds.) ICCSA 2016. LNCS, vol. 9790, pp. 386–397. Springer, Cham (2016). https://doi.org/10.1007/978-3-319-42092-9_30

University Web-Environment Readiness for Online Learning During COVID-19 Pandemic: Case of Financial University

Mikhail Eskindarov⬝, Vladimir Soloviev(✉) ⬝, Alexey Anosov⬝, and Mikhail Ivanov⬝

Financial University Under the Government of the Russian Federation, 38 Shcherbakovskaya, Moscow 105187, Russia

{MEskindarov,VSoloviev,AAAnosov,MNIvanov}@fa.ru

Abstract. The paper is devoted to the Financial University experience of transforming the learning process fully online during the COVID-19 pandemic. The architecture and principles of the web-environment were rethought significantly. It is shown that the competence in digital technologies formed in universities creates new challenges and tasks: continuity of digital footprints and technologies for learning process organization and implementation; openness and publicity of learning; professional development of teachers, and ethics in the digital space.

Keywords: Online learning · COVID-19 · Web

1 Introduction

The COVID-19 pandemic, which has covered almost all regions of the globe in a short time, has radically changed society's usual way, relations between people and countries. More than a hundred million people fell ill, several million died. In epidemiologically difficult areas of the world, severe quarantine restrictions have been and are still being introduced, depending on the situation, including the movement and communication of people, their participation in mass culture, sports, and other events. Many concerts and performances, sports competitions, including the 2020 Summer Olympics, world and continental championships in selected sports, have been canceled or postponed indefinitely. Various international and national conferences, scientific forums, and professional meetings have switched to online.

Significant damage has been done to the economies of different countries. The production and sales of numerous goods and services are declining. The standard of living of many people is falling. Transport, tourism, and other sectors of the economy are suffering severe losses. The sphere of education has undergone significant changes, having switched to a distance format. It caused many problems and difficulties that needed to be promptly resolved with the least cost and loss.

According to UNESCO data from March 18, 2020, due to the COVID-19 pandemic, educational institutions in 107 countries of the world were closed, which affected, in

O. Gervasi et al. (Eds.): ICCSA 2021, LNCS 12951, pp. 708–717, 2021.
https://doi.org/10.1007/978-3-030-86970-0_49

particular, 862 million schoolchildren, or about half of the school-age population world-wide [1]. A significant number of college and university students were forced to switch to a remote learning format. According to the Minister of Science and Higher Education of Russia Valery Falkov, in Russia, as of mid-December 2020, about 3 million students studied remotely, which was about 64% of the total number of all Russian students [2].

The transition to distance learning was aimed at reducing contacts between people. It contributed to a decrease in the number of cases, reducing the peak of tension due to the shortage in hospital beds, qualified medical personnel, and medicines at the beginning of the pandemic spread.

Nevertheless, this raises reasonable questions. How to measure the efficiency of distance learning? Are there any relevant methodological developments for organizing such training? Are there technological platforms and technical tools that support the remote educational format? Is it possible to quickly deploy them in sufficient numbers? What are the financial costs of the transition? How long will it take to prepare professors and students for remote learning? How effective will online learning be? Will there be a decline in the level of education? How can online learners be identified? Will the remote format contribute to falsifications and substitutions? How should you control the assimilation of educational material? How to assess the competencies acquired by students, theoretical knowledge, practical skills, and abilities? Moreover, many more other questions can probably be formulated when a vital necessity requires an urgent transition to distance learning.

2 Literature Review

Distance learning is a platform that educational institutions use to meet their learners' needs [1, 3, 4]. It has been established that distance learning is quite efficient when used in colleges and universities. Simultaneously, distance learning for elementary school students, when children must stay at home, as in the COVID-19 pandemic, causes some anxiety in many parents and creates many problems for them, including financial ones. Parents should help their primary school children in the most effective use of educational technologies, but not all of them have the appropriate academic and information technology skills [3]. Besides, not every family has computers with multimedia capabilities. Somewhere there is no Internet, or it works slowly or unstably, someone has several children of school age, and their classes are scheduled simultaneously [1].

Many works note that the course on digitalization of education previously adopted at universities played a positive role: by the beginning of the pandemic, most teachers already had digital competencies, and electronic teaching aids were developed in many academic disciplines. At the same time, the arrival of the pandemic in many educational institutions was fraught with many problems, difficulties, and shortcomings, the elimination of which required additional funding, time, the search for new methods and forms of conducting classes, and the improvement of the digital qualifications of individual teachers.

In [5], based on the analysis of the questionnaires of students and teachers of the Faculty of Pedagogical Sciences of the University of Granada (Spain), it was indicated the dissatisfaction of many students with the virtual educational process, to which they had to switch urgently. Many professors in their questionnaires noted the need to improve and generalize experience in conducting classes in a virtual environment and develop their digital literacy competencies and skills. The requirements were also recorded:

- on building up the bank of e-learning materials, first of all, audiovisual courses and banks of test items for the comprehensive assessment of students' knowledge and competencies;
- on the need to purchase modern multimedia and network equipment and specialized software;
- on teaching students and professors current technological capabilities for video and graphics processing.

Some university courses, in principle, cannot be entirely carried out in a distance form. In [6], the experience of using a mixed five-component online teaching strategy for first-year chemistry students at the University of Santo Tomas (Philippines) during a pandemic is presented. The strategy is called DLPCA (Discover, Learn, Practice, Collaborate, Evaluate). The asynchronous part of this hybrid strategy involves pre-recording lectures on video and posting on YouTube so that students can master the material at their own pace, as needed – repeatedly. The synchronous part of the training is conducted online using Zoom or Google Meetings video conferencing platforms. In the laboratory, students' knowledge and skills are assessed based on studying the lecture material. Practical experiments are carried out under a teacher's guidance by invited instructors – senior students, implementing actions proposed by a student at home, outside the laboratory, and answering the teacher's question. Questionnaires assessed the mixed learning strategy. Both students and teachers rated it positively. Among the negative points were the instability of the Internet and some students' constraint to turn on the video camera if other people nearby, such as parents.

In [7], it is noted that knowledge is a crucial resource of production, a source of sustainable competitive advantage in the creation of values and wealth. Consequently, knowledge must be effectively managed using intelligent methods of analysis and processing of various indicators that directly or indirectly affect the result: the knowledge and competencies of university graduates, in other words, their human capital. Using various models, approaches, and methods of knowledge formation in an educational organization, for example, in the period before and during the COVID-19 pandemic, it is possible, by use of machine learning, to obtain effective models of an individual, defining indicators, positively influencing the learning results [8, 9].

In [10], the teachers' and students' perception of the use of open digital innovations in leading universities in Spain, Italy, and Ecuador, which were severely affected by the pandemic in the spring of 2020, was assessed. It is noted that the digital environment has given rise to new learning paradigms associated with IT advances – multimedia, cloud services, wireless communications. Depending on the digital platforms and technological methods used, different learning approaches are used: online, mixed, and face-to-face. Thanks to the Internet, many educational processes are implemented online. The pandemic also contributed to this. The educational process is actively moving into online mode.

Moreover, there is no need to be afraid of possible deception on the part of careless students. Modern technological platforms used in online learning, for example, Teams, Skype, Zoom, in conjunction with video stream analysis modules based on intelligent machine learning methods, allow real-time tracking of students' emotions [11, 12] and their behavior when preparing for exams [13]. It all depends on the student's technical equipment, teacher's location during the exam, and organizational issues.

Further, if we want to prepare competitive graduates that meet the realities and needs of the 21st century, we, teachers, as noted in [14], need to better assess our students, not just by exam marks associated with answers to individual questions and solutions of specific, narrowly focused tasks. It is necessary to evaluate the student comprehensively, based on the results of solving urgent, complex tasks related to real-world problems and practical needs, requiring research and experiments, based on knowledge and competencies in several courses.

3 Online Learning in the Context of COVID-19 Pandemic

Many educational institutions in the world, including the Financial University, have switched to distance learning, both when conducting lectures and seminars and when taking tests and exams, coursework, and final qualification works. Such a transition is a forced measure aimed at reducing coronavirus spread, preserving students' and their teachers' health and lives.

According to many testimonies of teachers, students, and their parents in Russia, the learning process transfer to a distance format was painless. However, it was not devoid of some roughnesses and overlaps. Educational institutions had to quickly ensure the availability of information technology resources that allow working remotely. It was required to purchase licenses for specialized software for all participants in the educational process, purchase and update network equipment, multimedia tools, and increase network bandwidth. It was necessary to provide teachers and students with visual presentation materials and video instructions to master the technology of online training sessions independently or with the help of volunteers or participate in them.

It was also required to organize in a pandemic the uninterrupted functioning and interconnection of the university's structural units, their well-coordinated and efficient work, and operational availability to solve emerging urgent problems, when most of the employees were forced to be outside the university, in isolation. The generalized IT landscape supporting the educational process is shown in Fig. 1.

For the teaching staff, distance learning technologies are characterized by some more features. Changes in pedagogical methods of presenting educational material are required. Compared with traditional teaching technologies, it is necessary to work out new ways of organizing feedback with students and monitoring the assimilation of the teacher's questions. The forms of consultations, acceptance of tests and examinations, and other control means should be improved.

The quality of conducting distance learning largely depends not only on the teacher's professional and methodological skill and the students' activity but also on the capabilities of specialized packages that provide remote work, from stationary computers and mobile devices operating on various operating systems on the Internet.

At the Financial University, most lectures, seminars, and other classes are conducted on the Microsoft Teams platform (Fig. 2).

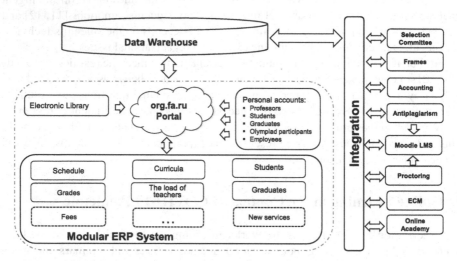

Fig. 1. Generalized IT landscape supporting the educational process.

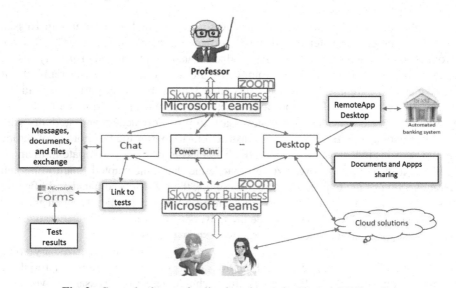

Fig. 2. General scheme of online learning at the Financial University.

Microsoft Teams allows professors and students to demonstrate presentations and different software during meetings and videoconferences. While all the standard classes in the Financial University during the pandemic were conducted on the Microsoft Teams, we also used Skype for business for staff internal meetings and Zoom for some research conferences.

Assignments to students can be issued during classes using standard LMS tools or OneNote Class notebooks. The use of videoconferencing in some cases requires an addition. Thus, studying Banking information systems, or Accounting, involves using an automated banking system or accounting software. We use Microsoft RemoteApp technology to give each student the ability to use remote applications with the same look and feel as local applications.

4 Discussion

The forced transition to the widespread use of online learning tools and distance educational technologies in 2020 caused a profound and complete rethinking of the university digitalization agenda issues. Understanding the strengths, weaknesses, opportunities, and threats associated with digital transformation at the present stage has radically changed.

Many challenging questions that had previously been heatedly discussed faded into the background or changed their meaning:

- the required level of IT competencies of teachers and students to participate in the educational process using distance learning technologies;
- equipping teachers and students with technical means and communication channels;
- choosing a platform for conducting the educational process, etc.

In the context of a real uncontested transition to the use of distance learning technologies, all these issues turned out to be solvable and not crucial.

At the same time, there was a rethinking of the value of distance learning technologies' widespread introduction. We can state that, contrary to the forecasts of ardent supporters of the use of digital tools, such a general use of distance educational technologies did not entail revolutionary changes in the university process management system. The formed competence in the field of digital technologies creates new challenges and tasks.

4.1 Digital Footprint Continuity

During the period of widespread use of distance technologies, the practice of forming a digital footprint about training has developed. Data about all classes conducted using Microsoft Teams fell into a single registry. Now you can set the time of the classes, the participants, the end time. Video materials have been generated for many lessons. Operationally, the management staff of departments and faculties could connect to classes and control the conduct.

Obviously, it is required, on the one hand, to maintain the level of managerial accessibility of the educational process and continue the accumulation of factual data on the educational process. On the other hand, we need to organize systematic analytical work aimed at finding hidden growth points. Thus it is necessary to preserve and expand digital tools for managing the educational process.

Analytical tasks need to be transformed from a narrow function of a small group of people into daily functions of each employee responsible for making any decisions, both in the learning process and in the university as a whole. In the target model, each decision-maker must be provided with independent direct access to the data necessary for analysis and subsequent decision-making and must also have the competencies required to work with such data.

Unfortunately, nowadays, significant amounts of information remain not analyzed yet for technical and organizational reasons. That, of course, is a growth point for the university both in managerial and academic terms.

4.2 Continuity of Technologies for Learning Process Organization and Implementation

Using webinars to conduct teaching during the pandemic has been very often criticized. Undoubtedly, this tool's full use cannot wholly replace the educational process in its classical understanding. At the same time, the infrastructure created at the university already makes it possible to combine webinars and classical classes effectively. The Financial University effectively operates a toolkit that allows you to control the online and hybrid learning process. The use of webinars in the conditions of professors' full methodological readiness for many classes may be preferable. The universities now have all the tools and skills for organizing, coordinating, and monitoring webinars. Now it would be advisable to find a new meaning for the experience gained by such labor.

4.3 Openness and Publicity of Learning

The learning process during the pandemic became public. University leadership, students, faculty, and staff observed training and exams previously limited by the classroom door. It became possible to compare formal educational content and the methodology of its presentation by professors. It creates new requirements for teachers in terms of preparation for online classes.

When classes can be explicitly or hidden recorded, additional work is required to develop and improve the presentation and teaching materials. The openness of online classes can and should be transferred into our classrooms, allowing us to guarantee the educational process's quality and effectively combine distance and classical technologies.

In this regard, the task of total equipping all classrooms with video cameras and sensitive microphones is of particular relevance. Such a solution was developed in the Financial University in 2017 and installed in one of the university buildings [13]. Still, in the last months of the pandemic, it was decided to implement it everywhere. It will also allow, in the event of new episodes of infectious diseases, to organize the learning process with less human and financial costs.

Of course, the learning process's new needs also require activating the teaching staff's advanced training based on digital tools.

4.4   Ethics in the Digital Space

During the pandemic, we also faced problems in terms of ethics of relations in the digital learning environment. We managed to effectively and quickly solve the sensational issues related to pranking in online classes.

However, at the same time, it is necessary to revise the regulations that separate digital freedom from digital hooliganism. We need to develop well-established mechanisms for representing each person in the digital walls of the universities.

Now we are forced to attend to what we previously considered entertainment or irrelevant issues: the image on the "avatars," the rights of access to publishing content in lecture materials. Now, in calm conditions, it is necessary to develop operating mechanisms for years to come.

5   Conclusion

The COVID-19 pandemic has forced educational institutions worldwide to switch to remote education urgently. In many universities, in which the learning process transformation strategy into the digital environment has been actively developing over the past years, this process was almost painless. At least at first, some educational institutions faced specific difficulties and problems – lack of appropriate technical means, insufficiently high IT qualification of individual teachers to work in a virtual environment, lack of audiovisual content in the courses taught, Internet failures, and low bandwidth. Nevertheless, over time, the problems were resolved positively. Ultimately, in most universities worldwide, the learning process in the distance format was assessed positively both by the teaching staff and the students.

There is a need to conduct a serious analysis of what happened to us in 2020, but today we can state the following:

- one of the most important conclusions is that any sane person comes to is the indispensability of personal communication, which is the basis of the university's activities; whatever changes take place, higher education will remain a school of interaction between the professor and the student, students with each other, as well as professors with each other;
- education in the future will be as a complex system that combines artificial intelligence, virtual reality, online courses, and much more; today, students can take all kinds of online classes on different learning platforms, and then they can choose online courses from teachers of their choice; an automatic test system can put assessments;
- universities without online courses will become uncompetitive and may subsequently disappear since they will not be in demand;
- companies and their employees are interested in new technologies to increase profits and, accordingly, bonuses, and employees of educational institutions will not receive any bonuses from the introduction of innovations; the introduction of new technologies in universities, on the contrary, can lead to staff cuts, which does not stimulate the active introduction of those very innovations; hence the unwillingness to take risks and change the methods of work that have been developed over the years;

- the experience of conducting two exam sessions led to the conclusion that the existing assessment system, transferred to existing technologies, showed ineffectiveness; intermediate or final testing is now challenging, if not impossible, so we need to look for new forms of knowledge assessment;
- there will be a reduction in classroom activities, primarily a reduction of lectures, which should lead to an active transition to project work based on real facts.

As a result of the Covid-19 pandemic, there has been a large-scale transition to online learning. As it is mentioned in [15], educational institutions started with establishing the instructional continuity, and then started to develop new instructional design and AI-enabled innovations.

References

1. Sawsan, A.: Barriers to distance learning during the COVID-19 outbreak: a qualitative review from parents' perspective. Heliyon **6**(11), e05482 (2020)
2. Russian Ministry of Education and Science Named the Number of Distance Learning Students (in Russian). https://news.rambler.ru/community/45498875. Accessed 06 Mar 2021
3. Hannum, W.H., Irvin, M.J., Lei, P.-W., Farmer, T.W.: Effectiveness of using learner-centered principles on student retention in distance education courses in rural schools. Distance Educ. **29**(3), 211–229 (2008)
4. Irvin, M.J., Hannum, W.H., de la Varre, C., Farmer, T.W., Keane, J.: Factors related to rural school administrators' satisfaction with distance education. Distance Educ. **33**(3), 331–345 (2012)
5. Torres, M., Acal, C., El Homrani, M., Mingorance, E.A.: Impact on the virtual learning environment due to COVID-19. Sustainability **13**(2), 582 (2021)
6. Lapitan, L.D.S., Tiangco, C.E., Sumalinog, D.A.G., Sabarillo, N.S., Diaz, J.M.: An effective blended online teaching and learning strategy during the COVID-19 pandemic. Educ. Chem. Eng. **35**, 116–131 (2021)
7. Velasquez, R.M.A., Lara, J.V.M.: Knowledge management in two universities before and during the COVID-19 effect in Peru. Technol. Soc. **64**, 101479 (2021)
8. Zolotaryuk, A.V., Zavgorodniy, V.I., Gorodetskaya, O.Y.: Intellectual prediction of student performance: opportunities and results. Adv. Econ. Bus. Manag. Res. **81**, 555–559 (2019)
9. Tarik, A., Aissa, H., Yousef, F.: Artificial intelligence and machine learning to predict student performance during the COVID-19. Procedia Comput. Sci. **184**, 835–840 (2021)
10. Tejedor, S., Cervi, L., Pérez-Escoda, A., Tusa, F., Parola, A.: Higher education response in the time of coronavirus: perceptions of teachers and students, and open innovation. J. Open Innov.: Technol. Mark. and Complex. **7**(1), 43 (2021)
11. Sebyakin, A., Zolotaryuk, A.: Tracking emotional state of a person with artificial intelligence methods and its application to customer services. In: 2019 Twelfth International Conference "Management of Large-Scale System Development" (MLSD), Moscow, Russia, pp. 1–5 (2019)
12. Sebyakin, A., Soloviev, V., Zolotaryuk, A.: Spatio-temporal deepfake detection with deep neural networks. In: Toeppe, K., Yan, H., Chu, S. K.W. (eds.) Diversity, Divergence, Dialogue, iConference 2021. LNCS, vol. 12645, pp. 78–94 (2021). https://doi.org/10.1007/978-3-030-71292-1_8
13. Soloviev, V.: Machine learning approach for student engagement automatic recognition from facial expressions. Sci. Publ. State Univ. Novi Pazar, Ser. A: Appl. Math. Inform. Mech. **10**(2), 79–86 (2018)

14. Butler-Henderson, K., Crawford, J.: A systematic review of online examinations: a pedagogical innovation for scalable authentication and integrity. Comput. Educ. **159**, 104024 (2020)
15. Krishnamurthy, S.: The future of business education: a commentary in the shadow of the Covid-19 pandemic. J. Bus. Res. **117**, 1–5 (2020)

Author Index

Printed in the United States
by Baker & Taylor Publisher Services